Topley and Wilson's

Principles of Bacteriology, Virology and Immunity

Seventh Edition in four volumes

Volume 1

W. W. C. Topley, 1886–1944

Topley and Wilson's
Principles of Bacteriology, Virology and Immunity

Seventh Edition in four volumes

Volume 1

WILLIAMS & WILKINS
Baltimore

General Editors

Sir Graham Wilson
MD, LLD, FRCP, FRCPath, DPH, FRS

Formerly Professor of Bacteriology as Applied to Hygiene, University of London, and Director of Public Health Laboratory Service, England and Wales.

Sir Ashley Miles CBE
MD, FRCP, FRCPath, FRS

Deputy Director, Department of Medical Microbiology, London Hospital Medical College, London.
Emeritus Professor of Experimental Pathology, University of London, and formerly Director of the Lister Institute of Preventive Medicine, London.

M. T. Parker
MD, FRCPath, Dip Bact

Formerly Director, Cross-Infection Reference Laboratory, Central Public Health Laboratory, Colindale, London.

Volume 1

General microbiology and immunity

Edited by

**Sir Graham Wilson
and
Heather M. Dick**

© G. S. Wilson, A. A. Miles and M. T. Parker 1983

First published 1929

Seventh edition in four volumes 1983 and 1984

Library of Congress Cataloging in Publication Data

Topley, W. W. C. (William Whiteman Carlton), 1886–1944.
  Topley and Wilson's Principles of bacteriology, virology, and immunity.
    Bibliography: p.
    Includes index.
    Contents: v. 1. Introduction to microbiology and immunity – v. 2. Systematic bacteriology.
  1. Medical microbiology – Collected works. 2. Immunology – Collected works. I. Wilson, Graham S. (Graham Selby), Sir, 1895–   . II. Miles, A. Ashley (Arnold Ashley), Sir, 1904–   . III. Parker, M. T. (Marler Thomas) IV. Title. V. Title: Principles of bacteriology, virology, and immunity.
[DNLM: 1. Bacteriology. 2. Immunity. 3. Viruses. QW 4 T675p]
QR46.T6 1983   616'.01   83-14501
ISBN 0-683-09064-X (v. 1)

All Rights Reserved. No part of this publication may be reproduced, stored in a retrieval system, or transmitted in any form or by any means, electronic, mechanical, photocopying, recording or otherwise, without the prior permission of the publisher.

Whilst the advice and information in this book is believed to be true and accurate at the date of going to press, neither the authors nor the publisher can accept any legal responsibility or liability for any errors or omissions that may be made.

To EM, BRP and to the memory of JW

Printed in Great Britain

Volume Editors

Sir Graham Wilson MD, LLD, FRCP, FRCPath, DPH, FRS
   Formerly Professor of Bacteriology as Applied to Hygiene, University of London, and Director of Public Health Laboratory Service, England and Wales.

Heather M. Dick MD, FRCP(Glas), FRCPath
   Professor and Consultant in Clinical Immunology, Royal Infirmary, Glasgow.

Contributors

J. P. Arbuthnott BSc(Glas), PhD(Glas), MIBiol
   Professor of Microbiology, Moyne Institute, Trinity College Dublin:
   Fellow of Trinity College, Dublin.
Naomi Datta MD(Lond)
   Professor of Microbial Genetics, University of London, Royal Postgraduate Medical School, London.
Heather M. Dick MD, FRCP(Glas), FRCPath
   Professor and Consultant in Clinical Immunology, Royal Infirmary, Glasgow.
Eve Kirkwood BSc, MB, ChB, MRCP
   Senior Lecturer in Bacteriology and Immunology, Western Infirmary, Glasgow.
Sir Ashley Miles CBE, MD, FRCP, FRCPath, FRS
   Deputy Director, Department of Medical Microbiology, London Hospital Medical College, London.
J. Gareth Morris DPhil, FIBiol
   Professor of Microbiology, University College of Wales, Aberystwyth.
Marilyn E. Nugent BSc, PhD
   Research Investigator, G. D. Searle & Co. Ltd, High Wycombe, Bucks.
Peter Owen BSc, PhD
   Lecturer in Microbiology, Trinity College Dublin.
Stephen Powis BSc
   St. John's College, Oxford

D. A. Ritchie BSc, PhD, FIBiol, FRSE
   Professor of Genetics and Head of Department of Genetics, Liverpool University, Liverpool.
Howard J. Rogers PhD
   Professor and Head of the Division of Microbiology, National Institute for Medical Research, Mill Hill, London.
R. J. Russell BA(MOD), MA, PhD
   Lecturer in Microbiology, University of Dublin, Department of Microbiology, Moyne Institute, Trinity College Dublin.
Frank Sheffield MB, ChB
   Head of Division of Bacterial Products, National Institute for Biological Standards and Control, London.
J. Douglas Sleigh MB, ChB, MRCP(Glas), FRCPath
   Senior Lecturer, Department of Bacteriology, University of Glasgow, Royal Infirmary, Glasgow.
J. David Williams BSc, MD, MRCPath DCP
   Professor of Medical Microbiology The London Hospital Medical College, London.
Peter C. Wilkinson MD
   Titular Professor in the Department of Bacteriology and Immunology Department, University of Glasgow (Western Infirmary), Glasgow.
Sir Graham Wilson MD, LLD, FRCP, FRCPath, DPH, FRS
   Formerly Professor of Bacteriology as Applied to Hygiene, University of London, and Director of Public Health Laboratory Service, England and Wales.

General Editors' Preface to 7th edition

After the publication of the 6th edition in 1975 we had to decide whether it would be desirable to embark on a further edition and, if so, what form it should take. Except for the single-volume edition of 1936, the book had always appeared in two volumes. We hesitated to alter this arrangement but reflection made us realize that a change would be necessary.

If due attention was to be paid to the increase in knowledge that had occurred during the previous ten years two volumes would no longer be sufficient. Not only had the whole subject of microbiology expanded greatly, but some portions of it had assumed a disciplinary status of their own. Remembering always that our primary concern was with the causation and prevention of microbial disease, we had to select that part of the newer knowledge that was of sufficient relevance to be incorporated in the next edition without substantial enlargement of the book as a whole.

One of the subjects that demanded consideration was virology, which would have to be dealt with more fully than in the 6th edition. Another was immunology. Important as this subject is, much of it is not directly concerned with immunity to infectious disease. Moreover, numerous books, reviews and reports were readily available for the student to consult. What was required by the microbiologist and allied workers was a knowledge of serology, and by the medical and veterinary student a knowledge of the mechanisms by which the body defends itself against attack by bacteria and viruses. We resolved, therefore, to provide a plain straightforward account of these two aspects of immunity similar to but less detailed than that in the 6th edition.

The book we now present consists of four volumes. The first serves as a general introduction to bacteriology including an account of the morphology, physiology, and variability of bacteria, disinfection, antibiotic agents, bacterial genetics and bacteriophages, together with immunity to infections, ecology, the bacteriology of air, water, and milk, and the normal flora of the body. Volume 2 deals entirely with systematic bacteriology, volume 3 with bacterial disease, and volume 4 with virology.

To this last volume we would draw special attention. It contains 27 chapters describing the viruses in detail and the diseases in man and animals to which they give rise, and is a compendium of information suitable alike for the general reader and the specialist virologist.

The first two editions of this book were written by Topley and Wilson, and the third and fourth by Wilson and Miles. For the next two editions a few outside contributors were brought in to bridge the gap that neither of us could fill. For the present edition we enlisted a total of over fifty contributors. With their help every chapter in the book has been either rewritten or extensively revised. This has led to certain innovations. The author's name is given at the head of each chapter; and each chapter is prefaced by a detailed contents list so as to afford the reader a conspectus of the subject matter. This, in turn, has led to a shortening of the index, which is now used principally to show where subjects not obviously related to any particular chapter may be found. A separate but consequently shorter index is provided for each of the first three volumes and a cumulative index for all four volumes at the end of volume 4. Each volume will be on sale separately. As a result of these changes we shall no longer be able to ensure the uniformity of style and presentation for which we have always striven, or to take responsibility for the truth of every factual statement.

We are fortunate in having Dr Parker, who has been associated with the 5th and 6th editions of the book, as the third general editor of all four parts of this edition and as editor of volume 2. Dr Geoffrey Smith with his extensive knowledge of animal disease has greatly assisted us both as a contributor and as editor of volume 3. Dr Fred Brown, of the Animal Virus Research Institute, has organized the production of volume 4, and Professor Heather Dick the immunity section of volume 1.

Two small technical matters may be mentioned. Firstly, in volume 2 we have retained many of the original photomicrographs and added others at similar magnifications because they portray what the student sees when he looks down an ordinary light microscope in the course of identifying bacteria. Electronmicrographs have been used mainly to illustrate general statements about the structure of the organisms under consideration. Secondly, all temperatures are given in degrees Celsius unless otherwise stated.

Apart from those to whom we have just expressed our thanks, and the authors and revisers of individual chapters, we are grateful to the numerous workers who have generously supplied us with illustrations; to Dr N. S. Galbraith and Mrs Hepner at Colindale for furnishing us with recent epidemiological information; to Dr Dorothy Jones at Leicester for advice on the *Corynebacterium* chapter and Dr Elizabeth Sharpe at Reading for information about *Lactobacillus*; to Dr R. Redfern at Tolworth for his opinion on the value of different rodent baits; to Mr C. J. Webb of the Visual Aids Department of the London School of Hygiene and Tropical Medicine for the reproduction of various photographs and diagrams; and finally to the Library staff at the London School and Miss Betty Whyte, until recently chief librarian of the Central Public Health Laboratory at Colindale, for the continuous and unstinted help they have given us in putting their bibliographical experience at our disposal.

GSW
AAM

Volume Editors' Preface

The decision to publish the present edition in four volumes instead of two, for reasons explained in the General Editors' Preface, has necessitated alterations in the content and sequence of the various chapters. A glance at the contents will show that the present volume covers a wide field corresponding fairly closely to that in Part I of the 6th edition. Most of the chapters have been written afresh by authors who have provided a comprehensive view of their subject embodying the advances made during the past ten years. It will be noticed, however, that the Growth and Death chapter of the 6th edition has been incorporated in the chapter on General Metabolism. A further innovation has been the transference of the five chapters on Ecology, Normal Flora and the Bacteriology of Air, Water and Milk from the end of Volume 2 of the 6th edition, where they were inaptly placed, to form part of the section on General Microbiology to which they really belong. For logical reasons they have been amalgamated so as to form only two chapters. As is made clear in Chapter 8, the subject of Bacteriocines has been tentatively added to that on the Normal Flora instead of ranking with the bacteriophages, as it did formerly.

The greatest change, however, has been the transference of the subject of Immunity from Volume 2 of the 6th edition to the present volume. Rather than attempting to give the complete experimental evidence supporting modern theories of immunity, we have concentrated on those aspects that pertain most closely to infectious disease. In a way, we have returned to the viewpoint of earlier microbiologists, but with the incorporation of modern experimental findings, often derived from work apparently far removed from the study of infection. Much of the detailed description of principles and techniques included in earlier editions has been omitted; this body of knowledge may still be consulted with profit, but has been sacrificed in the present edition to make way for a more wide-ranging discussion. Some of the more recent work on immunoglobulins, on lymphocyte functions, and on the role of the major histocompatibility complex (MHC) in Chapters 10 and 14 may seem strange to those who are not familiar with this rapidly advancing field of knowledge, but of their relevance to infection there can be no doubt. The generation of antibody diversity, the 'fine-tuning' of the controlling influences, and the disastrous consequences of disordered immunological responses (Chapters 10, 15, 16) can be explained only at the cellular level. Details of antigenic structure and the exploitation of this knowledge have also been brought up to date (Chapter 13). The role of antibody in infection (Chapter 12) and the importance of lymphocyte and antibody mediated tissue damage are examined in broad outline (Chapters 11 and 16). The underlying principles of the various problems of infection in patients with defective immunological responses are discussed in Chapter 17. Some readers may feel that we have omitted useful information, both old and new; but we would point to the existence of numerous specialized texts dealing with individual aspects of immunology and to the more important reviews and papers to which references have been made.

We have tried in this volume to provide a firm foundation for the general microbiologist who intends to work on bacterial chemistry, molecular biology, genetic engineering, or the prevention and control of infectious disease, as well as for those engaged in other branches of biology or in clinical medicine who need it for purposes of consultation or reference.

1983

GSW
HD

Contents of volume 1

General microbiology and immunity

1 History	1
2 Bacterial morphology	16
3 The metabolism, growth and death of bacteria	39
4 Bacterial resistance, disinfection and sterilization	70
5 Antibacterial substances used in the treatment of infections	97
6 Bacterial variation	145
7 Bacteriophages	177
8 Bacterial ecology, normal flora and bacteriocines	220
(i) Bacterial ecology	220
(ii) The normal bacterial flora of the body	230
(iii) Bacterial antagonism: bacteriocines	247
9 The bacteriology of air, water and milk:	251
(i) Air	251
(ii) Water	260
(iii) Milk	279
10 The normal immune system	296
11 Antigen-antibody reactions—*in vitro*	319
12 Antigen-antibody reactions—*in vivo*	330
13 Bacterial antigens	337
14 Immunity to infection—immunoglobulins	374
15 Immunity to infection—complement	384
16 Immunity to infection—hypersensitivity states and infection	389
17 Problems of defective immunity. The diminished immune response	402
18 Herd infection and herd immunity	413
19 The measurement of immunity	430
Index	445

Contents of volumes 2, 3 and 4

Contents of volume 2
Systematic bacteriology

20 Isolation, description and identification of bacteria
21 Classification and nomenclature of bacteria
22 *Actinomyces, Nocardia* and *Actinobacillus*
23 *Erysipelothrix* and *Listeria*
24 The mycobacteria
25 *Corynebacterium* and other coryneform organisms
26 The Bacteroidaceae: *Bacteroides, Fusobacterium* and *Leptotrichia*
27 *Vibrio, Aeromonas, Plesiomonas, Campylobacter* and *Spirillum*
28 *Neisseria, Branhamella* and *Moraxella*
29 *Streptococcus* and *Lactobacillus*
30 *Staphylococcus* and *Micrococcus*; the anaerobic gram-positive cocci
31 *Pseudomonas*
32 *Chromobacterium, Flavobacterium, Acinetobacter* and *Alkaligenes*
33 The Enterobacteriaceae
34 Coliform bacteria; various other members of the Enterobacteriaceae
35 *Proteus, Morganella* and *Providencia*
36 *Shigella*
37 *Salmonella*
38 *Pasteurella, Francisella* and *Yersinia*
39 *Haemophilus* and *Bordetella*
40 *Brucella*
41 *Bacillus*: the aerobic spore-bearing bacilli
42 *Clostridium*; the spore-bearing anaerobes
43 Miscellaneous bacteria
44 The spirochaetes
45 *Chlamydia*
46 The rickettsiae
47 The Mycoplasmatales: *Mycoplasma, Ureaplasma* and *Acholeplasma*

Contents of volume 3
Bacterial diseases

48 General epidemiology
49 Actinomycosis, actinobacillosis and related diseases
50 *Erysipelothrix* and *Listeria* infections
51 Tuberculosis
52 Leprosy, rat leprosy, sarcoidosis and Johne's disease
53 Diphtheria and other diseases due to corynebacteria
54 Anthrax
55 Plague and other yersinial diseases, pasteurella infections and tularaemia
56 Brucella infections of man and animals: vibrionic abortion
57 Pyogenic infections, generalized and local
58 Hospital-acquired infections
59 Streptococcal diseases
60 Staphylococcal diseases
61 Septic infections due to gram-negative aerobic bacilli
62 Infections due to gram-negative, non-sporing anaerobic bacilli
63 Gas gangrene and other clostridial infections of man and animals
64 Tetanus
65 Bacterial meningitis
66 Gonorrhoea
67 Bacterial infections of the respiratory tract
68 Enteric diseases: typhoid and paratyphoid fever
69 Bacillary dysentery
70 Cholera
71 Acute enteritis
72 Food-borne diseases: botulism
73 Miscellaneous diseases. Granuloma venereum, soft chancre, cat-scratch fever,

Legionnaires' disease, *Bartonella* infection and Lyme disease
74 Spirochaetal and leptospiral diseases
75 Syphilis, rabbit syphilis, yaws and pinta
76 Chlamydial diseases
77 Rickettsial diseases of man and animals
78 Mycoplasmal diseases of animals and man

Contents of volume 4
Virology

79 The nature of viruses
80 Classification of viruses
81 Morphology: virus structure
82 Virus replication
83 The genetics of viruses
84 The pathogenicity of viruses
85 Epidemiology of viral infections
86 Vaccines and antiviral drugs
87 Poxviruses
88 The herpes viruses
89 Vesicular viruses
90 *Togaviridae*
91 *Bunyaviridae*
92 *Arenaviridae*
93 Marburg and Ebola viruses
94 Rubella
95 Orbiviruses
96 Influenza
97 Respiratory disease: rhinoviruses, adenoviruses and coronaviruses
98 *Paramyxoviridae*
99 Enteroviruses: polio-, ECHO-, and Coxsackie viruses
100 Other enteric viruses
101 Viral hepatitis
102 Rabies
103 Slow viruses: conventional and unconventional
104 Oncogenic viruses
105 African swine fever

1

History
Ashley Miles

Introductory	1	Chemical analysis of bacterial products:	
Bacteriology and virology	1	Virulence factors	9
Pasteur's work	2	Antimicrobial agents	9
Putrefaction and fermentation	3	Immunology	10
Spontaneous generation	3	Metchnikoff's work: cellular defences	10
Tyndallization	4	Humoral defences: complement	11
Lister's work	4	Hypersensitivity: anaphylaxis	12
Microbial diseases	4	Antigens and antibodies: nature and combination	13
Koch's work	5	Antibody immunity	13
Fruition of medical microbiology	6	Cellular immunity	13
Classification of microbes: numerical taxonomy	6	B and T cells	14
Viruses	7	Monoclonal antibodies	14
Bacterial genetics	8	The immune response to tissues	14
		Immune tolerance	14

Introductory
Bacteriology and virology

In the study of any branch of science, an acquaintance with the historical development of knowledge is an important element in a clear understanding of our present conceptions. To the student of bacteriology such a basis is essential. From Pasteur onwards, for 60 years or more, the great majority of investigators were more interested in what bacteria do than in what they are, and much more interested in the ways in which they interfere with man's health or pursuits than in the ways in which they function as autonomous living beings. The relations of bacteria to disease, to agriculture, and to various commercial processes, presented problems which pressed for solution; and, as a result, we have seen the development of an applied science of bacteriology, or rather its application along many divergent lines, without the provision of any general basis of purely scientific knowledge.

In much the same way, the sister science of immunology grew from attempts to solve problems of immunity in man and animals. Nevertheless the last half of the nineteenth century and the first few decades of the twentieth saw, *pari passu* with the technical mastery of microbial technology, the emergence of the science of microbiology.

It is customary, in summarizing the history of bacteriology, at least in relation to medicine, to refer to the conception advanced by Fracastorius of Verona (1546), of a *contagium vivum* as the cause of infective disease, and to the views advanced by von Plenciz (1762) on the specificity of disease, based on a belief in its microbial origin. A concrete science is, however, seldom advanced by arguments, however ingenious, which are propounded without appeal to experiment, or to wide and detailed observation; and such views have acquired their main significance from knowledge gathered by later generations, rather than from their inherent fertility. The construction and use of the compound microscope was an essential prerequisite to the study of microbial forms. To van Leeuwenhoek (1683) must be ascribed the credit of placing the science of microbiology on the firm basis of direct observation (Dobell 1932). This Dutch maker of lenses devised an apparatus and technique (Cohen 1937) which enabled him to observe and describe various microbial forms with an accuracy and care that still serve as a model for all workers in this field. He observed, drew, and measured large numbers of minute living organisms, including bacterial and protozoal forms. This striking advance was not followed by further rapid progress in our knowledge of bacteria and their activities. Such progress was, however, impossible without further developments in technique. The world of minute living things, opened to morphological study by van Leeuwenhoek, was seen to be peopled by a multitude of dissimilar forms, whose interrelationships it was impossible to determine without preliminary isolation; and, so far as bacteria were concerned, this isolation was not accomplished until the problem of artificial cultivation was solved, almost two hundred years later.

Pasteur's work

The real development of bacteriology as a subject of scientific study dates from the middle of the nineteenth century, and is the direct outcome of the work of Louis Pasteur (1822-95). Isolated observations of microbial parasites, by Brassi, Pollender, Davaine and others, have priority in particular instances, just as Schultze, Schroeder and Dusch and others initiated technical methods which Pasteur applied to his own researches. But it was Pasteur and his pupils who settled the fundamental questions at issue, and developed a technique which made possible the cultivation and study of bacteria.

Trained as a chemist, Pasteur was led to the study of microscopic organisms by his observations on fermentation. His early studies on molecular asymmetry, had led him to believe that the property of optical activity possessed by certain organic compounds was characteristic of substances synthesized by living things. It was known that small amounts of an optically active substance, amyl alcohol, were formed during the fermentation of sugar. Since it was impossible to regard the molecule of amyl alcohol as derived from the molecule of sugar by any simple break-down process, he concluded that the optically active sugar was first broken down to relatively simple substances without optical activity, and that from such inactive substances the optically active amyl alcohol was synthesized. For Pasteur this was evidence of the presence of living things, and he therefore started on his study of fermentation with a strong *a priori* leaning towards the microbial theory of fermentation, and away from the then dominant hypothesis of Liebig. He was prepared to adopt the theories already propounded by Cagniard-Latour in 1836, and by Schwann in 1837, concerning the living nature of the yeast globules, which were always to be found in sugar solutions undergoing alcoholic fermentation, and which had been described by van Leeuwenhoek in 1683.

Since, however, the production of amyl alcohol had been observed in fermenting brews in which lactic acid also appeared, Pasteur first selected lactic fermentation for experimental study. Van Helmholtz (1843) had already indicated that alcoholic fermentation was due to the yeast itself or to some other organized material. He had shown that the substance, responsible for initiating alcoholic fermentation would not pass through membranes that allowed the passsage of

Fig. 1.1 Louis Pasteur (1822-1895).

organic substances in solution but held back particles in suspension. This experiment, successful with alcoholic fermentation, failed with many other ferments and fermentable liquids. Pasteur's mind was naturally addicted to generalization, and his interest lay in the phenomenon of fermentation as a general type of reaction. It was therefore natural that he should at first neglect the field in which the battle was more evenly balanced between the purely chemical conceptions of Liebig, and the biological theories of Cagniard-Latour, Schwann and Helmholtz, and turn to the field in which Liebig's views had never been successfully attacked. In Pasteur's first memoir published in 1857, he declared the lactic ferment to be a living organism, far smaller than the yeast cell, but which could be seen under the microscope, could be observed to increase in amount when transferred from one sugar solution to another, and had very decided preferences as regards the character of the medium in which it was allowed to develop; so that, for instance, by altering the acidity of the medium one could inhibit or accelerate its growth and activity. In this memoir Pasteur laid the first foundations of our knowledge of the conditions to be fulfilled for the cultivation of bacteria.

He also showed that the fermentation of various

organic fluids was always associated with living cells, and that different types of fermentation were associated with microscopic organisms that could be distinguished from one another by their morphology and their cultural requirements. Thus, at this early stage, the idea of specificity entered into bacteriology.

Putrefaction and fermentation

Spontaneous generation

It was impossible for Pasteur to pursue these studies without facing the problem of the origin of these minute living organisms, which he regarded as the essential agents of all fermentations. At this time (1859) there were two opposed schools of thought with regard to the genesis of microbial forms of life. One school, deriving from the great naturalists of antiquity, believed in the *spontaneous generation* of living things from dead, and especially from decomposing organic matter; but, as Pasteur astutely noted, the species of animals or plants believed to arise by spontaneous generation were diminishing in number, and the average size of those organisms still included in this category was getting smaller and smaller. However the discovery by Leeuwenhoek of the world of microbial organisms gave a powerful stimulus to the somewhat decadent theory. Here, at all events, were living things which obeyed no known law of reproduction, and whose existence seemed to lend support to a belief which had long been accepted by eminent authorities, and which had thereby acquired a natural prestige.

From the start of his inquiry, Pasteur assumed that these microscopic organisms, like other living things, were reproduced in some way from similar pre-existing cells. He had already convinced himself that these organized cells were the active agents of **fermentation**. Clearly then they could not arise *de novo* during the changes for which they were themselves responsible, but must have been introduced from without. Their striking specificity, maintained through repeated transferences from one specimen of fermentable fluid to another of the same kind, was strong evidence in favour of their autonomous reproduction. Here again Pasteur had tentatively adopted the correct solution before starting his experimental inquiry, but the main interest of his part in the controversy lies in the consummate skill with which he developed methods which enabled him to give clear demonstrations where others had left doubt and confusion, and which determined the main rules of a technique which made possible the cultivation and study of bacteria. Needham (1745) described spontaneous generation of microbes in closed flasks of heated putrescible fluid; but Spallanzani (1769) found no such generation after longer heating. Among the non-living substances suggested as the cause of **putrefaction,** oxygen occupied a prominent place at the beginning of the nineteenth century. Apperts' (1810) success with the preservation of foodstuffs, by heating and hermetical closure of the containing vessels, followed by a weighty expression of opinion by Gay-Lussac, had led to a general belief that the exclusion of this gas was the essential factor in ensuring the absence of fermentation. Schultz (1836) renewed the air in a heat-sterilized flask of putrescible fluid with air drawn through solutions of strong potash or concentrated sulphuric acid, and Schwann (1837) with air drawn through a heated tube; and could detect no putrefaction. But other treatments of the air were unsuccessful. A real advance was made by Schroeder and Dusch (1854), who sterilized the air by passing it through cotton wool, thereby avoiding exposure to strong chemicals. But their results were equivocal with some fluids owing, it appears, to their use of too short periods of initial heating.

This, then, was the position when Pasteur began his investigations in 1859. In a series of admirable memoirs, starting in 1860 and continuing for more than four years, he went over the ground already covered, added new and illuminating experiments of his own devising, and terminated the controversy by clear and decisive demonstrations. He showed that the material removed from air by passage through cotton-wool contained organized particles which were similar in appearance to the spores of moulds. Introduced into flasks of sterilized organic material, they were capable of giving rise to the growth of numerous kinds of living organisms. By other methods, he showed that these germs were numerous in the streets of cities, less numerous in the air of country uplands, rare in the quiet air of closed and uninhabited rooms or cellars, where the dust had deposited and remained undisturbed, and very rare in the pure air of the high Alps, above the level of human habitation. He showed that the failures of Schroeder and Dusch were due to the inadequate sterilization of their material, and that certain animal fluids, such as blood or urine, known to be eminently liable to undergo putrefaction, could be collected in such a way as to remain permanently unaltered (see Vallery-Radot 1919).

In 1876 Bastian published a communication controverting an early statement by Pasteur that urine, sterilized by boiling, remained free from growth on subsequent incubation. Bastian declared that, if the urine were made alkaline at the start, growth often ensued. Pasteur, on repeating the experiment, was forced to admit the truth of Bastian's statement. A careful retracing of all his steps resulted in the demonstration that fluids with an acid reaction, after sterilization at 100°, might remain apparently sterile because certain organisms, which remained alive, were unable to develop, while in an alkaline medium they might grow freely. It was found also that ordinary water frequently contained organisms which were not killed by heating to 100°, and that organisms which had become deposited on the surface of glassware in the dry state might withstand far higher temperatures. We know now that it is especially for those bacteria which form spores that these conditions hold true. As a result of this controversy Pasteur established the practice of heating fluid material to 120° under pressure for the purpose of sterilization, thus introducing the autoclave into the laboratory, and the practice of sterlizing glassware by dry heat at 170°.

Tyndallization An important advance was made by Tyndall who, observing that actively growing bacteria are easily destroyed by boiling, and that a certain amount of time is required for bacteria in culture in the resistant, inactive phase to pass into the growing phase in which they are heat-sensitive, introduced the method of sterilization by repeated heatings, with appropriate intervals for germination between them. This method is still known as *Tyndallization* (see Bulloch 1938).

Lister's work Pasteur's demonstration that both fermentation and putrefaction were initiated by airborne microbes prompted Joseph Lister's (1827–1912) work on wound sepsis. Lister was deeply concerned with the post-operative sepsis that exacted such a terrible toll on the lives of hospital patients. Assuming that putrefying wounds were analogous to putrefying organic matter, he became convinced that the key to the prevention of sepsis lay in denying access to the wound by the microbes of the environment, particularly of the air. This he achieved by the antiseptic dressing, first described in 1867, with strikingly successful results. With his antiseptic technique, the scope of surgery, hitherto limited by the fear of sepsis, was enormously enlarged. That this technique has largely been replaced by the aseptic technique in no way detracts from the merit of Lister's discovery, nor from the debt we owe him for fighting the usual battle against ignorance and prejudice. Although Lister did not directly prove that his success was due to the destruction of potentially infective microbes, in the light of contemporary work by French and German bacteriologists on the relation of microbes to disease, it became evident that his achievement was one of the first great triumphs of applied bacteriology in medicine.

While investigating fermentation Pasteur had used very various kinds of natural organic fluids and solutions, and had succeeded in growing micro-organisms on simple media. As a result he had realized that a medium eminently suitable for the growth of one bacterium or mould may be ill adapted for the growth of another, and that a prime necessity for the successful cultivation of any species of micro-organism is the discovery of a suitable growth medium. He had learned the need for the scrupulous sterilization of everything that came into contact with material to be examined bacteriologically; and had devised the necessary methods of sterilization in the steamer, in the autoclave, in the hot-air oven, or by direct flaming. He had proved the serviceableness of the cotton-wool plug for protecting media in flasks or tubes. He had realized the importance of the constitution of the nutrient material offered to a given bacterium, of the acidity or alkalinity of that medium, and of the oxygen pressure to which it was subjected. Armed with this knowledge, he proceeded to break new ground.

Pasteur was before all else a scientist, intensely curious, and loving knowledge for its own sake, but he was also a convinced utilitarian, and a Frenchman. He desired greatly that his discoveries should benefit mankind in general, France in particular, and, if possible, his neighbours in the first place. Thus we find him investigating the troubles of the local vintners, brewers, and vinegar-makers, and many of his memoirs are devoted to the diseases of wines or of beers, and the methods of preventing them. Here he faced the question whether one species could change into another, in particular whether mycoderma vini could change into the ordinary yeast of wine. Deceived on this point at first, he resorted as usual to rigorous and repeated experiments, and not only demonstrated that this mutation did not occur, but indicated clearly the conditions which led to its apparent occurrence, and the care which must be exercised before accepting any reported variation of this kind.

Microbial diseases It is evident from his memoirs on fermentation and spontaneous generation (see Vallery-Radot, P., 1922-1933) that the possibility of applying this new knowledge to the elucidation of infective disease was already in Pasteur's mind. A request from Dumas to investigate the disease then ruining the silkworm industry in the South of France, turned him permanently towards the study of infective processes. We cannot follow here, even in outline, Pasteur's researches into pébrine, anthrax, chicken cholera, or hydrophobia. Some of them are referred to in later chapters. But certain contributions are noteworthy. It was Pasteur who showed, in the case of anthrax, that a culture of a pathogenic organism could be passed through successive subcultures, in such a way as to dilute beyond possibility of significant action, any other material introduced with it into the primary culture from the blood or tissues, and still produce the disease when inoculated into a susceptible animal; though it is to Koch that priority must be given for much of our knowledge on the nature and mode of action of the anthrax bacillus. It was Pasteur who introduced into bacteriology the conception of virulence and of attenuation, and who demonstrated that an attenuated bacterial culture will act as a vaccine, that is, will confer immunity against subsequent infection with a virulent strain of the same bacterium. For Pasteur, indeed, a vaccine was synonymous with an attenuated culture, as opposed to a virulent culture on the one hand and to a dead culture on the other. It was Pasteur who, in the case of rabies, showed that it was possible to study the virus of an infective disease by animal passage when the organism could not be cultivated, and even to prepare an apparently effective vaccine by using suitably treated animal tissue.

Thus, throughout a long scientific life, Pasteur was largely concerned with the practical application of knowledge gained during his studies on fermentation; and with the problems which occupied the last thirty years of his life, the solution of which made his name a household word. But we shall miss the real significance of his work if we fail to realize that his fertile generalizations were of infinitely more importance for the progress of science than were his successful attacks on these isolated problems.

One further point must be noted. Pasteur and his colleagues had shown how to obtain cultures of micro-organisms, and propagate them indefinitely in the laboratory; but the methods which they employed were not well suited to the isolation of pure strains of bacteria from an originally mixed culture, except in those relatively rare instances in which it was possible to use a highly selective medium. Since all media were employed in the fluid state, the only method of purifying a culture was to make successive transfers with very small amounts of material, in the hope that only a few bacteria, all of one kind, would be carried over. Such a technique was very uncertain in its results.

Koch's work

Pasteur, starting as a chemist, founded bacteriology and revolutionized medicine. At about the time when he was propounding his germ theory of disease, a young German physician, some twenty years his junior, was turning from clinical medicine to bacteriology. Robert Koch (1843–1910), at that time a practising physician at Wollstein, attacked the problem of anthrax, and produced, as his first contribution to science, a demonstration of the character and mode of growth of the causative bacillus, which opened a new era in bacteriological technique. This memoir he published in 1876. In the following year he published his methods of preparing, fixing, and staining film-preparations of bacteria, using the aniline dyes introduced into histology by Weigert. In 1878 he published his memoir on traumatic infectious diseases, which remains a classical example of the study of experimental infections in laboratory animals. In 1881 he described his method of preparing cultures on solid media, a technical advance of the first importance, since it made possible the isolation of pure strains of bacteria from single colonies. Solid media prepared from naturally occurring material such as pieces of potato, had previously been used for the isolation of micro-organisms, particularly by mycologists, and the general principles to be observed in the preparation of pure cultures had been clearly enunciated by Brefeld, who had suggested the solidification of a nutrient medium by the addition of gelatin. The media and methods available for the cultivation of fungi were

Fig. 1.2 Robert Koch (1843–1910).

not, however, well suited for bacteria; and it was left for Koch to devise, in the form of his nutrient gelatin, and later, at the suggestion of Frau Hesse, of nutrient agar, a solid, transparent medium, easy to sterilize and handle, and thus admirably adapted for obtaining isolated colonies of bacteria (see Bulloch 1938). In 1882 and 1884 he published his classical papers on the bacillus of tuberculosis. In 1883 he discovered the vibrio of cholera. Already, Koch had enlisted the services of Loeffler and of Gaffky as his assistants. Later came Pfeiffer, Kitasato, Welch and many others, and, with his growing fame, he began to gather round him a group of keen and able young men, who were destined to introduce the methods he devised into the laboratories of many lands. In 1885 he was appointed Professor of Hygiene and Bacteriology in Berlin, and in 1891 he was made Director of the newly founded Institute for Infectious Diseases. His later years were devoted almost entirely to the investigation of bacteriological problems in their relation to the prevention and cure of disease, and many of his contributions to our knowledge will be considered in later chapters. Koch was, above all, an able and careful technician. He was greatly aided by the vigour and initiative of the large German chemical and optical firms, and the advances which he made in staining methods, in the use of the microscope for the observation of bacteriological preparations, and in the technique of cultivating bacteria, revolutionized this branch of science.

Fruition of medical microbiology

The fruits of this revolution appeared with surprising rapidity. During the last quarter of the nineteenth century a succession of discoveries was reported, bearing on the relation of bacteria to human and animal disease, which opened a new era in medicine.

In 1874 Hansen described the bacillus of leprosy, and Neisser, in 1879, the gonococcus. In 1880 Pasteur recorded the isolation of the bacillus of fowl cholera, and Eberth observed the bacillus of typhoid fever. In 1881 Ogston published an adequate description of the staphylococcus. In 1882 Koch discovered the tubercle bacillus, and Loeffler and Schütz the bacillus of glanders. In 1883 Koch isolated the cholera vibrio and Fehleisen the streptococcus of erysipelas; and Klebs described, but did not isolate, the bacillus of diphtheria. In 1884 Loeffler isolated the diphtheria bacillus, and Gaffky the typhoid bacillus, which Eberth had observed four years previously. In 1885 Loeffler discovered the bacillus of swine erysipelas, Kitt the bacillus of haemorrhagic septicaemia of cattle, and Salmon and Smith the bacillus associated with hog cholera. In the same year Nicolaier observed the tetanus bacillus in soil, inoculation of which produced the disease in animals.

The next five years saw the isolation by European, North American, and Japanese bacteriologists of the bacillus of swine erysipelas, the bacillus of haemorrhagic septicaemia of cattle, the bacillus associated with hog cholera, the pneumococcus, the colon bacillus, the meningococcus, the bacillus of Malta fever, the first of the many bacilli associated with gastro-enteritis, and the tetanus bacillus. This impressive list continues in the last decade of the century, with the bacillus then believed to be the cause of influenza, one of the bacilli of gas gangrene, the plague bacillus, the bacillus of fowl plague, the bacillus causing the variety of food poisoning known as botulism, the bacillus of bovine abortion, a dysentery bacillus and the cause of infectious pleuropneumonia of cattle.

Thus, by the close of the nineteenth century a great variety of micro-organisms had been identified as occurring in definite association with human or animal disease. In many instances it was clear that the association was one of cause and effect; in others, it was probably so. In others, again, there was doubt whether the bacterium present played any more important rôle than that of a secondary invader. Beyond dispute, however, the scientific investigation of infectious disease had become the province of the bacteriologist.

But the revolution inaugurated by Pasteur and extended by Koch spread far beyond the field of medicine. Agriculturists had long been puzzling over the problem of soil fertility, without arriving at any very helpful conclusions. One curious phenomenon was the reaccumulation of nitrates in the soil, in spite of their constant removal by the washing action of the rain. It was suspected that these nitrates might be derived in some way from the decomposition of organic material, and in 1877 Schloesing and Müntz, acting on a suggestion made by Pasteur in 1862, showed by experiment that the formation of nitrates was due to the action of living organisms. Warington, at Rothamsted, confirmed these results in 1878 and 1879, and showed that two stages were concerned, a preliminary conversion of ammonia to nitrites, and a subsequent oxidation of nitrites to nitrates. He believed that these two stages were carried out by different organisms, but failed to isolate or identify them. This problem was solved by Winogradsky in 1890, who isolated and described both the nitrite- and nitrate-forming organisms. In 1888 Hellriegel and Willfarth described the nitrogen-fixing bacteria which caused the formation of nodules on the roots of leguminous plants. Later Winogradsky described a free-living anaerobic organism which was able to fix atmospheric nitrogen, and Beijerinck, some ten years later, described a large, free-living, nitrogen-fixing aerobic bacterium, which he named *Azotobacter*, and which has since been extensively studied. The bacteriology of the soil thus became an important part of agricultural science.

In the early years of the bacteriological revolution it had been demonstrated that bacteria attacked plants, as well as animals. In 1878 Burrill described the organism of pear blight, and in 1883 Wakker discovered the bacillus which causes the 'yellows' of the hyacinth. This branch of bacteriology has since been pursued energetically.

The demonstration by Pasteur of the essential nature of fermentation led, as a natural consequence, to the entry of the bacteriologist into the industrial sphere. His help was required in dairy farming, in brewing, in the preservation of foods, and in all those commercial processes in which bacterial activity was desired or feared.

For more detailed history, the reader is referred to the writings of Bulloch (1938), Clark (1961) and Lechevalier and Solotorovsky (1965).

This brief summary will indicate with sufficient clearness to how great an extent the microbiologist has been occupied with applied problems.

Classification of microbes: numerical taxonomy

Description of microbes was usually limited to characters sufficient to distinguish them from those with which they might be confused. Thus fermentation reactions predominate in the classification of the coliform organisms; and specific toxins in that of the spore-bearing clostridia. The soil bacteriologist employs yet other methods. Systematic bacteriology was very generally neglected; only in more recent years was any real attempt made to survey bacterial groups as a whole, and to bring some order out of chaos.

Each addition to the characters by which bacteria can be recognized was a step towards precise definitions of bacterial species and genera; and the earlier years of the twentieth century saw a great refinement in the recognition and naming of the bacterial strains that had been incompletely described by their pioneer discoverers from 1880 onwards. Even so the characters used and the weight given to each of them tended to differ with the discipline—e.g. medical, agricultural or marine—concerned, thus limiting the scope of comparative microbiology. The introduction into microbiology by Sneath of **numerical taxonomy** in the late 50s, whereby the degree of relationship between organisms is expressed as a figure, which is based on a large number of characters all given equal weight in the calculations, has done much to minimize the use of the more subjectively determined criteria of earlier classifications.

Viruses The first two decades of this century witnessed no such striking advances in our knowledge of the bacteriology of disease as occurred between 1875 and 1900. Nevertheless, the technical innovations needed for a further decisive advance were being made. As early as 1892 Ivanovsky had reproduced mosaic disease in the tobacco plant by extracting the leaves of the diseased plant and applying the extract to healthy plants after it had passed through a filter that held back bacteria; and six years later Beijerinck fully confirmed this result. With the demonstration by Loeffler and Frosch, in 1898, of a similarly filtrable agent that could transmit foot-and-mouth disease, the filter-passing virus was established as a possible cause of disease in plants and animals. Viruses were at first defined in physico-chemical terms, like size as determined by filtrability and sedimentation in centrifuges, as well as by their action in the living animal. The larger viruses were visible in the light microscope, and the introduction in 1934 of the electronmicroscope by Ruska soon opened up the direct morphological study of smaller viruses. In the meantime, viruses were grown *in vitro* in non-proliferating suspensions of mammalian cells; and in the 1930s the technique of growing them in the bacteria-free, living chick embryo, introduced by Goodpasture in the United States, was perfected. By 1940, growth in tissue cultures of susceptible mammalian cells was established, a technique which greatly facilitated the study of many viruses as a substitute for the laborious methods of growing them in the living animal.

One class of viruses, the **bacteriophages** that infect

Fig. 1.3 Bacteriological Section, Congress of Hygiene and Demography, London 1891.

bacteria, was discovered independently by Twort in 1915 and d'Herelle in 1917. The comparative ease of studying the interaction of bacteriophage and bacteria invited an intensive study of the phenomenon, which has had profound implications for our understanding of the nature of virus infections in general; for bacteriophage proved to be an organism whose genetic apparatus was incorporated into that of the host cell, with either lethal results or the establishment of a heritable state of inapparent infection. The phage resembled plant and animal viruses, whose infective parts are either deoxyribonucleic acid (DNA) or ribonucleic acid (RNA) within a protein coat.

Bacterial genetics

At the beginning of the twentieth century it was generally recognized that the characters of bacterial strains were not fixed, but could vary, both physiologically and morphologically, sometimes according to their environment, and sometimes in a more permanent fashion that was independent of their environment. The facts pointed to a genetic apparatus in bacteria analogous to that in larger organisms; but valid genetic proof was lacking in the absence of any known form of sexuality in bacteria. In 1928 Griffith described the permanent transformation of the capsular material of a living pneumococcus by material derived from dead pneumococci of another type; and in 1944 Avery and his colleagues made a further advance, fundamental indeed for genetics in general, when they identified the transforming material as DNA. Here was an example of a permanent modification of the genetic apparatus of one organism by presumed genetic material from another. The unequivocal demonstration of bacterial sexuality was made by Tatum and Lederberg in 1947, when they recombined two biochemical mutants of *Esch. coli*, each lacking certain characters present in the parent strain, to produce a strain with all the original parental characteristics. This work, which laid the foundation of modern bacterial genetics, was made possible by advances in microbial chemistry during the previous 25 years, particularly as regards the action of non-proliferating cells on chemically defined substrates; definition of the growth requirements of bacteria, both the main energy sources and the bacterial vitamins, and the investigation of nutritional antagonisms and the action of chemotherapeutic agents; and the action of cell-free enzymes and coenzymes.

Two further important discoveries in bacterial genetics were made in 1952. In America, Zinder and Lederberg found that bacteriophages, whose genes, as noted above, are incorporated into the chromosomes of the bacteria they infect, can carry part of the bacterial chromosome with them when they leave one bacterium and infect another; and thus introduce into it alien bacterial genes. This is the phenomenon of transduction. In the same year, in Great Britain, Hayes discovered that recombination in *Esch. coli* was not only mediated by a sex factor, but that the transfer of genetic material was a one-way process, involving movement of some of the donor's genes to a recipient. Thus, although there is no full sexuality in bacteria, such as we recognize in more complex organisms, it has since proved possible, as we relate in Chapters 6, 7, to make detailed studies of the genetics of a number of bacterial species by exploiting the three phenomena of transformation, transduction and recombination.

The study of the genetic mechanisms in bacteria, together with the intensive work on the constitution of the nucleic acids, played a major role in Watson and Crick's formulation in 1953 of their hypothesis of the genetic code; according to which genetic information is embodied in paired linear strands of deoxyribonucleotides, whose pattern in a given region constitutes a gene. Bacterial genes are mainly chromosomal but some information is embodied in the extra-chromosomal plasmids, which are closed circles of DNA. Hayes' sex factor, e.g. is a plasmid, which facilitates the substitution of a region of a recipient chromosome by the corresponding region in the donor cell. In transformation there is recombination with an alien plasmid. In transduction, the phage particles contain fragments of host DNA, which may be transferred during infection of a recipient by the phage. These rather haphazard processes have in recent years been refined by recombinant techniques in which desired pieces of DNA are joined to a plasmid, a phage particle, or an animal virus. Certain endonucleases cleave double-stranded DNA at particular sequences of nucleotides, giving pieces with ends that are highly cohesive with the ends of the plasmid or phage DNA cloven with the same endonuclease. The inserted piece is then stabilized enzymically with a ligase and bacteria transformed with the resulting DNA. By this and related processes of cloning, bacteria may thus be made to synthesize, e.g. viral antigens and antiviral interferon of animal origin, making possible the bulk production of substances like these which it is impracticable to produce by conventional methods.

These are some of the immediately microbiological aspects of a new field of genetics, with its immense potential in exploiting the techniques of genetic engineering. One other property of DNA is of practical value to the microbiologist, the ready re-association of the two single strands of DNA produced by separation of the double helix. One such strand of unknown constitution may be hybridized with a strand of a known sequence of bases, and the constitution of the unknown deduced from the degree of association observed; or, as may be done in mammalian cells, the presence in the DNA of infected tissue of base sequences characteristic of a given virus may be inferred from the degree to which hybridization with a known virus occurs.

Chemical analysis also provided for genetic work microbial characters other than the enzymes responsible for biochemical activity.

Chemical analysis of bacterial products

One field of work that has expanded greatly since the 1920s is that of chemical analysis of bacterial products. In the United States Avery and Heidelberger from 1923 onwards were pioneers in the chemical definition of the carbohydrate capsule of the various types of pneumococci. Since that time capsules and cell walls of bacteria and the various exotoxins have been specified chemically with increasing precision. With the techniques at their disposal the earlier workers in this field were hampered by the fact that their methods were laborious and painstaking. In the last 20 years or so our understanding of the chemistry of biological macromolecules has advanced at a great rate, notably as a result of the refinement of the techniques of isolation—such as in gel filtration—and of characterizing complex molecules.

Virulence factors. It has been the aim of medical and veterinary microbiologists, as they have devised new methods by which to distinguish pathogenic bacteria, to relate pathogenicity to bacterial products or activities, sometimes referred to as virulence factors. With the protein exotoxins such as those of the tetanus bacillus, cholera and the diphtheria bacillus, and of certain streptococci and staphylococci, all are clearly virulence factors in that they are associated with virulent strains and in many respects mimic the natural disease when given separately. But with most organisms the factors which determine that one organism is pathogenic and another nonpathogenic have proved elusive. A microbe may be pathogenic because it has the power to invade tissue; or is resistant to a particular host defence; or because of any of a number of factors acting in concert. Of recent years the capacity to adhere to mucous surfaces, such as those of the respiratory alimentary and genital tract—sometimes associated with the possession of bacterial appendages—has been recognized as a feature of certain bacteria associated with the promotion of the early stages of infection. Virulence factors that can be relatively easily identified—like the production of exotoxins—are in some cases demonstrably under the control of genes, which occur mainly on plasmids; and a beginning has been made in experimental animals in locating chromosomal genes responsible for resistance to specific pathogens; though it is still not clear how these genes determine their effect.

Antimicrobial agents

As far as bacterial infections of man and animals are concerned, one of the most fruitful advances of the twentieth century was the introduction of effective agents, either synthetic or natural products, that kill bacteria in the body of the host. Ehrlich is justly described as the father of chemotherapy. He clearly saw the problem as one of finding agents that would interfere with the metabolism of the parasite in concentrations that did not affect the metabolism of the host; and, if possible, to use agents that acted on metabolic characters peculiar to the parasite. In the 1900s he cured one form of trypanosomiasis in rats with the dye trypan red, and another form in mice with an organic arsenic compound, atoxyl. By 1910 another of Ehrlich's organic arsenicals was successfully used to treat syphilis. A major advance in antibacterial therapy was made in 1935 when Domagk in Germany reported the successful treatment of streptococcal infections of mice with the dye Prontosil. It is a measure of the degree to which the study of bacterial nutrition had advanced by this time, that the active principle of the Prontosil molecule, sulphonamide, was soon proved to be antibacterial because it blocked the utilization of the closely related compound, para-aminobenzoic acid, which was at the same time established as an essential nutrient of the coccus. There followed a rapid exploitation of various sulphonamide drugs in antibacterial therapy.

Antibacterial effects by substances produced by microbes themselves had been observed since the end of the last century (Florey *et al.*, 1960), but had never been effectively exploited in therapy of man until 1940, when Chain and his colleagues made stable preparations of penicillin, the antibiotic derived from a *Penicillium* mould described by Fleming in 1929. With penicillin, and streptomycin, described in 1944 by Waksman and his colleagues, the antibiotic era was launched.

These discoveries had had consequences reaching far beyond that of providing for the treatment of infection. Other antibiotics—and chemotherapeutic agents also—were looked for and chemical manipulation applied to already known antibiotics, all with the idea of extending the range of diseases that could be treated successfully. But this increasing mastery of the treatment of a wide range of diseases was offset by a phenomenon that soon became a serious problem, namely the emergence of resistant strains of various pathogenic bacteria. It was in fact the investigation of the inheritance of resistance to antibiotics which provided much of the evidence for the discovery of plasmids and for some of the genetic mechanisms in bacteria that we have described above. The emergence of resistant strains in a general population meant not only that a particular disease became insusceptible to what had hitherto been an effective antibiotic but that resistant strains might themselves flourish. It was early realized that the selection of resistant strains by heavy antibiotic use, and their subsequent transmission between persons was a major factor in the accumulation of resistant strains in a community, especially in hospitals and like institutions. As a result, since the resistance is easily detectable with some accuracy *in vitro*, there was a renewed interest in the routes of transmission of bacteria.

Although they have an undoubted claim to be considered as a feature of non-specific immunity, the protein *interferons* are conveniently mentioned here, because, besides their protective role in mammalian natural viral infections, they are actively antiviral when given as preparations isolated from a variety of vertebrate cells. But in spite of extensive work ever since their discovery in 1957 by Isaacs and Lindenmann, attempts to produce them in therapeutic quantities for general use have so far had limited success.

We conclude this account of the more recent advances in microbiology with a mention of the speeding up and intensification of all forms of investigation by automation in the laboratory. It is evident not only in handling and processing of material, and identification of microbes, but also in computerized analysis of the resulting data. Adansonian classification, for example, and methods of identification of microbes have been very greatly facilitated and made more reliable, to the benefit of both practical and theoretical microbiology.

Immunology

Immunology as a scientific discipline was born of the study of how animals, by natural or artificial means, became immune to microbial infections and toxins of various kinds.

For as long as there are written accounts of infective disease it seems to have been realized that recovery is accompanied by an acquired resistance; the person who is convalescent is unlikely to suffer from the same disease again. There are many reports through the centuries of attempts to exploit this observation by inducing infections artificially. The eighteenth century provides the most remarkable example, in the work of the Gloucester physician, Edward Jenner, on prophylactic immunization against smallpox.

In his time, prophylaxis was attempted by inoculation of human smallpox material. The procedure was highly hazardous because of the occasional and sometimes fatal generalization of the disease, and its protective value was disputed. The first step to a safer procedure was the substitution of material derived from the lesions of cowpox, which is a much milder disease in man, as practised, for example, in the 1770s by a Dorsetshire farmer, Benjamin Jesty. Following up the postulate on which this practice was based Jenner made the first methodical investigation of the popular opinion that cowpox prevented subsequent smallpox. In 1798 he published privately his memoir 'An enquiry into the causes and effects of the variola vacciniae'. In this memoir Jenner included the first description of what we now recognize to be an example of 'hypersensitivity' associated with a state of immunity. 'It is remarkable that variolous matter, when the system is disposed to reject it should excite inflammation on the part to which it is applied more speedily than when it produces the Small Pox. Indeed it becomes almost a criterion by which we can determine whether the infection will be received or not. It seems as if a change, which endures through life, had been produced in the action, or disposition to action, in the vessels of the skin; and it is remarkable too, that whether this change has been effected by the Small Pox or the Cow Pox, that the disposition to sudden cuticular inflammation is the same on the application of variolous matter.'

Pasteur was the outstanding figure at the end of the nineteenth century to apply the new knowledge of bacteria to practical measures for prophylaxis of infective disease. But, although he was so successful in immunization against chicken cholera, veterinary anthrax and rabies, the mechanism whereby such protection arose remained obscure. In 1880 Pasteur put forward his 'exhaustion hypothesis', that resistance was brought about by the disappearance from the body of some necessary foodstuff which was used up during the first attack by the infective agent.

Metchnikoff's work: cellular defences

But the study *in vitro* of what happened to bacteria when exposed to the white cells—the leucocytes—or to the fluid part of blood soon led to more readily testable hypotheses about the mechanisms of resistance to infections. Élie Metchnikoff (1845-1919) was the first to recognize the general significance of the phenomenon of phagocytosis in animal tissues, whereby organic foreign particles are taken up by phagocytic cells and digested by them. Since living microbes so ingested may be killed in this way, Metchnikoff came to regard the phagocytic reaction as the prime defence against microbial invasion of the tissues. This concept of immunity as primarily cellular, promulgated in the 1890s, was at the time perhaps too sweeping; and although the importance of phagocytosis in defence was recognized early in the present century, it was not until some fifty years after Metchnikoff's first publication that cellular immunity once more became the subject of extensive study. The reason for the delay lies in the work of Metchnikoff's contemporaries,

Fig. 1.4 Élie Metchnikoff (1845–1916). From a photograph of 1907 by Nicola Perscheid. By courtesy of 'The Wellcome Trustees'.

whose striking discoveries turned the attention of bacteriologists to the concept of antimicrobial defences as predominantly due to substances in the blood serum; that is, humoral as distinct from cellular defences.

Humoral defences: complement

Nuttall in 1888 had shown that some species of bacteria were killed by defibrinated blood of certain animals, and one year later Buchner showed not only that cell-free serum was bactericidal but that it ceased to be so after being heated to 55° for one hour. The labile substance assumed to be the cause of the bactericidal effect he called 'alexine'. Another humoral defence was described by von Behring and Kitasato in 1890, who found that the serum of animals which had received a series of injections of non-lethal doses of tetanus toxin had the power to neutralize tetanus toxin specifically, and to protect normal animals from otherwise lethal doses of toxin. Yet another humoral defence was announced three years later by Pfeiffer; when living cholera vibrios were introduced into the peritoneal cavity of guinea-pigs previously inoculated with killed cultures of vibrios, they dissolved; i.e., they underwent bacterilysis. That the lysis of the vibrios was independent of cells was shown by Bordet in 1895, who defined two agents, both present in the serum of the immunized guinea-pig, necessary for the effect; one was relatively heat-stable and specific for the cholera vibrio, the other easily destroyed by heating to 55°. The first, not always present in normal serum, appeared in large amounts in the 'immune serum' in response to immunization, and the other was present in any normal serum. The first, Bordet's *substance sensibilatrice,* sensitized the vibrios to the lytic action of the second, which he assumed was identical with Buchner's 'alexine'; or in modern terminology, the vibrios were sensitized by specific antibody present in the antiserum, and lysed by complement. Bordet further noticed that the antiserum would clump the vibrios before lysis occurred. This was the phenomenon of agglutination by antibody, first studied in detail by Gruber and Durham (1896).

By 1897 Kraus had established that filtrates of cultures of the plague bacillus or of the cholera vibrio formed a precipitate with antisera from an immunized animal, and that the reaction was specific.

Thus by the early years of the present century a number of important *in vitro* effects of antibody were known—precipitation, agglutination, neutralization of toxins and the sensitization of microbes to the action of complement—which were available for the investigation of immunity, in both the normal and the specifically immunized animal. By this time too, the relation between the phagocytes and the antibacterial substances in the blood serum had been clarified. Denys and Leclef (1895) and later Mennes (1897), working on antistreptococcal immunity, had shown that phagocytosis by leucocytes was promoted by blood serum acting on the streptococci. In 1903, Wright and Douglas demonstrated that the effect was due to a thermolabile substance acting directly on the bacteria; they called the substance opsonin. And Neufeld and Rimpau (1904, 1905) found in the serum of animals immunized against streptococci and pneumococci thermostable substances which acted specifically on these bacteria, increasing the readiness with which they were ingested by phagocytic cells. To these substances they gave the name bacteriotropins. It soon became clear that the bacteriotropin effect was another manifestation of specific antibody activity, and the doctrine gradually became accepted that, whatever basic capacity the phagocytes of normal animals had for antibacterial defence, the high degree of specific immunity that could be induced by immunization was largely due to the formation of specific antibody, which in the infected animal would both neutralize any microbial toxins formed and render the microbes themselves much more susceptible to phagocytosis. As a result, specific immunity came to be studied largely in terms of antibody formation, and therapy and prophylaxis in infectious disease to be conceived in terms of administering to the host antibody of the right specificity for the infecting microbe, or of immunization designed to induce the formation of that antibody by the host itself.

Early in the century it was recognized that complement was not a single substance but a name covering substances responsible for a sequence of effects of which the starting point was the reaction of one component with an antigen-antibody complex and the other the final end point such as lysis. With more detailed analysis of both the direct chain of reactions and various side effects during the successive activation of the various components it became clear that the end result might be beneficial—as in the death of microbes or deleterious—as in certain allergic reactions. But over the years many complement-like effects that were independent of antibody, including death of microbes in fresh plasma, were observed—a process of activation by what is now known as the 'alternative pathway'; whereby complement is activated by a host of substances, including microbial substances, by-passing the stage of combination of specific antibody and antigen.

The first two decades of the twentieth century were concerned largely with the exploitation of specific immunization for the prophylaxis and therapy of infectious diseases (see Parish 1965), and attempts to determine the nature of the multitude of substances that stimulated the formation of specific antibodies—which came to be known by the generic name of antigens; and the nature of the antibodies themselves.

Hypersensitivity: anaphylaxis

The introduction of microbes, or soluble antigens like toxins, into the animal body did not always induce increased immunity to the microbes or the toxin. Jenner had observed that persons immunized with cowpox responded to a second inoculation of the skin with the cowpox material by an exaggerated inflammatory reaction. In 1890, Koch described an analogous form of this hypersensitivity, in the phenomenon known by his name, in which guinea-pigs infected with tubercle bacilli responded with an exaggerated inflammatory reaction to a local inoculation with more tubercle bacilli. A similar reaction, taking some 12-24 hours to reach maximum intensity, occurred when soluble material from the tubercle bacillus, later named tuberculin, was injected; the same amount of tuberculin was not toxic for normal animals. In this instance the hypersensitivity to substances from a bacillus was induced by the tubercle bacillus itself. Another form of hypersensitivity, first observed in 1894, occurred during immunization with certain non-living antigens. When many doses of the antigen were injected at intervals of several days, one of the later injections sometimes induced severe and often fatal reactions in the animal. Unlike the delayed response of the tuberculous guinea-pig to tuberculin, the response of these animals occurred immediately after the injection of the antigen. This paradoxical effect, whereby procedures intended to ensure prophylaxis proved to have an opposite effect, was studied in detail by Richet and Portier (1902) and by them named **anaphylaxis**. The two types of hypersensitive reaction, the one delayed in time, the other immediate, were distinguishable in another respect. *Delayed hypersensitivity* could not be induced in normal animals by the serum of hypersensitive animals; but anaphylactic hypersensitivity was transferable in this way. This operational distinction—in general the passive transfer from hypersensitive animals by lymphoid tissue and serum respectively—proved to be of major importance in defining the processes of immunization at the cellular level.

The outstanding feature of both immunization and hypersensitization was, of course, the *specificity* of the result. In general, an antibody induced by a particular antigen reacted most strongly and often, it appeared, exclusively with the inducing antigen. The multiplicity of different substances and different microbes which could act as antigens posed several questions. First, what made a substance antigenic and determined its specificity; second, what was the nature of antibody, and what determined its specificity; third, how did antigen and antibody react to produce the results like precipitation, agglutination, sensitization of cells to lysis, neutralization of toxin and so forth; and lastly, how did the animal body recognize the differences between antigens, and wherein lay its capacity to respond to them with antibodies of so many specificities?

Generally speaking, soluble antigens, such as proteins, and antigenic materials extracted from microbes and other cells, appeared to be large-molecular substances which, to be effective had as a rule to be 'foreign'; to the animal being immunized. Thus a rabbit would produce antibodies to, say, human serum proteins, but not to rabbit serum proteins; but both man and rabbit produced antibodies to the substances in, say, the typhoid bacillus, which was clearly 'foreign' to them both. It was evident that proteins were among the most powerful antigens, but the chemical configurations that determined immunological specificity were far from clear. The first indication of how sensitively the antibody-forming mechanism responded to structural detail in antigens came in 1906 from the experiments of Obermayer and Pick, who modified the surface of serum protein antigens by introducing nitro-groups or iodine, and found that the modified antigens induced the formation of antibodies which were to a large extent specific for the nitro or iodine groups. This work was extended from 1914 onwards by Landsteiner, who established that a wide variety of artificially introduced chemical groupings would alter the specificity of an antigen. An equally important step was the analysis of a natural antigen, made by Heidelberger and Avery, who in 1923 began their classical demonstration that the antigenic specificity of different types of pneumococcus was determined by differences in the composition of the polysaccharide substance of the capsular material that surrounded the coccus.

As regards antibody, many workers from 1907 onwards had found that antibody activity was associated with the serum globulins, and within the next twenty years, as the techniques of protein chemistry advanced, antibodies were firmly established as globulins.

Antigens and antibodies: nature and combination

The nature of the combination of antibody with antigen was only slowly elucidated. Ehrlich, in the 1890s, demonstrated its quantitative nature in the reaction of diphtheria toxin and antitoxin, and Ramon in 1923 recorded that optimum flocculation and neutralization of the toxin took place with a fixed proportion of the antitoxin. Three years later, Dean and Webb showed that optima of this kind indicated the equivalence of an antibody and the corresponding antigen, but that the two might combine in varying proportions.

From Heidelberger's pioneer chemical analysis of specific precipitates in 1929 it became evident that molecules of both antigen and antibody were multivalent and, as Marrack first proposed, precipitates of the two were primarily bound in a lattice of interlinked antibody and antigen molecules.

Antibody immunity

As to the mode of formation of antibodies, Ehrlich in 1897 suggested that antigens combined with the chemical 'side-chains' of what he called receptors present on the surface of the tissue cells. Combination with antigen then stimulated the cell to over-produce similar receptors, which were the antibodies, and release them into body fluids. Ehrlich apparently thought that the receptors, normally concerned with attaching nutrient substances to the cell, could also attach a wide variety of different chemical substances which, if antigenic, would elicit from the animal a corresponding variety of different antibodies. An opposite notion, that antibody-forming cells had a general synthetic mechanism which, in the presence of an antigen, used that antigen as a template, and so produced antibody with a configuration complementary to and therefore specific for it, was first proposed formally in 1930 by Breinl and Haurowitz; and for some years this 'template' hypothesis, that antigens 'instructed' the cell about the kind of antibody it was to make, held the field. In 1955, however, Jerne proposed a mechanism which was a return to Ehrlich's notion of an already formed combining site for each of the wide variety of antigenic configurations found in nature; only in Jerne's view these preformed sites were on the serum globulin molecules. An injected antigen combined with its appropriate globulin, and the resulting complex was carried to the antibody-forming cells, which were then stimulated to produce the globulins with the same configuration.

Two years later Talmage suggested that the antibody-forming cells themselves carried the preformed antigen-specific sites. On injection, the antigen selected the appropriate cell, and stimulated it to further synthesis of antibody. Soon after, Burnet extended the suggestion by postulating that the attachment of the required diversity of combining sites, and therefore of the cells carrying them, arose as the result of somatic mutations of the antibody-forming cells; and that the attachment of antigen to an appropriate site stimulated the cell to replicate, and so give rise to a population of descendants—a clone—all of which produced antibody of the same specificity. Of the two rival hypotheses of antibody formation, the instructive and the selective, the selective hypothesis as currently refined is now firmly established.

Cellular immunity

As early as 1898, Pffeifer and Marx suggested that antibody was formed by lymphoid tissues, but little progress towards substantiating the suggestion was made until the 1940s when, by a variety of approaches, lymphoid tissues were established as the source of antibody. The participation of lymphoid tissue in immune reactions was emphasized by Landsteiner and Chase in 1942, in their work on delayed hypersensitivity. They showed conclusively that the lymphoid cells, the lymphocytes in particular, of an animal made hypersensitive to a given antigen, would, on transfer to a normal animal, make that animal hypersensitive to the same antigen. This, the first proof that a form of specific immune reactivity could be transferred passively to an animal by a cell, as distinct from an antibody, was one decisive advance towards the rehabilitation of Metchnikoff's general thesis of the overriding importance of cells in immune reactions. The same year saw an equally decisive advance in the same direction, when Lurie proved that the phagocytic macrophages from rabbits immunized against tubercle bacilli, in circumstances that excluded the participation of antibody, had a greater power than normal macrophages to kill the tubercle bacilli they ingested; in other words, quite independently of antibody formation, the antimicrobial power of individual cells was increased as the result of immunization. In the subsequent elucidation of the interdependent rôles of lymphoid tissue and of macrophages and other phagocytes in specific immunity to infection much was owed to refinements of serological techniques, whereby the reactions of antibodies and antigens were observable with a precision and sensitivity far greater than those of the classical antigen-antibody reactions. These included precipitation in gels; the electrophoretic separation of reacting systems; and the detection of minimal or microscopic reactions by labelling one or other of the reagents with fluorescent dye, a radio-isotope or a readily detected enzyme.

B and T cells

A major advance of the last 30 years was the distinction, functional and to some extent topographical, in lymphoid tissue of two cellular systems; namely bursacytes (B cells) which responded with the production ultimately of antibody, and thymocytes (T cells) which on stimulation by antigen initiated the reaction of delayed hypersensitivity and the activation of macrophages to greater microbicidal activity; and the readiness with which both systems of cells, already sensitized by a particular antigen, responded vigorously to further experience of that antigen resided in cells endowed with a specific memory for making that response. The B and T systems have proved to be more complex than is indicated by the simple distinction outlined above; each has subsidiary functions, including inhibitory functions, operative in the regulation of the immune response.

By the 1950s antibodies had been recognized as belonging to one of five classes of the γ-globulins described by Tiselius in 1927. Of these, three were abundant: immunoglobulins G and A with two combining groups, and the macroglobulin M with five and perhaps ten such groups. Immunoglobulin E is associated with one form of hypersensitivity. Immunoglobulin A, as secreted at the mucous surfaces of respiratory, genital and alimentary tracts, constitutes an important defence against entry into the body of foreign antigens, including those of pathogenic microbes and their toxins, through these tracts.

Monoclonal antibodies

The discovery that B cells may give rise to myelomata which are monoclonal with respect to the type of globulin they secrete has opened up an entirely new field in the study of antibodies. Variants of a particular myeloma cell selected so as to produce no immunoglobulin are hybridized with the lymphocytes from the spleen of an animal already stimulated with antigen and the hybridoma selected which, when grown in tissue culture or in the peritoneal cavity of an animal, will give rise to one type antibody. Clones may be selected to yield antibody specific for a single antigenic determinant or for as much of an antigen molecule as is desired; and monoclonal antibodies are now exploited in the marking and purification of antigens, the discovery of new antigens, the investigation of cell functions inhibited by antibody and in the study of allergy.

Although it has been plausibly argued that the chief function of the specific immune response in the higher animals is defence against parasitic microbes of the environment, it was nevertheless recognized early in the study of immunity that it might have other equally biologically important functions.

The immune response to tissues

As a rule, an animal responds only to cells of substances 'foreign' to it. Its immune system is inert towards its own cells and tissue substances—Ehrlich's 'horror autotoxicus.' If, however, an aberrant cell or substance arose in the body, which so differed from the norm as to be 'foreign' to the animal in this sense, the immune response it stimulated might, as happens with infecting microbes, result in the destruction or elimination of the aberration. In the earlier part of this century, this aspect of the immune system, as a monitor of intrinsic deviations from the norm, received considerable attention, especially in relation to cancer. In retrospect, however, it can be seen to have been fruitless because it was thought of largely in terms of antibody immunity. Landsteiner and Chase's demonstration of the carriage by lymphocytes of specific immune reactivity had far reaching consequences in this and related fields. An animal may be immunized artificially with its own tissue but, although the animal reacts with acute inflammation in the organ or tissues used for the immunization, the reaction is not obviously associated with antibody formation. As Freund first demonstrated convincingly in 1953, the reaction closely resembles that of delayed hypersensitivity.

The same kind of response occurs in tissue transplantation. Most of an animal's tissues or organs cannot be successfully grafted on to another animal. The grafts are rejected by a mechanism which was shown by Medawar, first in 1958, to be another instance of the delayed hypersensitivity reaction, being immunologically specific in relation to the antigens of the grafted tissue, and being mediated by lymphocytes carrying the immune reactivity.

Immune tolerance

Medawar's studies of graft rejection also brought into prominence the phenomenon of *immune tolerance*. He took advantage of the fact that the lymphoid tissue of an animal does not begin to mature into a system capable of immune responses until about the time of birth. When cells from a given strain of donor mouse, for example, were injected into fetal mice of a different genetic constitution, these animals on reaching maturity accepted grafts from the donor mice which they would otherwise have rejected. On an extension of Burnet's hypothesis of antibody formation by *clonal selection* any somatic mutant cell with specificity for the donor cell, that arises in the maturing lymphoid tissue, combines with the donor antigens present and is suppressed before it can grow into a substantial reactive clone of cells with that specificity; so that the recipient mouse remains permanently indifferent to foreign antigens introduced at this stage of its development. In the same manner, it was argued that all the mouse's own tissue substances with potentialities as antigens that have access to the maturing lymphoid system would suppress any mutant lymphoid cells with corresponding specificities; so that the adult animal is indifferent to its own potential antigens—the state of affairs implied in Ehrlich's 'horror autotoxicus'.

The concept of immune tolerance had practical as well as theoretical importance, especially for the task of transplanting tissues and organs from one human being to another; since the problem of making a graft acceptable was clearly seen as one of inducing, temporarily at least, an indifference to the foreign antigens of the grafted tissue.

The implications of the diverse manifestations of the immune response are now evident to biologists in general. The last 25 years has seen an explosive development of immunology into a separate discipline, with a place in many biological sciences. Modern research on the cellular basis of the immune response, and on the genetics, development and aberrations of the system; on the genetics and nature of the specific response, and of the inter-relations of the B and T systems of lymphocytes; on the relation of tissue antigens to resistance to microbial infection; and on other beneficial and nocuous consequences of the response to antigens of the animal itself or its environment—is gradually establishing the concept of a system of which an important function is that of maintaining the *status quo* of the animal against disruptive internal change. It is a function that may be more important in the evolution of the immune system than that of ensuring the survival of the animal in the face of external attack, whether microbial or other. The antibacterial and antiviral immunity with which we are concerned in this book must now be placed in perspective as a facet, albeit an important one, of a type of response of a greater biological significance than was dreamed of by the microbiologists who laid the foundations of the science of immunology.

References

Bulloch, W. (1938) *The History of Bacteriology*. Oxford University Press, London.
Clark, P. F. (1961) *Pioneer Microbiologists of America*. University of Wisconsin Press, Madison.
Cohen, B. (1937) *J. Bact.* **34,** 343.
Dobell, C. (1932) *Antony van Leeuwenhoek and his 'Little Animals.'* John Bale, Sons and Danielsson, London.
Duclaux, E. (1920) *Pasteur, the History of Mind*. Engl. Transl. by E. F. Smith and Florence Hedges. Saunders, Philadelphia and London.
Florey, H. W. *et al.* (1960) *Antibiotics*. Oxford University Press, London.
Knight, B. C. J. G. (1962) *J. gen. Microbiol.* **27,** 357.
Lechevalier, H. A. and Solotorovsky, M. (1965) *Three Centuries of Microbiology*. McGraw-Hill, New York.
Parish, H. J. (1965) *A History of Immunization*. Livingstone, Edinburgh and London.
Vallery-Radot, P. (1922–33) *Oeuvres de Pasteur*. 6 vols. Masson, Paris.
Vallery-Radot, R. (1919) *The Life of Pasteur*. Engl. Transl. by Mrs R. L. Devonshire. Constable, London.

2

Bacterial morphology

Howard J. Rogers

Introductory	16	Cell walls and membranes of bacteria	23
Shape and size of bacteria	16	The cell wall	23
Arrangement of cells	18	The outer membrane	25
Differential staining of bacteria	18	The patterned outermost layers	26
Gram stain	18	The cytoplasmic membrane	27
Acid-fast stain	19	Mesosomes	27
Techniques for studying bacterial morphology	19	Endospores	28
Light microscopes	19	Activation	29
Phase contrast microscope	20	Germination	30
Transmission electron microscope	20	Outgrowth	30
Scanning electron microscope	20	Flagella	30
General plan of the bacterial cell	21	Fimbriae and pili	32
The nuclear apparatus	21	Sex pili	33
The bacterial cytoplasm	22	Protoplasts and spheroplasts	33
Intracellular granules	22	L-forms	34

Introductory

The majority of bacteria commonly encountered as mammalian parasites and saprophytes divide by binary fission and have rather simple shapes and ultrastructures. This must not lead us, however, to forget that more complex forms can be encountered such as, for example, in the genera *Cyanobacteria* (Stanier and Cohen-Bazire 1977) or *Rhodomicrobia* (Whittenbury and Dow 1977), but even these have simple ultrastructures compared with say a yeast or an amoeba. Such morphological simplicity of many bacteria does not extend to their biochemical behaviour which is at least as complicated as that of other cells. Thus the few structures visible have each to assume many metabolic functions. Our knowledge about bacterial morphology has mostly been derived from the study of rapidly growing laboratory cultures. After bacteria have entered the stationary growth phase many changes of degeneracy can occur, including those associated with cell death. Even rapidly multiplying bacteria show variation in their size, shape and ultrastructure according to the growth rates and the media used. It is, therefore, important in assessing morphological observations, particularly of claims for aberrant forms, to take note of the conditions used for growing the bacteria. In some genera such as *Arthrobacter* and with conditional morphological mutants in which gross changes in morphology occur when the bacteria are grown under varied conditions such as, for example, at different temperatures or with different glucose concentrations, precise details of the growth conditions become essential parts of the descriptions of morphology. A group of organisms which does not fit into this simple picture are the Actinomycetales. These grow as long frequently twisted filaments which form entangled aggregates. The lengths of the filaments are extremely variable.

The shape and size of bacteria

Light microscope observations allow the recognition of three main shapes for individual bacteria, namely spheroidal, cylindrical and spirillar. Many cocci appear to be approximately true spheres with, for streptococci and staphylococci, diameters of 0.75 to 1.25 μm. The higher magnification obtainable with the electron microscope, however, shows that these organisms are often not indeed true spheres, but have very characteristic shapes (see Fig. 2.1) made of cones and raised bars or indentations at the points of division and future division.

The cylindrical forms vary greatly in length from species to species. For example, *Bacillus anthracis* is between 3 and 8 μm, *Escherichia coli* from 2 to 3 μm and *Haemophilus influenzae* 0.7 to 1.5 μm long. Likewise diameters can be equally variable from 1 to 1.25 μm for *Bac. anthracis*, to *Clos. tetani* which is a long, very thin rod of 0.3–0.4 μm in diameter. It is probable from the careful studies of a few species that bacterial dimensions generally are dependent upon the growth rates of the organisms. For example, *Esch. coli* varies in mean length from 2.2 μm at a growth rate of about 0.5 generations per hour to 3.99 μm at 3.0 generations per hour (Donachie, Begg and Vicente 1976, Woldringh 1976, Woldringh et al 1977, Grover et al 1977). Likewise in some species the cell diameter varies in a similar sense according to the growth rate. Much effort over the years has been devoted, for various purposes, to the exact measurement of bacterial dimensions. All methods have their limitations; none can be guaranteed to give absolute answers. Methods dependent on light microscopy meet with limitations in accuracy due to inadequate resolution, particularly in measuring diameters of rod and coccal shaped organisms. Use of the split image eye-piece with a phase-contrast microscope is probably the method of choice. When electron-microscopy is used cell shrinkage occurs during preparation for transmission microscopy (Bayer and Remsen 1970) and can be as great as 30 per cent when the freeze-etch replica method is used as a criterion for the absolute size. The latter method even though very

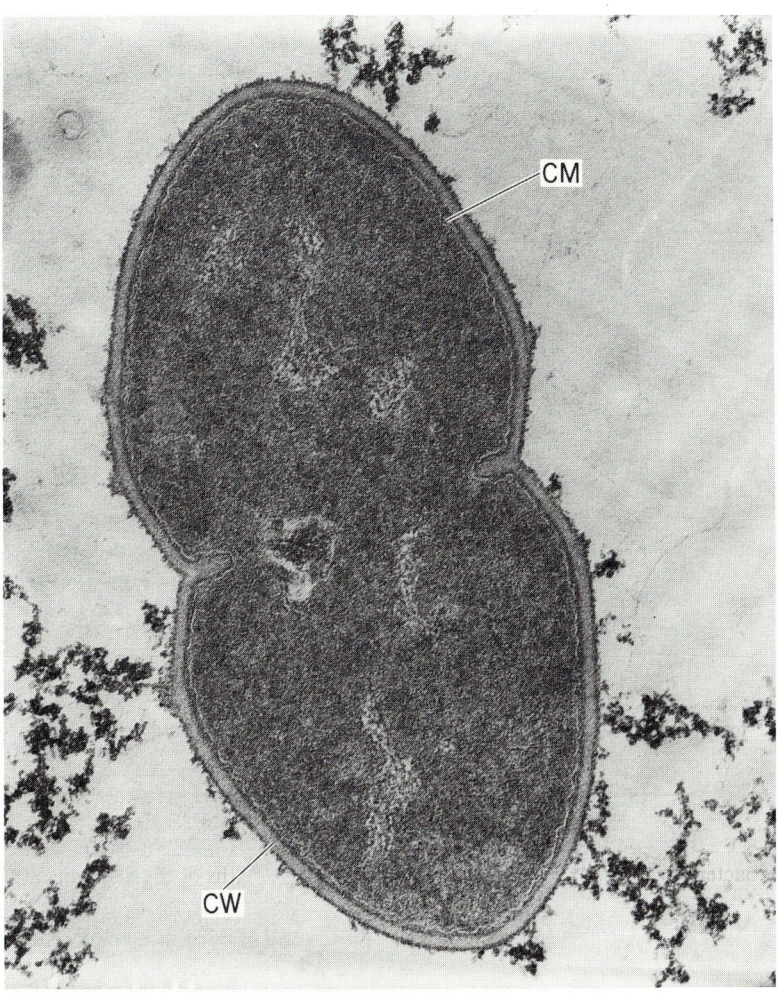

Fig. 2.1 Transverse section of *Streptococcus faecalis* showing the very distinctive shape of the organism, the cell wall (CW) and the cytoplasmic membrane (CM). Magnification 58 950 (Picture kindly supplied by Dr Michael Higgins, Department of Microbiology, Temple University, Philadelphia, USA.)

laborious in operation probably provides the correct ultimate standard. Electronic particle counters are now available with attachments which measure the population mean volumes of bacteria with accuracy, providing no changes in cell electrical resistance take place during the cell cycle and cells separate as individuals. Exponentially growing cultures are mixtures of cells at all stages of division and therefore the coefficient of variation of any length measurements of rod-shaped organisms can be expected to be high, of the order of 15–20 per cent. The widths of rods growing at a steady rate vary much less and have a coefficient of variation of 8–12 per cent.

Apart from the length and width of rod-shaped organisms the shapes of the ends often show features that are of differential value. They may be rounded, square cut or sharply pointed. Spiral shaped organisms are distinguished by the bending or twisting of the cells so that they can assume curved or helical forms. Well-known organisms showing a single curve are the vibrios. The spirilla are distinguished by a series of curves or twists and assume an approximately helical appearance.

The arrangement of cells

Almost as important to the bacteriologist as the shapes of individual cells are the constellations in which they are arranged. A streptococcus is a streptococcus, for example, because the cocci occur in chains rather than irregular groups. Some understanding of the mechanisms which lead to these characteristic arrangements of bacteria is beginning to be available. The seminal work of Cole and Hahn (1962) and the subsequent studies by Shockman and his colleagues (e.g. Higgins and Shockman 1976) of the surface growth and division of streptococci, has made clear the mechanisms determining the occurrence of linear arrays or chains. Likewise, studies of staphylococci (Tzagoloff and Novik 1977) and of sarcinae (Canale-Parola 1970; Chapman 1962) have shown that successive division planes occur perpendicularly to each other. This leads to the well known packets of cells characteristic of sarcinae and by a subsequent distortion, possibly caused by the action of autolysins, to grape-like clusters of staphylococci (Tzagoloff and Novick 1977). Indeed the autolysins, readily detected in almost all bacteria, play a major part in the separation of individual bacteria (Rogers 1979). Bacteria deprived of autolysins, whether by manipulation of the growth conditions in pneumococci (Tomasz 1968) or by genetic means in *Bac. subtilis* (Fein and Rogers 1976), grow as almost endless chains of unseparated cells. Some staphylococci so deprived grow as regular three-dimensional packages (Koyama, Yamada and Matsuhashi 1977). How far the growth forms of different species of bacteria with the same or similar individual cell morphology can be generally related to their autolytic activities it is not yet possible to say.

The differential staining of bacteria

Many combinations of different dyes and mordants followed by decolorizing agents have been applied to bacteria in attempts to find valuable means of distinguishing species and studying ultrastructure. With the common availability of electron microscopes, most of these attempts have been abandoned. The present almost universal use of the phase-contrast microscope makes it no longer necessary to dry, fix and stain bacteria in order to see them by light microscopy. Nevertheless, the importance of two methods for differentially staining groups of bacteria has remained undiminished.

The Gram stain

Gram (1884; see also Lautrop 1981) described a staining method which has been of the greatest service in distinguishing bacterial species, and which has been extensively studied, and frequently modified, by subsequent workers. This reaction depends on the fact that, when bacteria are stained with certain aniline dyes, such as gentian-violet, methyl-violet and others, and are subsequently treated with a solution of iodine in potassium iodide, a mordanting action occurs with some which prevents the subsequent decolorization of the bacteria on treatment with alcohol. Other bacteria, after similar treatment, are readily decolorized.

In employing Gram's method the distinction is not absolutely sharp and specific as regards a given bacterium at all stages of its growth, or as regards all bacterial species. Those organisms which are completely gram-negative never retain the stain, but those organisms which are gram-positive frequently fail to do so when preparations are made from old cultures. This is presumably because such cultures consist largely of dead, dying or degenerate cells, the physical and chemical properties of which are altered. Certain bacterial species show intermediate reactions, with the result that they are extremely sensitive to small changes in technique, sometimes appearing to be gram-positive and at others gram-negative. Due allowance must be made for this in determining the reaction of any given species.

The difference between the retention of the stain by gram-positive bacteria and its loss by gram-negative forms is correlated with certain other characters. Bartholomew and Mittwer (1951) record a large number of properties associated with reaction to the gram stain. For example, gram-positive bacteria are in general less susceptible to the ill effects of proteolytic enzymes, alkalies, azides, tellurites, and oxidizing agents, and to lysis by complement in the presence of antibody; they are more susceptible to acids, detergents, alkyl sulphates, organic solvents, and iodine, and they have a lower isoelectric point than gram-negative cells.

The difference between gram-positive and gram-negative organisms is particularly evident in their susceptibility to various groups of antibacterial agents, both inorganic and organic. The evidence about the mechanism of the gram reaction at the present time is compatible with the explanation that it reflects differences in the permeability of the cell wall to the dye-iodine complex formed in the cell. It has long been known that the retention of gram-positivity depends on the integrity of the cell wall (Benians 1920).

More recent studies indicate, however, that the reaction of gram-positive cells is in part due to their low permeability to alcoholic iodine and alcohol (Kaplan and Kaplan, 1933; Mittwer et al, 1950). Wensinck and Boeve (1957) found that the uptake of dye and iodine was much the same for gram-positive and gram-negative organisms. However, extraction of the complex from gram-positive and gram-negative cells differed depending on the concentration of aqueous alcohol used, at concentrations of 90 per cent (v/v) or greater the complex tended to be retained by gram-positive cells, whereas at this concentration it could easily be extracted from gram-negative cells. Salton (1963) labelled gram-positive and gram-negative cells with ^{32}P and measured the release of label when the cells were suspended in varying concentrations of alcohol and water. The pattern of release varied with the concentration of alcohol and, in general, agreed with that observed by Wensinck and Boeve. Release of ^{32}P from gram-positive cells dropped sharply in alcoholic concentration above 75 per cent (v/v), but remained high in gram-negative cells. Salton suggests that exposure of gram-positive cells to the usual decolorizers causes dehydration of the cell wall, which in turn results in a decrease in the permeability of the wall to the dye-iodine complex within the cell.

Acid-fastness

Certain bacteria, after being stained with warm solutions of fuchsin, resist the decolorizing action of strong mineral acids as was observed by Ehrlich (1882) and confirmed by Ziehl (1882, 1883), who modified the technique of staining. This reaction is characteristic of the mycobacteria, a group of acid-fast bacilli including the tubercle bacillus.

The mycobacteria are peculiarly resistant to the action of such solvents as strong solutions of alkalies, or mixtures of alkalies and sodium hypochlorite (antiformin), and also to the action of hydrolytic enzymes. Acid-fast bacilli are also gram-positive, though the majority of gram-positive organisms are not acid-fast. The property of acid-fastness in the tubercle bacilli and allied forms has been attributed to a lipid or a waxy substance (Klebs 1896, Koch 1897, Bulloch and Macleod 1904, Tamura 1913). Others find that it is the property of intact bacilli (see Sordelli and Arena 1934, Yegian and Vanderlinde 1947) and must therefore depend on the integrity of a membrane with selective permeability to movement of dye and decolorizing agent (see also Porter and Yegian 1945, Lamanna 1946 and Barksdale and Kim 1977). Acid-fastness is lost when tubercle bacilli are exposed under certain conditions to the antituberculous drug isoniazid (isonicotinic acid hydrazide)—an interesting effect not usually produced by other antituberculous drugs.

Techniques for studying bacterial morphology

In bacterial morphology we are mainly concerned with the individual cell and its component parts or with the arrangements of small groups examined by various microscopic techniques.

Individual cells may be studied at both cellular and subcellular levels. For whole cells light and electron microscopes are used. In light microscopy it is necessary or useful to examine living unstained bacteria with techniques which must be safe where pathogenic organisms are being studied. Wet preparations or specially mounted hanging drop preparations are used with light microscopy for the study of bacterial motility and with dark-ground or phase-contrast microscopy. But for most diagnostic purposes special methods of fixation are necessary; and these kill the bacterial cells. For the light microscope after fixation, the preparation is suitably stained; for the electron microscope various methods are used such as treatment with glutaraldehyde, osmium tetroxide (OsO_4), metal shadowing, negative staining, or thin sectioning.

At the subcellular level, mechanical disintegration of bacteria combined with high-speed centrifugation, sometimes through gradients of sucrose or caesium chloride, yields various cell fragments which can be studied. Much of our knowledge comes from the work of biochemists and physiologists interested in the location of macro-molecules including enzymes in different parts and fragments of the bacterial cell. The study of structure and function have contributed much to each other.

Light microscopes

There are six main techniques of light microscopy—ordinary bright field, dark-ground, ultraviolet, phase-contrast, interference, and fluorescence microscopy. The ordinary light microscope gives a theoretical resolving power of about 0.2 μm under optimal conditions; in practice it is of the order of 0.3 μm and objects smaller than this, though 'seen' in the sense of being visible, do not form images that reveal their real size or structure.

The resolving power of an optical system is greatly increased by using light of shorter wavelength. Barnard (1919, 1925, 1930) developed *ultraviolet-light microscopy* with a system of quartz lenses. With a wavelength of 257 nm he resolved and photographed particles with a diameter of 0.075 μm. Beyond this point a limit is again reached, because there is no refracting material to transmit light of shorter wavelength.

By transmitted light, little detail can be seen in bacteria unless they are selectively stained. Rather more can be seen in unstained bacteria by *dark-ground illumination* whereby the illuminating beam passes through a special sub-stage condenser which directs light along a path such that only the rays refracted, diffracted or scattered by the object reach the eye of the observer. Bacteria so examined appear as bright images on a dark background. This method is vastly superior to the former as a means of examining living, unstained organisms, but the smallest particle that can be resolved has a diameter of about 0.35 µm. Greater resolution is obtained by using ultraviolet light in dark-ground microscopy.

The phase contrast microscope

An unstained bacterium viewed by transmitted light is usually a featureless semi-transparency. The various structures it contains, however, differ in refractive index, and will slightly change the phase of light passing through them. The *phase microscope* is a means whereby differences in refractive index are made visible by translation into intensity differences. The theory of phase microscopy has been summarized by Bennett et al. 1951. An image of much greater contrast is produced and, in micro-organisms, much that could only be inferred from killed and stained specimens is revealed. As a result, the phase contrast microscope has become the workhorse of many bacteriological laboratories. From its use the general shape and arrangement of bacteria can be recognized without recourse to the use of stains. The interference microscope which, in principle, is one variation of the phase contrast instrument, is also capable of quantitatively measuring differences of density within cells by the use of interference fringes. It has not, however, been developed to give a sufficiently high resolution to apply to bacteria with confidence. Indeed, the main limitation of light microscopy generally in examining anything but the general shape of bacteria is that the maximum resolution is too poor. As a result almost all modern morphological work aiming at understanding ultrastructure is done with the electron microscope.

The transmission electron microscope

The electron microscope, devised by Ruska (1934) and by Marton (1934, 1941), makes use of the fact that an electron beam passing through a magnetic field behaves in a manner closely analogous to that of a beam of light passing through a refracting medium. This permits the construction of a microscope, generally similar to an optical microscope, that can be described in the terminology of light optics. The 'lenses' are circular electromagnets whose focus varies as the strength of the applied magnetic field; and the microscope has a 'condenser' coil, an 'objective' coil, and a 'projective' coil, the last being equivalent to the eyepiece. The 'wavelength' is a function of the speed of the electrons; those used in electron microscopy are equivalent to a wavelength about 1/100 000th that of visible light. Resolving powers commensurable with this wavelength are unattainable in practice, since 'chromatic' aberration in the magnetic lens cannot be corrected, and the 'spherical' aberration is over a thousand times that of a glass lens. With these limitations a theoretical resolution of 0.1 nm is possible. In practice, and with suitably thin specimens, resolutions of 0.5 to 2 nm may be obtained; with thicker specimens and with whole bacteria they are seldom better than 80 nm (Hillier 1950). For biological purposes, magnifications up to 200 000 are employed. The different opacities in an object recorded on the photographic plate are due to the scattering of the electron beams. The intensity of the transmitted beam, for a given electron speed, is roughly proportional to the thickness and density of the material examined. In a manner analogous to staining of optical microscopic objects, certain biological materials may be increased in density, i.e. electron scattering power, by impregnation with salts of heavy metals. Williams and Wyckoff (1945) introduced the technique of shadowing, whereby a stream of vaporized metal is directed obliquely on to the object. Where it accumulates, the resulting electron-opaque film appears light in the photographic image, and dark where the object has protected the supporting membrane from the vapour. The resulting photograph exactly resembles objects illuminated obliquely by an intense beam of light. Metal shadowing and negative staining are used for whole bacteria and are well fitted for showing surface structure and appendages, such as flagella and fimbriae. The value of these methods is limited by the necessity for completely dry specimens for examination, since the microscope works only in a high vacuum. The freeze-fracture and freeze-etching techniques, developed by Steere (1957) and Moor and his collaborators (Moor et al. 1961, Moor and Mühlethaler 1963), allow one to examine 'living' cells and eliminate the need for chemical fixation and dehydration. The technique consists essentially in quick freezing and cleaving of the specimen at low temperatures. 'Etching' is produced by differential sublimation of surface ice, a metal shadow is cast and a carbon replica of the frozen-etched specimen is made which can be viewed in the electron microscope. It has proved particularly useful in visualizing the spatial organization of some cellular components of bacteria.

The scanning electron microscope

For a number of purposes it is important to be able to examine three-dimensional pictures of bacteria including their superficial appearance. The principles of a microscope capable of building three-dimensional pictures by means of a beam scanning the surface of specimens were suggested by Knoll (1935). The first working model of such a machine is usually attributed to von Ardenne (1938). Further developments which eventually led to the commercial production of models by Cambridge physicists have been reviewed by Oatley (1972). Whereas in the transmission electron microscope the electrons from the source are focused so that they pass through the whole specimen, in the scanning instrument the beam is 'demagnified' so that it forms a probe of about 10 µm or less. The instrument is then built so that this probe scans the surface of the specimen building up a picture much in the same way as is the picture on a television screen. When the concentrated probe of electrons hits the specimen, secondary electrons are emitted as the result of events arising from electron-electron interactions in the surface layers. These secondary electrons are collected by a positively charged collector and accelerated into an aluminium coated scintillator. This generates photons which are conducted down a light pipe to give a picture on a fluorescent screen or photographic film. The resolution obtainable is inversely related to the depth to which the probe electrons penetrate the surface which is usually arranged to be about 10 nm. Surface charges on the specimen also effect the emission of secondary electrons and, to overcome this, specimens are usually covered with a thin layer of a conductor such as gold, gold/palladium or carbon. The resolution obtainable with the scanning electron microscope commonly available is of the order of about 10 nm or slightly less. Although very important for obtaining pictures of the surface of tissues and larger cells this resolution is rather too low to give important information about bacterial surfaces. These instruments, however, have a conventional tungsten hairpin filament as the source of the electron beam. The most recent microscopes have a field-emission electron source, using a single tungsten crystal, which by optical analogy approaches a source of infinite brightness. Such instruments are already commercially available having a resolution better than 5 nm. The development of electron microscopes is still being actively pursued, including the development of Scanning Transmission Electron Microscopes (STEM) with resolution at atomic level. A simple account of the principles and use of the scanning electron microscope, as applied to biological specimens, has been published by Hayat (1978). A more detailed account of the principles can be found in Black (1974) and papers on the developments of the field emission source and the STEM in a book edited by Venables (1976).

The general plan of the bacterial cell

Although there is dispute about some of the details, there is broad agreement about the main components of most bacterial cells. The interior of the cell—the protoplast—is differentiated into cytoplasm and nuclear material. Unlike the cells of higher organisms, there is no membrane separating the nucleus from the cytoplasm; neither is there any evidence of an endoplasmic reticulum or of structures corresponding to mitochondria. Membranous organelles—mesosomes—are seen in many bacteria after fixing and are particularly prominent in gram-positive organisms. Outside the cytoplasm are a cytoplasmic or plasma membrane, and the cell wall. The latter gives the bacterium its shape and rigidity. Outside the cell wall many bacteria have a slime layer or capsule which may have a well defined limit or merge gradually into the surrounding medium. Many bacteria have flagella associated with motility and in some species have fimbriae.

The nuclear apparatus

It is paradoxical that our knowledge of the genetic functions, structure, and replication of the DNA in bacteria should be so advanced, while we understand so little about its arrangement in the living microorganism. That the chromosome is circular we know from both genetic and physical examinations of DNA from a variety of species (Cairns 1962, 1963, Kleinschmidt et al. 1961). It, together with a variety of plasmids or episomes, which are also circular, account for all the DNA in bacteria. Despite this we have little trustworthy information as to how the chromosome, some 1000 μm in length, is folded up and packaged into the bacterium.

Early attempts to demonstrate 'nuclei' by conventional staining techniques were generally unsuccessful. A great advance was made by Stille (1937) and Piekarski (1937, 1940), who subjected bacteria to mild hydrolysis, and found regions that gave the positive Feulgen reaction typical of deoxyribonucleic acid.

Much of the affinity of bacteria for stains is due to ribonucleates in the cytoplasm (Vendrely and Lipardi 1946), which can be removed either by acid hydrolysis or by ribonuclease (Tulasne and Vendrely 1947). Robinow developed and extended these staining techniques (Robinow 1945, 1956a, 1956b). The general conclusion was that bacteria contain regions which, like the chromosome of eukaryotic organisms, are Feulgen-positive and stain bright red with Giemsa's stain. These regions correspond both in disposition and number to transparent regions seen when bacteria are examined by dark-phase-contrast microscopy (Mason and Powelson 1956), and also to those regions emitting the appropriate fluorescence, after staining with a variety of dyes that interact with DNA, and irradiating with ultraviolet light.

In thin sections of bacteria, fixed and stained by the Ryter and Kellenberger (1958) technique, the nuclear apparatus is seen as areas of low electron absorption usually scattered quite widely through the cytoplasm (see Fig. 2.3). In some circumstances concerned with the growth and species of the organism these lighter areas may be more compact, and are crossed by fibres which are probably DNA. Undoubtedly, as shown by autoradiography, after labelling the cell with radioactive thymine, the relatively transparent regions correspond to those rich in DNA (Caro et al 1958). Serial sections of bacilli have been thought (Ryter and Jacob, 1964; van Iterson et al. 1975) to show an intimate relationship between the nuclear material as defined above and the mesosomes. In view, however, of our uncertainty about the status of mesosomes, it would be well to withhold judgement on the importance of such an association. However, biochemical evidence would suggest some physical connection between the cytoplasmic membrane and the DNA, in the form of the chromosome origin, both in gram-positive and gram-negative bacteria (see Kleppe et al. 1979 for references). Were such an association proved beyond doubt to be true it would provide powerful support for a very seductive suggestion (Jacob et al. 1963) for the segregation of nuclei after DNA replication, by extension of the intervening envelope to which the DNA is attached.

The isolation of the nuclear apparatus from bacteria in an undamaged, preferably functional form, is an attractive proposition and it is not surprising that much research has grown up around such endeavours (Stonington and Petitjohn 1971, Worcel and Burgi 1972, 1974, Petitjohn 1976). Bodies containing the DNA of the bacterium together with some protein and nascent ribonucleic acid, either encased by envelope components or not, have been obtained from *Esch. coli* by several different methods. Examination by the electron microscope of such materials (e.g. Delius and Worcel 1973, Griffith 1976, Kavenoff and Brown 1976, Kavenoff and Ryder 1976) shows a large number, estimated at 140, of supercoiled loops of DNA extending from a central region containing clearly distinguishable material which can be removed by treatment with RNA'ase. The contour length of the longest loop is 20 μm. These pictures (see Fig. 2.2) suggest that the isolated nucleoids consist of separate supercoiled loops of DNA held together centrally by RNA. Other evidence, however, would make this unlikely to be a picture of the nuclear apparatus existing in the living bacterium since RNA synthesis has been shown to occur in the cytoplasm

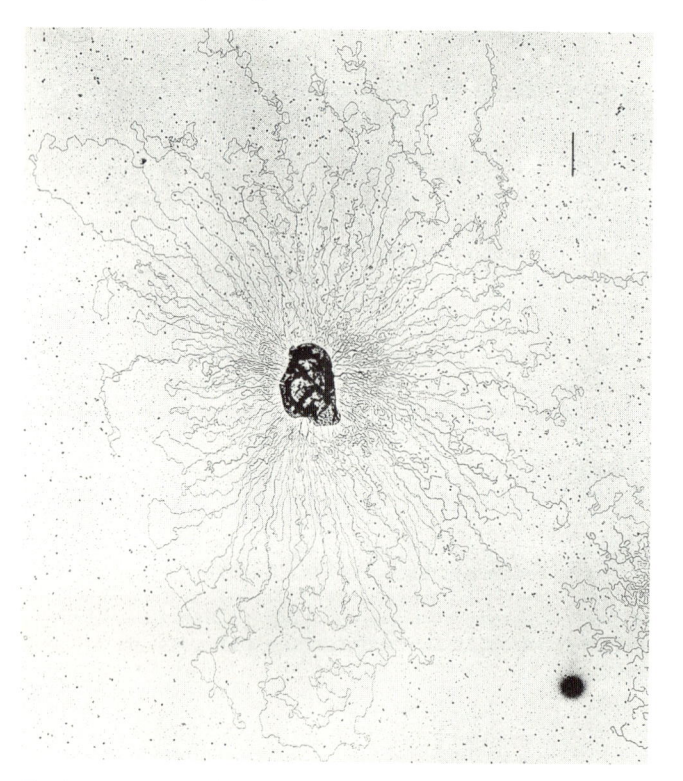

Fig. 2.2 The nuclear material from *Escherichia coli* spread out on salt solution showing the complex organization of the DNA into loops (picture kindly provided by Dr Ruth Kavenoff, Department of Chemistry, University of California, San Diego, La Jolla, California, USA).

rather than within the nuclear region. It has been proposed that such preparations of nuclear apparatus or nucleoids may have been turned inside out by the procedures used for their isolation. A different suggestion has, therefore, been put forward for the organization of the DNA in living cells (Kleppe, Ovrebö and Lossius 1979).

The bacterial cytoplasm

Application of the available methods for preparing and examining bacteria has yielded remarkably little information about the organization of the cytoplasm. In thin sections of material stained and fixed by the orthodox procedures, the cytoplasm appears to be packed with particles known from biochemical studies to be ribosomes, made of RNA's and a number of recognized proteins. Suitable preparations of isolated DNA can be obtained along with its associated messenger RNA to which ribosomes are seen to be attached. Although a high proportion of the ribosomes in growing bacteria is thought to be so attached, such organization cannot be seen in sections of whole bacteria. Freeze-fracture planes passing through the cytoplasm likewise show no particular organization of the ribosomes. (For a review of the structure and evolution of ribosomes, see Wittmann 1982.)

Intracellular granules

A variety of intracellular bodies have been seen. Some are called metachromatic granules, from their property of staining reddish purple with dyes such as polychrome methylene blue. Examination has shown that they are often surrounded by a type of membrane which is very much thinner than normal—being only 2 nm thick. Where these have been isolated and examined chemically they have been shown to be constructed predominantly of a single protein. Particularly valuable work has been done on the membranes of the gas vacuoles that are present in the cyanobacteria (blue green algae) (Walsby 1978).

Packaged within the granules is a variety of substances including polymetaphosphate, polysaccharides, sulphur, β-hydroxybutric acid, and ribulose-1,5-diphosphate carboxylase (Shively 1974). The function of the various granules is poorly understood, apart perhaps from that of gas vacuoles which enable the photosynthetic cyanobacteria in their natural habitats to float to variable depths with different intensities of illumination (Walsby 1978). Crystalline deposits in bacteria also occur, a particularly well-known class being the rhapidosomes. These intracellular structures, appearing as long rigid tubules, were first observed in lysates of strains of *Saprospira* by Lewin (1963). Similar structures have been described in strains of *Proteus* (van Iterson et al. 1967), *Azotobacter* (Pope and Jurtschuk 1967), *Pseudomonas* (Yamamoto 1967), and *Cl. botulinum* (Iida and Inoue 1968). Lewin distinguishes two types—one, a long hollow cylinder, 200 × 30 nm, and the other, a solid rod of the same dimensions, with projecting tails, circa 300 nm in length. Cog-wheel structures, reminiscent of certain phage tail components, are present in some preparations. The rhapidosomes found in *Saprospira* consist of RNA and protein in an approximate ratio of 1:2 (Correll and Lewin 1964). It has been suggested that rhapidosomes represent structural components of defective phages, but proof of their viral origin is lacking.

The cell walls and membranes of bacteria

Traditionally the outer layers of bacteria are classified into cell walls and the cytoplasmic membranes. This is satisfactory for gram-positive species but not for gram-negative organisms in which the cell envelope is sufficiently complex to provide difficulties in accepting these terms without qualification. For species such as the halobacteria and methanogenic bacteria the situation is worse and exact distinction between components of the surface layers becomes difficult, partly because of the rather poorly understood molecular organization of the wall. Acholeplasms and L-forms of bacteria have no wall that can be distinguished either morphologically or chemically. They are usually regarded as surrounded by a membrane having most of its properties in common with the cytoplasmic membrane of bacteria.

The cell wall

The presence of an outer relatively stiff coating around bacteria was recognized as early as the end of the nineteenth century. When motile rod-shaped organisms bumped into objects they did not bend, and when gram-negative organisms were suspended in strong solutions of neutral substances the cytoplasm shrank away from an outer layer that still maintained the original shape of the bacterium. Many attempts were made to find stains that would distinguish walls from the cytoplasm but the results were unsatisfactory owing to the limited resolution of the light microscope and the intense staining of the cytoplasm. It was not until the development of electron microscopes and of methods for preparing bacteria for examination had progressed some way that detailed structure and substructure of bacterial walls could be seen. The walls of all species that stain positively by the gram method appear as featureless layers outside the cytoplasmic membrane (see Fig. 2.1). In exactly transverse sections of rapidly growing bacteria, the wall is about 20–30 nm thick, but can be much thicker especially in old cultures. This is easily understood when it is realized that wall synthesis can continue unabated even when protein and nucleic acid synthesis cannot occur. This extra wall is added by increasing thickness more or less evenly over the whole cell. When bacteria are fixed and stained by techniques such as that introduced by Ryter and Kellenberger (1958), employing osmium tetra-oxide and uranium salts, the wall frequently appears with darker inner and outer layers (see Fig. 2.1). These can be seen even when unstained material is examined; several explanations have been put forward to account for them in terms of wall chemistry. The consensus of opinion would now be, however, that they do not correspond to any difference in the distribution of the polymers making up the wall (Rogers et al. 1980). Sub-structures have been claimed to be present in the walls after various staining and fixing treatments, their reality however is not generally accepted. When partly hydrolysed by treating with enzymes, such as lysozyme, the walls of some rods appear to disintegrate into ribbon-like material.

The region immediately outside the cytoplasmic membrane of gram-negative species in contrast to gram-positive ones is multilayered (see Fig. 2.3). The most prominent feature is an outer membrane having the characteristic trilaminar appearance of all membranes seen in sections of organisms. It is estimated to range in thickness from 5 to 10 nm according to the bacterial species (Glauert and Thornley 1969). Underlying this outer membrane is a relatively electron transparent region of a thickness again varying with the species. Then comes the so-called rigid or strengthening layer, usually 3–5 nm in thickness, to which in carefully fixed dehydrated and stained material, the cytoplasmic membrane is tightly fixed. Nevertheless, gaps have frequently been described between the cytoplasmic membrane and the strengthening layer. These have sometimes been referred to as a morphological realization of the otherwise physiologically defined periplasmic space. After organisms have been treated with chelating agents and shock treatments a large variety of enzymes, such as ribonucleases, phosphatase and β-lactamases, are rendered soluble (Heppel 1967, 1971). In growing bacteria they remain closely associated with the cells. They are referred to as periplasmic enzymes, and probably arise from between the rigid layer and the outer membrane rather than from beneath the former.

The rigid layer maintains the shape of all the bacteria except the halophiles and methanobacteria and contains peptidoglycan, also commonly known as mucopeptide or murein. Among gram-positive bacteria its proportion of the dry weight of the walls varies from about 40–60 per cent and exceptionally up to as much as 80–90 per cent in organisms like *Micrococcus luteus*. Among gram-negative species, it forms a very much lower proportion from about 5 to 15 per cent of the wall. The role of these polymers in providing the strength and shape maintaining properties of the wall is illustrated by the effect on bacteria of enzymes that hydrolyse them specifically, such as lysozyme and

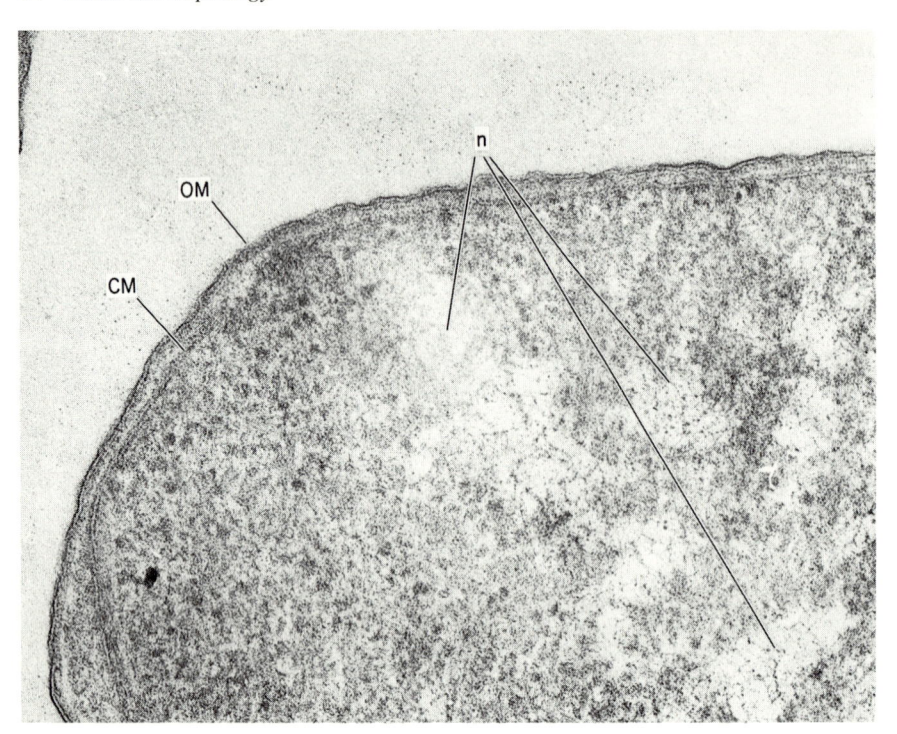

Fig. 2.3 A transverse section of *Escherichia coli* showing the outer membrane (OM), the cytoplasmic membrane (CM) and the light areas crossed by strands which is the nuclear material (n) of the bacterium. (Magnification 92 000) (Picture kindly supplied by Dr Ian Burdett of the Division of Microbiology, National Institute for Medical Research, Mill Hill, London.)

bacterial autolysins. Thus when bacteria are suspended in solutions containing high concentrations of sucrose or salts, all species are converted to spherical protoplasts or spheroplasts (Weibull 1953). Further evidence is that isolated walls have shapes similar to those of the bacteria from which they came, and are readily dissolved by enzymes that specifically attack identified chemical bonds in peptidoglycan.

Cell walls can be isolated from gram-positive bacteria by mechanical disintegration, removal of the insoluble material by centrifugation and washing successively with hot anionic detergent solutions, buffers and water. The isolation of the supporting layer from gram-negative bacteria requires more vigorous treatment with anionic detergent solutions and differential centrifugation.

The peptidoglycans from a very wide range of bacteria with one or two exceptions are built with a common general structure. They all have polysaccharide strands made of substituted glucosamine and muramic acid joined together by 1→4 β-linkages. These polysaccharide chains are linked together by short peptides, often heptapeptides containing both L and D enantiomorphs of the amino acids. Thus it is presumed a three-dimensional network completely covering the bacterium is constructed. The amino groups of the sugars are usually *N*-acetylated although in *Mycobacterium* and *Nocardia N*-glycolyl groups occur instead of *N*-acetyl and in *Bacillus cereus* a high proportion of unsubstituted glucosamine is present. The structure of the peptidoglycan present in all gram-negative bacteria and in bacilli such as *Bac. subtilis* is shown in Fig. 2.4 which serves as an illustration of the general structure of the class of compounds. A very limited number of amino acid types is found in each peptidoglycan; L and D amino acids alternate in the peptides, which are attached to the glycan chains by pseudo-peptide linkages to the carboxyl groups of the *N*-acetylmuramic acid residues. They are further joined together, or cross-linked, as it is known, through the D-alanyl terminus of one peptide and a free amino group, such as the 6-amino group in the L centre of the 2,6 meso diaminopimelic acid, of an amino acid residue in another. Two or three different cross-linkages have been found in peptidoglycans from various species of bacteria. Although the types of amino acid in any one peptidoglycan are very limited, the total number of amino acids now recognized in materials from the wide range of species is large. On the basis of the natures of the cross-linkages Ghuysen (1968) recognized five general types of peptidoglycan. A more detailed classification has been proposed by Schleifer and Kandler (1972).

Peptidoglycans from the majority of sources have other polymer molecules linked to them. In gram-positive species teichoic, teichuronic acids and polysaccharides are linked, sometimes by special small linker molecules to the 6-hydroxy positions of the *N*-acetylmuramic acid residues in the glycan chains. The teichoic acids (Archibald 1974) are a group of phosphate-containing polymers, the commonest of which consist of polyglycerol or polyribitol phosphate with carbohydrates glycosidically attached to one of the hydroxyl groups. The amino acid D-alanine is also attached to the polyol chain by a very alkali-labile ester linkage. A wide

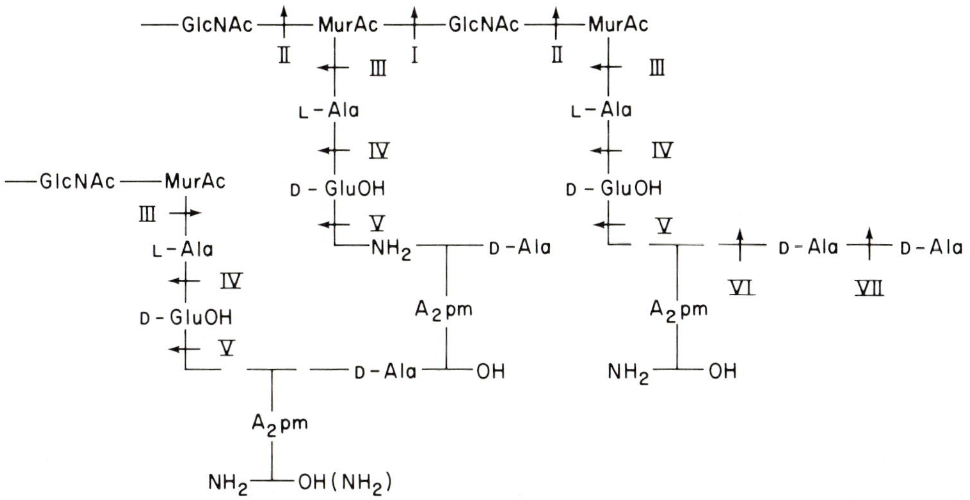

Fig. 2.4 The structure of peptidoglycan as seen in many bacilli and most gram-negative species. The Roman numerals represent points of attack of autolytic enzymes some of which are present in all organisms.

variety of teichoic acids has been discovered, including those with alternating residues of glycerol and carbohydrate in the main chain of the polymer, one made of N-acetylamino sugar phosphate alone and another based on mannitol phosphate (Anderton and Wilkinson 1980). The teichuronic acids are acidic polysaccharides containing uronic acids. Although they occur in small amounts in bacterial walls in ordinary batch grown cultures they often become the predominant polymers linked to peptidoglycans when organisms are grown in continuous cultures that are limited in their growth rate by the supply of inorganic phosphate. Teichoic acids are then almost completely replaced by teichuronic acids (Ellwood and Tempest 1969; Forsberg et al. 1973, Rogers et al. 1980. The teichoic acids in the walls act as immunogens and in some species as bacteriophage receptors; various speculations exist about their possible role in regulating the supply of cations to bacteria. Several organisms have been found to have more than one type of teichoic acid.

No gram-negative bacterium examined has been found indubitably to form teichoic or teichuronic acids. Instead, they have a unique polypeptide linked to the diaminopimelic acid residues in the peptidoglycan layer, as well as being present in the outer membrane (Braun and Rehn 1969, Inouye 1971, Inouye et al. 1972; Braun et al. 1970). This polypeptide has an unusual amino acid composition containing only ten different species, and at its amino terminus there is a cystine-containing lipid (Braun 1975). On average there is about one molecule of the lipoprotein for each ten residues of peptidoglycan. One of the functions of these molecules appears to be to form linkages between the peptidoglycan layer and the outer membrane. In mutants of Esch. coli that are deficient in making lipoprotein the outer membrane swells away from the cell particularly at the points at which the bacteria are dividing (Weigand et al. 1976).

The outer membrane

The composition and functions of the outer membrane of gram-negative bacteria, in particular Esch. coli, have been intensely investigated and a number of unique properties have been discovered. It can be isolated by methods many of which are modifications of that described first by Miura and Mizushima (1968). The organisms are first treated with lysozyme and EDTA in sucrose solution, the resulting spheroplasts are burst in a solution of 5 mM $MgCl_2$, the mixed membranes are washed, dialysed against 3 mM EDTA, and fractionated on a linear gradient of sucrose. Three bands of membrane result, the heaviest consisting of outer membrane, the lightest the cytoplasmic membrane and the intermediate one a mixture of both. A radically different method is that introduced by Schnaitman (1970) which depends on the insolubility of the outer but not the cytoplasmic membrane in detergents.

The outer membrane is distinguished from the cytoplasmic membrane by its absence of enzymic activity apart from a protease and a phospholipase (Albright et al. 1973, White et al. 1971, Bell et al. 1971) and by the presence of lipopolysaccharide. The lipid A of this latter material almost entirely replaces normal phospholipids in the outer leaflet of the membrane bi-layer (Smit et al. 1975; Mühlradt and Golecki 1975). Four or five unique proteins are present in the membrane in very much larger amounts than any of the others (see Osborn and Wu 1980 for a review of the nature of the outer membrane proteins). Although the molecular weights of these major proteins differ among species, the presence of a few dominating ones seems likely to be universal among gram-negative species. The known functions of the outer membrane are (1) to exclude hydrophilic molecules above a certain molecular size—in Esch. coli and Salmonella typhimurium, those above about 600 daltons, although this limit is likely to be higher in organisms other than the Enterobacteriaceae (2) to exclude hydrophobic compounds (3) to harbour specific receptors for bacteriophages and colicines in Esch. coli and similar lysogens in other species (4) to harbour proteins responsible for the passage of specific groups of substances such as nucleotides and Fe^{3+}-chelate compounds. Individual outer membrane proteins frequently appear to be able to perform both functions (3) and (4).

26   Bacterial morphology

Two of the major outer membrane proteins, or in some species one, are organized to form pores, these proteins are known as porins. The diameter of the pores sets limits to the size of hydrophilic molecules that can pass through the membrane. Another of the major proteins is the lipoprotein already described as covalently attached to the peptidoglycan. The interrelations between the lipoprotein attached to the peptidoglycan and that in the outer membrane are not fully understood. The limited permeability to hydrophilic substances and the impermeability to hydrophobic ones probably explain the relative resistance of gram-negative species to antibiotics such as some of the β-lactams, despite the sensitivity of the target proteins to them. These aspects of bacterial walls are more fully discussed in Rogers, Perkins and Ward (1980).

Patterned outermost layers

Examination of negatively stained or freeze-etched preparations of many bacteria of both gram-positive and -negative species shows that their surfaces are covered by a very regular patterned layer: Glauert and Thornley (1969), Sleytr (1978), Thorne (1977); (see Fig. 2.5). Gentle treatment with aqueous solutions of guanidine hydrochloride or urea will remove the layers, which can then be examined chemically.

The patterned layers consist of single proteins of molecular weights of, from 67 000 to 150 000 daltons. Suitable treatment of solutions of them leads to re-aggregation of the sub-units into sheets of material with the same patterns as those seen on the surface of the organisms. If the protein solutions are applied to bacteria from which the layers have already been stripped, the proteins will re-associate with the organisms under defined conditions to give the patterned outer layers again. The conditions for obtaining aggregation into sheets and re-association with the bacteria are closely defined but are not always the same. The function of these layers is not at all clear but in *Bac. sphaericus* they appear to be related to bacteriophage adsorption (Howard and Tipper 1973).

Fig. 2.5   The superficial patterned layer of *Desulfotomaculum nigrificans* as seen after freeze-etching magnification × 88 000. (Picture kindly supplied by Dr U. B. Sleytr, Center for Ultrastructure Research, University of Agriculture, Wien, Austria.)

The cytoplasmic membrane

The presence of a distinct membrane surrounding the cytoplasm of bacteria that has many properties in common with other cellular membranes was implied by the isolation and properties of protoplasts (Weibull 1953, Corner and Marquis 1969). The predicted membrane can now be clearly seen in transverse sections of both gram-negative and -positive species of bacteria as a trilaminar structure with a thickness of 7-8 nm. In freeze-fracture replicas it appears as a structure studded with particles on the convex face of the fracture and pits on the concave fracture face. Since the work of Branton and his colleagues (Branton 1966, Deamer and Branton 1967, Pinto da Silva and Branton 1970) it has become accepted that the fracture plane in membranes runs within the hydrophobic heart of the membrane splitting the lipid bi-layer. The particles seen as studs in such preparations are almost certainly proteins buried in the lipid bi-layer. The cytoplasmic membrane can be prepared from gram-positive species free from wall constituents by first forming protoplasts, washing and then bursting them by diluting the sucrose solution in which they are suspended. The membranes can be deposited subsequently by centrifuging. The complete removal of cytoplasmic constituents, however, such as ribosomes, is much less easy and criteria for such contamination are difficult to establish. Membranes have been described as a lipid sea with proteins floating in it. At a conservative estimate some 30-40 proteins can be demonstrated in membrane preparations by unidirectional polyacrylamide electrophoresis and 200 by the application of iso-electric focusing and polyacrylamide electrophoresis, used in a bidirectional fashion. The detection of contamination by small amounts of cytoplasmic proteins is, therefore, a problem of some magnitude which has never been completely solved.

Chemically the lipids of bacterial cytoplasmic membranes are distinguished from those from other sources by the following characteristics: (1) the absence of sterols except in those from the wall-less acholeplasmas; (2) the almost universal absence of the phosphatidylcholine (lecithin) and sphingomyelins; (3) the presence of a large range of carbohydrate-containing lipids and phospholipids (Shaw 1970, 1975); (4) the scarcity of polyunsaturated fatty acids and the abundance of branched chain and cyclopropane fatty acids. As would be expected, cytoplasmic membrane preparations show a multitude of enzymic functions such as, for example, a wide range of dehydrogenases, phosphatases, proteases and the ability to oxidize succinate and NADH by using molecular O_2 as electron acceptor. A complement of unique cytochromes is present except in those organisms not using them for electron-transport. Apart from energy supply to the cell the cytoplasmic membrane is also responsible for regulating the passage of metabolites into and out of the cytoplasm, including that by active transport which allows bacteria to take up molecules against a steep concentration gradient between the surrounding medium and the cytoplasm. This process is highly specific, and the membranes have a variety of proteins fixed within them that can combine specifically with metabolites such as amino acids, sugars and sugar derivatives. Some of these proteins have been isolated and studied. The so-called 'sided-ness' of cytoplasmic membranes is now well recognized, most of the enzymes they contain are not functionally expressed on both sides of the membranes. By comparing enzymic attack on substrates that are permeable and others that are not, a distinction can be made between inward (cytoplasmic) and outer (medium) facing enzymes. More general immunological methods have also been designed and used to study the problem (see Rogers et al. 1980).

Mesosomes

Sections of bacteria fixed and stained by any of the usual procedures show intrusions of the cytoplasmic membrane filled with either whorls or circles of further membranous material (see Fig. 2.6). Corresponding structures can be seen by negatively staining bacteria. They are very much more prominent in gram-positive than in gram-negative bacteria, although their presence has been described in most species. These structures have been called peripheral bodies (Chapman and Hillier 1953), chondroids (van Iterson and Leene 1964) or mesosomes (Fitz-James 1960). The latter name is the one that has persisted. Much was hoped for the functions of these structures but so far few or none have been proved. Indeed it would appear rather unlikely that they exist at all, in the form described, in living growing bacteria. Nevertheless, material corresponding in morphological form to the internal membrane of mesosomes can readily be isolated and shown to differ from the cytoplasmic membrane in its enzymic functions, in the antibodies it can provoke and in the proteins it contains. The isolation procedure simply involves making protoplasts, removing these at low centrifugal forces, and then centrifuging the resulting supernatant fluids at high speeds. Further purification on gradients of sucrose or caesium chloride is desirable to separate mesosomal membrane from flagella and vesiculated cytoplasmic membrane. Many of the electron transport functions of the cytoplasmic membrane are either missing or present as very low activities in the mesosomal membrane. Among these may be mentioned succinic dehydrogenase, and many of the cytochromes (Reaveley and Rogers 1969, Fernandez Fréhèl and Chaix 1970, Owen and Freer 1972). These observations are sufficient to make one hypothetical function of mesosomes very dubious, namely that they are equivalent in function to mitochondria in eukaryotes, since, of course, many of the special properties of mitochondria are missing. Vital enzymes concerned with wall synthesis are also missing from mesosomal membrane, an absence that rules out a hypothetical role of mesosomes in cell wall formation.

When bacteria that are usually seen to contain mesosomes are not treated with fixing or staining reagents but instead are frozen immediately to $-70°$ and studied by the

28 *Bacterial morphology* Ch. 2

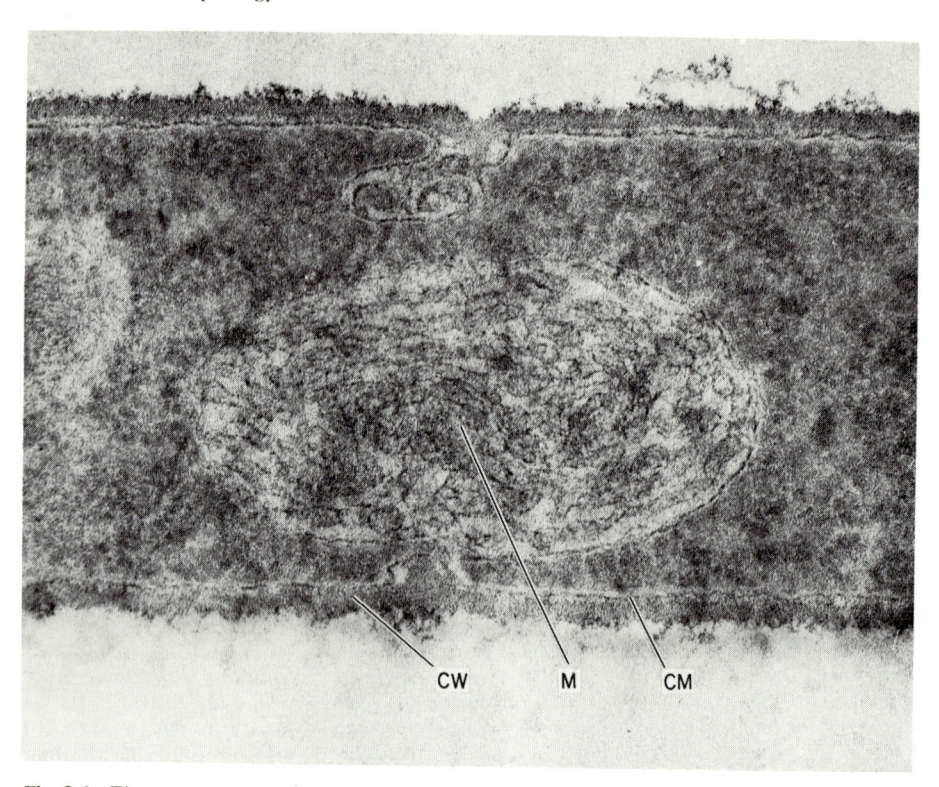

Fig. 2.6 The mesosome (m) of *Bacillus subtilis* as seen in longitudinal section of bacteria pre-fixed with glutaraldehyde and fixed with O_sO_4. Cytoplasmic membrane (cm), wall (cw). Magnification ×100 800. (Picture kindly supplied by Dr Ian Burdett, Division of Microbiology, National Institute for Medical Research, Mill Hill, London.)

freeze-fracture technique, no mesosomes are seen (Nanninga 1969, 1970). When the organisms have been fixed then mesosomes can be seen—observations suggesting that the process of fixing itself in some way causes their appearance. An experiment confirming this was done by Higgins, Tsien and Daneo-Moore (1976), who added glutaraldehyde, a common fixing agent, to cultures of *Streptococcus faecium* at 37° and then removed the bacteria at various intervals of time, freezing them to −70° and freeze-fracturing. Mesosomes appeared at the same rate as a radioactively labelled amino acid added at the same time as the glutaraldehyde became combined with the bacteria in such a way that it could not be removed by washing. It appears that the cross-linking reaction brought about by the glutaraldehyde both fixed the amino acid and caused the mesosomes to appear. However, the mesosomes always occurred at the same place in the cells, that is at the point where the bacteria divide. Thus, some special property or a cloud of imperfectly formed membrane material is likely to be in this area and is condensed by glutaraldehyde into the visible mesosomes (see Fig. 2.6).

Endospores

The formation of spores by bacteria was first observed by Cohn (1875), and first studied in detail by Koch (1876) in the anthrax bacillus. Spore formation is a distinguishing character of certain bacterial groups. The situation of the spore, which may be terminal, subterminal, or approximately equatorial, lends a distinctive morphology to many bacterial species. The spore is often wider than the bacillus which contains it, so that the cell is distorted. According to the position and diameter of the spore, 'drum-stick', 'barrel', or other irregular forms may be presented.

The mature spore is a multilayered structure, the innermost part—the spore protoplast or core—contains DNA, RNA and protein. One completed chromosome is present, much of the protein is ribosomal but a number of enzyme activities have been detected. The protoplast is surrounded by a membrane, which in turn is surrounded by the *germ cell wall* containing peptidoglycan of a similar structure to that in the vegetative cell wall (Tipper and Gauthier 1972). Outside this layer is the thick cortex (see Fig. 2.7) which has a large proportion of a highly modified peptidoglycan and makes an important contribution to spore properties (Pearce and Fitz-James 1971, Murrell and Warth 1964, Frese *et al.* 1970). Cortical peptidoglycan is distinguished from that in the wall of the vegetative cells of *Bac. sphaericus* and *Bac. subtilis* by (1), the presence of muramic anhydride which replaces every third muramic acid residue in the glycan, (2) the presence of incompleted peptide chains, which consist simply of L-alanine, and (3) a very low degree of cross-linking of its peptides (Warth and Strominger 1969; Tipper and Gauthier 1972). No teichoic acid or other polymers appear to be attached to this peptidoglycan. Outside the cortex is a membrane, and outside this the *spore coats* and *exo-sporium*,

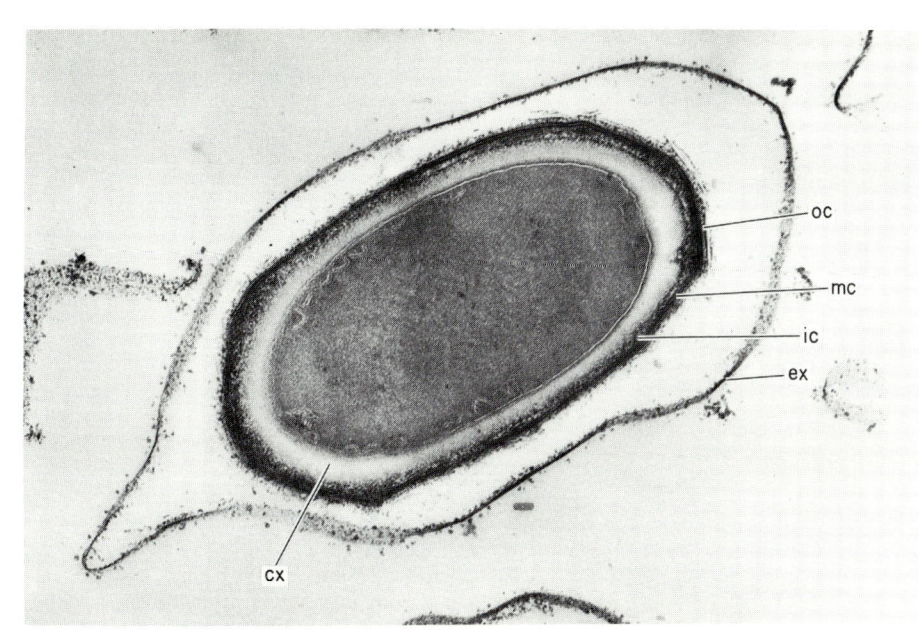

Fig. 2.7 Cross-section of a mature spore of *Bacillus cereus* showing the exosporium (ex), the outer (oc), Middle (mc) and inner coats (ic). The thick cortex is shown as (cx). Magnification 60 000. (Picture kindly provided by Professor W. G. Murrell, CSIRO Division of Food Research, Box 52, N. Ryde, N.S.W. Australia.)

the former of which contains, in *Bac. cereus*, *Bac. megaterium* and other bacilli, several very hydrophobic proteins (Murrell 1969, Aronson and Horn 1972, Kondo and Foster 1967, Kondo and Nishihara 1970). The outer layers of the spore are frequently sculptured into elaborate patterns (Bradley and Williams 1957, Holt and Leadbetter 1969).

Spores are dehydrated (Murrell and Scott 1957) and many consider that this has an important bearing on their well-known heat resistance. Mechanisms by which the dehydration may occur have been proposed and tested (Gould and Dring 1975). A further factor likely to be of importance to heat resistance, judging from the behaviour of mutants unable to synthesize it (Black et al. 1960, Wise et al. 1967) is dipicolinic acid (pyridine-2,6-dicarboxylic acid). This compound was found uniquely present in spores by Powell (1953); it forms insoluble calcium salts. Spores deficient in dipicolinic acid mature but are not heat resistant. Another of the unique characteristics of spores is their ability to concentrate Ca^{2+}, but not Mg^{2+} ions; the reverse is true for the vegetative cell (Eisenstadt and Silver 1972, Ellar 1978). Certain peptidases capable of hydrolysing the main peptide chain in peptidoglycans are also formed during sporulation, and in *Bacillus sphaericus* the enzymes concerned with 2,6-diaminopimelic acid formation and its incorporation appear. This is the amino acid present in the cortical peptidoglycan whereas the wall of the vegetative bacterium contains lysine. The enzymes concerned with lysine incorporation into peptidoglycan disappear during the onset of sporulation.

The morphological stages in spore formation have been clearly defined. In the first stage the DNA of the cell assembles into an axial filament. An asymmetric double membranous septum is then formed. Stage II is associated with a ring-like intrusion of the wall. In *Bacillus*, where the process of sporulation has been most thoroughly studied, the double membrane is associated in fixed and stained material with a large mesosome. Ballooning of the inner membrane then occurs towards the longitudinal centre of the cell. Eventually this membrane becomes detached from the vegetative cytoplasmic membrane and wall to form a separate prespore which is, as it were, subsequently engulfed by the cytoplasmic membrane of the vegetative cell (Stage III). The thick spore cortex forms between the two membranes (Stage IV) whilst the outer membrane forms the spore coats. The validity of the definition of five morphological stages in sporulation is supported by detailed genetic studies (Piggot and Coote 1976). A large number of mutants disturbed in each stage have been isolated; altogether it has been predicted that some 30-40 operons are likely to be involved in the process in *Bac. subtilis*, (Hranueli, Piggot and Mandelstam 1974, Piggot and Coote 1976). Recent ingenious studies of the transformation of sporulating cells suggest that expression of genes both in the mother cell and in the forming spore are necessary to effect sporulation (De Lencastre and Piggot 1979). A number of enzymes such as, for example, alkaline phosphatase, are greatly increased in activity at specific stages in sporulation but it seems unlikely that biochemical deductions about their direct involvement in the process can yet be drawn (Mandelstam 1976).

The total process whereby dormant spores revive and give rise to a new vegetative cell can be divided into three recognizable stages: activation, germination and outgrowth.

Activation. Freshly harvested spores are reluctant to germinate even under the most favourable circumstances. Germination rate is a function of ageing but can be greatly accelerated by heating, such as at 60° for one hour, by lowering the pH, or by treatment with mercaptoethanol. Metabolic inhibitors do not prevent activation and it is thought that metabolic activity is not necessary. It has been speculated that conformational changes in the spore macro-molecules may occur (Keynan and Evenchick 1969). No morphological changes resulting from activation have been described.

Germination. When activated spores are exposed to certain substances, such as amino acids, nucleosides or sugars, which have a degree of species or even strain-specificity, the process of germination starts. It consists in the loss of refractility and heat resistance and results in a body otherwise looking like a spore but which can be stained readily.

This stage, unlike activation involves active metabolism. It can be halted by metabolic poisons (see Vinter 1970) but not by inhibitors of protein synthesis, such as chloramphenicol, or of ribonucleic acid synthesis, such as actinomycin D (Steinberg, Halvorson, Keynan and Weinberg 1965). The cortex of the spore is mostly removed, possibly by the hydrolytic action of a β-Nacetyl glucosaminidase (Brown et al. 1978). Indeed an early observation was the excretion of peptidoglycan fragments by germinating spores of bacilli (Powell and Strange 1953, Strange and Powell 1954, Strange and Dark 1957a, & b). Germination-specific proteases are also formed which in the very short period of 3 min hydrolyse two specific spore proteins to amino acids (Setlow 1975a 1975b 1976). It has been calculated that the total process of converting the inert but activated, refractile, spore to a metabolically active, stainable, phage dark body occupies no more than 50 seconds (Vary and Halvorson 1964). In some ways the germinated spore can be looked upon as a protoplast but of spore origin and still sheathed in a spore-coat. Nevertheless, a number of metabolic differences exist between it and a vegetative protoplast. For example, there is evidence that in species such as *Bacillus*, electron transport is flavine-mediated instead of involving cytochromes as in the vegetative cell (see Keynan 1973).

Outgrowth. Under optimal conditions outgrowth of a new vegetative cell from the germinated spore of bacilli will occur in about 50 to 60 minutes (Vinter 1965). The first visible change is swelling of the cell due to water uptake so that the volume increases two to three times. DNA synthesis begins much later than the sequential formation of either protein or RNA, which start promptly. In *Bacillus* sp., DNA synthesis is delayed 120–160 minutes after germination (Keynan 1973). Examination (Cleveland and Gilvarg 1975) of wall formation during outgrowth suggests that the germ cell wall may be preserved during germination for use during outgrowth in the vegetative wall of the new cell.

Flagella

A large number of bacteria, including a few coccal forms, many strains of *Bacillus*, and most known spirilla and vibrios, are motile by means of flagella. These are long, thread-like processes, of uniform diameter, arranged in various ways on the bacterium. They vary greatly in length but are often longer than the organism to which they are attached. They may be sheathed or unsheathed. The type of flagellum is characteristic of a strain rather than a species; for example, some strains of *Pr. vulgaris* have unsheathed flagella (Astbury et al. 1955), whereas in other strains they are sheathed (Lowy and Hanson 1964, 1965). Unsheathed flagella have diameters ranging from 12 nm in *Pr. vulgaris* to 19 nm in a motile diphtheroid organism (Starr and Williams 1952). Sheathed flagella can have diameters as great as 55 nm, as in *Bdellovibrio bacteriovorus* (Seidler and Starr 1968), and consist of a core surrounded by a sheath which can be removed by treatment with autolysins, acid, or urea (Follett and Gordon 1963). Flagella have a cylindrical helical shape with a wavelength characteristic of the species, although in some organisms they are of two different wavelengths (biplicity) (see also Leifson 1959). These thread-like processes are arranged in one of four ways on the individual. There may be a single process, this arrangement being called monotrichate; one at either end, the amphitrichate condition; a bunch of flagella at one end, or more rarely at both ends, the lophotrichate condition, Lastly, the flagella may be arranged indiscriminately over the whole bacterial cell, the peritrichate form.

The flagellar bodies of *Esch. coli* and *Bac. subtilis* have been subject to intense investigation and three parts have been recognized. The filament which is readily seen, a hook-like body, to which it is attached, and a basal structure (Fig. 2.8). The filament is built from sub-units of a single protein, flagellin, which consists of sub-units of molecular weight ranging from about 30 000 daltons to 60 000 daltons in different species (Iino 1977), and in negatively stained preparations they appear as spherical bodies of 4.5 nm in diameter (Kerridge et al. 1962). These sub-units are arranged in helices to form the fibres (Finch and Klug 1972, Kerridge et al. 1962) of the flagella which are themselves helical. In section their packing appears hexagonal (Kerridge et al. 1962). In preparations subjected to ultrasonic treatment the sub-units are arranged around a hollow space but when ultrasonic treatment is avoided this space appears to be filled (Abram and Koffler 1964) with material which is presumably rather easily removed during ultrasonic treatment. The amino acid composition of flagellins from different species differs but cysteine and tryptophan are always absent. The unusual amino acid N-methyl-lysine is present in the flagellin from some serotypes of salmonellae but not others. The flagellar fibres are rather readily dissociated into soluble flagellin by a variety of treatments including those with acid, alkali, sonic oscillation, detergents, phenol, heat and agents like strong solutions of urea that weaken hydrogen bonds (Iino 1969a). Soluble flagellin can then be reassociated to form flagella-like fibres under a variety of conditions. For example, the flagellin from *Bac. pumilus* reassociates to fibres at temperatures lower than 26° giving either straight fibres at pH 4.0–4.9 or flagella-like helical fibres at pH 5.3–6.2. Flagellin from salmonellae will give flagellar fibres when high salt concentrations are present at neutral pH. At low salt concentrations (e.g. 0.05M phosphate) fragments of undissociated flagella must be

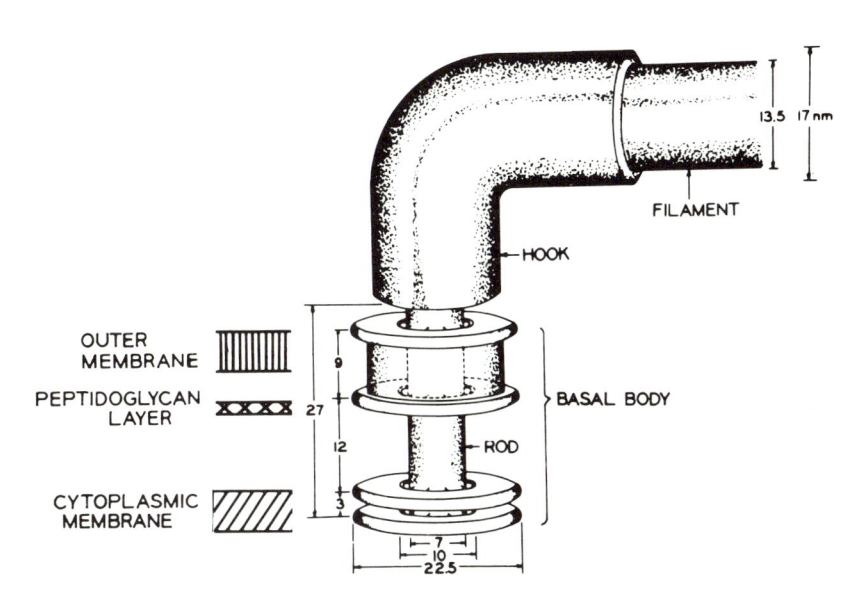

Fig. 2.8 Diagram of the flagellar apparatus of *Escherichia coli*. The figures represent nanometres.

added which then extend in length by a process akin to crystallization (Askura *et al.* 1964), appearing to occur from one end only (Abram, Koffler and Vattee 1966, Askura *et al.* 1968). The *in-vitro* growth of fibres from one end is likely to reflect the situation *in vivo*. The presence of *p*-fluorophenylalanine in growth media is known to lead to the formation of abnormal curly flagella. If *p*-fluorophenylalanine is added to cultures of *Salmonella typhimurium* for short times the distal tips of the flagellar fibres become curly. In other words, the new abnormal flagellin produced by incorporation into it of *p*-fluoro phenylalanine, is present in the distal tips of the fibres only. Also, the rate of growth of incompleted flagella in normal media can be shown to be inversely proportional to the length of the fibres (Iino 1969*a*), a result consistent with the feeding of the distal tip by flagellin sub-units passing from the cytoplasm of the cells up the hollow centre of the fibres. This mechanism also presumably limits the length of flagella *in vivo*. Other consistent evidence (Iino 1969*b*) strongly suggests that the flagellin sub-units do not leave the bacteria during flagellar growth.

The hooks and basal structures of flagella were first isolated from *Esch. coli* (De Pamphlis and Adler 1971) and *Bac. subtilis* (Dimmitt and Simon 1971) and although similar in construction show one clear difference. Whereas the basal structure of *Esch. coli* consists of a short rod, bearing two pairs of discs, that from *Bac. subtilis* has just one pair. The upper pair in the gram-negative organism appears to be associated with the outer membrane or with the lipopolysaccharide layer of the organism, whereas the lower pair is related to the peptidoglycan and cytoplasmic membrane. In the gram-positive *Bac. subtilis*, the lower ring appears to be associated with the cytoplasmic membrane, but exactly where the upper ring is situated is unknown.

The hook structure is slightly larger in diameter than the fibre and is about 900 nm long. In *Esch. coli* and *Salmonella typhimurium* it is composed of a single protein with a molecular weight of 42 000 daltons (Silverman and Simon 1972, Kagawa, Owaribe, Asakura and Takahashi 1976). In *Bac. subtilis*, it has a slightly smaller size of 33 000 daltons (Dimmitt and Simon 1971). Dissociation of the hook structures into soluble proteins is rather more difficult than for the fibres. The basal structures can also be dissociated but the picture is complex and 10–13 proteins can be recognized. Studies of the genetic control of flagella formation are well developed in *Salmonella typhimurium* and to a lesser extent in *Bac. subtilis* (Silverman and Simon 1977, Simon *et al.* 1978, Iino 1977). In general, three groups of genes control the formation and function of flagella, excluding those *che* genes that appear to relate to the flow or information from chemotactic stimuli to flagellar activity. The three groups of mutants directly concerned with flagella formation or action are *hag*, *mot* and *fla*. The *hag* genes appear to control the formation of flagellin, and may result either in bacteria without flagella or with abnormal ones, but with hook and basal structures present; the *fla* strains have either no flagellar structures at all or only partial ones, whilst *mot* strains have normal looking but non-motile flagella. It is estimated that for the enterobacteria more than 25 gene products are required for the assembly and function of the flagellar apparatus (Silverman and Simon 1977). Whether a specific sequence of expression of these genes is required for properly assembled and functioning flagella is not yet sufficiently clear. It is of interest that synthesis of the flagellar apparatus is subject to catabolite repression (Adler and Templeton 1967) and that this is reversed by cAMP (Dobrogosz and Hamilton 1971), thus linking the regulation of the formation of proteins of the flagella apparatus to that of the formation of enzymes such as β-galactosidase. Flagella are not formed in autolytic deficient strains of *Bacillus* (Fein and Rogers 1976, Fein 1979) but flagellin is still present in the cytoplasm probably of these and certainly of other lytic deficient mutants of *Bac. subtilis*. (Ayusawa *et al.* 1975) suggesting that the autolytic enzymes assist in penetrating the wall to allow flagellar assembly. It is now clear that the flagella drive bacteria forward by rotating, for if the tips of the flagella are fixed to a microscope slide by flagella-specific antibody, then the bacteria themselves rotate. Bacteria do not simply swim forward all the time, but after short distances reverse their direction, the so-called tumbling motion. This is caused by reversal of flagellar rotation. The energy that drives this 'flagellar motor' comes not directly from ATP as might be expected, but from the flow of protons giving rise to the proton-motive force predicted by the chemiosmotic hypothesis (Berg 1975, Berg and Anderson 1973, Manson *et al.* 1980, Kahn and McNab 1980).

Fimbriae and pili

The resolution possible by the light microscope is quite inadequate to visualize these structures which are attached to the surface of many, especially gram-negative, bacteria, because their diameter commonly ranges between only 3 and 14 nm. Soon after the electron microscope became commonly available they were discovered independently by Anderson (1949) and Houwink (1949) as fine hair-like appendages extruding from bacterial surfaces. A number of different terms have been used to describe them, but common usage has by now decided that the less numerous rather thicker ones concerned with DNA transfer between bacteria should be called 'sex pili', whilst those not so concerned are either simply, 'pili' or 'fimbriae'. Here they will be called fimbriae.

Fimbriae are formed by many groups of gram-negative bacteria including the Enterobacteriaceae, Pseudomonadaceae, *Neisseria, Proteus, Caulobacter,* *Rhizobium, Chromobacterium, Moraxella,* and *Vibrio* species. Whether exactly similar structures are found in gram-positive bacteria is not clear, although they have been described as occurring in *Corynebacterium renale* (Yanagawa and Otsuki 1970, Yanagawa, Otsuki and Tokui 1968), and *Actinomyces naeslundii* (Ellen *et al.* 1978). Among the enterobacteria fimbriae are usually peritrichously arranged and individual cells may carry anything from a few to 1000 of the appendages (see Fig. 2.9) although in other groups such as, for example, some pseudomonads, they may be polar or bipolar. Their length can vary from 0.2 μm to 2 μm. A number of attempts have been made to classify the various types of fimbriae (see Ottow 1975), partly on the basis of morphology, and partly on their one known function, namely that of causing adhesion of bacteria to surfaces (Duguid, Anderson and Campbell 1966). For example, type 1 are responsible for adhesion and haemagglutination by the organisms (Duguid *et al.* 1955); both properties are inhibited by D-mannose or methyl-α-D-mannoside (Duguid 1959, Duguid and Gillies 1958, Old 1972). They are about 7 nm in diameter. Another type found on some salmonellae seem to have no adhesive properties (Duguid 1968, Duguid *et al.* 1966, Old and Payne 1971); another, specific for *Klebsiella* and *Serratia marcescens* (Duguid 1959, 1968), shows a strong mannose-resistant adhesion to non-living surfaces such as glass and cellulose, as well as to the walls of fungi and plants. These latter fimbriae are thinner than the others—being 4–5 nm in diameter.

Fimbriae are built, like flagella, of sub-units of a single protein called 'fibrilin', which has a molecular weight of

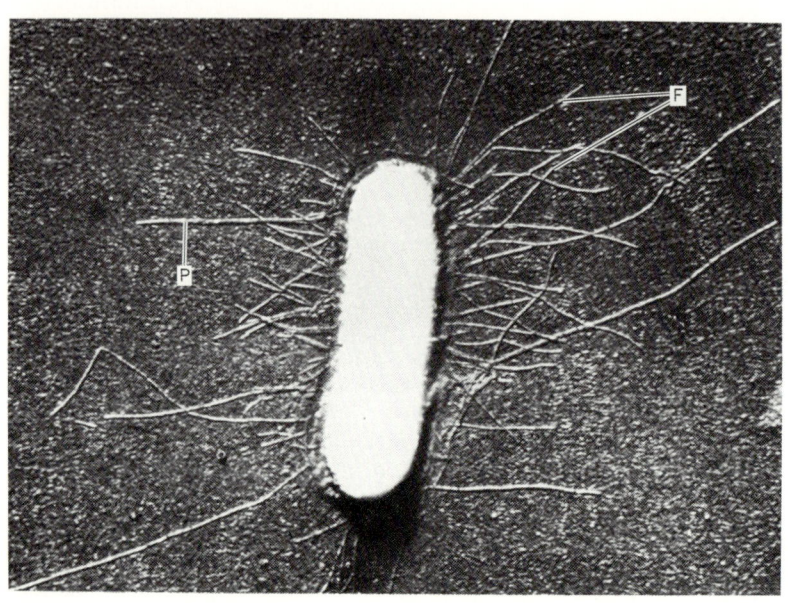

Fig. 2.9 A shadowed preparation of *Escherichia coli* showing the fimbriae (F) and pili (P).

approximately 16 000 (Brinton 1965). Methionine is absent from fibrilin which has a very high proportion of hydrophobic amino acids presumably accounting for the aggregation of isolated fimbriae and their ability to stick to animal cells. Examination by x-ray diffraction and image analysis suggests that the sub-units of fibrilin are so arranged in the structures as again to give helices, similar to flagellin in flagella (Brinton 1965, Mayer 1969). The biosynthesis of fimbriae occurs by first building a pool of fibrilin monomer molecules within the bacteria which is subsequently organized into the filamentous appendages. If, for example, defimbriated *Esch. coli* is treated with chloramphenicol, thus stopping protein synthesis, fimbriae are still re-formed, even after several further defimbriations (Brinton 1965, Brinton and Beer 1967, Martinez and Gordee 1966, Meynell and Lawn 1967). Genetic analysis of the control of the formation of fimbriae (or type 1 pili) identified a chromosomal gene, at first called '*fim*', but subsequently, '*pil*' (Bachmann, Low and Taylor 1976, Maccacaro and Hayes 1961). Complementation analysis of a large number of mutants has defined three closely linked cistrons *pil A*, *pil B* and *pil C* (Swaney *et al.* 1977*a*). Apart from this type of direct chromosomal control of formation however, phase variation also happens and two colony types occur, the bacteria in one being predominantly fimbriated the other not. Bacteria from either type can give rise to both sorts of colony. Although well recognized and studied (Maccacaro and Hayes 1961, Swaney *et al.* 1977*b*) the genetical basis of the phenomenon has not been fully elucidated.

Sex pili

Sex pili, first described by Crawford and Gesteland, (1964), are distinguished from fimbriae by a number of characteristics. They are many fewer and are larger than type 1 fimbriae, being about 9.0 nm in diameter. They adsorb male specific RNA bacteriophages, such as MS2 or R17 along their length and male specific filamentous DNA bacteriophages, like M13, by their tips. They are specified by the transfer factors in plasmids, e.g. F, drug transfer factors, or Col 1, specifying F-like or I-like pili. All sex pili are involved in some way in the passage of DNA from one bacterium to another. Their structure is, in general terms, similar to that of the fimbriae.

The F-specified pili consist of sub-units of a single protein, 'pilin', of molecular weight about 11 000 daltons (Brinton 1971, Tomoeda *et al.* 1975). Cystine, proline, arginine and histidine are all absent from the protein, whilst valine, alanine, glycine, serine and lysine account for 70–80 per cent of its total composition. One mole of glucose and two of phosphate per mole of protein are present (Brinton 1971). The sub-units are likely to be organized helically to give pili, although it has been suggested that they are not so arranged but are in a rod-like form. An axial canal can be seen after negative staining as a dark central line (Brinton 1965, 1971, Valentine and Strand 1965, Lawn 1966). Terminal knobs have also been described (Lawn 1966). The pili are dissociated into monomers by 10^{-3} M sodium dodecyl sulphate but not by reagents that weaken hydrogen bonds such as 8 M urea or 7 M guanidine hydrochloride (Tomoeda *et al.* 1975). There would seem no doubt that the sex pili in F^+ bacteria are in some way connected with the process of conjugation between bacteria; no transfer of DNA has been shown between organisms in their absence. When pili are removed from F^+ cells of *Esch. coli* by blending, mating ability is lost but recovers at the same rate as pili are formed and the appropriate RNA or DNA bacteriophages are adsorbed (Brinton 1965, Novotny *et al.* 1969). The mechanism by which they act in conjugation is quite unclear but two hypotheses have been formulated that can be summarized as: (1) the pilus serves as a simple hollow tube down which the DNA passes from an F^+ cell to a F^- one (Brinton 1965, Brinton *et al.* 1964). In favour of such a hypothesis is the observation (On and Anderson 1970) of recombination in loosely connected pairs although the frequency was higher between cells in close contact. 2) The pili serve only to connect male and female cells and then retract. In doing this they bring the two bacteria close together (Curtis 1969, Marvin and Hahn 1969). DNA is then transferred as a result of the wall-to-wall contact. Evidence in favour of the ability of sex pili to retract has been obtained by showing that sex pili disappear in the presence of cyanide or arsenate and that they do not appear as free structures in the culture supernatant. This disappearance is prevented when an RNA bacteriophage or antibody is first attached to the pili (Novotny and Fives-Taylor 1974). Achtman and Skurray (1977) however, point out that the concentration by workers on mating-pairs may be misleading, since tight groups of cells are much more common in conjugating populations of bacteria than are pairs.

Genetic control of the formation of sex pili has been intensively studied, mostly through analysis of transfer deficient mutants (Tra$^-$). Nineteen *tra* genes on the F-*lac* plasmid in *Esch. coli* have been identified, twelve of them possibly functioning as an operon under the control of the product of the thirteenth, the *tra J* gene (Finnegan and Willetts 1973, Helmuth and Achtman 1975, Willetts and Skurray 1980). Of the twelve mutations, those in cystrons 5–12 lead to absence of sex pili as do mutations in the locus *tra J*. It seems that tra A (cystron 12) is a likely candidate for the structural gene controlling pilin protein formation (Willetts 1971).

Protoplasts and spheroplasts

These forms are all, to varying degrees, sensitive to the osmolality of the medium that surrounds them and can be lysed by sudden lowering of its strength. Their surfaces are devoid of the strengthening polymers, the peptidoglycans. The distinction between spheroplasts and protoplasts as defined by Brenner *et al.* (1958) is that the surface of protoplasts is free of wall components, whereas residual wall materials are still present in spheroplasts. This frequently relates to whether the original organism stained gram-positively or gram-negatively. If the former, the appropriate wall-lytic enzyme is capable of removing the whole wall from the organism, leaving a spherical body surrounded only by the cytoplasmic membrane, as was first discovered by Weibull (1953). A gram-negative organism, such as *Esch. coli*, on the other hand, cannot be attacked directly by wall lytic enzymes, presumably because these are excluded from the sensitive peptidoglycan layer by the outer membrane of the wall. If, however, bacteria are treated with a solution of a substance capable of chelating divalent cations, such as ethylene diaminetetraacetic acid, the lytic enzymes can then break down the peptidoglycan layer (Repaske 1956, Birdsell and Cota-Robles 1967). Not all of the components of the wall, however, are liberated from the resulting osmotically sensitive bodies which are known as spheroplasts. When gram-positive cells are subjected to limited action of lytic enzymes so that some wall is left associated with the bacteria, these have also occasionally been called spheroplasts.

Protoplasts or spheroplasts can also be made by allowing the organism's own autolytic enzymes to remove the wall during incubation of the bacteria in buffer of a suitable pH containing 0.5 M sucrose (Joseph and Shockman 1974); those from *Streptococcus faecalis* have been called 'autoplasts'. They can also be made by treating sensitive organisms with antibiotics that act by inhibiting wall synthesis, such as the β-lactams. In this method too, the bacteria's own autolytic enzymes probably remove the existing wall. Other methods of interfering with wall synthesis such as, for example, the omission of an amino acid present in the peptidoglycan from the medium used for a suitable auxotroph, or the inclusion of certain amino acids, like glycine, in high concentrations in the medium, have also been used to produce protoplasts and spheroplasts.

Protoplasts and spheroplasts are capable of most of the respiratory and biosynthetic capabilities of the original bacteria (McQuillen 1960), but unless specially treated they will not divide, although they can grow in size. The solution commonly used to support them is sucrose at strengths varying from 0.5 to 1.0 M according to the species of bacterium and the investigator. Study, however, of a wide range of solutes (Corner and Marquis 1969, Marquis and Corner 1976) has shown that volume and shape are important molecular properties and that the simple molar concentration, even of very poorly ionized substances, is not alone adequate to predict their ability to protect protoplasts of bacilli.

L-forms

L-forms were discovered and investigated by Klieneberger-Nobel (1935, 1942) in cultures of *Streptobacillus moniliformis*.

She isolated them and maintained them in pure culture and, because of their likeness to mycoplasmas, at first believed them to be a separate species of bacterium, symbiotic with the streptobacillus. She named them L-forms, from the Lister Institute in which she was working. Dienes (1939, 1942), on the other hand, contended that they were part of a life cycle of the streptobacillus, and induced similar forms in *Esch. coli, H. influenzae*, flavobacteria, gonococci, and *Fusobacterium necrophorum*. Klieneberger-Nobel (1949) later adopted this view. L-forms have now been isolated from a wide variety of bacteria, and their properties have been the subject of intensive investigations.

In a few species, e.g. *Streptobacillus moniliformis*, L-forms arise spontaneously, but in most bacteria an inducing agent is needed. The discovery by Pierce (1942) that L-forms could be induced by exposure to penicillin made it possible to isolate them from a wide variety of bacteria; since then a number of inducing agents have been described. In general, those agents that induce spheroplast or protoplast formation can be used to induce L-forms, although lysozyme-induction is sometimes ineffective in that the resulting forms are often incapable of subsequent multiplication. Under the best conditions conversion to the L-form can be almost quantitative (Gooder and Maxted 1958)—thus ruling out the possibility that L-forms thus derived represent a mutant population. Successful isolation and cultivation of L-forms depends on the choice of inducer, osmotic stabilizer, and a suitable culture medium. Primary isolation is made on solid medium containing a low concentration of agar, inactivated serum, and an osmotic stabilizer, which is normally sucrose or sodium chloride. Once the organisms are stabilized, the composition of the medium is less critical, and often the L-forms can be trained to grow in liquid medium. L-form colonies on solid medium show a characteristic 'fried egg' appearance, with a dense centre and a lace-like periphery. The appearance is similar to that produced by mycoplasmas, but the two can be distinguished by staining reactions (Madoff 1960). The organisms owe their peculiar colonial morphology to their burrowing into the interstitial spaces in the agar; the support provided by the agar fibres is apparently essential for their development.

Microscopically, the L-form colony is seen to be composed of many highly pleomorphic elements, differing in both size and shape; each is bounded by a single or double membrane. Dienes (1967) distinguishes three basic elements: the large body, the granule, and the elementary body. Each of these shows great variation in size, e.g. the large bodies range from 1 μm to 50 μm, and may contain fairly homogeneous cytoplasm or may be highly vacuolated; the granules range in size from 0.1 to 1.0 μm and occur inside or outside the large bodies; the elementary corpuscles have a diameter of 0.05 to 0.5 μm. Electronmicrographs have confirmed the existence of these three 'basic' elements (Ryter and Landman 1964, Wyrick and Rogers 1973).

The method of reproduction of the L-forms is still not certain. There are two views. One suggests a reproductive cycle in which the smallest elements multiply within a large body and eventually cause its disintegration, with the release of many small bodies which then begin another cycle of reproduction. The other considers reproduction to occur by fission of the large bodies. In the absence, however, of a rigid cell wall, division often proceeds unequally so that many elements of varying size and complexity, not all of which are viable, are thrown off.

Studies made with filters of varying porosity to measure the diameter of the smallest viable unit (Klieneberger-Nobel 1956, Kellenberger *et al.* 1956, Tulasne and Lavillaureix 1958, Williams 1963, van Boven *et al.* 1968a, Wyrick and Gooder 1971) have given widely varying estimates, ranging from 0.1 to 0.2 μm for *Proteus* L-forms (Tulasne and Lavillaureix 1958) to 0.7 μm or greater for staphylococcal L-forms (Williams 1963). The use of the filtration method has been justifiably criticized by several workers on the grounds that L-form elements are plastic bodies and may be able to pass through the pores of a filter by deformation (Lederberg and St. Clair 1958, Weibull and Lundin 1962, Roux 1960). Van

Boven and his colleagues (1968a) estimated that the size of the smallest viable unit in streptococcal L-forms lay between 0.46 and 0.65 μm; but when the filtrate was examined by phase-contrast microscopy it was seen to contain a few large bodies with diameters from 0.8 to 1.7 μm; these were apparently the only elements in the filtrate that could form colonies (van Boven et al. 1968b). Microscopical studies of developing L-forms of *Vibrio cholerae* (Roux 1960), *Proteus species*, staphylococcal and diphtheroid L-forms (Weibull 1963), and streptococcal L-forms (van Boven et al. 1968b) lead to the conclusion that the small bodies are non-viable, but that the majority of large bodies are able to multiply. Small elements with a diameter less than 0.3 μm were separated by differential centrifugation and found to contain protein and RNA but little, if any, DNA (Weibull and Beckman 1961). On balance, the evidence favours the view that the large bodies with a diameter of approximately 0.7 μm or greater are the viable units, and that growth occurs by unequal fission of these bodies.

One of the characteristics of freshly isolated L-forms from several species of bacteria is that they can revert to the bacterial form. Reversion can sometimes be effected by removal of the inducing agent, e.g. penicillin, from the growth medium. Depending on their ability to revert, L-forms are often referred to as stable or unstable. As a rule, freshly isolated L-forms have a greater tendency to revert than old established lines. Landman (1967) showed that in newly isolated L-forms of a number of species the amount of retained cell wall material varied, and this was directly related to their ability to revert to the parent form. Recent work (Elliot et al. 1975a, b) however, has made any rôle for residual wall much less likely. Reversion can occur in protoplasts and L-forms that lack any detectable cell wall material. Landman and his associates (1963, 1968, 1969) obtained quantitative reversion of protoplasts and L-forms of *Bac. subtilis* by growing them at 26° in medium containing 25 per cent gelatin. Gooder (1967) found that streptococcal protoplasts could be induced to revert by the same treatment, but he was unable to induce reversion of streptococcal L-forms. Landman and Forman (1969) suggest that reversion takes place when cells are placed in an environment in which cell wall material excreted by the organism is prevented from diffusing away from the cells and can polymerize on the exterior of the cells. Reverted strains are usually identical with the parent strain.

A high proportion of protoplasts of *Bacillus licheniformis* with no remaining wall associated with them has been shown to revert when spread on the agar DP medium described by Landman, Ryter and Frèhèl (1968) and incubated at 35° (Elliot, Ward and Rogers 1975a, Elliot, Ward, Wyrick and Rogers 1975b). When, however, such protoplasts are spread on soft agar medium containing a wall-inhibiting antibiotic and repeatedly subcultured they grow as L-forms which eventually will no longer revert (Wyrick and Rogers 1973). DNA from such 'stable' L-forms of *Bacillus subtilis* will transform the original strain of the bacillus to give L-form colonies with a frequency similar to that found for other auxotrophic markers (Wyrick, McConnell and Rogers 1973). This suggests that 'stable' L-forms may have genetic lesions in the pathway for wall formation. Subsequent examination of a number of such L-forms from *B. subtilis* and *B. licheniformis* showed lesions leading to deficiency in specific enzymes involved in wall biosynthesis (Ward 1975). A similar deficiency has been reported in a stable L-form of *Streptococcus pyogenes* (Reusch and Panos 1976).

References

Abram, D. and Koffler, H. (1964) *J. molec. Biol.* **9**, 168.
Abram, D., Koffler, H. and Vattee, A. E. (1966) *J. Bact.* **101**, 250.
Achtman, M. and Skurray, R. (1977) In: *Microbial Interactions*, p. 235. Ed. by J. L. Reissig, Chapman and Hall, New York.
Adler, J. and Templeton, B. (1967) *J. gen. Microbiol.* **46**, 175.
Albright, F. R., White, D. A. and Lennarz, W. J. (1973) *J. biol. Chem.* **248**, 3968.
Anderson, T. F. (1949) In: *The Nature of the Bacterial Surface*. Ed. by A. A. Miles and N. W. Pirie. Oxford University Press, Oxford.
Anderton, W. J. and Wilkinson, S. G. (1980) *J. gen. Microbiol.* **118**, 353.
Archibald, A. R. (1974) *Advanc. appl. Microbiol* **11**, 53.
Ardenne, M. von (1938) *Z., Tech. Physik* **109**, 553; (1938) *Z. Tech. Physik* **19**, 407.
Aronson, A. I. and Horn, D. (1972). In: *Spores V*, p. 19. Ed. by H. O. Halvorson, R. Hansen and L. L. Campbell. *Amer. Soc. Microbiol.*
Askura, A., Eguchi, G. and Iino, T. (1964) *J. molec. Biol.* **10**, 42; (1968) *J. molec. Biol.*, **35**, 227.
Astbury, W. T., Beighton, E. and Weibull, C. (1955) Proceedings of the Ninth Symposium, Society Experimental Biology, p. 282.
Ayusawa, D., Yoneda, Y., Yamane, K. and Maruo, B. (1975) *J. Bact.*, **124**, 459.
Bachmann, B. J., Low, K. B. and Taylor, A. L. (1976) *Bact. Rev.* **40**, 116.
Barksdale, L. and Kim, K-S. (1977) *Bact., Rev. 41*, 217.
Barnard, J. E. (1919) *J. R. micro. Soc* pl; (1925) *Lancet* ii, 117; (1930) In: *A System of Bacteriology*. Vol 1, p. 115. Medical Research Council, London.
Bartholomew, J. W. and Mittwer, T. (1951) *Bact. Rev.* **16**, 1.
Bayer, M. E. and Remsen, C. C. (1970) *J. Bact.* **101**, 304.
Bell, R. M., Mavis, R. D., Osborn, M. J. and Vagelos, P. R. (1971) *Biochem. biophys. Acta.* **249**, 628.
Benians, T. H. C. (1920) *J. Path. Bact.* **23**, 401.
Bennett, A. H., Jupnik, H., Osterberg, H. and Richards, O. W. (1951) In: *Phase Microscopy*. Chapman and Hall, London.
Berg, H. C. (1975) *Nature, Lond.* **254**, 389.
Berg, H. C. and Anderson, R. A. (1973) *Nature, Lond.* **245**, 380; (1974). *Nature*, **249**, 77.
Birdsell, D. C. and Cota-Robles, E. H. (1967) *J. Bact.* **93**, 427.
Black, J. T. (1974). In: *The Principles and Technique of Scanning Microscopy*, Vol. 9, p. 1. Ed. by M. A. Hayat. van Nostrand Reinhold, New York.
Black, S. H., Hashimoto, T. and Gerhardt, P. (1960) *Canad. J. Microbiol* **6**, 213.
Boven, C. P. A. van., Ensering, H. and Hijmans, W. (1968a) *J. gen Microbiol.* **52**, 403; (1968b) *J. gen. Microbiol.*, **52**, 413.

Bradley, D. E. and Williams, D. J. (1957) *J. gen. Microbiol.* **17**, 75.
Branton, D. (1966) *Proc. nat. Acad. Sci. Wash.* **55**, 1048.
Braun, V. (1975) *Biochim. biophys. Acta.* **415**, 335.
Braun, V. and Rehn, K. (1969) *Eur. J. Biochem.* **10**, 426.
Braun, V., Rehn, K. and Wolff, H. (1970) *Biochemistry* **9**, 5041.
Brenner, S. *et al.* (1958) *Nature, Lond.* **181**, 1713.
Brinton, C. C. (1965) *Trans. N.Y. Acad. Sci.* **27**, 1003; (1971) *Crit. Rev. Microbiol.* **1**, 105.
Brinton, C. C. and Beer, H. (1967). In: *The Molecular Biology of Viruses*, p. 251. Ed. by J. S. Colter and W. Parnchych. Academic Press, New York.
Brinton, C. C., Geinski, P. and Carnaham, J. (1964) *Proc. nat. Acad. Sci., Wash.,* **52**, 776.
Brown, W. C., Velloun, D., Schnepf, E. and Greer, C. (1978) *FEBS Letts.* **3**, 247.
Bulloch, W. and Macleod, J. J. R. (1904) *J. Hyg. Camb.* **4**, 1.
Cairns, J. (1962) *J. molec. Biol.* **4**, 407; (1963) *J. Molec. Biol.* **6**, 208.
Canale-Parola, E. (1970) *Bact. Rev.* **34**, 82.
Caro, L. G., Tubergen, R. P. van and Farre, F. (1958) *J. biophys. biochem. Cytol.* **4**, 491.
Chapman, G. B. (1962) *J. Bact.* **79**, 132.
Chapman, G. B. and Hillier, J. (1953) *J. Bact.* **66**, 362.
Cleveland, E. F. and Gilvarg, C. (1975) In: *Spores VI*. Ed. by P. Gerhardt, R. N. Costelow and H. L. Sadoff, American Society of Microbiology.
Cohn, F. (1875) *Beitr. Biol. Pflanz.* **1**, 2.
Cole, R. M. and Hahn, J. J. (1962) *Science, N.Y.* **135**, 722.
Corner, T. R. and Marquis, R. E. (1969) *Biochim. biophys. Acta.* **183**, 544.
Correll, D. L. and Lewin, R. A. (1964) *Canad. J. Microbiol.* **10**, 63.
Crawford, E. M. and Gesteland, R. F. (1964) *Virology* **22**, 165.
Curtis, R. (1969) *Annu. Rev. Microbiol.* **23**, 69.
Deamer, D. W. and Branton, D. (1967) *Science, N.Y.* **158**, 655.
Delius, H. and Worcel, A. (1973) *J. molec. Biol.* **82**, 107.
Deussen, E. (1918) *Z. Hyg. InfektKr.* **85**, 235
Dienes, L. (1939) *J. infect. Dis.* **65**, 24; (1942) *J. Bact.* **44**, 37; (1967) In: *Microbial Protoplasts, Spheroplasts and L-forms*, p. 94. Ed. by L. B. Guze. Williams and Wilkins, Baltimore.
Dimmitt, K. and Simon, M. I. (1971) *J. Bact.* **108**, 282.
Dobrogosz, W. J. and Hamilton, P. B. (1971) *Biochem. biophys. Res. Commun.* **42**, 202.
Donachie, W. D., Begg, K. J. and Vicente, M. (1976) *Nature, Lond.* **264**, 328.
Duguid, J. P. (1959) *J. gen. Microbiol.* **21**, 271; (1968) *Arch. Immunol. Ther. exp.* **16**, 173.
Duguid, J. P., Anderson, E. S. and Campbell, I. (1966) *J. Path. Bact.* **92**, 107.
Duguid, J. P. and Gillies, R. R. (1958) *J. Path. Bact.* **75**, 519.
Duguid, J. P., Smith, I. W., Dempster, G. and Edmunds, P. N. (1955) *J. Path. Bact.* **70**, 335.
Ehrlich, P. (1882) *Dtsch. med. Wschr.* **8**, 269.
Eisenstadt, E. and Silver, S. (1972). In: *Spores V*, p. 180. Ed. by H. O. Halvorson, R. Hanson and L. L. Campbell. American Society for Microbiology.
Ellar, D. J. (1978) *Symp. Soc. gen. Microbiol.* **28**, 295.
Ellen, R. P., Walker, D. L. and Chan, K. H. (1978) *J. Bact.* **134**, 1171.

Elliot, T. S. J, Ward, J. B. and Rogers, H. J. (1975a) *J. Bact.* **124**, 623.
Elliot, T. S. J., Ward, J. B., Wyrick, P. B. and Rogers, H. J. (1975b) *J. Bact.* **124**, 905.
Ellwood, D. C. and Tempest, D. W. (1969) *Biochem. J.* **111**, 1.
Fein, J. E. (1979) *J. Bact.* **137**, 933.
Fein, J. E. and Rogers, H. J. (1976) *J. Bact.* **127**, 1427.
Fernandez, B., Fréhèl, C. and Chaix, P. (1970) *Biochim. biophys. Acta* **223**, 292.
Finch, J. T. and Klug, A. (1972). In: *The Generation of subcellular structures*, p. 167. Ed. by R. Markham and J. B. Bancroft. North Holland Publishers, Amsterdam.
Finnegan, D. and Willetts, N. (1973) *Molec. gen. Genetics.* **127**, 307.
Fitz-James, P. C. (1960) *J. biophys. biochem. Cytol.* **8**, 507; (1962) *J. Bact.* **84**, 104; (1971) *J. Bacteriol.* **105**, 1119.
Follett, A. E. C. and Gordon, J. (1963) *J. gen. Microbiol.* **32**, 235.
Forsberg, C. W., Wyrick, P. B., Ward, J. B. and Rogers, H. J. (1973) *J. Bact.* **113**, 929.
Freese, E. B., Cole, R. M., Klofat, W. and Freese, E. (1970) *J. Bact.* **101**, 1046.
Glauert, A. M. and Thornley, M. J. (1969) *Annu. Rev. Microbiol.* **23**, 159.
Ghuysen, J-M. (1968) *Bact. Rev.* **32**, 425–64.
Gooder, H. (1967) In: *Microbial Protoplasts, Soheroplasts and L-forms*, p. 40. Ed. by L. B. Guze. Williams and Wilkins, Baltimore.
Gooder, H. and Maxted, W. R. (1958) *Nature, Lond.* **183**, 808.
Gould, G. W. and Dring, C. J. (1975) *Nature, Lond.* **258**, 402.
Gram, C. (1884) *Fortschr. Med.* **2**, 185.
Griffith, J. (1976) *Proc. nat. Acad. Sci., Wash.* **73**, 3872.
Grover, N. B., Woldringh, C. L., Zaritsky, A. and Rosenberger, R. F. (1977) *J. theoret. Biol.* **67**, 181.
Hayat, M. A. (1978). Introduction to: *Biological Scanning Electron Microscopy*. University Park Press.
Helmuth, R. and Achtman, M. (1975) *Nature, Lond.* **257**, 652.
Heppel, L. A. (1967) *Science, N.Y.* **156**, 1451; (1971). In: *Structure and Function of Biological Membranes*. Ed. by L. I. Rothfield. Academic Press, New York.
Higgins, M. L. and Shockman, G. D. (1976) *J. Bact.* **127**, 1346.
Higgins, M. L., Tsien, H. C. and Daneo-Moore, L. (1976) *J. Bact.* **127**, 1519.
Hillier, J. (1950) *Annu. Rev. Microbiol.* **4**, 1.
Holt, S. C. and Leadbetter, E. R. (1969) *Bact. Rev.* **33**, 346.
Houwink, A. L. (1949) In: *The Nature of the Bacterial Surface*, p. 92. Ed. by A. A. Miles and N. W. Price. Oxford University Press.
Howard, L. and Tipper, D. J. (1973) *J. Bact.* **113**, 1491.
Hranueli, D., Piggot, P. J. and Mandelstam, J. (1974) *J. Bact.* **119**, 684.
Iida, H. and Inoue, K. (1968) *Jap. J. Microbiol.* **12**, 353.
Iino, T. (1969a) *Bact. Rev.* **33**, 454; (1969b) *J. gen. Microbiol.* **56**, 227; (1977) *Annu. Rev. Genetics.* **11**, 161.
Inouye, M. (1971) *J. biol. Chem.* **246**, 4834.
Inouye, M., Shaw, J. and Shen, C. (1972) *J. biol. Chem.* **247**, 8154.
Iterson, W. van, Hoeniger, J. M. F. and Zanten, E. V. van. (1967) *J. Cell Biol.* **32**, 1.
Iterson, W. van and Leene, W. (1964) *J. Cell Biol.* **20**, 361.

Iterson, W. van, Michels, P. A. M., Vyth-Dresse, F. and Aten, J. A. (1975) *J. Bact.* **121,** 1189.
Jacob, F., Brenner, S. and Cuzin, F. (1963) *Cold. Spr. Harb. Symp. quant. Biol.* **28,** 329.
Joseph, R. and Shockman, G. D. (1974) *J. Bact.* **118,** 735.
Kagawa, H., Owaribe, K., Asakura, S. and Takahashi, N. (1976) *J. Bact.* **125,** 68.
Kahn, S. and McNab, R. M. (1980) *J. molec. Biol.* **138,** 563, 599.
Kaplan, M. L. and Kaplan, L. (1933) *J. Bact.* **25,** 309.
Kavenoff, R. and Brown, O. A. (1976) *Chromosoma* **59,** 89.
Kavenoff, R. and Ryder, O. A. (1976) *Chromosoma* **55,** 13.
Kellenberger, E., Liebermeister, K. and Bonifas, V. (1956) *Z. Naturf.* **11B,** 206.
Kerridge, D., Horne, R. W. and Glauert, A. M. (1962) *J. molec. Biol.* **4,** 227.
Keynan, A. (1973) 23rd Symposium of the Society for General Microbiology. p. 55.
Keynan, A. and Evenchick, Z. (1969) In: *The Bacterial Spore*, p. 359. Ed. by G. W. Gould and A. Hurst. Academic Press, N.Y.
Klebs, E. (1896) *Zbl. Bakt.* **20,** 488.
Kleinschmidt, A., Lanig, D. and Zahn, R. K. (1961) *Z. Naturf.* **166,** 730.
Kleppe, K., Övrebö, S. and Lossius, I. (1979) *J. gen. Microbiol.* **112,** 1.
Klieneberger-Nobel, E. (1935) *J. Path. Bact.* **40,** 93; (1942) *J. Hyg., Camb.* **42,** 485; (1949) *J. gen. Microbiol.* **3,** 434; (1956) *Zbl. Bakt., I. Abt. Orig.* **165,** 329.
Knoll, M. (1935) *Z. tech. Physik.* **16,** 467.
Koch, R. (1876) *Cohn's Beitr. Biol. Pflanz.* **2,** 277; (1897) *Drsch. med. Wschr.* **23,** 209.
Kondo, M. and Foster, J. W. (1967) *J. gen. Microbiol.* **47,** 257.
Kondo, M. and Nishihara, T. (1970) *Jap. J. Bact.* **25,** 215.
Koyama, T., Yamada, M. and Matsuhashi, M. (1977) *J. Bact.* **129,** 1518.
Lamanna, C. (1946) *J. Bact.* **52,** 99.
Landman, O. E. (1967) In: *Microbial Protoplasts, Spheroplasts and L-forms*. Ed. by L. B. Guze. Williams and Wilkins, Baltimore.
Landman, O. E. and Forman, A. (1969) *J. Bact.* **99,** 576.
Landman, O. E. and Halle, S. (1963) *J. molec. Biol.* **7,** 721.
Landman, O. E., Ryter, A. and Fréhél, C. (1968) *J. Bact.* **96,** 2154.
Lautrop, H. (1981) *Amer. Soc. Microbiol. News* **47,** 44.
Lawn, A. M. (1966) *J. gen. Microbiol.* **45,** 377.
Lederberg, J. and St Clair, J. (1958) *J. Bact.* **75,** 143.
Leifson, E. (1959) In: *Atlas of Bacterial Flagellation*. Academic Press, N.Y.
Lencastre, H. de and Piggot, P. (1979) *J. gen. Microbiol.* **114,** 377.
Lewin, R.A. (1963) *Nature, Lond.* **198,** 103.
Lowy, J. and Hanson, J. (1964) *Nature, Lond.* **202,** 538; (1965) *J. molec. Biol.* **11,** 293.
McQuillen, K. (1960) In: *The Bacteria*, Vol I, p. 249. Academic Press, New York.
Maccacaro, G. A. and Hayes, W. (1961) *Genetic Res.* **2,** 394; (1961) *Ibid.* **2,** 406.
Madoff, S. (1960) *Ann. N.Y. Acad Sci.* **79,** 383.
Mandelstam, J. (1976) *Proc. Roy. Soc. London. B.* **193,** 89.
Manson, M. D., Tedesco, P. M. and Berg, H. C. (1980) *J. molec. Biol.* **138,** 541.

Marquis, R. E. and Corner, T. R. (1976). In: *Microbial and Plant Ptotoplasts*. p. 1. Ed. by J. F. Peberdy, A. H. Rose, H. J. Rogers and E. C. Cocking. Academic Press, London.
Martinez, R. J. and Gordee, E. Z. (1966) *J. Bact.* **91,** 870.
Marton, L. (1934) *Bull. Acad. Belge. Cl. Sci.* **20,** 439; (1941) *J. Bact.* **41,** 397.
Marvin, D. A. and Hahn, B. (1969) *Bact. Rev.* **33,** 172.
Mason, D. J. and Powelson, D. M. (1956) *J. Bact.* **71,** 474.
Mayer, F. (1969) *Arch. Mikrobiol.* **68,** 179.
Meynell, E. and Lawn, A. M. (1967) *Genet. Res.* **9,** 359.
Mittwer, T., Bartholomew, J. W. and Kallman, B. J. (1950) *Stain Technol.* **25,** 196.
Miura, T. and Mizushima, S. (1968) *Biochim. biophys. Acta.* **150,** 159.
Moor, H. and Mühlethaler, K. (1963) *J. cell. Biol.* **17,** 609.
Moore, H., Mühlethaler, K., Waldner, H. and Frey-Wyssting, A. (1961) *J. biophys biochem Cytol.* **10,** 1.
Mühlradt, P. F. and Golecki, J. R. (1975) *Europ. J. Biochem.* **51,** 343.
Murrell, W. G. (1969) In: *The Bacterial Spore*. p. 215. Ed. by G. W. Gould and A. Hurst. Academic Press, New York.
Murrell, W. G. and Scott, W. J. (1957) *Nature, Lond.,* **179,** 481.
Murrell, W. G. and Warth, A. D. (1964). In: *Spores II*. Ed. by L. L. Campbell and H. O. Halvorson. American Society of Microbiology.
Nanninga, N. (1969) *J. cell. Biol.* **42,** 733; (1970) In: *Proceedings of the 7th Internat. Congress de Microscop. électronique*, Grenoble, III, 349.
Novotny, C. P. and Fives-Taylor, P. (1974) *J. Bact.* **117,** 1306.
Novotny, C., Raizen, E., Knight, W. S. and Brinton, C. C. (1969) *J. Bact.* **98,** 1307.
Oatley, C. W. (1972). In: *The Scanning Electron Microscope*. Cambridge University Press, England.
Old, D. C. (1972) *J. gen. Microbiol.* **71,** 149.
Old, D. C. and Payne, S. (1971) *J. med. Microbiol.* **4,** 215.
On, J. T. and Anderson, T. F. (1970) *J. Bact.* **102,** 648.
Osborn, M. J. and Wu, H. C. P. (1980) *Annu. Rev. Microbiol.* **34,** 369.
Ottow, J. C. G. (1975) *Annu. Rev., Microbiol.* **29,** 79.
Owen, P. and Freer, J. H. (1972) *Biochem. J.* **129,** 907.
Pamphlis, M. L. de and Adler, J. (1971) *J. Bact.* **105,** 384.
Pearce, S. M. and Fitz-James, P. C. (1971) *J. Bact.* **105,** 339.
Petitjohn, D. E. (1976) *CRC. Critical Rev. Biochem.* **4,** 175.
Piekarski, G. (1937) *Arch. Mikrobiol.* **8,** 428; (1940) *Ibid.* **11,** 406.
Pierce, C. H. (1942) *J. Bact.* **43,** 780.
Piggot, P. J. and Coote, J. G. (1976) *Bact. Rev.* **40,** 908.
Pinto da Silva, P. and Branton, D. (1970) *J. cell. Biol.* **45,** 598.
Porter, K. R. and Yegian, D. (1945) *J. Bact.* **50,** 563.
Pope, L. M. and Jurtshuk, P. (1967) *J. Bact.* **94,** 2062.
Powell, J. F. (1953) *Biochem. J.* **54,** 210.
Powell, J. F. and Strange, R. E. (1953) *Biochem. J.* **54,** 205.
Reaveley, D. A. and Rogers, H. J. (1969) *Biochem. J.* **113,** 67.
Repaske, R. (1956) *Biochim. biophys. Acta.* **22,** 189.
Reusch, V. M. and Panos, C. (1976) *J. Bact.* **126,** 300.
Robinow, C. F. (1945). In: Addendum to *The Bacterial Cell*. Ed. by R. J. Dubos. Harvard; University Press, Cambridge, Mass. (1956a). *Symp. Soc. gen. Microbiol.* **6,** 181; (1956b) *Bact. Rev.* **20,** 207.
Rogers, H. J. (1979). In: *Microbial Polysaccharides and*

Polysaccharases. p. 237. Ed. by R. C. W. Berkeley, G. W. Gooday and D. C. Ellwood. Society of General Microbiology, Academic Press, London and N.Y.

Rogers, H. J., Perkins, H. R. and Ward, J. B. (1980) In: *Microbial Walls and Membranes.* Chapman and Hall, London.

Roux, J. (1960) *Ann. Inst. Pasteur.* **99,** 286.

Ruska, E. (1934) *Z. Physik.* **87,** 580.

Ryter, A. and Jacob, F. (1964) *Ann. Inst. Pasteur* **107,** 384.

Ryter, A. and Kellenberger, E. (1958) *Z. Naturforsch.* **13B,** 597.

Ryter, A. and Landman, O. E. (1964) *J. Bact.* **88,** 457.

Salton, M. R. J. (1963) *Bact. Rev.* **25,** 77.

Schleifer, K. H. and Kandler, O. (1972) *Bact. Rev.* **36,** 407.

Schnaitman, C. (1970) *J. Bact.* **104,** 890.

Seidler, R. J. and Starr, M. P. (1968) *J. Bact.* **95,** 1952.

Setlow, P. (1975*a*) *J. biol. Chem.* **250,** 8159; (1975*b*) *Ibid.* **250,** 8168; (1976) *J. biol. Chem.* **251,** 7853.

Shaw, N. (1970) *Bact. Rev.* **34,** 365; (1975) *Advanc. Microbiol. Physiol.* **12,** 141.

Shively, J. M. (1974) *Annu. Rev. Microbiol.* **28,** 167.

Shively, J. M., Ball, F., Brown, D. H. and Saunders, R. E. (1973) *Science, N.Y.* **182,** 584.

Shockman (see Higgins and Shockman, 1976).

Silverman, M. R. and Simon, H. I., (1972) *J. Bact.* **112,** 986; (1977) *Annu. Rev. Microbiol.* **31,** 397.

Simon, H. I., Silverman, M. R., Matsumura, P., Ridgway, H., Korneda, Y. and Hilmen, M. (1978) *Symp. Soc. gen. Microbiol.* **28,** 272.

Sleytr, U. B. (1978) *Int. Rev. Cytol.* **53,** 1.

Smit, J., Kamio, Y. and Nikaido, H. (1975). *J. Bact.* **124,** 942.

Smith, H. (1977) *Bact. Rev.* **41,** 475.

Sordelli, A. and Arena, A. (1934) *C. R. Soc. Biol.* **117,** 63.

Stanier, R. Y. and Cohen-Bazire, G. (1977) *Annu. Rev. Microbiol.* **31,** 225.

Starr, M. P. and Williams, R. C. (1952) *J. Bact.* **63,** 701.

Steere, L. (1957) *J. biophys. biochem. Cytol.* **3,** 45.

Steinberg, W., Halvorson, H. O., Keynan, A. and Weinberg, E. (1965) *Nature, London* **208,** 710.

Stille, B. (1937) *Arch. Mikrobiol.* **8,** 125.

Stonington, O. G. and Petitjohn, D. E. (1971) *Proc. nat. Acad. Sci., Wash.* **68,** 6.

Strange, R. E. and Dark, F. A. (1957*a*) *J. gen. Microbiol.* **16,** 236; (1957*b*) *Ibid.* **17,** 525.

Strange, R. E. and Powell, J. F. (1954) *Biochem. J.* **58,** 80.

Swaney, L. M., Liu, Y-P., Ippen-Ihler, K. and Brinton, C. C. (1977*a*) *J. Bact.* **130,** 495; (1977*b*) *Ibid.* **130,** 506.

Tamura, S. (1913) *Z. physiol. Chem.* **87,** 85.

Thorne, K. (1977) *Biol. Rev.* **52,** 219.

Tipper, D. J. and Gauthier, J. J. (1972) In: *Spores V.* Ed. by H. O. Halverman,. R. Hanson and L. L. Campbell. American Society of Microbiology.

Tomasz, A. (1968) *Proc. nat. Acad. Sci, Wash.* **59,** 86.

Tomoeda, M., Inuzaka, M. and Date, T. (1975) *Progr. biophys. molec. Biol.* **30,** 23.

Tulasne, R. and Lavillaureix, J. (1958) *C.R. Acad. Sci.* **246,** 2396.

Tulasne, R. and Vendrely, R. (1947) *Nature, Lond.* **160,** 225.

Tzagoloff, H. and Novik, R. (1977) *J. Bact.* **129,** 343.

Valentine, R. C. and Strand, M. (1965) *Science, N.Y.* **148,** 511.

Vary, J. C. and Halvorson, H. O. (1964) *J. Bact.* **89,** 1340.

Venables, J. A. (1976) In: *Developments in Electron Microscopy and Analysis.* Academic Press, London.

Vendrely, R. and Lipardi, J. (1946) *C.R. Acad. Sci.* **223,** 342.

Vinter, V. (1965) *Folia Microbiol.* **10,** 288; (1970) *J. appl. Microbiol.* **33,** 50.

Walsby, A. E. (1978) *Symp. Soc. gen. Microbiol.* **28,** 327.

Ward, J. B. (1975) *J. Bact.* **124,** 668.

Warth, A. D. and Strominger, J. L. (1969) *Proc. nat. Acad. Sci., Wash.* **64,** 528.

Weibull, C. (1953) *J. Bact.* **66,** 688.

Weibull, C. (1963) *Proc. Soc. exp. Biol., N.Y.* **113,** 32.

Weibull, C. and Beckman, H. (1961) *J. gen. Microbiol.* **24,** 379.

Weibull, C. and Lundin, B-M. (1962) *J. gen. Microbiol.* **27,** 241.

Weigand, R. A., Vinci, K. D. and Rothfield, L. I. (1976) *Proc. Nat. Acad. Sci., Wash.* **73,** 1882.

Wensinck, F. and Boeve, J. J. (1957) *J. gen. Microbiol.* **17,** 401.

White, D. A., Albright, F. R., Lennarz, W. J. and Schnaitman, C. A. (1971) *Biochim. biophys. Acta.* **249,** 636.

Whittenbury, R. and Dow, C. S. (1977) *Bact. Rev.* **41,** 754–808.

Wilkinson, S. G. (1977) In: *Surface Carbohydrates of the Carbohydrates of the Prokaryotic Cell.* pp. 97–175. Ed. by I. W. Sutherland. Academic Press, London.

Willetts, N. (1971) *Nature (New biol.).* **230,** 183.

Willetts, N. and Skurray, R. (1980) *Annu. Rev. Genetics* **14,** 41.

Williams, R. E. O. (1963) *J. gen. Microbiol.* **33,** 325.

Williams, R. C. and Wyckoff, R. W. G. (1945) *Proc. Soc. exp. Biol. N.Y.* **59,** 265.

Wise, J., Swanson, A. and Halvorson, H. O. (1967) *J. Bact.* **94,** 2075.

Wittmann, H. G. (1982) *Proc. roy. Soc. Lond. B* **216,** 117.

Woldringh, C. L. (1976) *J. Bact.* **125,** 248.

Woldringh, C. L. deJong, M. A., van den Berg, W. and Koppes, L. (1977) *J. Bact.* **131,** 270.

Worcel, A. and Burgi, E. (1972) *J. molec. Biol.* **71,** 127; (1974) *Ibid.* **82,** 91.

Wyrick, P. B. and Gooder, H. (1971) *J. Bact.* **105,** 284.

Wyrick, P. B., McConnell, M. and Rogers, H. J. (1973) *Nature, Lond.* **244,** 505.

Wyrick, P. B. and Rogers, H. J. (1973) *J. Bact.* **116,** 456.

Yamamoto, T. (1967) *J. Bact.* **94,** 1746.

Yanagawa, R. and Otsuki, K. (1970) *J. Bact.* **101,** 1063.

Yanagawa, R., Otsuki, K. and Tokui, T. (1968) *Jap. J. vet. Res.* **16,** 31.

Yegian, D. and Vanderlinde, R. J. (1947) *J. Bact.* **54,** 777.

Ziehl, F. (1882) *Dtsch. med. Wschr.* **8, 451;** (1883) *Ibid.* **9,** 247.

3

The metabolism, growth and death of bacteria

J. Gareth Morris

Bacterial metabolism	39	Bacterial nutrition and the design of culture	
Conservation of free energy by bacteria	40	media	55
Enzymes in bacterial metabolism	41	Culture pH	56
Regulatory mechanisms in metabolism	42	Culture oxidation-reduction potential E_h	56
Energy-yielding metabolism in		Gaseous requirements	57
chemoheterotrophic bacteria	44	(a) Oxygen	57
Fermentation	44	(b) Carbon dioxide	57
Respiration	46	Biochemical tests in diagnostic bacteriology	58
Transport of solutes across the bacterial cell		**Growth of bacteria**	58
membrane	48	Growth of a bacterial cell	58
Permeation of the outer membrane of the gram-		Growth in batch culture	59
negative bacterium	50	The lag phase	60
Catabolic pathways	50	The exponential phase	60
Anabolic pathways	51	The stationary phase	61
Biosynthesis of small molecules	51	Culture growth rates and nutrient limitation	61
(a) Amino acids	51	Continuous flow culture	61
(b) Purines and pyrimidines	52	(a) Turbidostat	61
(c) Fatty acids and lipids	52	(b) Chemostat	62
Biosynthesis of macromolecules	53	Bacterial growth on solid surfaces	63
Biosynthesis of periodic macromolecules	53	Temperature and growth rates	63
Biosynthesis of informational		**Survival and death of bacteria**	64
macromolecules	54	Survival of bacteria as resting forms—	
(a) Biosynthesis of DNA	54	dormancy	64
(b) Biosynthesis of RNA	54		
(c) Biosynthesis of protein	54		

Bacterial metabolism

The metabolism of the bacterial cell is designed to enable it to maintain its viability over as wide a range of environmental conditions as possible and to respond to appropriate conditions by growth leading to reproduction. The component metabolic processes thus consist of sequences of enzyme-catalysed chemical reactions so co-ordinated as to enable the living organism to abstract from its environment both the ingredients required to construct its constituents and the free energy necessary to accomplish this and other life-sustaining tasks. These functions are performed via metabolic routes that are broadly divisible into three major types (Kornberg 1965). *Catabolic pathways* serve primarily to fragment foodstuffs and supply the consequent limited range of product-molecules to certain *central metabolic pathways*; these conserve free energy by the direct or indirect production of adenosine triphosphate (ATP), and also yield 'reducing power', often in the form of reduced pyridine dinucleotides (NADH and NADPH). Certain of the intermediates on the central metabolic pathways are withdrawn into *anabolic pathways*, which create from them the building blocks from which are then constructed the macromolecular components of the cell. The central metabolic pathways thus play a dual, *amphibolic* role in the supply of both the energy and the precursor molecules consumed in biosynthesis (Davis 1961). Intermediates drained from a central pathway for anabolic purposes will have to be replaced either from catabolic pathways or, under certain circumstances when this is not feasible, by the operation of an ancillary pathway having a specific replenishment function, i.e. an *anaplerotic pathway* (Kornberg 1966).

The chemical reactions which feature in these metabolic pathways are catalysed by enzymes whose synthesis and activities must be carefully controlled if the purposes of metabolism are to be satisfactorily fulfilled and the free energy harnessed by the cell is not to be squandered. So numerous and diverse are the metabolic pathways exploited by bacteria, and so varied and cunning are the devices that regulate their operations, that in this chapter we shall be able only to outline some of the more important features of bacterial metabolism, with special reference to heterotrophic bacteria of the types that are of medical or veterinary importance. We shall therefore have to omit many of the most spectacular examples of biochemical ingenuity practised by bacteria which are able to grow at extremes of pH, temperature or salinity (Kushner 1978) or are able to accomplish photosynthesis, fixation of dinitrogen, or the utilization of unusual organic and inorganic substrates (Stanier *et al.* 1977). Necessary details of metabolic routes, their component enzyme-catalysed reactions and mechanisms of control of gene expression and enzyme activity are provided in comprehensive textbooks of biochemistry (e.g. Lehninger 1975, Metzler 1977), molecular and bacterial genetics (e.g. Watson 1976, Stent and Calendar 1978), and bacterial physiology (e.g. Rose 1976, Gottschalk 1979, Moat 1979, Mandelstam *et al.* 1982).

Conservation of free energy by bacteria

It is characteristic of all growing and reproducing living cells that they fabricate highly ordered structures from a variety of low-molecular weight foodstuffs. This spontaneous creation of order from disorder—decrease in entropy—can be accomplished only at the expense of some input of free energy. The living cell in this sense represents an open thermodynamic system through which energy spontaneously flows from an external source to some external sink, conserving a high degree of order in the system through which it passes. Sustained by this flow of energy the living cell maintains a dynamic steady state with its environment; bereft of it, death inevitably supervenes. Thus the continued viability of any bacterium is dependent on its continuous exploitation of some exogenous source of free energy, save in those special situations in which an organism may survive in a 'resting state' e.g. an endospore, wherein all metabolism is shut down.

Bacteria of clinical or veterinary importance are chemoheterotrophs deriving free energy from the chemical transformation of organic compounds supplied in their growth media. In such organisms chemical reactions that would, in isolation, not occur spontaneously, i.e. endergonic reactions, are coupled to the highly exergonic processes of fermentation or respiration (Lehninger 1971, Morris 1974). A variety of coupling devices are used, which fall into two main categories: (a) chemical coupling mediated by so-called 'energy-rich' metabolites i.e. compounds characterized by high group-transfer potentials (Lipmann 1941), (b) chemiosmotic coupling via ion currents established across the cytoplasmic membrane of the bacterial cell (Mitchell 1966, 1976; Garland 1977; see Haddock and Hamilton 1977, p.1; Harold 1979). The crucial mediator in chemical coupling is adenosine 5'-pyrophosphate (ATP), whose production is the chief purpose of fermentation and respiration and whose consumption in anabolic and other cell processes, directly or indirectly, enables these to be accomplished. It is for this reason that ATP has been termed the 'currency of energy exchange in living systems', since its generation and expenditure is the basis of the cell's energy economy (Atkinson 1977).

Equilibrium thermodynamics tells us that in a closed system, at constant temperature and pressure, the spontaneous reaction will be that whose occurrence is associated with a loss of free energy by the system (ΔG is negative). This exergonic reaction will cease when chemical equilibrium is reached i.e. when the system has no further capacity for spontaneous change (ΔG is zero). An endergonic reaction (ΔG is positive) will not occur spontaneously. Thus thermodynamic calculations based on changes in free energy content measured in J mol^{-1} or cal. mol^{-1} enable one to predict which reactions in a given system will be exergonic and which will be endergonic. It is important to note that such thermodynamic considerations tell us nothing of the likely speed of a reaction or of the mechanism whereby it is accomplished. Such information can be obtained only from kinetic studies (Morris 1974).

In many metabolic pathways some of the key steps are oxidation-reduction reactions in which electrons are transferred by a reductant (electron donor) to an oxidant (electron acceptor). As a consequence of its donation of electrons the reductant is converted into its conjugative oxidant, and *vice versa* for the oxidant whose acceptance of electrons produces its conjugate reductant:

$$\text{reductant} \rightleftharpoons \text{conjugate oxidant} + \text{n electrons}$$

Such a conjugate pair is termed a *redox couple* so that, in essence, an oxidation-reduction reaction occurs between two redox couples which differ in their affinity for electrons. The couple that displays the greater electron affinity assumes the oxidizing role, and its oxidant member will be the electron acceptor in the reaction. The couple with the lesser electron affinity plays the reducing role and its reductant member will serve as the electron donor. The intensity of the affinity that a redox couple displays for electrons is termed its *redox potential* or electrode potential, E_h. The standard hydrogen redox couple—H_2 at 1 atmosphere: H^+ ions at unit activity—is stipulated to display zero redox potential, and thus to form the null point on a scale of redox potentials upon which all other redox couples may be assigned experimentally determined E_h values. The E_h value of any redox couple will vary within limits about a mid value depending on the ratio of concentrations in which oxidant and conjugate reductant forms are present. The mid value, midpoint potential, E_m, is the redox potential displayed by the redox couple when reductant and oxidant are present in equal activities—approximately, concentrations. Electron transfer from one redox couple to another of more positive E_h value will be a spontaneous process associated with a loss of free energy ($-\Delta G$) whose magnitude will be proportional to the difference between the E_h values of the interacting couples i.e. to the electromotive force, ΔE_h (see Morris 1974).

Enzymes in bacterial metabolism

These specialized catalytic proteins play a crucial role in every aspect of bacterial metabolism, for only by their intervention can the component chemical reactions be performed with the necessary precision and rapidity. Enzymes accelerate only reactions that are thermodynamically feasible, i.e. spontaneous, causing them to reach more swiftly the equilibria to which they are destined to proceed even in the absence of any catalyst. Although an enzyme must participate in the reaction it catalyses, it will be regenerated concurrently with the formation of the reaction products. Even though some enzymes are catalytically active only when associated with a co-factor of low molecular weight (prosthetic group and/or coenzyme), the explanation of the specificity of action of every enzyme must be sought in its protein structure (Fersht 1977). A portion of the enzyme protein i.e. the *catalytic site* is so constructed that it displays an affinity for the substrate with which it interacts to create an enzyme-substrate complex. This is so instrinsically unstable that it readily breaks down to yield the reaction product(s) whilst regenerating the enzyme molecule with an unoccupied catalytic site. The acceleration in reaction rate is thus attributable to the reaction proceeding by a novel mechanism, via the enzyme-substrate complex, which is associated with a lower energy of activation than that attending the uncatalysed reaction. The specificity of enzyme action is attributable to the specific recognition of the substrate by the catalytic site. The presence at this site of special functional groups borne on amino acid side chains and prosthetic molecules, held in appropriate juxtapositions by the tertiary structure of the protein molecule, not only ensures that the appropriate substrate(s) are brought together and bound in proper orientation, but is also responsible for facilitating the desired reaction. Enzymes are naturally vulnerable to all treatments that destroy or disrupt protein molecules and to attack by reagents that interact with their amino side chains. It follows that the catalytic activity of an enzyme can be inhibited by a wide variety of compounds. Some of these are irreversible in their action (Rando 1974), but the inhibition caused by some other compounds is reversible (Morris 1974, Wynn 1979). Amongst the latter category one finds substrate analogues which unproductively bind to the catalytic site, thus excluding the substrate and preventing creation of the productive enzyme-substrate complex.

As increasing concentrations of a substrate are presented to an enzyme, the velocity of the enzyme-catalysed reaction will increase until, at substrate concentrations sufficient to saturate the available catalytic sites, the rate of reaction will be sustained at some maximum value (V_{max}). Thus a plot of reaction rate (v) versus substrate concentration ([S]) frequently approximates to a rectangular hyperbola. An enzyme displaying these kinetic properties is said to demonstrate Michaelis-Menten kinetics; and the relation between the reaction rate and substrate concentration is adequately described by two constants viz. the maximum velocity V_{max} and the Michaelis constant K_m, which is that substrate concentration at which the rate of the enzyme-catalysed reaction is $0.5\ V_{max}$. From the several ways in which the observed relation between v and [S] may differ from the simple hyperbolic plot and be affected by the presence of various inhibitors and reagents, the skilled enzyme kineticist can deduce much information about the nature of the enzyme-substrate complex (Segel 1975). Enzymes will inevitably be inhibited by treatments which would denature proteins (e.g. heat, extremes of pH), but within limits will respond to changes in temperature and pH in a more subtle manner. Thus involvement of ionizable groups at the enzyme catalytic site will inevitably mean that the rate of the enzyme-catalysed reaction will vary with pH in a distinctive manner; generally a plot of V_{max} versus pH yields a bell-shaped curve whose crown indicates the optimal pH range for activity of that enzyme.

In some instances profound changes in the catalytic activity of an enzyme can follow interaction of the enzyme molecule with a low molecular weight metabolite that bears little or no structural resemblance to the substrate, and hence does not bind at the catalytic site (Monod *et al.* 1965) These *effector* molecules bind to a second site on the enzyme molecule (an *allosteric* site), resulting in some conformational change in the enzyme (Koshland 1969, Wyman 1972). Repercussions at the catalytic site either cause enhanced activity (evoked by an *allosteric activator*) or diminished activity (produced by an *allosteric inhibitor*). Though a monomeric enzyme protein can carry an allosteric site and be subject to regulation by allosteric effectors, very often the allosteric enzyme is an oligomeric protein composed of two, four, six or more subunits which need not be identical. In certain cases such a multimeric enzyme displays co-operativity of binding of substrate molecules. In the absence of substrate the catalytic sites on the enzyme subunits would all be in some low-affinity configuration. Binding of substrate to the active site on one subunit could cause the catalytic site(s) on neighbouring subunits to adopt an altered high-affinity configuration. Such behaviour, homotropic substrate activation, would be mirrored in a sigmoidal relation between the reaction rate and increasing substrate concentration. A negative effector, i.e. inhibitor, bound to the allosteric site on one subunit could prevent adoption of the high-affinity configuration by the catalytic site on this and neighbouring subunits. A positive effector, i.e. allosteric activator, could have the opposite effect, forcing all of the catalytic sites into their most active configuration and thus causing transformation of what, in the absence of activator, would have been a sigmoidal rate-substrate concentration relation into a rectangular hyperbolic plot reflecting much higher reaction rates at low substrate concentrations. A number of enzymes are known to contain catalytic subunits, which bear the catalytic sites, and additional regulatory subunits, which carry the allosteric effector binding sites. Control of the activities of key enzymes by various metabolites acting as allosteric effectors indeed forms the basis of many aspects of metabolic regulation (Sanwal 1970).

Regulatory mechanisms in metabolism

The control and co-ordination of metabolic events is chiefly accomplished through mechanisms that regulate the location, amounts, and catalytic activities of enzymes.

Bacterial cells being devoid of specialist subcellular organelles, enzyme activities are not compartmentalized in the elaborate manner that they may be in the mitochondria or lysosomes of eukaryotic cells. Even so, many enzymes and transport proteins are necessarily incorporated into the bacterial cytoplasmic membrane (Salton 1974); in gram-negative bacteria some may be located in the periplasmic space. Within the cytoplasm itself there may exist multienzyme complexes i.e. associations of several non-covalently linked enzymes whose aggregation evidently facilitates their co-ordinated function (Gaertner 1978).

The rates of synthesis of enzymes, as of other bacterial proteins, can be regulated 'at source' by controls exercised on the transcription of those structural genes which encode the information to be translated at the ribosomes into newly synthesized enzyme molecules. Within a bacterial cell some of these structural genes are continuously transcribed at a high rate, resulting in the constitutive synthesis of the corresponding enzymes in relatively high quantities regardless of the prevailing growth conditions. Other genes which specify *inducible* enzymes remain unexpressed until their transcription is evoked by an appropriate inducer molecule. This 'response to need' is an economic device ensuring that the bacterium does not continue to produce unnecessary batteries of enzymes and other proteins (Clarke 1972).

The β-galactosidase of *Escherichia coli* is one example of an inducibly synthesized enzyme which is present in an exceedingly low concentration in cells grown in a glucose-minimal medium, but is produced in large quantity when the organism is grown with lactose as sole source of carbon and energy. Monod and his colleagues (Jacob and Monod 1961, 1970) found that synthesis of β-galactosidase was accompanied by the production of two other proteins (a lactose transport protein and a galactoside transacetylase), and that co-ordinate synthesis of the three proteins could be triggered by any of several synthetic β-galactosides, some of which were not recognized as substrates by the β-galactosidase enzyme. Thus the enzyme itself was not the agent whereby the inducer exerted its influence. They proposed that the three structural genes—*lac*Z, *lac*Y and *lac*A—which specified the three inducibly synthesized proteins were closely linked and subject to co-ordinate control by a single repressor, whose negative, inhibitory control of their transcription was relieved only when it specifically bound an inducer molecule. Furthermore, they proposed that the genes formed an operational linkage group—an operon—which not only shared the gene specifying the repressor substance (assumed to be a protein), but also a second regulatory locus termed the *operator*. This operator, lying at the proximal end of the operon, was the target of the repressor protein, which by binding to it prevented transcription into mRNA of all of the structural genes of the operon. The repressor protein-inducer complex could not combine with the operator so that in the presence of inducer this would remain open, allowing initiation of transcription of the operon. All of these predictions have been validated by subsequent biochemical and genetic studies (Fig. 3.1).

The *lac* repressor protein has been isolated (Gilbert and Müller-Hill 1966) and its amino acid sequence determined (Beyreuther *et al.* 1973). It is a tetrameric protein which combines specifically and very tightly with the operator (*lac*O). The *lac*O locus (Gilbert and Maxam 1973) includes the mRNA transcriptional initiation site for the lac operon (Müller-Hill 1975). The repressor protein is specified by a *lac*I regulatory gene which is separated from *lac*O by 85 base pairs, which constitute another regulatory element termed the promoter—*lac*P. This promoter locus includes two functional sites, (a) a CAP binding site lying proximally to *lac*I, and (b) a RNA polymerase binding site (Beckwith and Rossow 1974, Dickson *et al.* 1975).

The novel finding made in these subsequent investigations was that the *lac* operon was subject to dual control, the second being a positive, i.e. enabling, control mediated by a requirement for a general catabolite activator protein CAP, specified by a distant *crp* gene which, to be effective, had first to bind cyclic AMP (cAMP). By binding to the appropriate site

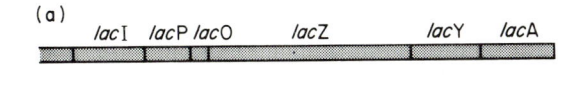

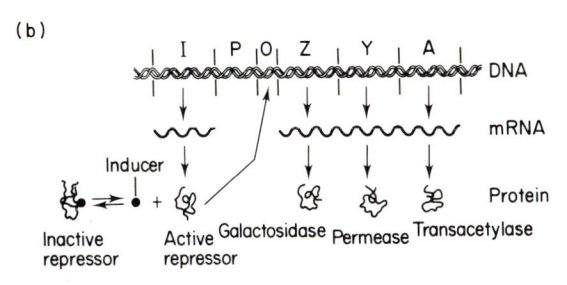

Fig. 3.1 The *lac* operon in *Esch. coli*. (a) Relative order of genes coding for enzymes of lactose fermentation and of genes implicated in the control of synthesis of these enzymes. (b) The original operon model of Jacob and Monod, as proposed for the regulation of the *lac* genes of *Esch. coli* (from Stent and Calendar 1978).

within the *lac*P promoter locus, the cAMP-CAP complex enables RNA polymerase to bind to its template as a prelude to transcription. In *Esch. coli*, intracellular cAMP concentration serves as an indicator of the general availability of catabolites (Pastan and Adhya 1976). When glucose is provided in excess in the culture medium, such catabolites will be in plentiful supply and the intracellular concentration of cAMP is low. Under these conditions the *lac* operon is poorly transcribed, even when lactose is also provided in the medium. Should the supply of glucose fail, the cAMP concentration within the cell would rise sharply and the formation of the cAMP-CAP complex would bring the *lac* operon to a state of readiness to be transcribed. Only if lactose were available, however, would it be profitable for the organism to synthesize lactose transport protein and β-galactosidase; hence the additional restraint on transcription which is released by the inducer.

Catabolite repression of synthesis of inducible catabolic enzymes is a common phenomenon in bacteria (Magasanik 1961, Paigen and Williams 1970), although its relief is not always mediated via cAMP (Ullmann 1974), which is *Esch. coli* plays additional metabolic roles (Ullmann and Danchin 1980).

In the *lac* system an inducer alleviates the negative control exerted by a repressor protein. The converse is also found, for example in arabinose utilization by *Esch. coli* (Engelsberg and Wilcox 1974), where a regulator gene specifies an activator protein facilitating transcription only when it combines with an inducer molecule, i.e. positive control. An alternative to the induction of enzyme synthesis by a potential substrate is repression of synthesis of an enzyme by a product of the reaction it catalyses. Thus, tryptophan exerts co-ordinate control over the expression of five contiguous structural genes that constitute the *trp* operon in *Esch. coli*, and specify the five enzymes whose sequential operation produces tryptophan from chorismate (Bertrand et al. 1975). Again, two quite separate control elements work in series to regulate the rate of synthesis of the *trp* enzymes. The first involves a repressor protein which binds tryptophan and only then blocks transcription by attaching to the *trp* O operator locus. The second is a control exercised not on the initiation but on the termination of mRNA synthesis. This *attenuator control* ensures that only in the presence of tryptophan is there premature termination of transcription of the *trp* operon, mediated by a termination rho factor. Such attenuator control is of primary importance in the regulation of synthesis of the enzymes of histidine biosynthesis in *Salm. typhimurium* (Artz and Broach 1975).

All these means of regulating gene transcription are relatively slow acting in that they effect changes in the amounts of enzymes present in growing organisms. They have therefore been termed 'coarse controls' to distinguish them operationally from the more rapid 'fine controls' exerted by devices which regulate the catalytic activities of existing enzyme molecules. In some instances the activity of an enzyme is dramatically altered by covalent modification of the enzyme molecule accomplished by ancillary enzymes (Segal 1973). An example is the adenylylation and deadenylylation of the glutamine synthetase of *Esch. coli*, which regulates the activity of this enzyme in response indirectly to the supply of ammonium ions (Magasanik 1977). More commonly, fine control of enzyme activity is effected by allosteric interactions with metabolites that may act as activators or inhibitors. In *feed-back inhibition* the accumulated end product of a metabolic sequence, or the product when provided exogenously in excess, inhibits an enzyme which is specific to its synthesis (Umbarger 1969). Less commonly, in precursor activation a metabolite serving as an allosteric activator stimulates an enzyme which acts either upon it or on some product formed a little further ahead in the metabolic sequence. Quite often the effector molecule is neither a substrate nor a direct product in the pathway in which the enzyme participates, but is a metabolic indicator whose concentration mirrors some more general physiological condition e.g. availability of energy reflected in the ATP/ADP ratio (Knowles 1977; see Haddock and Hamilton 1977, p. 241) or availability of reducing power, evidenced by the NADH/NAD$^+$ ratio (Weitzman and Danson 1976). The enzymes most prone to fine control are those which serve as pacemakers in that they control rate-limiting steps in metabolic pathways. Such enzymes catalyse reactions that may (a) determine the rate of generation of some crucial metabolite such as ATP, (b) initiate attack on some substance (the committal step), when all subsequent steps are accomplished so rapidly that no intermediate

Energy-yielding metabolism in chemoheterotrophic bacteria

Fermentation

The term 'fermentation' has been abused by having been indiscriminately applied to any process wherein microbial growth results in the accumulation in the culture medium of some useful organic product. In fact, fermentation has a much more restricted meaning. It is a mode of metabolism wherein an exogenous organic compound is utilized to generate ATP solely via substrate level phosphorylation reactions (SLP reactions). The end-products of the fermentation which are excreted into the medium or are liberated as gases (e.g. H_2, CO_2) are substances the organisms are incapable of using to generate still more ATP. Most fermentations are balanced oxidation-reduction processes, and the major organic end-products are reduced compounds formed from the intermediates which serve as the terminal electron acceptors in the fermentation. The homolactic fermentation which is accomplished during anaerobic utilization of glucose as energy source by bacteria such as *Str. pyogenes* well illustrates this feature. In this fermentation the glucose molecule is converted to two molecules of pyruvate by the Embden-Meyerhof-Parnas (EMP) pathway of glycolysis. The terminal reduction of the pyruvate to yield lactate is rendered essential by the necessity to regenerate NAD^+ from the NADH produced in the earlier dehydrogenation of 3-phosphoglyceraldehyde. The homolactic fermentation is thus a balanced oxidation-reduction in which one molecule of glucose ($C_6H_{12}O_6$) yields two molecules of lactic acid ($C_3H_6O_3$), with 3-phosphoglyceraldehyde serving as the primary electron donor and pyruvate as the terminal electron acceptor. The course of glucose metabolism via the EMP pathway is so contrived that among the intermediates formed are two compounds, namely 1,3-diphosphoglycerate and phosphoenolpyruvate, whose phosphate transfer potentials are such that they can phosphorylate ADP to yield ATP in exergonic reactions. It is these two reactions, catalysed respectively by phosphoglycerylkinase and pyruvate kinase, that are the key 'free energy-conserving' SLP reactions of the fermentation (Fig. 3.2).

Despite the enormous variety of substances such as carbohydrates, amino acids, purines, pyrimidines, and fatty acids that can be fermented by different bacteria, all make use of a surprisingly limited range of SLP reactions. Table 3.1 lists the six most common substrates of these ATP generating reactions. Of these, carbamoyl phosphate and N^{10}-formyltetrahydrofol-

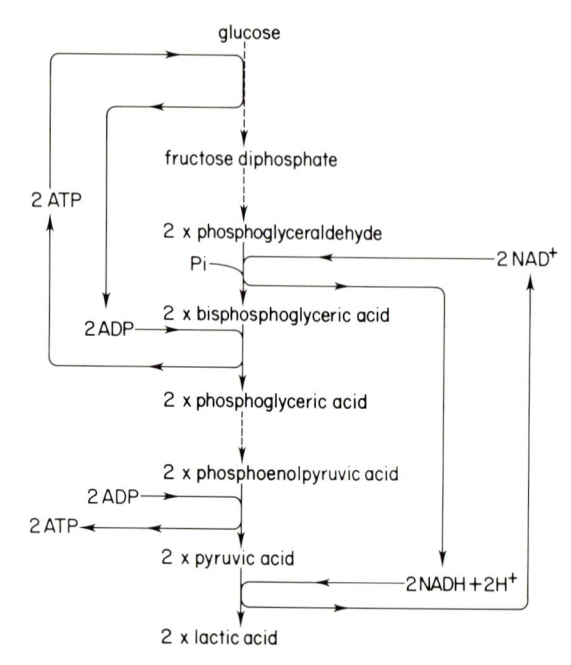

Fig. 3.2 Outline pathway for the fermentation of glucose by a homolactic fermentative bacterium.

Table 3.1 'Energy-rich' compounds implicated in substrate level phosphorylation reactions in bacteria (after Thauer *et al.* 1977)

Type of compound	Energy-rich compound	$-\Delta G^{o\prime}$ of hydrolysis kcal mol^{-1}	kJ mol^{-1}
Phosphoacyl anhydride	Acetyl phosphate	10.7	44.8
	Butyryl phosphate	10.7	44.8
	Bisphosphoglycerate	12.4	51.9
	Carbamoyl phosphate	9.4	39.3
Phosphoenol ester	Phosphoenolpyruvate	12.3	51.6
Acyl anilide	N^{10}-formyltetrahydro-folate	5.6	23.4

The ascribed values of $\Delta G^{o\prime}$ refer to modified standard state conditions of pH 7, ionic strength 0.25 and a free Mg^{2+} concentration of 10 mol dm^{-3}.

ate are of importance in only a few fermentations. Carbamoyl phosphate is generated during fermentations of arginine (Deibel 1964, Mitruka and Costilow 1967, Venugopal and Nadkarni 1977) and allantoin (Valentine and Wolfe 1960, Tigier and Grisolia 1965); N^{10}-formyltetrahydrofolate is an intermediate in certain clostridial fermentations of purines and histidine (Barker 1961). The remaining four most common SLP reactions are kinase reactions which utilize as phosphate donor to ADP either a phosphoacylanhydride (acetyl phosphate, bisphosphoglycerate, butyryl phosphate) or the phosphoenolester, phosphoenolpyruvate.

The necessity that one or more of these compounds must be generated in the course of a fermentation has led to the exploitation by fermenting bacteria of many biochemical reactions that have no parallel in the metabolism of respiring organisms. A good example is provided by the various routes of glutamate utilization by bacteria. In aerobic bacteria glutamate would most likely be oxidized to α-oxoglutarate, which would then be consumed via the tricarboxylic acid cycle. In contrast, various anaerobic bacteria ferment glutamate to yield ammonia, hydrogen and carbon dioxide, together with acetic and butyric acids. Yet, although they yield the same end products,*Cl. tetanomorphum* and *Fusobacterium fusiforme* employ quite different routes for the fermentation of glutamate i.e. they have evolved different solutions to the same problem posed by the requirement that substrates of SLP reactions should be produced in the course of glutamate catabolism. Whereas the clostridium exploits a vitamin B_{12}-dependent pathway via methylaspartate, the fusobacterium degrades the glutamate via a route in which α-hydroxyglutarate and possibly glutaconyl CoA and crotonylCoA are intermediates (Buckel and Barker 1974). One can learn from this the important lesson that the fermentative production of the same products from the same initial substrate must not be taken as evidence that the same metabolic pathway is being followed.

Many more examples can be given of the folly of making such an assumption. Thus *Cl. propionicum* ferments lactate to propionate via acrylylCoA, whilst *Propionibacterium shermanii* also ferments lactate to propionate but by a different route via methylmalonylCoA (Anderson and Wood 1969). Many fermentations of carbohydrates yield reduced end-products from pyruvate, yet the EMP pathway is by no means the only route for the anaerobic conversion of glucose to pyruvate e.g. *Zymomonas mobilis* uses the Entner-Doudoroff pathway, whereas heterofermentative lactic bacteria such as *Leuconostoc mesenteroides* exploit a phosphoketolase (pentose phosphate) pathway (Wood 1961, Anderson and Wood 1969). Again, acetate is a very common end-product of bacterial fermentation, its occurrence generally being indicative of a key role for acetyl phosphate as a SLP reaction substrate in the fermentation. Yet the acetyl phosphate may have been generated in any of several ways. In *Lacto. delbruecki* acetyl phosphate is produced directly from pyruvate by the action of a pyruvate dehydrogenase which is CoA-independent but reliant on FAD as electron acceptor (Hager and Lipmann 1961). In most instances, however, the acetyl phosphate will have been formed from acetyl CoA (by a phosphotransacetylase) and the acetyl CoA will in turn have been produced from pyruvate. But though in a *Clostridium* species this will have been accomplished by the action of a pyruvate-ferredoxin oxidoreductase, in anaerobically growing enterobacteria acetyl CoA is produced by a pyruvate formate lyase. In still other fermentations of sugars, by some lactic bacteria, acetyl phosphate is not produced from pyruvate at all, but by a prior lytic reaction viz. cleavage of fructose 6-phosphate or xylulose 5-phosphate by phosphoketolases (Gottschalk and Andreesen 1979).

In the many bacterial oxidation-reduction fermentations of single substrates the primary electron donor and terminal electron acceptor are manufactured by the organism as intermediates on the pathway of fermentation. In some cases however the bacterium must be supplied with a pair of organic compounds, one of which is oxidized at the expense of reduction of the other. The most notable example of this requirement is provided by those proteolytic clostridia which ferment pairs of amino acids via the Stickland reaction (Stickland 1934) e.g. the fermentation of alanine plus proline by *Cl. sticklandi* (Stadtman 1954, Schwartz et al. 1979).

The lactic fermentation of glucose illustrated in Fig. 3.2 is an example of a simple homofermentative process in which a linear unbranched pathway generates a single electron acceptor. Quite often, however, branched fermentation pathways are exploited, each branch being associated with the production of a distinctive SLP reaction substrate or different electron acceptor. In consequence, the fermentation yields a mixture of end-products whose proportions can vary in response to changes in metabolic and growth requirements. This argues for the existence of control mechanisms which can regulate the flow of metabolites through the various branch routes. Though comparatively little is known of these regulatory processes in most branched pathway fermentations, they probably operate by allosteric controls exerted by key metabolites on the activities of branch-point enzymes. By this means the efficiency of ATP generation can be optimally adjusted to the demands made by the growth conditions in a manner that is not open to the organism which operates a linear fermentation pathway of invariant ATP yield (Thauer et al. 1977). In some fermentations there is latitude for variable discharge of excess reducing power either in the form of H_2 gas or by over-reduction of the normal end-products of fermentation, e.g. production of ethanol in place of acetate or of butanol in place of butyrate. Nor should we overlook the co-operation that can be established between different species of bacteria in mixed culture—anaerobic consortia—where the H_2 liberated by one species may avidly be seized upon by another species which employs it as a reductant.

Not all fermentations are oxidation-reduction processes, for some exploit lytic reactions to yield the required SLP substrate(s). Examples are the fermentation of arginine and agmatine by *Str. faecalis* (Bauchop and Elsden 1960, Roon and Barker 1972) and the fermentation of xanthine by *Cl. cylindrosporum* (Barker 1961, Vogels and van der Drift 1976).

A fermentation is an oxygen-independent and hence anaerobic means of generating ATP. That it does not proceed in aerated cultures of most facultative anaerobes is due to the fact that for such organisms oxygen is the preferred terminal electron acceptor, and aerobic respiration therefore supervenes. A portion of the fermentative pathway may still be exploited in the respiring organism. For example, the EMP pathway continues to be employed to catabolize glucose to pyruvate in aerobically growing *Esch. coli*, though the yield of ATP from the associated SLP reactions is now small in comparison with that derived from the aerobic respiration of NADH chiefly derived from the subsequent complete oxidation of acetylCoA via the tricarboxylic acid cycle. In some bacteria key enzymes of fermentation pathways are not synthesized under aerobic conditions e.g. arginine deimidase in *Str. faecalis* (Abdelal 1979). On occasion, even traces of an alien electron acceptor can perturb the normal course of an oxidation-reduction fermentation e.g. suppression of butyrate formation in favour of acetate production when *Cl. acetobutylicum* is exposed to non-lethal concentrations of oxygen (O'Brien and Morris 1971) or when *Cl. perfringens* is supplied with nitrate (Hasan and Hall 1975). Reducing power generated by anaerobically fermenting organisms can in this way often be diverted to effect specific reductions of exogenous substances; there is a suggestion that reductive formation of carcinogens by the activity of the anaerobic microflora of the human gut might be a factor in the aetiology of cancer of the large bowel (Miraglia 1974).

The range of compounds that can be fermented anaerobically is frequently characteristic of the species. Carbohydrate fermentation tests conducted in media containing a pH indicator have long been among the most generally useful of biochemical tests employed in a diagnostic bacteriology laboratory (Hugh and Leifson 1953, McDade and Weaver 1959, MacFaddin 1976, Mitruka 1976). The tests are usually performed qualitatively, with the investigator observing production of acid with or without gas evolution during anaerobic incubation. The availability of simple gas chromatographic procedures which enable samples of the growth medium to be rapidly and quantitatively analysed for their contents of fermentation end-products has greatly extended the potential usefulness of such tests (Mitruka 1976). Chromatographic procedures are similarly useful in the identification of proteolytic anaerobes e.g. species of *Clostridium*, by indicating which amino acids are used and what products are formed (Elsden et al. 1976, Elsden and Hilton 1979).

All fermentations are catalysed by soluble enzymes located in the bacterial cytoplasm. Thus part of the ATP that is generated by SLP reactions must, in an obligately fermentative bacterium, be consumed at the cytoplasmic membrane in order to generate the transmembrane electrochemical gradient of protons that is required to maintain the ionic integrity of the cell and to effect selective import of nutrients and export of undesirable products and ions. This protonmotive force is created by a membrane-associated, proton-exporting ATP phosphohydrolase (Harold 1977, Clarke et al. 1979), whose activity may dispose of a substantial part of the total ATP synthesized by a fermenting bacterium (Stouthamer 1979).

Respiration

Cellular respiration is invariably a membrane-associated process which in the bacterium occurs at the cytoplasmic membrane (Jones 1976, Haddock and Jones 1977). Within this membrane are a series of electron carriers—specific redox couples—arranged in such a fashion that electrons donated by some primary reductant, which may be an inorganic or organic compound, are transferred along the chain of carriers in order of their ascending midpoint redox potentials until finally they are employed to reduce a terminal electron acceptor—the respiratory oxidant—supplied by the environment. Simultaneously with this electron transport, protons are pumped (exported) across the membrane; by this means the interior of a respiring bacterial cell is made electronegative and alkaline with respect to the external medium. In this situation protons would tend to return along their electrochemical gradient, and by channeling their return via a proton-translocating membrane ATP synthetase the protic current is harnessed to generate ATP. Note that this is the converse of the situation in the obligately fermentative bacterium in which ATP is consumed to generate the transmembrane protonmotive force (Fig. 3.3). In the respiring organism, free energy deriving from an oxidation-reduction process is thus initially conserved as a protonic potential difference, which is then transduced to yield ATP via a H^+-ATP synthetase (Postma and van Dam 1976).

This chemiosmotic explanation of the mechanism whereby respiratory electron transport is coupled to ATP synthesis was initially postulated by Mitchell (1961) on purely theoretical grounds, but has since been fully justified by a wealth of experimental evidence (Mitchell 1966, Garland 1977; see Haddock and Hamilton 1977, p. 1).

The reductant that is most often employed as the primary electron donor in bacterial respiration is NADH generated by metabolism in the cytoplasm. Thus the first component of many bacterial respira-

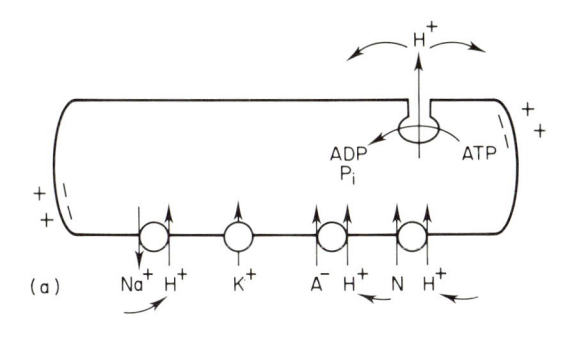

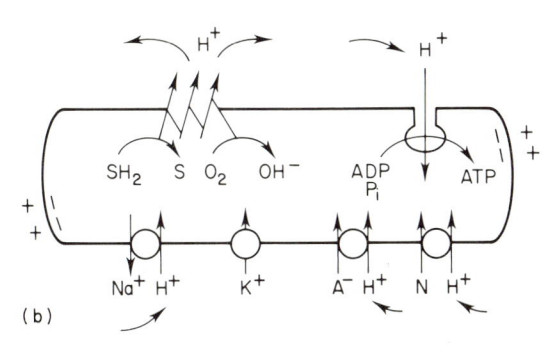

Fig. 3.3 Free energy transduction by the transmembrane proton circulation in bacteria (modified from Harold 1977). (a) Obligately fermentative bacterium in which the protonmotive force is generated by hydrolysis of ATP via a proton-exporting ATP phosphohydrolase. (b) Aerobically respiring bacterium in which the sequence of membrane-located electron transport reactions (which oxidizes substrate SH_2 by O_2) is associated with the export of protons whose return via the proton-translocating ATP synthetase impels the synthesis of ATP. In both types of organism the protonmotive force (external medium acid and electro-positive) is harnessed to accomplish ion and substrate transport via specific membrane-integrated 'porters' viz. cation uniporters (K^+), anion (A^-) and neutral molecule (N) proton symporters and cation proton antiporters (Na^+).

Several other compounds (known as uncoupling agents), though allowing electron transport to proceed and even accelerating it, prevent the associated ATP synthesis by abolishing the transmembrane protonmotive force. These uncoupling agents, in effect, dissociate oxidation from phosphorylation (Hamilton 1975). Specific inhibitors of the membrane H^+-ATP synthetase will act more directly as phosphorylation inhibitors (West 1974). Details of the constitution and operation of bacterial respiratory electron transport chains have been revealed both by the discriminating use of such inhibitors and by the examination of mutant strains blocked at various stages in oxidative phosphorylation (Haddock 1977; see Haddock and Hamilton 1977, p. 95).

Bacteria differ greatly in the detailed composition of their respiratory chains; and variations in redox carrier patterns, particularly in the quinones and terminal oxidases, can be induced even in a single species by changes in its growth conditions (Haddock and Jones 1977). All bacterial respiratory systems show some degree of branching. This is likely to be most extensive at the level of the primary dehydrogenases, since electrons can be fed into the respiratory chain from a number of reductants in addition to NADH. Terminal branching of the chain is also not uncommon, with more than one cytochrome oxidase being present. In a few bacteria the branching is even more complex, involving b or c-type cytochromes as well as multiple cytochrome oxidases. The extent of such branching can depend on the concentration of oxygen supplied during growth, which can determine what proportions of various redox carriers are synthesized. The major benefit to the organism of a branched respiratory chain is undoubtedly to allow some flexibility in the route of electron transfer, thus enabling it to minimize the potentially deleterious effects of certain growth environments and to take maximum advantage of others. Bacterial respiration in any event is subject to both coarse and fine controls (Jones 1976; see Haddock and Hamilton 1977, p. 23).

The types of cytochrome possessed by various bacteria are of interest to the bacterial taxonomist as well as to the physiologist (Meyer and Jones 1973, Jones and Meyer 1976). The qualitative oxidase test of Kovács (1956) or some variant of it (MacFaddin 1976) is taken to denote the presence of a distinctive component of the respiratory system in bacteria that can aerobically oxidise the test reagent (Stanier et al. 1966). This test is particularly helpful in the identification of the genus *Neisseria*, though it is also used to distinguish the Pseudomonadaceae from oxidase-negative members of the Enterobacteriaceae (Steel 1961). It is relevant to note however that in *N. meningitidis* a portion of this oxidase activity is located in the outer membrane of the cell envelope (Devoe and Gilchrist 1976).

tory chains is a NADH dehydrogenase complex (containing non-haem iron sulphur proteins and flavoproteins). Electrons are donated by this complex to a quinone—ubiquinone or menaquinone—and thence via haem-containing cytochromes to the terminal component of the electron transport chain, which then reduces the terminal oxidant (Fig. 3.4). In aerobic respiration the last member of this chain consists of one or more cytochrome oxidases (Jurtshuk et al. 1975).

A number of compounds interact with specific components of the respiratory chain, blocking electron transport and halting ATP generation. Among these potentially lethal agents are three compounds which react with cytochrome oxidase: carbon monoxide competes with oxygen for its binding site on the reduced cytochrome oxidase, whilst both azide and cyanide combine with the oxidized form of this molecule.

Until quite recently it was thought that *anaerobic respiration* could be accomplished only by somewhat specialized bacteria using a few inorganic oxidants in place of molecular oxygen. Denitrifying bacteria when growing anaerobically with nitrate reduce this oxidant to dinitrogen gas (Payne 1973, Stouthamer 1976).

```
Succinate ⋯⋯⋯⋯⋯⋯⋯⋯⋯↘              Cytochrome b₅₅₆ ────→ Cytochrome o
                                                                  oxidase
NADH ─────→ Ubiquinone ↗
                       ↘                                  (Cytochrome a₁)
α glycerophosphate ⋯⋯⋯↗              Cytochrome b₅₅₈ ────→ Cytochrome d
                                                                  oxidase
```

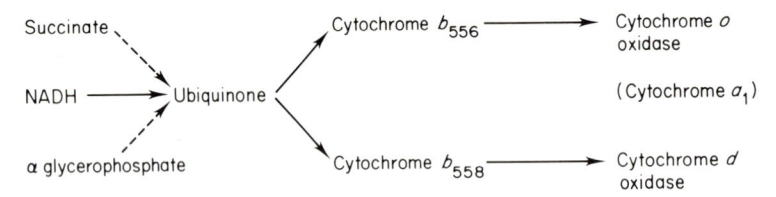

Fig. 3.4 Branched respiratory chain of *Esch. coli* grown under conditions of oxygen limitation. Cytochrome oxidase *o* is cyanide-sensitive and cytochrome oxidase *d* is cyanide-resistant. Though present, cytochrome a₁ does not appear to serve as a terminal oxidase (after Jones 1977).

Sulphate-reducing bacteria such as *Desulfovibrio desulfuricans* are obligate anaerobes which respire by reducing sulphate to sulphide ions (Le Gall and Postgate 1973, Postgate 1979). The cell membranes of these bacteria contain electron transport couples, including quinones and cytochromes, which resemble those exploited by aerobically respiring bacteria. The methane-generating bacteria, on the other hand, are strict anaerobes which can use hydrogen as primary electron donor and carbon dioxide as terminal electron acceptor in a specialized route of membrane-associated respiration which exploits entirely novel electron transport carriers (Wolf 1979). Though *Des. desulfuricans* was the first example of a non-photosynthetic strict anaerobe which contained cytochrome (Ishimoto et al. 1954), other anaerobes have now been discovered to contain both quinones and cytochromes e.g. *Cl. formicoaceticum* and *Cl. thermoaceticum*; Gottwald et al. 1975. Some anaerobes e.g. species of *Bacteroides* (Macy et al. 1975) that are unable to synthesize haem can, however, synthesize and exploitcytochromes when provided with haem or haematin in their growth medium. These and other discoveries indicate that anaerobic respiration is probably of much more widespread occurrence among bacteria than was formerly realized. It is now evident that a suitable organic oxidant can serve as terminal electron acceptor in anaerobic respiration, fumarate having been identified to serve this role in many obligate and facultative anaerobes (Kroger 1977; see Haddock and Hamilton 1977, p. 61).

Indeed, some capacity to supplement ATP generation, and hence growth yields, by fumarate respiration appears to be fairly widespread among facultatively anaerobic, gram-negative bacteria e.g. species of *Haemophilus*, *Pasteurella*, and *Vibrio*, and also in a lesser number of gram-positive bacteria, including some species of *Bacillus* and *Staphylococcus*. Anaerobic respiration with fumarate as terminal oxidant explains the ability of *Cl. formicoaceticum* to grow on fumarate and of *V. succinogenes* to use hydrogen or formate plus fumarate as the source of energy for growth. Even *Esch. coli* can similarly grow anaerobically on a mixture of hydrogen plus fumarate; and there is reason to believe that during its anaerobic growth on glucose, ATP generation by fermentation is supplemented by fumarate respiration (Boonstra and Konings 1978). *Str. faecalis* growing on glucose plus fumarate significantly produces succinate plus acetate in place of the lactate that it forms from glucose alone (Deibel and Kvetkas 1964). The ability to undertake anaerobic respiration with fumarate as oxidant appears to be resricted to organisms possessing menaquinone or desmethylquinone in their cell membranes; *b* cytochromes may also be required for the respiratory electron transfer to the terminal, membrane-located fumarate reductase.

In respiration, and especially aerobic respiration, disposal of excess reducing power generated during energy-yielding catabolism, poses as a rule no problem to the bacterium. In aerobic respiration, water is the accumulated end-product of reduction of the terminal electron acceptor, disposal of the hydrogen peroxide produced incidentally being accomplished by catalase or peroxidases or both. Respiring organisms therefore generally operate catabolic pathways in which the supplied source of the respired reductant(s) is completely oxidized (Kornberg 1959). Thus the tricarboxylic acid cycle which effects complete oxidation of acetylCoA—or of any of the cycle intermediates when these are supplied—is frequently a key amphibolic pathway in respiring organisms, supplying much of the NADH which is reoxidized at the cell membrane. The associated high yield of ATP (approximately 3 mol ATP per mol of NADH respired) explains why total oxidation, associated with respiration, of an organic substrate is a much more effective means of ATP generation than is fermentative metabolism of the same substrate.

Transport of solutes across the bacterial cell membrane

Before any component of the culture medium can be metabolized it must first be taken up by the bacterial cell. The need to acquire certain substances from the culture medium and to exclude others presents bacteria not only with a problem in discrimination but also with the task of translocating a large number of hydrophilic and possibly ionized compounds across the hydrophobic osmotic barrier presented by the cytoplasmic membrane (Tristram 1978, Rosen 1978). It was evident to Monod and his colleagues (Cohen and Monod 1957) that only the synthesis of some specific protein component of the inner cell membrane of *Esch. coli* could explain the acquisition by cells that had been exposed to lactose of the means of taking up this sugar. They postulated that the transport process required transitory formation of a specific complex between this inducibly synthesized membrane protein and the β-galactoside; and they coined the term 'permease' to draw attention to the function of this protein. The vital role of the β-galactoside permease was evident from the fact that it was possible to obtain 'cryptic' mutant strains of *Esch. coli* which were unable to grow on lactose as a source of carbon and energy, even though they possessed the means of metabolizing the disaccharide once it had gained entry to the cytoplasm—as was demonstrable when brief treatment of the cells with toluene breached the permeability barrier posed by the cytoplasmic membrane. Although it is no longer fashionable to denote such transport proteins as permeases, the concept of functionally specialized transport systems mediating the transfer of compounds across the bacterial cell membrane has been wholly vindicated. We now know that, when *Esch. coli* is grown on lactose, one of the three proteins whose synthesis is evoked is a transport protein (*lac*Y protein) responsible for the uptake of β-galactosides. It is a small molecular weight intrinsic membrane protein capable of being extracted in its native state from the cytoplasmic membrane of the lactose-induced cell (Altendorf et al. 1977).

The same solute may be transported by different mechanisms in different bacteria and by multiple means in the one organism. It is not only the specificity and variety of these transport systems that is impressive, but also the fact that many can translocate the substrate against its concentration gradient. Such 'uphill' translocation requires the transport process to be coupled to some source of free energy. In practice this means either (a) utilization of ATP or of some other 'high-energy' group transfer agent such as phosphoenolpyruvate, or (b) exploitation of the transmembrane electrochemical gradient of protons generated by respiration or by the hydrolysis of ATP supplied by fermentation.

The mechanisms whereby active transport across the bacterial cell membrane can be sustained by the protonmotive force have been expertly reviewed (Harold 1972, 1977a, 1979; Hamilton 1975, 1977; see Haddock and Hamilton 1977, p. 185).

Briefly, the protonic potential difference (Δp) across the membrane has both electric ($\Delta\psi$) and thermodynamic activity (ΔpH) components, both of which tend to impel spontaneous return of protons to the cell cytoplasm. This can take place only via certain specific channels plugged through the otherwise proton-impermeable insulating membrane. We have seen how proton return via the H^+-ATP synthetase is associated with the phosphorylation of ADP in respiring bacteria; but the protonic potential difference can also be exploited to propel other ions and even uncharged solutes across the membrane via intrinsic membrane proteins (porters) which act as *specific* carriers. Cations other than protons might be imported into the cell by unifunctional porters in response to the electrical potential difference ($\Delta\psi$, internal negative), or conversely could be exported in exchange for imported protons via a proton antiporter. Anions might enter on bifunctional carriers (proton symports) in response to the proton activity gradient (ΔpH), whilst neutral solutes could also be taken into the cell via proton symports, the movement being motivated by both $\Delta\psi$ and ΔpH. These possibilities are illustrated in Fig. 3.3. So important is the protonic potential difference that even in an obligately fermentative bacterium it may be dispensed with only under certain extraordinary conditions (Harold and van Brunt 1977).

Our knowledge of individual membrane-transport systems has often been acquired by kinetic measurements of the rates of flux of permeant substrates across the membranes of whole cells and protoplasts. Additional information has been obtained by studies performed upon mutant strains defective in their transport properties, and with suspensions of closed membrane vesicles which can be prepared from bacterial cell membranes in such a way that they retain the ability to undertake energy-dependent transport processes characteristic of the parent membrane (Kaback 1971, Konings 1977).

The translocation of some, notably hydrophobic, substances through the bacterial cell membrane may occur by passive diffusion along their normal concentration gradients e.g. excretion of undissociated fatty acid end-products of fermentation. However, in all other cases substrate transport occurs by one of three mechanisms:

(1) *Facilitated diffusion.* In this process an intrinsic membrane protein (the facilitator) interacts stereospecifically with the solute and accelerates its passage across the membrane along its normal concentration gradient. Theoretically, at least, this process should not require input of free energy, since it represents merely an acceleration of spontaneous movement towards the equilibrium reached when equal activities of solute are present on both sides of the membrane.

(2) *Active transport.* A membrane carrier is again responsible for the accelerated and specific translocation of the solute, but the process is dependent on a second component which supplies the free energy necessary for its operation against the prevailing electrochemical gradient of that solute, so that ultimately it is concentrated either within the cell or the medium, depending on whether the system functions to effect influx or efflux of the solute.

(3) *Group translocation.* Whereas in facilitated diffusion and active transport the solute is translocated unchanged, group translocation results in the transported solute appearing on the other side of the membrane in a modified form i.e. as some derivative. Thus the membrane component functions as an enzyme with vectorial properties.

All three types of transport system can be illustrated by known mechanisms for the uptake of carbohydrates by *Esch. coli* (Silhavy et al. 1978; see Rosen 1978,

p. 127). Entry of glycerol is mediated by a facilitator, but it is not concentrated within the cell and the process appears to be one of facilitated diffusion (Lin 1976). The active transport systems for sugars fall into two classes depending on whether or not a periplasmic substrate-binding protein is implicated. Such periplasmic proteins are lost from organisms subjected to severe cold osmotic shock (Neu and Heppel 1965). Among the 'shock-resistant' active transport systems involving no periplasmic component is the lactose-uptake system mediated by the *lac*Y protein, which presumably functions as a proton symport lactose carrier. 'Shock-sensitive' transport systems are concerned in the uptake of galactose, arabinose, maltose and ribose, with a specific sugar-binding, periplasmic protein being implicated in each case. This sugar plus binding-protein complex probably interacts with the membrane-located component, the active transport drawing its energy directly from ATP, though a transmembrane proton potential may be necessary to keep the membrane component in its active state. Finally, as an example of group translocation, we find in *Esch. coli* the phosphoenolpyruvate phosphotransferase (PTS) system (Kundig *et al.* 1964). This consists of a complex of enzymic and carrier proteins which convert hexose outside the membrane to hexose phosphate inside the membrane by a series of phosphoryl transfer reactions (Saier 1979). It is of considerable significance that the synthesis and activities of all such transport systems are regulated. We have noted the induced synthesis of the *lac*Y protein. But even when a mixture of sugars is provided to a bacterium which could be expected to transport them all, a preference can be displayed which can only be explained by regulatory interaction between the various transport systems. Thus in *Esch. coli* (Kornberg and Jones-Mortimer 1977; see Haddock and Hamilton 1977, p. 217) and in *V. cholerae* (Bag 1974) rapid uptake of glucose via the PTS system has an inhibitory effect on the concurrent transport of other sugars both via PTS and non-PTS systems.

Bacterial transport systems for amino acids (Anraku 1978; see Rosen 1978, p. 171) are of various degrees of specificity. Several effect symport of the amino acid with protons (Hirata *et al.* 1976), but others implicate a periplasmic binding protein and ATP (Berger 1973). The transport systems for inorganic anions and cations are highly specific and hence numerous (Silver 1978; see Rosen 1978, p. 221). Of especial interest are the uptake systems for potassium ions (Harold and Altendorf 1974, Epstein and Laimins 1980) and phosphate ions (Rosenberg *et al.* 1977), and the excretory systems for sodium ions (West and Mitchell 1974), and calcium ions (Rosen and Brey 1979). Under conditions of iron starvation the bacterial uptake of iron frequently necessitates secretion into the medium of iron-chelating agents of high affinity—generally catechols or hydroxamates. After binding with iron, the ferrichelate is transported into the bacterium by a highly specific uptake system (Neilands 1974).

Permeation of the outer membrane of the gram-negative bacterium

The major components of the outer membrane are proteins, phospholipids, and lipopolysaccharides. The lipopolysaccharide in particular presents a permeability barrier to hydrophobic substances, including certain bacterial dyes, detergents, and antibiotics; this property has been exploited in the formulation of selective media for the gram-negative enteric bacteria e.g. deoxycholate agar or eosin methylene blue agar. Significantly, 'deep rough' mutant strains of *Esch. coli* or *Salm. typhimurium* which produce lipopolysaccharide with much shortened polysaccharide side chains (O antigen) are as sensitive as are gram-positive bacteria to these agents and to antibiotics such as actinomycin D, erythromycin, or novobiocin. Wild-type, LPS complete, strains of gram-negative enteric bacteria are, however, sensitive to antibiotics such as ampicillin, cephalothin, cycloserine, and neomycin which apparently diffuse across the outer membrane by some hydrophilic route (Nikaido and Nakae 1979).

The outer membrane of *Esch. coli* is generally permeable to hydrophilic molecules smaller than about 600 daltons, acting in this respect rather like a molecular sieve. The aqueous pores or channels through which these solutes passively diffuse are provided by certain acidic proteins, the porins, whose channel-forming ability has been demonstrated by using reconstituted vesicles (Nakae 1976) or planar lipid bilayer membranes (Benz *et al.* 1978, Schindler and Rosenbusch 1978). Porins from the outer membrane of *Ps. aeruginosa* allow the passage of solutes of significantly higher molecular weights (Hancock and Nikaido 1978); yet other factors such as ionic charge and degree of hydrophobicity as well as molecular size determine passage through porin pores. Additionally the outer membrane contains various intrinsic proteins which serve as specific transporters for otherwise non-permeant solutes, e.g. in *Esch. coli*, a maltose carrier, a vitamin B_{12} transport protein, several carriers for ferrichelates (Kadner and Bassford 1978; see Rosen 1978, p. 413). Thus there exists a multiplicity of routes for the passage of various materials across the outer membrane, including macromolecules which may occasionally gain access to the periplasmic space via transient rupture and resealing of the membrane.

Catabolic pathways

These are the metabolic pathways which effect the degradation and chemical transormation of substances provided to the organism, to produce therefrom metabolites which can be used by the central metabolic pathways. The catabolic routes and the enzymes that service them are therefore almost as diverse as the range of organic compounds bacteria use as sources of carbon and energy for growth. Of course no species of bacterium is capable of catabolizing more than a restricted range of substances. Their abilities in this regard vary considerably, from the bacterium which is capable of using a single compound only, e.g. *Bac. fastidiosus* which can use only uric acid or its degradation products allantoin and allantoate as energy source for growth (Vogels and van der Drift 1976), to an organism such as *Ps. multivorans* which can use more than a hundred different organic compounds as carbon and energy sources (Stanier *et al.* 1966). The latter organism acquires its spectacular versatility by exploiting inducibly synthesized catabolic enzymes and, as in other genera also but especially in *Pseudomonas*, may be aided by its carriage of plasmids which encode for unusual degradative enzymes (Broda 1979). The inducer may be the first substrate of the catabolic pathway, and may control the synthesis of a number of sequentially acting enzymes. Alternatively, some later enzymes of the pathway may be induced by one of the catabolic intermediates. The advantage to the bacterium of such *sequential induction* is that, if the first substrate is supplied the synthesis of all enzymes required for its catabolism is assured, whereas if the intermediate inducer is supplied in the medium the early enzymes, which would now be redundant, are not synthesized (Stanier and Ornston 1973). Another variant is control by *product induction* e.g. urocanate as the inducer of the enzymes for histidine degradation in *Ps. aeruginosa*, which ensures that histidine is not directed to the catabolic pathway unless its concentration exceeds that required for protein synthesis. Many catabolic pathways are additionally subject to catabolite repression (Clarke, 1978).

Aerobic organisms frequently exploit oxygen as a reactant in catabolic processes wherein aromatic rings are opened by oxygenation (Dagley 1978). Although obligately anaerobic bacteria are denied this ability, they can use hydration and dehydration reactions to accomplish similar feats (Morris 1975), so that catabolic versatility in this and other instances is not the prerogative of aerobes. Examples of specific catabolic pathways are given in reviews describing the bacterial breakdown of carbohydrates (Fraenkel and Vinopal 1973, Mortlock 1976), aromatic compounds (Dagley 1971, 1978; Ornston 1971), amino acids (Massey *et al.* 1976), and purines and pyrimidines (Vogels and van der Drift 1976). Even when a given compound fails to support growth of a bacterium, it may nevertheless be degraded, if only partly, alongside the metabolism of another substance which acts as a carbon and energy source—a phenomenon termed *co-metabolism* (Horvath 1972).

In their natural environment bacteria are often poorly provided with readily assimilable substrates and are forced to use large-molecular weight materials which they must break down extracellularly to components that are transportable into the cell.. This they accomplish by the secretion of extracellular hydrolytic enzymes. In a clinical context the macromolecules likely to be present in the environment include proteins, nucleic acids, collagen, mucopolysaccharides, lipids etc. Against these, many invasive pathogenic bacteria produce one or more of the appropriate extracellular enzymes viz. proteinases, peptidases, deoxyribonuclease, ribonuclease, collagenase, hyaluronidase, lipases (Rogers 1961, Glenn 1976). Such enzymes are recognizable also as biologically active antigens (toxins) whose identification by immunological methods frequently forms the basis of strain-typing procedures. For example, the lethal α toxin of *Cl. perfringens* is a Ca^{2+}-dependent lecithinase; and minor antigens produced by *Cl. perfringens* type A include a collagenase, a hyaluronidase, and a deoxyribonuclease (Sterne and Batty 1975). Bacterial extracellular enzymes may also be implicated in detoxification processes e.g. bacterial production of penicillinase (Lampen 1978; see Stanier, Rogers and Ward 1978, p. 231).

Anabolic pathways

The metabolic pathways employed to synthesize cellular components differ from those that are used to degrade them. In part, this is an inevitable consequence of thermodynamic constraints e.g. use of 'activated precursors' to bring about spontaneous biosynthetic reactions, or of the need to copy template molecules when fabricating informational macromolecules. Yet the employment of quite distinct anabolic and catabolic reaction sequences also facilitates their independent regulation and harmonious cooperation.

Biosynthesis of small molecules

(a) Amino acids Familial relationships are discernible amongst groups of amino acids whose members share a common precursor provided by one of the central metabolic pathways (Cohen 1967, Umbarger 1978).

Six groups are recognizable:

1. glutamate group, consisting of glutamate, glutamine, arginine, proline, and hydroxyproline—stems from α-oxoglutarate

2. aspartate group, consisting of aspartate, threonine, methionine, isoleucine, and lysine—originates in oxaloacetate

3. alanine, valine and leucine, share a common origin in pyruvate

4. serine, glycine, cysteine and cystine may all originate in 3-phosphoglycerate, though in some bacteria

glycine may first be synthesized from glyoxylate or threonine and thereafter be hydroxymethylated to yield serine in a process involving N^5-formyltetrahydrofolate

5. histidine is produced from ATP plus phosphoribosylpyrophosphate

6. aromatic amino acids viz. phenylalanine, tyrosine, and tryptophan stem from shikimate, which has its origin in the condensation of erythrose 4-phosphate with phosphoenolpyruvate.

The pathways of biosynthesis generate the α-oxo acid precursor of the amino acid, which is then formed by a terminal transamination reaction. The biosynthesis of each amino acid may be regulated both by controls exercised on the synthesis of the enzymes specifically concerned with its production, and also by fine controls exercised on the activities of certain of these enzymes (Umbarger 1969). In multifunctional biosynthetic pathways leading to the synthesis of more than one amino acid from a common precursor, the patterns of enzymic control may be complex, but are nicely adjusted to the requirements of the bacterium under different growth conditions (Umbarger 1978).

Though L-amino acids predominate, D-amino acids are to be found in certain bacterial products e.g. D-alanine and D-glutamate in cell wall peptidoglycan, D-glutamate in the capsular material of *Bac. anthracis* and D-amino acids in several cyclic polypeptide antibiotics such as the gramicidins and tyrocidins. When produced, the D-amino acid is generally formed by enzyme-catalysed racemization of the L-amino acid. Thus cycloserine is toxic to many bacteria because, having a similar structure to alanine, it competitively inhibits alanine racemase and D alanine: D alanine synthetase prevents normal cell wall synthesis.

(b) Purines and pyrimidines These two types of heterocyclic nitrogenous bases are synthesized by quite different routes. The purine ring has as its starting material ribose 5-phosphate; the pyrimidine ring, on the other hand, originates in aspartate plus carbamoyl phosphate. The pathways of biosynthesis, which in the first instance yield inosine 5'-phosphate (IMP) and uridylic acid (UMP) respectively, are complex (Moat 1979), as are the details of the interconversion of the various purines and pyrimidines and their derivatives. The end result is to supply the variously phosphorylated deoxyribonucleosides of adenine, guanine, cytosine, and thymine, and phosphorylated ribonucleosides of adenine, guanine, cytosine, and uracil, that are required for the synthesis of DNA and RNA, the activation of various biosynthetic precursors, and the manufacture of coenzymes such as FMN, FAD and NAD(P) (e.g. Foster and Moat 1980).

In bacteria such as *Esch. coli*, deoxyribonucleoside diphosphates are produced by the reduction of the corresponding ribonucleoside diphosphates in a reaction catalysed by a ribonucleotide reductase, using reduced thioredoxin as the reductant. In a few species of bacteria vitamin B_{12} is required for the reduction, which now takes place at the nucleoside triphosphate level. Methylation of the uracil ring to produce thymine occurs at the level of dUMP with methylenetetrahydrofolate generally serving as donor of the C1 group and as the reductant. In bacteria which operate this route—and some obligate anaerobes use an alternative—the drug 5-flurodeoxyuridine is bactericidal since, after its conversion to 5-fluorodeoxyuridylic acid, it inhibits the thymidylate synthetase. Similarly, such bacteria are particularly susceptible to inhibitors of dihydrofolate reductase (e.g. trimethoprim).

Regulation of purine and pyrimidine biosynthesis again involves control of gene expression and regulation of enzyme activities (Gots 1971). In purine biosynthesis, feed-back inhibition by AMP and GMP of the amination of PRPP by glutamine, and feed-back regulation of the first branch-point enzymes after IMP are particularly important. In pyrimidine biosynthesis, feed-back regulation of the first enzyme specific to this pathway i.e. aspartate transcarbamylase provides an informative example of the operation of this type of regulatory mechanism (Kantrowitz et al. 1980).

(c) Fatty acids and lipids Except for the rather special homopolymer of β-hydroxybutyrate that is produced by certain species (Dawes and Senior 1973), bacteria do not accumulate lipids as reserve materials. The major lipids are therefore membrane constituents including phospholipids, lipoproteins, and the glycolipids found in many gram-positive bacteria (Goldfine 1972, O'Leary 1975). Lipopolysaccharide is a major component of the outer membrane of gram-negative bacteria; and a few bacteria are noted for the high and somewhat unusual lipid content of their cell walls e.g. *C. diphtheriae Myco. tuberculosis*.

The precursor of fatty acids is acetylCoA, yet the intermediates in the manufacture of the chief fatty acids in a bacterium are not CoA esters. Instead, bacteria employ an acyl carrier protein, ACP with a prosthetic group of 4'-phosphopantetheine, synthesis being accomplished by the enzymic addition of C2 units from malonyl-ACP to the growing acyl chain attached to ACP. The final acyl residue may be transferred from the ACP to glycerophosphate to yield a phosphatidic acid. This may then interact with CTP to produce an 'activated' CDP-derivative susceptible to esterification with alcohols such as serine which displace CDP. Decarboxylation of phosphatidylserine yields phosphatidylethanolamine, which may be methylated by S-adenosylmethionine to yield phosphatidylcholine.

Introduction of a double bond into a fatty acid molecule to yield a monounsaturated fatty acid can be accomplished in one of two ways. In bacteria operating the *anaerobic mechanism*, e.g. *Esch. coli*, *Salm. typhimurium*, *Staph. aureus*, and *Cl. butyricum*, at the C_{10} stage of fatty acid synthesis the βhydroxydecanoyl-ACP is acted upon by a special dehydratase to yield a *cis* double bond between carbon atoms 3 and 4; the resulting compound does not serve as a substrate for the normal enoyl-ATC reductase, but instead continues to be elongated by the accretion of C2 units until the required C_{16} or C_{18} mono-unsaturated acyl-ACP is produced. In the *aerobic mechanism* operated by other bacteria, preformed long-chain fatty acids are suitably hydroxylated before desaturation by an oxygenation reaction using molecular oxygen and NADPH. Since multiply-unsaturated fatty acids can only be produced by such an oxygenative mechanism, it follows that anaerobes and other bacteria operating the anaerobic mechanism do not contain such compounds.

There are numerous variants of the basic pattern of lipid synthesis in bacteria that give rise to very diverse products, including branched chain- and ring-containing fatty acids, and fatty acids bound in ether linkages to glycerol, as in the plasmalogens found in certain anaerobes (Goldfine 1972). Furthermore, the conditions under which the bacterium is grown—composition of the growth medium, availability of oxygen, temperature of incubation, prevailing pH—can substantially affect the nature of the lipids it contains. The changes so wrought are frequently explicable by the need for the organism to preserve a fluid membrane (Cronan and Gelmann 1975). The bacterium may even make opportunistic use of unusual fatty acids present in the growth medium (Finnerty and Makula 1975, Cronan 1978, Finnerty 1978, Raetz 1978).

Biosynthesis of macromolecules

The bacterium is required to synthesize two categories of macromolecule:

(1) Periodic macromolecules which generally serve as reserve materials or have a structural role. They may be homopolymers, e.g. glycogen, poly β-hydroxybutyrate, or heteropolymers e.g. cell wall peptidoglycan, outer membrane lipopolysaccharide.

(2) Information macromolecules i.e. DNA, RNA and protein, wherein the precise order in which a small selection of monomers are linked together is of vital importance to their function.

Synthesis of both types of macromolecule poses the requirement that stable covalent linkages must be forged between the unit components by a mechanism rendered spontaneous by the provision of the necessary free energy. In all cases, serial coupling of the components is accomplished by suitable enzymes catalysing the synthesis of the requisite bond—glycosidic, peptide, sugar phosphate ester etc.; spontaneity is assured by the provision of one of the reactants in a suitably activated form. Invariably it is the small-molecular weight unit that is to be attached to the growing macromolecule which is appropriately activated. In this way monosaccharide nucleotides are used in the synthesis of polysaccharides, amino acids are delivered to the site of protein synthesis as amino acyl tRNAs, and the purine and pyrimidine nucleosides which are to be employed for the synthesis of DNA and RNA are supplied as their 5′-triphosphates.

In one respect, however, the fabrication of informational macromolecules is accomplished quite differently from that of other products. The requirement that their structures must be free from error makes it necessary for the components to be joined together in a sequence dictated by information pre-encoded on a template molecule. The key to how this is done with the necessary fidelity lies in the structure of the genetic material i.e. DNA.

Biosynthesis of periodic macromolecules

It would be wrong to suppose that, because synthesis of these molecules does not require provision of a template, they can be constructed in a haphazard fashion. Even with homopolymers the manner in which the units are joined together, e.g. degree of branching of the molecule, will determine the distinctive nature of the product. Thus ADP-glucose may be used to synthesize glycogen in *Esch. coli* and an amylopectin, granulose, in butyric clostridia, the synthesis being subject to both coarse and fine controls (Preiss 1969, Dawes and Senior 1973). The structural macromolecules of the bacterial cell wall are complex heteropolymers (Schleifer and Kandler 1972, Braun and Hantke 1974, Schleifer *et al.* 1976). The peptidoglycan, murein, component is synthesized from UDP-*N*-acetyl-muramyl pentapeptide and UDP-*N*-acetylglucosamine, the former first being transferred to the membrane lipid carrier undecaprenyl phosphate (Rogers 1979, Tonn and Gander 1979). The penicillins and cephalosporins interfere with subsequent cross-linking (transpeptidation) reactions and possibly with other steps in this sequence (Spratt 1980, Yocum *et al.* 1980), resulting in the inhibition of cell-wall synthesis in sensitive growing bacteria (Blumberg and Strominger 1974, Tomasz 1979). The lipopolysaccharide of the outer membrane of gram-negative bacteria has a complex structure (Nikaido 1973, Ørskov *et al.* 1977); the nature and sequence of sugar units in the repeating oligosaccharide units of the exposed side chain determine the somatic (O) antigen specificity of these bacteria (Roantree 1967).

Bacterial exopolysaccharides, though of simple composition (repeating units of two to six sugars to which various acyl groups are attached), differ in their functions and physical properties (Powell 1979). Their synthesis from sugar nucleotides again requires their assembly on the lipid phosphate used in the synthesis of peptidoglycan and lipopolysaccharide, followed by extrusion from the cell (Sutherland 1979*a, b*; Troy 1979).

Biosynthesis of informational macromolecules

(a) Biosynthesis of DNA The inspired deduction by Watson and Crick in 1953 that the DNA molecule consisted of two complementary polynucleotide chains twisted about each other in the form of a regular double helix proved so germinal a concept that it is well worthwhile to read the original statement of their hypothesis and to place this in the context of previous speculations (Watson 1968, Crick 1974). The twin chains, or strands, are joined together by hydrogen bonds between specific pairs of bases. For sound stereochemical reasons adenine is invariably paired with thymine, and guanine with cytosine. This specific base pairing ensures the complementarity that enables the double helix to be faithfully replicated. When the strands are separated, each may specify the order in which nucleotides must be sequentially linked together to form the unique new strand which will complement that serving as its template. In this way DNA provides the template for its own reproduction i.e. DNA replication is semiconservative with each of the two daughter double helices being composed of one parental and one copied new polynucleotide strand.

The replication of chromosomal DNA is in fact a highly complex process even in bacteria (Szekely 1980), requiring the cooperative action of many enzymes, including DNA unwinding enzymes (Abdel-Monem and Hoffmann-Berling 1980), and, very possibly, association of the chromosome with specific membrane sites (Kornberg 1974, Matsushita and Kubitschek 1975, Kolter and Helsinki 1979). One of the reasons for this complexity is the requirement for the utmost fidelity in the strand-copying process. In consequence, the error rate is probably only 1 in every 10^9 to 10^{10} base pairs replicated in *Esch. coli*. Fidelity is further ensured by the ability of the DNA polymerases to erase their own errors in a 'proof-reading' fashion (Alberts and Sternglanz 1977). The situation is complicated by the possession by bacteria of additional enzymes that participate in DNA recombination (Radding 1977, Venema 1979) and repair processes (Moseley and Williams 1977) whereby DNA strands may be broken and reunited—involving both nucleases and ligases—and potentially mutagenic damage can be repaired by various mechanisms, some of which are inducible (Hanawalt *et al.* 1978, 1979, Smith 1978). DNA may be further modified by the bacterium, which protects its own DNA from the degradative action of the restriction enzymes it employs to degrade certain categories of alien DNA. These restriction enzymes are now being employed by genetic engineers to incorporate genes into suitable vectors as a preliminary to cloning in host bacteria (Sinsheimer 1977, Glover 1978, Old and Primrose 1980).

(b) Biosynthesis of RNA There are three classes of RNA, namely ribosomal RNA (rRNA), transfer RNA (tRNA) and messenger RNA (mRNA). The last has only a short life span in bacterial cells, being degraded to its nucleoside monophosphates; these can be used again for RNA synthesis after their rephosphorylation to the corresponding triphosphates. The template employed for RNA biosynthesis is the appropriate segment of a strand of DNA, with the same rules of base-pairing, though with uracil in place of thymine in the product RNA, determining the base sequence in the novel polyribonucleotide. The transcribing enzyme (RNA polymerase) is a multi-subunit protein that is capable of recognizing the encoded signals on the DNA strand which indicate where transcription should be initiated (Losick and Chamberlain 1976, Talkington and Pero 1978). In many instances an accessory protein (rho factor) indicates where transcription should be terminated (Roberts 1969). We have already noted how control of DNA transcription to produce a mRNA message can regulate, at source, genetic expression e.g. in the examples provided by the *lac* and *trp* operons. In *Esch. coli* 3 to 5% of the RNA is mRNA, 15 to 25% is tRNA and 70 to 80% is rRNA (Norris and Koch 1972). There is an interesting correspondence between the level of RNA synthesis in *Esch. coli* and the intracellular concentration of guanosine tetraphosphate (ppGpp) which is accumulated during conditions e.g. starvation of an essential amino acid when protein and RNA synthesis are simultaneously halted. This phenomenon, termed *stringent control* is not displayed by *relaxed* control mutants having a lesion in the regulatory *rel* A gene (Reiness *et al.* 1975, Molin *et al.* 1977).

(c) Protein synthesis Protein biosynthesis is accomplished at the ribosome where it is directed by mRNA templates whose nucleotide codon sequences are recognized by the corresponding complementary (anticodon) sequences of aminoacyl tRNAs (Szekely 1980). A polypeptide chain is then generated by coupling of aminoacyl tRNA and peptidyl tRNA molecules by a simple enzymic reaction (peptidyltransferase activity). The linear polypeptide chain so formed will then, by spontaneous folding, acquire its secondary and tertiary structure. In some instances, however, the structure may be altered by post-translational modification e.g. removal of a leader peptide fragment from proteins secreted through the cell membrane (Inouye *et al.* 1977, Ambler and Scott 1978, Lampen 1978; see Stanier, Rogers and Ward 1978, p. 231). Thus a triplet of bases on one of the DNA strands determines the nature of a complementary triplet of bases in the mRNA, and these in turn determine which amino acid should be inserted into the growing polypeptide chain. This 'co-linearity of gene and protein' accounts for the fact that a change in the nature of a single base in the DNA—a point mutation—may cause the substitution of one amino acid for another at a particular location in the finally synthesized protein. With four different types of purine and pyrimidine base in mRNA but twenty different amino acids available for protein synthesis, assuming unidirectional reading of the mRNA, there will be 64 distinct triplets of bases specifying the insertion of 20 amino acids. This means that the genetic code is in part 'degenerate', with most amino acids being specified by several alternative base triplets. In fact, 61 of the 64 available triplets (or mRNA codons) code for individual amino acids, the remaining three codons being concerned with signalling the end of synthesis of a particular polypeptide chain (Clark 1977).

The process of protein synthesis at the ribosome is therefore one of translating the mRNA-encoded message.

To be precise, the actual translation is accomplished by the aminoacyl tRNA synthetases, which assign the appropriate amino acid to the tRNA carrying the complementary anticodon. The ribosome's main function is to supervise the interaction between the two types of RNA i.e. mRNA and aminoacyl tRNA plus peptidyl tRNA, and it is thus cast in the role of reader. The tRNAs are particularly fascinating molecules. Although they consist of few nucleotides—75 to 90, some of them unusual—their structure has evidently been highly conserved during evolution. It can be represented in two dimensions as an 'extended clover leaf' and it provides (a) an attachment site for the amino acid, (b) a recognition site for the ligase which attaches this amino acid, (c) the anticodon, and (d) the means of ribosome recognition (Rich 1978). In a real sense the tRNA delivers the apposite amino acid at the appropriate stage in the course of peptide elongation and in a suitably activated condition. The ribosome is a much more complex structure. In bacteria it is 70s in size, where s is the Svedberg Unit, and is composed of two parts (30s and 50s). In the ribosome of *Esch. coli* the smaller subunit contains 16s RNA molecules and 21 proteins whilst the larger subunit contains 5s and 23s RNA molecules plus 34 proteins (Brimacombe 1978; see Stanier, Rogers and Ward 1978, p. 1).

When one considers the translation machinery as a whole (Grunberg-Manago *et al.* 1978; see Stanier, Rogers and Ward 1978, p. 27), consisting as it does not only of mRNA, aminoacyl RNAs and ribosomes but also of numerous ions, nucleotides, and proteins transiently associated with the ribosome at the different steps of protein synthesis, one can well comprehend how it is that a wide selection of antibiotics act by interfering with the mechanism of protein synthesis e.g. chloramphenicol, streptomycin, erythromycin, tetracyclines (Gale *et al.* 1972). The synthesis of some small and rather specialist polypeptides is accomplished non-ribosomally and without a mRNA linear template (Kleinkauf and Koischwitz 1978, Lipmann 1980). We have already encountered this in the synthesis of the peptide cross-links of peptidoglycan, but an even more striking example is provided by the biosynthesis of several polypeptide antibiotics (Katz and Demain 1977). Nisin and subtilin are thought to be copied from mRNA templates with later enzymic modification (Ingram 1970), but others are produced by non-ribosomal systems e.g. bacitracins, gramicidins, colistins, and polymyxins (Kleinkauff 1979).

Bacterial nutrition and the design of culture media

The bacteria of interest to the clinical microbiologist are chemoheterotrophs which grow only when supplied with some organic compound(s). For certain bacteria a single organic compound will suffice; for example *Esch. coli* will grow aerobically on acetate, glucose, or succinate. Such an organism must be able to synthesize all of its substance from materials and free energy supplied by the metabolism of this single carbon source, so long as it is simultaneously provided with a suitable source of nitrogen plus a number of salts which supply essential ions such as potassium, magnesium, iron, phosphate, and sulphate. Other elements will be required in lesser amounts e.g. calcium and manganese, and still others in minor (trace) concentrations only e.g. chlorine, cobalt, copper, molybdenum, nickel, selenium, sodium, tungsten, vanadium, and zinc (Hutner 1972). Yet many more bacteria grow only in complex media fortified with several organic nutrients. These may include specific amino acids, purines, pyrimidines and, at lesser concentrations, B vitamins. Occasionally, other special substances serve as growth stimulants, as, for example, erythritol for *Brucella abortus* (Smith 1978). A bacterium with such requirements evidently lacks the means of synthesizing these substances and hence is dependent on their provision preformed in the culture medium. Not surprisingly, this behaviour is particularly common among bacteria in whose normal habitat these nutrients are invariably present. Knowledge of the various growth factors required by a bacterium enables one to formulate a growth medium of defined composition; though for many purposes it may be sufficient to employ an undefined medium containing ingredients such as peptone, yeast extract, blood, casein hydrolysate, brain infusion (Bridson and Brecker 1970, Lapage *et al.* 1970). The choice of such components is frequently suggested by the nature of the menstruum from which the bacterium was initially isolated (Meynell and Meynell 1970). In the early days of bacteriology the formulation of such ill-defined nutrient media was more of an art than a science and many famous medical bacteriologists are now commemorated by the media which they devised e.g. Löffler's serum medium for the cultivation of *C. diphtheriae*, Dorset's egg medium for the growth of laboratory strains of *Myco. tuberculosis*, Robertson's meat medium for the growth of species of *Clostridium*. In analysing the preferences shown by bacteria for various media, much new information of general biochemical importance was amassed, including the nature of several novel growth factors (Guirard and Snell 1962). There now remain only a few pathogenic bacteria for which no medium for in-vitro cultivation has been devised e.g. *Myco. leprae* and some spirochetes.

The nutritional foibles of a bacterial species can be exploited to prepare an elective medium which will preferentially encourage growth of that organism (Veldkamp 1970). The efficacy of such media can further be enhanced by the incorporation of selectively inhibitory compounds which are substantially less toxic to the desired isolate than to its normal competitors. The inhibitors chosen for this purpose are many and varied, having for the most part been discovered empirically e.g. tellurite in media designed to isolate *C. diphtheriae* from throat swab cultures, or bismuth sulphite for the isolation of salmonellae, including *Salm. typhi*, from faecal material. A selective and diagnostic medium may combine the effects of several agents. For example, MacConkey medium which besides nutrients (peptone and lactose) contains sodium chloride, bile salts, and neutral red, favours the growth of lactose-positive, gram-negative bacteria, especially enterobacteria, while discouraging that of many gram-positive bacteria which are more sensitive to the combined toxic action of detergent plus hydrophilic bactericidal dyes. Similar use can be made of combinations of antibiotics for repressing the growth of sensitive bacteria but not of resistant ones.

In culture media, most bacteria rely on pre-reduced nitrogen in the form of ammonium ions or amino acids, sometimes better supplied as peptides (Payne 1976). Nitrate may sometimes be assimilated by reduction to ammonium ions—a process not to be confused with the reductive use of nitrate in anaerobic respiration (Stouthamer 1976). Although aspartate and alanine may in some bacteria act as early products of ammonium-ion assimilation and as potential reservoirs of transferable amino groups, most bacteria assimilate ammonium ions via glutamate (Brown *et al.* 1974). Synthesis of glutamate from α-oxoglutarate is accomplished either (a) directly by a NAD(P)H-dependent glutamate dehydrogenase, or (b) indirectly by the prior formation of glutamine from glutamate—catalysed by glutamine synthetase and consuming 1 mol ATP per mol glutamine synthesized—followed by a NAD(P)H-dependent transfer of the δ-amino group from glutamine to α-oxoglutarate, catalysed by glutamate synthase. In bacteria which contain both of these systems the former is usually concerned with the assimilation of ammonium ions when these are abundant, the latter with much lower ammonium ion concentrations. Other bacteria augment the available supply of ammonium ions by the controlled catabolism of exogenous nitrogenous organic compounds. Only infrequently, as with some bacteria that degrade urea or use amino acids as major energy sources, will the generated concentration of ammonium ions/ammonia so exceed what is assimilable that the resultant accumulation of ammonia causes alkalinization of the growth medium.

Thus far we have considered only the nutrients provided in a culture medium. Whether it will support growth of a given bacterium will depend on additional factors such as moisture, pH, and oxidation-reduction potentials. These are often influenced by the nature of the atmosphere in the incubator. Indeed, the gaseous components of a culture, especially oxygen, carbon dioxide, and sometimes hydrogen, can be quite as important as any of the substances initially incorporated into the medium.

Culture pH

Although some alkalophilic bacteria require media of high pH e.g. pH 8 to 10, and others (acidophiles) express a preference for media of pH 4 or less (Langworthy 1978), most bacteria of clinical significance grow best in media whose pH is close to neutrality (pH 7). Many organisms by their metabolic creation of acids or bases change the pH of their media in the course of growth, sometimes so drastically as to lead to their premature death. It is for this reason that culture media generally contain fairly high concentrations of suitable pH buffers; these may be supplied by potassium phosphate buffer in minimal media, and by the amino acids and polypeptides in complex undefined media. The pH of the culture can be continually monitored with a glass pH electrode and microvoltmeter and, if necessary, automatically stabilized (Munro 1970). Alternatively when some visual indication only of change in culture pH is required, a suitable pH indicator can be incorporated into the medium. Care must be taken to add the dye at a low, non-toxic concentration, and to choose an indicator that will not be decolorized by reduction in anaerobic cultures (Munro 1970, MacFaddin 1976).

Culture oxidation-reduction potential: E_h

Many obligately anaerobic bacteria will not grow in media that are not pre-reduced or until, as a consequence of their metabolic activity, the organisms in the inoculum have led to the desired reduction of the medium. The oxidation-reduction potential of the medium, its redox potential, E_h, can be measured, and thereafter monitored potentiometrically by means of a platinum electrode. It is not easy to interpret the E_h value so measured, but in crude terms it appears to indicate the state of reduction of the culture (Morris 1975). Thus most species of *Clostridium* grow only in media whose E_h value is less than -100mV; vigorously growing cultures would have redox potentials as low as -400mV (Morris and O'Brien 1971). As with pH measurements, various dyes can be used as visual indicators of the prevailing E_h of the culture, subject to similar caution being exercised over their possible toxicity and simultaneous action as pH indicators (Jacobs 1970).

The presence of dissolved oxygen is incompatible with the maintenance of a low redox potential in a culture medium;

hence for anaerobic cultivation all dissolved oxygen must be expelled and replaced by inert gases (Finegold *et al.* 1977). Reducing agents may be incorporated into the medium to poise it at a suitably low E_h value. Thiols such as mercaptoethanol, cysteine and thioglycollate have been used for this purpose; a mixture of cysteine with dithiothreitol has been commended as a particularly effective combination (Moore 1968). When thiols are inappropriate, ascorbate has been employed as a reductant, or more simply iron filings. The presence of hydrogen and CO_2 in the anaerobic atmosphere is frequently helpful.

Gaseous requirements

(a) Oxygen Though oxygen is an absolute requirement for the growth of obligate aerobes, it is lethal to obligate anaerobes. Facultative bacteria can grow either in the presence or in the absence of oxygen. In aerobic metabolism, molecular oxygen is used for two purposes: (1) as a terminal electron acceptor in aerobic respiration; and (2) as a substrate in metabolic, oxygenase-catalysed reactions in which oxygen atoms are incorporated into organic compounds. Oxygen is only slightly soluble in water and even less soluble in culture media. Nutrient broth equilibrated with air at 30° contains only about 8 mg dm^{-3} of dissolved oxygen (i.e. 0.25 mmol). The respiratory rates of aerobic bacteria are so great that dissolved oxygen is quickly exhausted by a growing culture. Thus to increase the rate of transfer of oxygen from the atmosphere to the medium, the maximum surface of medium may be exposed to the air, as in shake-flask cultures; or air may be bubbled through a well stirred medium which is further agitated by turbulence. Even so, the growth of a dense culture of aerobic bacteria can be limited by the shortage of oxygen (Wilson 1930, Harrison 1976).

Despite the oxygen-dependence displayed by obligately aerobic organisms, oxygen is potentially toxic to all living cells. The tolerance displayed by bacteria ranges from the most exacting, extremely oxygen-sensitive (EOS), obligate anaerobes, through moderate anaerobes which are able to survive brief exposure to low concentrations of oxygen, to facultative and obligate aerobes (Loesche 1969, Morris 1976). Within this range of organisms some bacteria, the microaerophiles, are apparently able to grow only when molecular oxygen is supplied but at an atmospheric concentration less than the normal 20% v/v. It is now known that certain of these organisms, e.g. *Campylobacter fetus*, can grow anaerobically if supplied with an appropriate respiratory electron acceptor such as fumarate (Stouthamer *et al.* 1979).

The biochemical basis of oxygen toxicity is ill understood. It seems certain that it originates in the consumption of oxygen by the living cell when partial reduction of the dioxygen molecule invariably accompanies its harmless complete reduction to water. The by-products of such partial reductions are variously toxic substances among which hydrogen peroxide and superoxide anion might initially be the most important. In fact, although superoxide anion is a free radical, it may not be so harmful as some of the products to which it can give rise, especially the hydroxyl free radical which is an indiscriminately reactive and hence damaging reagent. Living cells which normally consume oxygen i.e. are oxybiontic are protected from the lethal effects of these by-products by various agencies including enzymes such as catalase and peroxidases which dispose of hydrogen peroxide, and superoxide dismutases. The intolerance of oxygen by obligate anaerobes has been ascribed to their total lack of catalase (McLeod and Gordon 1923) or of superoxide dismutase (McCord *et al.* 1971). In fact, this is an oversimplification. Thus both catalase (Sherman 1926) and superoxide dismutase (Hewitt and Morris 1975) are present in some obligate anaerobes and though it is true that, in general, the greater their contents of these protective enzymes the more aerotolerant, oxyduric, are these organisms (Tally *et al.* 1977), other factors also contribute to the wide range of oxygen sensitivities displayed by aero-intolerant bacteria. It is therefore helpful to view the degree of aerotolerance that is displayed by any bacterium as the resultant of two opposing sets of tendencies (1) those that predispose to oxygen tolerance and (2) those that make for oxygen sensitivity. This equilibrium concept of oxygen tolerance (Morris 1979) accommodates the somewhat capricious distribution of oxygen protective enzymes amongst the moderate anaerobes and helps to explain why the apparent oxygen sensitivity of an organism can vary, amongst other things, with the nature of the medium in which it is grown, the interaction of whose constituents with oxygen might additionally generate potentially damaging compounds such as organic peroxides. Indeed, addition of catalase to a culture medium can sometimes prove helpful to an anaerobic bacterium and aid the inception of culture growth from a small inoculum (Harmon and Kautter 1977).

(b) Carbon dioxide Although heterotrophic bacteria are not able to use carbon dioxide as the sole source of carbon for growth, they generally require to be provided with some carbon dioxide which they assimilate in many key carboxylation reactions. Thus ^{14}C-labelled carbon dioxide can be shown to be incorporated into such end products as amino acids, purines and pyrimidines (Wood and Stjernholm 1962). It is therefore not surprising that the demand for CO_2 may be lessened by the addition to a minimal medium of such compounds as aspartate, glutamate, adenine, guanine, and possibly arginine and fatty acids, as well as biotin whose prime role is as cofactor in carboxylation reactions.

Carbon dioxide is much more soluble in water than is oxygen and in culture media will be in equilibrium with bicarbonate ions (HCO_3^-) whose concentration will also be pH dependent. Furthermore, although the concentration of carbon dioxide in air is comparatively low (0.03% v/v), all aerobically respiring organisms generate carbon dioxide and hence elevate its concentration in the immediate environment. The same is frequently true of anaerobically fermenting bacteria though when, as in the homolactic fermentation of glucose, carbon dioxide is not formed it may have to be supplied. This is true of staphylococci, which fail to grow in the absence of any exogenous source of carbon dioxide (Gladstone 1937). Some (capneic) bacteria e.g. *Brucella abortus* (Huddleson 1921), *Neissseria meningitidis*, and *N. gonorrhoeae* (Chan *et al.* 1975), especially when freshly isolated, require an elevated concentration of carbon dioxide (5 to 10% v/v); such organisms presumably have a lower affinity rather than a special use for carbon dioxide. On the other hand, even if care is taken to control the culture pH, excess atmospheric carbon dioxide can prove inhibitory e.g. carbon dioxide concentrations above 50% v/v at pH 6.6 inhibited very many bacterial species (Coyne 1933).

Biochemical tests in diagnostic bacteriology

Biochemical tests are employed in diagnostic bacteriology or bacterial systematics to identify different bacterial species or to distinguish between different strains of the same species. Nutritional and other growth requirements may be used for the same purpose. Either enzymes or transport proteins, or both, needed for the utilization of some substrates may have to be induced; thus, to ensure reproducible results, the composition of the test medium and the 'condition' of the organisms used as the inoculum must be most carefully controlled. Similarly, care must be taken when washed suspensions are used that the organisms have been grown under suitable conditions. For example, to determine whether an organism is capable of catabolizing tryptophan to yield indole, it would be unwise to employ a test medium containing a high concentration of glucose, since the inducible enzyme tryptophanase which catalyses the reaction is in many organisms susceptible to catabolite repression i.e. to the glucose effect (Clarke and Cowan 1952). Or again, when testing for the production of an alkaline phosphatase e.g. in *Pseudomonas* sp., or in enterobacteria, the medium must contain not more than a minimum of free phosphate ions, else a false-negative result will be obtained (Torriani 1960). That the composition of the growth medium and other cultural conditions influence the yield of a desired end-product of metabolism is a lesson well learned by industrial microbiologists (Demain *et al.* 1979, Neijssel and Tempest 1979, Malik 1980), whose 'genetically improved', high-producing strains have frequently acquired faults in the normal regulatory processes.

Growth of bacteria

In a bacterial culture each of the growing organisms is becoming progressively larger until, at some critical size, it divides to give rise to two identical daughter cells. Since the members of the bacterial population are dividing asynchronously, whether growth of the culture is followed by measurement of the biomass concentration or by monitoring the increase in the number of organisms in a given culture volume, similarly continuous growth curves will be obtained. As direct counting procedures, based on microscopic observations or automated particle counting, do not discriminate between living and dead organisms, these methods have to be augmented by procedures which enumerate only viable cells. Although approximate viable counts can be performed in liquid media by the dilution series method, it is generally more satisfactory to employ agar-solidified nutrient media—pour plates, spread plates or roll tubes—or membranes on which the organisms have been collected by filtration and which are then incubated on the surface of a nutrient-rich medium. The variety of techniques for undertaking such total and viable counts and also for estimating biomass concentration were fully described in the 6th edition of this work (pages 116 to 120). Other standard accounts are available of these basic procedures (e.g. Meynell and Meynell 1970, Mallette 1970, Postgate 1970), of automated or mechanized variants of these methods (Sharpe 1973, Goldschmidt and Fung 1979), and of techniques for the rapid detection and assay of sparse bacterial populations (Strange 1972, Hartog and Heerink 1979) which include electrometric (Ur and Brown 1975) and microcalorimetric procedures (Russell *et al.* 1975).

Growth of a bacterial cell

When examining the growth of a bacterium from its origin by parental cell division through the course of enlargement to its own division, we are considering the temporarily ordered sequence of biochemical and morphological events that constitute the bacterial cell cycle (Helmstetter *et al.* 1979). The time taken to accomplish the complete cell cycle—*the generation time*—is determined by several factors including the temperature and the composition of the culture medium. Furthermore, the composition and size of the bacterium at division will vary greatly with its growth rate. For example, the faster its growth rate, the larger will be the cell before it divides and the greater will be its proportional content of RNA (Nierlich 1978). On the other hand, bacteria differ greatly in their maximum possible growth rates. Thus, in a nutrient-rich medium at 37° *Salm. typhimurium* might divide every 25 min, but at 37° in a minimal medium containing proline as the sole source of carbon and energy it would divide only every 300 min; and *Myc. tuberculosis* not for six hours. It follows that although the environment can greatly influence the growth rate of a bacterium, its maximum growth rate is ultimately genetically determined and is undoubtedly dependent on many interrelated factors.

Studies with *Esch. coli* and various of its mutants have shown that, independent of the duration of the cell cycle (e.g. 20 to 60 min), the replication of its genome takes a constant 40 min at 37°. This means that in cells growing with a doubling time of less than 40 min new rounds of DNA replication have to be started before the previous round of DNA replication is complete. Consequently in the most rapidly growing cells, genes that are located close to the point of initiation of replication of the circular chromosome, and on both sides of it, since the replication proceeds bidirectionally, will be present in more than one copy per cell, while genes close to the terminus of chromosome replication are still present in only single copies. Such selective gene amplification can affect the properties of the resultant organisms. Indeed, the bacterium exploits this phenomenon to its advantage; for one of the ways by which it ensures that the rate of protein synthesis matches the demands of accelerated growth is by grouping the genes encoding for ribosomal RNAs and proteins close to the origin of chromosome replication. The attainment of balanced growth at greatly disparate rates has, of course, implications beyond adjustments of rates of DNA and RNA and protein synthesis (Lutkenhaus *et al.* 1979); in particular, how is the synthesis of the cell envelope controlled to keep pace with requirements? These and related problems have been discussed in several accounts of the bacterial cell cycle, which consider the coupling between DNA replication and cell division (Donachie *et al.* 1973, Koch 1977, James 1978, Cooper 1979), and surface extension during the bacterial cell cycle (Sargent 1978).

To undertake biochemical studies of the bacterial cell cycle one must in some way synchronize the division of all the viable cells in a large culture volume. Many of the procedures devised for this purpose depend on the imposition of some temporary stress that interferes with the physiology of the organism. The best of these are based on size-selection whereby the smallest, i.e. newborn, cells in a normally growing culture are differentially harvested (Helmstetter 1967, Poole *et al.* 1973) and used to inoculate a medium in which they are able to grow with minimal lag. In this way synchronous cell division can be obtained over at least two cell cycles—as evidenced by a stepped culture growth curve with total cell numbers doubling at the times of division.

Growth in batch culture

In a routine bacteriology laboratory organisms are inoculated into liquid media which are then incubated until growth ceases. In such a batch culture the organisms are dividing asynchronously and, because it is a closed system i.e. with no net input or output, the environment changes progressively during incubation

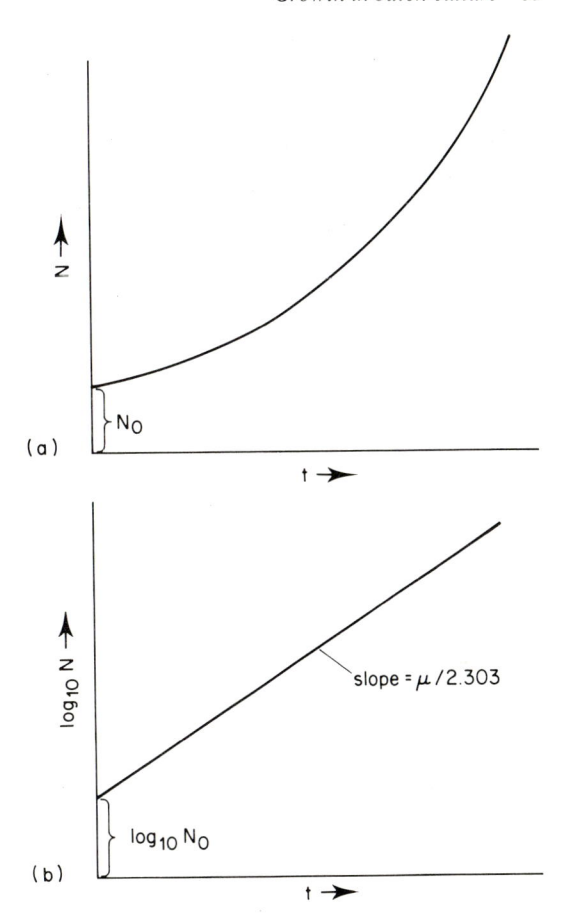

Fig. 3.5 Theoretical growth curves illustrating instantaneous exponential growth in a bacterial batch culture. Total cell numbers (N; i.e. number of bacterial cells per ml) plotted against time (t). N_0 = inoculum cell concentration: μ = specific growth rate.

as nutrients are consumed and end-products accumulate. It follows that bacteria harvested at different times during the course of a batch culture differ in their properties.

If all the bacteria in an inoculum were viable and were immediately able to grow and multiply at a constant rate in the new medium then the bacterial concentration in the culture—its cell number, N—would increase with time (t) in the manner shown in Fig. 3.5a. This exponential rate of culture growth is the consequence of the bacteria multiplying by binary fission. Thus, were we to replot the data using a logarithmic scale for cell numbers, we should obtain the relation shown in Fig. 3.5b (it would be most appropriate to use logarithms to the base 2 as described by Finney *et al.* 1955, but it is generally more convenient to employ logarithms to the base 10). This is the linear semi-logarithmic plot we would expect to obtain with bacteria growing at a constant rate i.e. at any time (t) the rate of increase in cell numbers dN/dt is directly proportional to the concentration of cells in the culture at that time (i.e. N_t). This simple relation $dN/dt \propto N_t$ enables us to define an important index of the rate of growth viz. the proportionality constant (μ) i.e. $dN/dt = \mu N_t$. This constant (μ) is known as the *growth-rate constant* or the *specific growth rate*, and has the dimension time^{-1}. It can easily be derived from the plot of log N versus t wherein its value will be given by the slope of the linear plot. When, as in Fig. 3.5b, the ordinate scale is $\log_{10} N$ the measured slope of the line will equal $\mu/2.303$, since $\log_2 N = 2.303 \log_{10} N$. When exponential growth is occurring in the culture in a balanced fashion, i.e. with a proportional increase in concentration of each cell component, then irrespective of whether we choose to measure the increase in concentration in terms of cell number, cell mass—biomass—DNA, protein or any other cellular component, the rate of increase would be determined by the value of μ. Incidentally, it follows that in this exponentially growing culture the duration of the cell cycle (doubling time, mean generation time) will equal $0.693/\mu$.

In practice, a bacterial inoculum when introduced into fresh culture medium does not immediately start to grow in an exponential manner. In fact, the behaviour of the culture is much more nearly that represented in Fig. 3.6. This culture growth curve, though somewhat idealized, indicates the sequential phases of batch culture development that are generally recognizable.

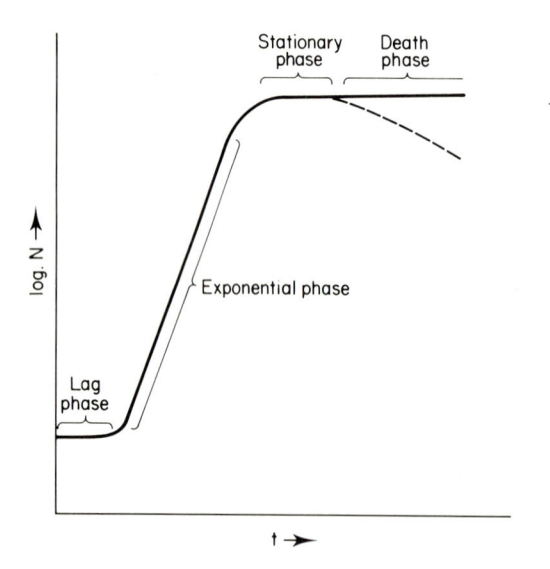

Fig. 3.6 Idealized growth curve of a bacterial batch culture. Total cell count (—); viable cell count (- - -).

The lag phase The duration of this initial phase can vary considerably, depending on the composition of the medium, it being generally shorter in a nutrient-rich than in a minimal medium. Its length may also be determined by the history of the organisms employed as the inoculum e.g. what proportion are viable and whether or not they were harvested from a medium identical in composition with that into which they are now introduced. Furthermore the lag phase is generally much more evident when the increase in cell number is followed rather than the increase in cell mass; the cells are metabolizing and even increasing in size though not in number. It appears therefore that the lag phase is one of adaptation of the organisms to their new environment, where their enzymic machinery and their other macromolecular components, co-factor and metabolite concentrations are adjusted to conform with the demands made by the novel growth conditions. In this sense the lag phase must be regarded as a period not of rest but of intense metabolic activity. In extreme circumstances, growth after a prolonged lag may be the consequence of selection of a mutant strain which, alone of all the organisms in the inoculum, can multiply in the new culture conditions.

The exponential phase The lag phase is succeeded by a brief acceleration phase of unbalanced growth which quickly gives way to the phase of exponential, balanced growth. Although the growth rate remains constant throughout the duration of this phase, culture conditions at its start—early exponential phase—will be different from those near its end—late exponential phase. This is because nutrient concentrations will have declined, end-product concentrations will have increased, the pH may have changed, and the cell density will have risen, with consequential increases in demands for nutrients e.g. in aerobes, for oxygen. For a multitude of reasons, therefore, the exponential phase is of limited duration. Then follows a brief phase of deceleration, before the culture passes into the stationary phase.

It is wise, indeed necessary, when employing suspensions of bacteria for biochemical purposes to specify at what stage during growth in batch culture they had been harvested, since the differences between organisms taken from stationary and exponential phases of growth are likely to be great. Yet, though it is legitimate to refer to organisms harvested early and late in the exponential phase of batch culture it is not correct to designate such organisms as 'young' or 'old' respectively, for these terms are meaningless when applied to bacteria taken from asynchronously dividing cultures.

Given a surfeit of nutrients, and other optimal conditions, the rates of growth of bacteria in batch culture approach near maximum values (μ_{max}) far in excess of the growth rates displayed by the same organisms in their natural habitats (Mason 1935, Brock 1971). An extreme example is provided by a marine bacterium *Vibrio natriegens* which at the elevated temperature of 37° and in a nutrient-rich medium displayed a generation time of less than 10 min (Eagon 1962). Even the rates of growth of pathogenic bacteria are considerably lower in their infected hosts than in culture (Smith 1978). For example, Maw and Meynell (1968) found that in the spleen of a mouse infected with *Salm. typhimurium* the bacteria grew at a rate only 5 to 10 per cent of that displayed in an artificial medium. When the prime cause of termination of the exponential phase of growth is the accumulation of toxic metabolites then, if desired, steps may be taken to prolong such growth and, by postponing the onset of the stationary phase, obtain a greater final biomass concentration e.g. control of pH or removal of toxic substances by dialysis (Schulz and Gerhardt 1969).

The stationary phase This phase ensues when the culture conditions are so changed that further balanced growth and cell division cannot be sustained. The total cell number will stay constant unless lysis supervenes, but the viable cell count may quickly decline—the phase of death or decline. Increase in cell mass does not always cease with stasis in total cell numbers. It may occur as the result of unbalanced growth, such as may be seen when a carbon and energy source remains in excess after some other nutrient, e.g. the nitrogen source, has been exhausted (Dawes and Senior 1973). The kind and extent of continued anabolic activity will vary according to the nutrient which is limiting. Some of the nutrients exuded from dead organisms may be used by survivors in the population in what has been termed cryptic growth (Ryan 1959, Postgate 1967).

Culture growth rates and nutrient limitation

When a key nutrient is supplied in limiting concentration, the specific growth rate (μ) sustainable in the culture is diminished. Monod empirically discovered (see Monod 1949) that the relation between μ and the prevailing concentration of a growth rate determining substrate ([S]) could generally be expressed as a rectangular hyperbola. This relation can therefore be represented by the equation for such a curve i.e.

$$\mu = \mu_{max} \frac{[S]}{(K_s + [S])}$$

where μ_{max} and K_s are constants, μ_{max} being the maximum value of μ approached as [S] is increased; K_s is that value of [S] which supports a specific growth rate which is one half of μ_{max}. By plotting $1/\mu$ versus $1/[S]$ we obtain a straight line whose intercept on the $1/\mu$ ordinate gives the value of $1/\mu_{max}$ and whose intercept on the $-1/[S]$ abscissa, obtained by extrapolation, gives the value of $-1/K_s$. Since bacteria may differ greatly in their affinities for a given nutrient, reflected in their K_s values for that nutrient, and also in the values of μ_{max} when the nutrient is provided in excess, the values of K_s and μ_{max} are frequently helpful in explaining the nature of the competition between the component organisms in a mixed bacterial population, and in predicting the likely outcome at various values of [S] i.e. which organism will triumph in the competition for the scarce resources (Meers 1973, Veldkamp 1976).

In batch cultures all nutrients are usually supplied in excess, so that during the exponential phase the specific growth rate is very close to μ_{max}. When by design or mischance the concentration of some key nutrient falls below that capable of sustaining μ_{max}, the exponential phase will be prematurely terminated, but may be restored and extended by the addition of more of that nutrient. The technique may be used to prolong the exponential phase when this is in danger of being abbreviated by other causes, e.g. by the potentially lethal accumulation of toxic end-products of metabolism. A substantial proportion of the culture may be removed in its mid-exponential phase and an equal volume added of fresh culture medium, pre-equilibrated to the temperature of incubation. By this means a quasi-steady state may be established in which the culture grows perpetually at a specific growth rate close to μ_{max}. Extension of this principle to a continuous flow culture system gives us the *turbidostat*.

Continuous flow culture

Continuous flow cultures may be established in a growth vessel of the type illustrated in Fig. 3.7 wherein the inflow rate of fresh medium is compensated for by an automatically equal outflow rate of culture from the vessel via some constant-level device.

(a) **Turbidostat** In a turbidostat the inflow rate of a complex medium with all components in adequate but non-toxic excess is automatically adjusted so that the cell density, i.e. biomass concentration, in the culture is kept constant. This can be done by continuous spectrophotometric monitoring of the absorbance of the culture; any significant decrease in the required value serves as a trigger for the addition of a pulse of fresh sterile medium. In such a medium the culture

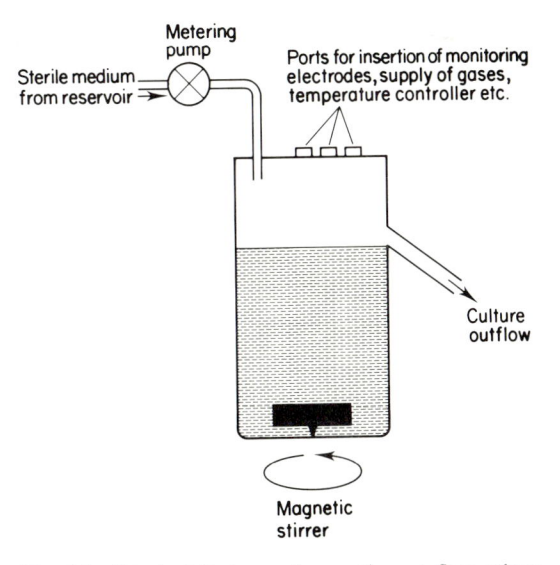

Fig. 3.7 Principal features of a continuous flow culture apparatus.

remains in a continuous exponential phase, with a specific growth rate close to μ_{max}. The turbidostat therefore provides a means of maintaining cultures for many generations in a constant environment of excess nutrients (Munson 1970, Watson 1972). The selection pressure it imposes may, however, favour the ultimate emergence and predominance of some mutant strain with an improved value of μ_{max} (Bryson and Szybalski 1952).

(b) Chemostat Chemostat culture has emerged as a powerful analytical tool in microbial ecology and physiology. The principles that underly the procedure and mathematical derivations of the relations that are established between the various growth characteristics of the culture have been fully described in many articles (Herbert et al. 1965, Kubitschek 1970, Tempest 1970a, b; 1978, Dean et al. 1972, Pirt 1975).

Chemostat culture exploits the fact that, when a bacterial culture contains a limiting suboptimal concentration of an essential nutrient, its specific growth rate must be less than its μ_{max}, and will in fact retain a steady value ($\bar{\mu}$) so long as the sustrate concentration in the culture ([s̄]) remains constant.

The simplest way in which such a continuous flow system may be regulated is by extrinsic control of the limiting substrate concentration. The culture medium may be supplied at an appropriate rate, and a solution of the limiting substrate added by means of a pump whose operation is controlled by signals fed back from an appropriate monitor of the resident substrate concentration in the culture e.g. a substrate-selective electrode. Such an extrinsic control system is essential when one wishes to limit the substrate to bacterial growth rate by virtue of its toxicity (see Veldkamp 1976). More usually, however, control of the resident substrate concentration and hence of the growth rate is brought about by the culture itself i.e. control is intrinsic and the chemostat culture is self-regulating. A culture medium is supplied which contains a potentially limiting concentration of the substrate—generally designated as the reservoir substrate concentration s_r. An appropriate flow rate (F litre h^{-1}) is chosen for the introduction of this medium into the culture, whose volume is maintained at V litre. When a bacterial inoculum is now introduced into the culture vessel, the medium being aerated or maintained anaerobic, and its pH and/or redox potential being monitored and controlled as necessary, growth will ensue and the biomass concentration will increase. This leads to progressively greater demands on the limited supply of substrate and a decrease in its concentration. Concurrently with this, the specific growth rate falls until this equals the imposed dilution rate at which the fresh medium is being supplied (dilution rate D h^{-1} equals F/V). At this point the culture enters a steady state (when $\bar{\mu}=D$), any tendency to increase on the part of the biomass concentration ($\bar{x}$) being self-corrected by the consequent temporary decrease in the concentration of s below the value s̄. Similarly any converse tendency to decrease in biomass concentration will immediately be countered by the consequent temporary increase in the concentration of s above the value s̄; this would support an increase in growth rate which would very quickly restore the biomass concentration to $\bar{x}$.

The most important feature therefore of the chemostat is that it enables the investigator to impose a chosen growth rate upon the bacteria he is studying, at any value less than μ_{max}, merely by suitably adjusting the dilution rate in his culture. Simultaneously, this growth occurs in a medium specifically limited by a single nutrient, thus enabling the biochemical consequence of such limitation to be investigated at different rates of growth. Studies of the relation between growth rate and the composition and metabolic activities of bacteria have thus greatly benefited from the introduction of the technique of chemostat culture (Ellwood and Tempest 1972). The sudden changes in growth rate which in batch culture (Schaechter et al. 1958) had only been rendered possible by transfer of bacteria from minimal to nutrient rich media—'shift-up' or conversely 'shift-down'—could now be accomplished merely by altering the dilution rate in a single chemostat culture (Tempest 1970, Koch 1971). Values of μ_{max} may, somewhat paradoxically, also be more accurately measured in chemostat culture, for μ_{max} will approximately equal the critical dilution rate D_c. This is the threshold value of D above which the specific growth rate cannot keep pace with the continuous flow rate through the culture with the consequence that the culture density falls precipitately and irretrievably, a condition known as *wash-out*.

Chemostat culture has also been employed to select mutant strains of bacteria, particularly those displaying a greater affinity (lower K_s) for the limiting substrate than does the parent organism; and to enrich from natural mixed inocula bacteria which, although growing slowly, are capable of thriving on exceptionally low concentrations of nutrients in their environment. Such organisms would be rapidly overgrown in batch culture by faster growing competitors (Veldkamp 1976). The use of chemostat culture procedures by microbial ecologists has shed new light on the competition between species for nutrients; the mechanisms whereby stable consortia of bacteria may be established and consume otherwise poorly degradable substrates; and even on predator-prey relationships in microbial communities (Jannasch and Mateles 1974). In the medical and veterinary fields we find such diverse applications of continuous or semi-continuous flow culture techniques as the 'model mouth', in which may be studied the colonization of teeth by bacteria contributing to dental plaque and implicated in dental caries (Russell and Coulter 1975); or anaerobic chemostat culture of the mixed bacterial flora of the rumen or large intestine (Hobson and Summers 1967). The plug-flow culture procedure, which spatially separates zones of culture corresponding to different phases of batch culture (Pirt 1975), may prove an even more helpful model of the interactions of the microflora of the intestinal tract.

A word of caution should be entered at this point. Just because an experimental apparatus provides for continuous flow, and nutrient(s) are in short supply, it does not necessarily follow that the system will operate as a chemostat. It would be misleading if the equations appropriate only to the self-regulating chemostat were applied to systems that do not establish a true steady state. For this reason care has to be taken to ensure perfect mixing in a chemostat culture, and to avoid conditions, such as clumping of the organisms or their adhesion as a pellicle or film on the walls of the culture vessel.

Studies of bacterial growth on various energy-yielding substances have particularly benefited from the applications of chemostat culture techniques. It was at one time thought that the yield of bacterial biomass per unit mass of energy-yielding substrate, i.e. the molar growth yield, should prove to be predictable from knowledge of the number of mol of ATP generated per mol of substrate used by the bacterium (Bauchop and Elsden 1960). However, the relation has proved not to be one of direct proportionality, for a part of the free energy, ATP, conserved by the bacterium is thereafter expended in essential tasks other than biosynthesis. Even this *maintenance energy* fraction is not constant, but varies with the growth rate of the bacterium (Neijssel and Tempest 1976, Stouthamer 1979). A bacterium may have to pay a high price, in energetic terms, (a) to acquire essential nutrients from its environment when these are present in low concentration, and (b) to conserve high intracellular concentrations of key ions and solutes while expelling others (Stouthamer and Bettenhausen 1977, Hueting et al. 1979). Conversely, bacteria may have several means at their disposal for ridding themselves of excess ATP when this is being generated at too high a rate e.g. the so-called slip reactions and futile cycles (Neijssel and Tempest 1979, Stouthamer 1979).

Theoretically it should be possible to perpetuate the steady state initially established in a chemostat bacterial culture. However, problems of contamination, inhomogeneity caused by wall growth etc. would in time defeat this objective. Even more important is the fact that the intense selection pressure imposed by the chemostat culture favours the emergence of a mutant strain which may displace the parent organism in a surprisingly short time (Hartley 1974).

Bacterial growth on solid surfaces

Colonization of the surfaces of materials and of host tissues is a common and important feature of microbial growth in nature. The factors which contribute to the adhesion of bacteria to such solid surfaces are of considerable interest both to microbial ecologists and to medical microbiologists (Ellwood et al. 1979, Savage 1977, Swanson 1977).

Colonial growth of pure clones of bacteria on agar-solidified media forms the basis of most procedures for their isolation. Moreover, the constancy of shape and appearance of the colonies of different bacterial species on various solid media aids in their identification. A number of factors determine colony size and appearance, including motility of the organism, e.g. swarming of colonies of *Proteus* species, the presence or absence of capsular material surrounding the bacterium, or of other cell surface components, as evidenced in smooth or rough textures of their colonies. The rate of growth of the bacterial colony is the resultant of a combination of unrestricted plus diffusion-limited growth (Wimpenny 1979); the diffusion gradients of nutrients, including oxygen, and of end-products of metabolism that are established, can result in bacteria located in different positions within the final colony displaying different physiological characteristics (Pirt 1975, Wimpenny and Parr 1979).

Temperature and growth rates

Barber (1908), who made an exhaustive study of the rate of growth of *Esch. coli* in batch culture, reported the findings illustrated in Fig. 3.8. In this, it will be seen that at 15° the doubling time was 180 min, at 25° 44 min, at 35° 22 min and at 40° 17 min. The maximum growth rate was at about 37°, but between 36° and 46° there was little change in the doubling time. Interestingly, the growth rate increased with the temperature of incubation until the optimal temperature was reached. The relation between temperature and growth rate is similar in other genera of bacteria, differing chiefly in displacement of the curve along the temperature scale to give lower or higher, minimum, optimum and maximum growth temperatures

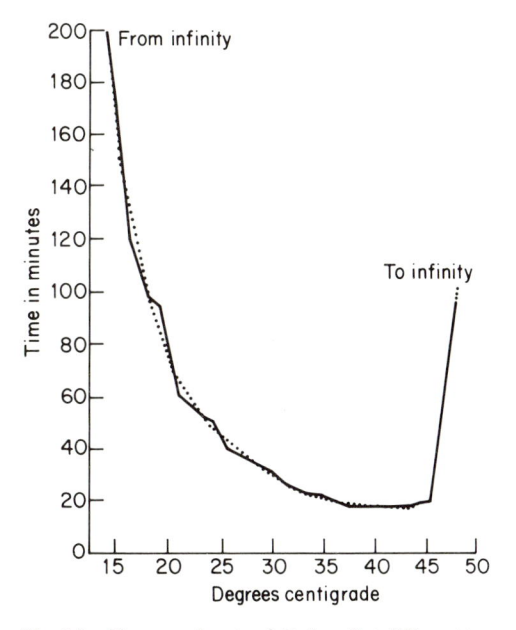

Fig. 3.8 The growth rate of *Esch. coli* at different temperatures. Curve plotted from actual observations (—); smoothed curve (- - -). (After Barber.)

(Neilson et al. 1959, Ingraham 1962). The range of temperature over which bacteria will grow is very wide. Species are known that can multiply below minus 10° and others at over 90°. We have no space to describe the metabolism of these organisms, but those who are interested should consult the writings of Kushner (1978) and Gould and Corry (1980), who discuss the growth and survival of bacteria not only at such different temperatures but also in other extreme natural environments.

Survival and death of bacteria

As pointed out by Postgate (1976; see Gray and Postgate 1976, p. 1), the death of a microbe can be discovered only retrospectively. To ascertain this a population is incubated in a recovery medium, and those individuals which do not form progeny are taken to be dead. Bacteria which undergo binary fission do not grow old and die, they vanish as the parent is replaced by two equally young individuals. Death results only from some environmental stress; so that by survival we mean maintenance of viability in adverse circumstances. A proportion of most bacterial populations will prove to be non-viable. The cells may have succumbed to any of a variety of ill-defined stress conditions (Quesnel 1963). The study of death and survival in a bacterial population is therefore an examination of changes in the rate of decrease in viability, expressed as a ratio of viable to total cell counts, in response to defined stresses e.g. freezing and thawing, elevated temperatures, desiccation, shortage of oxygen, various forms of radiation, or exposure to toxic chemical agents (Gray and Postgate 1976). An important though more insidious stress is that of starvation, wherein the nature of the material whereof the organism is deprived, i.e. energy source, carbon or nitrogen source, essential ion, will determine the nature and intensity of the stress. Some bacteria, by controlled endogenous metabolism and utilization of reserve materials, are better able to cope with 'feast and famine' fluctuations than are others (Koch 1971, Dawes 1976; see Gray and Postgate, p. 19). In natural situations many motile bacteria can seek out fresh nutrients—positive chemotaxis—or escape sources of toxicity (Berg 1975). Maintenance of a protonmotive force at the cytoplasmic membrane may be important in the sustenance of cellular viability (Konings and Veldkamp 1980); but, of course, any irreversible lesion in a key process in bacterial growth and division would be scored as lethal, even if the cell were to continue to metabolize and otherwise function normally for a prolonged period of time. Various methods have been devised for assessing the damage caused by different agencies to bacterial cells (Russell et al. 1973).

Since it is usual to regard the phase of batch culture following the stationary phase as that of death or decline, it should be understood that this follows no predictable course, but may take almost any form depending on the multiplicity of stresses to which the population is responding.

Recovery media supply the organism with conditions congenial to growth and division tending to stimulate metabolism; the converse is attempted when it is desired to maintain the bacteria in a state of suspended animation over long periods of time. Refrigeration is an obvious means whereby to decrease the metabolic rate, but even more effective is desiccation, which lowers the water activity within the organism below the threshold value compatible with sustained enzymic activities. The chief aim of the procedures and menstrua employed in the storage of bacterial cultures by lyophilization (i.e. freeze drying) is to effect such desiccation in a non-harmful manner so that the organism may subsequently be revived by rehydration in a suitable resuscitation medium (Lapage et al. 1970).

Survival of vegetative bacteria in animals requires that the organisms avoid or counter the numerous antimicrobial defence mechanisms that can be mobilized by the host. The fate of a bacterial population in such circumstances will be determined by the dynamic relation established between the organisms that succumb to and those that resist the host defences. The interaction between bacterial aggressins and the host defences is dealt with in the Immunity section of this book. Relatively little is known, however, of the mechanisms of long term survival of bacteria in chronic and carrier states of animal hosts (Smith 1976; see Gray and Postgate 1976, p. 299).

Survival of bacteria as resting forms: Dormancy

Although various forms of 'resting cell' are produced by different bacterial species (Sudo and Dworkin 1973), that of greatest interest to the medical microbiologist is the endospore produced by species of *Bacillus* and *Clostridium* (Gould and Hurst 1969). This is a differentiated cell produced as a consequence of a sequence of biochemical and structural changes directed by sporulation genes in response to an imposed environmental stress e.g. impending starvation (Mandelstam 1976, Young and Mandelstam 1979). The mature spore has a unique structure, the central protoplast containing an entire chromosome and being surrounded in turn by a plasma membrane, germ-cell wall, cortex, spore coat and frequently an exosporial integument (Warth 1978; see Chapter 2). In its mature condition the spore is unusually resistant to desiccation, radiation, heat, oxygen and various solvents and chemicals. Fortunately, although spores of different bacterial species exhibit a wide range of heat resistance, nearly all can be killed by heating at 160° for 90 min in a hot air oven or by autoclaving at 120° for 30 min. Though the basis of the heat resistance and the metabolic dormancy of the spore is not fully understood, it is believed to be attributable to the greatly diminished water activity of the spore protoplast. The high content of dipicolinic acid in the protoplast may be a contributory factor, but the spore cortex plays a key role in the dehydration of the protoplast and in sustaining this desiccation (Gould and Dring 1974). The dormant, cryptobiotic spore may be activated by various agencies, both physical and chemical, with resultant germination occurring in suitable media to give rise again to the vegetative form of the bacterium (Gould and Hurst 1969). For microbial growth and survival in extremes of environment, see Gould and Corry (1980).

Mixed cultures Until recently bacteriologists have studied mainly pure cultures. In nature, however, a pure culture is the exception rather than the rule. Normally, bacteria occur in mixed communities in which the fate of any one species depends on a host of different factors such, for example, as the nature and number of other bacteria present; its growth requirements and the extent to which these are met by the substrates available; the antagonistic effect of bacteriocines and antibiotics excreted by bacteria and fungi; ingestion by protozoa; and the temperature and acidity of the medium. For these reasons increasing attention is being paid to the study of mixed cultures and microbial communities. Knowledge so obtained is of particular value in the fields of agriculture and biotechnology. For further information on this subject the student is referred to the papers given at a special meeting in 1981 at the Royal Society edited by Quayle and Bull (1982).

References

Anderson, R. L. and Wood, W. A. (1969) *Annu. Rev. Microbiol.* **23**, 539.
Artz, S. and Broach, J. (1975) *Proc. nat. Acad. Sci. USA* **72**, 3453.
Atkinson, D. E. (1977) *Cellular Energy Metabolism and its Regulation.* Academic Press, New York.
Bag, J. (1974) *J. Bact.* **118**, 764.
Barber, M. A. (1908) *J. infect. Dis.* **5**, 379.
Abdelal, A. T. (1979) *Annu. Rev. Microbiol.* **33**, 139.
Abdel-Monem, M. and Hoffmann-Berling, H. (1980) *Trends in biochem. Sci.* **5**, 128.
Alberts, B. and Sternglanz, R. (1977) *Nature* **269**, 655.
Altendorf, K., Müller, C. R. and Sandermann, H. Jr (1977) *Europ. J. Biochem.* **73**, 545.
Ambler, R. P. and Scott, G. K. (1978). *Proc. nat. Acad. Sci. USA* **75**, 3732
Barker, H. A. (1961) In: *The Bacteria,* Vol. 2, p. 151. Ed. by I. C. Gunsalus and R. Y. Stanier, Academic Press, New York; (1978) *Annu. Rev. Biochem.* **47**, 1.
Bauchop, T. and Elsden, S. R. (1960) *J. gen. Microbiol.* **23**, 457.
Beckwith, J. and Rossow, P. (1974) *Annu. Rev. Genet.* **8**, 1.
Benz, R., Janko, K., Boos, W. and Laüger, P. (1978) *Biochim. biophys. Acta* **511**, 305.
Berg, H. C. (1975) *Sci. Amer.,* August Issue, p. 36.
Berger, E. A. (1973) *Proc. nat. Acad. Sci. USA* **70**, 1514.
Bertrand, K., Korn, L., Lee, F., Platt, T., Squires, C. L., Squires, C. and Yanofsky, C. (1975) *Science* **189**, 22.
Beyreuther, K., Adler, K., Geisler, N. and Klemm, A. (1973) *Proc. nat. Acad. Sci. USA* **70**, 3576.
Blumberg, P. M. and Strominger, J. L. (1974) *Bact. Rev.* **38**, 291.
Boonstra, J. and Konings, W. N. (1978) Abst. XII Int. Cong. Microbiol. München, P. 115.
Braun, V. and Hantke, K. (1974) *Annu. Rev. Biochem.* **43**, 89.
Bridson, E. Y. and Brecker, A. (1970) In: *Methods in Microbiology,* Vol. 3A, p. 229. Ed. by J. R. Norris and D. W. Ribbons, Academic Press, London.
Brock, T. D. (1971) *Bact. Rev.* **35**, 39.
Broda, P. (1979) *Plasmids.* Freeman, Oxford.
Brown, C. M., Macdonald-Brown, D. S. and Meers, J. L. (1974) *Advanc. microb. Physiol.* **11**, 1.
Bryson, V. and Szybalski, W. (1952) *Science* **116**, 45.
Buckel, W. and Barker, H. A. (1974) *J. Bact.* **117**, 1248.
Chan, K., Wiseman, G. M. and Caird, J. D. (1975) *Brit. J. ven. Dis.* **51**, 382.
Clark, B. F. C. (1977) *The Genetic Code.* Studies in Biology No. 83. Edward Arnold, London.
Clarke, D. J., Fuller, F. M. and Morris, J. G. (1979) *Europ. J. Biochem.* **98**, 597.
Clarke, P. H. (1972) *Sci. Prog. Oxf.* **60**, 245; (1978) In: *Essays in Microbiology,* p. 8/1. Ed. by J. R. Norris and M. H. Richmond. John Wiley, Chichester.
Clarke, P. H. and Cowan, S. T. (1952) *J. gen. Microbiol.* **6**, 187.
Cohen, G. N. (1967) *Biosynthesis of Small Molecules.* Harper and Row, New York.
Cohen, G. N. and Monod, J. (1957) *Bact. Rev.* **21**, 169.
Cooper, S. (1979) *Nature* **280**, 17.
Coyne, F. P. (1933) *Proc. Roy. Soc. B.* **113**, 196.
Crick, F. H. C. (1966) *J. mol. Biol.* **19**, 548; (1974) *Nature* **248**, 766.
Cronan, J. E. Jr (1978) *Annu. Rev. Biochem.* **47**, 163.
Cronan, J. E. Jr and Gelmann, E. P. (1975) *Bact. Rev.* **39**, 232.
Dagley, S. (1971) *Advanc. microb. Physiol.* **6**, 1; (1978) In: *The Bacteria,* Vol. 6 p. 305., Ed. by L. N. Ornston and J. R. Sokatch. Academic Press: New York.
Davis, B. D. (1961) *Cold Spring Harbor Symp. quant. Biol.* **26**, 1.
Dawes, E. A. and Senior, P. J. (1973) *Advanc. microb. Physiol.* **10**, 136.
Dean, A. C. R., Pirt, S. J. and Tempest, D. W. (Eds), (1972) *Environmental Control of Cell Synthesis and Function.* Academic Press, London.
Deibel, R. H. (1964) *J. Bact.* **87**, 988.
Deibel, R. H. and Kvetkas, M. J. (1964) *J. Bact.* **88**, 858.
Demain, A. L., Kennel, Y. M. and Aharonowitz, Y. (1979) In: *Microbial Technology: Current State and Future Prospects,* p. 163. Soc. Gen. Microbiol. Symposium 29. Ed. by A. T. Bull, D. R. Ellwood and C. Ratledge, Cambridge University Press: Cambridge.
Devoe, I. W. and Gilchrist, J. E. (1976) *J. Bact.* **128**, 144.
Dickson, R. C., Abelson, J., Barnes, W. M. and Reznikoff, W. S. (1975) *Science* **187**, 27.
Donachie, W. D., Jones, N. C. and Teather, R. (1973) In: *Microbial Differentiation,* p. 9. Ed. by J. M. Ashworth and J. E. Smith, Symp. 23. Soc. Gen. Microbiol. Cambridge University Press: Cambridge.

Eagon, R. G. (1962) *J. Bact.* **83**, 736.
Ellwood, D. C., Melling, J. and Rutter, P. R. (Eds) (1979) *Adhesion of Microorganisms to Surfaces.* Spec. Publ. Soc. Gen. Microbiol, Vol. 2. Academic Press, London.
Ellwood, D. C. and Tempest, D. W. (1972) *Advanc. microbial Physiol.* **7**, 83.
Elsden, S. R. and Hilton, M. G. (1979) *Archiv. Microbiol.* **123**, 137.
Elsden, S. R., Hilton, M. G. and Waller, J. M. (1976) *Arch. Microbiol.* **107**, 283.
Engelsberg, E. and Wilcox, G. (1974) *Annu. Rev. Genet.* **8**, 219.
Epstein, W. and Laimins, L. (1980) *Trends biochem Sci.* **5**, 21.
Fersht, A. (1977). *Enzyme Structure and Mechanism.* Freeman, Oxford.
Finegold, S. M., Shepherd, W. E. and Spaulding, E. H. (1977) *Practical Anaerobic Bacteriology.* Cumitech 5. Amer. Soc. Microbiol., Washington DC.
Finnerty, W. R. (1978) *Advanc. microbial Physiol.* **18**, 177.
Finnerty, W. R. and Makula, R. A. (1975) *CRC Crit. Rev. Microbiol.* **4**, 1.
Finney, D. J., Hazelwood, T. and Smith, M. J. (1955) *J. gen. Microbiol.* **12**, 222.
Foster, J. W. and Moat, A. G. (1980) *Microbiol. Rev.* **44**, 83.
Fraenkel, D. G. and Vinopal, R. T. (1973) *Annu. Rev. Microbiol.* **27**, 69.
Gaertner, F. H. (1978) *Trends biochem. Sci.* **3**, 63.
Gale, E. F., Cundliffe, E., Reynolds, P. E., Richmond, M. H. and Waring, M. J. (1972) *The Molecular Basis of Antibiotic Action.* Wiley, New York.
Gilbert, W. and Maxam, A. (1973) *Proc. nat. Acad. Sci. USA* **70**, 3581.
Gilbert, W. and Müller-Hill, B. (1966) *Proc. nat. Acad. Sci. USA* **56**, 1891.
Gladstone, G. P. (1937) *Brit. J. exp.. Path.* **18**, 322.
Glenn, A. R. (1976) *Annu. Rev. Microbiol.* **30**, 41.
Glover, S. W. (1978) *Advanc. microbial Physiol.* **18**, 235.
Goldfine, H. (1972) *Advanc. microb. Physiol.* **8**, 1.
Goldschmidt, M. C. and Fung, D. Y. C. (1979) *Food Technol.* **33**, 63.
Gots, J. S. (1971) In: *Metabolic Pathways,* 3rd edn., Vol. 5, p. 225, Ed. by H. J. Vogel. Academic Press, New York.
Gottschalk, G. (1979) *Bacterial Metabolism* Springer-Verlag, New York.
Gottschalk, G. and Andreesen, J. R. (1979) In: *Microbial Biochemistry, p. 85. Int. Rev. Biochem. Vol. 121.* Ed. by J. R. Quayle. University Park Press, Baltimore.
Gottwald, M., Andreesen, J. R., Le Gall, J. and Ljungdahl, L. G. (1975) *J. Bact.* **122**, 325.
Gould, G. W. and Corry, Janet E. L. (1980) *Microbial Growth and Survival in Extremes of Environment. Soc. appl. Microbiol. Tech. Science No. 15* Academic Press, London.
Gould, G. W. and Dring, G. J. (1974) *Advanc. microbial Physiol.* **11**, 137.
Gould, G. W. and Hurst, A. (1969) *The Bacterial Spore.* Academic Press, London.
Gray, T. R. G. and Postgate, J. R. (Eds) (1976) *The Survival of Vegetative Microbes, Soc. Gen. Microbiol.* Symposium 26, Cambridge University Press, Cambridge.
Guirard, B. M. and Snell, E. E. (1962) In: *The Bacteria,* Vol. 4, p. 33 (Ed. by I. C. Gunsalus and R. Y. Stanier) Academic Press, New York.
Haddock, B. A. and Hamilton, W. A. (Eds) (1977) *Microbial Energetics,* Soc. Gen. Microbiol. Symp. 27. University Press, Cambridge.
Haddock, B. A. and Jones, C. W. (1977) *Bact. Rev.* **41**, 47.
Hager, L. P. and Lipmann, F. (1961) *Proc. nat. Acad. Sci. USA* **47**, 1768.
Hamilton, W. A. (1975) *Advanc. microbial Physiol.* **12**, 1.
Hanawalt, P. C., Cooper, P. K., Ganesan, A. K., Smith, C. A. (1979) *Annu. Rev. Biochem.* **48**, 783.
Hanawalt, P. C., Friedberg, E. C. and Fox, C. F. (Eds) (1978) *Repair Mechanisms.* Academic Press, New York.
Hancock, R. E. W. and Nikaido, H. (1978) *J. Bact.* **136**, 381.
Harmon, S. M. and Kautter, D. A. (1977) *Appl. environ. Microbiol.* **33**, 762.
Harold, F. M. (1972) *Bact. Rev.* **36**, 172; (1977*a*) *Annu. Rev. Microbiol.* **31**, 181; (1977*b*) *Curr. Top. Bioenerg.* **6**, 83; (1979) In: *Microbiology—1979.* p, 42, Ed. by D. Schlessinger. Amer. Soc. Microbiol., Washington DC.
Harold, F. M. and Altendorf, K. (1974) *Curr. Top. Memb. Trans.* **5**, 1.
Harold, F. M. and van Brunt, J. (1977) *Science* **197**, 372.
Harrison, D. E. F. (1976) *Advanc. microb. Physiol.* **14**, 243.
Hartley, B. S. (1974) In: *Evolution in the Microbial World.* p. 151. Soc. Gen. Microbiol. Symposium 24. Ed. by M. J. Carlile and J. J. Skehel. Cambridge University Press, Cambridge.
Hartog, B J. and Heerink, G. J. (1979) *Antonie v. Leeuwenhoek,* **45**, 327.
Hasan, S. M. and Hall, J. B. (1975) *J. gen. Microbiol.* **87**, 120.
Helmstetter, C. E. (1967) *J. mol. Biol.* **24**, 417.
Helmstetter, C. E., Pierucci, O., Weinberger, M., Holmes, M. and Tang, M-S. (1979) In: *The Bacteria,* Vol. 7, p. 517. Ed. by J. R. Sokatch and L. N. Ornston, Academic Press, New York.
Herbert, D. (1961) *Rep. Soc. Chem. Industr., Monograph* No. 12, p. 21.
Herbert, D., Phipps, P. J. and Tempest, D. W. (1965) *Lab. Pract.* **14**, 1150.
Hewitt, J. and Morris, J. G. (1975) *FEBS Lett.* **50**, 315.
Hirata, H., Sone, N., Yoshida, M. and Kagawa, Y. (1976) *Biochem. biophys. Res. Commun.* **69**, 665.
Hobson, P. N. and Summers, R. (1967) *J. gen. Microbiol.* **47**, 53.
Horvath, R. S. (1972) *Bact. Rev.* **36**, 146.
Huddleson, I. F. (1921) *Cornell Vet.* **11**, 210.
Hueting, S., de Lange, T. and Tempest, D. W. (1979) *Arch. Microbiol.* **123**, 183.
Hugh, R. and Leifson, E. (1953) *J. Bact.* **66**, 24.
Hutner, S. H. (1972) *Annu. Rev. Microbiol.* **26**, 313.
Ingraham, J. L. (1962) In: *The Bacteria,* Vol. 4, 265. Ed. by I. C. Gunsalus and R. Y. Stanier. Academic Press, New York.
Ingram, L. C. (1970) *Biochem. biophys. Acta* **224**, 263.
Inouye, S., Wang, S., Sekizawa, J., Halegoua, S. and Inouye, M. (1977) *Proc. nat. Acad. Sci. USA* **74**, 1004.
Ishimoto, M., Koyama, J. and Nagai, Y. (1954); *J Biochem, Tokyo,* **41**, 763; (1954) *Bull. chem. Soc. Japan* **27**, 565.
Jacob, F. and Monod, J. (1961) *J. molec. Biol.* **3**, 318; (1970) In: *The Lactose Operon.* Cold Spring Harbor Symp. quant. Biol.
Jacob, H-E. (1970) In: *Methods in Microbiology,* Vol. 2. Ed. by J. R. Norris and D. W. Ribbons. Academic Press, London.
James, R. (1978) In: *Companion to Microbiology:* p. 59. Ed by A. T. Bull and P. M. Meadow. Longman, London.

Jannasch, H. W. and Mateles, R. I. (1974) *Advanc. microbial Physiol.* **11,** 165.
Jones, C. W. (1976) *Biological Energy Conservation.* Chapman and Hall, London.
Jones, C. W. and Meyer, D. J. (1976) In: *Handbook of Microbiology,* Vol. 2. Ed. by H. Le Chevalier. Cleveland Rubber Co. Press.
Jones, J. G. (1979) *A Guide to Methods for Estimating Microbial Numbers and Biomass in Fresh Water.* Freshwater Biological Association, UK.
Jurtshuk, P., Mueller, T. J. and Acord, W. C. (1975) *C.R.C. critic. Rev. Microbiol.* **3,** 399.
Kaback, H. R. (1960) *Fed. Proc.* **19,** 130; (1971) *Meth. in Enzymol.* **22,** 99.
Kantrowitz, E. R., Pastra-Landis, S. C. and Lipscomb, W. N. (1980) *Trends in biochem. Sci.* **5,** 124; **5,** 150.
Katz, E. and Demain, A. L. (1977) *Bact. Rev.* **41,** 339.
Kleinkauf, H. (1979) *Planta medica* **35,** 1.
Kleinkauf, H. and Koischwitz, H. (1978) In: *Progress in Molecular and Subcellular Biology,* Vol. 6, p. 59. Ed. by F. E. Hahn. Springer-Verlag, Berlin.
Koch, A. L. (1971) *Advanc. microbial Physiol.* **6,** 147; (1977) *Advanc. microbial Physiol.* **16,** 49.
Kolter, R. and Helinski, D. R. (1979) *Annu. Rev. Genet.* **13,** 355.
Konings, W. N. (1977) *Advanc. microbial Physiol.* **15,** 175.
Konings, W. N. and Veldkamp, H. (1980) In: *Contemporary Microbial Ecology.* P. 161. Ed. by D. C. Ellwood, J. N. Hedger, M. J. Latham, J. M. Lynch and J. H. Slater. Academic Press, London.
Kornberg, A. (1974) *DNA Synthesis.* W H. Freeman, San Francisco.
Kornberg, H. L. (1959) *Annu. Rev. Microbiol.* **13,** 49; (1965) In: *Function and Structure in Micro-organisms,* p. 8. Soc. Gen. Microbiol. Symposium 15. Ed. by M. R. Pollock and M. H. Richmond. Cambridge University Press, London; (1966) In: *Essays in Biochemistry,* Vol. 2, p. 1. Ed. by P. N. Campbell and G. D. Greville. Academic Press, London.
Koshland, D. E. Jr (1969) *Curr. Top. Cell. Regul.* **1,** 1.
Kovács, N. (1956) *Nature* **178,** 703.
Kubitschek, H. E. (1970) *Introduction to Research with Continuous Cultures.* Prentice-Hall, Englewood Cliffs, N.J.
Kundig, W., Gosh, S. and Roseman, S. (1964) *Proc. nat. Acad. Sci. USA* **52,** 1067.
Kushner, D. J. (Ed.) (1978) *Microbial Life in Extreme Environments.* Academic Press, New York.
Langworthy, T. A. (1978) In: *Microbial Life in Extreme Environments.* p. 279. Ed. by D. J. Kushner. Academic Press, New York.
Lapage, S. P., Shelton, J. E. and Mitchell, T. G. (1970) In: *Methods in Microbiology,* Vol. 3A, p. 1. Ed. by J. R. Norris and D. W. Ribbons. Academic Press, London.
Le Gall, J. and Postgate, J. R. (1973) *Advanc. microb. Physiol.* **10,** 81.
Lehninger, A. L. (1971) *Bioenergetics,* 2nd edn. W. A. Benjamin, Menlo Park, Calif.; (1975) *Biochemistry,* 2nd edn. Worth Publishers, New York.
Lin, E. C. C. (1976) *Annu. Rev. Microbiol.* **30,** 535.
Lipmann, F. (1941) *Advanc. Enzymol.* **1,** 99; (1980) *Advanc. microb. Physiol.* **21,** 228.
Loesche, W. J. (1969) *Appl. Microbiol.* **18,** 723.
Losick, R. and Chamberlain, M. (Eds) (1976) *RNA Polymerase.* Cold Spring Harbor Laboratory, Cold Spring Harbor, New York.

Lutkenhaus, J. F., Moore, B. A., Masters, M. and Donachie, W. D. (1979) *J. Bact.* **138,** 352.
McCord, J. M., Keele, B. B. and Fridovich, I. (1971) *Proc. nat. Acad. Sci. USA* **68,** 1024.
McDade, J. J. and Weaver, R. H. (1959) *J. Bact.* **77,** 65.
MacFaddin, J. F. (1976) *Biochemical Tests for Identification of Medical Bacteria.* Williams and Wilkins, Baltimore.
McLeod, J. W. and Gordon, J. (1923) *J. Path 'act.* **26,** 332.
Macy, J., Probst, I. and Gottschalk, G. (1975) *J. Bact.* **123,** 436.
Magasanik, B. (1961) *Cold Spring Harbor Symp. quant. Biol.* **26,** 249;(1977) *Trends biochem. Sci.* **2,** 9.
Magasanik, B., Prival, M., Brenchley, J., Tyler, B., Deloe, A., Streicher, S., Bender, R. and Paris, C. (1974) *Curr. Top. Cell. Regul.* **8,** 118.
Malik, V. S. (1980) *Trends in biochem. Sci.* **5,** 68.
Mallette, M. F. (1970) In: *Methods in Microbiology,* Vol. 1, p. 522. Ed. by J. R. Norris and D. W. Ribbons. Academic Press, London.
Mandelstam, J. (1976) *Proc. Roy. Soc. Lond. B.* **193,** 89.
Mandelstam, J., McQuillen, K. and Dawes, I. W. (1982) *Biochemistry of Bacterial Growth,* 3rd edn. Blackwell, Oxford.
Mason, W. W. (1935) *J. Bact.* **29,** 103.
Massey, L. K., Sokatch, J. R. and Conrad, R. S. (1976) *Bact. Rev.* **40,** 42.
Matsushita, T. and Kubitschek, H. E. (1975) *Advanc. microbial Physiol.* **12,** 247.
Maw, J. and Meynell, G. G. (1968) *Brit. J. exp. Pathol.* **40,** 597.
Meers, J. L. (1973) *C.R.C. Crit. Rev. Microbiol.* **1,** 139.
Metzler, D. E. (1977) *Biochemistry. The Chemical Reactions of Living Cells.* Academic Press, New York.
Meyer, D. J. and Jones, C. W. (1973) *Int. J. syst. Bact.* **23,** 459.
Meynell, G. G. and Meynell, E. (1970) *Theory and Practice in Experimental Bacteriology,* 2nd Edn. Cambridge University Press, Cambridge.
Miraglia, G. J. (1974) *C.R.C. Crit. Rev. Microbiol.* **10,** 81.
Mitchell, P. (1961) *Nature* **191,** 144; (1966) *Biol. Rev. Cambridge Phil. Soc.* **41,** 445; (1976) *Biochem. Soc. Trans.* **4,** 399.
Mitruka, B. M. (1976) *Methods of Detection and Identification of Bacteria.* C.R.C. Press, Cleveland, Ohio.
Mitruka, B. M. and Costilow, R. N. (1967) *J. Bact.* **93,** 295.
Moat, A. G. (1979) *Microbial Physiology.* Wiley, New York.
Molin, S., von Meyenburg, K., Malløe, O., Hansen, M. T. and Pato, M. L. (1977) *J. Bact.* **131,** 7.
Monod, J. (1949) *Annu. Rev. Microbiol.* **3,** 371.
Monod, J., Wyman, J. and Changeux, J. D. (1965) *J. mol. Biol.* **12,** 88.
Moore, W. B. (1968) *J. gen. Microbiol.* **53,** 415.
Morris, J. G. (1974) *A Biologist's Physical Chemistry,* 2nd edn. Edward Arnold, London; (1975) *Advanc. microb. Physiol.* **12,** 169; (1976) *J. appl. Bact.* **40,** 229; (1979) In: *Strategies of Microbial Life in Extreme Environments.* p. 149. Ed. by M. Shilo. Dahlem Konferenzen, Berlin.
Morris, J. G. and O'Brien, R. W. (1971) In: *Spore Research 1971.* p. 1. Ed. by A. N. Barker, G. W. Gould and J. Wolf. Academic Press, London.
Mortlock, R. P. (1976) *Advanc. microb. Physiol.* **13,** 1.
Moseley, B. E. B. and Williams, E. (1977). *Advanc. microb. Physiol.* **16,** 100.

Müller-Hill, B. (1975) *Prog. biophys. molec. Biol.* **30**, 227.
Munro, A. L. S. (1970) In: *Methods in Microbiology*, Vol. 2, p. 39. Ed. by J. R. Norris and D. W. Ribbons. Academic Press, London.
Munson, R. J. (1970) In: *Methods in Microbiology*, Vol. 2, p. 349. Ed. by J. R. Norris and D. W. Ribbons. Academic Press, London.
Nakae, T. (1976) *J. biol. Chem.* **251**, 2176.
Neijssel, O. M. and Tempest, D. W. (1976) *Arch. Microbiol.* **110**, 305; (1979). In: *Microbial Technology: Current State, Future Prospects*, p. 53. Soc. Gen. Microbiol. Symposium 29 Ed. by A. T. Bull, D. R. Ellwood and C. Ratledge. Cambridge University Press, Cambridge.
Neilands, J. B. (Ed.) (1974) *Microbial Iron Metabolism*. Academic Press, New York.
Neilson, N. E., MacQuillan, M. F. and Campbell, J. J. R. (1959) *Canad. J. Microbiol.* **5**, 293.
Neu, H. C. and Heppel, L. A. (1965) *J. biol. Chem.* **240**, 3685.
Nierlich, D. P. (1978) *Annu. Rev. Microbiol.* **32**, 393.
Nikaido, H. (1973) In: *Bacterial Membranes and Walls*. p. 132. Ed. by L. Leive. Marcel Dekker, New York.
Nikaido, H. and Nakae, T. (1979) *Advanc. microb. Physiol.* **20**, 163.
Norris, T. E. and Koch, A. L. (1972) *J. molec. Biol.* **64**, 635.
Novick, A. (1955) *Annu. Rev. Microbiol.* **9**, 97.
O'Brien, R. W. and Morris, J. G. (1971) *J. gen. Microbiol.* **68**, 307.
Old, R. W. and Primrose, S. B. (1980) *Principles of Gene Manipulation*. Blackwell, Oxford.
O'Leary, W. M. (1975) *CRC crit. Rev. Microbiol.* **4**, 41.
Ørskov, I., Ørskov, F., Jann, B. and Jann, K. (1977) *Bact. Rev.* **41**, 667.
Ornston, L. N. (1971) *Bact. Rev.* **35**, 87.
Paigen, K. and Williams, B. (1970) *Advanc. microb. Physiol.* **4**, 251.
Pastan, I. and Adhya, S. (1976) *Bact. Rev.* **40**, 527.
Payne, W. J. (1973) *Bact. Rev.* **37**, 409; (1976) *Advanc. microb. Physiol.* **13**, 55.
Payne, W. J. and Balderston, W. L. (1978) In: *Microbiology—1978*, p. 339. Ed. by D. Schlessinger. Amer. Soc. Microbiol., Washington DC.
Pirt, S. J. (1975) *Principles of Microbe and Cell Cultivation*. Blackwell, Oxford.
Polk, H. C. and Miles, A. A. (1973) *Brit. J. exp. Path.* **54**, 99.
Poole, R. K., Lloyd, D. and Kemp, R. (1973) *J. gen. Microbiol.* **77**, 209.
Postgate, J. R. (1967) *Advanc. microb. Physiol.* **1**, 2; (1970) In: *Methods in Microbiology*, Vol. 1, p. 611. Ed. by J. R. Norris and D. W. Ribbons. Academic Press, London;(1979) *The Sulphate-Reducing Bacteria*. Cambridge University Press, Cambridge.
Postma, P. W. and van Dam, K. (1976) *Trends in biochem. Sci.* **1**, 16.
Powell, D. A. (1979) *Spec. Publ. Soc. Gen. Microbiol.* **3**, 117. Academic Press, London.
Preiss, J. (1969) *Curr. Top. Cell. Regul.* **1**, 125.
Quayle, J. R. and Bull, A. T. (1982) *New dimensions in bacteriology. Phil. Trans.* Royal Society, London. *B* **297**, 445.
Quesnel, L. B. (1963) *J. appl. Bact.* **26**, 127.
Radding, C. M. (1977) *Annu. Rev. Genetics* **7**, 87.
Raetz, C. R. H. (1978) *Microbiol. Rev.* **42**, 614.
Rando, R. R. (1974) *Science* **185**, 320.

Reiness, G., Yang, H., Zubay, G. and Cashel, M. (1975) *Proc. nat. Acad. Sci. USA* **72**, 2881.
Rich, A. (1978) *Trends in biochem. Sci.* **3**, 34.
Roantree, R. J. (1967) *Annu. Rev. Microbiol.* **21**, 443.
Roberts, J. (1969) *Nature* **224**. 1168
Rogers, H. J. (1961) In: *The Bacteria*. Vol. 2, p. 257. Ed. by I. C. Gunsalus and R. Y. Stanier. Academic Press, New York; (1979) *Advanc. microb. Physiol.* **19**, 1.
Roon, R. J. and Barker, H. A. (1972) *J. Bact.* **109**, 44.
Rose, A. H. (1976) *Chemical Microbiology. An Introduction to Microbial Physiology*, 3rd edn. Butterworths, London.
Rosen, B. P. (Ed.) (1978) *Bacterial Transport* Marcel Dekker, New York.
Rosen, B. P. and Brey, R. N. (1979) In; *Microbiology—1979*. Ed. by D. Schlessinger. *Amer. Soc. Microbiol.*, Washington DC.
Rosenberg, H., Gerdes, R. G. and Chegwidden, K. (1977) *J. Bact.* **131**, 505.
Russell, A. D., Morris, A. and Allwood, M. C. (1973) In: *Methods in Microbiology*, Vol. 8, p. 96. Ed. by J. R. Norris and D. W. Ribbons. Academic Press, London.
Russell, C. and Coulter, W. A. (1975) *Appl. Microbiol.* **29**, 141.
Russell, W. J., Farling, S. R., Blanchard, G. C. and Boling, E. A. (1975) In: *Microbiology—1975*, p. 22. Ed. by D. Schlessinger. *Amer. Soc. Microbiol.*, Washington D.C.
Ryan, F. J. (1959) *J. gen. Microbiol.* **21**, 530.
Saier, M. H. Jr (1979) In: *Microbiology—1979*, p. 72. Ed. by D. Schlessinger. *Amer. Soc. Microbiol.*, Washington D.C.
Salton, M. R. J. (1974) *Advanc. microb. Physiol.* **11**, 213.
Sanwal, B. D. (1970) *Bact. Rev.* **34**, 20.
Sargent, M. G. (1978) *Advanc. microb. Physiol.* **18**, 105.
Savage, D. C. (1977) In: *Microbiology—1977*, p. 422. Ed. by D. Schlessinger. *Amer. Soc. Microbiol.*, Washington, D.C.
Schaechter, M., Maaløe, O. and Kjeldgaard, N. (1958) *J. gen. Microbiol.* **19**, 592.
Schindler, H. and Rosenbusch, J. P. (1978) *Proc. nat. Acad. Sci. USA* **75**, 3751.
Schleifer, K. H., Hammes, W. P. and Kandler, O. (1976) *Advanc. microb. Physiol.* **13**, 245.
Schleifer, K. H. and Kandler, O. (1972) *Bact. Rev.* **36**, 407.
Schulz, J. S. and Gerhardt, P. (1969) *Bact. Rev.* **33**, 1.
Schwartz, A. C., Quecke, W. and Brenschede, G. (1979) *Z. für allg. Mikrobiol.* **19**, 211.
Segal, H. L. (1973) *Science* **180**, 25; (1975) *Enzyme Kinetics*. Wiley, New York.
Sharpe, A. N. (1973) *Soc. Appl. Bacteriol. Technical Series*, Vol. 7, p. 197. Academic Press, London.
Shaw, N. and Baddiley, J. (1968) *Nature* **217**, 142.
Sherman, J. M. (1926) *J. Bact.* **11**, 417.
Sinsheimer, R. L. (1977)) *Annu. Rev. Biochem.* **46**. 415.
Smibert, R. M. (1978) *Annu. Rev. Microbiol.* **32**, 673.
Smith, H. (1978) In: *Essays in Microbiology*. Ed. by J. R. Norris and M. H. Richmond p. 13/1 John Wiley and Sons, Chichester.
Smith, K. C. (1978) *Photochem. Photobiol.* **28**, 121.
Spratt, B. G. (1980) *Phil. Trans. R. Soc. Lond. B.* **289**, 273.
Springer, M. S., Goy, M. F. and Adler, J. (1979) *Nature* **280**, 279.
Stadtman, T. C. (1954) *J. Bact.* **67**, 314.
Stanier, R. Y., Adelberg, E. A. and Ingraham, J. L. (1977) *General Microbiology*, 4th edn. Macmillan, London.
Stanier, R. Y. and Ornston, L. N. (1973) *Advanc. microb. Physiol.* **9**, 89.

Stanier, R. Y., Palleroni, N. J. and Doudoroff, M. (1966) *J. gen. Microbiol.* **43**, 159.
Stanier, R. Y., Rogers, H. J. and Ward, J. B. (Eds) (1978) *Relations Between Structure and Function in the Prokaryotic Cell.* Soc. Gen. Microbiol. Symposium 28, Cambridge University Press, Cambridge.
Steel, K. J. (1961) *J. gen. Microbiol.* **25**, 297.
Stent, G. S. and Calendar, R. (1978) *Molecular Genetics. An Introductory Narrative.* Freeman, San Francisco.
Sterne, M. and Batty, I. (1975) *Pathogenic Clostridia.* Butterworths, London.
Stickland, L. H. (1934) *Biochem. J.* **28**, 1746.
Stock, J. and Roseman, S. (1971) *Biochem. biophys. Res. Commun.* **44**, 132.
Stouthamer, A. H. (1976) *Advanc microb. Physiol.* **14**, 315.
Stouthamer, A. H. and Bettenhausen, C. W. (1977) *Arch. Microbiol.* **113**, 185.
Stouthamer, A. H., de Vries, W. and Niekus, H. G. D. (1979) *Ant. van Leeuw.* **45**, 5.
Strange, R. E. (1972) *Advanc. microb. Physiol.* **8**, 105.
Sudo, S. Z. and Dworkin, M. (1973) *Advanc. microb. Physiol.* **9**, 153.
Sutherland, I. W. (1979a) *Trends biochem. Sci.* **4**, 55; (1979b) *Spec. Publ. Soc. Gen. Microbiol.* **3**, 1. Academic Press, London.
Swanson, J. (1977) In: *Microbiology—1977,* p. 427. Ed. by D. Schlessinger. *Amer. Soc. Microbiol.,* Washington, D.C.
Szekely, M. (1980) *From DNA to Protein. The Transfer of Genetic Information.* Macmillan, London.
Talkington, C. and Pero, J. (1978) *Proc. Nat. Acad. Sci. USA* **75**, 1185.
Tally, F. P., Goldin, B. R., Jacobus, N. V. and Gorbach, S. L. (1977) *Infect. Immun.* **16**, 20.
Tempest, D. W. (1970a) *Advanc. microb. Physiol.* **4**, 223; (1970b) In: *Methods in Microbiology,* Vol. 2, P. 259. Ed. by J. R. Norris and D. W. Ribbons. Academic Press, London; (1978) In: *Essays in Microbiology,* p. 7/1. Ed. by J. R. Norris and M. H. Richmond John Wiley, Chichester.
Tempest, D. W., Meers, J. L. and Brown, C. M. (1970) *Biochem. J.* **117**, 405.
Thauer, R. K., Jungermann, K. and Decker, K. (1977) *Bact. Rev.* **41**, 100.
Tigier, H. and Grisolia, S. (1965) *Biochem. biophys. Res. Commun.* **19**, 209.
Tomasz, A. (1979) *Annu. Rev. Microbiol.* **33**, 113.
Tonn, S. J. and Gander, J. E. (1979) *Annu. Rev. Microbiol.* **33**, 169.
Torriani, A. (1960) *Biochim. biophys. Acta* **38**, 460.

Tristram, H. (1978) In: *Companion to Microbiology,* p. 297. Ed. by A. T. Bull and P. Meadow. Longman, London.
Troy, F. A. (1979) *Annu. Rev. Microbiol.* **33**, 519.
Ullmann, A. (1974) *Biochem. biophys. Res. Commun.* **57**, 348.
Ullmann, A. and Danchin, A. (1980) *Trends biochem. Sci.* **5**, 95; **38**, 323; (1978) *Annu. Rev. Biochem.* **47**, 533.
Umbarger, H. E. (1969) *Annu. Rev. Biochem.,* **38**, 323.
Ur, A. and Brown, D. F. J. (1975) *J. med. Microbiol.* **8**, 19.
Valentine, R. C. and Wolfe, R. S. (1960) *Biochem. biophys. Acta.* **45**, 389.
Veldkamp, H. (1970) In: *Methods in Microbiology,* Vol. 3A, p. 305. Ed. by J. R. Norris and D. W. Ribbons. Academic Press, London; (1976) *Continuous Culture in Microbial Physiology and Ecology.* Meadowfield Press, Durham.
Venema, G. (1979) *Advanc. microb. Physiol.* **19**, 245.
Venugopal, V. and Nadkarni, G. B. (1977) *J. Bact.* **131**, 693.
Vogels, G. D. and van der Drift, C. (1976) *Bact. Rev.* **40**, 403.
Warth, A. D. (1978) *Advanc. microb. Physiol.* **17**, 1.
Watson, J. D. (1968) *The Double Helix.* Atheneum Press, New York; (1976) *Molecular Biology of the Gene,* 3rd edn. W. A. Benjamin, Menlo Park, California.
Watson, J. D. and Crick, F. H. C. (1953) *Nature* **171**, 737 and 964.
Watson, T. G. (1972) In: *Environmental Control of Cell Synthesis and Function.* Ed. by A. C. R. Dean, S. J. Pirt and D. W. Tempest. Academic Press, London.
Weitzman, P. D. J. and Danson, M. J. (1976) *Curr Top. Cell. Regul.* **10**, 161.
West, I. C. (1974) *Biochem. Soc. Spec. Publ.* **4**, 27.
West, I. C. and Mitchell, P. (1974) *Biochem. J.* **144**, 87.
Wilson, G. S. (1930) *J. Hyg. Camb.* **30**, 433; (1931) *Brit. J. exp. Path.* **12**, 88.
Wimpenny, J. W. T. (1979) *J. gen. Microbiol.* **114**, 483.
Wimpenny, J. W. T. and Parr, J. A. (1979) *J. gen. Microbiol.* **114**, 487.
Wolfe, R. S. (1979) *Ant. van Leeuw.* **45**, 353.
Wood, H. G. and Stjernholm, R. L. (1962) In: *The Bacteria,* Vol. 3, p. 41. Ed. by I. C. Gunsalus and R. Y. Stanier. Academic Press, New York.
Wood, W. A. (1961) In: *The Bacteria,* Vol. 2, p. 59. Ed. by I. C. Gunsalus and R. Y. Stanier. Academic Press, New York.
Wyman, J. (1972) *Curr. Top. Cell Regul.* **6**, 209.
Wynn, C. H. (1979) *The Structure and Function of Enzymes,* 2nd edn. Edward Arnold, London.
Yocum, R. R., Waxman, D. J. and Strominger, J. L. (1980) *Trends in biochem. Sci.* **5**, 97.
Young, M. and Mandelstam, J. *Advanc. microb. Physiol.* **20**, 103.

4

Bacterial resistance, disinfection and sterilization

Graham Wilson

Introductory	70	Alcohols	83
Disinfection and sterilization	71	Ethers	83
Physical agencies	71	Phenols and cresols	83
Ultraviolet rays	71	Chloroxylenols, chlorhexidine, and	
Photodynamic sensitization	73	hexachlorophane	84
Ionizing radiations	73	Halogens	84
Sonic and ultrasonic methods	74	Dyes	85
Desiccation	75	Essential oils	85
Cold	75	Dynamics of disinfection	85
Heat, dry and moist	76	Reaction velocity	85
Thermal death point of bacteria	78	Concentration coefficient	87
Pressure	78	Temperature coefficient	88
Filtration	78	Standardization of disinfectants	88
Chemical agencies	78	Rideal-Walker test, Chick-Martin test	89
Distilled water	79	Gaseous disinfectants	90
Acids, alkalies, and aldehydes	79	Sulphur dioxide	90
Salt action	80	Methyl bromide	90
Mode of action of salts	80	Propylene oxide	90
Antagonistic effect of salts	81	Formaldehyde	90
Soaps	81	Liquid disinfectants, use of in practice	91
Synthetic detergents	82	Solid disinfectants	91

Introductory

The use of aromatic vapours and of smoking, salting and drying for the prevention of putrefaction and for warding off infection goes back to the days of ancient Egypt and the Old Testament. Pringle's observations in 1750 constituted the first scientific approach to the study of disinfection. Early in the nineteenth century hypochlorites came into prominence, and by 1847 Semmelweis (1861) in Vienna had shown conclusively that chloride of lime was able to counteract the agent responsible for puerperal infection. Though their mode of action was not understood at the time, these substances were employed probably in the belief that infectious diseases arose either from miasmata or from putrefying material. Substances with a strong deodorant property that could cloak miasmata and substances that could prevent putrefaction were both accepted as disinfectants.

To the hypochlorites was soon added carbolic acid, which was used so successfully by Lister in 1865 and onwards in the control of wound infection (see Godlee 1924) and advocated by Budd (1873) for preventing the spread of typhoid fever. These practices were employed empiricially for many years before their bacteriological justification was established. The influence of organic matter in diminishing the activity of disinfectants was demonstrated by Baxter (1875).

Robert Koch (1881) was the first to make a serious study of the comparative germicidal effect of different disinfectants. By drying anthrax spores on silk threads of the same length, immersing them in a solution of the substance to be tested, and subsequently transferring them to a nutrient medium in order to ascertain if the bacteria were still alive, he collected a considerable quantity of information on the relative activity of different disinfectants. His work was criticized and his methods improved by Geppert (1889, 1891a, b). In 1897 Krönig and Paul published their classical paper, describing a new method for the quantitative study of disinfection, and demonstrating that in a culture submitted to the influence of a germicidal agent bacteria die, not simultaneously but in an orderly sequence. To Madsen and Nyman (1907) and to Chick (1908, 1910, 1912) must be ascribed the merit of analysing the various factors upon which disinfection depends, and of showing that the law underlying the death of bacteria is similar to that underlying a simple unimolecular chemical reaction.

In the present chapter we deal with the principles rather than with the practice of disinfection and sterilization. Our main endeavour is to discover as far as possible the underlying principles of disinfection, to discuss the laws governing the killing of bacteria, and to point out the importance of a thorough knowledge of these laws and principles to anyone who, whether engaged in medicine, hygiene, dairy-farming, food-preservation, or agriculture, is confronted with the problem of controlling bacterial activity.

Disinfection and sterilization

Many terms are used in relation to the destruction of bacteria and viruses. Most of them have been given different connotations (see Sykes 1958). For our purpose we shall regard *sterilization* as the complete destruction of all living matter. In medical parlance the term is often used in a restricted sense to refer to the destruction of pathogenic organisms only, but though this is convenient it is inaccurate and misleading.

Disinfection, on the other hand, has this latter meaning, indicating the destruction or removal of all organisms capable of giving rise to infection. Through long usage the term is closely associated with chemical substances, though there is no reason why articles suitably treated by heat or other physical agents should not be referred to as disinfected.

The term *antisepsis* is most conveniently used to indicate the prevention of infection, usually by inhibiting the growth of bacteria. Most disinfectants when suitably diluted have an antiseptic action.

Agents that destroy bacteria, without any special degree of destruction being defined, are often said to be *bactericidal* or *germicidal*, and those that restrain or prevent bacterial growth as *bacteristatic*.

The complete destruction of a bacterium or virus is not always easy to ascertain. Some organisms that appear to be dead may be revived by appropriate methods; and, as is made clear in Chapter 3, organisms that are incapable of multiplying may nevertheless retain some enzymic activity.

Physical agencies

Irradiation

The kinds of radiation used for sterilization and disinfection fall into two groups—ionizing and non-ionizing (see Glasser *et al.* 1952). We shall deal with this second group first.

Non-ionizing radiation. Ultraviolet rays Downes and Blunt (1877, 1878) showed that sunlight was able to destroy bacteria in a putrescible fluid, and Duclaux (1887) found that it killed vegetative bacilli more rapidly than spores. Roux (1887) observed that nutrient media exposed to sunlight in the presence of air were rendered bacteristatic; and Burnet (1925) many years later brought evidence to show that this effect was due to the production of hydrogen peroxide—a substance so powerful that its inhibitory action may be noticeable in a dilution of even 1/40 000.

That sunlight owed its germicidal effect to its actinic rather than to its heat waves was demonstrated by Ward (1892). By throwing a spectrum across an agar plate, Ward further showed that blue light was more germicidal than red. This observation was confirmed and amplified by Barnard and Morgan (1903), who found that the bactericidal radiations were confined mainly to those in the ultraviolet region included between the wavelengths 328 and 226 nm, that is the light between the visible violet and the extreme ultraviolet. Browning and Russ (1917) defined this range more closely as that region within the wavelengths 296 and 210 nm. Since the limit of retinal sensibility on the lower side is reached at 397 nm, it follows that the most germicidal rays are in a part of the spectrum which is invisible to the human eye (Fig. 4.1). Visible rays are not completely devoid of germicidal power, but they require very much longer to produce their lethal effect (Thiele and Wolf 1907, Laroquette 1918). Later work by Gates (1930), Ehrismann and Noethling (1932), Buchholz and von Jeney (1935), Prudhomme (1937), and Rouyer and Servigne (1938) showed that two of the most active wavelengths in the ultraviolet region were 265 and 253 nm.

The time necessary for destruction of micro-organisms by ultraviolet light depends on the intensity of the light, the distance of the source of illumination, and the nature of the medium in which the organisms are exposed. In general, the Bunsen-Roscoe law holds true; that is, within given limits the product of the intensity of irradiation and the length of exposure is constant. The temperature of the organisms at the time of exposure, provided that it is within the normal limits of viability, seems to have no effect on the action of ultraviolet light (Rentschler *et al.* 1941), though with some air-dried organisms the relative humidity

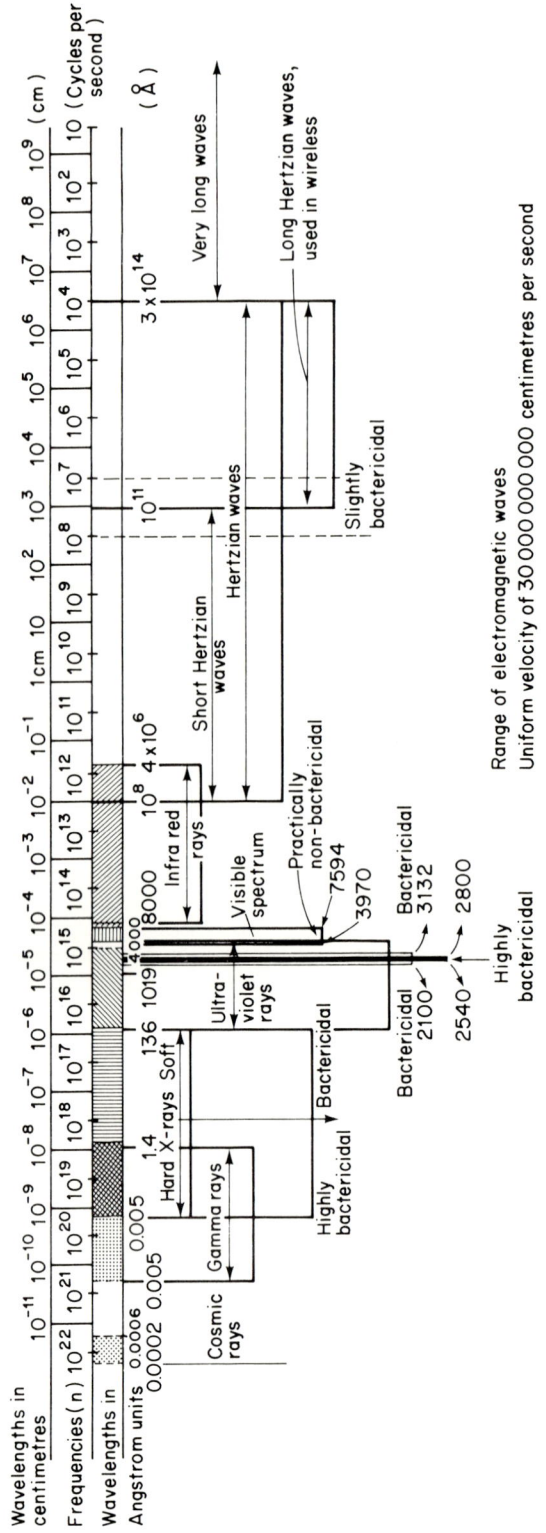

Fig. 4.1 Diagram showing the range of electromagnetic waves with the bactericidal action of rays of different wavelength.
Note.—Little work has been done on some parts of this range and none on others, so that the figure must be regarded as affording no more than an indication, incomplete and probably in part erroneous, of the bactericidal action of these waves. (10 Angström units = 1 nm)

The scale of this chart is taken from a diagram prepared by Vivian T. Saunders, M.A., of Uppingham School, and published by John Murray, London, W.1, to whom we are indebted for permission to reproduce it.

of the atmosphere may influence the result (Webb 1963).

Under natural conditions the germicidal effect of sunlight varies according to circumstances. In southern lands the combined effect of the ultraviolet and the heat rays is often considerable; thus Semple and Greig (see Hewlett 1909) in India found that typhoid bacilli exposed to the sun on pieces of white drill cloth were killed in two hours, whereas controls kept in the dark were still alive after six days. In the smoke-covered towns of the northern hemisphere, the ultraviolet rays are often very weak, being largely filtered off by the impurities in the atmosphere, thus depriving sunlight of most of its activity. That light under these conditions is not entirely without effect, however, was shown by Garrod (1944), who found that in hospital wards hæmolytic streptococci could be readily demonstrated in dust from the darker portions of the ward, but not from dust on window sills and other parts exposed to diffuse daylight.

Ultraviolet light generated by a Cooper Hewitt Mercury vapour lamp has been used for the sterilization of drinking water. Foulds (1911) found that it was quite easy to ensure a 99 per cent reduction in the bacterial count, including all *Esch. coli*, by this method. (For further information see Taylor 1949.) It has also a limited usefulness for the destruction of bacteria and viruses in the air (see Chapter 9).

Ultraviolet light has little penetrating power; it is readily filtered off by thin layers of glass or of protein. Bacteria, however, are probably sufficiently small for the rays to reach their interior; and toxins (Kitasato 1891, Fermi and Pernossi 1894), serum complement (Sellards 1918, Brooks 1920) and other proteins can be inactivated if they are exposed in thin layers. The rays appear to act by coagulating the proteins (Dreyer and Hanssen 1907) and causing irreversible damage to the DNA (Davis *et al.* 1968). Though the production of hydrogen peroxide or ozone was at one time thought to be responsible for the germicidal effect (D'Arcy and Hardy 1894, Wesbrook 1896), later work showed that atmospheric oxygen was not essential (Thiele and Wolf 1906, 1907, Blum 1932). The mechanism by which the colloidal change in the protoplasm is brought about and the quanta of energy required for killing bacteria are discussed by Wyckoff (1932), Buchholz and von Jeney (1935), Lea and Haines (1940), and Silverman (1966). (For a discussion of the photochemical effects of ultraviolet light see Smith and Hanawalt 1969; and for the inactivation of viruses, see Booth 1979.)

Kelner (1949) described the phenomenon of *photoreactivation*. He found that if, after a certain length of treatment with ultraviolet light, a suspension of coliform bacilli was exposed for 45–60 minutes to visible light from a 500-watt tungsten projection lamp, a high proportion of the apparently dead organisms were able to form colonies on incubation. Recovery has also been observed as a result of heating cells either during or after irradiation with substances such as pyruvate, acetate, or citric acid cycle metabolites, or by making suitable changes in the growth medium (Roberts and Aldous 1949, Heinmets *et al.* 1954, Ellison, Erlanger and Allen 1955, Wainwright and Nevill 1955).

The effect appears to be due to an enzyme that combines specifically with thymine dimers but is inactive in the dark. When the complex is irradiated with ultraviolet light the enzyme cleaves the dimer, restoring the original thymine residue and releasing the enzyme (Davis *et al.* 1968).

Photodynamic sensitization Though the visible rays of the spectrum have only a weak germicidal action on bacteria, they can be rendered almost as active as ultraviolet waves by passing them through a dye that causes them to fluoresce (see Raab 1900, von Tappeiner 1900). In one of his original experiments, for example, von Tappeiner projected a spectrum across a table and exposed a culture of a paramœcium, suspended in a 1/800 solution of eosin, to the different rays. The culture exposed to the green rays, which fluoresced in the presence of eosin, was killed in 2 to 4 hours; cultures exposed to the other rays, which did not fluoresce, were unharmed. For sensitization to occur, the paramœcia had to be in close contact with the fluorescing particles; if they were merely exposed to light that had been filtered through eosin, they were unaffected. The phenomenon of photodynamic sensitization has been observed not only with protozoa but with toxins, antitoxins, and bacteria (von Tappeiner and Jodlbauer 1904), bacteriophages (Clifton 1931), and filtrable viruses (Perdrau and Todd 1933*b*). It is a process in which molecules of biological importance are oxidized in the presence of oxygen or a suitable hydrogen donor, a dye, and visible light. The oxygen and the substrate are consumed, but the dye is re-used (Smith and Hanawalt 1969). The damage can to some extent be counteracted by carotenoid pigments. This may partly account for the prevalence of pigmented strains among airborne bacteria and bacteria in natural brines (Harrison 1967). In some instances the reaction is completely reversible by simple dilution of the dye-substrate mixture. Enteroviruses are exceptional in not being readily photosensitized by direct mixture with cationic dyes. For purposes of vaccine production, photodynamically inactivated virus suspensions may prove of value, since the dyes bind to the nucleic acid with little or no degradation of the surface antigens (Booth 1979). For a review of the subject, see Bayliss 1924, Blum 1932, Harrison 1967, Booth 1979.)

Ionizing radiations In practice only two kinds of ionizing radiation are used for sterilization or disinfection, namely (*a*) electromagnetic γ-rays and (*b*) particulate high-energy electrons sometimes referred to as β-rays. Other kinds, such as α-particles, neutrons, protons and x-rays, are unsatisfactory either because they have poor penetrating power, or induce radioactivity in exposed materials, or present difficulties in the regulation of their dosage. Both γ-rays and high-energy electrons are used extensively in industry, mainly for the sterilization of instruments and dressing packs, food substances, pharmaceutical products and a miscellany of objects that cannot conveniently be dealt with in other ways (see Report 1964, 1967, Silverman 1966). By virtue of their penetrating power they can be used for the treatment of pre-packed articles, so that the risk of contamination after processing is avoided. Though radioresistant mutants may develop when organisms are exposed repeatedly to sublethal doses of radiation with intervening periods of outgrowth, they are of little practical importance in the industrial field.

Gamma radiation has a narrow range of wavelength and results from the nuclear disintegration of certain radioactive substances, such as the isotopes Cobalt-60 or Caesium-137. High-speed electrons used in industry are produced either by an electrostatic generator of the van de Graaff type in which a high voltage is applied directly, or by a linear accelerator in which the charged particles emanating from a heated cathode are given successive increments of energy.

Gamma rays are more penetrating than fast electrons of the same energy and do not induce secondary radiation in exposed materials (Powell 1961); they require, however, a longer time and their source cannot be switched off. With high-speed electrons, on the other hand, the beam can be accurately directed and shielded, sterilization is complete within a few seconds, and the machine can be switched on and off as required. However, deep penetration can be obtained only by the use of 5–10 million electron volts (MeV), and with this dosage radioactivity may be induced in the treated objects. X-rays, which are produced by a special machine, do not differ in principle from γ-rays, but they extend over a wide range of wavelengths, each with different penetrating power, and are therefore less convenient to use for industrial purposes.

The chemical and physical changes produced in matter by ionizing radiation depend on the amount of energy absorbed, and on the nature of the atoms in the matter, as well as on the temperature, the partial pressure of oxygen, the amount of organic material, the presence of protective compounds such as aliphatic alcohols and sulphydryl-containing substances, the species of organism, the physiological state of the organisms, and their water content (Silverman 1966). Spores are, in general, more resistant than vegetative organisms, but some non-sporing organisms, e.g. *Micrococcus radiodurans*, have D_{10} and LD_{90} values 52 and 200 times greater, respectively, than those for such an organism as *Serratia marcescens*; and are said to be as resistant as spores.

The energy of ionizing radiation is measured in electron volts (eV), but the energy absorbed by unit mass of the irradiated material is expressed in rads—the rad being an arbitrary unit corresponding to the absorption of 100 ergs per gram of material. Most articles are sterilized by the absorption of 2 to 5 megarads (M rad), the megarad being 1 000 000 rads.

The effect of ionizing radiation on matters depends on the amount of energy imparted. With radiation of high energy the nuclei of the atoms may be so excited as to become unstable and emit a neutron. Such a neutron may enter the nucleus of another atom. An atomic nucleus that loses or gains a nuclear particle may become radioactive; and in this way ionizing radiation may induce radioactivity in matter that previously possessed none. Gamma radiation, however, from the isotope sources used for industrial purposes is of too low energy to have this effect. In biological materials, such as food, the radiation acts mainly on the molecules of water, leading to the formation of highly chemically reactive free H and OH radicals which serve as powerful reducing and oxidizing agents. Apart from this indirect method, the work of Wyckoff and Rivers (1930), Lea, Haines and Bretscher (1941), Pollard and Forro (1949) and others suggests that the radiation acts directly on the micro-organisms themselves, giving up during its passage through them one or more quanta of energy which lead to the disruption of the internal economy including the degradation of DNA (Goldblith 1967). Whether the one-quantum-hit-to-kill explanation, first put forward by Crowther (1926), is correct, or whether several quanta are required is still in doubt. So also are the number and the size of the sensitive areas or 'targets' contained in the organism (see Brown and Melling 1971).

It may be added that sublethal doses of various forms of radiation may bring about the changes in the morphology and growth characters of bacteria and induce heritable mutations. The subject of ionizing radiations is reviewed by Goldblith (1967, 1971, and Davies, 1976).

Electricity Observations by numerous workers (for references see 4th edition) have shown that electricity in the form of galvanic, low-frequency, or high-frequency currents has little destructive action on bacteria apart from the heat or the chemical substances generated.

Radiofrequency induction heating, which depends on the use of a special generator and coil, may be used for the sterilization of vessels *in situ* and for other purposes (Trotman 1969).

Mechanical, sonic and ultrasonic methods of breakage Violent shaking, with or without the addition of small inert particles such as glass ballotini, may disrupt and kill both vegetative bacteria and spores (Curran and Evans 1942, Nossal 1953, Cooper 1953). Bacterial cells may also be broken and killed by mechanical grinding as in a mortar, by crushing as in the Hughes (1951) press, by mechanical shearing forces as in the extrusion press of Milner, Lawrence and French (1950), and by sonic or ultrasonic vibrations (Wood and Loomis 1927). The disintegrative effect of mechanical agitation alone is very variable; generally it is much more effective when combined with the use of small particles. Curran and Evans (1942), working with spores of two *Bacillus* species, obtained the best results by shaking in a reciprocating beaker with tiny glass beads.

Sonic waves, i.e. waves of audible frequency, of about 8900 cycles per second, produced, for example, by a nickel tube vibrating in a strong electromagnetic field in resonance with a 2000-volt oscillating power circuit, may kill coliform and other bacteria exposed to them for a sufficient length of time, but the results are variable (Chambers and Flosdorf 1936). (For earlier references see Chambers and Gaines 1932.)

Ultrasonic waves,[1] or waves with a frequency of vibration greater than that appreciable by the human ear, i.e. >20 000 cycles per second, produced by connecting a piezo-electric crystal with a high-frequency oscillator, are also credited with bactericidal power. The observations of Takahashi and Christensen (1934) and others suggest that bacteria, and even filtrable viruses, may be largely destroyed by exposure to these waves for an hour or so. Païc and his colleagues (1935*a*, *b*), however, found that ultrasonic waves of a frequency of 280 000 cycles per second had no destructive action in 2 hours on certain toxins, a coli bacteriophage, the herpes virus, or a number of different micro-organisms, though completely sterilizing a culture of *Paramœcium* in 5 minutes. These methods are used mainly for obtaining materials of different kinds, such as enzymes, cell walls, or nucleic acids from bacterial cells, which are disrupted in the process. None of them can be relied on for sterilization. (For further information see Sykes 1958, Rodgers and Hughes 1960.)

Desiccation When dried on silk threads or glass slips, the proportion of organisms surviving for any given length of time varies with a great number of factors, such as the species of bacterium, the initial numbers present, the nature of the suspending medium, the rapidity of drying, and the temperature and gaseous nature of the environment (see Ficker 1898). Anthrax spores dried on silk threads may survive for over 20 years, but many of the pathogenic non-sporing bacteria die in a few hours. Paul, Birstein, and Reusz (1910*a*), working with staphylococci dried on garnets, found that the velocity of disinfection was equal to the square root of the oxygen concentration. The lower the temperature at which the organisms were kept after being dried in this way, the smaller was the proportion that succumbed. Owing to the numerous factors that affect the result, drying cannot be relied upon for the destruction of bacteria. It performs a useful purpose, however, in preservation of decomposable matter, such as food. Little or no growth of bacteria is possible in less than 15 per cent of moisture and of moulds in less than 5 per cent, so that properly dried food may remain free from putrefaction for a long time.

When stored after drying, various species of bacteria survive best in a dry atmosphere, but viruses differ greatly, some surviving best at a high and others at a low relative humidity (Songer 1967). The drying of bacteria in gelatin ascorbic acid discs is a simple and convenient method for preserving many common organisms (Stamp 1947; see also McCracken 1964). Dried in the type of gum used for postage stamps, some bacteria and viruses may remain viable for weeks or months (Selwyn 1965). For a study of bacterial survival on different surfaces—glass, tiles, stainless steel, silk—at different degrees of temperature and humidity see and McDade, Hall and Street (1964), and on wool and cotton Pressley (1967).

Several workers have studied the rate of death of bacteria and viruses in aerosol particles atomized from distilled water or some other liquid medium. So many variables are concerned in determining the result that generalization is difficult. It would appear that at a high relative humidity oxygen plays no part in the survival of *Esch. coli*, death being due to an effect on RNA synthesis, but that at a low relative humidity oxygen is toxic, possibly owing to its action on flavin-linked enzymes (Benbough 1967, Cox and Baldwin 1967). The death-rate is high—80 per cent or so—in the first second, after which the organisms die off slowly (Webb 1959); Poon (1966) found it to be proportional to the rate of evaporation of water, and to increase exponentially with a rise of temperature in the environment.

Cold The early qualitative experiments of Macfadyen (1900) and Macfadyen and Rowland (1900) showed that organisms grew readily after exposure for several hours to liquid air at $-182°$ to $-190°$, or even to liquid hydrogen at $-252°$. Paul and Prall (1907) found that staphylococci dried on garnets and exposed to liquid air remained viable for several months.

Like desiccation, the effect of cold on bacteria is determined by numerous factors. In general, low temperatures arrest growth but are not bactericidal. The minimum temperature for growth varies greatly, not only with the species of bacterium, but with such variables as nutrient requirements, stage of growth, water activity (a_w), acidity, salt concentration, and cell damage (see Ingram and Mackey 1976). Growth of many organisms below freezing point may occur down to $-7°$, but rarely below $-10°$. Freezing of water within the cell occurs slowly and may not be complete till the temperature falls to $-50°$ or even $-70°$. It is probably this unfrozen water that enables bacteria to grow below freezing point. A sudden change in temperature—*thermal shock*—may destroy organisms that survive when the change is made more slowly.

The combination of *freezing and drying* is widely made use of in the preservation of bacteria. Numerous

---
[1] The term supersonic is an aeronautical expression used to indicate speeds above the speed of sound.

methods may be used for this purpose. The bacterial suspension may be frozen before and during drying by immersing the container in a mixture of ice and salt or of CO_2 snow; or, as in the lyophile process, the moisture may be drawn off by a high-vacuum pump and trapped by condensation in a vessel immersed in CO_2 or by a chamber containing anhydrous calcium sulphate (Flosdorf and Mudd 1935, 1938). During the freeze-drying process the immediate death-rate is often high (Fry and Greaves 1951). Further organisms may perish during the process of storage, particularly between temperatures of 0° and −30°. It was thought at one time that the organisms were mechanically crushed by extracellular ice crystals, or disrupted at low temperature by the intracellular formation of ice. It seems more probable, however, that at low temperatures the damage to the organisms is caused by a rise in concentration of electrolytes within the cells when the water separates out as ice. Protection against this action is afforded by the addition of 10 per cent or so of glycerol in the suspending medium, by glucose, or by some protein such as gelatin, serum, or milk. Most bacteria survive well when subjected to temperatures between −30° and −60°, but for indefinite survival a temperature below −70° is best. For this purpose storage in liquid nitrogen at −180° is the most satisfactory method. At such a temperature all metabolic activity is suspended. Deep-frozen bacteria have, in fact, been cultivated after having been embedded for many millennia in the frozen soil and rocks of Antarctica (see Report 1974).

(For a discussion on the effect of freezing bacteria, see Audrey Smith 1961, Ingram and Mackie 1976; for a description of various methods of freeze-drying, see Busby 1967, Lapage and Redway 1974, Bousfield and MacKenzie 1976, and for a simple method for maintaining fastidious organisms, see Park 1976.)

Dry heat Like desiccation, the resistance of micro-organisms to heat depends on a number of factors, such as the nature of the heat itself—dry or moist; the temperature of exposure; the species of bacterium; the nutrient composition of the medium on which the organisms are grown; the presence of protective substances such as sucrose, glucose, sorbitol, or glycerol in the medium; the stage of growth, organisms in the stationary phase being generally more resistant than those in the logarithmic phase; the H-ion concentration of the suspending medium—heat, as Pasteur found, being more destructive in an acid than in a neutral or alkaline medium. The degree of injury caused by heat seems to depend on (a) damage to the cytoplasmic membrane allowing leakage of intracellular components and ingress of substances to which the membrane is normally impermeable; (b) denaturation of the ribosomes and degradation of their contained RNA; and (c) rupture of the DNA strands. Recovery from damage to cells heated to a sub-lethal temperature depends on resumption of the integrity of the cytoplasmic membrane followed by synthesis of lipids and ribosomal protein, and repair of the broken strands of DNA (see Tomlins and Ordal 1976). (For the effect of water activity (a_w) on heat survival of bacteria, see Horner and Anagnostopoulos 1975.) Koch and Wolffhügel (1881) were the first to make exact measurements of the effect of heat on micro-organisms. They found that vegetative bacteria were killed by a temperature of just over 100° in $1\frac{1}{2}$ hours; many, of course, succumbed well within this interval, but this was the time necessary for complete sterilization. Spores, on the other hand, were much more resistant, requiring a temperature of 140° for 3 hours. Their heat resistance is associated with the presence of calcium dipicolinate. It is also related to their lower water content, since there is evidence to show that desiccation can raise the time-temperature limit necessary to cause coagulation of proteins (see Hewlett 1909, Cameron 1930). The resistance of both vegetative bacteria and spores varies considerably with the different species, some being killed much more rapidly than others. The spores of moulds are intermediate in resistance between the vegetative and sporing bacteria; they require a temperature of 110–115° for $1\frac{1}{2}$ hours for their destruction.

The higher the temperature to which the organisms are submitted, the shorter is the survival time. Thus, if the temperature is raised from 140° to 160° spores are killed in 1 to $1\frac{1}{2}$ hours. At 400° they are killed in 20–30 seconds (Oag 1940).

Koch did not regard dry heat as an effective method of disinfection. Though satisfactory when dealing with naked bacteria, it is quite ineffectual within the times usually employed when the bacteria are protected by textile or other relatively non-conducting material; this is due to the low power of penetration of hot air. Thus, when a bundle of tow measuring 55 × 50 cm was exposed to a temperature of 140–150°, the interior after 3 hours had only reached the temperature of 74.5°—a temperature quite inadequate to kill the spores enclosed in the bundle. Moreover a temperature of 140° is sufficient in a short time to ruin most cloth fabrics.

Flaming is a useful method of surface disinfection for non-inflammable substances; its efficacy appears to depend on the temperature to which the exposed surface is raised (Mayser 1925).

The apparatus most commonly employed for sterilization by dry heat is the hot-air oven. Laboratory glassware is usually exposed to a temperature of 160° for 1 to $1\frac{1}{2}$ hours. Every part of the material must be held at the required temperature for a sufficient time. Unless provided with a fan or some satisfactory method of air circulation, hot-air ovens cannot be relied upon. Temperature variations of 30–40° may occur in different parts of a loaded oven. A fan greatly decreases this variation. Care must always be taken not to pack the load in the oven too tightly (Darmady and Brock 1954).

Another apparatus is the moving-belt infrared sterilizer. This consists of an insulated tunnel in which are sited a series of infrared heating elements; objects to be sterilized are passed through the tunnel on a conveyor belt. The amount of heat, the heating period, and the rate of movement of the conveyor must be carefully controlled (Darmady et al. 1957, Report 1962).

For the sterilization of small numbers of syringes, conducted heat may be used. The syringes are held in holes bored through an aluminium block which is heated by a thermostatically controlled electric hot plate (Darmady et al. 1958).

Moist heat Koch (Koch et al. 1881), in conjunction with Gaffky and Loeffler, was the first to make a quantitative study of the germicidal action of moist heat. He found that the temperature required for sterilization of spores was much lower than with dry heat. Thus anthrax spores were killed in 10 minutes at 95°, and spores present in garden earth in less than 10 minutes at 105°. He also showed that steam under pressure was more effective than steam at atmospheric pressure. For the disinfection of clothes, too, he found moist heat to be preferable to dry heat, as it had a greater penetrating power. Thus after 4 hours' dry heat at 140-150° the temperature inside a roll of flannel was only 83°, and the contained spores germinated freely, whereas after 1½ hours of moist heat at 120°, the temperature inside was 117°, and all the spores were dead. The higher the temperature of exposure, the more rapid is the destruction of the bacteria. Bigelow and Esty (1920), for example, exposed the spores of thermophilic organisms, suspended in oil baths, with the following results:

Temperature.°	Killed in minutes.
100	1320
110	225
120	23
130	3.5
140	1.0

It will be noted that at 100° sterilization took nearly a day.

In the hot springs of Yellowstone Park bacteria are said to be abundant at a temperature of 100° provided the water is neutral or alkaline; but below pH 4 the numbers fall off, and at pH 2 to 4 they are absent at 100° though present at 60°. Thus bacteria can withstand heat and acid separately but not together (Brock and Darland 1970).

Mündel (1937) found that the addition of 2 per cent washing soda (Na_2CO_3) greatly increased the disinfecting power of boiling water, as well as reducing the tendency of metal instruments to rust. Thus, spores in a 0.35 per cent suspension of earth in water resisted boiling for about 10 hours, but were killed in a 2 per cent solution of soda at 98° in 10 to 30 minutes.

Spores vary greatly in their heat resistance. Working with clostridial spores, Perkins (1965) found the D_{10} value, i.e. the time taken to destroy 90 per cent of spores, to range from a minimum of 1.8 minutes at 77° with Cl. botulinum type E to a maximum of 110 minutes at 121° with Cl. thermosaccharolyticum. The lower their water content, the more alike do spores behave. Thus, under very moist conditions the spores of B. stearothermophilus were 50 000 times more heat resistant than those of Cl. botulinum E, but under dry conditions the ratio fell to about 10:1 (Murrell and Scott 1966).

Tyndallization Because of the resistance of spores to boiling water the autoclave has largely displaced the steamer in laboratory practice. Steam is still employed at atmospheric pressure for sterilizing media the physical or chemical composition of which would be altered by steam under pressure, but when this is necessary we take advantage of Tyndall's observation, and submit the medium to steaming for 30 minutes on 3 successive days; any sporing organisms that have not been killed on the first day germinate, and thus become susceptible to exposure on the second day. It must be realized that tyndallization can be successful only if the nature of the medium and the conditions to which it is subjected between successive heatings are such as to enable all the spores to germinate. It is quite inapplicable, for instance, to the sterilization of bacterial suspensions in a non-nutrient fluid. Similarly, it is unsuitable for the destruction of anaerobic spore-bearing bacteria the spores of which are unable to germinate aerobically, or of thermophilic spore-bearing organisms which fail to grow at a temperature below 50-55°.

Autoclaving In the autoclave steam is heated under pressure and should remain saturated with moisture. Saturated steam condenses on objects cooler than itself and, by giving up its latent heat, quickly raises them to its own temperature. Superheated steam, on the other hand, behaves like a gas and condenses very slowly. All air must therefore be expelled from an autoclave before the pressure is allowed to rise; otherwise the temperaure developed by a given pressure of steam will be less than that reached by saturated steam (see Sooner and Turnbull 1942, and Table 4.1). Though there are a few exceptions, it is safe to say that saturated steam under pressure of 15 lb

Table 4.1 Autoclave temperatures according to degree of steam saturation and gauge pressure (Report 1942).

Gauge pressure in pounds per sq. inch	Amount of air discharged before closing the steam outlet tap				
	All (pure steam)	2/3 (20-in. vacuum)	1/2 (15-in. vacuum)	1/3 (10-in. vacuum)	None
	°	°	°	°	°
5	109	100	94	90	72
10	115	109	105	100	90
15	121	115	112	109	100
20	126	121	118	115	109
25	130	126	124	121	115
30	135	130	128	126	121

per square inch, i.e. with a temperature of about 120°, is sufficient to sterilize any medium in 30 minutes.

(For general information on sterilization by steam, see Underwood 1934, Konrich 1938; on the sterilization and testing of *dressings*, in particular, and the modern forms of autoclave, Report 1942, Bowie 1958*a, b*, Alder and Gillespie 1957, Kelsey 1958, Report 1960*a*, Howie 1961, Knox and Pickerill 1964.)

Thermal death point of bacteria Chick and Martin (1910) showed that heat coagulation of proteins was an orderly process occurring in two stages: in the first, known as denaturation, the water reacts with the protein; in the second, known as agglutination, the altered protein separates out in a particulate form. Chick (1910) made similar observations on bacteria, pointing out that coagulation of the protein was a function of time and temperature. Though different species of bacteria vary in their heat susceptibility, there is no one temperature which can be strictly spoken of as the thermal death point of any particular species.

For example, the typhoid bacillus in a suspension of 100 000 organisms per ml is destroyed at 47° in 2 hours, at 49° in 48 minutes, at 51° in 18 minutes, at 53° in 7 minutes, at 55° in 2½ minutes, and at 59° in 21 seconds. If, therefore, a suspension was gradually heated, death might apparently take place suddenly at 55°; but this is merely a point, of no special significance, near the upper end of a series of temperatures each of which might legitimately be regarded as a thermal death point.

Effect of heat on subsequent multiplication. Heat reactivation In his studies on disinfection Koch noticed that spores which had been heated but not quite killed required longer to germinate than unheated spores. Eijkman (1908) made similar observations on vegetative organisms. Working with *Esch. coli* heated at 52°, he found that the longer the organisms were heated, the longer the survivors took to grow. Allen (1923) likewise found that the generation time of non-sporing organisms submitted to pasteurization in milk was longer than that in raw milk, indicating the damage caused to them by heat.

Though sub-lethal heat may delay germination of spores, the opposite effect has been recorded by some workers. Christian (1931*a, b*), for example, working with an aerobic spore-bearing bacillus isolated from tainted milk, found that germination appeared to be stimulated by heating the spores to 100° for 30 minutes at the time of inoculation. Evans and Curran (1943) also observed that heating spores of some species of aerobic spore-bearing bacilli for 10 minutes at 85° often stimulated germination. In general, mild heating followed by 3 hours' incubation led to about the same degree of germination as 24 hours' incubation without pre-heating (see p. 29). Heat is not the only method of aiding germination. Ionizing radiation and certain chemical agents, such as L-alanine and calcium dipicolinate, have a similar effect (Gould and Ordal 1968).

Pressure Numerous studies, mainly of marine organisms, have been made of the effect of hydrostatic pressure on bacteria. Most organisms are unaffected by a pressure of 100 atmospheres, but in the vegetative state most of the well-known organisms are affected in one way or another by a pressure of 300 atmospheres, and are completely inactivated by one of 600 atmospheres—the pressure existing at ocean depths of 6000 metres. The effect on organisms cultivated under increasing pressures is to bring about the formation of long filaments devoid of septation and incapable of division. Alternatively, gross pleomorphism may be seen, accompanied as a rule by thickening and sometimes convolution of the cell wall. Mobility is inhibited. Several other effects are noticeable. For example, bacteria will often grow at high pressure when the temperature is increased; the pH range normally tolerated by bacteria growing at 1 atmosphere becomes restricted as the hydrophilic pressure is increased; and the activity of certain antibiotics is raised.

(For reviews of this imperfectly explored subject, see Zimmerman 1970, Zobell and Kin 1972, Dring 1976.)

(For a review of physical agencies in disinfection, see Skinner and Hugo 1977.)

Filtration Though filtration to remove bacteria from a liquid suspension or from air may be regarded as a mechanical method of disinfection, we do not propose to consider it in this chapter, since reference is made to it elsewhere (Chapter 9).

Chemical agencies

In the following pages attention will be focused mainly on groups of chemical disinfectants. So far as it is known, the mode of action of each group will be discussed in relation to that particular group, but it may be well to preface this section with some general remarks.

Disinfectants act in a great variety of ways, both physical and chemical. Factors such as the degree of adsorption to the cell wall, alteration in electrophoretic mobility, modification of permeability of the cell wall with leakage of the constituents or abnormal penetration of substances from without, cellular lysis, irreversible coagulation of the cell proteins, and interference with the metabolism and enzyme systems of the cell—are all concerned to a varying extent (Hugo 1967, 1977)

. The points of attack, or so-called targets, are the cell wall, the cytoplasmic membrane with its associated enzymes, and the cytoplasm. The ribosomes, though a target for many antibiotics, are not a specific target for chemical disinfectants (Hugo 1977). Many chemical agents that are bactericidal in high concentrations are merely bacteristatic in low concentrations; and many disinfectants when used in a concentration slightly higher than the minimal bacteristatic concentration and added to bacteria in the logarithmic phase of growth cause lysis (Pulvertaft and Lumb 1948). (For the survival of microbes exposed to chemical stress, see Hugo 1976.)

In general, chlorination or nitration of the aromatic nucleus increases the bactericidal action of a disinfectant such as phenol; so also does alkyl substitution. A rise in pH is said to be detrimental to the activity of phenols, organic acids, and compounds containing chlorine, but to increase that of the acridines, crystal violet and quaternary ammonium compounds (Bean 1967). The higher the bactericidal activity of a compound, the more is it affected by the presence of organic matter (Bean 1967).

Distilled water Observations on the ability of bacteria to survive in distilled water have yielded conflicting results. There are numerous reasons for this. When distilled from copper vessels the water may contain traces of the highly toxic metallic ion. Thus water containing 1 in 50 million parts of copper sulphate will kill cholera vibrios in one hour (Ficker 1898); and water containing 1 part of copper in 10 million will sterilize a suspension of *Ps. aeruginosa* in two hours (Hoder 1932). Organisms may perish rapidly in glass-distilled water when inoculated in small numbers but remain viable for months when inoculated in large numbers, presumably owing to the nutrient material carried over by the organisms (Ficker 1898). Moulds are said to survive almost indefinitely in distilled water, and so may some of the enterobacteria (Castellani 1967).

Numerous other factors may affect the results, such as the H-ion concentration, the presence of traces of alkali absorbed from the glass, traces of unsaturated fatty acids distilled from heated cotton-wool plugs, the amount of CO_2 absorbed from the air, the quantity of dissolved oxygen, and the temperature at which the suspension is maintained.

There is no evidence that distilled water acts by disrupting bacteria as it does many unicellular organisms; bacteria are too resistant to changes in osmotic pressure for this to be probable.

Acids Krönig and Paul (1897) were the first to show that the disinfectant action of acids in general was proportional to their degree of electrolytic dissociation i.e. to the H-ion concentration of their solutions. This was confirmed by Winslow and Lochridge (1906). It may be noted, however, that some mineral acids have an additional action depending on their oxidizing effect. Thus Krönig and Paul (1897) drew attention to the high germicidal activity of nitric acid, dichromic acid, chloric acid, persulphuric acid, and permanganic acid, and found that their oxidative and germicidal powers ran parallel to each other in this order of decreasing activity.

The acid-tolerance of micro-organisms varies with different species. In Winslow and Lochridge's experiments, already referred to, the parts per million of dissociated hydrogen necessary to sterilize a suspension of *Esch. coli* in 40 minutes were 12.80; to sterilize a suspension of *Salm. typhi* only 4.85 were required.

The effect of the H-ion concentration of the medium on bacteria suspended in it is rather complex. There is, first of all, an optimum concentration for growth; for *Esch. coli* this is about pH 7.6. There is, secondly, an optimum concentration for survival; for *Esch. coli* this is about pH 6.0. Thirdly, there is a point at which the acid-tolerance of the organism fails; this for *Esch. coli* is about pH 4.6. And lastly, there is evidence that the H-ion concentration most suitable for certain fermentative processes is different from the optimum pH for growth (Cohen and Clark 1919, Gale 1940).

Organic acids, which dissociate much less than mineral acids, have a disinfectant action that appears to depend on the nature of their anionic component or of their undissociated molecule. Thus Winslow and Lochridge (1906) found that, to produce a 99 per cent reduction in the number of *Esch. coli* in 40 minutes, a 0.0812 N solution of acetic acid or a 0.0097 N solution of benzoic acid was required. The degree of dissociation of each acid at these concentrations is only about 1 per cent, so that the amount of dissociated hydrogen in the acetic acid was 1.2 parts per million and in the benzoic acid 0.1 part per million. Since 7.49 parts per million were required when HCl was used, it follows that their H-ion concentration alone will not explain the disinfectant activity of these organic acids.

There is some evidence that the bactericidal activity of the monobasic series of organic acids increases with increase in molecular weight and decrease in surface tension, whereas with the dibasic organic acids the reverse holds true (Reid 1932).

Aldehydes Among the aldehydes, formaldehyde and glutaraldehyde stand out as powerful disinfectants. According to Rubbo, Gardner and Webb (1967), a 2 per cent aqueous alkaline solution of **Glutaraldehyde** rapidly destroys both vegetative and sporing bacteria. Its activity towards viruses is intermediate between that to vegetative bacteria and spores. Its antimicrobial activity increases with concentration, pH, and duration of contact, but is diminished by organic matter, and is low against the tubercle bacillus. It has an agglutinating action owing to the formation of intercellular bonds caused by a preferential action for the outer layers of the cells (Navarro and Monsan 1976). It irritates the abraded skin, corrodes rubber tubing but not metals, and has no deleterious effect on the cement used for lenses. In the medical field its main value is in the sterilization of equipment that cannot be sterilized by physical methods. In the industrial field it is used extensively for tanning leather. Its maximum activity is at pH 7.5 to 8.5, but at this pH the solution deteriorates. It is therefore best to keep it at an acid pH and dilute it with a buffer solution to a slightly alkaline pH just before use (Relyfeld 1977). Its mode of action is discussed by Navarro and Monsan (1976), and by Gorman, Scott and Russell (1980), to whose review the reader is referred. (For the action of formaldehyde, see p. 90.)

The disinfectant action of the higher fatty acids will be more conveniently considered with that of the soaps (see p. 81).

Alkalies Krönig and Paul (1897) showed that the

disinfectant action of alkalies was dependent on their degree of dissociation, and hence on their concentration of OH-ions.

Thus, of the bases KOH, NaOH, LiOH, and NH$_4$OH, KOH shows the highest degree of dissociation, and is hence the most actively germicidal; NH$_4$OH is dissociated the least and is the least actively germicidal.

Levine (1952), on the other hand, working with a spore suspension of *Bac. metiens*, found that, though the pH appeared to be the controlling factor in the germicidal activity of hypochlorites, it was not so of alkalies.

For example, the killing time in minutes for 4 per cent NaOH and for concentrations of Na$_2$CO$_3$ and Na$_3$PO$_4$ equivalent to 4 per cent NaOH was: 0.6 minute at pH 13.4 for NaOH, 5 minutes at pH 12.15 for Na$_3$PO$_4$, and 118 minutes at pH 11.35 for Na$_2$CO$_3$. But when the concentrations were all adjusted to give a pH of 11.35, the time was: 220 minutes for NaOH, 163 minutes for Na$_3$PO$_4$, and 118 minutes for Na$_2$CO$_3$.

Levine suggests that the concentration of undissociated NaOH which penetrates the cells, rather than the OH-concentration, may determine the rate of disinfection with alkalies.

When we turn to other bases, we find exceptions. Thus, Ba(OH)$_2$ is less dissociated than KOH, but is very much more toxic; similarly with the hydroxides of the other alkaline earths. The reason for this is that the metallic ion is often highly toxic, and assists the hydroxyl ion in its germicidal activities.

Just as bacteria possess a limit of acid-tolerance, so they possess a limit of alkali-tolerance. Cohen (1922) found that for *Salm. typhi* this was about pH 8.7. It is of interest to note that H-ions appear to be more toxic than OH-ions in similar concentration.

Salt action The disinfectant action of salts is influenced by many different factors, such as the nature of the salt itself, the strength in which it is used, the composition, hydrogen-ion concentration, and temperature of the medium in which it is dissolved, the bacterial group—or even species—on which it acts, and the simultaneous presence of other salts or germicides. In general, it may be said that (*a*) the salts of heavy metals are more toxic than those of lighter metals; (*b*) in pure solution all salts are germicidal to some extent in very low concentration; (*c*) protein in solution weakens the germicidal effect of salts by combining with the cations to form an insoluble albuminate, thus diminishing the number of free ions in the medium (Behring 1890); (*d*) within limits, the more nutritive the medium is in which the bacteria are suspended, the higher is the salt concentration required to exert a germicidal action; (*e*) in nutrient media, very low concentrations of salts may stimulate growth, higher concentrations inhibit growth, and higher concentrations still destroy the organisms; (*f*) gram-positive and gram-negative organisms often react differently to the same salts; (*g*) salts with bivalent cations tend to neutralize the toxic action of salts with monovalent cations; (*h*) the presence of another germicide in solution may increase or decrease the germicidal action of the salt itself; (*i*) in pure solution the germicidal action of salts increases with temperature.

Mode of action of salts (1) *The osmotic effects*. Bacteria differ from practically all other living cells in being resistant to changes in osmotic pressure, except of high degree. Salts may, however, affect them indirectly by dehydrating the proteins on which they are growing. This action is relied on in many processes of food preservation.

(2) *Oxidation*. Salts and certain allied substances that contain a high proportion of oxygen, or that are able to liberate oxygen from other compounds, have long been known to be germicidal. The dichromates, perchlorates, persulphates, and permanganates seem to act mainly by virtue of their oxidizing activity. In a weakly acid solution they are very strong disinfectants. Bleaching powder owes its germicidal effect to its ability, when acted on by weak acids, to give rise to nascent oxygen; this combines to form hypochlorous acid, which in a 0.01 per cent concentration is said to kill anthrax spores in 30 seconds (Andrewes and Orton 1904). In the presence of ammonia, or of organic substances containing the =NH group, the hypochlorites combine to form compounds known as chloramines, which on account of their prolonged action are of use in water purification.

$$\begin{array}{c} R^1 \\ > NH + HOCl \rightarrow \\ R^2 \end{array} \begin{array}{c} R^1 \\ > NCl + H^2O \\ R^2 \end{array}$$

According to Dakin (1915), all bodies containing the NCl group are strongly antiseptic. Various other powerful disinfectants, such as chlorine, bromine, iodine, H$_2$O$_2$, and ozone, seem to owe their germicidal effect to their oxidizing power.

(3) *Reduction*. Certain salts, such as the sulphites and the ferrous compounds, appear to act by virtue of their reducing power.

(4) *Molecular action*. In a previous section we saw that certain acids, such as acetic and benzoic, and in fact most of the organic acids, act not by virtue of their H-ion concentration but by virtue of the undissociated molecule.

(5) *Ionic action*. As the salts of mineral acids are electrolytically dissociated in solution, it follows that their germicidal action may be due either to the undissociated molecule, to the anion, to the cation, or to all three in combination. To assess the importance of each of these factors, comparative tests have been made with salts of one metal combined with different anions, and of one anion combined with different metals.

(*a*) *Cations*. Working with *Esch. coli*, Winslow and Hotchkiss (1922) arranged the cations in order of increasing toxicity as follows: K, Na, NH$_4$, Li, Sr, Mg, Ca, Ba, Mn, Ti$^{···}$, Sn, Ni, Ti$^{·}$, Zn, Cu, Fe$^{··}$, Fe$^{···}$, Co, Pb, Al, Ce, Cd, Hg. Though there are exceptions, metals of low atomic weight are less toxic than those of high atomic weight, and bivalent cations tend to be more toxic than monovalent cations.

Koch (1881) was very impressed with the high germicidal activity of salts of the heavy metals, such as mercury and silver. Geppert (1889), however, pointed out that, if the excess mercury or silver was removed at the end of the test by bubbling H$_2$S through the suspension, considerably higher concentrations of the heavy metal were required to destroy the organisms than had been found by Koch. A similar reactivation of anthrax spores in a sublimate solution could also be effected by inoculation of the mixture into guineapigs (Geppert 1891b). Geppert showed that minute concentrations of heavy metals were sufficient to inhibit growth, but were unable to kill the organisms. Fildes (1940) brought evidence to show that mercury acted by combining with the −SH groups of the bacterial cell, which are essential for metabolism. If the mercury is neutralized by the addition of −SH compounds, like glutathione, cysteine, or thioglycollate, bacteria are able to grow after treatment with strong mercury solutions which previous workers have regarded as being actively germicidal. In virtue of this property, the bacteristatic power of mercury is high, but its bactericidal effect is comparatively low. In estimating the bactericidal effect in practice, it is necessary to remove the excess of mercury from the suspension at the end of the test period by treatment with H$_2$S or ammonium sulphide, and then to cultivate the organisms in a liquid medium containing 1 per cent thioglycollate in order to provide an adequate concentration of −SH groups. Records of the germicidal effect of mercury solutions not based on the use of this method must be regarded as unreliable. The same criticism applies to the organic salts of mercury, such as phenyl mercuric nitrate and several proprietary preparations. If thioglycollate is added to the broth used for subculture, none of these substances is able to destroy *Staph. aureus* or *Esch. coli* in a 1/1000 dilution in 10 minutes at room temperature (Hoyt, Fisk and Burde 1942). A similar observation was made for streptococci by Morton, North and Engley (1948).

(*b*) *Anions*. Less work has been done on the germicidal action of anions than of cations. The most extensive study is that of Eisenberg (1919), who arranged the anions in order of increasing, toxicity from SO$_4$ at the bottom to TeO$_3$ at the top.

Various explanations have been put forward for the action of electrolytes on living matter. They are supposed to act through the electrical charges they bear; through the coagulation of colloids either directly (see Hofmeister 1888, 1889) or indirectly through altering the properties of the solvent—the so-called lyotropic effect of Freundlich (1903); or through some specific chemical property such as must characterize sodium and potassium ions, which, though having the same electrical charge and exercising similar lyotropic effects, are yet quite different in their action on heart muscle.

The different reaction of different bacteria to given salts seems to be determined sometimes by the nature of the anion and sometimes by that of the cation. Many of the most successful selective media make use of the differential effect of the anionic radicle. Tellurites (Fleming 1932, Fleming and Young 1940) and azides (Snyder and Lichstein 1940, Mallmann *et al.* 1941) are of special value in the suppression of gramnegative bacteria, and some of the polyphenols, particularly cacotheline (nitrobruciquinone hydrate), inhibit organisms of the *Proteus* group (Jones and Handley 1945).

Antagonistic effect of salts Ficker (1898) was the first to make observations on the effect of physiological saline on bacteria. He found that, far from being harmless, it killed the cholera vibrio very rapidly. The same was shown to be true for the meningococcus by Flexner (1907), who made the additional observation that the action of the monovalent sodium ion could be neutralized by the bivalent calcium ion. Similar results were obtained by other workers (see von Eisler 1909). The mechanism of this action is unknown.

Attempts have been made to devise a balanced solution in which the proportions of the different salts are so ordered as to neutralize their individual toxic effects. Ringer's solution is of this type, and has the following composition:

NaCl	0.9 g
KCl	0.042 g
CaCl$_2$	0.048 g
NaHCO$_3$	0.02 g
Glass-distilled water	100.00 ml

Though greatly superior to simple saline solutions, Ringer's solution is not perfect; coliform bacteria, for example, begin to die off in it at room temperature after 2–3 hours. Moreover, a variable amount of the salt content is precipitated on autoclaving, so that it is not easy to standardize completely. The addition of a protein, such as 0.1 per cent gelatin, serves to neutralize the slight toxic effect of the mixture, and is often useful as a diluent solution for the more delicate organisms.

For further information on salt action, reference may be made to the admirable review of Falk (1923), and the general discussion by Bayliss (1924).

Soaps Long-chain unsaturated fatty acids, such as oleic, linoleic, and linolenic, either free or in the form of soaps, have a bactericidal or bacteristatic effect on many bacteria. Lamar (1911a, b) found that the sodium or potassium soaps of these acids in a concentration of 1/500 to 1/5000 killed and dissolved pneumococci, and that sodium oleate in a dilution as high as 1/20 000 greatly accelerated the autolysis of pneumococci, and favoured their lysis by normal or 'immune' serum. Many other bacteria, such as hæmolytic streptococci, meningococci, gonococci, diphtheria bacilli, *Bord. pertussis* (Pollock 1947), lactobacilli (Kodicek and Worden 1945), tubercle bacilli (Dubos and Middlebrook 1947), and *Erysipelothrix* (Hutner 1942) resemble pneumococci in their sensitivity to soaps of the unsaturated fatty acids, though it should be mentioned that the growth of some of these organisms is actually improved by the incorporation in the medium of these acids in very low concentration. The gram-negative bacilli of the coli-typhoid group are fairly susceptible to soaps of the saturated fatty acids, but are resistant to soaps of the unsaturated fatty-acids series (Reichenbach 1908, Walker 1926, Belin and Ripert 1937). In contact with soap most organisms die quickly, though they differ in their susceptibility. For example, the vegetative cells of *Bac. mesentericus* are dead within 10 seconds, *Staph. aureus* not for 20 minutes (Eschment and Lutz-Dettinger 1970). The risk of infection being carried over from one person to another by bar soap is said to be slight (Bannan and Judge 1965, Heiss 1969).

There is a suggestion that the germicidal activity of soaps increases with increase in molecular weight in the saturated fatty-acid series, but decreases in the unsaturated fatty-acid series (Bayliss 1936). Of commercial soaps, Nichols (1920) found yellow or brown bar soap, such as was used in washing dishes, to be effective in a 1/200 concentration in killing pneumococci and streptococci (see also Colebrook and Maxted 1933). Soap prepared from coconut oil, such as salt-water soap, is more germicidal than any other soap to the typhoid bacillus (Walker 1926). If a stiff lather is made on the hands, even *Esch. coli* is killed within a minute. The germicidal effect of soaps is increased by rise in temperature. In practice, thorough washing with yellow bar or ivory bar soap in a minimum of hot water can be relied upon to kill a high proportion of pathogenic organisms on the hands, with the exception of *Staph. aureus*. The transient flora is almost completely removed, but only about half of the resident flora. To remove the remainder, stronger disinfectants are required (Lowbury 1966).

The mode of action of soaps in destroying bacteria is far from clear. It is certainly not due entirely to free alkali, since this may be present in much too small an amount to have any deleterious effect at all. Lamar (1911a, b) was of the opinion that soap acted on the lipid moiety of the cells, rendering them more permeable to germicidal substances in the solution. There is some evidence that soaps and long-chain fatty acids are mainly responsible for the auto-disinfecting action of the skin (Burtenshaw 1942, Ricketts *et al.* 1951).

Reference may be made here to the liberation of unsaturated fatty acids from *cotton-wool* during exposure to dry heat. Wright (1934) observed that the fatty distillate from cotton-wool plugs inhibited the growth of pneumococci, and Drea (1942) showed the same for tubercle bacilli. The identification of the inhibitory principle with unsaturated fatty acids was made by Pollock (1948). He found that a greasy film was formed on the inside of tubes that were plugged with cotton-wool and sterilized for $1\frac{1}{2}$ hours in the hot-air oven at 180°. If the cotton-wool was extracted with methanol before heating no film appeared and the characteristic effect on growth was not observed. In practice, it is advisable when culturing susceptible organisms to replace cotton-wool by aluminium caps.

Synthetic detergents In 1935 Domagk drew attention to the germicidal activity of a new class of surface-active agents. Since then, these substances have assumed an important place in industry by virtue of their wetting, detergent, or emulsifying properties, and to a less extent in surgery and in food hygiene (Johns 1947) on account of their cleansing and bactericidal powers.

Chemically, they are diverse. There is a large group of anionic detergents including the sulphated alcohols, the paraffin sulphonates, the alkylaryl sulphonates, and the petroleum sulphonates, many of which are made from by-products of the coal-tar and petroleum industries. Another group, mainly of cationic detergents, consists of quaternary ammonium compounds in which the four hydrogen atoms are replaced by organic radicles. Non-ionic detergents consisting of polyesters and ethers, formed as condensation products from fatty acids, alcohols and phenols, constitute a smaller group. So-called amophoteric or ampholytic detergents, which are compounds of mixed cationic and anionic structure, are also prepared (Glassman 1948).

Of these four groups, the cationic quaternary ammonium compounds alone possess appreciable germicidal activity. They are good wetting and cleaning agents. They are odourless, non-corrosive to metals, practically non-irritant to the skin and raw surfaces, and they destroy most vegetative bacteria in dilutions ranging from 1/100 to 1/16 000 or so. They are more effective against gram-positive than gram-negative bacteria. Their germicidal action is often greatly diminished by the presence of organic matter and of phospholipids; their penetrating power is usually low; some organisms such as the tubercle bacillus are resistant to them (Smith *et al.* 1950); and others, such as *Ps. æruginosa*, may not only survive but may actually multiply in them (Dold and Gust 1957). On the hands they are said to be more bacteristatic than bactericidal, and to be neutralized by saline though not by distilled water (Skadhauge 1958). For removal of microbes from the skin they should be combined with a disinfectant, such as chlorhexidine (Lilly *et al.* 1979). Their mode of action is doubtful, but interference with the cellular enzymic processes (Knox *et al.* 1949) and alteration in the permeability of the cell wall (Hotchkiss 1946, Gale and Taylor 1947, Salton 1951) may partly account for their germicidal property. Ampholytic surface-active compounds, such as dodecyldi(aminoethyl)glycine contained in Tego, are not noticeably affected by serum, but on the skin their bactericidal power is slight (Kuipers and Dankert 1970).

Since some of the quaternary ammonium compounds may clump bacteria and prove bacteristatic in high dilution, special precautions must be taken to neutralize these effects when estimating their bactericidal power (see McCulloch 1947, Davies 1949, Davis 1961). (See the review by Glassman 1948, and papers by Baker *et al. 1941a, b*, Albert 1942; Williams *et al.* 1943, Iland 1944, Guiteras and Shapiro 1946, van Eseltine and Hucker 1948, Curran and Evans 1950, Hobbs *et al.* 1960.)

Alcohols Epstein (1897) found that absolute ethyl alcohol was not a germicide, but that when diluted it became germicidal. Minervini (1898) confirmed this, and showed in addition that alcohol had little or no action on spores. For the destruction of vegetative bacteria the optimal strength depends on the degree of moisture present. A final concentration of 50–70 per cent appears to be most effective. Thus, an equal amount of absolute alcohol should be added to an aqueous suspension of bacteria, whereas for dry bacteria a solution of alcohol already diluted to 50–60 per cent should be used. For disinfection of moist hands 80–96 per cent alcohol is recommended; for the disinfection of dry hands 70–80 per cent alcohol is better. Dry vegetative bacteria are destroyed less rapidly than moist—presumably because the penetration of alcohol takes longer. Viruses vary greatly in their susceptibility (see below).

The presence of protein increases the disinfection time of alcohol, but not to any considerable extent. The addition of a dilute mineral acid or alkali greatly increases its activity, enabling alcohol to kill spores. Thus Coulthard and Sykes (1936) found that a solution of 70 per cent alcohol containing 1 per cent sulphuric acid destroyed spores of *Bac. subtilis* in less than 24 hours, and a solution of 70 per cent alcohol containing 1 per cent sodium hydroxide in 24–48 hours. Alcohol lowers the germicidal effect of some substances, such as the heavy metal salts, phenol and formaldehyde, that are dissolved in it (Koch 1881, Krönig and Paul 1897), but is said to raise the germicidal effect of others, such as iodine. In fact, a strong tincture of iodine—4.5 per cent iodine in 70 per cent alcohol, together with 2 per cent potassium iodide to stabilize the iodine—is one of the best skin disinfectants known. Alcohol should not be relied upon for the disinfection of syringes (see Report 1945, 1962). Though the bactericidal activity of alcohol is negligible below 10–20 per cent, it may prove bacteristatic to many organisms in concentrations as low as 1 per cent (Wirgin 1902). As an antiseptic for the preservation of vaccines, 25 per cent alcohol has been found to be rather more potent than 0.5 per cent phenol (see Cruickshank *et al.* 1942). Commercial alcohol as a rule contains spores, so that for surgical or biological use it should be filtered through a Berkefeld or similar candle (not a Seitz, which is effective only in the presence of water) or distilled. Viruses vary in their susceptibility to alcohol. Some, such as the rotaviruses and the herpes simplex virus, readily succumb, whereas others, such as the enteroviruses and astroviruses, are resistant to all but high concentrations (J. B. Kurtz, Oxford 1980).

Ritchie (1899) showed that the germicidal action of different alcohols increased with their molecular weight, ethyl alcohol being more potent than methyl, propyl than ethyl, and butyl than propyl alcohol. This has been confirmed by subsequent workers (Wirgin 1904, Lockemann, Bär and Totzeck 1941). For rapid disinfection of the skin 70 per cent isopropyl alcohol is particularly useful (Lee, Schoen and Malkin 1967); it is more bactericidal than ethyl alcohol, it is a better fat solvent, and it is not so volatile. (For a detailed review of the disinfectant action of alcohol, see Sobernheim 1943, and for its value as a hand disinfectant, see Ahlfeld and Vahle 1896, Neufeld and Schiemann 1943, Archer 1945.) *Acetone* is said to kill tubercle bacilli in under 5 minutes, and in 50 per cent concentration in under 3 hours (Weiszfeiler *et al.* 1968).

Ethers These are possessed of some degree of germicidal activity. Cultures of non-sporing bacteria incubated in an atmosphere saturated with the vapour of diethyl ether—$C_2H_5OC_2H_5$—exhibited no growth; subcultures showed that the organisms had been killed in a period varying from about 1 to 48 hours (Topley 1915). Direct immersion of *Esch. coli* in 50 per cent either proved fatal in about 3 minutes at room temperature. On the other hand, exposure of *Cl. septicum* to pure ether failed to destroy the spores in 24 hours. According to Krönig and Paul (1897) ethereal solutions of disinfectants are almost without effect on anthrax spores. Among the vegetative organisms, gram-negative bacteria are said to be more susceptible than gram-positive (Pearce 1946). The animal viruses tend to be either very sensitive or very resistant.

Phenols and cresols Under this heading we shall consider the action of those bodies that are obtained from the destructive distillation of coal, and that pass over between the temperatures of 170° and 270°. Phenol itself in certain proportions is soluble in water, but most of the compounds in this group are not so; when mixed with water they form emulsions of varying degrees of fineness. Their mode of action is therefore different from the action of the germicides which we have so far considered. The phenols and cresols have a fairly high germicidal activity when employed in solutions above a given concentration; but it requires quite a low degree of dilution to deprive them entirely of this activity. In this respect they differ from the saline disinfectants (see p. 84).

The mode of action of phenol is open to discussion.

In low concentration it is thought to disrupt the cytoplasmic membrane and allow leakage of the cellular constituents, and in high concentration to coagulate the cellular proteins (Whittet, Hugo and Wilkinson 1965, Hugo and Bloomfield 1971). Srivastava and Thompson (1966) suggest that, as its activity is greatest immediately after cell division—a time at which the layer of phospholipid around the cell is very thin— the phenol penetrates the cell through what is virtually an open hole. Organisms differ in their susceptibility to phenol; *Esch. coli*, for example, is readily killed, whereas *Str. faecalis* is much more resistant (Slonim et al. 1969).

The emulsified disinfectants, such as the cresols, probably act in much the same way as phenol, but their germicidal activity is usually somewhat higher. By virtue of their emulsoid state, their particles are adsorbed on to the surface of suspended matter, and hence their concentration is increased in the immediate neighbourhood of the bacteria. This action is interfered with by the presence of other suspended organic matter, which serves to adsorb the germicide, and thus lower its effective concentration around the bacteria. Emulsoids of the cresol group are generally most active when freshly made up in solution; after a day or two, probably because of an alteration in their colloidal state, their activity diminishes. Some of the cresols can be employed in true solution, but their solubility in water is very low. Para-chlor-meta-cresol, for example, has a solubility of about 1/300; towards naked bacteria it is approximately ten times as active as phenol (see Withell 1942*a*).

The **chloroxylenols,** many of which form a clear solution in water, came into prominence mainly for use in skin disinfection. They are comparatively non-irritant, but their bactericidal power, on the whole, is considerably less than that of phenol; unless employed in 30, 50, or even 100 per cent concentration, they cannot be relied upon to destroy staphylococci on the skin (see Colebrook 1941). Their germicidal action is also weak, and some gram-negative aerobic bacilli can multiply in them at use-dilution.

The biguanide compound 1:6-bis-*p*-chlorophenyl-diguanidohexane (**Chlorhexidine,** Hibitane) is widely used for skin disinfection. Lowbury, Lilly and Bull (1960) found that a 0.5 per cent solution in 70 per cent alcohol gave roughly similar results in the disinfection of skin to 1 per cent iodine in 70 per cent ethanol or to 70 per cent ethanol alone. Caution has to be used, because lower concentrations, such as 0.05 per cent, will not destroy pseudomonads; these organisms may survive even in 0.1 per cent solution (Burdon and Whitby 1967). It appears to act by causing leakage of the cellular constituents and precipitation of the bacterial cytoplasm (Hugo and Longworth 1964, 1966).

Several diphenylalkane compounds, all of which are 2:2-dihydroxychloro-derivatives, are highly bactericidal. One of the commonst is **hexachlorophane,** which will kill staphylococci at 20° in 1 to 10 minutes in a concentration of between 1/100 and 1/500 (see Sykes 1958). It is made up in a 2–3 per cent concentration for skin disinfection in the form of a cream, a liquid soap, or a solid soap. Its action is cumulative and for the best effects it should be used on the skin several times daily (Gezon et al. 1964, Lowbury 1966). Liquid soap containing 3 per cent hexachlorophane and 0.3 per cent chlorocresol is particularly valuable for disinfection of the hands (Lilly and Lowbury 1971). Hexachlorophane is toxic and if absorbed through the skin in sufficient quantity may prove fatal to infants. At necropsy, spongy degeneration of the brain is found (see Kimbrough 1973, Mullick 1973). (For a historical account of the development of phenols as antimicrobial agents, see Hugo 1979.)

Thymol is said by Wahl and Blum-Emerique (1949) to be bacteristatic and mildly bactericidal; when added in crystalline form to a culture of dysentery bacilli it killed all the organisms in 5 hours at 37°. Though destroying bacteria under these conditions, it did not destroy particles of bacteriophage, which could therefore be obtained free from the organisms against which they were active.

According to Jesenská (1966) *toluol* is without effect on staphylococci, streptococci and corynebacteria but does act on *Proteus* and some other gram-negative bacteria.

(For information on the use of tar derivatives in practice see Report 1942, 1944, Sykes 1958, Dunklin and Lester 1959, Whittet et al. 1965.)

Of the aromatic amidines, *propamidine* has good germicidal activity against gram-positive organisms; it is not much affected by serum and does not inhibit phagocytosis. Halogenation increases the germicidal power of the amidines (Tetsumoto 1937) but renders them destructive to muscle and connective tissue in wounds (Selbie and McIntosh 1943). The introduction of a nitro group into position 5 of the furan ring enhances germicidal activity (Whittet et al. 1965). *Nitrofurazone,* which is only slightly soluble, is used for the treatment of superficial wounds. *Furazolidone* acts on both gram-positive and gram-negative bacteria and has been used in the treatment of intestinal infections in man and animals.

Halogens Enough has just been said, and previously (p. 80), to indicate the high bactericidal activity of *chlorine* alone or in combination with other disinfectants. It owes its germicidal action to its hydrolysis in aqueous solution to form HOCl. This substance is most active in acid solutions, in which it is undissociated, as at pH 4.0. As the pH rises, so does the degree of ionization; the disinfectant activity, however, falls. ClO_2, on the other hand, does not ionize, is much less affected by pH, and has $2\frac{1}{2}$ times the oxidizing action of chlorine (Bernarde et al. 1967). Chlorine is readily neutralized by protein and, when used for the disinfection of objects such as dirty glassware, crockery, or dairy equipment and utensils, it must be combined with a detergent, or used after fat and protein have been removed by soap detergent treatment. In aqueous solution it is unsuitable for the disinfection of polished surfaces because it fails to spread evenly over them. It has the disadvantage of attacking metals and is therefore unsuitable for surgical instruments. It is the most satisfactory agent for the destruction of bacteria in water and is widely used for this purpose. For the rapid disinfection of hands a 1 per cent solution may be recommended. The residual smell may be removed by a solution of thiosulphate.

Iodine as a 1 per cent solution in 70 per cent ethanol or in potassium iodide is a good disinfectant for the skin before operation, comparing favourably in this respect with other substances (Lowbury *et al.* 1960). Mixtures of iodine with surface active agents sold under the name of *Iodophors* are used, particularly in dairy hygiene and for disinfection of the skin. They have a wide range of activity and are lethal to gram-negative as well as to gram-positive bacteria (Lowbury 1966). An additional advantage is that they do not stain and are practically odourless. *Bromine* is very irritant and has no advantages over chlorine. *Fluorine* is restricted in its use mainly to protection against fungi.

Dyes The triphenylmethane dyes, such as crystal violet, methyl violet, and brilliant green, have a strong bacteristatic and a weak bactericidal action against bacteria, particularly the gram-positive group (Churchman 1912). Staphylococci are more susceptible than streptococci (Garrod 1933).

Browning and Gilmour (1913) drew attention to the bactericidal action of the amninoacridines. Unlike most other disinfectants, their activity is not inhibited by serum, and they have therefore been used in the treatment of wounds. For this purpose proflavine is to be preferred to acridine, since it is less irritating to the tissues (Browning 1943, Russell and Falconer 1943, Gordon *et al.* 1947).

According to Albert (1966), whose monograph should be consulted, the antibacterial action of the aminoacridines depends not on the nature of the substituents (amino-methyl-, chloro-), but on the degree of cationic ionization, which increases with basic strength. Thus 3,6 diaminoacridine (proflavine), which is completely ionized at pH 7.3 and 37°, has a minimal bacteristatic concentration for *Streptococcus pyogenes* of 1 in 160 000, whereas the corresponding concentration of 2,7 diamonoacridine, which is only 4 per cent ionized under these conditions, is 1 in 20 000. Lerman (1961) suggested that the planar molecules of the acridines might be intercalated between adjacent base pairs of the double DNA helix. The effect of this would be to distort the DNA molecule with consequential effects on the normal pattern of growth, cell division, and enzymic activity. There is evidence that this interpretation is correct (see Waring 1966).

The non-exacting species of bacteria—*Pseudomonas, Proteus, Escherichia*—are the most resistant to the aminoacridines; the somewhat exacting species—*Salm. typhi, Vibrio, Staphylococcus*—are less resistant; and the exacting species—*Streptococcus, Neisseria, Haemophilus, Brucella, Clostridium* are the least resistant of all (Albert 1966).

Essential oils Many of the essential oils have a weakly bactericidal effect killing vegetative bacilli, sometimes within an hour, but more often not for several hours or even days (Chamberland 1887, Cadéac and Meunier 1889, Böcker 1938). They are valued most for their antiseptic action; hence their use in embalming (see Risler 1936). Of the animal oils, seal and tuna oils are said to give off germicidal vapours (Harris, Bunker and Milas 1932); others, such as cod-liver and sardine oils, become germicidal only after exposure to ultraviolet light.

(For a review of chemical agents in disinfection, see Skinner and Hugo 1977 and Collins *et al.* 1981; and for a short account of sublethal damage to bacteria by antimicrobial agents and its repair, see Busta 1978.)

The dynamics of disinfection

Reaction velocity

The figures obtained by Krönig and Paul (1897) in their work on the disinfection of anthrax spores by $HgCl_2$ were submitted by Madsen and Nyman (1907) to a mathematical analysis, with the result that the reaction velocity of disinfection was found to be similar to that obtaining in a unimolecular reaction. Madsen and Nyman themselves made fresh experiments, using the garnet method, and were able to confirm the findings of Krönig and Paul. In the following year Chick (1908), working independently, reached the same conclusions with regard to the analogy between disinfection and a unimolecular reaction (Fig. 4.2).

We may adapt the definition of a unimolecular reaction to the process of disinfection by saying that at any moment the reaction velocity is proportional to

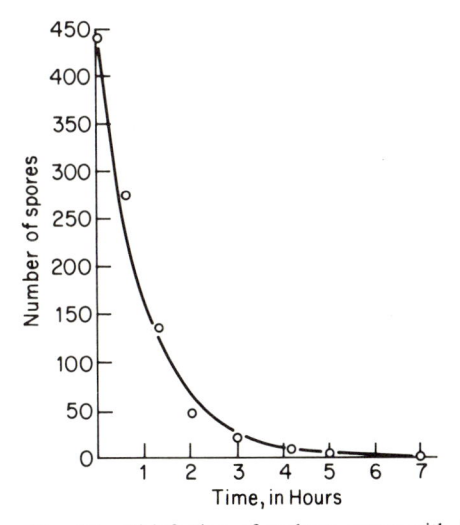

Fig. 4.2 Disinfection of anthrax spores with 5 per cent phenol at 33.3°. The curve is drawn through a series of calculated points; the circles represent the experimental observations. (After Chick.)

86 *Bacterial resistance, disinfection and sterilization* Ch. 4

the number of surviving bacteria per unit volume. Supposing that B represents the initial number of living organisms, and b the final number, then the reaction velocity may be expressed by the equation:

$$k = \frac{1}{t} \log \frac{B}{b}$$

Chick, using the drop method, studied the disinfection of anthrax spores by 5 per cent phenol. Her results are given in Table 4.2 and Figs. 4.2 and 4.3.

Table 4.2 Anthrax spores; 5 per cent phenol. 33.3°

Time (hours)	Mean no. of bacteria per drop	Value of k*
0	439.0	—
0.5	275.5	0.40
1.25	137.5	0.40
2.0	46.0	0.49
3.0	15.8	0.48
4.1	5.45	0.46
5.0	3.6	0.42
7.0	0.5	0.42

* Values of k are calculated to base 10.

It will be seen from Table 4.2 that k has a mean value of 0.44; from Fig. 4.2 that the velocity of the reaction becomes slower and slower till it is almost negligible (in theory it never reaches completion), and from Fig. 4.3 that numbers of surviving organisms plotted against time in hours fall along a descending straight line.

On the other hand, Eijkman (1908), Henderson Smith (1921), and several more recent observers (Hobbs and Wilson 1942, Withell 1942*a*, Jordan and Jacobs 1944*a, b*, 1945, Berry and Michaels 1947, Moats 1971), working mainly with vegetative organisms exposed to chemical disinfectants, failed to obtain a constant value for k throughout the experiment. The value tended to be low at the start, to rise to a maximum, and then to fall off again towards the close. The resulting sigmoid curve obtained by plotting the logarithms of the surviving organisms was clearly depicted by Eijkman, and is regarded by Jordan and Jacobs as being more characteristic than the linear exponential curve.

Further observations have shown that survivor curves, other than logarithmic, may be of three or four different types (Fig. 4.4, taken from Moats 1971).

Various explanations have been offered for the difference in form of survivor curves. It is assumed that the linear or logarithmic form occurs when the death or survival of any given bacterium during any interval of time is determined by a multitude of small and independent causes—by 'chance' in the statistical sense—the chance of survival at any given moment being weighted equally against each bacterium. The non-logarithmic curves are assumed to be due mainly to differences in resistance of the individual bacteria

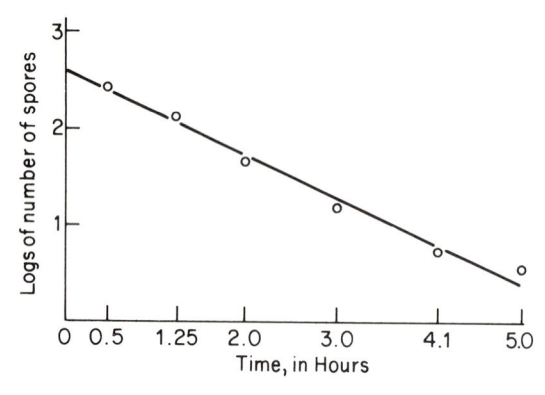

Fig. 4.3 Disinfection of anthrax spores with 5 per cent phenol at 33.3°. The curve is drawn through a series of calculated points; the circles represent the experimental observations. This curve is constructed from the same observations as those used for Fig. 4.2, but the numbers of organisms are expressed logarithmically. (After Chick.)

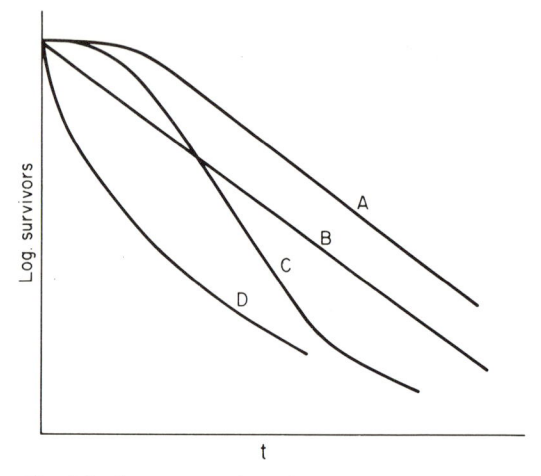

Fig. 4.4 Some types of survivor curves observed. (Taken from Moats 1971.)
A Convex (Shoulder) with initial lag phase followed by logarithmic death-rate
B Logarithmic death-rate
C Like curve A but with a tail
D Convex downwards the whole way, suggestive of a heterogeneous population

in the suspension. The particular form of the curve, however, will vary with the conditions of the experiment. Thus Henderson Smith (1921), working with the spores of *Botrytis cinerea*, found that in a 0.4 per cent solution of phenol a sigmoid curve with a pronounced shoulder at the beginning and tail at the end was transformed, as the concentration of phenol was increased, to a linear form in a 0.7 per cent solution (Fig. 4.5). Much the same transformation was noted when, with a given strength of phenol, the temperature was raised from 27° to 39°. It may be noted

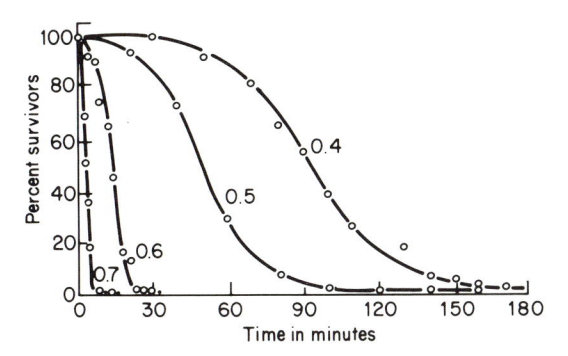

Fig. 4.5 Destruction of spores of *Botrytis cinerea* by varying strengths of phenol. (After Henderson Smith 1921.)

that, whatever form the survival curve takes, there is generally a linear stage in the middle. Whether with non-logarithmic curves the difference in resistance of members of the bacterial population is physiological or genetic is still doubtful. Similarly, whether the resistance of subcultures of the treated organisms is greater than that of the parent strain is a question on whose answer various observers have failed to agree (Eddy 1953, Nagy 1969, Corry and Roberts 1970, Moats 1971, Duitschaever and Jordan 1974, Han 1975).

(For a further discussion of this subject see Eijkmans 1908, Hewlett 1909, Reichenbach 1911, Loeb and Northrop 1917, Brooks 1919, Cohen 1922, Jordan and Jacobs 1944*a*, Berry and Michaels 1947, Sykes 1958, Whittet *et al.* 1965, Moats 1971, Tomlins and Ordal (1976).)

Concentration of disinfectant

Chick (1908) found that the relationship between the concentration of a disinfectant and the time taken for disinfection was not a simple but an exponential one, the exponent of the concentration being a factor varying with each disinfectant. That is to say, doubling

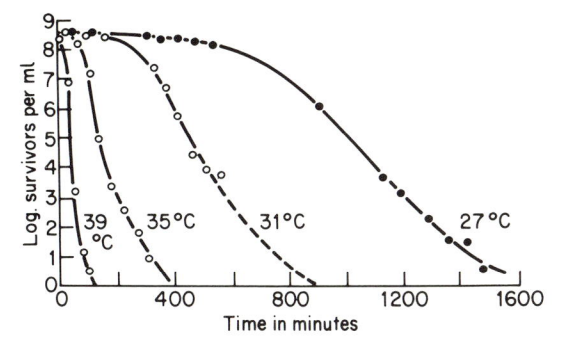

Fig. 4.6 Destruction of *Esch. coli* by 0.52 per cent phenol at various temperatures. The numbers of organisms are expressed logarithmically. (After Jordan and Jacobs 1945.)

the concentration of phenol does not halve the time necessary for the completion of the reaction as might be expected, but diminishes it to a far greater extent. Watson (1908), working on Chick's figures, found that the relation could be expressed by the formula

$$C^n t = \text{a constant}$$

where C is the concentration, n a constant varying with each disinfectant, and t the time necessary for disinfection. This equation represents the relation when one molecule of one substance reacts with an excess of molecules of a second. For purposes of calculation it may be written

$$n \log C + \log t = \text{a constant},$$

that is, the relation between $\log C$ and $\log t$ is a linear one.

An example will make this clear (Table 4.3).

Table 4.3 Disinfection of paratyphoid bacilli by phenol at 20°

Parts of phenol per 1000	Time taken for disinfection (min)	$5.5 \log C + \log t$
8.0	45	6.62
7.5	75	6.69
7.0	105	6.67
6.5	125	6.58
6.0	225	6.64
5.5	440	6.71
5.0	690	6.68

In this table the value of n is taken as 5.5. It will be seen that the values of the constant are closely similar. If the logarithms of the concentrations are plotted against the logarithms of time, the resulting curve is found to be linear.

For dealing with the salts of the heavy metals, a slight modification of the formula is required, owing to the fact that these salts are dissociated in solution, and their action depends not on their molecular but on their ionic concentration. If the concentration of Hg-ions is substituted for concentration of $HgCl_2$, for example, then the formula holds good.

Paul, Birstein and Reusz (1910*b*) found that the value of the velocity constant k for aqueous solutions of HCl was approximately proportional to the square root of its concentration.

To calculate the value of the exponent n, observations are made on the reaction velocities k_1 and k_2 of the disinfectant at two different concentrations, and the values substituted in the formula

$$KC^n = \frac{1}{t} \log \frac{B}{b}$$

(For the arithmetic concerned, see 4th edition, pp. 154-5.)

The behaviour of any disinfectant is influenced by the value of n. Thus, for phenol let us take $n = 6$, and for HgCl$_2$, $n=1$. Then a doubling of the concentration of HgCl$_2$, i.e. C^n or 2^1, will halve the time taken for completion of the reaction; doubling the concentration of phenol, i.e. C^n or 2^6, will diminish it 64 times. Conversely, halving the concentration of HgCl$_2$ doubles the time of the reaction; halving the concentration of phenol increases it 64 times.

A substance with a high value of n is germicidal above a given concentration; it requires, however, but a low degree of dilution to abolish its germicidal activity entirely. In contrast, a substance with a low value of n, though germicidal in solutions above a given concentration, exercises an inhibiting effect on the growth of bacteria even when employed in high dilution.

One further point may be dealt with here, namely the question of whether the numbers of bacteria present in a suspension affect the reaction velocity. Working with HgCl$_2$ and anthrax spores, Madsen and Nyman (1907) found the numbers of spores to be immaterial; a suspension containing 124 800 was sterilized as rapidly as one containing only 7750. Eisenberg and Okolska (1913), however, divided the disinfectants into 3 classes: in the first, comprising alcohol, phenol, and formaldehyde, the numbers of bacteria had but little effect, i.e. a concentration of disinfectant that would destroy a given number of bacteria would also destroy 100 times that number. In a second group, comprising acetone, HgCl$_2$ and K$_2$Mn$_2$O$_8$, the numbers of bacteria affected the issue; thus a concentration of disinfectant that destroyed a given number of bacteria failed to destroy 10 times that number. A third group comprising HCl, H$_2$SO$_4$, oxalic acid, KOH, and other bodies, occupied an intermediate position between the first two classes.

Temperature coefficient

As the temperature increases in arithmetical progression, the velocity of the reaction increases in geometrical progression, or mathematically expressed

$$\frac{k'}{k} = \theta^{(T'-T)}$$

in which k' and k are the velocity constants of the reaction at temperatures T′ and T respectively, and θ is the temperature coefficient.

In the disinfection of paratyphoid bacilli by 0.6 per cent phenol at 20° and at 30°, Chick (1908) obtained the figures shown in Table 4.4.

Most observers have obtained higher values of θ for phenol—generally about 7 or 8 for each rise in temperature of 10°. For HgCl$_2$ the temperature coefficient is lower, generally 2 to 4 for 10°, and for chloramine 4.5 (Butterfield and Wattie 1946).

An alternative method of estimating the value of θ is to start with a suspension of organisms the number of which need not be estimated, and determine how long it takes for com-

Table 4.4 Paratyphoid bacilli. Phenol 6 per 1000

Time in minutes		Average no. of survivors	k
At 20°	1	539	—
	2	276.6	0.29
	3	137.5	0.30
	4	80.1	0.28
	6	42	0.22
At 30°	1	1 368	—
	2	162	0.93
	3	65.5	0.66
	4.1	15.1	0.63
	6	1.5	0.59

The mean value of k at 20° = 0.27.
The mean value of k' at 30° = 0.7.
The value of $\theta = 1.1$ for 1°
$\qquad = 1.1^{10}$ or 2.6 for 10°.

plete sterilization at two or more different temperatures. Since the time taken for the completion of a reaction may be considered as inversely proportional to the velocity, there is no need to estimate the value of k. Thus:

	Paratyphoid bacilli.
Phenol 6 per 1000.	Time elapsing in minutes
11°C	2.2
21°C	1.0

Then $\theta = \dfrac{2.2}{1.0} = 2.2$ for 10°, or $2.2^{0.1} = 1.08$ for 1.0°.

In disinfection by hot water a very much higher value is obtained for θ. Thus Chick (1910), working with *Salm. typhi* at temperatures of 49° and 54.1°, found that the velocity constant of the reaction was increased 13.1 times for the 5° rise in temperature, i.e. $\theta = 1.67$ for 1°, or about 170 for 10°. (For variations in the value of θ in different temperature ranges and at different concentrations of disinfectant, see Jordan and Jacobs 1946, Jordan, Jacobs and Davies 1947.)

It must be pointed out that when an organism is suspended in a favourable medium, the effect of temperature on inhibition of growth may be opposite to that on disinfection. Thus Behring (1890) found that the growth of anthrax bacilli at room temperature was inhibited in the presence of 1/400 000 HgCl$_2$; at 37° a concentration of 1/100 000 was necessary to inhibit growth. It appears that the rise of temperature favours the growth of the bacilli more than it does the action of the germicide. Similar results were obtained by Chick (1908) with paratyphoid bacilli and phenol.

Standardization of disinfectants

In the practical application of disinfection, it is desirable to have some measure of the relative germicidal activity of different disinfectants. The first method devised for this purpose was Koch's (1881) thread method. Suitable organisms, such as anthrax spores, were dried on silk threads, submitted to the action of the disinfectant, and subsequently washed, and transferred to a solid medium in order to ascertain whether all the spores had been killed. This method is open to the objection that traces of disinfectant remaining in the interstices of the thread are liable to prevent the growth of surviving organisms. This is particularly likely to happen with disinfectants having a low concentration exponent, such as mercury salts (Geppert 1889, 1891a, b; see p. 81). Krönig and Paul (1897), therefore, replaced the threads by garnets and rendered the test quantitative by plating out the washings from the garnets and counting the number of colonies that developed.

After these tests came the Rideal-Walker drop method (1903), which is still employed for some purposes, and the Chick-Martin test (1908). In the *Rideal-Walker Test* similar quantities of organisms are submitted to the action of varying concentrations of phenol and of the germicide to be tested. Subcultures are made into broth every 2½ minutes up to 15 minutes and the tubes incubated at 37° for 3 days. That dilution of disinfectant X which sterilizes the suspension in a given time is divided by that dilution of phenol which sterilizes the suspension in the same time, and a phenol coefficient obtained (Table 4.5).

test (Kelsey and Maurer 1966). The terminology of these tests is a little confusing. A use-dilution test is intended to estimate the dilution of a disinfectant required for a particular purpose. An in-use test is intended to estimate the number of living micro-organisms in a vessel of disinfectant in actual use. Bergan and Lystad (1971) in Norway strongly favour the Kelsey-Sykes capacity use-dilution test in a slightly modified form. They (1972) give a list of inactivators for various types of disinfectant that are required for the process of standardization, as does also Reybrouck (1979). More recently, van Klingeren and Mossel (1978) developed a shake-release test; and Beck and his colleagues (1977) have reviewed a wide range of tests for evaluating disinfectant action *in vitro* under real-life conditions. (See also a report on the use, testing and evaluation of disinfectants in hospitals Report 1965b) and a comparison of three in-use tests by Christensen *et al.* 1982).

Some of the more important information we need might be obtained by following up Phelps's (1911) suggestion of determining the value of k, n, and θ for each disinfectant. The determination would have to be made in the presence and absence of organic matter and with three or four different organisms. The task

Table 4.5 *Salm. typhi*, 24 hours' broth culture at 37°. Temperature at which test was conducted 60° F

Disinfectant	Dilution	2½′	5′	7½′	10′	12½′	15′
Phenol	1–110	+	−	−	−	−	−
	1–120	+	+	−	−	−	−
	1–130	+	+	+	−	−	−
	1–140	+	+	+	+	−	−
X	1–225	+	−	−	−	−	−
	1–250	+	+	−	−	−	−
	1–275	+	+	+	−	−	−
	1–300	+	+	+	+	+	−

Time in minutes of exposure of suspension to disinfectant

+ = growth.
− = no growth.

The phenol coefficient of X is therefore $\dfrac{275}{130} = 2.1$.

(For a full discussion of the Rideal-Walker method, see *Lancet*, 1909, ii, 1516, and for the technique used by the Association of Official Agricultural Chemists of the USA, see Report 1960b.)

In the *Chick-Martin test* the disinfectant acts, not in pure solution, but in water containing a suspension of 3 per cent dried human faeces. The time, moreover, for subculture is fixed at 30 minutes. By this method the phenol coefficient is lower than that given by the Rideal-Walker method.

Both of these tests have been strongly criticized for their artificiality. In the last few years numerous other tests have been proposed, such as the use-dilution tests in Germany (Report 1958b) and the United States (Report 1965a); the capacity use-dilution tests in Great Britain (Kelsey and Sykes 1969); and the in-use

would be a formidable one and would not be made easier by the fact that the reaction velocity, k, often varies during the progress of disinfection. Should we take k to be the value observed in the middle state of the reaction (Hobbs and Wilson 1942), or over the first half (Withell 1942a, b), or over the whole of the reaction (Jordan and Jacobs 1944b)? Again, what organisms should be selected to display the greatest differences in their reaction to a given disinfectant? Even if we had all the information yielded by Phelps's method, we should still be a long way from knowing exactly how the compound would behave under practical conditions or whether it might safely be applied to the tissues.

Clearly, no one method of standardization can give us all the information we need. The only logical solution is to adopt some form of in-use test carried out under the conditions the disinfectant is expected to meet with in practice.

For deciding whether a given disinfectant is suitable for application to living tissue its toxicity for tissue cells in culture may be compared with its disinfectant power. A toxicity index may then be calculated by dividing the highest dilution of germicide killing the tissue cells by the highest dilution killing the bacteria. The smaller the index, the more suitable is the compound likely to be for clinical use (Salle 1961).

Gaseous disinfectants

The gaseous disinfectants most commonly employed are sulphur dioxide, chlorine, formaldehyde, ethylene oxide, and β-propiolactone. To destroy vegetative organisms sulphur dioxide should be present in the air in a concentration of 2–3 per cent and chlorine in 1 per cent. Both are active only in a moist atmosphere, sulphur dioxide combining with water to form sulphurous acid and chlorine hypochlorous acid. Formaldehyde and ethylene oxide are said to owe their virtue as disinfectants to their alkylating properties, with consequent blocking of reactive groups needed in metabolic reactions (Phillips 1966). *Sulphur dioxide* is used extensively in the soft-drink industry (see Hammond and Carr 1976); and *Methyl bromide* for the fumigation of cereals and other foods, ships' holds, cargo areas, and the decontamination of furs and hides (Kereluk 1971). For the fumigation of rooms, sulphur dioxide and chlorine have been largely replaced by formaldehyde (see Beeby *et al.* 1967). *Ozone* is sometimes used for the sterilization of water, and for preventing the growth of moulds on shell eggs (Ingram and Haines 1949). It is a powerful oxidizing agent, unstable, difficult to generate in high concentration in air, and particularly damaging to rubber (Phillips 1966). It has little penetrating power and, though strongly bactericidal, is of value for the destruction only of organisms unprotected by colloidal or other material (Elford and van den Ende 1942). *Propylene oxide* gas acts best at a relative humidity of 30 to 60 per cent, has a moderate penetrating and bactericidal power, and is used chiefly for decontamination.

Formaldehyde gas may be generated by (a) evaporation of formalin, which consists of a 40 per cent solution of formaldehyde in water, plus 10 per cent methanol to prevent polymerization; (b) addition of formalin to potassium permanganate; (c) volatilization of paraformaldehyde, which is a polymer of formaldehyde and which, when heated, depolymerizes rapidly to liberate formaldehyde. It acts best at a relative humidity of 60–90 per cent. A linear relation exists between its concentration and its killing rate, but increase in temperature between 10° and 30° makes little difference to its activity. Its effect is seriously impaired by the presence of protein matter, and its penetration into fabrics is poor. In a concentration of 1–2 mg per litre of air at R.H. 80 per cent and at 20° it will destroy most vegetative organisms within 6 hours, but it cannot be relied on for the destruction of acid-fast bacilli, spores, or smallpox virus. For the disinfection of blankets it is best used in combination with free steam in a special chamber (Report 1958*b*). Its chief value lies in the disinfection of plain surfaces, such as rooms from which all fabrics, books, and similar articles have been removed, and of objects, such as silks, furs, and leather goods, that are damaged by steam. It is affected, however, by so many different factors that the results are often unpredictable (see Nordgren 1939, Maassen and Dörffler 1953, Report 1958*a*). In liquid solution, on the other hand, as in 1 per cent formalin, it is a powerful and reliable disinfectant. According to Spicher and Peters (1976), spores vary in their susceptibility; and many that appear to be dead can be reactivated by heating at 80° or 95°. In any test of its activity, the residual formaldehyde must be neutralized by sulphite. For further information on its use in room disinfection, see Beeby, Kingston and Whitehouse (1967).

Ethylene oxide is a poisonous epoxide with a faint sweet smell and a boiling-point of 10.7°. It is freely soluble in water and oil, and to some extent in rubber and certain plastics. In concentrations over 3 per cent, it forms an explosive mixture with air, and is therefore generally mixed with CO_2 or with one of the Freons, such as dichlorodifluoromethane. Unlike formaldehyde, whose saturated vapour concentrations is 2 mg per litre of air at 20°, its vapour pressure is over 2 g per litre. It acts best at a low relative humidity, 10–40 per cent. It has the advantage over formaldehyde of being highly diffusible, and of being suitable, therefore, for the disinfection of fabrics, of heart-lung machines, catheters, cystoscopes, bronchoscopes, and of various plastic articles. It is best used in a closed chamber where its concentration can be controlled. The rate of killing is slow and disinfection is usually not complete for several hours (see Phillips 1949, Ernst and Doyle 1968, Kereluk *et al.* 1970). Spores and acid-fast bacilli are killed almost as readily as ordinary vegetative organisms (Kelsey 1967). Its activity increases up to a temperature of 55°, but not further (Weymes 1966). According to Dadd and Daly (1980) resistance to ethylene oxide does not run parallel with that to heat, irradiation, or other chemical disinfectants.

Beta-propiolactone is a colourless liquid with a sweetish odour and a boiling-point of 163°. In gaseous form it acts best at a relative humidity of 75 per cent or over. Its disinfectant activity increases 2–3 times with each 10° rise of temperature. Its diffusibility is poor. It is said to be 25 times as active as formaldehyde and 4000 times as active as ethylene oxide (Hoffman and Warshowsky 1958). In a concentration of 5 mg per litre at 80 per cent R.H. and 24° it is said to kill spores of *Bac. subtilis* in 2 hours. It leaves no non-volatile residue, as formaldehyde does, and is therefore recommended as a substitute for formaldehyde for the decontamination of large rooms and buildings. For this purpose the use of 1 gallon to every 12 000–25 000 cft of space is recommended, according to the presence or absence of furnishings (Barbeito 1966, Hoffman *et al.* 1966). It is also useful for the disinfection of delicate instruments, such as those used in ophthalmology (Allen and Murphy 1960, Spiner and Hoffman 1960). (For an account of the inactivation of non-sporing bacteria by gases, see Russell 1976.)

The use of aerosols and vaporized disinfectants for the air, as apart from fumigation, is considered in Chapter 9.

Liquid disinfectants

We have no space here to discuss the practical application of liquid disinfectants. All we can do is to draw attention to certain general rules and precautions to be observed in their use.

There are not many purposes for which these substances are essential, and the number of different compounds required is really very small. As investigation has shown (Report 1965*b*), there is in British hospitals a large and quite unnecessary expenditure on disinfectants. Far more different compounds are used than are necessary; many are used regardless of concentration, temperature, or time of action; others are employed for purposes for which they are quite useless; and expensive disinfectants are often used when cheap ones would be equally good or better. In some situations cleaning with a good detergent is all that is required, and disinfection by special compounds is superfluous (see Ayliffe *et al.* 1967).

Except when a specific organism is in question, it is generally best to use a disinfectant that has a wide range of antibacterial action, such as the hypochlorites, and most, though not all, of the phenolic compounds. It must be remembered, however, that the germicidal activity of chlorine compounds and of the emulsified phenolic disinfectants is greatly diminished by the presence of organic matter. Some modern plastics have a similar effect (Leigh and Whittaker 1967). When disinfection is really called for, as much organic matter as possible should first be removed by mechanical or detergent means so as to lessen the amount of work the disinfectant has to do. For the disinfection of varnished or greasy surfaces, aqueous solutions, as of the hypochlorites, are of little value; an emulsified disinfectant must be used or a wetting agent added to ensure complete coverage of the whole surface. For the destruction of spores and of acid-fast bacilli that have a high content of lipid, heat is always to be preferred to chemical disinfectants. The resistance of viruses to chemical disinfectants has been little studied. Present experience, however, would suggest that, when treatment by heat or ionizing radiations is impracticable, the halogens are generally to be preferred; alternatively pure phenol may be used (Report 1965*b*). Dry-cleaning of clothes and blankets, it may be remarked, may bring about a 90 per cent reduction in their bacterial content, but cannot be relied upon to destroy all organisms, even of the vegetative group (Pressley 1966).

The following references may be of help to those wanting more specific information.

General use of disinfectants in hospitals: Report (1944, 1965*b*), Sykes (1958), Whittet *et al.* (1965), Schmidt (1966), Beauchamp (1967), Kelsey and Maurer (1967, 1972), Report (1968), Spicher (1970), Report (1981).

Disinfection of the hands and skin: Ahlfeld and Vahle (1896), Colebrook (1941), Neufeld and Schiemann (1943), Gardner and Vincent (1948), Grün and Schopner (1957), Skadhauge (1958), Lowbury, Lilly and Bull (1964), Lowbury (1966), Kirk (1966), Pazdiora (1966), Lee *et al.* (1967), van der Hoeven and Hinton (1968), Lilly and Lowbury (1971), Lilly, Lowbury and Wilkins (1979), Ojajärvi (1980).

Disinfection of syringes: Hughes (1946), Baumann (1948), Evans and Spooner (1950), Darmady *et al.* (1957), Report (1962). Dry heat is best.

Disinfection of rubber gloves: Report (1959), Oliver and Tomlinson (1960), and of *glove powder* Kelsey (1962).

Disinfection of dressings (p. 74); heat-sensitive instruments (Alder and Mitchell 1972).

Disinfection of plastics: Mackenzie (1966), Plester (1967).

Disinfection in the dairy industry: Clegg (1967), Cousins and Allan (1967), Cousins (1976).

Selective toxicity: Albert (1950).

Disinfection of cosmetics and articles for the toilet: Malcolm (1976).

Tabulated information on the *survival of different bacteria* in human blood, urine, fæces, and tissues, in culture media and in food: (see Engley 1956).

Detailed information on *individual disinfectants:* Savolainen (1948), Sykes (1958), Lawrence and Block (1968), Hugo (1971).

Solid disinfectants These are generally made up in the form of powders, with a basis of lime, silicious matter, or vegetable fibre. Phenol is the commonest disinfectant incorporated. To destroy bacteria they must pass into solution; in the dry state they act merely as deodorants.

(For a description of various solid disinfectants that were used in the last century, see Kuchenmeister 1860; for the bacteristatic action of onions and garlic, see Wit *et al.* 1979; for the resistance of bacteria to various inimical agencies, see Gray and Postgate 1976; for the effect of sodium chloride on disinfectants, see Heinzel (1982); and for a general review of the use and effectiveness of disinfectants, see Collins *et al.* 1981.)

References

Ahlfeld, F. and Vahle, F. (1896) *Dtsch. med. Wschr.* **22**, 81.
Albert, A. (1942) *Lancet,* **ii** 633; (1950) *Nature, Lond.,* **165**, 12; (1951) *The Acridines,* Edward Arnold, London; (1966) *The Acridines,* 2nd edn.
Alder, V. G. and Gillespie, W. A. (1957) *J. clin. Path.* **10**, 299.
Alder, V. G. and Mitchell, J. P. (1972) *Soc. appl. Bact. Tech. Ser.* No. 8, p. 229.
Allen, H. F. and Murphy, J. T. (1960) *J. Amer. med. Ass.* **172**, 1759.
Allen, P. W. (1923) *J. Bact.* **8**, 555.
Andrewes, F. W. and Orton K. J. P. (1904) *Zbl. Bakt.* **35**, 645, 811.
Archer, G. T. L. (1945) *Brit. med. J.* **ii**, 148.
D'Arcy, R. F. and Hardy, W. B. (1894) *J. Physiol.* **17**, 390.
Ayliffe, G. A. J., Collins, B. J., Lowbury, E. J. L., Babb, J. R. and Lilly, H. A. (1967) *J. Hyg., Camb.* **65**, 515.
Baker, Z., Harison, R. W. and Miller, B. F. (1941*a*) *J. exp. Med.* **73**, 249; (1941*b*) *ibid.* **74**, 611, 621.
Bannan, E. A. and Judge, L. F. (1965) *Amer. J. publ. Hlth* **55**, 915.
Barbeito, M. S. (1966) *Hospitals* **40**, 100.
Barnard, J. E. and Morgan, H. de R. (1903) *Proc. roy. Soc.* **72**, 126.
Baumann, E. (1948) *Sterilisation und sterile Aufbewahrung von Spritzen und Hohlnadeln,* 2nd ed. Benno Schwabe & Co. Basle.
Baxter. (1875) *Rep. loc. Govt. Bd. publ. Hlth.* New Ser. No. 5, Appendix p. 216.
Bayliss, M. (1936) *J. Bact.* **31**, 489.
Bayliss, W. M. (1924) *Principles of General Physiology,* 4th edn. London.
Bean, H. S. (1967) *J. appl. Bact.* **30**, 6.
Beauchamp, I. L. (1967) *Amer. ind. Hyg. Ass. J.* **28**, 31.
Beck, E. G. *et al.* (1977). *Zbl. Bakt., 1 B,* **165**, 335.
Beeby, M. M., Kingston, D. and Whitehouse, C. E. (1967) *J. Hyg., Camb.* **65**, 115.
Behring. (1890) *Z. Hyg. InfektKr.* **9**, 395.
Belin, M. and Ripert, J. (1937) *C. R. Soc. Biol.* **124**, 612.
Benbough, J. E. (1967) *J. gen. Microbiol.* **47**, 325.
Bergan, T. and Lystad, A. (1971) *J. appl. Bact.* **34**, 741; (1972) *Acta path microbiol. scand.* B, **80**, 507.
Bernarde, M. A., Snow, W. B. and Olivieri, V. P. (1967) *J. appl. Bact.* **30**, 159.
Berry, H. and Michael, I. (1947) *Quart. J. Pharm.* **20**, 348.
Bigelow, W. D. and Esty, J. R. (1920) *J. infect. Dis.* **27**, 602.
Blowers, R. and Wallace, K. R. (1955) *Lancet* i. 1250.
Blum, H. F. (1932) *Physiol. Rev.* **12**, 23.
Böcker, O. E. (1938) *Z. Hyg. InfektKr.* **121**, 166.
Booth, J. C. (1979) *Lab-Lore,* **8**, 563.
Bousfield, I. J. and Mackenzie, A. R. (1976) *Inhibition and Inactivation of Vegetative Microbes,* p. 329. Ed. by F. A. Skinner and W. B. Hugo, Academic Press, London.
Bowie, J. H. (1958*a*) *Hlth. Bull., Edin.* **16**, 36; (1958*b*) *Hosp. Engng.* **12**, 158, 182.
Brock, T. D. and Darland, G. K. (1970) *Science* **169**, 1316.
Brooks, S. C. (1919) *J. gen. Physiol.,* **1**, 61; (1920) *J. med. Res.* **41**, 411.
Brown, M. R. W. and Melling, J. (1971) See Hugo 1971, p. 1.
Browning, C. H. (1943) *Brit. med. J.* **i**, 341.
Browning, C. H. and Gilmour, W. (1913) *J. Path. Bact.* **18**, 144.
Browning, C. H. and Russ, S. (1917) *Proc. roy. Soc.,* B **90**, 33.
Buchholz, J. and Jeney, A. V. (1935) *Zbl. Bakt.* **133**, 299.
Budd, W. (1873) *Typhoid Fever. Its Nature, Mode of Spreading, and Prevention.* Longmans, Green & Co., London.
Burdon, D. W. and Whitby, J. L. (1967) *Brit. Med. J.* **ii**, 153.
Burnet, F. M. (1925) *Aust. J. exp. Biol. med. Sci.* **2**, 65.
Burtenshaw, J. M. L. (1942) *J. Hyg., Camb.* **42**, 184.
Busby, D. W. G. (1967) In Collins' *Progress in Microbiological Techniques,* p. 36. Butterworths, London.
Busta, F. F. (1978) *Advance. appl. Microbiol,* **23**, 195.
Butterfield, C. T. and Wattie, E. (1946) *Publ. Hlth. Rep., Wash.* **61**, 157.
Cadéac and Meunier, A. (1889) *Ann. Inst. Pasteur* **3**, 317.
Cameron, A. T. (1930) *Trans. roy. Soc. Can.,* 3rd Series **24**, Section V, 53.
Castellani, A. (1967) *J. trop. Med. Hyg.* **70**, 181.
Chamberland, M. (1887) *Ann. Inst. Pasteur* **1**, 153.
Chambers, L. A. and Flosdorf, E. W. (1936) *Proc. Soc. exp. Biol., N.Y.* **34**, 631.
Chambers, L. A. and Gaines, N. (1932) *J. cell. comp. Physiol.* **1**, 451.
Chick, H. (1908) *J. Hyg., Camb.* **8**, 92; (1910) *Ibid.* **10**, 237; (1912) *Ibid.* **12**, 414.
Chick, H. and Martin, C. J. (1908) *J. Hyg., Camb.* **8**, 698; (1910) *J. Physiol.* **40**, 404.
Christensen, E. A., Jepsen, O. B., Kristensen, H. and Steen, G. (1982) *Acta path. microbiol scand,* B, **90**, 95.
Christian, M. I. (1931*a*) *J. Dairy Res.* **3**, 113; (1931*b*) *Nature, Lond.* **127**, 558.
Churchman, J. W. (1912) *J. exp. Med.* **16**, 221; (1923) *Ibid.* **37**, 1.
Clegg, L. F. L. (1967) *J. appl. Bact.* **30**, 117.
Clifton, C. E. (1931) *Proc. Soc. exp. Biol., N.Y.* **28**, 745.
Cohen, B. (1922) *J. Bact.* **7**, 183.
Cohen, B. and Clark, W. M. (1919) *J. Bact.* **4**, 409.
Colebrook, L. (1941) *Bull. War Med.* **2**, 73.
Colebrook, L. and Maxted, W. R. (1933) *J. Obstet. Gynæc.* **40**, 966.
Collins, C. H., Allwood, M. C., Bloomfield, S. F. and Fox, A. (1981) *Disinfectants: their Use and Evaluation of Effectiveness.* Academic Press, London.
Collins, C. H., Allwood, M. C., Bloomfield, Sally, F. and Fox, A. (1981) *Disinfectants: their Use and Evaluation of Effectiveness.* Soc. appl. Bact., Tech. Series No. 16. Academic Press, London.
Cooper, P. D. (1953) *J. gen. Microbiol.* **9**, 199.
Corry, J. E. L. and Roberts, T. A. (1970) *J. appl. Bact.,* **33**, 733.
Coulthard, C. E. and Sykes, G. (1936) *Pharmaceut. J.* **83**, 79.
Cousins, C. M. and Allan, C. D. (1967) *J. appl. Bact.* **30**, 168.
Cousins, Christina M. (1976) *Inhibition and Inactivation of Vegetative Microbes,* p. 13. Ed. by F. A. Skinner and W. B. Hugo, Academic Press, London.
Cox, C. S. and Baldwin, F. (1967) *J. gen. Microbiol.* **49**, 115.
Crowther, J. A. (1926) *Nature, Lond.* **118**, 86.
Cruickshank, J. C., Hobbs, B. C., McFarlan, A. M. and Maier, I. (1942) *Brit. med. J.* **ii**, 182.
Curran, H. R. and Evans, F. R. (1942) *J. Bact.* **43**, 125; (1950) *J. Dairy Sci.* **33**, 1.
Dadd, A. H. and Daly, G. M. (1980) *J. appl. Bact.,* **49**, 89.
Dakin, H. D. (1915) *Brit. med. J.* **ii**, 318.
Darmady, E. M. and Brock, R. B. (1954) *J. clin. Path.* **7**, 290.

Darmady, E. M., Hughes, K. E. A. and Jones, J. D. (1958) *Lancet* **ii**, 766.

Darmady, E. M., Hughes, K. E. A. and Tuke, W. (1957) *J. clin. Path.* **10**, 291.

Davies, G. E. (1949) *J. Hyg., Camb.* **47**, 271.

Davies, R. (1976) *Inhibition and Inactivation of Vegetative Microbes*, p. 239. Ed. by F. A. Skinner and W. B. Hugo, Academic Press, London.

Davis, B. D., Dulbecco, R., Eisen, H. N., Ginsberg, H. S. and Wood, W. B. (1968) *Principles of microbiology and immunology.* Harper and Row, New York.

Davis, J. G. (1961) *Lab. Pract.* **10**, 639.

Dold, H. and Gust, R. (1957) *Arch. Hyg. Berl.* **141**, 321.

Domagk, G. (1935) *Dtsch. med. Wschr.* **61**, 829.

Downes, A. and Blunt, T. P. (1877) *Proc. roy. Soc.* **26**, 488; (1878) *ibid.* **28**, 199.

Drea, W. F. (1942) *J. Bact.* **44**, 149.

Dreyer and Hanssen. (1907) *C. R. Acad. Sci.* **145**, 234.

Dring, G. J. (1976) *Inhibition and Inactivation of Vegetative Microbes*, p. 257. Ed. by F. A. Skinner and W. B. Hugo, Academic Press, London.

Dubos, R. J. and Middlebrook, G. (1947) *Amer. Rev. Tuberc.* **56**, 334.

Duclaux, E. (1887) *Ann. Inst. Pasteur* **1**, 88.

Duitschaever, C. L. and Jordan, D. C. (1974) *J. Milk Technol.*, **37**, 382.

Dunklin, E. W. and Lester, W. (1959) *J. infect. Dis.* **104**, 41.

Eddy, A. A. (1953) *Proc. roy. Soc., B*, **141**, 137.

Eggerth, A. H. (1926) *J. gen. Physiol.* **10**, 147; (1929a) *J. exp. Med.* **49**, 53; (1929b) *Ibid.* **50**, 299; (1931) *Ibid.* **53**, 27.

Ehrismann, O. and Noethling, W. (1932) *Z. Hyg. InfektKr.* **113**, 597.

Eijkman, C. (1908) *Biochem. Z.* **11**, 12.

Eisenberg, P. (1919) *Zbl. Bakt.* **82**, 69.

Eisenberg, P. and Okolska, M. (1913) *Zbl. Bakt.* **69**, 312.

Eisler, M. v. (1909) *Zbl. Bakt.* **51**, 546.

Elford, W. J. and Ende, J. van den. (1942) *J. Hyg., Camb.* **42**, 240.

Ellison, S. M., Erlanger, B. R. and Allen, P. (1955) *J. Bact.* **69**, 536.

Engley, F. B. (1956) *Texas Rep. Biol. Med.* **14**, 10, 114, 313.

Epstein, F. (1897) *Z. Hyg. InfektKr.* **24**, 1.

Ernst, R. R. and Doyle, J. E. (1968) *Biotechnol. Bioengng* **10**, 1.

Eschment, R. and Lutz-Dettinger, U. (1970) *Off. Gesundh Dienst.* **32**, 527.

Eseltine, W. P. van and Hucker, G. J. (1948) *N.Y. St. agric. Exp. Sta., Tec. Bull.* No. 282.

Evans, F. R. and Curran, H. R. (1943) *J. Bact.* **46**, 513.

Evans, R. J. and Spooner, E. T. C. (1950) *Brit. med. J.* **ii**, 185.

Falk, I. S. (1923) *Abst. Bact.* **7**, 33, 87, 133.

Fermi, C. and Pernossi, L. (1894) *z. Hyg. InfektKr.* **16**, 385.

Ficker, M. (1898) *Z. Hyg. InfektKr.* **29**, 1.

Fildes, P. (1940) *Brit. J. exp. Path.* **21**, 67.

Fleming, A. (1932) *J. Path. Bact.* **35**, 831.

Fleming, A and Young, M. Y. (1940) *J. Path. Bact.* **51**, 29.

Flexner, S. (1907) *J. exp. Med.* **9**, 105.

Flosdorf, E. W. and Mudd, S. (1935) *J. Immunol.* **29**, 389; (1938) *Ibid.* **34**, 469.

Foulds, M. (1911) *J. R. Army med. Cps.* **16**, 167.

Freundlich, H. (1903) *Z. phys. Chem.* **44**, 129.

Fry, R. M. and Greaves, R. (1951) *J. Hyg., Camb.* **49**, 220.

Gale, E. F. (1940) *Bact. Rev.* **4**, 135.

Gale, E. F. and Taylor, E. S. (1947) *J. gen. Microbiol.* **1**, 77.

Gardner, A. D. and Vincent, E. (1948) *Lancet* **ii**, 760.

Garrod, L. P. (1933) *St, Bart's Hosp. med. Rep.* **66**, 203; (1944) *Brit. med. J.* **i**, 245.

Gates, F. L. (1930]) *J. gen. Physiol.* **14**, 31.

Geppert, J. (1889) *Berl. klin. Wschr.*, **26**, 789, 819; (1891a) *Dtsch. med. Wschr.* **17**, 797, 825, 855; (1891b) *Ibid.* **17**, 1065.

Gezon, H. M., Thompson, D. J., Rogers, K. D., Hatch, T. F. and Taylor, P. M. (1964) *New Engl. J. Med.* **270**, 379.

Glasser, O., Quimby, E. H., Taylor, L. S. and Weatherwax, J. L. (1952) *Physical Foundations of Radiology* pp. 36, 41. Hoeber, New York.

Glassman, H. N. (1948) *Bact. Rev.*, **12**, 105.

Godlee, R. J. (1924) *Lord Lister.* Clarendon Press, Oxford.

Goldblith, S. A. (1967) *Proc. Symph. B'pest*, p. 3. *int. atom. Energy Agency, Vienna;* (1971) See Hugo 1971, p. 285.

Gordon, J., McLeod, J. W., Mayr-Harting, A., Orr, J. W. and Zinnemann, K. (1947) *J. Hyg., Camb.* **45**, 297.

Gorman, S. P., Scott, E. M. and Russell, A. D. (1980) *J. appl. Bact.*, **48**, 161.

Gould, G. W. and Ordal, Z. J. (1968) *J. gen. Microbiol.* **50**, 77.

Gray, T. R. G. and Postgate, J. R. (1976) *The Survival of Vegetative Microbes*, Cambridge University Press, Cambridge.

Grün, L. and Schopner, R. (1957) *Z. Hyg. InfektKr.* **143**, 521.

Guiteras, A. F. and Shapiro, R. L. (1946) *J. Bact.* **52**, 635.

Hammond, S. M. and Carr, J. G. (1976) *Inhibition and Inactivation of Vegetative Microbes*, p. 89. Ed. by F. A. Skinner and W. B. Hugo, Academic Press, London.

Han, Y. W. (1975) *Canad. J. Microbiol.* **21**, 1464.

Harris, R. S., Bunker, J. W. M. and Milas, N. A. (1932) *J. Bact.* **23**, 429.

Harrison, A. P. (1967) *Annu. Rev. Microbiol.* **21**, 143.

Heinmets, F., Lehman, J. J., Taylor, W. W. and Kathan, R. H. (1954) *J. Bact.* **67**, 511.

Heinzel, M. (1982) *Arch. Lebensmitt. Hyg.* **33**, 84.

Heise, R. (1917) *Arb. ReichsgesundhAmt.* **50**, 204, 418.

Heiss, F. (1969) *Arch. Hyg. Bakt.* **153**, 247.

Hewlett, R. T. (1909) *Lancet* **i**, 741, 815, 889.

Hobbs, B. C., Emberley, N., Pryor, H. M. and Smith, M. E. (1960) *J. appl. Bact.* **23**, 350.

Hobbs, B. C. and Wilson, G. S. (1942) *J. Hyg., Camb.* **42**, 436.

Hoder, F. (1932) *Z. ImmunForsch.* **74**, 455.

Hoeven, E. van der and Hinton, N. A. (1968) *Canad. med. Ass. J.* **99**, 402.

Hoffman, R. K., Buchanan, L. M. and Spiner, D. R. (1966) *Appl. Microbiol.* **14**, 989.

Hoffman, R. K. and Warshowsky, B. (1958) *Appl. Microbiol.* **6**, 358.

Hofmeister, F. (1888) *Arch. exp. Path. Pharmak.* **24**, 247; (1889) *Ibid.* **25**, 1.

Horner, K. J. and Anagnostopoulos, G. D. (1975) *J. appl. Bact.* **38**, 9.

Hotchkiss, R. D. (1946) *Ann. N.Y. Acad. Sci.* **46**, 479.

Howie, J. W. (1961) *J. clin. Path.* **14**, 49.

Hoyt, A., Fisk, R. T. and Burde, G. (1942) *Surgery* **12**, 786.

Hughes, D. E. (1951) *Brit. J. exp. Path.* **32**, 97.

Hughes, R. R. (1946) *Br. med. J.* **ii**, 685.

Hugo, W. B. (1967) *J. appl. Bact.* **30**, 17; (1971) *Inhibition and Destruction of the Microbial Cell*. Academic Press, London.

Hugo, W. B. (1976) In Gray and Postgate *The Survival of Vegetative Microbes*, p. 383. Ed. by T. R. G. Gray and J. R. Postgate Cambridge University Press, Cambridge;

(1977) In *Inhibition and Inactivation of Vegetative Microbes*, p. 1. Ed. by F. A. Skinner and W. B. Hugo, Academic Press, London; (1979) *Microbios*, **23**, 83.
Hugo, W. B. and Bloomfield, S. F. (1971) *J. appl. Bact.* **34**, 557, 569, 579.
Hugo, W. B. and Longworth, A. R. (1964) *J. Pharm. Pharmac.* **16**, 655, 751; (1966) *Ibid.* **18**, 569.
Hutner, S. H. (1942) *J. Bact.* **43**, 629.
Iland, C. N. (1944) *Lancet* **i**, 49.
Ingram, M. and Haines, R. B. (1949) *J. Hyg., Camb.* **47**, 146.
Ingram, M. and Mackey, B. M. (1976) In: *Inhibition and Inactivation of Vegetative Microbes*, p. 111. Ed. by F. A. Skinner and W. B. Hugo, Academic Press, London.
Jesenská, Z. (1966) *Čslká Epidem. Mikrobiol. Immunol.* **15**, 107.
Johns, C. K. (1947) *Canad. J. Res.*, F., **25**, 76.
Jones, R. E. and Handley, W. R. C. (1945) *Mon. Bull. Minist. Hlth. Lab. Serv.* **4**, 107.
Jordan, R. C. and Jacobs, S. E. (1944a) *J. Hyg., Camb.* **43**, 275; (1944b) *Ibid.* **43**, 363; (1945) *Ibid.* **44**, 210; (19467) *ibid.* **44**, 243, 249.
Jordan, R. C., Jacobs, S. E. and Davies, H. E. F. (1947) *J. Hyg., Camb.* **45**, 333.
Kelner, A. (1949) *J. Bact.* **58**, 511.
Kelsey, J. C. (1958) *Lancet* **i**, 306; (1962) *Mon. Bull. Minist. Hlth. Lab. Serv.* **21**, 17; (1967) *J. appl. Microbiol.* **30**, 92.
Kelsey, J. C. (1967) *J. appl. Bact.*, **30**, 92.
Kelsey, J. C. and Maurer, I. M. (1966) *Mon. Bull. Minist. Hlth* **25**, 180; (1967) *Ibid.* **26**, 110; (1972) *PHLS Monogr. Ser.* No. 2.
Kelsey, J. C. and Sykes, G. (1969) *Pharm. J.* **202**, 607.
Kereluk, K. (1971) In; *Progress in Industrial Microbiology*, Vol. 10. Ed. by D. J. D. Hockenhull, Churchill, Edinburgh.
Kereluk, K., Gammon, R. A. and Lloyd, R. S. (1970) *Appl. Microbiol.* **19**, 146, 152, 157, 163.
Kimbrough, R. D. (1973). *Pediatrics* **51**, No. 2, Pt ii, p. 391.
Kirk, J. E. (1966) *Acta derm.-vener., Stockh.* **46**, Suppl. No. 57.
Kitasato, S. (1891) *Z. Hyg. InfektKr.* **10**, 267.
Klingeren, A. van and Mossel, D. A. A. (1978) *Zbl. Bakt., 1st Abt., Orig. B* **166**, 540.
Knox, R. and Pickerill, J. K. (1964) *Lancet* **i**, 1318.
Knox, W. E., Auerbach, V. H., Zarudnaya, K. and Spirtes, M. (1949) *J. Bact.* **58**, 443.
Koch, R. (1881) *Mitt. ReichsgesundhAmt.* **1**, 234.
Koch, R., Gaffky and Loeffler. (1881) *Mitt. Reichsgesundh Amt.* **1**, 322.
Koch, R. and Wolffhügel, G. (1881) *Mitt. ReichsgesundhAmt.* **1**, 301.
Kodicek, E. and Worden, A. N. (1945) *Biochem. J.* **39**, 78.
Konrich, F. (1938) *Die bakterielle Keimtötung durch Wärme*. Ferd. Enke Verlag, Stuttgart.
Krönig, B. and Paul, T. (1897) *Z. Hyg. InfektKr.* **25**, 1.
Küchenmeister, G. F. M. (1860) *Dtsch. Klin.* **12**, 123.
Kuipers, J. S. and Dankert, J. (1970) *J. Hyg.. Camb.* **68**, 343.
Kurtz, J. B. (1950) Unpublished Communication, Oxford.
Lamar, R. V. (1911a) *J. exp. Med.* **13**, 1, 380; (1911b) *Ibid.* **14**, 256.
Lapage, S. P. and Redway, K. F. (1974) *Preservation of Bacteria with Notes on other Micro-organisms. Publ. Hlth Lab. Service, Monograph Series*, No. 7 H.M.S.O., London.
Laroquette, M. de. (1918) *Ann. Inst. Pasteur* **32**, 170.
Lawrence, C. A. and Block, S. S. (1968) *Disinfection, Sterilization, and Preservation*. Henry Kimpton, London.
Lea, D. E. and Haines, R. B. (1940) *J. Hyg., Camb.* **40**, 162.
Lea, D. E., Haines, R. B. and Bretscher, E. (1941) *J. Hyg., Camb.* **41**, 1.
Lee, S., Schoen, I. and Malkin, A. (1967) *Amer. J. clin. Path.* **47**, 646.
Leigh, D. A. and Whittaker, C. (1967) *Brit. med. J.* **iii**, 435.
Lerman, L. S. (1961) *J. molec. Biol.* **3**, 18.
Levine, M. (1952) *J. Bact.* **16**, 117.
Lilly, H. A. and Lowbury, E. J. L. (1971) *Brit. med. J.* **iii**, 674.
Lilly, H. A., Lowbury, E. J. L. and Wilkins, M. D. (1979) *J. Hys., Camb.,* **82**, 89.
Lockemann, G., Bär, F. and Totzeck, W. (1941) *Zbl. Bakt.* **147**, 1.
Loeb, J. and Northrop, J. H. (1917) *J. biol. Chem.* **32**, 103.
Lowbury, E. J. L. (1966) *Prescribers J.* **6**, 78.
Lowbury, E. J. L., Lilly, H. A. and Bull, J. P. (1960) *Brit. med. J.* **ii**, 1039; (1964) *Ibid.* **ii**, 531.
Maasen, W. and Dörffler, P. (1953) *Z. Hyg. InfektKr.* **137**, 492.
McCracken, A. W. (1964) *Mon. Bull. Minist. Hlth* **23**, 17.
McCulloch, E. C. (1947) *Science* **105**, 480.
McDade, J, J, Hall, L. B. and Street, A. R. (1964) *Amer. J. Hyg.* **80**, 184, 192.
Macfadyen, A. (1900) *Proc. roy. Soc. B* **66**, 180, 339.
Macfadyen, A. and Rowland, S. D. (1900) *Proc. roy. Soc. B* **66**, 488.
Mackenzie, S. (1966) *Brit. Hosp. J.* 16 Sept. p. 1733.
Madsen, T. and Nyman, M. (1907) *Z. Hyg. InfektKr.* **57**, 388.
Malcolm, S. A. (1976) In: *Inhibition and Inactivation of Vegetative Microbes*, p. 305. Ed. by F. A. Skinner and W. B. Hugo, Academic press, London.
Mallmann, W. L., Botwright, W. E. and Churchill, E. S. (1941) *J. infect. Dis.* **69**, 215.
Mayser, H. (1925) *Zbl. Bakt.* **94**, 238.
Milner, H. W., Lawrence, N. S. and French, C. S. (1950) *Science* **111**, 633.
Minervini, R. (1898) *Z. Hyg. InfektKr.* **29**, 117.
Moats, W. A. (1971) *J. Bact.* **105**, 165.
Morton, H. E., North, L. L. and Engley, F. B. (1948) *J. Amer. med. Ass.* **136**, 37.
Mullick, F. G. (1973) *Pediatrics* **51**, No. 2, Pt ii, p. 395.
Mündel, O. (1937) *Z. Hyg. InfektKr.* **120**, 267.
Murrell, W. G. and Scott, W. J. (1966) *J. gen. Microbiol.* **43**, 411.
Neufeld, F. and Schiemann, O. (1943) *Z. Hyg. InfektKr.* **124**, 751.
Nagy, E. (1969) *Wien tierärztl. Mschr.,* **56**, 410.
Navarro, J. M. and Monsan, P. (1976) *Ann. Microbiol., Paris,* **127**B, 295.
Nichols, H. J. (1920) *J. Lab. clin. Med.* **5**, 502.
Nordgren, G. (1939) *Acta path. microbiol. scand.* Suppl. XL.
Nossal, P. M. (1953) *Biochim. biophys. Acta* **11**, 596.
Oag, R. K. (1940) *J. Path. Bact.* **51**, 137.
Ojajärvi, J. (1980) *J. Hyg., Camb.,* **85**, 193.
Oliver, R. and Tomlinson, A. H. (1960) *J. Hyg., Camb.* **58**, 465.
Païc, M., Deutsch, V. and Borcila, I. (1935a) *C. R. Soc. Biol.* **119**, 1063.
Païc, M., Haber, P., Voet, J. and Eliasz, A. (1935b) *C. R. Soc. Biol.* **119**, 1061.
Park, C. H. (1976) *Amer. J. clin. Path.,* **66**, 927.
Paul, T., Birstein, G. and Reusz, A. (1910a) *Biochem. Z.* **25**, 367; (1910b) *Ibid.* **29**, 202.

Paul, T. and Prall, F. (1907) *Arb. ReichsgesundhAmt.* **26,** 73.
Pazdiora, A. (1966) *ČslkáEpidem. Mikrobiol. Immunol.* **15,** 102.
Pearce, R. (1946) *Lancet* **i,** 347.
Perdrau, J. R. and Todd, C. (1933a) *Proc. roy. Soc. B,* **112,** 277; (1933b) *Ibid.* **112,** 288.
Perkins, W. E. (1965) *J. appl. Bact.* **28,** 1.
Phelps, E. B. (1911) *J. infect. Dis.* **8,** 27.
Phillips, C. R. (1949) *Amer. J. Hyg.* **50,** 280; (1966) *Spacecraft Sterilization.* Natn Aeronautics and Space Administration, Washington, D.C.
Plester, D. W. (1967) *Trans. J. Plastics Inst.* Aug., p. 579.
Pollard, E. C. and Forro, F. (1949) *Science* **109,** 374.
Pollock, M. R. (1947) *Brit. J. exp. Path.* **28,** 295; (1948) *Nature Lond.* **161,** 853.
Poon, C. P. C. (1966) *Amer. J. Epidem.* **84,** 1.
Powell, D. B. (1961) *Recent Developments in the Sterilization of Surgical Materials,* pp. 9–16. Pharmaceutical Press, London.
Pressley, T. A. (1966) *Med. J. Aust.* **ii,** 1033; (1967) *ibid.* **i,** 550.
Pringle. (1750) *Philos. Trans.,* **46,** 480, 525.
Prudhomme, R. O. (1937) *C. R. Soc. Biol.* **126,** 289.
Pulvertaft, R. J. V. and Lumb, G. D. (1948) *J. Hyg., Camb.,* **46,** 62.
Raab, O. (1900) *Z. Biol.* **39,** 524.
Reichenbach, H. (1908) *Z. Hyg. InfektKr.* **59,** 296; (1911) *Ibid.* **69,** 171.
Reid, J. D. (1932) *Amer. J. Hyg.* **16,** 540.
Relyfeld, E. H. (1977) *Ann. Microbiol., Paris,* **128** b, 495.
Rentschler, H. C., Nagy, R. and Mouromseff, G. (1941) *J. Bact.* **41,** 745.
Report. (1942) *The Prevention of 'Hospital Infection' of Wounds.* Med. Res. Coun. Lond., War Memo., No. 6; (1944) *The Control of Cross Infection in Hospitals.* Med. Res. Coun., Lond., War Memo., No. 11; (1945) *The Sterilization, Use and Care of Syringes.* Med. Res. Coun., Lond., War Memo., No. 15; (1958a) *J. Hyg. Camb.* **56,** 488; (1958b) *Zbl. Bakt.* **173,** 307; (1959) *Lancet* **i,** 425; (1960a) *Lancet* **ii,** 1243; (1960b) *Official Methods of Analysis of the Association of Official Agricultural Chemists,* p. 63, 9th ed. Wash., D. C.; (1962) *The Sterilization, Use and Care of Syringes.* Med. Res. Coun., Lond., Memo., No. 41; (1964) Report of the Working Party on Irradiation of Food. Minist. Hlth, Lond., H.M.S.O.; (1965a) *Official Methods of Analysis.* 10th edn. Ass. agric. Chemists. Washington, D.C.; (1965b) *Br. med. J.* **i,** 408; (1967) Radio-Agency, Tech. Rep. Ser., No. 72: (1968) *Lancet* **i,** 705, 763, 831.
Report (1974) *New Scientist,* **62,** 296;(1981) *Disinfectants: their Use and Evaluation of Effectiveness.* Ed. by C. H. Collins, M. C. Allwood, S. F. Bloomfield and A. Fox. *Soc. appl. Bact., Technical Series,* No. 16.
Reybrouck, G. (1979) *Zbl. Bakt., 1 Abt., B,* **168,** 480.
Ricketts, C. R., Squire, J. R., Topley, E. and Lilly, H. A. (1951) *Clin. Sci.* **10,** 89.
Rideal, S. and Walker, J. T. A. (1903) *J. R. sanit. Inst.* **24,** 424.
Risler, J. (1936) *C. R. Acad. Sci.* **203,** 517.
Ritchie, J. (1899) *Trans. path. Soc. Lond.* **50,** 256.
Roberts, R. B. and Aldous, E. (1949) *J. Bact.* **57,** 363.
Rodgers, A. and Hughes, D. E. (1960) *J. biochem. microbiol. Technol. Engng.* **2,** 49.
Roux, E. (1887) *Ann. Inst. Pasteur* **1,** 445.
Rouyer, M. and Servigne, M. (1938) *Ann. Inst. Pasteur* **61.** 565.

Rubbo, S. D., Gardner, J. F. and Webb, R. L. (1967) *J. appl. Bact.* **30,** 78.
Russell, A. D. (1976) *Inhibition and Inactivation of Vegetative Microbes,* p. 61. Ed. by F. A. Skinner and W. B. Hugo. Academic Press, London.
Russell, D. S. and Falconer, M. A. (1941) *Brit. J. Surg.* **28,** 472; (1943) *Lancet* **i,** 580.
Salle, A. J. (1961) *Arch. Mikrobiol* **39,** 116.
Salton, M. R. J. (1951) *J. gen. Microbiol.* **5,** 391.
Savolainen, T. (1948) *Studies on the Growth-Inhibition of Certain Anaerobic Bacterial Strains by Organic Compounds.* Ann. Med. exp. Biol. Fenn., Suppl.
Schmidt, B. (1966) *Gesundheitswesen in Desinfektion* **58,** 89.
Selbie, F. R. and McIntosh, J. (1943) *J. Path. Bact.* **55,** 477.
Sellards, A. W. (1918) *J. med. Res.* **38,** 293.
Selwyn, S. (1965) *J. Hyg., Camb.* **63,** 411.
Semmelweis, I. P. (1861) *Ätiologie, Begriff und Prophylaxis des Kindbettfiebers.* J. A. Barth, Leipzig (Reprinted 1912).
Silverman, G. (1966) *spacecraft Sterilization Technology.* Natn Aeronautics and Space Administration, Washington, D.C.
Skadhauge, K. (1958) *Acta path. microbiol. scand.* **43,** 211.
Skinner, F. A. and Hugo, W. B. (1977) *Inhibition and Inactivation of vegetative microbes.* Academic press, London.
Slonim, D., Diawara, S. M. and Brazdova, R. (1969) *J. Hyg. Epidem. Praha.* **13,** 313.
Smith, Audrey U. (1961) *Biological Effects of Freezing and Supercooling.* Edward Arnold, London.
Smith C. R., Nishihara, H., Golden, F., Hoyt, A., Guss, C. O. and Kloetzel, M. C. (1950) *Publ. Hlth. Rep., Wash.* **65,** 1588.
Smith, J. H. (1921) *Ann. appl. Biol.* **8,** 27.
Smith, K. C. and Hanawalt, P. C. (1969) *Molecular photobiology. Inactivation and Recovery.* Academic Press, New York and london.
Snyder, M. L. and Lichstein, H. C. (1940) *J. infect. Dis.* **67,** 113.
Sobernheim, G. (1943) *Schweiz. med. Wschr.* **73,** 1280, 1304, 1333.
Songer, J. R. (1967) *Appl. Microbiol.* **15,** 35.
Spicher, G. (1970) *Path. et Microbiol.* **36,** 259.
Spicher, G. and Peters, J. (1976) *Zw. Bakt., Ite Abt. B,* **163,** 486.
Spiner, D. R. and Hoffman, R. K. (1960) *Appl. Microbiol.* **8,** 152.
Spooner, E. T. C. and Turnbull, L. H. (1942) *Bull. War Med.* **2,** 345.
Srivastava, R. B. and Thompson, R. E. M. (1966) *Brit. J. exp. Path.* **47,** 315.
Stamp (Lord) (1947) *J. gen. Microbiol.* **1,** 251.
Sykes, G. (1958) *Disinfection and Sterilization.* Spon, London.
Takahashi, W. N. and Christensen, R. J. (1934) *Science* **79,** 415.
Tappeiner, H. v. (1900) *Münch. med. Wschr.* **47,** 5.
Tappeiner, H. v. and Jodlbauer, A. (1904) *Münch. med. Wschr* **51,** 737.
Taylor, E. W. (1949) Thresh, Beale and Suckling's *The Examination of Waters and Water Supplies,* 6th ed. J. & A. Churchill Ltd., London.
Tetsumoto, S. (1937) *Jap. J. exp. Med.* **15,** 1.
Thiele, H. and Wolf, K. (1906) *Arch. Hyg.* **57,** 29; (1907) *Ibid.* **60,** 29.
Tomlins, R. I. and Ordal, Z. T. (1976) *Inhibition and*

Inactivation of Vegetative Microbes, p, 153. Ed. by F. A. Skinner and W. B. Hugo, Academic Press, London.
Topley, W. W. C. (1915) *Brit. med. J.* **i,** 237.
Trotman, R. E. (1969) *J. appl. Bact.* **32,** 297.
Underwood, W. B. (1934) *Textbook of Sterilisation.* Amer. Sterilizer Co., Eerie, Pa.
Wahl, R. and Blum-Emerique, L. (1949) *Ann. Inst. Pasteur* **77,** 561.
Wainwright, S. D. and Nevill, A. (1955) *J. gen. Microbiol.* **12,** 1.
Walker, J. E. (1926) *J. Infect. Dis.* **38,** 127.
Ward, H. M. (1892) *Proc. roy. Soc.* **52,** 393.
Waring, M. J. (1966) *Biochemical Studies of Antimicrobial Drugs,* p. 235. Symposium 16. *Soc. gen. Microbiol.*
Watson, H. E. (1908) *J. Hyg., Camb.* **8,** 536.
Webb, S. J. (1959) *Can. J. Microbiol.* **5,** 649; (1963) *J. appl. Bact.* **26,** 307.
Weiszfeiler, G. J., Czanik, P. and Bátkai, L. (1968) *Zbl. Bakt.* **207,** 517.
Wesbrook, F. F. (1896) *J. Path. Bact.* **3,** 70.
Weymes, C. (1966) *Brit. Hosp. J.* 16 Sept.
Whittet, T. D, Hugo, W. B. and Wilkinson, G. R. (1965) *Sterilisation and Disinfection.* Heinemann Med. Books, London.
Williams, R., Clayton-Cooper, B., Duncan, J. M. and Miles, E. M. (1943) *Lancet,* **i,** 522.
Winslow, C.-E. A. and Hotchkiss, M. (1922) *Proc. Soc. exp. Biol., N.Y.* **19,** 314.
Winslow, C.-E. A. and Lochridge, E. E. (1906) *J. infect. Dis.* **3,** 547.
Wirgin, G. (1902) *Z. Hyg. InfecktKr.* **40,** 307; (1904) *Ibid.* **46,** 149.
Wit, J. C. de, Notermans, S. H. W. and Gorin, N. (1979) *Antonie v. Leeuwenhoek,* **45.** 156.
Withell, E. R. (1942a) *J. Hyg., Camb.* **42,** 124; (1942b) *Ibid.* **42,** 339.
Wood, R. W. and Loomis, A. L. (1927) *Phil. Mag.* VII, **4,** 417.
Wright, H. D. (1934) *J. Path. Bact.* **38,** 499.
Wyckoff, R. W. G. (1932) *J. gen. Physiol.* **15,** 351.
Wyckoff, R. W. G. and Rivers, T. M. (1930) *J. exp. Med.* **51,** 921.
Zimmerman, A. M. (1970) *High Pressure Effects on Cellular Processes.* Academic Press, New York.
Zobell, C. E. and Kim, J. (1972) *The Effects of Pressure on Organisms.* Ed. by M. A. Sleigh and A. G. MacDonald, Cambridge University Press, London.

5

Antibacterial substances used in the treatment of infections

J. David Williams

Introductory	98	Mechanisms of resistance to aminoglycosides	116
The development of chemotherapy	98	Pharmacological properties and toxicity	119
Resistance to antibiotics	99	The polymyxins	119
Epidemiological aspects of resistance	100	Nitrofurans	121
Assessment of activity of antibiotics against micro-organisms	102	Nalidixic acid and analogues	121
Minimum inhibitory concentrations (MIC)	102	Broad spectrum antibiotics	122
Minimum bactericidal concentrations (MBCs)	102	Sulphonamides	122
		Trimethoprim	124
Antibiotics in clinical use	103	Cephalosporins and related compounds	125
The beta-lactam antibiotics	103	Thienamycin	131
Antibiotics mainly against gram-positive organisms	105	Beta-lactamase inhibitors	131
		Clavulanic acid	131
Benzyl penicillin and related acid-stable penicillins	105	Sulbactam	131
6APA and the antistaphylococcal penicillins	105	Tetracyclines	132
The macrolides and lincomycin	107	Chloramphenicol	133
Erythromycin	109	Antibiotics active against anaerobes	133
Lincomycin	109	Metronidazole and related nitroimidazoles	133
Fusidic acid	109	Antituberculous drugs	134
Rifampicin	109	Ethambutol, cycloserine, isoniazid, para-aminosalicylic acid (PAS), thiosemicarbazones	134
Vancomycin	111		
Bacitracin	111	Anti-fungal drugs	136
Antibiotics active mainly against gram-negative rods	111	The polyenes: nystatin, amphotericin B	136
Beta-lactam antibiotics	111	5-Fluorocytosine	138
Ampicillin and related compounds	111	Imidazoles	138
The acylureido penicillins	114	Griseofulvin	138
Amidino penicillins	114	Antiviral agents	139
Temocillin	115	Amantadine	139
Monocyclic beta-lactams	115	Compounds active against Herpesvirus	139
Aminoglycoside antibiotics	116	Nucleoside analogues	139
Streptomycin	116	Arabinosides	140
Neomycin and kanamycin	116	Acyclo-guanosine	140
Gentamicin	116	Phosphono compounds	140
		Methisazone	140

Introductory

Antibiotics are substances that are obtained from micro-organisms and are able, even in high dilution, to inhibit or kill other micro-organisms. Chemotherapeutic substances have similar activity in high dilution but they are obtained by chemical processes rather than derived from biological sources. Benzyl penicillin is an antibiotic, sulphadimidine a chemotherapeutic agent. This simple distinction has been complicated by two facts. Firstly an antibiotic with a simple chemical formula such as chloramphenicol may be easier to produce chemically than to obtain by fermentation. Secondly, many substances now in use are derived initially from fermentation and are then chemically modified to increase activity or modify the behaviour *in vivo*. These semi-synthetic substances are therefore part antibiotic, part chemotherapeutic; consequently their nomenclature is rather imprecise. The term 'anti-microbial agent' covers all substances whether naturally occurring, synthesized or semi-synthetic but despite its length is still not very precise as it includes other substances such as disinfectants in remit. Antibiotic is now often used as a trivial name to cover all the substances except disinfectants and will be so used in this chapter. The term chemotherapy is used to cover treatment whether this is with a chemotherapeutic or antibiotic substance.

Antibiotics differ from disinfectants in being precisely directed against specific targets in the micro-organisms and are therefore active in very low concentrations, i.e. mg per litre concentrations. Because many of the targets are absent from mammalian cells the direct toxicity of antibiotics is low, whereas disinfectants are general poisons which affect host as well as parasite functions. There are only a few cases where the specific target has been identified and an antibiotic designed to attack that site. More usually a substance is discovered to have antibiotic action either incidentally or as a result of a deliberate search for such compounds and the exact targets are unravelled only after further intensive research. It is only in the last five years that the proteins which are bound by penicillin and lead to cell wall disruption have been recognized although penicillin has been in use for 40 years. In the search for antiviral agents much work has been carried out to find possible targets in the virus which might be open to attack by antibiotic. Even with viral chemotherapy however the alternative approach of screening extracts and compounds for antiviral action has so far produced more success than efforts carried out at the molecular level. After an organism producing substances with antibiotic action has been detected much effort is given to increasing the field of the agent by the producing strain, e.g. the series of Wisconsin mutants of *Penicillin crysogenum* and also to modifying the structure to increase activity.

The development of chemotherapy

Folk remedies against infection with protozoa such as amoebae (ipecachuana) and malaria (quinine bark) had been used for centuries before a more scientific approach to chemotherapy was instituted by Paul Ehrlich. He investigated a series of organic arsenical compounds in order to find one which was far more active on the infecting organisms than upon the host, i.e. had a high therapeutic index. This culminated in the discovery of salvarsan (compound 606) which was highly active against *Tr. pallidum* and other spirochaetes. The specific biochemical differences between bacteria and man were important factors in the selective toxicity of salvarsan for spirochaetes rather than the mammalian cells and this pointed the way towards a search for specific targets. Unfortunately the progress in this direction was slow and the next agents to be found were discovered more by accident than design. The early work with *Penicillium* cultures carried out by Lister is detailed by Selwyn (1980), and the work of Fleming by Hare (1970) and Abraham (1971). The first publication on the use of penicillin was on its value as a selective agent to isolate *H. influenzae* from sputum (Fleming, 1929) and over a decade elapsed before the therapeutic use in man was established (Chain *et al.* 1940).

Many other antibiotics were subsequently isolated by screening moulds for antibiotic production. Waksman isolated streptomycin from *Streptomyces griseus* and this was found to be active against *Myco. tuberculosis* (Schatz and Waksman 1944). Florey *et al.* (1960) should be consulted for other details of the early development of antibiotics.

The use of dyestuffs as killing agents for bacteria led to the use of dyes to treat experimental human infections. One such dye, prontosil rubrum, had a protective effect in mice fatally infected with beta-haemolytic streptococci. There was some delay in the knowledge of these findings becoming widespread because the dye itself was not active *in vitro* against streptococci. Apart from underlining the value of *in vivo* experimentation in conjunction with work in test-tubes, the difficulty in explaining the findings produced an unfortunate delay until Domagk's verbal presentations in London in 1936 and his publication in the previous year. Within a short period Tréfouël and Nitti (1935) had elucidated the splitting of prontosil *in vivo* to produce the active agent, sulphanilamide, and Woods (1940) had shown that interference with folate metabolism was caused by the similarity between sulphanilamide and para-aminobenzoic acid. Man uses preformed folate which is absorbed from the intestine whereas bacteria need to manufacture their own supply. The metabolism of the organism was therefore seriously compromised while man was unaffected in that particular way. One might wish that the metabolic difference between man and bacteria might have pointed the way to design of a folic acid antagonist such as sulphonamide but, as with almost all antibiotic substances, the observed biological effects preceded elucidation of the mode of action.

By 1940 both sulphonamides and penicillin had indicated that major therapeutic advances had occurred and since then the development of antibiotic substances has been rapid. There are now many thousands of such substances available for study and many have already established a place in treatment of infection in man, in animals and to some extent in plants. Classification of the heterogeneous collection of compounds now in use presents some difficulty. Many efforts have been made on a purely chemical basis and systems of increasing complexity have been developed. A chemical basis of classification is scientifically the most useful, particularly as the structure of a compound bears considerable relation to its activity both in range and intensity. For a microbiologist, however, the classification requires to be based mainly upon what types of micro-organisms are inhibited or killed, i.e. the classification is based on activity rather than structure. The classification used in this chapter will therefore also be mainly based on activity but whenever relevant the structures related to the variable activities of the compounds are presented. Therefore the classification of antibiotics shown in Table 5.1 is derived from the spectrum of activity of the agents and subdivided into the different chemical structures with that range of activity. In subsequent tables and figures the relation of the structure to the activity is made more apparent.

Resistance to antibiotics

Antibiotics have a defined spectrum of activity. Some bacteria are naturally resistant to some antibiotics and naturally sensitive to others. Table 5.2 shows examples of natural resistance to bacteria. The use of antibiotics will eliminate or at least reduce the numbers of sensitive bacteria and the ecological niche created will be filled by bacteria which are naturally resistant to that antibiotic. It is therefore fairly common to see the mouths of patients receiving tetracycline becoming colonized by *Proteus mirabilis*, those receiving ampicillin colonized by *Klebsiella aerogenes*, and those receiving cephalosporins colonized by *Ps. aeruginosa*. These organisms are common, especially in hospital, or can be transferred by contact with attendants or apparatus. Because of the antibiotic therapy they are at an advantage in replacing the normal oral microbial flora. It is not surprising that micro-organisms which are able to adapt to adverse conditions can acquire resistance to antibiotics to which they are usually

Table 5.1 Classification of antimicrobial agents based on their spectrum of activity

Group 1 Active against gram-positive bacteria and gram-negative cocci	Group 2 Active mainly against gram-negative bacilli	Group 3 Broad spectrum antibiotics	Group 4 Specific antibacterials	Group 5 Antifungal; antiviral agents

Standard penicillins
  Benzyl penicillin
  Phenoxymethylpenicillin
  Other oral penicillins
Antistaphylococcal penicillins
  Methicillin
  Cloxacillin
  Flucloxacillin
Erythromycin and other macrolides
Lincomycin ⎫ (not Neisseria)
Clindamycin ⎭
Rifampicin
Vancomycin
Bacitracin
	2a. *For systemic infections*
	Penicillins
	  Aminopenicillins
	  Carboxypenicillins
	  Acylureidopenicillins
	  Amidinopenicillins
	  Temocillin
	  Monobactams
	Aminoglycosides
	  Streptomycin
	  Neomycin
	  Kanamycin
	  Gentamicin
	  Tobramycin, etc.
	Polymyxins
	
	2b. *For urinary tract infections*
	Nitrofurantoin
	Nalidixic Acid
	Cinoxacin
	Mandelamine
	Hexamine
		Sulphonamides
		  Sulphadimidine
		  Sulphadiazine
		  Sulphafurazole
		  etc.
		Cotrimoxazole
		Cephalosporins
		  Cephaloridine
		  Cephalothin
		  etc.
		Thienamycin
		Beta-lactamase inhibitors
		  Clavulanic acid
		  Sulbactam
		Tetracycline
		Oxytetracycline
		etc.
		Chloramphenicol
			4a. *Anaerobic organisms*
			Lincomycin (clindamycin)
			Nitroimidazoles
			  Metronidazole
			  etc.
			
			4b. *Tuberculosis*
			Streptomycin
			PAS
			Isoniazid
			Rifampicin
			Ethionamide
			Pyrazinamide
			Ethambutol
			Cycloserine
			Thioacetazone
			
			4c. *Chlamydia; Rickettsia Mycoplasma*
			Erythromycin
			Rifampicin
			Tetracycline
			Chloramphenicol
				5a. *Antifungal agents*
				Polyenes
				  Nystatin
				  Amphotericin B
				  Candicidin
				5-Fluorocytosine
				Imidazoles
				  Clotrimazole
				  Miconazole
				  Econazole
				Griseofulvin
				
				5b. *Antiviral agents*
				Methisazone
				Idoxuridine
				Amantadine
				Arabinosides
				Acycloguanides
				Phosphono compounds

Table 5.2 The natural susceptibility of some groups of micro-organisms to the major groups of antibiotics

	Streptococci	Neisseria	Staphylococci	Enterobacteria	Bacteroides[4]	Haemophilic[5]	Pseudomonas	Mycobacteria	Mycoplasma; Chlamydia	Fungi
Benzyl penicillin	S	S	S[1]	R	R	R	R	R	R	R
'Gram-negative' penicillins	S	S	S[1]	S	R	S	R	R	R	R
Cephalosporins	S	S	S	S	R	R	R	R	R	R
Aminoglycosides	R	R	S	S	R	R	S	S[2]	R	R
Polymyxins	R	R	R	S	R	R	S	R	R	R
Macrolides	S	S[3]	S	R	S	S	R	R	S	R
Tetracycline	S	S	S	S	S	S	R	R	S	R
Chloramphenicol	S	S	S	S	S	S	R	R	S	R
Sulphonamides	S	S	S	S	S	S	R	R	R	R
Rifamycins	S	S	S	R	S	R	S	S	S	R
Nitroimidazoles	R	R	R	R	S	R	R	R	R	R
Polyenes	R	R	R	R	R	R	R	R	S	S

[1] Very few staphylococci are now in the natural state and most are resistant to benzyl penicillin and the anti-gram-negative penicillins
[2] Particularly streptomycin
[3] Not lincomycins
[4] Not oral strains
[5] Comprising Haemophilus, Bordetella, Legionella, Campylobacter

susceptible. The mechanisms of acquisition are dealt with elsewhere in detail. The result of the acquisition of additional genetic material, however, is that the organism now has additional biochemical mechanisms to disturb antibiotic action. The mechanisms that may be present are (1) alteration of the target organ in the microbial cell; (2) exclusion of the antibiotic from the cell; (3) elaboration of enzymes capable of destroying or modifying the antibiotic; (4) by-pass of the blocked reaction by a different pathway.

The mechanisms which operate against different antibiotics are considered in the section describing that antibiotic but examples of each mechanism are given here. Streptomycin acts by becoming bound on to proteins derived from 30s subunit of the ribosomes (Davies 1964, Ozaki et al. 1969). Cells that have acquired resistance produce the protein in an altered form no longer capable of binding streptomycin. Resistance of Esch. coli to tetracycline is associated with an ability of the cell to synthesize novel envelope proteins leading to accumulation of tetracycline in the envelope and exclusion from the ribosomes (Chopra et al. 1981). Many bacteria acquire enzymes capable of destroying antibiotics; the classical example is the wide range of beta-lactamases capable of opening the beta-lactam ring of cephalosporins and penicillins. All organisms produce a small amount of beta-lactamase which is chromosomally mediated and can be induced by the presence of a beta-lactam antibiotic to increase its production of beta-lactamase. These enzymes show differences between species and are relatively species-specific, so much so that they have been used as an identification marker for species (Matthew and Harris 1976). In some organisms production of a large amount of beta-lactamase is part of the constitution of the cell. Of more importance for antibiotic resistance are the various plasmid-mediated beta-lactamases which are transmissible between bacterial species and are generally produced in large enough amounts to confer a high degree of resistance. Aminoglycoside antibiotics and chloramphenicol are also capable of being modified enzymically by bacteria by a wide variety of bacterial enzymes. Resistance to trimethoprim can be brought about by modifying the metabolic pathway which is interrupted by trimethoprim.

Epidemiological aspects of resistance

Antibiotic resistance is increasing among almost all bacterial species. The epidemiological factors in this increase are complex and should be considered under three sections (Table 5.3). Firstly there are the acquisition episodes by which bacteria receive their resistance genes. These are considered elsewhere in this volume (Ch. 6). Secondly, the biochemical mechanisms the organisms use to display their resistance which have already been briefly mentioned. Thirdly there is the problem of spread of resistance in bacteria which might give rise to epidemic events in the com-

Table 5.3 Relevant factors in the epidemiology of antibiotic resistance

1. *Acquisition of resistance* *Chromosomal* 　Mutation 　Transduction 　Transformation *Extrachromosomal (plasmids)* 　Transduction 　Transfer 　(Transposition) 2. *Mechanisms of resistance* Barriers to penetration by antibiotics Antibiotic-modifying enzymes Alteration at target-site in the organism Bypass of the target-site	3. *Means of dissemination of resistant organisms* Case to case transfer Acquisition from carriers Acquisition from the inanimate environment (including food) 4. *Factors influencing the spread of resistant organisms* Virulence and transmissibility of the organism Use of antibiotics Opportunities for infection in hospital

munity and the hospital and the factors which influence this spread.

Two types of clinical problems may arise from antibiotic-resistant bacteria. There may be a major outbreak or epidemic of illness caused by a single antibiotic-resistant species which spreads widely in the community or hospital producing severe infection with high mortality. The second type of problem is the spread of resistance among organisms which are not responsible for epidemic disease. They may, however, cause large numbers of single episodes of infection creating treatment problems for individual patients. Organisms such as *Esch. coli, H. influenzae* and *N. gonorrhoeae* are examples of such organisms. In addition the acquisition of antibiotic resistance in the normal flora of man and animals provides a gene pool of resistance which can either transfer to more virulent species or may themselves give rise to infection in suitable immunologically defective patients.

Epidemic diseases

Acquisition of resistance by organisms of high virulence for man has given rise to major public health problems. Details of the epidemiology of *Sh. dysenteriae* 1 are given in Chapters 36 and 69, of chloramphenicol-resistant *Salm. typhi* in Chapters 37 and 68, and tetracycline-resistant *Vibrio cholerae* in Chapter 27.

Endemic diseases

After only a few years of benzyl penicillin penicillinase-producing *Staph. aureus* became a problem in hospital (Barber and Rozwadoska-Dawzenko 1948). After the introduction of other antibiotics such as chloramphenicol, tetracycline, erythromycin and novobiocin active against staphylococci, resistance appeared to these also. The incidence of resistant strains in hospital patients is often closely related to the usage of a particular agent (Knight and Holzer 1954, Alder and Gillespie 1967). Sometimes a lag phase may occur before the resistant strains appear; neomycin-resistant strains emerged only after nine years of neomycin use and gentamicin-resistant strains after a similar period (Shanson 1980). Further association of antibiotic use and antibiotic resistance are given by Ayliffe (1973) and Mouton (1976). A full account is given in Chapter 60.

Ampicillin has been one of the main antibiotics used against *H. influenzae* since 1960. In 1974 the first report of ampicillin resistance appeared (Gunn *et al.* 1974), a resistance which was later shown to be due to production of beta-lactamase (Williams *et al.* 1974) of a type similar to TEM-1. The proportion of beta-lactamase-producing strains has continued to increase from 1.5 per cent in 1977 to 6.5 per cent in 1981 (Howard *et al.* 1978, Philpott-Howard and Williams 1982). Hospital epidemics are unusual, although three children in an orphanage in Bangkok died of influenzal meningitis and the organism isolated in each case was beta-lactamase-producing and chloramphenicol resistant. The carriage rate of the strain in the other children was 17 per cent—38 out of 219 children (Simasathien *et al.* 1980). Low-level resistance which is found in beta-lactamase non-producers is unusual—less than 0.5 per cent of isolates.

With *N. gonorrhoeae* low-level resistance to benzyl penicillin has become fairly common and currently is found in approximately 10 per cent of isolates (Seth and Johnston 1980). Production of beta-lactamase was first detected in 1976 in returning visitors from West Africa and the Far East (Phillips 1976, Percival *et al.* 1976). In some countries in South-East Asia about half the strains of gonococci isolated are producers of beta-lactamase.

Resistance of *Str. pneumoniae* to tetracycline was first recognized by Evans and Hansman in 1963. The distribution of these strains in the UK is patchy with an incidence of 23 per cent of pneumococcal strains resistant to tetracycline in Liverpool (Percival 1972) whereas in more rural areas resistance occurs in only 2 per cent of strains.

Resistance of pneumococci to benzyl penicillin was reported from South Africa in 1977 (Appelbaum *et al.* 1977). The organisms were isolated from five children in Durban, three of whom died of meningitis and two of whom recovered after treatment with rifampicin and high dose ampicillin. Subsequent monitoring of other children in the community from which the index cases were drawn uncovered a large number of symptomless carriers of the penicillin-resistant strain.

Klebsiella spp. resistant to several antibiotics provide a major reservoir of resistant genes in the gastro-intestinal tract of patients in hospital. Many of the organisms which are isolated from these patients do not come from infected areas but merely indicate colonization of the patient by hospital strains. Of 50 patients who had coliform bacilli isolated from the sputum or bronchial washings while in an intensive care unit, none showed evidence of pneumonia (French and Horn 1979). Nevertheless faecal organisms have been involved in many outbreaks of infection (Casewell *et al.* 1977, Currie *et al.* 1978, Houang *et al.* 1979). Hospital food can provide a starting point for colonization of the gastro-intestinal tract (Montgomerie *et al.* 1970). The spread of *Klebsiella* sp. and other gram-negative rods can be partly explained by faecal contamination, but the organisms can also be found on the skin of patients' hands and body and colonizing urine drainage bags. Antibiotic therapy increases nasopharyngeal, hand and gut carriage rates of klebsiellae and other species (Rose and Schrier 1968, Pollock *et al.* 1972, Haverhorn and Michael 1979).

Antibiotic resistance is very closely linked to antibiotic use, possibly to some degree to use in animals but predominantly to use in man. The use of antibiotics in feed additives for growth-promotion of pigs, poultry and some other farm animals results in the emergence of resistant bacteria in the gut flora of animals. In some instances resistant strains may pass from animals to man (Anderson *et al.* 1961) but most resistant strains colonizing or infecting man do not have an animal origin. Therefore even the total prohibition of antibiotic use in animals would have little effect on the total sum of antibiotic resistances encountered in bacteria infecting man.

Assessment of activity of antibiotics against micro-organisms

There are many laboratory techniques for determining the *in-vitro* susceptibility of micro-organisms to antibiotics. Because of the large number of antibiotics now available and the wide range of micro-organisms, from viruses to fungi which are possibly susceptible, the techniques vary in complexity. Some features are common to all techniques however and in particular the attempts to express the activity of an antibiotic in numerical terms.

The **minimum inhibitory concentration** (MIC) of an antibiotic for a particular strain of micro-organisms is generally taken as that concentration which totally (or almost totally) inhibits that organism. When tests are carried out in doubling dilutions of agar or broth and an inoculum of 10^3 to 10^4 bacteria is used a sharp end point to the titration is obtained. Provided the same culture medium, culture conditions and inoculum size are used this end point is reproducible. One problem is that variations in technique, inoculum size, salt content, temperature or other conditions may bring about wide fluctuations in the MIC. There is no standard, agreed method for determining MICs. Many microbiologists now make up the antibiotic concentration in doubling dilutions based upon 1 (rather than 0.1 or 10, or 100) and this at least allows different workers to compare the results of tests with the same set of numbers. Antibiotics have an effect on micro-organisms at concentrations much lower than the total MIC. The lowest concentration at which they begin to act is sometimes called the MAC (minimum active concentration) and by others as the MIC(S)—subtotal to distinguish from the MIC(T)—total. Antibiotics acting *in-vivo* at sub-total MIC levels may have a role in eradication of infection in a person with normal host defences; on the other hand antibiotic activity at MIC(S) levels may favour the emergence of antibiotic-resistant bacteria.

Minimum bactericidal concentrations (MBCs) of antibiotics may be determined by subculturing organisms from liquid media containing an antibiotic to antibiotic-free media and determining the number of survivors. For some antibiotics, e.g. aminoglycosides, the bactericidal concentrations are almost the same as total MIC levels whereas for others such as chloramphenicol, bactericidal levels may be many-fold greater than inhibitory levels. Micro-organisms vary also in their susceptibility to cidal action of antibiotics. There are variations between species (e.g. *Str. faecalis* is more resistant to the bactericidal activity of penicillins than *Str. pneumoniae*) or within strains of the same species. Some strains of staphylococci for example, are killed at concentrations of methicillin only slightly higher than the MIC whereas other strains are tolerant of levels greatly in excess of the MIC (Sabath *et al.* 1977).

The methods and materials for determining inhibitory and bactericidal concentrations of antibiotics for micro-organisms vary according to the antibiotic and the organisms and the original papers should be consulted for details (Reeves *et al.* 1978, Lorian 1980, Waterworth 1981).

Antibiotic-impregnated paper discs are widely used in clinical microbiology laboratories for the determination of susceptibility of bacteria and in order to predict the outcome of therapy and guide the correct choice of antibiotic therapy. Several attempts have been made to establish an internationally recognized standard method for carrying out such tests (WHO Tech. Rep. Series 1978). Unfortunately there is no standard test which is suitable for testing all bacteria against all antibiotics. As the types of media and methods for disc testing increase in complexity it is clear that with some organisms and some antibiotics the differences between susceptible and resistant are so great that most types of simple test will give the 'correct' answer whereas other organisms and antibiotics are more difficult to assess. Many systems are in use—the International Collaborative Study (ICS)/Ericsson method, the comparative method, the Stokes method and the modified Kirby Bauer method. The four systems were compared by Brown and Blowers (1978) and were found to give similar results. All systems however have to be supplemented either by modification of the techniques or by employing other tests when evaluating the susceptibility of some organisms and some antibiotics. The books by Reeves et al. (1978) or Lorian (1980) should be consulted for detailed methodology.

Antibiotics in clinical use

The beta-lactam antibiotics

Fleming (1929) observed the suppression of the growth of *Staph. aureus* around a colony of a *Penicillium* that had contaminated a nutrient agar plate. The inhibitory substance, to which he gave the name penicillin, could be extracted from cultures of the organism and was highly active against certain groups of bacteria, especially the gram-positive cocci. The crude preparations of penicillin were non-toxic to leucocytes in antibacterial concentrations. However the amount of penicillin in culture filtrates was low and, owing to its instability, attempts at that time to concentrate it proved abortive (Fleming 1932). A few years later Chain and his colleagues (Chain et al. 1940, Abraham et al. 1941) re-investigated the properties of penicillin, and we owe to them its extraction, its production in bulk, its concentration in a stable, highly active form, the methods of its assay, and the definitions of its efficacy and limitations in the treatment of animal and human infections. For an early review of the properties of penicillin see Florey et al. (1960).

During the first 20 years which followed the clinical uses of penicillin, development of new compounds based on the beta-lactamases took place slowly but in the 20 years from 1960-80 many different types of penicillin became available. The cephalosporins and many other compounds also owe their activity to the beta-lactam structure. The several types of beta-lactam antibiotics are described throughout this chapter in an order relative to the spectrum of activity but some general classification of beta-lactam antibiotics is needed before describing the first members of the group. Three main subdivisions can be used based upon the chemical structure: (1) penicillins (p. 105)—general structures shown in Fig. 5.1; (2) cephalosporins (p. 125)—general structures shown in Fig. 5.13; and (3) for want of a generic name what I have termed low molecular weight beta-lactams (p. 103). The group includes the thienamycins (p. 131), beta-lactamase inhibitors (p. 131) and monocyclic beta-lactams (p. 115). The chemical structure of penicillins is based on a two-ringed nucleus of 6-aminopenicillanic acid—a thiazolidine ring attached to an unstable beta-lactam ring. In cephalosporins and related compounds a dihydrothiazine ring replaces thiazolidine in the central nucleus—7-aminocephalosporanic acid. A variety of ring structures or side chains is attached to the beta-lactam ring in the third group of beta-lactam antibiotics described here. An outline of the classification of beta-lactam antibiotics is shown in Table 5.4 and further details of classification of penicillins in Table 5.5.

Table 5.4 Classification of beta-lactam antibiotics

Penicillins	*Cephalosporins*	*Low molecular weight beta-lactams*
1. Standard penicillins Penicillin of Fleming	1. Oral group 2. Anti-gram-positive	1. Thienamycin 2. Beta-lactamase inhibitors

2. Antistaphylococcal penicillins
    (stable to staphylococcal
    beta-lactams)
	3. Anti-enterobacterial
	4. Anti-pseudomonal
	5. Cephamycins
		3. Monocyclic beta-lactams

3. Anti-'gram- negative' penicillins
    (a) destroyed by 'gram- negative'
        beta-lactamases
    (b) stable to some 'gram-negative'
        beta-lactamases		

In this classification general usage has been followed in dividing the compounds mainly on the basis of their antibacterial activity and chemical configuration. All cephalosporins show a broad spectrum of activity but it is becoming clear that there is considerable variation in the degree of activity against different groups of bacteria. For this reason the names of the groups of cephalosporins used in this classification are intended to highlight the most important antibacterial activity of that subgroup.

Table 5.5 Classification of penicillins

This grouping is based upon the activity of the penicillins against gram-positive and gram-negative bacteria and on their stability to gram-positive and gram-negative beta-lactamases. With each group (and subgroup of anti-gram-negative penicillins) there are oral and parenteral compounds, analogues, esters and complexes which have a similar antibacterial spectrum to the parent compound.

Groups of penicillins	Type of compound	Analogues, complexes and esters of similar antibacterial spectrum
Group I Standard penicillins. Highly active against cocci, both positive and negative, and against gram-positive rods. Susceptible to staphylococcal penicillinase (therefore 90% of clinical isolates of *Staph. aureus* are resistant). Very high activity against streptococci and *Neisseria*. Barrier type resistance rare in *Str. pneumoniae* and enzyme type resistance increasing in *N. gonorrhoeae*.	Penicillin G (benzyl penicillin) given by parenteral route	V analogues which are absorbed after oral administration include penicillin V. Procaine penicillin and benzathine penicillins are complexes which slow down absorption from intramuscular injection sites.
Group II Antistaphylococcal penicillins. Penicillins which are stable to staphylococcal penicillinase with MIC range 1–2 mg/l. Reduced activity against the range of organisms susceptible to standard penicillins and used only for treatment of staphylococcal infection. Methicillin resistance in staphylococci, show cross-resistance to all other members of the group.	(a) Methicillin. The original parenterally administered penicillinase-stable penicillin (b) Cloxacillin. Acid-stable isoxazole penicillin which is absorbed after oral administration	Analogue—nafcillin. Analogues—flucloxacillin, oxacillin, dicloxacillin
Group III Penicillins which are active against gram-negative bacteria but susceptible to hydrolysis by gram-negative and gram-positive beta-lactamases. There are four types of penicillins in this group which have different chemical substituents at position 6 of the penicillin nucleus. The substituent gives the name to the subgroup and conveys a characteristic anti-bacterial spectrum. Within each subgroup cross-resistance occurs with other members of that subgroup and in some instances to other Group III compounds. The analogues and ester modifications have marginal effects on antibacterial activity but alter the pharmacology compared to the parent compound.	(a) Aminopenicillins Ampicillin—activity similar to benzyl penicillin but additional activity against *Esch. coli, Proteus, Str. faecalis, H. influenzae* and salmonellae. Rather poorly absorbed orally. (b) *Carboxypenicillins* Carbenicillin—extends the antibacterial range of ampicillin to include *Ps. aeruginosa* and indole-positive *Proteus* sp. Not absorbed after oral administration. (c) Acylureidopenicillins Mezlocillin—extends the range of ampicillin to include Klebsiella and other resistant gram-negative rods. Higher activity than ampicillin. Not absorbed orally. (d) *Amidinopenicillins* Mecillinam—has similar anti-gram-negative spectrum to ampicillin but minimal activity against gram-positive organisms.	Esters of ampicillin which hydrolyse after absorption to release ampicillin including talampicillin, pivaloylampicillin and bacampicillin. Closely related analogues are amoxycillin and epicillin. Esters of carbenicillin which are hydrolysed after absorption include carfecillin and indanyl carbenicillin. Closely related analogue is ticarcillin which is not absorbed orally. No esters are available. One closely related analogue—azlocillin—shows fairly high activity against *Ps. aeruginosa*. Piperacillin is an analogue with some activity against *Bacteroides fragilis* and *Ps. aeruginosa*. One ester—pivaloylmecillinam—is absorbed after oral administration.
Group IV Penicillins which are stable to gram-negative beta-lactamases. Unlike the situation with cephalosporins, the development of such penicillins has been slow and only one agent has so far been described.	Temocillin—this compound shows antibacterial activity similar to ticarcillin against gram-negative rods without beta-lactamases but much higher activity than ticarcillin against beta-lactamase-producing strains.	None

Antibiotics active mainly against gram-positive organisms

Benzyl penicillin and related acid-stable penicillins

The structure of benzyl penicillin and other acid-susceptible or 'standard' penicillins is shown in Fig. 5.1. The nucleus is 6-aminopenicillanic acid (6APA), a dipeptide of valine and cysteine joined to give a four-membered beta-lactam ring containing nitrogen and a five-membered ring containing sulphur. The properties of penicillins depend on the side chain in the carbon atom in the 6-beta position. Benzyl penicillin is destroyed to a large extent by gastric acid and only a small proportion is absorbed after oral administration. The introduction of an oxygen link to the benzyl side chain confers acid-stability and produces compounds such as phenoxymethyl penicillin (penicillin V) which is absorbed when taken orally. Other modifications of the pharmacological properties of benzyl penicillin can be effected by complexing it with procaine or benzathine (Fig. 5.1) in order to slow its absorption from intramuscular deposits and so limit the number of injections which have to be given.

The antibacterial spectrum of all the above-mentioned penicillins is the same. They are active against gram-positive and gram-negative cocci and against gram-positive bacilli. They act by interfering with the synthesis of the cell wall peptidoglycan presumably because it is a structural analogue of the amino acids concerned in the synthetic processes. A series of enzymes acts in the process of cell wall synthesis and several of these have been found to become bound to penicillins, which then lower the effectiveness of the bacterial enzymes.

The range of activity of benzyl penicillin has been altered by the emergence of resistance in some bacterial species but the number of susceptible strains remains high. Furthermore the intrinsic activity of benzyl penicillin is higher than that of most other penicillins, so that organisms, if they are susceptible at all, are highly susceptible. Modification of the side chains to increase the spectrum usually reduces the antibacterial activity. *Staph. aureus* is naturally inhibited by approximately 0.03 mg of benzyl penicillin per litre. At the present time however 90 per cent of strains of *Staph. aureus* produce a diffusible beta-lactamase capable of destroying benzyl penicillin. It is now usual for *Staph. aureus* infections to be unresponsive to therapy with standard penicillins. Of the other gram-positive cocci, *Streptococcus faecalis* strains are moderately resistant but beta-haemolytic streptococci and *Str. pneumoniae* are susceptible to 0.03 mg/l. A few isolations of *Str. pneumoniae* resistant to penicillin have been made in South Africa and the United States but these are rare events. Other alpha-haemolytic streptococci are more variable in susceptibility. In persons not receiving penicillin therapy the oral streptococci are highly susceptible to benzyl penicillin but after a few days of therapy penicillin-resistant strains predominate in the oral flora. Of the gram-negative cocci *N. meningitidis* remains highly susceptible to benzyl penicillin but resistance has been encountered in *N. gonorrhoeae* and *Branhamella catarrhalis*. The resistance of *N. gonorrhoeae* can be due to a barrier to the penetration of penicillins raising the minimum inhibitory concentration (MIC) from 0.03 to about 0.5 mg/l, a level at which clinical infections remain curable, albeit at increased dosage. Other strains of *N. gonorrhoeae* have acquired plasmids that determine the production of large quantities of gram-negative beta-lactamase, which makes the infection unresponsive to all penicillins and cephalosporins that are susceptible to hydrolysis by gram-negative beta-lactamases. *Branhamella catarrhalis* has also acquired beta-lactamase production in recent years.

Gram-positive rods, including corynebacteria and clostridia, remain generally highly susceptible to benzyl penicillin. Occasional strains show increased resistance but beta-lactamase production has not been reported to date in these species. Gram-negative bacilli are inhibited only by high concentrations of standard penicillins which have little or no role in therapy of infection with these organisms.

Hypersensitivity

Penicillins show little direct toxicity unless very large doses are used. The occurrence of hypersensitivity reactions however is a prominent hazard of penicillin therapy. Some allergic manifestations are due to impurities in the preparations but the penicillin molecule itself and the derivatives can combine with protein to produce antigenic compounds (Idsoe *et al.* 1972). The most important derivative is penicilloyl which combines with protein to give the major antigenic determinant. Polymerization of penicillin can also occur after a period in solution and the polymers may also produce allergic reactions (Smith *et al.* 1971). The two most serious reactions are anaphylactic shock which occurs in about 0.004 per cent of patients (Porter and Jick 1977) and may prove fatal if not treated promptly with adrenaline, and serum sickness. Serum sickness is more common than anaphylaxis and is a typical type III reaction occurring 7–10 days after the start of treatment.

6APA and the antistaphylococcal penicillins

Two reports from Japanese workers (Sakaguchi and Murao 1950, Kato 1953) showed that standard penicillins could be deacylated to yield the nucleus 6APA. It was subsequently shown that a wide variety of micro-organisms were capable of carrying out this reaction (Rolinson *et al.* 1960*b*), and by the addition of selected side chains to the nucleus several penicillins

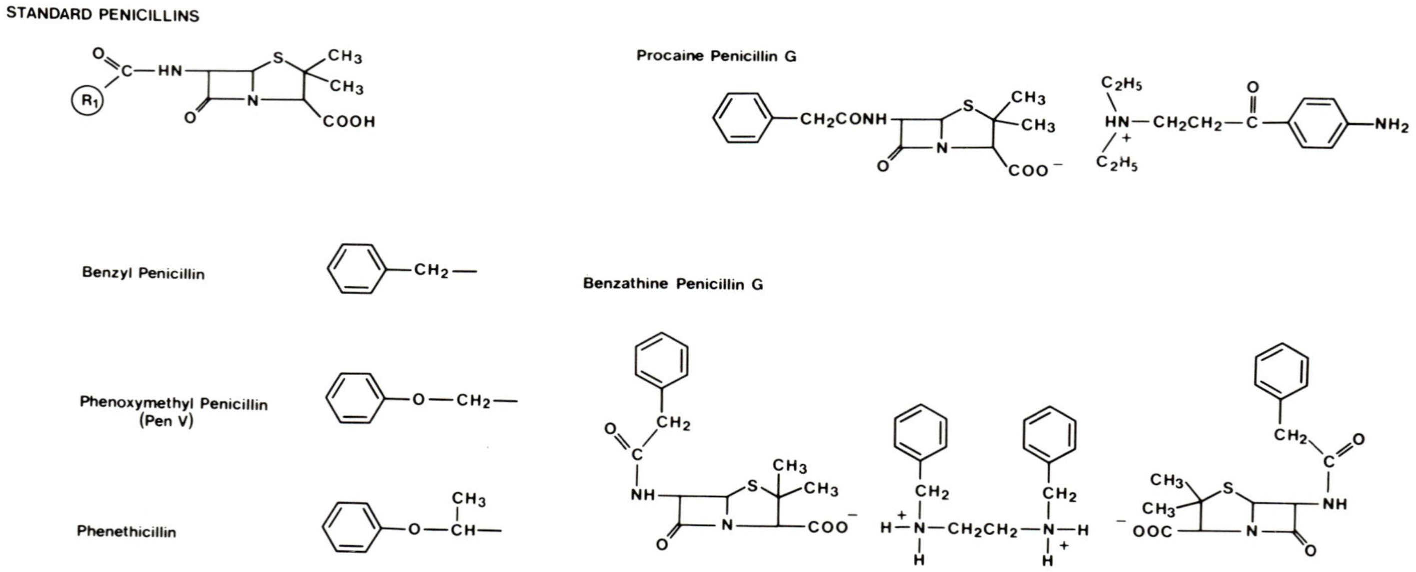

Fig. 5.1 Benzyl penicillin and the orally absorbed analogues of benzyl penicillin; complexes of benzyl penicillin.

ANTISTAPHYLOCOCCAL PENICILLINS

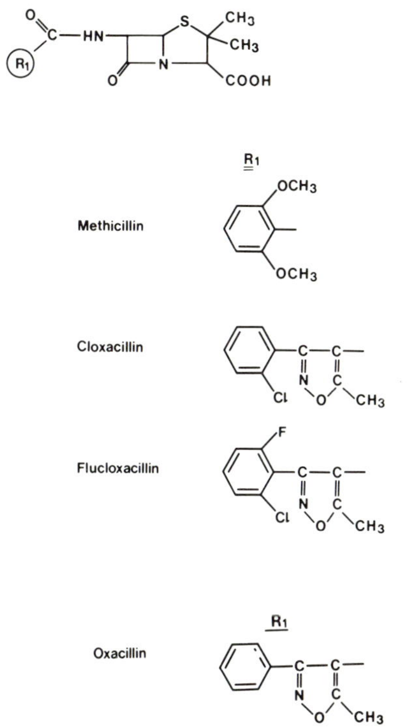

Fig. 5.2 Penicillins which are stable to staphylococcal penicillinases.

have been produced which are resistant to staphylococcal beta-lactamases (Fig. 5.2). The first compound produced was dimethoxy benzyl penicillin (methicillin) which was unstable to gastric acid and required parenteral administration (Rolinson et al. 1960a). Several other acid-stable compounds followed; some of these such as cloxacillin (Knudsen et al. 1962), oxacillin (Doyle et al. 1961) and flucloxacillin (Sutherland et al. 1970) are in widespread use. The side chains are bulky and are supposed to prevent attachment of beta-lactamases to the penicillin. All these compounds show a reduction of activity in comparison with benzyl penicillin; the MIC for a susceptible staphylococcus for example is 0.5 to 1 mg per litre compared to 0.03 mg/l for benzyl penicillin and the activity against other bacteria is similarly reduced. Against penicillinase-producing staphylococci however methicillin-like penicillins have a similar MIC (0.5–1 mg/l) to that exhibited against penicillinase-negative strains. Some strains of staphylococci do show an increased resistance to methicillin by mechanisms other than beta-lactamase production (Jevons 1961). For a short period after the introduction of methicillin it was feared that the emergence of such strains would place some limitation on the usefulness of this substance but, apart from some hospitals where such strains are prevalent, the incidence appears to be less than 2 per cent. This group of penicillins therefore retains its valuable role in treatment of staphylococcal infections.

Penicillins active against gram-negative bacilli are described in a later section (p. 111 and Fig. 5.5).

The macrolides and lincomycin

Although these compounds are chemically dissimilar they are often grouped together because of a similar antibacterial spectrum. The main microbiological difference is that the lincomycins are not active against *Neisseria* and other gram-negative cocci whereas the macrolides show some activity. They are both active against a wide range of gram-positive organisms including streptococci, staphylococci and clostridia. In addition they show activity against *Bacteroides* spp. (see later) and this activity is more predictable with the lincomycins than with the macrolides.

Some other differences in spectrum are seen with respect to small aerobic gram-negative rods against which erythromycin shows *in vitro* activity that is sometimes of clinical value. One problem with macrolides is the relatively poor absorption when they are not esterified, which results in blood levels usually below 2 mg/l. Therefore the *in vitro* activity which is shown against *Haemophilus, Pasteurella, Brucella,* mycoplasmas, rickettsiae and chlamydia is often not realized in clinical use. In the treatment of many of these infections tetracycline is usually preferred. Erythromycin is used however in chest infections which fail to respond to ampicillin and where an unusual organism is suspected, e.g. *Mycoplasma* or *Legionella*.

The macrolides are a group of antibiotics derived from *Streptomyces erythreus* (McGuire et al. 1952) and have the general structure shown in Fig. 5.3. They have a large lactone ring to which two or more sugars or aminosugars are glycosidically linked. Erythromycin A is the most active and most widely used compound.

The clinical use of **erythromycin** for treatment of staphylococcal infection has varied considerably over the years. When first introduced erythromycin was active against most clinical strains of staphylococci and, because its appearance preceded that of methicillin, erythromycin was widely used. Unfortunately resistance soon developed in hospital isolates (McCabe et al. 1961). The resistance is due to an alteration in the 50s ribosomal subunit (Tanaka et al. 1968, Taubman et al. 1966) and may be either constitutive, i.e. present at a high level, or inducible, i.e. promoted by exposure to subinhibitory concentrations of erythromycin. Induction is usually reversible in that the resistance declines when the drug is removed but the inducibility of the strain remains. Furthermore, some strains once induced may remain highly resistant to erythromycin and other macrolides, i.e. acquire a

Erythromycin A

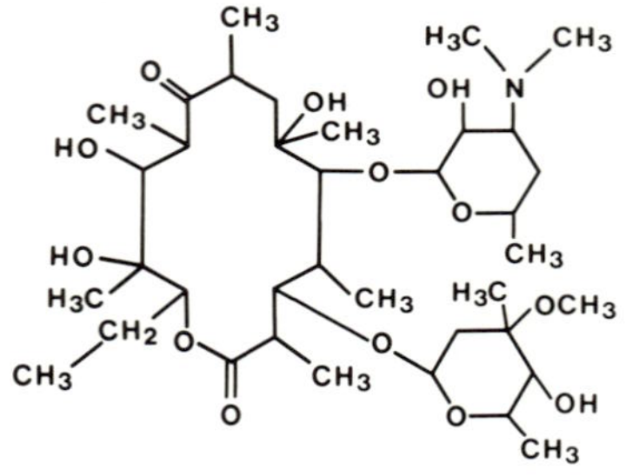

Erythromycin Ethyl Succinate

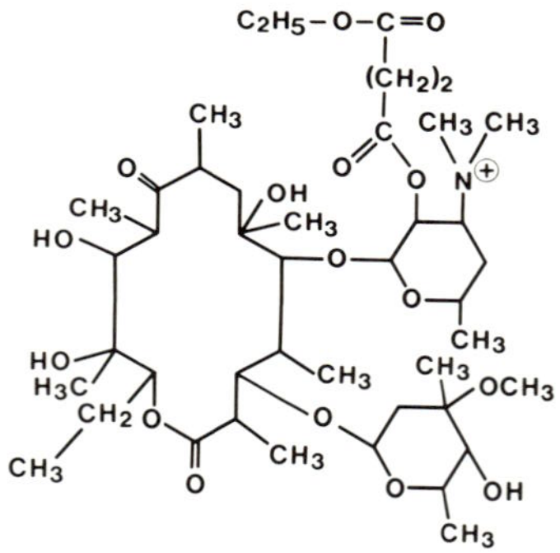

Erythromycin Stearate

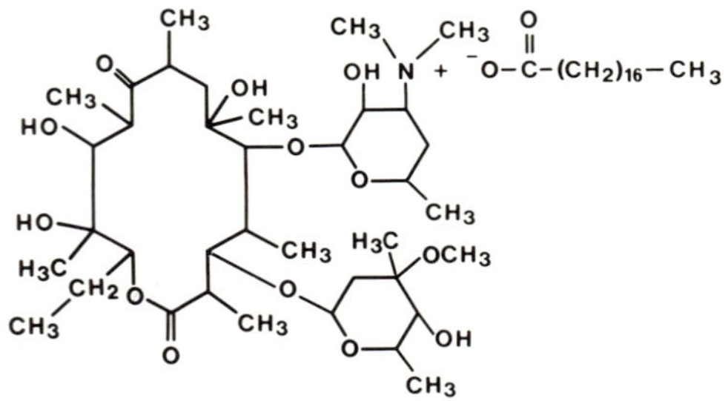

Erythromycin Lactobionate

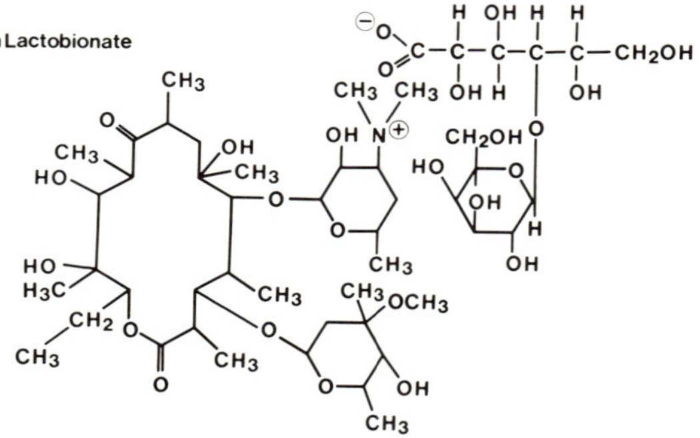

Fig. 5.3 Chemical structures of erythromycin A and its esters.

constitutive resistance. These strains also show the induction of resistance to lincomycin when grown in the presence of erythromycin (Garrod 1957). Strains of staphylococci showing this feature—often called dissociated resistance—may be less amenable to lincomycin therapy although only very few instances of this phenomenon have been reported to occur during therapy (Duncan 1967).

Streptococci (with the exception of *Str. faecalis*) are susceptible to 0.2 mg/l of less of erythromycin and *Staph. aureus* strains (Phillips *et al.* 1970), including those producing penicillinase, are susceptible to 2 mg/l or less. Very few strains of group A streptococci are resistant to erythromycin. Fluctuation in the prevalence of erythromycin-resistant strains of staphylococci has been a feature of hospital epidemiology. The emergence of erythromycin-resistant staphylococci was one factor which led to a sharp diminution in the use of erythromycin and this decline in use became greater after the introduction of lincomycin, fusidic acid and methicillin. The main clinical uses were firstly as alternative therapy in patients with hypersensitivity to penicillins who had an infection which required treatment with penicillin G (Shooter 1969) and secondly in a variety of infections in young children (Cohen *et al.* 1965, Sutherland 1962).

Campylobacter fetus var. *jejuni* is now recognized in many countries to be one of the most prevalent causes of diarrhoeal disease in man (Butzler *et al.* 1973, Skirrow 1977). It is highly susceptible to erythromycin (Butzler *et al.* 1974) which has become widely used in treatment of campylobacter infections. *Legionella pneumophila* causes severe pulmonary infection in man and erythromycin is one of the few antibiotics active against it (Thornsberry *et al.* 1978). There is no clear evidence that erythromycin has an ameliorating effect 50in established whooping cough but the organism is susceptible to 2 mg/l or less of erythromycin and several trials of efficacy are in progress. Erythromycin also shows some *in vitro* activity against *Chlamydia trachomatis* (Ridgway *et al.* 1976, Chan, Kuo Cho *et al.* 1977).

Erythromycin is poorly absorbed from the gut. The peak blood level of about 1.4 mg/l is not reached for 2–3 hours after administration of 500 mg on the fasting stomach (Nightingale *et al.* 1976). Most of the drug is excreted in the faeces, less than 5 per cent being found in the urine (Heilman *et al.* 1952). A proportion of the drug passes directly through the intestine unabsorbed and most of what is absorbed is excreted unchanged into the bile (Bilibin *et al.* 1967). Of considerable interest is the concentration of erythromycin which occurs in the prostatic fluid (Winningham *et al.* 1968, Stamey 1971).

Some esters of erythromycin result in higher blood levels of the antibiotic than is seen with the base (Griffith 1955, Hirsch and Finland 1959). There are two problems which stem from this; the first concerns the site at which hydrolysis of the ester occurs and the second relates to toxicity. After oral administration of the esters, hydrolysis has to occur for the base to be released to have its antibacterial effect (Tardrew *et al.* 1969). The possible sites for this hydrolysis are in the duodenum, in the wall of the small intestine, in the blood, or in the liver. Reports on where the hydrolysis actually occurs appear to be conflicting.

The second problem is the toxicity associated with the esters. Erythromycin is one of the safest of antibiotics (Dunlop and Murdoch 1960), but reports of cholestatic jaundice following administration of erythromycin estolate (Farmer *et al.* 1963, Etteldorf and Crawford 1967) have led to a fall in the use of the estolate and certain other esters.

Lincomycin has a much simpler structure than has erythromycin (Fig. 5.4). It was isolated from *Streptomyces lincolensis* (Mason *et al.* 1962) and has been widely used for treatment of staphylococcal and streptococcal infections. It has found a particular niche in the treatment of staphylococcal osteomyelitis. The activity against *Bacteroides* spp. will be considered later.

Both erythromycin and lincomycin have irregular absorption after oral administration. Many of the analogues of erythromycin which have been produced, e.g. oleandomycin, carbomycin, were only marginally better absorbed and were less active microbiologically. The only analogue of lincomycin which shows an improved absorption combined with increased antibacterial activity is clindamycin (Fig. 5.4). Unfortunately this compound came to be associated with pseudomembranous enterocolitis and the use of both lincomycin and clindamycin has declined considerably since 1974 (see Ch. 58).

Fusidic acid

This antibiotic derived from *Fusidium coccineum* has an unusual steroidal structure (Fig. 5.4) and has a similar spectrum of antibacterial activity to that of benzyl penicillin except in that it is also active against penicillinase-producing staphylococci (Barber and Waterworth 1962).

The main clinical uses have been against infections with penicillinase-producing staphylococci. It is often used in combination with other antistaphylococcal drugs such as lincomycin because of the rapidity with which resistant strains may emerge. These resistances may result from mutational change in the strain or by selection of previously resistant cells present in a lesion due to staphylococci. Fusidic acid inhibits protein synthesis *in vitro* by inhibition of ribosome translocation and the actions have been reviewed by Tanaka *et al.* (1968).

Rifampicin

This antibiotic is best known for its activity against *Myco. tuberculosis* (see p. 135 and Chapters 24 and

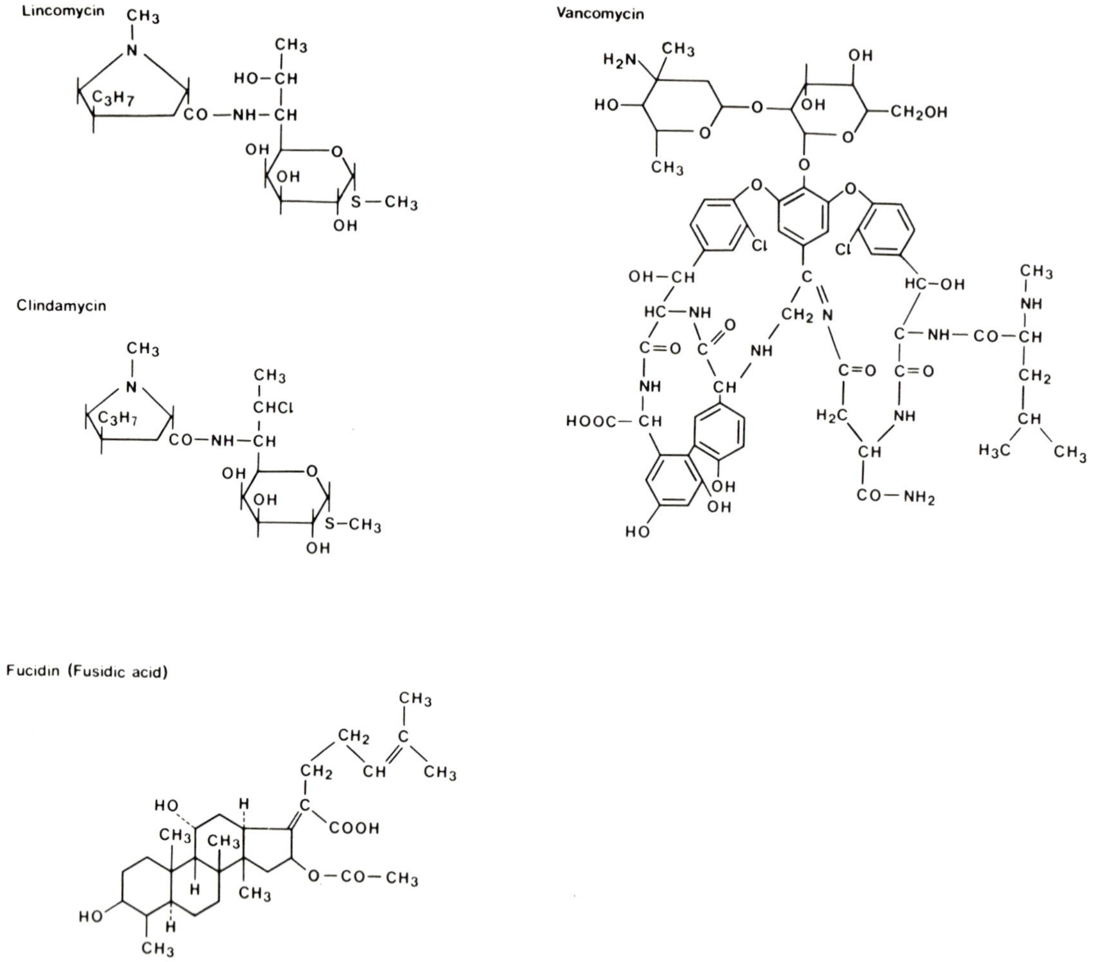

Fig. 5.4 The chemical structures of several antibiotics active against gram-positive cocci—lincomycin, clindamycin, fusidic acid and vancomycin.

51), but apart from this specialized use the antibacterial spectrum is similar to that of benzyl penicillin and is therefore included in the group of antibiotics active against cocci and gram-positive rods. In addition it shows activity against *Bacteroides* spp. Its activity against staphylococci is as high as that of benzyl penicillin, concentrations of 0.05 mg/l being bactericidal for most strains. As with fusidic acid, resistant strains may emerge rapidly *in vitro* (Kunin *et al.* 1969) and in patients during a course of treatment (Bouanchaud and Acar 1969). Rifampicin should therefore not be used alone but only in combination with another antistaphylococcal drug. The activity against *Neisseria meningitidis* is useful in the treatment of carriers of meningococci, especially in geographical areas where sulphonamide-resistant strains of meningococci are prevalent. The activity against gram-negative bacilli is low (20–50 mg/l) and systemic infections with gram-negative rods are not amenable to therapy with rifampicin; but, because of the high excretion in the biliary and urinary tracts, the compound is sometimes used for infections in those regions (Bevan and Williams 1971). Because of the value of rifampicin in the treatment of tuberculosis its use in non-tuberculous infections should be limited.

Rifampicin (Fig. 5.16) is the most widely used of a group of semi-synthetic derivatives of an antibiotic produced by *Streptomyces mediteranei*. These derivatives have greater activity and less toxicity than the parent compound (Sensi *et al.* 1960). One other derivative, rifamide, has the unusual character of being almost wholly excreted into the bile where it reaches a concentration of several thousand mg/l. This compound was used for treatment of biliary-tract infection, but has been replaced by rifampicin in most countries. The drug acts by interfering with RNA synthesis by becoming bound to DNA-dependent RNA polymerase. Resistance is due to an alteration in the enzyme that prevents this binding (Torchini-Valentini et al. 1968).

Rifampicin is well absorbed after oral administration. Rifamide was the original parenteral analogue of rifampicin with a similar antibacterial spectrum but in recent years other parenteral formulations of rifampicin have also become available. The toxicity is low but effects on the liver and immunological disorders have been reported. One unusual side effect is stimulation of the liver enzymes to metabolize oestrogens at a rapid rate, resulting in failure of contraception therapy on occasion.

Several other properties of rifampicin are worthy of note. It is a red, highly coloured compound like other members of the group. It has a wide distribution in the body and being fat-soluble shows remarkable ability to penetrate mammalian cells. It is the antibiotic most effective in attaining high levels inside cells. *In vitro* high concentrations will inhibit vaccinia virus (Subak-Sharpe et al. 1969, Katz et al. 1970) and although these findings have no therapeutic value in man or in animal infections (Engle et al. 1970) it is unique to find an antibiotic capable of inhibiting both bacteria and viruses.

Vancomycin

Vancomycin is a high molecular weight heterocyclic compound isolated from *Streptomyces orientalis* (Fig. 5.4). It is highly active against gram-positive bacteria particularly staphylococci and streptococci. It is active against almost all strains of staphylococci by inhibiting the incorporation of amino acids into the peptidoglycan. The absence of cross-resistance with penicillin indicates that different sites of action are concerned. Staphylococci may be trained to resistance by growth in increasing concentrations of vancomycin (Geraci et al. 1957) but resistance has not been observed to appear during therapy. Vancomycin is at present used chiefly for the prophylaxis and treatment of endocarditis due to gram-positive bacteria. Penicillin-resistant pneumococci are fully sensitive to it (CDC Report 1977); faecal streptococci are partially sensitive, and a combination of vancomycin and an aminoglycoside are bactericidal for them (Watanakunakorn and Bakie 1973, Harwick et al. 1974). The activity against gram-negative bacteria is small although some strains of *Neisseria* may be susceptible. A more recent indication for the use of vancomycin is in the treatment of pseudomembranous enterocolitis associated with *Clostridium difficile*.

The high molecular weight necessitates the drug being given intravenously. It is toxic for the eighth cranial nerve, leading to deafness, and also for the kidney and therefore is used only in short courses and with monitoring of serum levels particularly when there is pre-existing renal disease.

Bacitracin

Bacitracin (Fig. 5.9) is a large molecular weight polypeptide derived from *Bacillus subtilis* and is active mainly against gram-positive bacteria and *Neisseria* (Johnson et al. 1945, Eagle et al. 1947). Strains of haemolytic streptococci which are not group A are usually less susceptible than group A strains (Maxted 1953). Bacitracin inhibits mucopeptide synthesis leading to accumulation of phospholipids (Park 1952, Siewert and Strominger 1967). It is too toxic for parenteral administration and its main use is for topical application to infected wounds and for skin infections with *Staph. aureus*.

Antibiotics active mainly against gram-negative rods

Beta-lactam antibiotics

The use of bacterial enzymes to produce 6-aminopenicillanic acid opened the way not only for penicillins stable to staphylococcal beta-lactamases but also for the development of penicillins which are active against gram-negative aerobic bacteria. A classification of these compounds is set out in Table 5.5 and the chemical structures are shown in Fig. 5.5. These penicillins fall into four chemical groups, amino penicillins, carboxy penicillins, acylureido penicillins, which are derivatives of ampicillin, and amidino penicillins (Table 5.5). In addition there is one recent penicillin stable to gram-negative beta-lactamases and presumably the first of a new series of compounds, the 6-alpha-methoxy penicillins. Of the acyl penicillins, ampicillin (alpha-aminobenzylpenicillin) is the primary substance and of the carboxy penicillins, carbenicillin (alpha-carboxybenzylpenicillin) is the parent substance from which a variety of other compounds have been developed.

Ampicillin and related compounds

Ampicillin is acid-stable but the absorption is slower and more irregular than that of some of the later developed esters and analogues. It may also be given parenterally. It has four to eightfold greater activity than benzyl penicillin against gram-negative bacteria including *H. influenzae*, *Esch. coli*, *Proteus mirabilis*, salmonellae and shigellae. Other enterobacteria and pseudomonads are resistant. Ampicillin retains high activity against many strains of streptococci, including a higher activity than benzyl penicillin against *Str.*

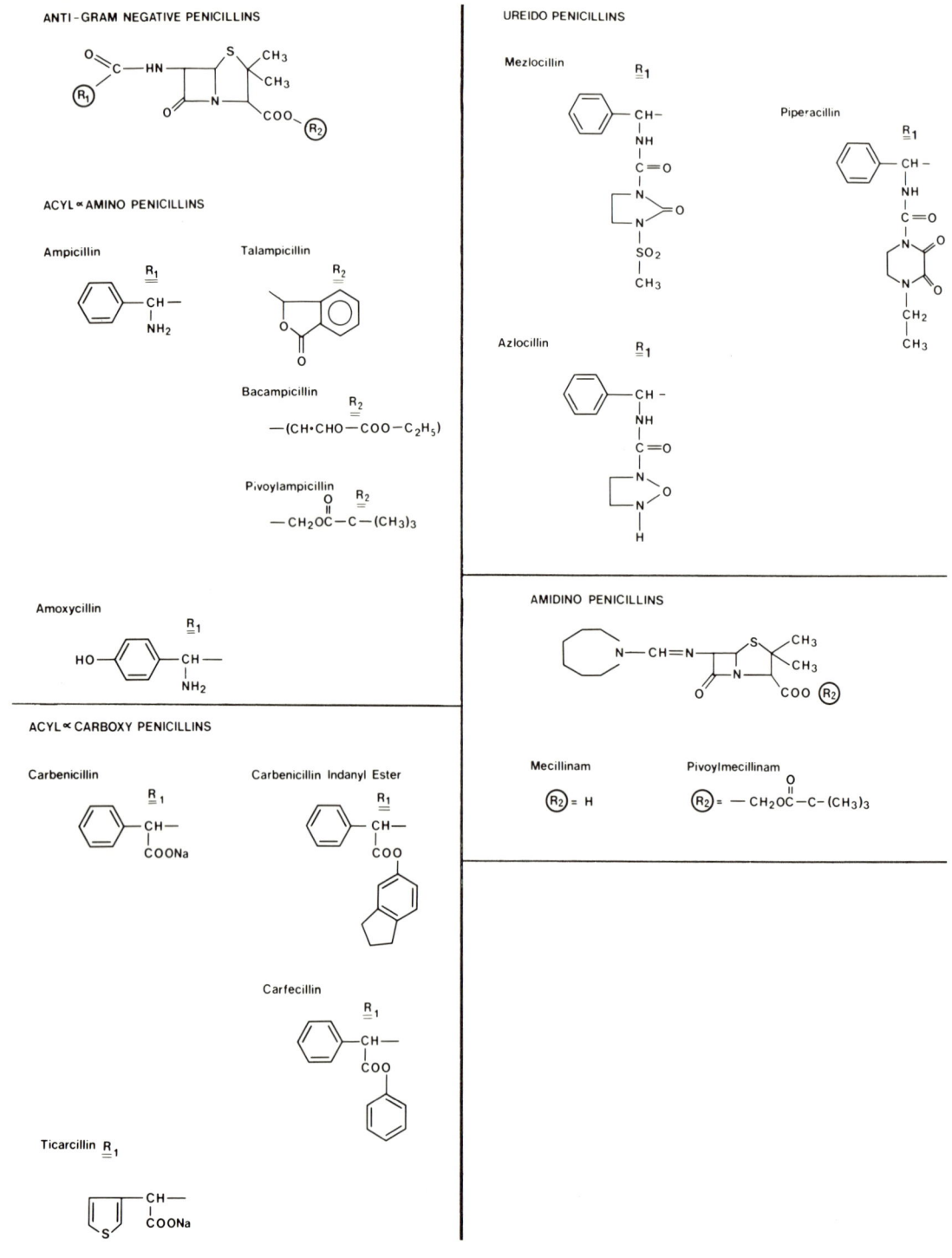

Fig. 5.5 Penicillins active against gram-negative rods together with their esters and analogues.

faecalis. Strains of staphylococci which produce beta-lactamase are also able to hydrolyse ampicillin and are therefore not susceptible.

The antibacterial spectrum of ampicillin is detailed in Table 5.6. Ampicillin is the antibiotic most frequently administered to man because (1) like other beta-lactam antibiotics its toxicity is low, (2) it is active on a wide-range of bacteria that cause common infections, including not only those sensitive to benzylpenicillin but also most strains of *Esch. coli* and *H. influenzae*, and (3) it can be given by the mouth. Such defects that ampicillin has are the allergic reactions resulting in a maculo-papular rash, which occurs ten days after the start of treatment in about 2 per cent of patients, the susceptibility to beta-lactamases produced by gram-negative bacteria, and the variability in absorption after oral administration. The further development of the esters and analogues of ampicillin has been carried out with these three defects in mind. Only the first objective, that is reducing the side effects, and the third, that is improving the absorption after oral administration, have been realized. All the esters and analogues of ampicillin share the susceptibility to gram-negative beta-lactamases. One analogue that is widely used is *amoxycillin* (Fig. 5.5) which produces blood levels of approximately twice those of ampicillin after oral administration. The serum level time course also follows the same pattern as that of ampicillin, peak levels of 4–7 mg/l being reached 1½–2 hours after administration. Amoxycillin is also marginally more active than ampicillin against *H. influenzae*.

Several esters of ampicillin are shown in Fig. 5.5. These compounds are hydrolysed by the esterases of the intestinal mucosa and of the liver to release ampicillin while in the process of absorption or shortly afterwards. The esters are absorbed very rapidly and the serum level time course is quite different from those of ampicillin or amoxycillin. Peak levels are reached in 20–30 minutes after taking an oral dose and the ampicillin is excreted rapidly into the urine. The therapeutic significance of the pharmacological differences between the analogues and the esters is not clear.

Carbenicillin and related compounds

Carbenicillin was the first penicillin to be produced which showed significant antibacterial activity against *Pseudomonas aeruginosa*. This activity is not high; concentrations of 32–128 mg/l being required to inhibit most strains of *Pseudomonas*. This means that doses in excess of 20–30 g per day may be needed for treatment of septicaemia due to these organisms. The mode of action of carbenicillin against *Pseudomonas* is complex. Probably it is better able to penetrate the envelope of *Pseudomonas aeruginosa* than other penicillins but, in addition, carbenicillin can inactivate the chromosomally-mediated type Id beta-lactamase which is produced by all *Pseudomonas* strains (Sabath and Abraham 1964).

Carbenicillin also shows a degree of activity against indole-positive *Proteus* and *Enterobacter* strains which are resistant to ampicillin whereas *Klebsiella* species are resistant both to ampicillin and to carbenicillin (Table 5.6). The enhanced gram-negative spectrum is

Table 5.6 Comparative activity of anti-gram-negative penicillins expressed as mode of the most sensitive strains (mg/l)

	Ampicillin Amoxycillin and esters	Carbenicillin Ticarcillin and esters	Mezlocillin	Azlocillin	Piperacillin	Mecillinam	Temocillin
Streptococci[1]	0.03	0.1	0.03	0.03	0.25	1–2	64+
Str. faecalis	2–4	25	8	16	2	64+	64+
Staph. aureus	0.03	2	0.25	0.25	1	5	64+
(beta-lactamase-positive)	32+	32+	32+	32+	32+	64+	64+
Esch. coli	0.5	8–16	8–16	8–16	1	0.03	4
Klebsiella sp.	32+	R	32+	32+	4	0.25	1
Enterobacter	32+	16–32	16–32	32+	4	0.25	4
Proteus indole-negative	0.5	4	2	2	0.25	0.25	1
indole-positive	32+	4–8	2	16	1	0.25	4
Pseudomonas sp.	32+	32–64	32	8–16	4–8	64+	32
Other gram-negative rods[2]	32+	32+	32+	32+	32+	64+	32
Haemophilus influenzae	0.25	0.5–1	0.12	0.06	0.1	16	1–2
(beta-lactamase-positive)	8–16	4–8	4–8	4	—	64+	1–2
N. gonorrhoeae	0.03	0.25	0.1	0.01	0.01	0.25	—
(beta-lactamase-positive)	8–16	32+	32+	32+	4–8	8–16	—

[1] Excluding *Str. faecalis*
[2] *Serratia, Acinetobacter, Alkaligenes* and related organisms

gained at the expense of reduced activity compared to ampicillin against gram-positive bacteria, especially *Str. faecalis* and penicillin-sensitive staphylococci, and against *H. influenzae*.

Carbenicillin is not absorbed orally and has to be given parenterally which in practice, because of the large doses required, means intravenous administration. As carbenicillin is usually a tri-sodium salt the large amount of sodium administered can have deleterious effects on renal function. The esters and analogues were developed to overcome these problems. Ticarcillin has an identical spectrum to that of carbenicillin with somewhat enhanced specific activity (Sutherland et al., 1970). Because it is given as the mono-sodium salt, the electrolyte disturbance is less than with carbenicillin. Bleeding due to alteration of platelet function has been observed with both the carboxy penicillins, possibly related to the very high blood levels which are a result of the high doses employed.

The esters of carbenicillin are shown in Fig. 5.5. As with the ampicillin esters, free carbenicillin is released on absorption of the ester. The high levels of carbenicillin reached in the urine after oral administration of the esters justify their use in treatment of urinary tract infection, particularly for infections caused by *Ps. aeruginosa* and indole-positive *Proteus* strains.

The acylureido penicillins

Three ureido penicillins are shown in Fig. 5.5 and the antibacterial spectrum is shown in Table 5.6. The ureido penicillins are derived from ampicillin by substitution of the alpha amino group with a heterocyclic compound. This has the effect of enhancing the antibacterial spectrum towards some micro-organisms. The effects on beta-lactamase stability are marginal and all ureido penicillins are unstable to gastric acid and have to be given parenterally. Azlocillin and mezlocillin are very similar in their activity, the former being generally less active than mezlocillin except against *Pseudomonas aeruginosa* where the activity is about four times that of carbenicillin. Apart from the activity against *Pseudomonas* and to some extent against indole-positive *Proteus* strains the spectrum is very similar to that of ampicillin; and they share the same susceptibility to hydrolysis by gram-negative beta-lactamases. *Pseudomonas aeruginosa* plasmid-mediated beta-lactamases do not inactivate ureido penicillins to the same extent as does carbenicillin, but the efficacy of ureido penicillins in *Pseudomonas* infections remains to be confirmed. The higher specific activity which mezlocillin and azlocillin show, compared to ampicillin, against gram-negative rods, *N. gonorrhoeae* and *H. influenzae* is due to more rapid penetration of the cell wall or higher activity against the carboxypeptidases rather than to beta-lactamase stability. Consequently organisms that are resistant to ampicillin because of beta-lactamase production are resistant also to azlocillin and mezlocillin. The same conditions apply to piperacillin, which is the most active of the ureido penicillins, against strains of *Esch. coli, Klebsiella, Enterobacter, Proteus, Pseudomonas, Neisseria* and *Haemophilus*. The range of MICs is shown in Table 5.6. The enhanced activity against gram-negative organisms is balanced by some reduction in activity against gram-positive cocci with the exception of *Str. faecalis* against which the activity is similar to that of ampicillin.

Esters of ureido penicillins have not been produced, so that oral equivalents of ampicillin are not available.

Amidino penicillins

Mecillinam and its pivoyalyl ester, pivmecillinam, are two amidino penicillins now available (Fig. 5.5). They show a marked difference in spectrum of activity to other penicillins active against gram-negative rods in that the activity against gram-positive bacteria is low; MICs of mecillinam even for penicillin-sensitive staphylococci, are 4–8 mg/l and *Streptococcus faecalis* is highly resistant. Almost all rapidly growing gram-negative rods (*Esch. coli, Enterobacter, Klebsiella, Salmonella, Shigella* and *Proteus* species) are susceptible to 1 mg or less per litre. *Pseudomonas* strains, *Bacteroides fragilis* and *Haemophilus influenzae* are resistant (Neu 1976b). The mode of action of mecillinam differs to some extent from other penicillins in that the main binding of this penicillin is to penicillin-binding-protein 2 rather than 1 and 3 which are the main targets for benzyl penicillin. This produces visible changes in cells exposed to mecillinam consisting of large distorted irregularly round cells which are slow to lyse (Greenwood and O'Grady 1973). Other penicillins and cephalosporins produce either distinct rapidly lysing protoplasts or grossly elongated cells (Spratt 1977a). Against the gram-positive organism *Bac. subtilis* however, mecillinam acts as a weak penicillin, binding to all the penicillin-binding-proteins; it is possible that mecillinam behaves similarly against other gram-positive organisms (Spratt 1977b).

Because of the slightly different mode of action of mecillinam and other penicillins synergy between mecillinam and other penicillins might be expected. This does occur against most types of gram-negative organisms exposed to ampicillin and mecillinam (Tybring and Melchior 1975); many other beta-lactam antibiotics have been shown to exert a similar effect (Neu 1976, Baltimore et al. 1976).

Acquired resistance can be readily demonstrated in the laboratory but so far has not been a prominent feature in patients treated with mecillinam for short periods. However four out of 12 carriers of *Salm. typhi* treated with mecillinam eventually excreted a mecillinam-resistant strain of the organism (Jonsson and Tunewall 1975). Cross-resistance with ampicillin

Temocillin

This compound is a new beta-lactam antibiotic which is of considerable interest because it is the first penicillin which shows complete stability to hydrolysis by most gram-negative beta-lactamases (Slocombe et al. 1981). The formula is shown in Figure 5.6. The activity against carbenicillin-sensitive gram-negative rods is of the same order as carbenicillin, azlocillin and piperacillin, but against carbenicillin-resistant gram-negative rods temocillin remains active whereas other penicillins show varying degrees of resistance. Similarly with *H. influenzae* strains which produce beta-lactamase the susceptibility is unchanged to temocillin, whereas the resistance to ampicillin and carbenicillin increases 4–8 fold (Chen and Williams 1982). It has lowered activity against gram-positive organisms in comparison with ampicillin and the ureido penicillins.

is related to the production of large amounts of beta-lactamase which will hydrolyse mecillinam. Strains resistant to ampicillin by reason of the inability of ampicillin to penetrate the cell envelope are generally susceptible to mecillinam (Tybring 1975).

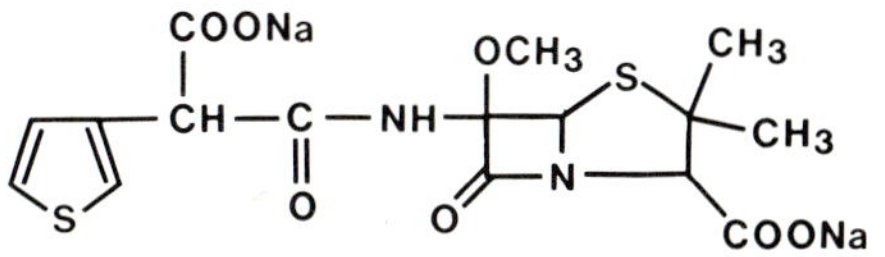

Fig. 5.6 Temocillin—a penicillin resistant to hydrolysis by beta-lactamases of gram-negative bacteria.

Monocyclic beta-lactams

A series of compounds has recently been described (Sykes et al. 1981) in which the beta-lactam ring is not fused with another heterocyclic structure as in

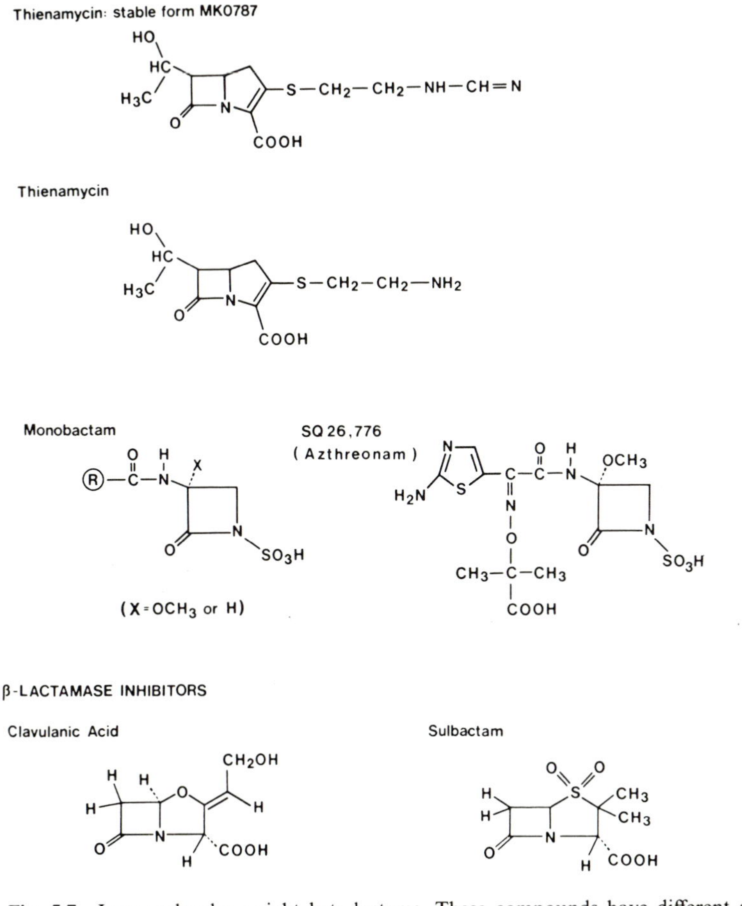

Fig. 5.7 Low molecular weight beta-lactams. These compounds have different activities—the monobactams are active mainly against gram-negative rods, thienamycin has antibacterial activity against a wide range of organisms; clavulanic acid and sulbactam are beta-lactamase inhibitors.

penicillins (Fig. 5.7). These monocyclic compounds, or monobactams, have a spectrum of activity very similar to temocillin in that they are active against a wide range of gram-negative bacilli regardless of whether or not they are producing beta-lactamases (Livermore and Williams 1981). *Pseudomonas aeruginosa* is relatively sensitive (MIC 4–8 mg/l) but *Esch. coli, Proteus, Klebsiella* and other related organisms are fully susceptible to 1–2 mg/l. Gram-positive organisms are generally resistant. The monobactams, one of which is currently undergoing trial, are stable to almost all beta-lactamases which have been tested except for slow hydrolysis by one of the *Pseudomonas* specific enzymes. According to available information they have the low toxicity of penicillins and like temocillin may well become widely used.

Aminoglycoside antibiotics

The aminoglycosides are a family of antibiotics which contain 2-deoxystreptamine or, as in streptomycin, a non-amino hexose as an essential structure with two or more other cyclic amino-sugars attached (Fig. 5.8). Streptomycin was the first member of the family and the other groups to be developed were the kanamycins, neomycins and gentamicins, all of which have been modified to produce analogues with enhanced antibacterial activity in an attempt to reduce toxicity. Because of the large number of different aminoglycosides available one becomes readily confused unless a clear system of classification is followed. Several of the aminoglycosides in use are mixtures of related chemical structure. A suggested classification of the aminoglycosides is shown in Table 5.7 and the structures in Figure 5.8.

Streptomycin

Streptomycin was the first of the major antibiotics to be discovered as a result of a systematic search among microbes for antibacterial agents. It was isolated from *Streptomyces griseus* (Schatz et al. 1944) as a water-soluble substance to which gram-negative and gram-positive organisms, including the tubercle bacillus, were susceptible (Schatz and Waksman 1944). It is moderately active in rickettsial (Huebner et al. 1948) and treponemal infections (Dunham and Rake 1946), and is effective in the treatment of plague and tularaemia.

Neomycin and kanamycin

Neomycin is an aminoglycoside from *Streptomyces fradiae* (Waksman and Lechevalier 1949) and consists of two similar biologically active components, neomycin B and C. Because of its high ototoxicity it is not used for systemic infections, but is given orally for the treatment of infantile enteritis due to enteropathogenic *Esch. coli* strains or for pre-operative sterilization of the gut. It is widely used in creams for topical application, but even here its use is not without risk of toxic effects (Editorial 1969). Kanamycin, produced by *Streptomyces kanamycelius*, is considerably less toxic than neomycin and was the first aminoglycoside to become widely used for the treatment of severe gram-negative rod infections. One of the derivatives, tobramycin, is not susceptible to some of the inactivating enzymes of *Pseudomonas aeruginosa* which attack gentamicin. A further derivative, amikacin, is stable to most aminoglycoside inactivating enzymes.

Gentamicin

Gentamicin was isolated not from a fungus but from *Micromonospora purpurea*, and is separable into three components C_1, C_{1A} and C_2. These have identical antibacterial activity, both *in vitro* and *in vivo*, and a similar toxicity (Weinstein et al. 1967). The activity of gentamicin against *Ps. aeruginosa* in addition to other gram-negative rods susceptible to kanamycin soon made gentamicin one of the most widely used of all parenteral antibiotics.

Mechanisms of resistance to aminoglycosides

The aminoglycosides have a rapid bactericidal action exerted by irreversible binding of the drug to the 30s portion of the bacterial ribosome. The passage of antibiotic into sensitive cells is very rapid and the substance may be concentrated many times inside the cell (Bryan and van den Elzen 1975). Aminoglycosides are active against a wide range of gram-negative rods—*Esch., coli, Klebsiella, Enterobacter, Proteus*, and several compounds are active against *Pseudomonas aeruginosa* also. MICs of aminoglycosides for gram-negative rods are usually less than 1 mg/l of gentamicin and 2 mg/l or less of kanamycin. Streptomycin is of major importance in treatment of infection due to mycobacteria and is now used almost entirely for that purpose. The others have some antimycobacterial action but less than that of streptomycin.

Among gram-positive species only staphylococci are naturally sensitive to aminoglycosides with MICs of 0.5 mg/l or less. Some bacterial species are naturally resistant to aminoglycosides; these include streptococci, pneumococci, bacteroides and clostridia and other gram-positive bacilli. Aminoglycosides, particularly gentamicin and related compounds, are often considered to be of very broad spectrum. In fact many important species are resistant and gentamicin is seldom used alone.

Resistance to aminoglycosides is usually due to enzymes which attack the side chains attached to the aminoglycoside nucleus at specific points in the molecule. Study of aminoglycoside-inactivating enzymes

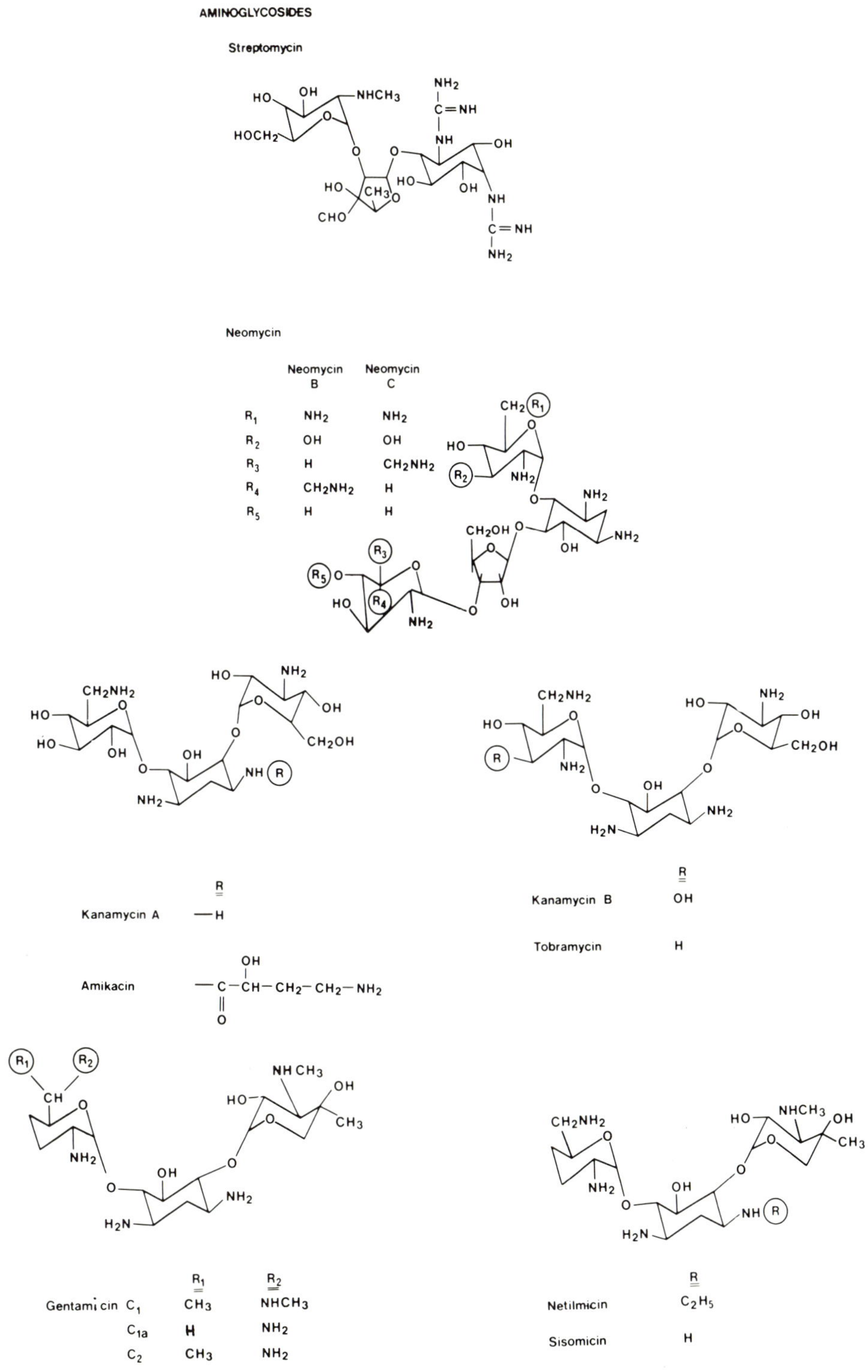

Fig. 5.8 Structures of the main aminoglycoside antibiotics.

Table 5.7 Relationships of the aminoglycoside (aminocyclitol) antibiotics

Group	Chemistry	Type of compound	Features	Derivatives
1	The central ring is streptone and the amino sugar is glucosamine	Streptomycin	The first of the aminoglycosides to be discovered. Of prime importance is the activity against *Myco. tuberculosis*. Resistance to streptomycin develops quickly in aerobic gram-negative rods due to mutation in a single step to produce ribosomes which will not bind streptomycin.	Dihydrostreptomycin—formed by dehydration of the free aldehyde group; has greater ototoxicity than streptomycin and no longer used.
2	The central sugar is deoxystreptone either a 4–6 substituent (2a) or 4–6 disubstituent (2b, 2c) and the amino sugar varies.			
	2a. Pentose and glucosamine	Neomycin B C	Neomycin is a combination of two closely related compounds. Too toxic for systemic use and applied only as a topical agent mainly against staphylococci.	Framycetin is neomycin B.
	2b. Kanosamine	Kanamycin A B	These are two closely related kanamycins of which kanamycin A is most commonly used. The compound was widely used in the 1960s as the most active agent then available for gram-negative rods (excluding *Pseudomonas*). Mainly of importance now because of its two derivatives.	Amikacin is a kanamycin A derivative which has a hydroxy butyric acid substituent which confers resistance to most inactivating enzymes from gram-negative rods. Tobramycin is a kanamycin B derivative which is dehydrogenated in the 3′ position conferring resistance to many phosphorylating enzymes from *Ps. aeruginosa*
	2c. Garosamine	Gentamicin C_1 C_{1a} C_2	Gentamicin was the first aminoglycoside used for treatment of infections with *Ps. aeruginosa* but is also active against a wide range of gram-negative rods. Gentamicin is a mixture of three related compounds derived from actinomycete therefore not spelled with a 'y'. More toxic than kanamycin.	Sissomicin is 4′–5″ dehydrogenated gentamicin C_{1a} and has a similar spectrum to gentamicin. Netilmicin is N-ethyl sissomicin and this substituent is resistant to many aminoglycoside-inactivating enzymes.
3.	Actinamide The ring is glycosylated at 2 points to give a hemitietal formation.	Spectinomycin	Differs from other aminoglycosides in its low oto- and nephrotoxicity. It only has moderate activity and its main use is against *N. gonorrhoeae* (including beta-lactamase-producing strains)	There are several derivatives, none of which is so far available for clinical use.

has enabled new aminoglycosides to be synthesized which are able to circumvent some of these enzymes.

The inactivating enzymes are of three main types—phosphotransferases (APH), acetyltransferases (AAC) and adenyltransferases (ANT). The phosphotransferases and adenyltransferases attack the hydroxyl (OH) groups attached to the carbon atoms and acetyltransferases attack the amino NH_2 groups (Table 5.8). The enzymes are named according to which carbon site they modify; there are at least three acetyltransferases capable of attacking various sites; they are distinguished as AAC 6′, AAC 2′ and AAC 3. At least 16 different modifying enzymes have now been described and most gram-negative rods and staphylococci are capable of producing them. Other mechanisms of resistance do occur however. Streptomycin resistance results from alteration of the target site in the ribosome which is no longer open to attack

Table 5.8 The reactions involved in the action of three types of aminoglycoside-inactivating enzyme

Acetyltransferase	
$AA-NH_2 + CH_3CoSCoA$	$AA-NH-CoCH_3 +$ CoASH
Adenyltransferase	
$AA-OH + ATP$	$AA-O-adenyl +$ pyrophosphate
Phosphotransferase	
$AA-OH + ATP$	$AA-O-phosphate + ADP$

AA = aminoglycoside antibiotic

by streptomycin. Some resistant strains have a barrier to penetration by the aminoglycoside and this results in cross-resistance across all members of the group. Where a specific enzyme is responsible for the resistance the organism remains susceptible to antibiotics which are not modified by that particular enzyme or group of enzymes.

An interesting development has been the attachment to carbon-1 on the 2-deoxystreptamine ring of short chain compounds which stoichiometrically interfere with the action of inactivating enzymes even though susceptible amino and hydroxyl groups are present on the other rings. These side chains are sometimes called 'harbour' groups; for example, amikacin possesses an aminohydroxyl butyric acid and netilmicin (Miller *et al.* 1976) has an ethyl group attached in this position. This has the effect in amikacin of distorting the action of the inactivating enzymes except for those acetylating in the 6' position. Netilmicin, a derivative of gentamicin C_{1a}, is susceptible to attack by only three enzymes.

Development of resistance in aminoglycosides

Three types of resistance mechanisms are found in bacteria resistant to aminoglycosides. Streptomycin-resistant mutants of gram-negative bacilli arise readily both in laboratory experiments and during treatment of patients (Finland *et al.* 1946). A second type of resistance to streptomycin and to other aminoglycosides may arise even when the ribosomes are normal. This type of resistance is generally of a low level producing MIC requirements of 8–32 mg/l, and is fairly common in clinical isolates and is probably due to a barrier to the penetration of the antibiotic. Permeability barriers usually produce cross-resistance to all members of the aminoglycoside group. The third type of resistance is mediated through the aminoglycoside-inactivating enzymes which confer a high degree of resistance. MICs of specific antibiotics attacked by the enzymes are generally 128 mg/l or more. Witchitz and Chabbert (1972) showed that the enzyme production was under control of plasmids. Gentamicin-resistant gram-negative rods are distributed widely. Several hospital epidemics have occurred caused by gentamicin-resistant *Staphylococcus aureus* (Shanson *et al.* 1976, Speller *et al.* 1976) or *Klebsiella* (Houang *et al.* 1979).

Pharmacological properties and toxicity

Aminoglycoides have similar pharmacological properties; they are absorbed to a minimal extent when given by mouth. After intramuscular administration a peak level is reached after 30–60 minutes and in patients with normal renal function the half-life is about 2 hours (Gyselynck *et al.* 1971). Ninety percent of the drug can be recovered from the urine. The excretion of the drugs is altered by changes in renal function which influence the plasma half-life. High plasma levels are toxic and require careful monitoring. The sites of toxic effects are the eighth cranial nerve and the kidney. Streptomycin, gentamicin and the analogues of gentamicin predominantly affect the vestibular branch while kanamycin and the derivatives, tobramycin and amikacin, preferentially affect the cochlea; the toxic effects are usually irreversible. Toxicity is most often seen in patients with disordered renal function. Elderly patients with low renal reserves and perhaps pre-existing auditory loss appear to be especially prone to suffer ototoxic effects. Nephrotoxicity is generally manifested after several days' treatment with aminoglycosides. Nephrotoxicity is related to prolonged moderately raised plasma levels of aminoglycoside and usually is reversible on stopping the drug. There is little room for manoeuvre between the toxic level and therapeutic level of aminoglycosides and the adjustment of dosage requires special care. Three points require particular attention: the selection of initial dosage; ensuring an adequate serum level; and the avoidance of accumulation.

Because the parenteral route of administration is necessary and the potential toxicity of aminoglycosides demands regular monitoring of serum levels one finds that aminoglycosides are usually given only for serious infections where gram-negative rods are prominent.

Synergy of penicillins and aminoglycosides is of considerable laboratory interest as it is the most clear cut of all antibiotic interactions. The synergy has found many clinical indications, for example, in the treatment of streptococcal endocarditis where benzyl penicillin is combined with streptomycin or gentamicin, or in treatment of pseudomonas infection where gentamicin or tobramycin is combined with carbenicillin, ticarcillin or a ureido penicillin.

The polymyxins

Polymyxins are cyclic polypeptides of high molecular weight (Fig. 5.9). They diffuse poorly in culture media so that zones of inhibition are small around discs

POLYPEPTIDES

Bacitracin

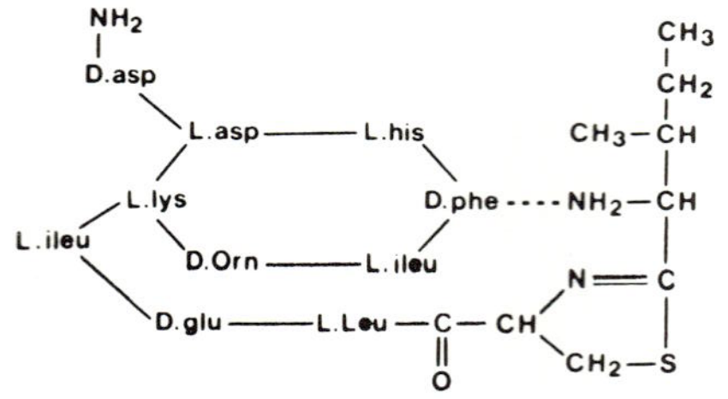

Colistin

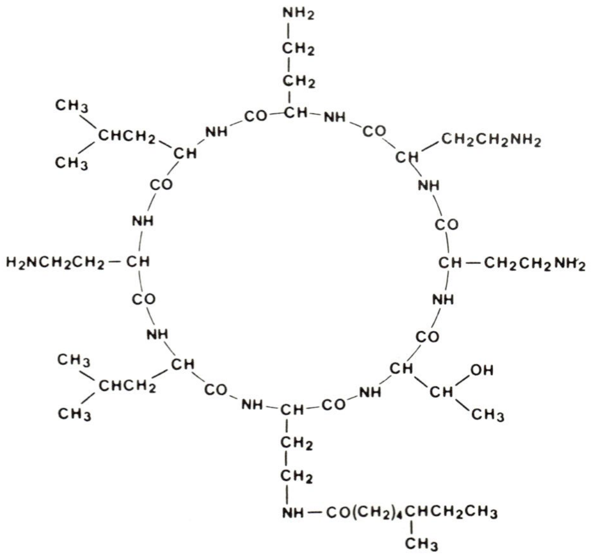

Polymyxin B

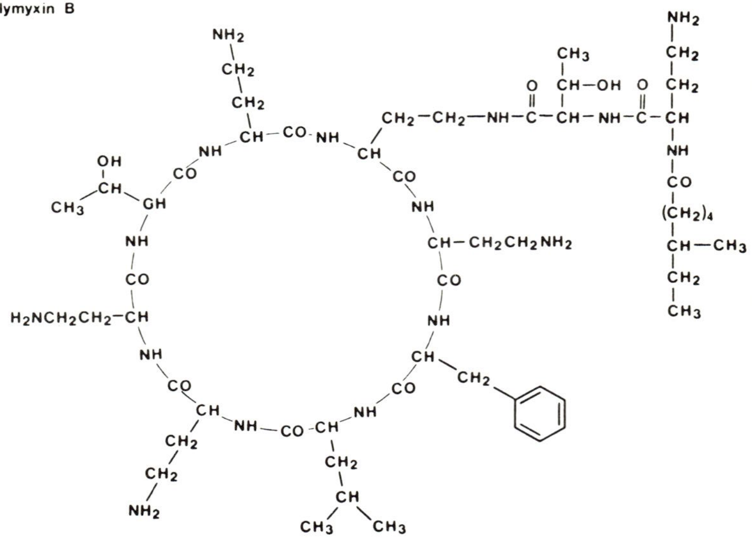

Fig. 5.9 The polypeptide antibodies—polymyxins and bacitracin.

containing these antibiotics. Similarly they diffuse poorly into the tissues and, because they are not absorbed after oral administration, have to be given by parenteral injection. Two compounds are generally available for use, polymyxins B and E; the latter is identical with colistin. The antibiotics are prepared as sulphate or sulphomethate salts, the former being more active *in vitro* although the degree of modification of the salts which occurs *in vivo* is not known. The sulphates are very painful on intramuscular injection. The antibacterial activity of polymyxins is directed solely against aerobic gram-negative rods. Strains of *Esch. coli*, *Klebsiella* and *Enterobacter* are uniformly susceptible to 1–2 mg/l. *Pseudomonas aeruginosa* is likewise highly susceptible, the polymyxins being the first antibiotic discovered which showed such activity. *Proteus* spp. are all resistant, as are gram-positive organisms. The polymyxins act on the cell membrane by increasing the permeability to electrolytes and larger molecules (including some antibiotics such as sulphonamides) which leads to loss of essential cell constituents and death of the bacterium (Newton 1956). Resistant mutants arise occasionally but are unusual. There is no evidence that transferable antibiotic resistance is associated with polymyxin antibiotics. At this time most strains of enterobacteria (excluding *Proteus* and *Pseudomonas*) are susceptible to polymyxins.

Excretion of the drug after injection is mainly into the urine where levels of up to 300 mg/l can be reached. The drugs show a predictable effect on reducing renal function even in normal subjects (Brumfitt *et al.* 1964). There is a rise of plasma urea and creatinine levels which revert to normal after treatment. This degree of nephrotoxicity was acceptable when colistin and polymyxin B were the only antibiotics available for treatment of severe *Pseudomonas* infections. More alarming are the paraesthesiae and sensory loss when blood levels are high (as may occur in renal impairment), but these also resolve on stopping the antibiotic. The polymyxins have been replaced in severe *Pseudomonas* infection by gentamicin, the carboxy penicillins and anti-pseudomonas cephalosporins, but they are still used on occasion for urinary tract infections caused by *Ps. aeruginosa* or multi-resistant klebsiellae, and for topical application to skin sites infected with *Pseudomonas aeruginosa*.

Nitrofurans

Many nitrofurans have been synthesized since 1940; they are widely used in both human and veterinary medicine. The most commonly used agent is nitrofurantoin, which along with other nitrofurans was reviewed by Chamberlain (1976). Nitrofurantoin acts by inhibiting bacterial enzymes involved in the synthesis of DNA (McCalla 1977). The main activity is directed against gram-negative rods—*Esch. coli*, *Klebsiella* and *Enterobacter* in particular. *In vitro* they also show some activity against *Proteus* sp. particularly at acid pH but in the alkaline milieu which *Proteus* produces the activity is much reduced (Brumfitt and Percival 1967). The gram-positive cocci found in urinary tract infection, i.e. *Str. faecalis* and coagulase-negative staphylococci, are also generally susceptible to nitrofurantoin. Resistance occurs in gram-negative rods and is thought not to be related to plasmid acquisition but the frequency of resistance in *Esch. coli* is less than 5 per cent in community strains. The MICs of nitrofurantoin for *Esch. coli* are relatively high—8–16 mg/l and for other gram-negative rods somewhat higher. *Pseudomonas*, *Serratia* and related species are resistant to nitrofurantoin.

The pharmacology of nitrofurantoin dictates the clinical uses of the drug. After oral administration the drug is excreted very rapidly into the urine with blood levels usually below 1–2 mg per litre, while urine levels are generally in excess of 100 mg/l despite some inactivation of nitrofurantoin in the liver and the kidneys. The clinical use of nitrofurantoin in man is primarily for *Esch. coli* urinary tract infection. The dose which is administered sometimes is quite large, up to 400 mg a day, which gives rise to nausea and other gastrointestinal effects. Lower doses are equally effective. Prolonged administration may lead to neurotoxicity.

Nalidixic acid and analogues

Nalidixic acid is chemically unrelated to nitrofurantoin (Fig. 5.10) although it has very similar pharmacological properties and is often grouped with nitrofurantoin as an agent primarily used in urinary tract infection. The mode of action is supposed to be directed at enzymes responsible for joining intermediate-sized fragments of DNA into high molecular weight DNA (Crumplin and Smith 1976). Higher concentrations of nalidixic acid also inhibit protein synthesis which modifies the more lethal effects on the DNA. Unlike nitrofurantoin, nalidixic acid acts only on gram-negative rods and has no action on streptococci or micrococci. Most of the common gram-negative urinary tract pathogens (*Klebsiella*, *E. coli*, *Enterobacter* and *Proteus*) are susceptible to nalidixic acid (8–16 mg/l) but *Pseudomonas* spp. are resistant. The incidence of resistant strains in the general population is low. Some bacterial species mutate at a high rate to give resistance to nalidixic acid. This seems to be particularly true of *Klebsiella* sp. and it is not uncommon to encounter resistant mutants emerging during a single course of treatment with nalidixic acid. *Klebsiella* strains are much more common in hospital urinary tract infections than in domiciliary practice and for this reason in hospital infections nalidixic acid is not commonly used, or if it is used requires to be combined with another agent such as a beta-lactam compound. Many antibiotics are antagonistic when

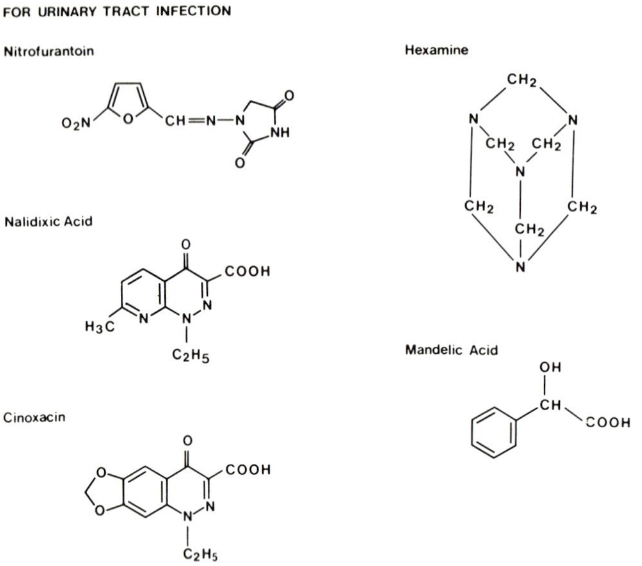

FOR URINARY TRACT INFECTION

Fig. 5.10 The chemical structures of agents used mainly for their activity against organisms in the urinary tract and with little or no effect on systemic infections.

combined with nalidixic acid and these include tetracyclines, chloramphenicol and nitrofurantoin.

After oral administration the blood levels that are reached with nalidixic acid are almost as low as nitrofurantoin although some subjects have considerably higher levels which are probably related to variation in the mechanism for metabolizing the drug. Some analogues of nalidixic acid such as oxolinic acid and cinoxacin produce more predictable blood levels but side effects such as sleeplessness and excitability may occur particularly with oxolinic acid, and these side effects may be related to high blood levels of the antibiotic. Much of the drug is metabolized to an inactive form before excretion into the urine but microbiologically active moieties reach levels of 100-200 mg/l in the urine after a dose of 1 gm. In man the compound is used solely for urinary tract infection. The structures of nitrofurantoin, nalidixic acid and some other compounds used solely for urinary tract infection are shown in Figure 5.10.

Broad spectrum antibiotics

Sulphonamides

Sulphonamides are simple chemicals with a confusing nomenclature. The different sulphonamide analogues have broadly similar antibacterial properties but they differ widely in their pharmacological actions. The main clinical uses today are in the treatment of community-acquired urinary tract infections and of meningococcal carriers but sulphonamides were formerly used for a wide variety of infections. The main toxic effects are seen on the skin.

The development of sulphonamides took place in Germany under the direction of Gerhardt Domagk (1935). A red dye, prontosil, was found to have a curative effect on mice infected with beta-haemolytic streptococci, and was later used to treat human infections with this organism. Trefouël et al. (1935) showed that the active agent was split off in vivo from prontosil, releasing the simple well known chemical sulphanilamide. Soon chemical modifications were being made all over the world, and compounds with higher antibacterial activity and, frequently, greater toxicity were produced. Figure 5.11 shows the structure of some agents that are still in common use.

The sulphonamide nucleus is substituted by a limited number of simple ring structures—pyrimidine, pyridazine, pyrazine, or isoxazole—and these are further substituted by methyl groups or oxymethyl groups. The names of sulphonamides are confusing, not only because there are so many, but because they all sound alike. The name stems from the chemical substituent. Sulphamethoxydiazine is just as correctly called sulphamethoxypyrimidine and quite distinct from sulphamethoxypyridazine or sulfametopyrazine.

Bacteria are generally unable to absorb preformed folate but need to make it for themselves. An essential constituent in folate metabolism is para-aminobenzoic acid (PABA), to which sulphonamides bear a close similarity; the enzyme carrying out the incorporation of PABA into a tetrahydropterydine prefers to incorporate sulphonamide. Vital phases of purine meta-

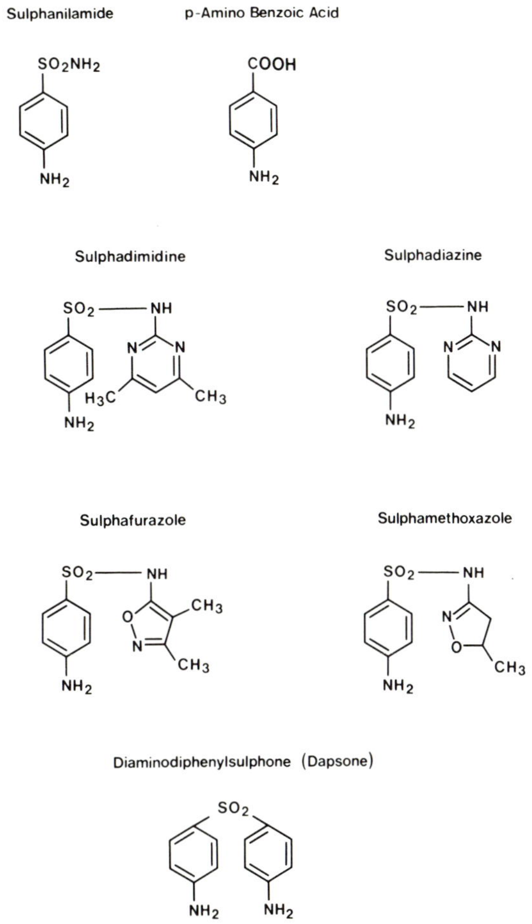

Fig. 5.11 Chemical formulae of sulphonamides and related compounds.

bolism are thus disrupted. Resistance can emerge by development of a folic acid synthetase which has a lower affinity for sulphonamide than for PABA or by increased production of PABA (Richmond 1966).

PABA is an almost universal constituent of culture media which interferes with laboratory tests for sensitivity testing; the size of the inoculum used in the laboratory also greatly influences the results. These two facts explain most of the discrepancies in reported figures for microbial susceptibilities to sulphonamides. For example, *Escherichia coli* has been reported to be sensitive to 0.8–1.6 mg/l (Schmidt *et al.* 1944) and 250 mg/l upwards (Schweinburg and Rutenburg 1949).

The main activity of sulphanilamide is directed against beta-haemolytic streptococci; sulphadiazine extended the range to include pneumococci. Gonococci and meningococci were also susceptible to sulphonamide therapy. The highest activity is shown by compounds with a dissociation constant (pKa) of 6–7, for example sulphadiazine (6.8), sulphamethoxydiazine (7.0) and sulfametopyrazine (7.0), with decreasing activity above and below these figures (Neumann 1968). The pKas of other commonly used agents are sulphamethoxazole 5.7, sulphacetamide 5.2, and sulphadimidine 7.2. Sulphanilamide has a pKa of 10.5 and shows low activity.

Resistant strains of *Neisseria gonorrhoeae* were isolated soon after sulphonamides were introduced, and sulphonamides are no longer used to treat gonorrhoea but sulphonamide resistance among meningococci was not seen until 1963 (Millar *et al.* 1963) and is still relatively uncommon in the UK (Abbott and Graves 1972, Fallon 1974). Differences also occur in the susceptibility of gram-negative bacilli isolated inside and outside hospital. Hospital strains of gram-negative bacilli show a high incidence of sulphonamide resistance; in the urinary tract this may affect 60 per cent of isolates (Williams and Leigh 1966).

Big differences occur in the pharmacological properties of sulphonamides (Table 5.9). Most sulphonamides are absorbed readily from the gut. A few such as sulphasuccidine are scarcely absorbed at all and are used solely for their action on intestinal flora. However, even well absorbed compounds can alter the gut flora by suppressing the sensitive coliforms (Reeves 1974).

After absorption, sulphonamides undergo varying degrees of protein binding; acetylation and conjugation with glucuronic acid occur in the liver. The acetyl derivatives are microbiologically inactive, more protein bound, more rapidly eliminated, and many are poorly soluble, which may give rise to crystaluria. Some persons are able to acetylate sulphonamides only at a slow rate and derive more advantage and less toxicity from sulphonamides. Protein binding affects excretion, the highly protein-bound components such as sulphamethoxydiazine producing prolonged blood levels. Compounds that are the least protein bound, such as sulphadiazine, diffuse most readily (Reider 1963) and reach their highest levels in CSF, amniotic fluid, and peritoneal fluid. Variations in renal handling also affect the blood level. The mechanism of excretion of a sulphonamide influences its rate of elimination and accounts for the wide variation in the half-lives of different compounds. Certain features of excretion are shared by the sulphonamides. Alkalization of the urine speeds up the excretion of almost all compounds; the lower the pKa the more pronounced is the effect; thus sulphanilamide with a high pKa of 10.5 is not affected.

Although sulphonamide therapy has been superseded by antibiotics in several areas of antibacterial therapy, there are still well defined indications for their use. Most urinary tract infections seen in general practice are caused by sulphonamide-sensitive *Esch. coli* (Mond *et al.* 1965, Slade and Crowther 1972). These infections respond to short courses of therapy but

Table 5.9 The sulphonamides in clinical use

| 1. Rapidly absorbed with different routes of excretion
(a) Half-life less than 6 hours
    Sulphadimidine
    Sulphafurazole
    Sulphathiazole
    Sulphamethizole
(b) Half-life up to 20 hours
    Sulphadiazine
    Sulphamethoxazole
(c) Half-life up to 40 hours
    Sulphamethoxydiazine
    Sulphamethoxypyridazine
    Sulphadimethoxine
(d) Half-life over 40 hours
    Sulphadoxine | 2. Poorly absorbed, excreted unchanged in faeces
    Succinylsulphathiazole
    Phthalylsulphathiazole | 3. Used for special purposes
    Sulphasalazine—ulcerative colitis
    Mafenide—burns |

occasionally long-term low-dose prophylactic therapy is necessary such as in children with urinary tract infection associated with vesico-ureteric reflux and non-correctable abnormalities, and patients with frequent symptomatic recurrences of urinary tract infections. Meningococci have remained sensitive to sulphonamides which are used along with benzyl penicillin in meningococcal meningitis. Other forms of meningitis in children, due to *Haemophilus influenzae* and *Str. pneumoniae*, are not usually susceptible to treatment with sulphonamides even though the organisms are inhibited *in vitro*.

Because of the effects of sulphonamides on the bowel flora pre-operative administration of sulphonamides before operation on the large bowel is carried out by many surgeons. Sulphacetamide has been widely used for eye infections because of its rapid penetration of ocular tissues when instilled into the conjunctivae (Robson and Tebrich 1942). *Nocardia* species are usually highly resistant to antibiotics but sensitive to sulphonamides (Strauss *et al.* 1951).

Many of the earlier sulphonamides produced cyanosis owing to methaemoglobinaemia but this complication, along with other haematological abnormalities (haemolytic anaemia, agranulocytosis), is rarely produced by modern agents. Skin rashes occur in approximately 1 per cent of patients receiving sulphonamide therapy. They usually appear in the first week of therapy and are generally morbilliform in type, involve the whole body surface, and are frequently accompanied by pruritus.

Trimethoprim

The enzyme dihydrofolate reductase is present in both prokaryotes and eukaryotes. The reductases from different species vary in their susceptibility to inhibition by a series of 2.4 diamino benzyl pyrimidines (Burchall and Hitchings 1965). This series of pyrimidines includes methotrexate and pyrimethamine which have uses in the inhibition of mammalian and plasmodial reductases. Trimethoprim (Fig. 5.12), another member of the series, preferentially inhibits the dihydrofolate reductases of prokaryotes and, apart from mycobacteria and *Pseudomonas aeruginosa*, most bacteria are inhibited by trimethoprim. Resistant mutants to the drug arise fairly readily and trimethoprim is usually combined with another antibiotic to suppress this resistance. Sulphamethoxazole is commonly combined as cotrimoxazole but other agents such as rifampicin or tetracycline may be used for specific purposes. MICs of trimethoprim for susceptible gram-negative rods are about 0.25–1 mg/l. Resistance among *Esch. coli* and *Klebsiella* was generally uncommon but since 1977 a rising incidence has been found (Brumfitt *et al.* 1977, Hart *et al.* 1977). This increase has also been observed in patients undergoing outpatient treatment, both in the UK and Finland, and may be related to increasing use of the drug as a single agent. Among the less susceptible species are *Neisseria*, but there is synergy between sulphonamides and trimethoprim against *N. gonorrhoeae*. Similar synergy is seen against *Bacteroides fragilis* (Phillips and Warren 1974). Resistance in *H. influenzae* is unusual (Howard *et al.* 1978) but is increasing (Philpott-Howard and Williams

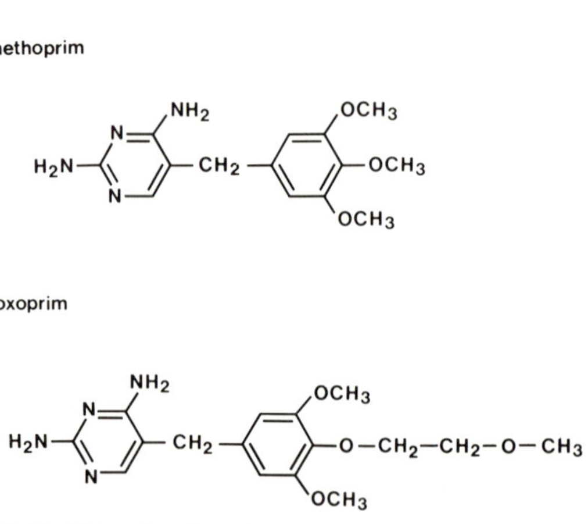

Fig. 5.12 Trimethoprin and tetroxaprim—the two dihydrofolate reductase inhibitors in clinical use.

1982). Only occasional resistant strains of *Salm. typhi* have been encountered. Gram-positive cocci are also somewhat less sensitive than coliform bacilli and in particular *Str. pneumoniae* should be considered to be on the borderline of susceptibility.

Blood levels of trimethoprim after a dose of 160 mg reach a peak of about 2 mg/l after 2 hours. The half-life of the compound is about 12 hours which is similar to that of sulphamethoxazole, the most usual partner for trimethoprim. The main clinical uses of trimethoprim and of cotrimoxazole are in urinary tract infection and chronic bronchitis although it has a valuable role in treatment of typhoid fever where it is almost as effective as chloramphenicol (Geddes *et al.* 1971).

Cephalosporins and related compounds

The development of the cephalosporin group of antibiotics stems from the isolation of cephalosporin C in 1955 (Newton and Abraham 1955). Some years previously Brotzu had examined the microbial flora of the seawater and isolated a fungus similar to *Cephalosporium acremonium*. This fungus appeared to have an inhibitory effect on certain bacteria *in vitro* and Brotzu used the crude products of its growth with some success for treating typhoid fever and brucellosis (Burton and Abraham 1951, Crawford *et al.* 1952). In 1949 two groups of workers (Newton and Abraham 1955, Crawford *et al.* 1952) in England started to examine the culture fluids from Brotzu's fungus and recognized two active substances; cephalosporin P and cephalosporin N. In 1952, cephalosporin N was shown to be a new type of penicillin and a purified sample yielded penicillamine on acid hydrolysis (Abraham and Newton 1954). It was later renamed penicillin N. Cephalosporin P was found to have a steroid structure similar to that of fusidic acid. The following year a second hydrophilic antibiotic was discovered in the metabolic products of the *Cephalosporium* sp. This was named cephalosporin C. Early studies showed that cephalosporin C could be separated from impure penicillin N but it was present in such small amounts in the culture fluids from the Sardinian *Cephalosporium* sp. that it could not be detected by antibacterial assay and had been discovered only because it was concentrated during the purification of penicillin N.

Cephalosporin C was shown to have a wide range of activity and to be equally effective against strains of penicillin-resistant and penicillin-sensitive staphylococci. By 1957, the potential importance of cephalosporin C, especially its resistance to penicillinase, had been recognized. In 1959 the structure of cephalosporin C was proposed and later confirmed by x-ray crystallography, and small amounts of 7-aminocephalosporanic acid were isolated in a relatively pure form. The N-phenylacetyl derivative was shown to be much more active than cephalosporin C itself against penicillinase-producing strains of *Staphylococcus aureus* (Loder *et al.* 1961). In 1960, a chemical procedure was discovered by Lilly Research Laboratories that produced 7-aminocephalosporanic acid in much higher yields and from then on chemical manipulations were undertaken to improve the performance of the cephalosporins. The identification and isolation of the active moiety from cephalosporium were slow, but the development of cephalosporin antibiotics has proceeded at an accelerating rate through the 1960s and 1970s (Fig. 5.13, see p. 128). The cephalosporin nucleus contains several positions where chemical modification can be made. The early modification efforts were directed at the 7-acyl group because this largely determined the kind and amount of antibacterial activity possessed by an analogue. Substitution at the 3 position affected mainly pharmacokinetic properties. This led to the evolution of several main groups of cephalosporins. A suggested classification of these compounds is shown in Table 5.10.

The cephalosporins are all broad spectrum compounds affecting gram-positive and gram-negative bacteria. However major differences in the spectrum exist and need emphasis because increased activity against one group of bacteria may be obtained at the cost of reduced activity against others. The five groups of cephalosporins we recognize (Table 5.10) are distinguishable partly by their pharmacodynamic properties, but mainly by their spectra of activity on various bacteria. In some instances these characteristics can be clearly attributed to their chemical structure but in others this is not at present possible.

Group I This group comprises orally absorbable cephalosporins with a small substituent radicle in the 3 position. Oral absorption of cephalosporins was due to both the alpha-amino group in the 7-acyl substituent and to this small uncharged group in position 3. The first of the orally-absorbable cephalosporins, cephalexin, was introduced in 1969. Later followed cephradine, and in 1978, cefaclor.

Group II This group comprises cephalosporins that are active mainly on gram-positive organisms. It includes cephalothin, a 3-acetoxymethyl compound, the first cephalosporin to become generally available for clinical use. It was found to be metabolically unstable and partly converted by esterases to the corresponding deacetyl component which has low antibacterial activity. This problem was overcome in those compounds in the second part of this group of cephalosporins by replacing the ester group at position 3. This led in 1964 to the introduction of cephaloridine and some time later, cefazolin. The activity of cephalosporins in group II against gram-positive cocci is as high as benzyl penicillin even if they are producers of beta-lactamase. Methicillin-resistant staphylococci, however, show increased resistance to cephalosporins although the MIC for cephaloridine and cephalothin is usually not more than 1-2 mg/l when tests are

Table 5.10 Classification of cephalosporins

This grouping is based upon the following features: 1) chemical structure; 2) antibacterial activity, beta-lactamase stability and metabolic stability. Cephalosporins have a wide antibacterial activity but there are significant variations in the degree of activity against groups of bacteria. Four of these groups are named after the group of bacteria which has particular importance.

	MICs of cephalosporins for 5 groups of organisms (mg/l)				
	Gram +ve	Entero-bacter	Pseudo-monas	Bacte-roides	Haemo-philic*

Group I: Oral group
Compounds with small radical at position 3 which are very well absorbed after oral administration. The antibacterial activity of these compounds is relatively low compared to parenteral cephalosporins. They are stable however to gram-positive beta-lactamases and to some gram-negative beta-lactamases.

Cephalexin	1–4	4–8	>128	>128	4–8
Cephradine	1–4	4–8	>128	>128	4–8

Cl replaces CH$_3$ at position 3 which confers some increase in antibacterial activity and some reduction in absorption.

Cefaclor	1–4	1	>128	>128	0.3–2

Group II: Gram-positive group
Group of cephalosporins which were originally introduced because of their high activity against penicillinase-producing staphylococci. The activity against gram-positive bacteria is much greater than other groups of cephalosporins. The activity against gram-negative rods is relatively low. Group IIa are metabolically stable but Group IIb are desacetylated with diminution of antibacterial activity.

(a) Very high activity against staphylococci (including penicillinase producers) and streptococci. Methicillin-resistant staphylococci show increased resistance.

Cephaloridine	0.03	1–2	>128	32–64	4–8
Cefazolin	0.12	1–2	>128	32	0.5–4

(b) As IIa but metabolism occurs *in vivo* which produces desacetyl derivatives which are much less active microbiologically.

Cephalothin	0.03	2–4	>128	64	0.5–1
Cephacetrile	—	—	>128	16	—

Group III: Enterobacteria group
Compounds with high activity against many aerobic gram-negative rods and stability to many beta-lactamases produced by these bacteria. Group IIIb is metabolically unstable (see IIb). Highest activity against *Haemophilus influenzae* is seen in this group.

(a) Active against most gram-negative rods with exception of *Pseudomonas* and *Enterobacter*. Cefuroxime is more stable to a greater range of beta-lactamases than cefamandole.

Cefamandole	0.5	0.5–1	>128	16–32	0.1
Cefuroxime	1–2	0.5–1	>128	8–16	0.25

(b) Higher specific activity than IIIa but metabolic changes as in IIb result in a derivative of less active metabolite (ceftizoxime).

Cefotaxime	1–2	0.06	16–32	8–32	0.01

Group IV: Pseudomonas group
Compounds which exhibit high activity against *Pseudomonas* strains in addition to activity against Enterobacteria. Group IVa are more stable to plasmid-mediated beta-lactamases.

(a) These compounds show high specific activity against *Pseudomonas* and stability to almost all gram-negative beta-lactamases.

Ceftazidime	4–32	0.25–0.5	1–2	16	0.1
Cefsulodin	2–8	>128	2–4	>128	128

(b) These compounds show instability to some plasmid-mediated beta-lactamases which are found in occasional strains of *Pseudomonas aeruginosa*. They both show differential activity against *Bacteroides* spp. Some, such as *B. fragilis* are sensitive and some, such as *B. thetaiotaomicron* are resistant.

Cefoperazone	2–8	0.25–2	4–8	4–8	0.25–0.5
Ceftriaxone	4–32	0.25–1	4–8	2–32	0.06

Group V: Cephamycins
Compounds with 7-alpha-methoxy group conferring high stability to beta-lactamases of aerobic and anaerobic bacteria.

Inhibits many beta-lactamases including *B. fragilis*.

Cefoxitin	2	2–4	R	4–8	2.0

Has an oxygen atom replacing sulphur in the nucleus (1-oxa-cephem). Marginal activity against *Ps. aeruginosa*.

Moxalactam	4–16	0.25–0.5	8–16	0.5	0.05

Slightly anomalous inclusion in that it is relatively inactive compared to cefoxitin against *Bacteroides* but resembles cefoxitin in all other respects including inhibition of beta-lactamases.

Cefotetan	—	0.06	>128	4–32	1–2

* includes *Haemophilus*, *Bordetella*, *Legionella*, *Campylobacter*

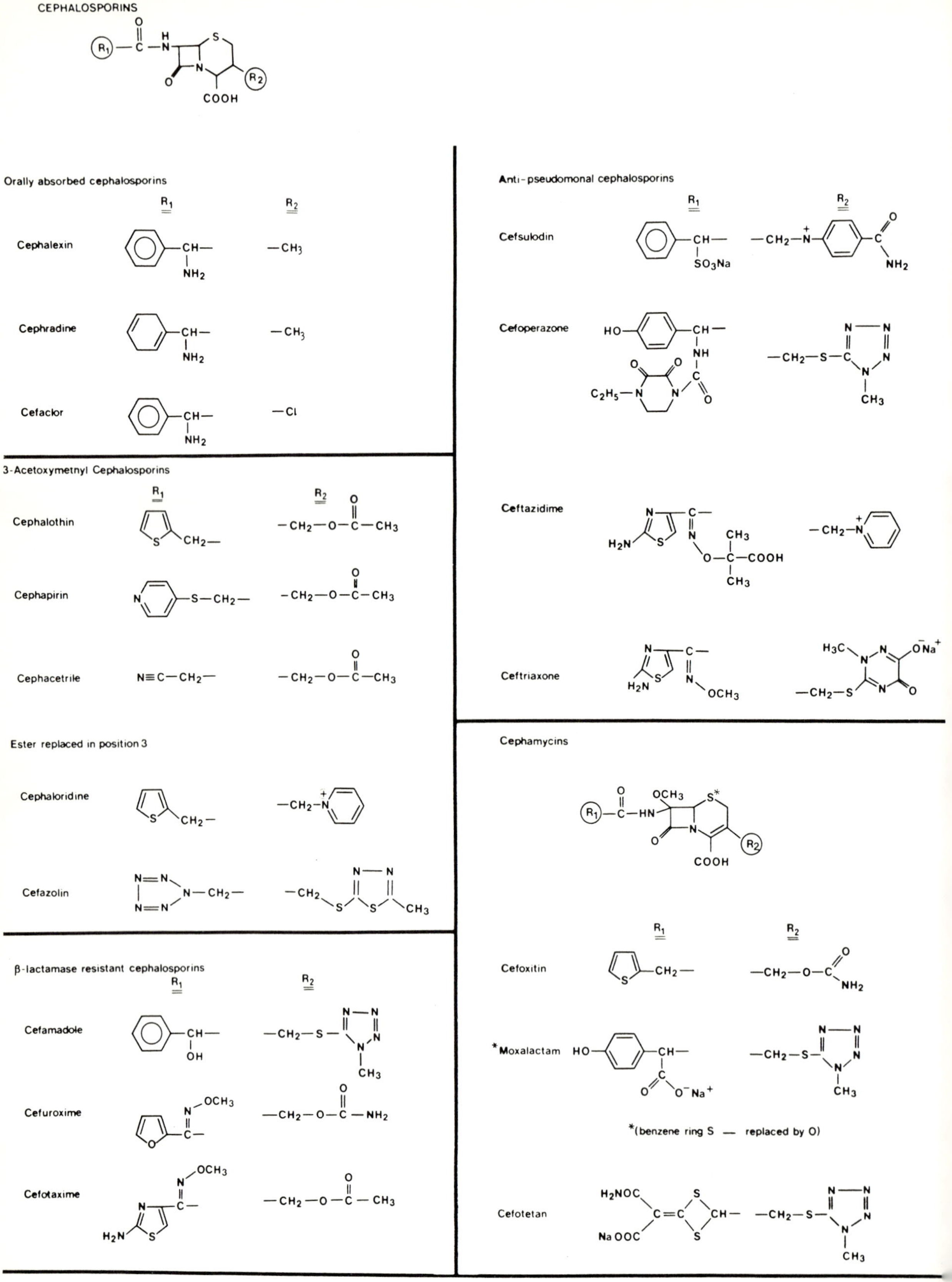

Fig. 5.13 The cephalosporin group of antibiotics and the related cephamycins.

performed at 37° in media not containing high salt concentrations.

Group III Cephalosporins of this group are active on many gram-negative organisms, mainly members of the Enterobacteriaceae, and usually exhibit substantial resistance to destruction by the beta-lactamases of these organisms. It proved difficult to modify the structure of beta-lactamases so as to render them insensitive to all beta-lactams while retaining high antibacterial efficacy, particularly against gram-positive organisms. The nature of the substituent at the 7-acyl position seems to have the primary influence on the susceptibility of the molecule to the beta-lactamase of gram-negative organisms. Once the 7-acyl group has introduced a degree of resistance to hydrolysis by beta-lactamase, alteration at the 3 position does not alter this resistance and may enhance it. The first of the beta-lactamase-resistant cephalosporins to be brought into clinical use was cefamandole, followed by cefuroxime. Both these cephalosporins have large radicles substituted at the 7-acyl position (group IIIa compounds). Cefotaxime has the highest activity of all cephalosporins against enterobacteria but is metabolically unstable and the desacetyl derivative, ceph-tizoxime, is less active than the parent.

Group IV In this group we include cephalosporins that show antibacterial activity against *Ps. aeruginosa*, as well as against most of the enterobacteria. The chemical relations of these compounds are unclear.

Group V The *cephamycins* form a distinct group of compounds with 7-alpha-methoxy substituents that are stable in the presence of a wide range of beta-lactamases. The first of them to be introduced clinically is cefoxitin, a naturally-occurring substance produced by *Streptomyces lactamdurans*. The additional 7-alpha-methoxy group renders it resistant to hydrolysis by most beta-lactamases of gram-positive and gram-negative bacteria including *Bacteroides*. Two other compounds of importance, cefotetan and a 1-oxa-cephamycin, are placed also in this group.

Like penicillins, resistance of bacteria to the cephalosporins depends on two major factors, a permeability barrier to the penetration of the antibiotic into the site of peptidoglycan synthesis and the production of enzymes that destroy the antibiotic. Many gram-positive organisms have fairly simple walls lying outside the cell membrane. Thus, the target for attack by cephalosporins is superficially placed on the cell and relatively easily accessible; therefore the permeability barrier was seen to be of little consequence in the resistance of gram-positive organisms to the earliest cephalosporins (Group II). As the succeeding cephalosporins have been produced the very high activity against gram-positive organisms has been diminished almost in line with the enhanced activity against gram-negative rods.

Gram-negative cells have an envelope of lipoprotein attached to the underlying peptidoglycan and outside this is the membrane which has a typical lipid bi-layer structure. Closely associated with the outer face of this membrane is the lipopolysaccharide. Therefore the target enzymes in gram-negative rods are less accessible than those of gram-positive organisms. The important determinants of permeability appear to be the outer membrane and the lipopolysaccharide which have the effect of restricting the rate at which the cephalosporin reaches its target sites. The barrier may be species-specific. For example, in some strains of *Esch. coli* the barrier excludes penicillins to a greater extent than cephalosporins.

One aim of *in vitro* susceptibility testing is to predict the possible *in vivo* behaviour of an agent. Cephalosporins provide peculiar difficulties in assessment because of their complex mode of action and because of the various mechanisms of resistance to them. Reference has been made to the components of cephalosporin activity—the penetration of the outer layers of the bacterial cell; the affinity for the target site inside the periplasmic space; and the stability of the cephalosporin to chromosomally- and plasmid-mediated beta-lactamases. It is difficult to determine which laboratory susceptibility tests most accurately reflect the combined effect of these three factors. If there is a readily penetrable barrier, the antibiotic may diffuse so rapidly into the cell that the latter is killed before any beta-lactamase present has had the opportunity to act. *In vivo* the antibiotic may be delivered to the cells more slowly than in a laboratory experiment and the beta-lactamase may be able to inactivate the antibiotic before the cell is killed.

The effect of the beta-lactamase contribution to the outcome of cephalosporin-organism interaction is difficult to assess. There is a shortage of information on antibacterial activity which is closely related to *in vivo* response in man.

In order to compare the activity of the cephalosporins against a variety of organisms, the MIC values expressed in Table 5.10 are the median MICs for the strains tested. In general the 'newer' cephalosporins tend to be slightly less active than cephalothin and cephaloridine. There is little difference between the MIC values for beta-lactamase-producing and non-producing strains of *Staph. aureus*. Gram-positive beta-lactamases have relatively little activity against cephalosporins compared with their activity against penicillins. *Str. faecalis* is generally resistant to cephalosporins. Beta-lactamase has a role in the resistance of *H. influenzae* to cephalosporins. The MICs of enzyme-sensitive cephalosporins for beta-lactamase-producing strains are approximately four times greater than those for beta-lactamase non-producers. The enzyme-resistant cephalosporins—cefamandole, cefuroxime and cefotaxime—and the cephamycins show equally good activity *in vitro* against beta-lactamase-producing and non-producing strains of *H. influenzae*.

Table 5.10 also shows the activity of parenterally- and orally-administered cephalosporins against aerobic gram-negative bacilli. The effects of the enzyme stability of group III cephalosporins can be seen clearly.

The cephamycin group has been found to be more active than any other cephalosporins tested against strains of *Bacteroides fragilis*. Many strains of *B. fragilis* are beta-lactamase producers but their resistance also depends on the possession of a permeability barrier.

Cefsulodin is a cephalosporin with activity directed primarily against *Ps. aeruginosa* and gram-positive cocci. It is less active than other cephalosporins against gram-negative bacilli other than *Ps. aeruginosa*. High antipseudomonal activity is related partly to a high affinity for the penicillin-binding-proteins and partly to the resistance to pseudomonal beta-lactamases (Tosch *et al.* 1977). Cefsulodin tested against 180 clinical isolates of *Ps. aeruginosa* inhibited 90 per cent of strains at a concentration of 16 mg/l or less.

Other antipseudomonal cephalosporins included in Table 5.10 are ceftazidime, ceftriaxone and cefoperazone. These compounds appear to have similar characteristics *in vitro*; they are active against *Ps. aeruginosa* and also against other gram-negative bacilli including *Proteus* spp., *Enterobacter* and *Serratia*. None shows particular activity against gram-positive cocci. Ceftazidime and ceftriaxone are very resistant to hydrolysis by plasmid-mediated gram-negative beta-lactamases which occasionally are found in *Pseudomonas aeruginosa*.

The relative advantage of one cephalosporin over another may depend primarily on differences in pharmacokinetic characteristics. Oral absorption of cefoxitin, cephradine and cefaclor result in serum concentrations ranging from 18–48 mg/l (Spyker *et al.* 1978, Neiss 1973). Urine recovery of the drugs is almost 100 per cent indicating the completeness of absorption of the drugs from the gastro-intestinal tract. As can be seen from Table 5.10 the activities of the oral cephalosporins are much less than those of the parenteral compounds. Cephalosporins manufactured for parenteral use can be given intravenously or intramuscularly. They are all well absorbed after intramuscular administration but pain is common with cephalothin. Cephradine, which is available for parenteral and oral administration, is better absorbed after oral than after intramuscular administration; the latter may result in unpredictable serum concentrations (Neiss 1973). Cephaloridine reaches peak serum concentrations approximately twice those of cephalothin after similar doses and is excreted more slowly, detectable activity being present in the serum eight hours after the dose (Andersson 1978).

Some cephalosporins undergo metabolic degradation in the body. Cephalothin is metabolized to desacetylcephalothin, probably in the liver and kidneys. Cefotaxime and cephacetrile are also desacetylated. The desacetyl derivatives retain some antibacterial activity but this is considerably lower than that of the parent compounds. Cefoxitin is metabolized to the descarbamyl form but only to an insignificant amount *in vivo*.

The main route of excretion of all cephalosporins is via the kidney. More than 90 per cent of cephaloridine is excreted by glomerular filtration, whereas cephalothin is excreted mainly by tubular excretion. Other cephalosporins are excreted by both routes to varying extents. The major part of a dose of any cephalosporin is excreted within 24 hours of administration and for several cephalosporins more than 50 per cent of the dose is excreted within 6 hours of administration. Cephalosporins may also be excreted via the biliary system. Cephalothin bile concentrations in patients undergoing cholecystectomy are approximately the same as serum concentrations one hour after a 1 g intravenous dose. Cefamandole is reported to reach concentrations 50 times those of cefazolin and 400 times those of cephalothin in the gall-bladder bile of patients with non-obstructive cholecystitis and cholelithiasis (Quinn *et al.* 1977).

Both penicillins and cephalosporins stimulate in man and experimental animals a specific immunological response but this is not paralleled by allergic responsiveness; adverse reactions of a hypersensitivity type are quite uncommon. Serum from persons who develop allergic reactions to cephalosporins contains antibodies detectable by passive haemagglutination with red blood cells coated with cephalothin (Levine 1973); the antibodies appear to be IgE in nature. Allergic reactions are possible with all cephalosporin preparations but the incidence of cephalosporin allergy is less than that of penicillin allergy (Thoburn *et al.* 1966). Not all cephalosporin allergies are found in patients who are penicillin-allergic; some are due to independent sensitization. To what extent do patients with penicillin allergy cross-react with cephalosporins? Dash (1975) suggested that 91–94 per cent of patients with a history of penicillin allergy did not react to cephalosporins. The probable clinical outcome of therapy with a cephalosporin in a patient allergic to benzyl penicillin is likely to be predicted by skin tests (Levine and Zolov 1969). The newer cephalosporins appear to be only rarely cross-allergic with penicillins. The incidence of hypersensitivity to cephalosporins is less than 10 per cent.

Administration of high doses of cephaloridine or cephalothin to experimental animals damages the proximal kidney tubules, resulting in acute tubular necrosis; it is probable that tubular transport mechanisms are involved. In an analysis of 36 cases of possible cephalosporin-associated nephrotoxicity, Foord (1975) showed that patients who developed renal failure during cephaloridine therapy had peak serum concentrations in the range 130–360 mg/l. Nephrotoxicity does not occur when peak serum concentrations are maintained in the range 20–80 mg/l and concentrations greater than 100 mg/l are unnecessary and unsafe. When renal function is normal a maximum dose of 6 g a day should be used and the dose reduced in renal impairment.

The cephalosporins are a group of antibiotics with antibacterial activity against many bacterial species. They are widely and increasingly used in the treatment of many types of infection. However, the precise place of each of the cephalosporins in the treatment of infections remains to be defined in relation to the alternative antibiotics available.

Thienamycin

The formulae of thienamycin and its formimidoyl derivative are shown in Figure 5.7 along with other low molecular weight beta-lactam compounds. The antibacterial activity of thienamycins is remarkably high and of broad spectrum (Weaver *et al.* 1979). Thienamycin is as active as standard penicillins against gram-positive cocci (Weaver *et al.* 1979), as gentamicin against *Pseudomonas aeruginosa* (Livermore and Williams 1981), as ampicillin against *H. influenzae* and as metronidazole against *Bacteroides* spp. (Nasu *et al.* 1981). Thienamycin binds to cell wall synthesizing proteins, so presumably the mode of action is similar to that of penicillins and cephalosporins.

The pharmacology of thienamycin shows some differences from that of penicillins. Thienamycin itself is unstable and derivatives are necessary to confer stability to the molecule. The derivative shown in Fig. 5.7 is microbiologically as active as thienamycin. Further instability is shown after injection as the molecule is subject to degradation by dipeptidases in the kidney so that the antibiotic is excreted in an inactive form in the urine. To reach its full potential for treatment of infection thienamycin derivatives may have to be combined with a dipeptidase inhibitor.

Beta-lactamase inhibitors

Beta-lactamase inhibitors have little antibacterial action at concentrations that are clinically attainable but may inhibit some organisms at high concentrations. Some strains of *Bacteroides* are inhibited by 8–32 mg/l, but more is needed to affect other organisms. The importance of these agents is their ability to prevent the action of a wide range of beta-lactamases of gram-positive and gram-negative bacteria. The inhibitors of antibiotic-modifying enzymes have been reviewed by Cole (1979).

Most of the beta-lactamase inhibitors are themselves beta-lactam compounds which compete with the beta-lactam antibiotic for the attention of the enzyme. Sometimes the enzyme and the inhibitor are only loosely bound together but with the majority of inhibitors an irreversible inactivation of the enzyme occurs. Isoxazole penicillins and methicillin are weak competitive inhibitors of beta-lactamases from gram-negative bacteria; concentrations of 500 mg/l of methicillin will protect ampicillin from hydrolysis (Rolinson *et al.* 1960, Hamilton-Miller and Smith 1964, Sabath and Abraham 1964). This high concentration of methicillin can be reached in man only in the urinary tract, and attempts have been made to utilize the effect of methicillin in treatment of urinary tract infection (Shirley and Moore 1965). Some of the more recently described beta-lactam antibiotics such as moxalactam, the monobactams and thienamycin also strongly inhibit beta-lactamase activity but, because they have high intrinsic antibacterial activity, they are not usually classed as beta-lactamase inhibitors *per se*.

Of those compounds with low antibacterial activity but high activity against beta-lactamases, two, clavulanic acid and sulbactam are shown in Figure 5.7.

Clavulanic acid is a product of *Streptomyces clavuligerus* (Reading and Cole 1977). It will inhibit the activity of staphylococcal beta-lactamase, most plasmid-mediated beta-lactamases of gram-negative rods and some chromosomal enzymes such as those from *Bacteroides fragilis* (Cole 1979, Wise *et al.* 1978). The effect on some chromosomal enzymes however is slight; for example the Id or Sabath and Abrahams enzymes of *Pseudomonas aeruginosa*. The effect of the low concentrations of clavulanic acid on the MICs of penicillinase-sensitive antibiotics for penicillinase-producing bacteria is dramatic. For example, the MIC of ampicillin for *Bacteroides fragilis* strains falls from 128 mg/l or more to less than 0.5 mg/l in the presence of 2 mg/l of clavulanic acid. In general the susceptibility is increased to that of non-penicillinase-producing strains. The effects have been described with *Staph. aureus* and benzyl penicillin, *Bacteroides* and penicillins (Wise *et al.* 1980), with *Esch. coli* and ampicillin (Cole 1979) and with plasmid-mediated beta-lactamase-producing *Ps. aeruginosa* and carbenicillin (Li *et al.* 1982). Many other penicillins and cephalosporins have been studied in combination with clavulanic acid including piperacillin (Neu and Fu 1980).

Clavulanic acid is well absorbed after oral administration. Clinical studies have shown that the combination with amoxycillin (augmentin) is effective in the treatment of urinary tract infection (Ball *et al.* 1980) and soft tissue infections (Leigh *et al.* 1980) caused by beta-lactamase-producing bacteria. At the present time if one wishes to use clavulanic acid with a beta-lactam compound other than augmentin it is necessary to give a small dose of amoxycillin in addition. Combinations of clavulanic acid or other beta-lactamase inhibitors are expected to be available soon.

Sulbactam Many beta-lactamase inhibitors other than clavulanic acid have been shown to have high activity in the laboratory. Some agents which are highly potent inhibitors such as the 6-halogenated penicillanic acids penetrate bacterial cells poorly and will probably not be of value for clinical use (Wise *et al.* 1981, Li *et al.* 1982). Sulbactam is penicillanic acid sulphone (Fig. 5.7) and, like clavulanic acid, has little intrinsic antibacterial activity (English *et al.* 1978). Like clavulanic acid it is active against many beta-lactamases of gram-negative and gram-positive bacteria. Sulbactam is rather less active than clavulanic acid and is considered to be less able to penetrate cells than clavulanic acid (Li *et al.* 1982). It is at present being developed for oral and parenteral use either as a pro-drug combined with ampicillin or alone (Wise *et al.* 1981).

Tetracyclines

The first tetracyclines to be introduced were oxytetracycline isolated from *Streptomyces aureofaciens* (Finlay *et al.* 1950), and chlortetracycline isolated from *Streptomyces rimosus* (Duggar 1948). The 'parent' compound tetracycline was not introduced until 1953. All the tetracyclines have a broad antibacterial spectrum. They are active against both gram-negative and gram-positive bacteria, aerobic and anaerobic. In addition *Mycoplasma, Chlamydia* and *Rickettsia* are also affected by these compounds. The structures are shown in Figure 5.14. Tetracyclines interrupt synthesis by preventing the attachment of transfer RNA to the ribosomes (Gale *et al.* 1972); more specifically, tetracycline binds to the S4 and S18 proteins of 70S ribosomes at a 1 to 1 molar ratio (Goldman *et al.* 1980). *In vivo* the effects of tetracycline are seen on the ribosomes of prokaryotes but not on those of eukaryotes although there is some activity on the latter *in vitro* (Franklin 1966).

Several analogues have been prepared by producing minor changes in the side chains which have important clinical implications. The three types of effects produced by modifying tetracycline molecules are as follows:

a) Improvement of pharmacological properties. Several compounds have been produced which are more completely absorbed than the three standard tetracyclines and which are also more slowly excreted. Examples of compounds such as these are dimethylchlortetracycline (Finland and Garrod 1960) and lymecyline-tetracycline methylene lysine (Cassano 1961).

b) Reduction of toxicity. Doxycycline is better absorbed than tetracycline (English 1967), but the main interest in this compound is in treatment of infections in patients with renal impairment; tetracyclines have an anti-anabolic effect and may increase uraemia in renal failure (Edwards *et al.* 1970). This effect is not seen with doxycycline (Little and Bailey 1970). Other side effects of tetracycline such as the deposition in developing teeth and bones are not diminished in newer analogues.

c) Activity against tetracycline-resistant organisms. All the tetracyclines mentioned above have an identical antibacterial spectrum and only minor differences are seen in their specific activity. Minocycline, in addition to improved absorption compared to tetracycline, is active against strains which have developed resistance to tetracycline notably *Staph. aureus* (Steigbigel *et al.* 1968), streptococci and *Haemophilus* (Wood *et al.* 1975), and *Nocardia* (Bach *et al.* 1973).

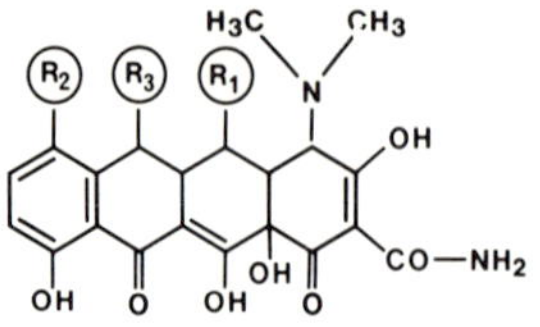

	R_1	R_2	R_3
Tetracycline	–H	–H	CH_3, OH
Oxytetracycline	–OH	–H	CH_3, OH
Chlortetracycline	–H	–Cl	CH_3, OH
Minocycline	–H	$–N(CH_3)_2$	–H
Doxycycline	–OH	–H	$–CH_3$

Chloramphenicol

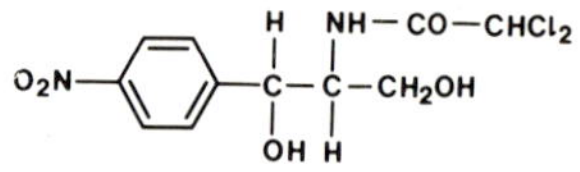

Fig. 5.14 Chemical structures of the tetracyclines and chloramphenicol.

At one time the tetracyclines were perhaps the most widely used of all antibiotics because of the wide spectrum of micro-organisms affected but other agents have gradually taken their place in the treatment of many of the common pathogens. Tetracyclines are now mainly used for infections which are relatively non-responsive to other agents such as chlamydial urethritis (Dunlop 1977), for rickettsial infections including Q fever endocarditis (Wilson *et al.* 1976) and uncommon infections such as brucellosis, yersiniosis, tularaemia, and melioidosis (Cahia and Woodward 1978). Although the indications for tetracycline therapy have been reduced, it is still used on a considerable scale even in the UK (Richmond and Linton 1980, Chopra et al. 1981).

Chloramphenicol

Chloramphenicol has a relatively simple chemical formula and is now a synthetic antibiotic (Fig. 5.14). Originally it was isolated from *Streptomyces venezuelae* (Ehrlich et al. 1947). When originally isolated it was active against many bacterial species but the emergence of resistance among staphylococci on a wide scale and the first reports of fatal aplastic anaemia due to chloramphenicol both occurred in the early 1950s and since that time chloramphenicol has only limited use.

Chloramphenicol interferes with protein synthesis (Gale and Folkes 1953) by binding to the 50S subunit of bacteria but not the 80S ribosome of eukaryotes (Vazquez 1966). The action is bacteristatic against most bacterial species so that elimination of pathogens from infected sites requires intact host defence systems. The spectrum of activity includes most streptococci and staphylococci, *Neisseria*, gram-positive rods including corynebacteria and clostridia, enterobacteria, small aerobic gram-negative rods, anaerobic organisms, including *Bacteroides fragilis*, spirochaetes, rickettsiae and chlamydiae. The organisms not affected by chloramphenicol include mycobacteria, *Pseudomonas aeruginosa*, *Nocardia* as well as fungi or viruses. It has probably the broadest spectrum of any antibiotic and is a popular antibiotic in countries where antibiotics may be purchased without prescription. The inhibitory concentrations of chloramphenicol for organisms against which it is active are usually in the region of 1-4 mg/l. Some of the antibiotic resistance problems which have arisen are of serious import. Chloramphenicol is mainly used in three infections, namely typhoid fever, *Haemophilus influenzae* meningitis and cerebral abscess. Chloramphenicol-resistant *Salm. typhi* have been isolated from several outbreaks in Central America and Asia (see Chapter 37). Many developing countries carry out some surveillance procedures as requested by WHO (WHO Technical Report Series 1978). Chloramphenicol resistance in *H. influenzae* has become more prevalent in recent years; very rare in 1977 (Howard et al. 1978), approximately 1 per cent of strains are now resistant to chloramphenicol in the UK (Philpott-Howard and Williams 1982). The resistance of anaerobic bacteria to chloramphenicol has not been reported to show increased prevalence.

Chloramphenicol is absorbed readily from the gastro-intestinal tract although in the past many formulations of chloramphenicol were irregular in their absorption (Kucers 1980). It is readily diffusible into body fluids including the CSF, and peak blood levels after a 1 g dose reach only about 5 mg/l. The drug is acetylated in the liver and much of the chloramphenicol excreted in the urine is microbiologically inactive.

One analogue of chloramphenicol, thiamphenicol, is less conjugated than chloramphenicol. High levels of active drug are found in the urine. The analogue is not available in many countries including the UK.

Antibiotics active against anaerobes

Some of the antibiotics described previously have activity against some obligate anaerobes. Anaerobes are often divided into (1) those susceptible to benzyl penicillin, which include the majority of anaerobic cocci, *Bacteroides* excluding the *B. fragilis* group, and most strains of clostridia; (2) those not susceptible to benzyl penicillin which include most strains of *Bacteroides fragilis* and related species such as *B. thetaiotaomicron*, *B. distasonis*, *B. vulgatus*, *B. ovatus* and *B. uniformis* (Sutter and Finegold 1976). In recent years resistance to benzyl penicillin in isolates of other species of *Bacteroides* has apparently increased (Nord and Olsson-Liljequist 1981). Of the other penicillins none show greater activity against those strains which are sensitive to benzyl penicillin. Some of the acylureido penicillins are active against the *B. fragilis* group (Thadepalli et al. 1979) but it is doubtful if this is sufficient to be of clinical importance. The cephamycin group of cephalosporins especially moxalactam have high activity against the *B. fragilis* group (Nasu et al. 1981). Cefoxitin is also active because even though the MICs of cefoxitin for the *B. fragilis* are relatively high—8-16 mg/l, the stability to hydrolysis by the beta-lactamases of these organisms (which cefoxitin shares with moxalactam) aids the activity (Leung and Williams 1978). Thienamycin and its stabilized derivative, N-formimidoyl thienamycin, have the highest activity of any beta-lactam compound against anaerobic bacteria, almost 100 per cent being inhibited by 0.5 mg/l (Martin et al. 1982).

Of the other antibiotics already described, erythromycin, the lincomycins, tetracycline and chloramphenicol have been widely used to treat anaerobic infections including those caused by the *B. fragilis* group. Clindamycin has been one of the most important because strains of all *Bacteroides* are almost uniformly susceptible to low concentrations (Sutter and Finegold 1976). The association, however, of clindamycin with pseudomembranous enterocolitis has limited the use of this group of antibiotics in the treatment of anaerobic infections.

Metronidazole and related nitroimidazoles

Azomycin (2 nitro imidazole) was originally described by Maeda et al. (1953) as a new antibiotic and the structure elucidated by Nakamura (1955). Further work in France on the nitroimidazole group finally produced metronidazole (Cosor et al. 1966) as an agent which was subsequently widely used for treatment of trichomoniasis, amoebiasis and protozoal diseases of animals. The activity of metronidazole against anaerobes was unrecognized until the mid-1960s and even then the activity was in the main regarded as an interesting side reaction to its antiprotozoal effects. The anti-anaerobe activity came into prominence when clindamycin was linked to enterocolitis. At that period in 1973–4 there was a paucity of effective drugs active against *Bacteroides fragilis* and metronidazole has since then held a leading place in the treatment of infections caused by obligate anaerobic bacteria.

Metronidazole acts only against obligate anaerobes and has no action on facultative anaerobes; even microaerophilic cocci are unaffected by this compound. Metronidazole passes freely into bacterial cells where it interferes with the anaerobic processes inherent in protein synthesis. The biochemical mechanisms are complex and involve electron transport proteins such as ferredoxin (Lindmark and Muller 1976).

The majority of *Bacteroides* strains including *B. fragilis* and related species are inhibited by 0.5 mg/l or less of metronidazole and anaerobic cocci by 0.25 mg/l or less. These levels are readily reached in the blood after oral administration of 200 mg or by 1–2 g rectally. Because many of the patients with anaerobic infections require parenteral therapy intravenous preparations containing 500 mg doses are also now available. When metronidazoles were used mainly for the treatment of trichomoniasis relatively small doses were given for short periods but higher doses appear to be given for anaerobic infections. The only untoward effect reported as a result of increased dosage or prolonged administration has been the development of peripheral neuritis. The structure of metronidazole and two structural analogues of approximately similar characteristics are shown in Figure 5.15.

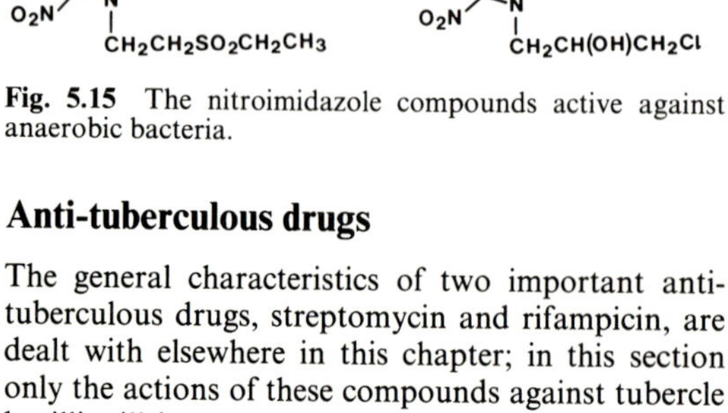

Fig. 5.15 The nitroimidazole compounds active against anaerobic bacteria.

Anti-tuberculous drugs

The general characteristics of two important anti-tuberculous drugs, streptomycin and rifampicin, are dealt with elsewhere in this chapter; in this section only the actions of these compounds against tubercle bacilli will be considered. The other compounds are iso-nicotinic acid hydrazide (INAH), ethambutol, pyrazinamide and, briefly, three drugs, cycloserine, para-aminosalycilic acid (PAS) and the thiosemicarbazones which are seldom used nowadays (Fig. 5.16). The activity of these compounds against mycobacteria is shown in Table 5.11 (see p. 136). The concentration required for bacteristatic or bactericidal activity varies considerably between the different agents used; therefore doses range for example from 450 mg daily doses for rifampicin to 15 g a day for PAS. The majority of the drugs have many side effects and are often used at levels close to those producing toxicity. Furthermore monotherapy with these drugs results in the production of resistant strains of *Myco. tuberculosis* within a few weeks or months of starting therapy. For these reasons the therapeutic regimens for tuberculosis are complex and follow-up has to be carried out for a long period. The regimen may need alteration in order to produce the best results and reduce toxicity to a minimum. One of the most widely accepted regimens is that recommended by the BTTA in 1979. Therapy is started with three drugs; rifampicin, ethambutol and INAH; ethambutol is discontinued after 2–3 months and the others after 12–18 months (for further information, see Chapter 51).

Ethambutol

Ethambutol was introduced in 1961 (Thomas *et al.* 1961). It acts only on growing mycobacteria by interfering with protein synthesis and nucleic acid, more particularly RNA, synthesis. Some polyamines and magnesium are antagonistic to ethambutol indicating that the mode of action involves the intracellular polyamines of mycobacteria (Forbes *et al.* 1966). High level one-step resistant mutants can be selected in the laboratory and during treatment (Tsukamura 1965) but there is no cross-resistance with other antituberculous drugs.

About 80 per cent of an orally administered dose of ethambutol is rapidly absorbed producing peak blood levels of 4.5 mg/l 2 hours after a dose of 25 mg/kg (Donamae and Yamamoto 1966). The greater part of the dose is excreted in the urine, partly as inactive metabolites (Peets *et al.* 1965).

Cycloserine

Cycloserine is produced by a variety of *Streptomyces* including *Str. lavendulae*. Cycloserine has a simple chemical formula and is now prepared synthetically. The activity of the compound is expressed against a wide range of gram-positive and gram-negative bacteria and, apart from the activity against mycobacteria, cycloserine has been used extensively for treatment of urinary tract infection. The site of action is on the cell wall where it acts as a competitive inhibitor of d-alanine. Against mycobacteria synergy occurs with

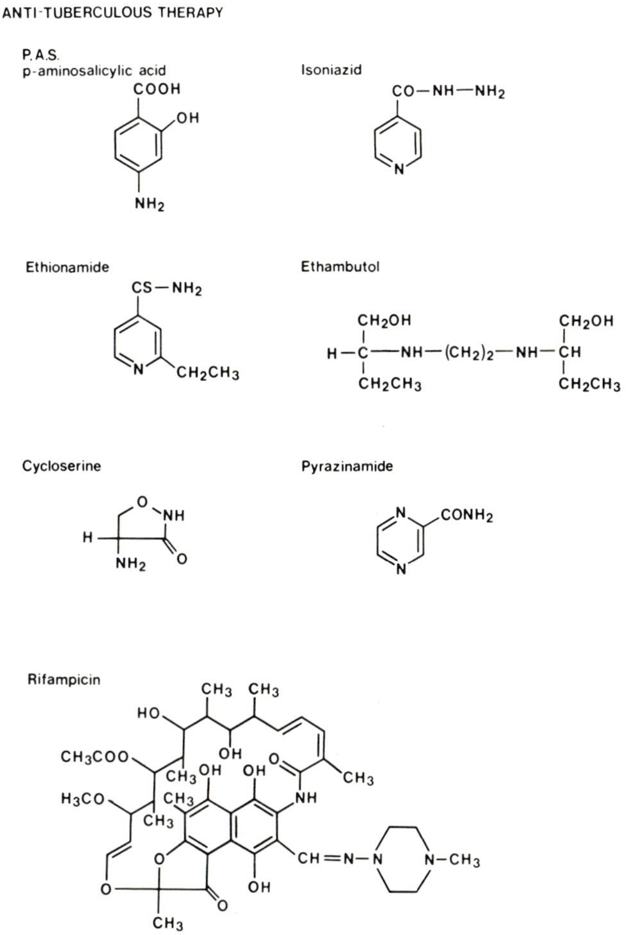

ANTI-TUBERCULOUS THERAPY

Fig. 5.16 Antibiotics used in treatment of tuberculosis.

isoniazide and against other bacteria synergy is seen with penicillin, streptomycin and other antibiotics (Neuhaus 1967). Resistance can emerge during therapy but cross-resistance with other anti-tuberculous agents is not recorded. Peak blood levels of 4 mg/l are reached about 2 hours after a dose of 0.25 g, and 6 mg/l after a dose of 0.5 g; 60 per cent of the dose is excreted in the urine within 24 hours (Krackhardt and Schweiz 1960).

Isoniazid (INAH)

Isoniazid was first reported by Bernstein et al. (1952). It is highly active against *Myco. tuberculosis* which is inhibited by 0.05 mg/l. Resistance develops rapidly *in vitro* and emerges within 1–2 months in patients treated with INAH alone, but primary resistance is rare (Miller et al. 1966). INAH interferes with nucleic acid synthesis in mycobacteria (Wimpenny 1967) and also with the production of mycolic acids. (For further information about the action of isoniazid, and for the mechanisms of resistance to it, see Chapter 24).

INAH is well absorbed from the gastro-intestinal tract after oral administration but the duration of the high blood levels attained is affected by the rate of acetylation of the compound. Some persons are able to acetylate rapidly, which lowers the blood levels more quickly than in slow acetylators of the drug. The slow acetylators are autosomal homozygous recessives (Evans et al. 1960) and acetylate sulphonamides at a similar slow rate. Ethionamide and pyrazinamide are both chemically related to INAH, and are discussed briefly in Chapter 24.

Para-aminosalycilic acid (PAS)

Most strains of *Myco. tuberculosis* are inhibited by 0.5–2 mg/l of PAS. Primary drug resistance is uncommon in *Myco. tuberculosis* (Miller et al. 1966, Hobby et al. 1974). Like sulphonamides, PAS interferes with

Table 5.11 The activity of antituberculous drugs against mycobacteria

Data relating to tests on solid media, mainly extracted from Walter and Heilmeyer (1969). Susceptibility tests on mycobacteria are not readily reproducible and results are usually expressed as a resistance ratio related to the susceptibility of a standard strain (H37R$_v$) tested in parallel to the test strains.

	MIC (mg/l)		
	Myco. tuberculosis	*Myco. bovis*	Other mycobacteria
Streptomycin	1–2	1–5	Rapid growers resistant other variable
Rifampicin	0.5–2.5	0.5–2.5	Rapid growers resistant, others sensitive
Isoniazid	0.05–0.1	0.05–0.2	Resistant
Ethambutol	1–2	1–2	Some 1–5, others resistant
Pyrazinamide	50–200	2000 (R)	500
Cycloserine	10–20	10–50	Rapid growers > 100, others 25–50
PAS	0.5–2	10–50	Resistant
Thioacetazone	2–5	1–5	Often sensitive except rapid growers
Ethionamide	10–20	10–20	Mostly resistant

folic acid metabolism as a competitive analogue of para-amino benzoic acid. Large doses need to be given, usually in the range of 9–12 g per day, in order to obtain satisfactory blood levels; this leads to some degree of gastro-intestinal disturbance. Most of the absorbed compound is excreted fairly rapidly into the urine and modification of the dosage is essential in renal impairment.

Thiosemicarbazones

Thiacetazone is the most active of the thiosemicarbazones against *Myco. tuberculosis*. The structure-activity relations of these compounds are described by Benisch *et al*. (1950). It was widely used as an alternative to PAS but it is fairly toxic and, like many anti-tuberculous agents, may produce liver damage or agranulocytosis.

Anti-fungal drugs

The polyene antibiotics

Of the anti-fungal antibiotics, polyenes are the most widely used in man. Chemically, they comprise a macrolide ring closed with an ester or lactone; the ring sizes vary along with the number of hydroxyl groups on the ring (Fig. 5.17). The first polyene to be discovered was nystatin (fungicidin) isolated from *Streptomyces albidus* (Hazen and Brown 1951). (The name nystatin is derived from the laboratory of the New York State Department of Health.) Many related compounds have subsequently been isolated, notably amphotericin B from *Streptomyces noclosus* (Gold *et al*. 1956), which is the only polyene which can be given by intravenous injection. The polyenes act by increasing permeability of the cell membrane, which causes the loss of cellular constituents resulting in lysis (Kinsky 1961). The polyenes increase cell permeability by interaction with the sterols of the membrane forming a complex and the cholesterol of the liposomes (de Kruijff *et al*. 1974). Amphotericin may be less toxic than some other polyenes because it binds more avidly to ergosterol, the main fungal sterol, than it does to cholesterol, the main membrane sterol in animal cells (Brajtburg *et al*. 1974).

Nystatin rapidly becomes bound to many fungal cells including the yeasts, *Candida* and *Cryptococcus*, to dimorphic fungi such as *Histoplasma*, to dermatophytes, and moulds such as *Aspergillus*. Anti-fungal activity increases with the number of double bonds in the macrolide ring although the MICs of polyenes which have been reported from different centres vary widely (Medoff and Kobayashi 1980). The median MICs for *Candida albicans* given by Medoff and Kobayashi are nystatin 3 mg/l, amphotericin B 0.5 mg/l, and candicidin 0.5 mg/l. The anti-fungal activity of amphotericin B against yeasts is increased by 5-fluorocytosine (Medoff *et al*. 1971). Mutants of yeasts resistant to nystatin may be produced in the laboratory during growth in the presence of the antibiotic but have not been isolated from clinical infections.

Both mystatin and amphotericin B are used for treating superficial infections of the skin and mucous membranes, as ointment, tablets, lozenges or pessaries; both compounds are very effective. Amphotericin is the only polyene used parenterally for systemic fungal infections. The side effects and toxicity of amphotericin vary from batch to batch. The antibiotic appears to be fixed fairly rapidly to the sterols in the cell membranes but little is known of the tissue distribution or metabolism of the drug.

Ch. 5

ANTI-FUNGAL AGENTS

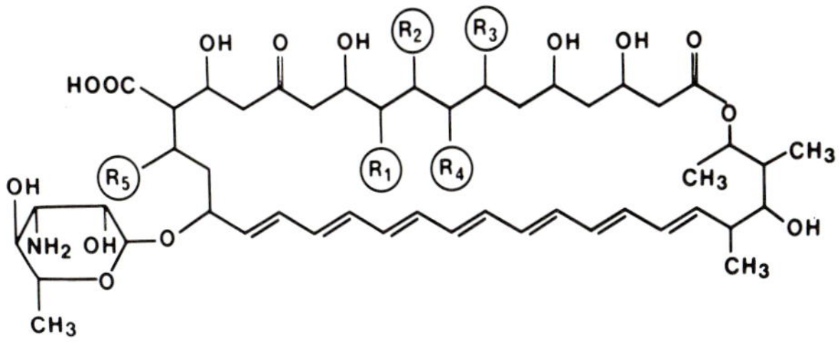

	R_1	R_2	R_3	R_4	R_5
Nystatin	—H	—H	—OH	—H	—OH
Amphotericin B	—H	—OH	—H	—OH	—H
Candidin	—OH	—H	=O	—H	—OH

Griseofulvin

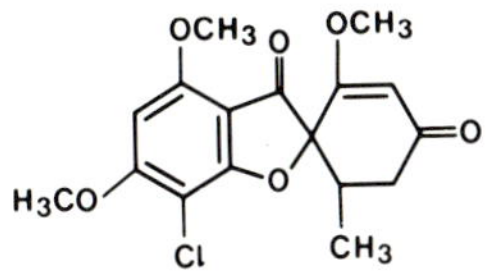

5-Fluorocytosine

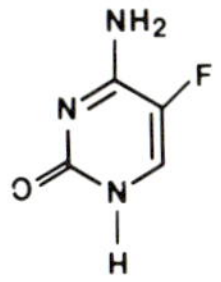

Clotrimazole

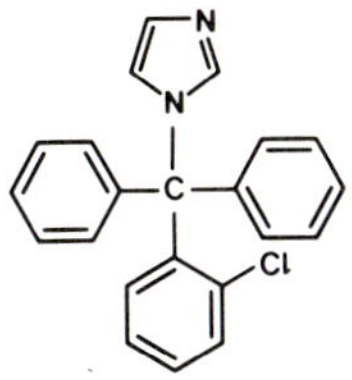

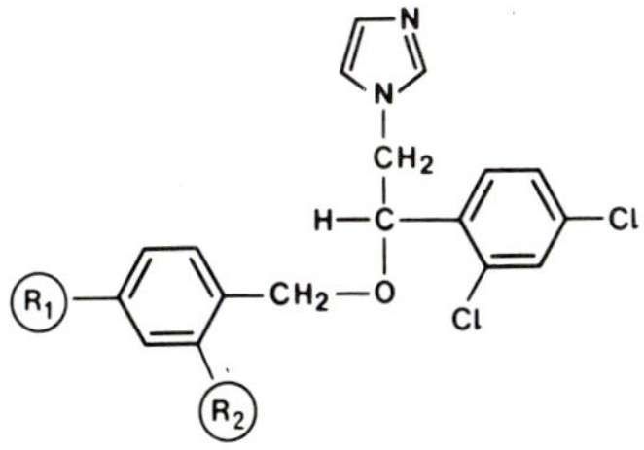

	R_1	R_2
Econazole	—Cl	—H
Miconazole	—Cl	—Cl

Fig. 5.17 Antibiotics used in treatment of fungal infections.

5-Fluorocytosine

This simple compound was originally produced as a possible cytotoxic agent but was found to be fungistatic rather than cytotoxic (Duschinsky et al. 1957, Berger and Duschinsky 1962). It aroused considerable interest as it was an orally administrable drug with its main antimicrobial activity directed against yeasts including *Candida* and *Torulopsis*. MICs of 5-fluorocytosine for sensitive strains of these species range from 0.1 to 1.0 mg/l (Hamilton-Miller 1972, Holt and Newman 1973). The MIC for *Aspergillus fumigatus* is 1.0–12.0 mg/l (Drouhet 1973, Scholer 1970). 5-Fluorocytosine acts as an abnormal analogue of uracil leading to incorporation of 5-fluorouracil into fungal RNA. Biosynthesis of fungal DNA is also impaired. The biochemical pathways are detailed by Scholer (1980).

Resistance to 5-fluorocytosine was found in 7.9 per cent of strains of *Candida albicans* reported in the literature between 1971 and 1976 (Scholer 1980). A similar percentage of resistant strains was found in *Torulopsis* but resistance was found in about 20 per cent in non-albicans strains of *Candida*. In addition to this natural resistance rate, mutation to high-level resistance may occur at fairly high frequency (Jund and Lacroute 1970). For this reason 5-fluorocytosine is generally used in combination with another drug. Synergy occurs with amphotericin B (Medoff et al. 1971) and other polyenes; this finding is utilized in combined therapy with 5-fluorocytosine and amphotericin B for the treatment of systemic candidiasis, cryptococcosis and aspergillosis. In man the tolerance of the drug is high with about 15 per cent of patients suffering from minor gastro-intestinal problems (Dufresne 1972); elevation of serum transaminases (Utz et al. 1969); partial bone marrow suppression occur occasionally (Davies and Reeves 1971). Leukopenic and thrombocytopenia occurred more often when 5-fluorocytosine was combined with amphotericin B (Utz et al. 1975).

Imidazoles

The first imidazoles to be used experimentally for inhibition of micro-organisms were the benzimidazoles which are structural analogues of adenine, in the hope that these agents would interfere with the amino acid pool and inhibit protein synthesis (Woolley 1944). In the event, the most effective of the early imidazoles was clotrimazole which acts on the cell membranes causing increased permeability and loss of intracellular phosphate and potassium (Iwata et al. 1973). The MIC of clotrimazole for *Candida* is 0.5 to 2 mg/l with a few strains requiring 10 mg/l (Holt 1974). Other fungi such as *Cryptococcus neoformans*, *Aspergillus fumigatus* and some dermatophytes show similar susceptibility (Holt 1974). *Torulopsis* is reported to be rather more resistant (Marks et al. 1971). Naturally occurring resistant strains are rare and resistance does not appear to emerge during treatment. The drug has been used for topical and systemic treatment of candidiasis and aspergillosis but only in the former were satisfactory results obtained (Holt 1980). The drug is given orally for both systemic and topical infections. Side effects are common, of which the most troublesome are the psychological disturbances such as disorientation and hallucination; the most frequent are the gastro-intestinal effects.

Two related compounds, miconazole and econazole, are more recently introduced imidazoles. The modes of action of these compounds are similar to those of clotrimazole and resistance to the action is very rare amongst those fungal genera usually sensitive to the drugs. Lower concentrations of miconazole than econazole are required for inhibition of *Candida* spp., *Aspergillus fumigatus* and *Trichophyton* spp. (Holt 1980).

The imidazoles are mainly used as topical application in cutaneous or muco-cutaneous fungal infections and have been most successful in the treatment of candidiasis. The effectiveness of the newer imidazoles in the treatment of systemic fungal infections, however, remains to be determined. The blood levels after oral administration are low.

Griseofulvin

Griseofulvin is quite different from all other anti-fungal agents in its mode of action, pharmacology, antibacterial activity and clinical uses.

Griseofulvin was isolated from *Penicillium griseofulvum* and characterized by Oxford et al. (1939). It was later seen to be the same as the curling factor of Brian (Grove and MacGowan 1947). This curling factor has been produced by *Penicillium janczewski* and has caused distortion of the germ tubes of *Botrytes alli* (Brian et al. 1945, 1946). The first use of griseofulvin was to treat fungal infection in tomatoes and lettuce; it was not used in mammals until Gentles (1958) demonstrated its effectiveness in the treatment of experimental dermatophyate infections in guinea-pigs. Within one year the compound had been widely used and shown to be highly effective in man (Ainsworth 1961). Griseofulvin is present in the stratum corneum within 12 hours of oral administration (Roth and Blank 1960) and appears to be concentrated in infected areas of the skin. It is rather slowly absorbed after oral administration with peak blood levels not being reached until 4 hours after a dose (Atkinson et al. 1962). The half-life is about 20 hours and the drug is mainly excreted in the urine. The anti-fungal activity is directed only against the three main species of dermatophyte, *Microsporum*, *Epidermophyton* and *Trichophyton*, which are all inhibited by concentrations of 2–5 mg/l (Davies et al. 1967). *Candida* and *Torulopsis* and other yeasts are insensitive. The action is expressed on the cell walls of growing fungi, the elongating hyphae being susceptible to the drug while dividing yeast cells are unaffected. It is suggested that the biosynthesis of chitin is antagonized (Brian 1960).

Antiviral agents

Virus diseases have remained much less amenable to antibiotic therapy than bacteria or fungi. Similar lines of research have been followed in the search for antiviral agents as have been used for antibacterial compounds, namely the screening of possible agents against virus cultures *in vitro* and in animals. However, considerable effort has also been expended on examining possible targets in the virus metabolic processes which might be open to disturbance, and the synthesis of agents which might interfere with these processes.

Antiviral agents show relatively high toxicity which is related to the intimate intracellular relation between the virus and the host. Antiviral agents are required to inhibit selectively those intracellular processes which are virus-specific without altering those of the host. The most effective therapy is given very early in the onset of the disease; even more effective results have been seen with some infections when the agents have been used prophylactically. The groups of virus which have shown some response to therapy are the Poxvirus, Herpesvirus and *Myxovirus influenzae*. At the present time the interferons have shown little or no therapeutic effects in man and will not be discussed further.

Amantadine is represented three-dimensionally in Figure 5.18. It is a tricyclic amine hydrochloride which has some activity against influenza A in preventing the penetration of susceptible cells by the virus (Davies *et al.* 1964). Because of this mode of action it clearly is most effective when used prophylactically or in the earlier stages of influenza infection. Clinical trials have indicated that this is the most effective use of the drug (Jackson and Stanley 1976). There is some variation in the susceptibility of strains of influenza to amantadine (Oxford 1975), but most of the recently isolated strains of influenza are susceptible. Amantadine is well absorbed after oral administration. The half-life of the drug is long and doses of 100 mg are given twice daily. The frequency of side effects is low (Couch and Jackson 1976). Amantadine appears to have a confirmed role as an antiviral agent.

Compounds active against Herpesvirus

Nucleoside analogues A number of nucleoside analogues have antiviral activity, notably iodo-deoxyuridine, cytosine arabinoside, adenine arabinoside (Fig 5.18) and trifluorothymidine. They are incorporated into viral DNA in place of the normal nucleotides (Easterbrook and Davern 1963, Kjellern *et al.* 1963). They are all active against herpes viruses.

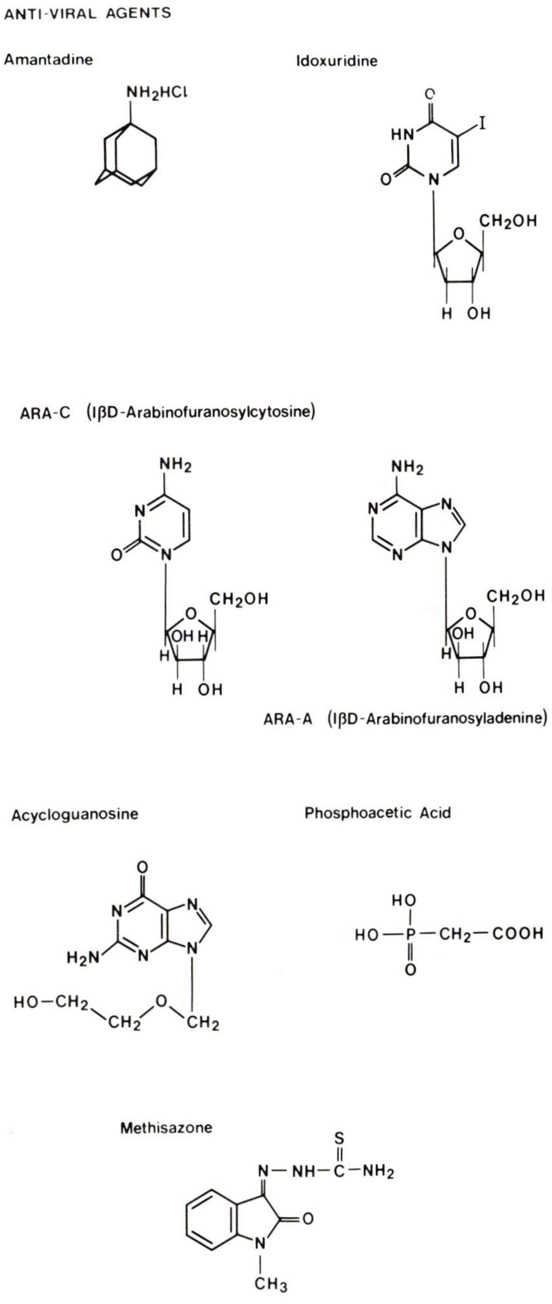

Fig. 5.18 Antibiotics used in treatment of viral infections.

Idoxuridine inhibits *Herpesvirus hominis*, the varicella-zoster group and vaccinia virus to different degrees in different test systems (Marks 1974). In man the main use is for topical therapy of herpes infection of the conjunctiva, particularly in the early stages of infection. Varicella-zoster infections of the eye have also been successfully treated (Pavan-Langston and McCulley 1973), but treatment of mucous membrane and skin infection has been less satisfactory. Skin lesions of herpes zoster however have shown some response to high concentrations of tropically applied idoxuridine (Juel-Jensen and MacCallum 1964).

Arabinosides Cytosine arabinoside and adenine arabinoside have a similar range of antiviral activity to idoxuridine (Underwood et al. 1964) but have somewhat higher activity (De Garile and De Rudder 1964, Walker et al 1970, Collins and Bauer 1977). Both agents have been used with some success topically for treatment of ocular infections with *Herpesvirus hominis*. Parenteral use of these compounds to treat systemic herpetic infections, i.e. herpes encephalitis and disseminated zoster, have been subject to some controversy owing to their doubtful clinical efficacy and to the subsequent immunosuppression of the host defences.

Acyclo-guanosine The anti-herpetic activity of this compound has created much current interest (Elion et al. 1977, Schaeffer et al. 1978). The mode of action appears to be very specifically related to virus metabolism. The antiviral moiety is the phosphorylated form of acyclo-guanosine and a virus-induced enzyme is responsible for bringing about phosphorylation. This means the active substance is produced more avidly in virus-infected cells than in uninfected cells. The phosphorylated compound has an inhibitory effect on DNA polymerases of both type I and type II *Herpesvirus hominis* (Oxford 1979).

The pharmacokinetics of the compound suggest that the agent could be used parenterally for treatment of disseminated herpes infection in man. The animal models of herpes infection both of skin and eye showed considerable improvement; and trials in man of acyclo-guanosine in eye infection with herpesvirus infection have also been encouraging (Jones et al. 1979). Trials of this compound in herpes encephalitis and related infections are at present in progress.

Phosphono compounds Phosphonoacetic acid has been reported to have high activity against herpesviruses (Shipkowits 1973) but has some toxic effect on the skin which limits its use. One derivative, phosphonoformate, does show some potential for treatment of superficial Herpes simplex infection as it has a lower toxicity (Helgstrand et al. 1978). The antiviral spectrum is also rather different in that it affects the DNA polymerase of Hepatitis B virus *in vitro* (Nordenfelt et al. 1979). Work in animal models, particularly in the guinea-pig skin model, has established its therapeutic activity (Alenius et al. 1978), and clinical trials in man have also shown some early success in treatment of herpes folialis (Wallin et al. 1980).

Methisazone Methisazone is an effective antiviral drug which has been rendered superfluous by the successful eradication of smallpox—the infection it was used to treat. Antiviral activity of thiosemicarbazones was shown to be present against Poxvirus by Hamre et al. (1950), and subsequently methisazone was shown to be the most effective thiosemicarbazone (Bauer and Apostolov 1966). They inhibit the synthesis of viral proteins (Appleyard et al. 1965). Trials of methisazone as a prophylactic agent were carried out on smallpox contacts in India and showed considerably efficacy (Rao et al. 1966, Bauer et al. 1969).

References

Abbott, J. D. and Graves, J. F. R. (1972) *J. clin. Pathol.* **25**, 528.
Abraham, E. P. (1971) *Biographical Memoirs of Fellows of the Royal Society* **17**, 255.
Abraham, E. P. et al. (1941) *Lancet* **ii**, 191.
Abraham, E. P. and Newton, G. G. F. (1954) *Biochem. J.* **58**, 266.
Ainsworth, G. C. (1961). *Nature (Lond.)* **191**, 12.
Alder, V. G. and Gillespie, W. A. (1967) *Lancet* **ii**, 1062.
Alenius, S., Dinter, Z. and Oberg, B. (1978) *Antimicrob. Agents Chemother.* **14**, 408.
Anderson, E. S., Galbraith, N. S. and Taylor, C. E. D. (1961) *Lancet* **i**, 854.
Andersson, K. E. (1978) *Scand. J. infect. Dis.*, (Suppl.) **13**, 37.
Appelbaum, P. C., Bhamjee, A., Scragg, J. M., Hallett, A. E., Brown, A. and Cooper, R. C. (1977) *Lancet* **ii**, 995.
Appleyard, G., Hume, V. B. M. and Westwood, J. C. N. (1965) *Ann. N.Y. Acad. Sci.* **130**, 92.
Atkinson, R. M., Bedford, C., Child, K. J. and Tomich, E. G. (1962) *Antibiotics Chemother.* **12**, 225.
Ayliffe, G. A. J. (1973) In: *Antibiotic Therapy*, p. 53. Churchill Livingstone, London.
Bach, M. C., Sabath, L. D. and Finland, M. (1973) *Antimicrob. Agents Chemother.* **3**, 1.
Ball, A. P., Geddes, A. M., Davey, P. G., Farrell, I. D. and Brookes, G. R. (1980) Proc. First Symp. on Augmentin, Clavulanate-potentiated amoxycillin. *Int. Congress Series* **544**, 132.
Baltimore, R. S., Klein, J. D., Wilcox, C. and Finland, M. (1976) *Antimicrob. Agents Chemother.* **9**, 701.
Barber, M. and Rozwadowska-Dowzenko, M. (1948) *Lancet* **ii**, 641.
Barber, M. and Waterworth, P. M. (1962) *Lancet* **i**, 931.
Bauer, D. J. and Apostolov, K. (1966) *Science*, **154**, 796.
Bauer, D. J., St. Vincent, L., Kempe, C. H., Young, P. A. and Downie, A. W. (1969) *Amer. J. Epidemiol.* **90**, 130.
Beach, M. W., Gamble, W. B., Zemp, C. H. and Jenkins, M. Q. (1955) *Pediatrics* **16**, 335.
Benisch, R., Mietzsch, F. and Schmidt, H. (1950) *Amer. Rev. Tuberc.* **61**, 1.
Berger, J. and Duschinsky, R. (1962) *U.S. Pat. Appl.*, Ser. No. 181, p. 822.
Bernstein, J., Lott, W. A., Steinberg, B. A. and Yale, H. L. (1952) *Amer. Rev. Tuberc.* **65**, 357.
Bevan, P. G. and Williams, J. D. (1971) *Brit. med. J.* **3**, 284.
Bilibin, A. F., Gracheva, N. M., Tarasova, A. P., Rubtsova, L. K. and Ermakova, E. T. (1967) *Antibiotiki* **12**, 846.
Bouanchaud, D. H. and Acar, J. F. (1969) *Path. et Biol. Paris* **17**, 763.
Brajtburg, J. K., Price, H. D., Medoff, G., Schlessinger, D. and Kobayashi, G. S. (1974) *Antimicrob. Agents Chemother.* **5**, 377.

Brian, P. W. (1960) *Trans. Brit. Mycol. Soc.* **43,** 1.
Brian, P. W., Curtis, P. J. and Hemming, H. G. (1946) *Trans. Brit. Mycol. Soc.* **29,** 173.
Brian, P. W., Hemming, H. G. and McGowan, J. C. (1945) *Nature (Lond.)* **155,** 637.
Brotzu, G. (1948) *Lavori dell'Istituto d'Igiene di Cagliare* **1,** 1.
Brown, D. F. J. and Blowers, R. (1978) In: *Laboratory Methods in Antimicrobial Chemotherapy.* p. 8. Churchill Livingstone, Edinburgh, London, New York.
Brumfitt, W., Black, M. and Williams, J. D. (1964) *Brit. J. Urol.* **38,** 495.
Brumfitt, W., Hamilton-Miller, J. M. T. and Grey, D. (1977) *Lancet* **ii,** 926.
Brumfitt, W. and Percival, A. (1967) *Ann. N.Y. Acad. Sci.* **145,** 329.
Bryan, L. E. and van den Elzen, H. M. (1975) *J. Antibiot.* **28,** 696.
Burchall, J. J. and Hitchings, G. H. (1965) *Molec. Pharmacol.* **1,** 126.
Burton, H. S. and Abraham, E. P. (1951) *Biochem. J.* **50,** 168.
Butzler, J. P., Dekeyser, P., Detrain, M. and Dehaen, F. (1973) *J. Paediatr.* **82,** 318.
Butzler, J. P., Dekeyser, P. and Lafontaine, T. H. (1974) *Antimicrob. Agents Chemother.* **5,** 86.
Cahia, F. M. and Woodward, T. E. (1978) *Bull. N.Y. Acad. Med.* **54,** 237.
Casewell, M. W., Dalton, M. T., Webster, M. and Phillips, I. (1977) *Lancet* **ii,** 444.
Cassano, A., Filice, A., Costa, C. and Trimarco, C. (1961) *Rif. Med.* **75,** 1383.
CDC (1977) *Morbidity and Mortality Weekly Report* **26,** 285.
Chain, E., Florey, H. W., Gardner, A. D., Heatley, N. G., Jennings, M. A., Orr-Ewing, J. and Sanders, A. G. (1940) *Lancet* **ii,** 226.
Chamberlain, R. E. (1976) *J. Antimicrob. Chemother.* **2,** 325.
Chan, Kuo Cho, Wang San Pin and Grayston J. Y. T. (1977) *Antimicrob. Agents Chemother.* **12,** 80.
Chen, H. Y. and Williams, J. D. (1982) *J. Antimicrob. Chemother.* **10,** 279.
Chopra, I., Howe, T. G. B., Linton, A. H., Linton, K. B., Richmond, M. H. and Speller, D. C. E. (1981) *J. Antimicrob. Chemother.* **8,** 5.
Cohen, J. R., Shuman, H. H., Blake, A. and Plotkin, L. H. (1965) *Clin. Med.* **72,** 341.
Cole, M. (1979) In: *Antibiotic Interactions.* p. 99. Academic Press, London.
Collins, P. and Bauer, D. J. (1977) *J. Antimicrob. Chemother.* (Suppl. A), **3,** 73.
Cosor, C., Crisan, C., Horclois, R., Jacob, R., Roberts, J., Tchelitcheff, S. and Vaupre, R. (1966) *Arzneimittel Forschung* **16,** 23.
Couch, R. B. and Jackson, G. G. (1976) *J. infect. Dis.* **134,** 516.
Crawford, K., Heatley, N. G., Boyd, P. F., Hale, C. W., Kelly, B. K., Miller, G. A. and Smith, N. (1952) *J. gen. Microbiol.* **6,** 47.
Crumplin, G. C. and Smith, J. T. (1976) *Nature, (Lond.)* **260,** 643.
Crimplin, G. C. and Smith, J. T. (1976) *Antimicrob. Agents Chemother.* **8,** 251.
Currie, K., Speller, D. C. E., Simpson, R. A., Stevens, M. and Cooke, D. I. (1978) *J. Hyg., Camb.* **80,** 115.
Dash, C. H. (1975) *J. Antimicrob. Chemother.* **1** (Suppl.), 107.

Davies, J. E. (1964) *Proc. Natl. Acad. Sci. USA* **51,** 659.
Davies, R. R., Everall, J. D. and Hamilton, E. (1967) *Brit. med. J.* **iii,** 464.
Davies, R. R. and Reeves, D. S. (1971) *Brit. med. J.* **i,** 577.
Davies, W. L. et al. (1964) *Science, N.Y.* **144,** 862.
Domagk, G. (1935) *Deutsche medizinische Wochenschrift* **61,** 250.
Donomae, I. H. and Yamamoto, K. (1966) *Ann. N.Y. Acad. Sc.* **135,** 849.
Doyle, F. P. et al. (1961) *Nature (Lond.)* **192,** 1183.
Drouhet, E. (1973) In: *Handbook of Microbiology.* **3,** p. 697. CRC Press, Cleveland, Ohio.
Dufresne, J. J. (1972) *Rev. Méd. Suisse romande* **92,** 597.
Duggar, (1948) *Ann. N.Y. Acad. Sci.* **51,** 177.
Duncan, I. B. (1967) *Antimicrob. Agents Chemother.* **1,** 723.
Dunham, W. B. and Rake, G. (1946) *Science* **103,** 365.
Dunlop, D. M. and Murdoch, J. (1960) *Brit. med. Bull.* **16,** 67.
Dunlop, E. M. C. (1977) *J. Antimicrob. Chemother.* **3,** 377.
Duschinsky, R., Pleven, E. and Heidelberger, C. (1957) *J. Amer. Chem. Soc.* **79,** 4599.
Eagle, H., Newman, E. V., Greif, R., Burkholder, T. M. and Goodman, S. C. (1947) *J. clin. Invest.* **26,** 919.
Easterbrook, K. B. and Davern, C. I. (1963) *Virology* **19,** 509.
Editorial (1969) *Brit. med. J.* **ii,** 181.
Edwards, O. M., Haskinsson, E. C. and Taylor, R. T. (1970) *Brit. med. J.* **i,** 26.
Ehrlich, J., Bartz, Q. R., Smith, R. M., Joslyn, D. A. and Burkholder, P. R. (1947) *Science* **106,** 417.
Elion, G. B., Furman, P. A., Fyfe, J. A., De Miranda, P., Beauchamp, L. and Schaeffer, J. H. (1977) *Proc. natl. Acad. Sci. USA* **74,** 5716.
Engle, C., Lasinski, E. and Gelzer, J. (1970) *Nature (Lond.)* **228,** 1190.
English, A. R. (1967) *Proc. Soc. exp. Biol. (NY)* **126,** 487.
English, A. R., Retsema, J. A., Girard, A. E., Lynch, J. E. and Borth, W. E. (1978) *Antimicrob. Agents Chemother.* **14,** 414.
Etteldorf, J. N. and Crawford, S. E. (1967) *S. Dakota J. Med.* **20,** 30.
Evans, W. and Hansman, D. (1963) *Lancet* **i,** 451.
Evans, D. A. S., Manley, K. A. and McKurich, V. A. (1960) *Brit. med. J.* **ii,** 485.
Fallon, R. J. (1974) *Brit. med. J.* **ii,** 272.
Farmer, C. D., Hoffman, H. N., Shorter, R. G., Thurber, D. L. and Bartholomew, L. G. (1963) *Gastroenterology* **45,** 157.
Finland, M., Brumfitt, W. and Kass, E. K. (1976) *J. infect. Dis.* **134**(S), 235.
Finland, M. and Garrod, L. P. (1960) *Brit. med. J.* **ii,** 959.
Finland, M., Murray, R., Harris, H. W., Kilhan, L. and Meads, M. (1946) *J. Amer. med. Assoc.* **132,** 16.
Finlay, A. C., Hobby, G. et al. (1950) *Science* **111,** 85.
Fleming, A. (1929) *Brit. J. exp. Path.* **10,** 226; (1932) *J. Path. Bact.* **85,** 831.
Florey, H. W., Chain, E., Heatley, N. G., Jennings, M. A., Sanders, A. G., Abraham, E. P. and Florey, M. E. (1960) In: *Antibiotics* Oxford University Press, London.
Foord, R. D. (1975) *J. Antimicrob. Chemother.* **1** (Suppl.), 119.
Forbes, M., Peets, E. A. and Kuck, N. A. (1966) *Ann. N.Y. Acad. Sci.* **135,** 726.

Franklin. T. J. (1966) in *16th Symposium Soc. gen. Microbiol.* p. 192, Cambridge University Press, Cambridge.
French, G. L. and Horn, J. (1979) *Critical Care Medicine* **7**, 487.
Gale, E. F., Cundliffe, E., Reynolds, P. E., Richmond, M. H. and Waring, M. J. (1972) in: *The Molecular Basis of Antibiotic Action* p. 315, Wiley, London.
Gale, E. F. and Folkes, J. P. (1953) *Biochem. J.* **53**, 493.
De Garile, W. P. and De Rudder, J. (1964) *C. R. Acad. Sci.* **259**, 2725.
Garrod, L. P. (1957) *Brit. med. J.* **ii**, 57.
Geddes, A. M., Fothergill, R., Goodall, J. A. D. and Dorken, P. R. (1971) *Brit. med. J.* **iii**, 451.
Gentles, J. C. (1958) *Nature (Lond.)* **182**, 476.
Geraci, J. E., Heilman, F. R., Nichols, D. R., Wellman, W. E. and Ross, G. T. (1957) *Antibiot. Annual* p. 90.
Godtfredsen, N. W., Roholt, K. and Tybring, L. (1962) *Lancet* **i**, 928.
Gold, W., Stout, H. A., Pagano, J. E. and Donovick, R. (1956) *Antibiotics A.* 579.
Goldman, R. A., Cooperman, B. S., Strychouz, W. A., Williams, B. A. and Tritton, T. R. (1980) *FEBS Letts.* **118**, 113.
Greenwood, D. and O'Grady, F. (1973) *J. clin. Path.* **26**, 1.
Griffith, R. S. (1955) *Antibiotic. Annu.* p. 269.
Grove, J. F. and MacGowan, J. C. (1947) *Nature (Lond.)* **160**, 574.
Gunn, B. A., Woodall, J. B. and Jones, J. F. (1974) *Lancet* **ii**, 845.
Gyselynk, A. M., Forrey, A. and Cutler, R. (1971) *J. infect. Dis.* **124**(S), 570.
Hamre, D., Bernstein, J. and Donovick, R. (1950) *Proc. Soc. exp. Biol. Med.* **73**, 275.
Hamilton-Miller, J. M. T. (1972) *Sabouraudia* **10**, 276.
Hamilton-Miller, J. M. T. and Smith J. T. (1964) *Nature* **201**, 999.
Hare, R. (1970) *The Birth of Penicillin*, Allen & Unwin, London.
Hart, C. A., Gibson, M. F., Mulvihill, E. and Green, H. T. (1977) *Lancet* **ii**, 1081.
Harwick, H. J., Kalmanson, G. M. and Guze, L. B. (1974) *J. infect. Dis.* **129**, 358.
Haverhorn, M. L. and Michel, M. F. (1979) *J. Hyg., Camb* **82**, 177.
Hazen, E. L. and Brown, R. (1951) *Proc. Soc. exp. Biol. Med.* **76**, 93.
Heilman, F. R., Herrell, W. E., Wellman, W. E. and Geraci, J. E. (1952) *Proc. Mayo Clinic* **27**, 285.
Helgstrand, E. *et al.* (1978) *Science* **201**, 819.
Hirsch, H. A. and Finland, M. (1959) *Amer. J. med. Sci.* **237**, 693.
Hobby, G. L., Johnson, P. M., Boytar-Papirnyik, V. (1974) *Amer. Rev. resp. Dis.* **110**, 95.
Holt, R. J. (1974) *Infection* **2** 95.
Holt, R. J. (1980) In: "*Antifungal Chemotherapy*" p. 107, Wiley, Chichester.
Holt, R. J. and Newman, R. L. (1973) *J. clin. Path.* **26**, 167.
Houang, E. T., Evans, M. A. L. and Simpson, C. N. (1979) *Lancet* **ii**, 205.
Howard, A. J., Hince, C. J. and Williams, J. D. (1978) *Brit. med. J.* **i**, 1657.
Huebner, R. J., Hottle, G. A. and Robinson, E. B. (1948) *Publ. Hlth. Rep., Wash.* **63**, 357.

Idsøe, O., Guthe, T. and Willcox, R. R. (1972) *Bull. Wld. Hlth. Org.* **47** (Suppl.), 1.
Iwata, K., Yamaguchi, H. and Hiratani, T. (1973) *Sabouraudia* **11**, 158.
Jackson, G. G. and Stanley, E. D. (1976) *J. Amer. med. Ass.* **235**, 2739.
Jevons, P. (1961) *Brit. med. J.* **i**, 124.
Johnson, B. A., Anker, H. and Meleney, F. (1945) *Science* **102**, 376.
Jones, B. R., Coster, D. J., Fison, P. N., Thompson, G. M., Cobo, L. M. and Falcon, M. G. (1979) *Lancet* **i**, 243.
Jonsson, M. and Tunevall, G. (1975) *Infection* **3**, 31.
Juel-Jensen, B. E. (1973) *Brit. med. J.* **i**, 406.
Juel-Jensen, B. E. and MacCallum, F. O. (1964) *Brit. med. J.* **ii**, 987.
Jund, R. and Lacroute, F. (1970) *J. Bact.* **102**, 607.
Kato, K. (1953) *J. Antibiotics (Japan)* **130**, 184.
Katz, E., Grimley, P. and Moss, B. (1970) *Nature (Lond.)* **227**, 1050.
Kinsky, S. C. (1961) *J.Bact.* **82**, 889.
Kjellern, L., Pereira, H. G., Valentine, R. and Armstrong, J. A. (1963) *Nature (Lond.)* **199**, 1210.
Knight, V. and Holzer, A. R. (1954) *J. clin. Invest.* **33**, 1190.
Knudsen, E. T., Brown, D. M. and Rolinson, G. N. (1962) *Lancet* **ii**, 632.
Krackhardt, H. and Schweiz, Z. (1960) *Zuberk. Arzt.* **17**, 403.
De Kruijff, B., Gerritsen, W. J., Oerlemans, A., Van Dijck, P. W. M., Demel, R. A. and Van De Enen, L. L. M. (1974) *Biochim. biophys. Acta* **339**, 30.
Kucers, A. and Bennet, N. McK. (1979) *The Use of Antibiotics: A Comprehensive Review with Clinical Emphasis.* Heinemann Medical Books, London.
Kunin, C. M., Brandt, D. and Wood, H. (1969) *J. infect. Dis.* **119**, 132.
Leigh, D. A., Marriner, J. M., Freeth, M., Bradnock, K. and Nisbet, D. (1980) In: Proc. 1st Symp. on Augmentin, clavulanate-potentiated amoxycillin. *International Congress Series* **544**, 222.
Leung, T. and Williams, J. D. (1978) *J. Antimicrob. Chemother.* **4**, 47.
Levine, B. B. (1973) *J. infect. Dis.* **128**(S), 364.
Levine, B. B. and Zolov, D. M. (1969) *J. Allergy* **43**, 231.
Li, J. T., Moosdeen, F. and Williams, J. D. (1982) *J. Antimicrob. Chemother,* **9**, 287.
Lindmark, D. G. and Muller, M. (1976) *Antimicrob. Agents Chemother.* **10**, 476.
Little, P. J. and Bailey, R. R. (1970) *New Zealand med. J.* **72**, 183.
Livermore, D. M. and Williams, J. D. (1981) *J. antimicrob. Chemother.* **9**, (Suppl. E), 29.
Livermore, D. M., Williams, R. J. and Williams, J. D. (1981) *J. antimicrob. Chemother,* **8**, 355.
Loder, P. B., Newton, G. G. F. and Abraham, E. P. (1961) *Biochem. J.* **79**, 408.
Lorian, V. (1980) *Antibiotics in Laboratory Medicine.* Williams & Wilkins, Baltimore, London.
McCabe, A. F., Gould, J G. and Forfar, J. O. (1961) *Lancet* **ii**, 7.
McCalla, D. R. (1977) *J. Antimicrob. Chemother.* **3**, 517.
McGuire, J. M. *et al.* (1952) *Antibiotics Chemother.* **2**, 281.
Mach, M. C., Gold, O. and Finland, M. (1973) *J. Lab. clin. Med.* **81**, 787.

Maeda, K., Osato, T. and Umezawa, H. (1953) *J. Antibiotics (Tokyo)* **6A,** 182.
Marks, M. I. (1974) *Antimicrob. Agents Chemother.* **6,** 34.
Marks, M. I., Steer, P. and Eickhoff, T. C. (1971) *Appl. Microbiol.* **22,** 93.
Martin, D. A., Sanders, C. V. and Marier, R. L. (1982) *Antimicrob. Agents Chemother.* **21,** 168.
Mason, D. J., Dietz, A. and De Boer, C. (1962) *Antimicrob. Agents Chemother*; **2,** 554.
Matthew, M. and Harris, A. M. (1976) *J. gen. Microbiol.* **94,** 55.
Maxted, W. R. (1953) *J. clin. Path.* **6,** 224.
Medoff, G., Comfort, M. and Kobayashi, G. S. (1971) *Proc. Soc. exp. Biol. Med.* **138,** 571.
Medoff, G. and Kobayashi, G. S. (1980) In: *Antifungal Chemotherapy.* p. 3, Wiley, New York.
Medoff, G., Kobayashi, G. S., Kwan, C. N., Schlessinger, D. and Venkov, P. (1972) *Proc. Nat. Acad. Sci. USA* **69,** 196.
Millar, J. W., Siess, E. E., Feldman, H. A., Silverman, C. and Frank, J. (1963) *J. Amer. med. Ass.* **186,** 139.
Miller, A. B., Fall, F., Fow, W., Gefford, M. J. and Mitchison, D. A. (1966) *Tubercle (Lond.)* **47,** 92.
Miller, G. H., Arcieri, G., Weinstein, M. J. and Waitz, J. A. (1976) *Antimicrob. Agents Chemother.* **10,** 827.
Mond, N. C., Percival, A., Williams, J. D. and Brumfitt, W. (1965) *Lancet* **i,** 514.
Montgomerie, J. Z., Doak, P. B., Taylor, D. E. M., North, D. J. K. and Martin, W. J. (1970) *Lancet* **ii,** 787.
Mouton, R. P., Glerum, J. H. and Van Loenen, A. C. (1976) *J. Antimicrob. Chemother.* **2,** 9.
Nakamura, B. (1955) *Pharmaceutical Bull. (Japan)* **3,** 379.
Nasu, M., Maskell, J. P., Williams, R. J. and Williams, J. D. (1981) *Antimicrob. Agents Chemother.* **20,** 433.
Neiss, E. S. (1973) *J. Irish med. Assoc.* **66** (Suppl.) 1.
Neu, H. C. (1976a) *Antimicrob. Agents Chemother.* **9,** 793; (1976b) *Antimicrob. Agents Chemother.* **10,** 535.
Neu, H. C. and Fu, K. P. (1980) *Antimicrob. Agents Chemother.* **18,** 582.
Neuhaus, F. C. (1967) In: *Mechanism of Action—Antibiotics I.* p. 40. Springer, New York.
Neumann, M. (1968) *Médicine et Hygiène* **846,** 1322.
Newton, B. A. (1956) *Bact. Rev.* **20,** 14.
Newton, G. G. F. and Abraham, E. P. (1955) *Nature* **175,** 548.
Nightingale, C. H., Dittert, L. W. and Tozer, T. N. (1976) *J. Amer. pharm. Ass.* **16,** 203.
Nord, C. E. and Olsson-Liljequist, B. (1981) *J. antimicrob. Chemother.* **8,** (Suppl. D), 33.
Nordenfelt, E., Helgstrand, E. and Oberg, B. (1979) *Acta path. microbiol. Scand.*, Sect. B, **87,** 75.
Oxford, A. E., Raistrick, H. and Simonart, P. (1939) *Biochem. J.* **33,** 240.
Oxford, J. S. (1975) *J. antimicrob. Chemother.* **1,** 7.
Oxford, J. S. (1979) *J. antimicrob. Chemother.* **6,** 223.
Ozaki, M., Mizushima, S. and Nomura, M. (1969) *Nature (Lond.)* **222,** 333.
Park, J. T. (1952) *J. biol. Chem.* **194,** 897.
Pavan-Langston, D. and McCulley, J. P. (1973) *Arch. Ophthalmol.* **89,** 25.
Peets, E. A., Sweeney, W. M., Place, V. A. and Buyske, D. A. (1965) *Amer. Rev. resp. Dis.* **91,** 51.
Percival, A. (1972) In: *Current Antibiotic Therapy,* p. 156. Ed. by A. M. Geddes and J. D. Williams, Churchill-Livingstone, London.
Percival, A., Corkhill, J. E., Arya, O. P., Rowlands, J., Alerganit, C. D., Rees, E. and Annels, E. H. (1976) *Lancet* **ii,** 1379.
Phillips, I. (1976) *Lancet* **ii,** 1379.
Phillips, I., Fernandes, R. and Warren, C. (1970) *Brit. med. J.* **ii,** 89.
Phillips, I. and Warren, C. (1974) *Lancet* **i,** 827.
Philpott-Howard, J. and Williams, J. D. (1982) *Brit. med. J.* **ii,** 1597.
Pollock, H., Charache, P., Nieman, R. E., Jeff, H. P., Reinhardt, J. A. and Hardy, P. H. Jr. (1972) *Lancet* **ii,** 668.
Porter, J. and Jick, H. (1977) *Lancet* **i,** 587.
Quinn, E. L. et al., (1977) In: *Current Chemotherapy,* **II.** p. 803, *Amer. Soc. Microbiol.* New York.
Rao, A. E., McFadzean, J. A. and Squires, S. (1965) *Ann. N.Y. Acad. Sci.* **130,** 118.
Rao, A. R., McKendrick, G. D. W., Velayudhan, L. and Kamalakshi, K. (1966) *Lancet* **i,** 1072.
Reading, C. and Cole, M. (1977) *Antimicrob. Agents Chemother.* **11,** 852.
Reeves, D. S. (1974) *Pers. Comm.*
Reeves, D. S., Phillips, I., Williams, J. D. and Wise, R. *Laboratory Methods in Antimicrobial Chemotherapy.* (1978) Churchill Livingstone, Edinburgh, London, New York.
Reider, J. (1963) *Arzneimittel Forschung* **13,** 81.
Richmond, M. H. (1966) In: *Biochemical Studies of Antimicrobial Drugs.* Cambridge University Press, Cambridge.
Richmond, M. H. and Linton, K. B. (1980) *J. antimicrob. Chemother.* **6,** 33.
Ridgway, G. L., Owen, J. M. and Oriel, J. D. (1976) *J. antimicrob. Chemother.* **2,** 71.
Robson, J. M. and Tebrich, W. (1942) *Brit. med. J.* **i,** 687.
Rolinson, G. N., Stevens, S., Batchelor, F. R., Cameron-Wood, J. and Chain, E. B. (1960a) *Lancet* **ii,** 2564.
Rolinson, G. N. et al. (1960b) *Nature (Lond.)* **187,** 236.
Rose, H. D. and Schrier, J. (1968) *A. J. med. Sci.* **255,** 228.
Roth, F. J. Jr. (1960) *Ann. N.Y. Acad. Sci.* **89** 247.
Roth, F. J. and Blank, H. (1960) *Archs. Derm.,* **81,** 662.
Sabath, L. D. and Abraham, E. P. (1964) *Nature (Lond.)* **204,** 1066.
Sabath, L. D., Wheeler, N., Laverdiere, M., Blazevic, D. and Wilkinson, B. J. (1977) *Lancet* **i,** 443.
Sakaguchi, K. and Murao, S. (1950) *J. agric. Chem. Soc. (Japan)* **23,** 411.
Schaeffer, H. J., Beauchamp, L., De Miranda, P., Elion, G. B., Bauer, D. J. and Collins, P. (1978) *Nature* **272,** 583.
Schatz, A., Bugie, E. and Waksman, S. A. (1944) *Proc. Soc. exp. Biol. N.Y.* **55,** 66.
Schatz, A. and Waksman, S. A. (1944), *Proc. Soc. exp. Biol. N.Y.* **57,** 244.
Schmidt, L. H., Sesler, C. L. and Hughes, H. B. (1944) *J. Pharmacol.* **81,** 43.
Scholer, H. J. (1970) *Mykosen,* **13,** 179; (1980) In *Antifungal Chemotherapy.* Wiley, Chichester.
Schweinburg, F. B. and Rutenburg, A. M. (1949) *J. Lab. clin. Med.* **34,** 1457.
Selwyn, S. (1980) In: *The Beta-lactam Antibiotics,* p. 2, Hodder & Stoughton, London.
Sensi, P., Timbal, M. T. and Maffu, G. (1960) *Experentia* **16,** 412.
Seth, A. D. and Johnston, N. A. (1980) *Lancet* **ii,** 531.
Shanson, D. C. (1981) *J. Hosp. Infect.* **2,** 11.

Shanson, D. C., Kensit, J. and Duke, R. (1976) *Lancet* **ii**, 1347.
Shipkowitz, N. L. et al., (1973) *Appl. Microbiol.* **26**, 264.
Shirley, R. L. and Moore, J. W. (1965) *New Engl. J. Med.* **273**, 282.
Shooter, R. A. (1969) *Prescribers J.* **9**, 74.
Siewert, G. and Strominger, J. L. (1967) *Proc. nat. Acad. Sci. Wash.* **57**, 767.
Simasathien Sriluck, Duangmani Chiraphun and Echeverria, P. (1980) *Lancet* **ii**, 1214.
Skirrow, M. B. (1977) *Brit. med. J.* **ii**, 9.
Slade, N. and Crowther, S. T. (1972) *Brit. J. Urol.* **44**, 105.
Slocombe, B. et al., (1981) *Antimicrob. Agents Chemother.* **20**, 38.
Smith, H., Dewdney, J. M. and Wheeler, A. W. (1971) *Immunology* **21**, 527.
Speller, D. C. E. et al., (1976) *Lancet* **i**, 464.
Spratt, B. G. (1977a) *Antimicrob. Agents Chemother.*, **11**, 161; (1977b) *Antimicrob. Chemother.* **3**, (Suppl. B), 13.
Spyker, D. A., Thomas, B. L., Sande, M. A. and Bolton, W. K. (1978) *Antimicrob. Agents Chemother.* **14**, 172.
Stamey, A. (1971) *Hosp. Practit.* **6**, 49.
Steigbigel, N. H., Reed, C. W. and Finland, M. (1968) *Amer. J. med. Sci.* **255**, 179.
Strauss, R. E., Kligman, A. M. and Pillsbury, D. M. (1951) *Amer. Rev. Tuber.* **63**, 441.
Subak-Sharpe, J. H., Timbury, M. C. and Williams, J. F. (1969) *Nature (Lond.)* **222**, 341.
Sutherland, J. M. (1962) *Postgrad. Med.* **32**, 127.
Sutherland, R., Burnett, J. and Rolinson, G. N. (1970) *Antimicrob. Agents Chemother.* **10**, 390.
Sutherland, R., Croydon, E. A. P. and Rolinson, G. N. (1970) *Brit. med. J.* **iv**, 455.
Sutter, V. L. and Finegold, S. M. (1976) *Antimicrob. Agents Chemother.* **10**, 736.
Sykes, R. B. et al., (1981) *Nature (Lond.)* **291**, 489.
Tanaka, K., Teraoka, H. and Tamaki, M. (1968) *Science* **162**, 576.
Tardrew, P. L., Mao, J. C. H. and Kenney, D. (1969) *Appl. Microbiol.* **18**, 159.
Taubman, S. B., Jones, N. R., Young, F. E. and Corcoran, J. W. (1966) *Biochim. biophys. Acta* **123**, 438.
Thadepalli, H., Roy, U., Bach, V. T. and Webb, D. (1979) *Antimicrob. Agents Chemother.* **15**, 487.
Thoburn, R., Johnson, J. E. and Cluff, L. E. (1966) *J. Amer. med. Ass.* **198**, 345.
Thomas, J. P., Baughn, C. O., Wilkinson, R. G. and Shepherd, R. G. (1961) *Amer. Rev. resp. Dis.* **83**, 891.
Thornsberry, C., Baker, C. N. and Kirven, L. A. (1978) *Antimicrob. Agents Chemother.* **13**, 78.
Tocchini-Valentini, G. P., Marino, P. and Colvill, A. J. (1968) *Nature (Lond.)* **220**, 273.
Tosch, W., Kradolfer, F., Konopka, E. A., Regos, J. Zimmermann, W. and Zak, O. (1977) In: *Current Chemotherapy*, p. 843. Amer. Soc. Microbiol. New York.
Tréfouël, J., Tréfouël, I., Nitti, F. and Bovet, D. (1935) *C. R. Soc. Biol.* **120**, 756.
Tsukamura, M. (1965) *Acta tuberc. scand.* **46**, 89.
Tybring, L. (1975) *Antimicrob. Agents Chemother.* **8**, 266.
Tybring, L. and Melchior, N. H. (1975) *Antimicrob. Agents Chemother.* **8**, 271.
Underwood, G. E., Wisner, C. A. and Weed, S. A. (1964) *Arch. Ophthal.* **72**, 505.
Utz, J. P., Garriques, I. L., Sande, M. A., Warner, J. F., McGehee, R. F. and Shadomy, S. (1975) *J. infect. Dis.* **132**, 368.
Utz, J. P., Tynes, B. S., Shadomy, H. J., Duma, R. J., Kannan, M. M. and Mason, K. N. (1969) In: *Antimicrobial Agents and Chemotherapy-1968*, p. 344. Amer. Soc. Microbiol. Ann Arbor.
Vazquez, D. (1966) *Soc. gen. Microbiol. Symp.* **16**, p.169.
Waksman, S. A. and Lechevalier, H. A. (1949) *Science* **109**, 305.
Walker, W. E., Waisbren, B. A., Martins, R. R. and Batayias, G. E. (1970) *Antimicrob. Agents Chemother.* **10**, 380.
Wallin, J., Lernestedt, J. O. and Lycke, E. (1980) *Proc. Int. Conf. human Herpes Viruses,* Atlanta, Georgia.
Watanakunakorn, C. and Bakie, C. (1973) *Antimicrob. Agents Chemother.* **4**, 120.
Waterworth, P. M. (1963) *Clin. Med.* **70**, 941.
Waterworth, P. M. (1981) *Antibiotic and Chemotherapy*, 5th edn., p. 464. Churchill Livingstone, Edinburgh.
Weaver, S. S., Bodey, G. P. and Leblanc, B M. (1979) *Antimicrob. Agents Chemother.* **15**, 518.
Weinstein, M. J., Wagman, G. H., Oden, E. H. and Marquez, J. A. (1967) *J. Bact.* **94**, 789.
WHO *Tech. Rep. Series* (1978) p. 624.
WHO *Weekly epidem. Rep.* (1974) Rec. 49. No. 37, 311.
Williams, J. D., Kattan, S. and Cavanagh, P. (1974) *Lancet* **ii**, 103.
Williams, J. D., and Leigh, D. A. (1966) *Brit. J. clin. Pract.* **20**, 177.
Williams, R. J. and Williams, J. D. (1980) *Antibiotics and Chemotherapy*, p. 63, MTP Press, Lancaster.
Wilson, H. G., Neilson, G. H., Galea, E. G., Stafford, G. and O'Brien, M. F. (1976) *Circulation* **53**, 680.
Wimpenny, J. W. T. (1967) *J. gen. Microbiol.* **47**, 389.
Winningham, D. G., Nemoy, N. J. and Stamey, T. A. (1968) *Nature (Lond.)* **219**, 139.
Wise, R., Andrews, J. M. and Bedford, K. A. (1978) *Antimicrob. Agents Chemother.* **13**, 389.
Wise, R., Andrews, J. M. and Bedford, K. A. (1980) *J. antimicrob. Chemother.* **6**, 197.
Wise, R., Andrews, J. M. and Patel, N. (1981) *J. antimicrob. Chemother.* **7**, 531.
Witchitz, J. L. and Chabbert, Y. A. (1972) *Ann. Inst. Pasteur* **122**, 367.
Wood, M. J., Farrell, W., Kattan, S. and Williams, J. D. (1975) *J. antimicrob. Chemother.* **1**, 323.
Woods, D. D. (1940) *Brit. J. exp. Path.* **21**, 74.
Woolley, D. W. (1944) *J. biol. Chem.* **152**, 225.

6

Bacterial variation

Naomi Datta and Marilyn E. Nugent

Introductory	145
Mutation and mutagenesis	146
Microlesions	147
Macrolesions	149
UV-induced mutation	150
Other types of mutation	150
Flip-flop mechanisms	150
Insertion sequences and transposons	151
Specificity and mechanism of insertion	153
Deletions and insertions	154
Transposon-encoded proteins	154
Ubiquity of insertion sequences	154
Acquisition of new genes	155
Transformation	156
Physiological competence	156
Artificial competence	156
Conjugation	157
Determined by F	157
Conjugative pili	157
Mating pair formation	158
Surface exclusion	158
Transfer of DNA	158
Hfr and F' formation	159
Mobilization by autonomous F	160
Intergeneric F-mediated conjugation	160
Conjugation other than F-mediated	160
Repression of transfer genes	160
Conjugation in other genera	161
Phage-mediated conjugation	161
Transduction	162
Phage conversion	162
Protoplast fusion	163
Mapping of chromosomal genes	163
Plasmids	163
Essential genes	164
Genes determining conjugation	164
Other plasmid genes	164
Antibiotic resistance	164
Resistance to other agents	166
Virulence factors	166
Metabolic characters	167
Phage inhibition	167
Mapping of plasmids	167
Evolution and classification of plasmids	168
Bacterial variation and epidemiology	172
Genetic engineering	173

Introductory

Among bacteria, the range of metabolic capabilities is immensely wide; even those that infect or colonize man and other animals vary greatly in their morphology, physiology and metabolism. Variation, as discussed here, refers not to variation amongst all bacteria but to changes that may take place in a bacterial cell line.

All the characteristics of bacteria, as of other forms of life, are genetically determined and encoded in DNA. These expressed characteristics are referred to as the **phenotype** of a strain and the genetic information encoding them as the **genotype**. The potential to vary the expression of characteristics, in response to environmental changes, is itself genetically determined; for example, the β-galactosidase of *Escherichia coli* is not synthesized continuously, but its production is turned on when its substrate, lactose, is present. The genes that determine repression of β-galactosidase synthesis when it is not required, induction of enzyme synthesis when it is required, and the polypeptide sequence of the enzyme itself are all encoded in one DNA sequence, the *lac* operon (chapter 3). This is just one of many and complex systems that allow bacteria to make best use of the nutrients available. Metabolic changes in response to changing environment, within the limits of the existing genetic make-up of the organism, are described as *phenotypic* variations. *Genotypic* variation, with which this chapter is concerned, depends upon a change in the genetic information, i.e. in the DNA, that determines the phenotype. Such changes can occur by mutations of various kinds that change the sequence of DNA or by the acquisition of new genetic information from an outside source.

In general, phenotypic change occurs in all the cells in a bacterial culture and is rapidly reversible. Genotypic change, when it occurs, affects an individual cell in a culture, and is irreversible. There are, however, exceptions to these generalizations, and it is not always easy to be sure into which category a variation fits. For example, lysozyme treatment of a culture of *Bacillus subtilis* converted the cells to protoplasts. Removal of lysozyme did not necessarily result in reversion to growth of normal bacilli. The culture either continued to grow as protoplasts or showed mass reversion to the bacilliary form, according to the cultural conditions. This was an example of a phenotypic change that, in some circumstances, could appear as heritable and permanent (Landman and Halle 1963). On the other hand, changes resulting from mutation may give the impression of being phenotypically determined because bacteria multiply so rapidly; a mutant clone, with some temporary selective advantage, can overgrow a population of unaltered cells within a few hours. If the conditions are reversed, the mutant may be at a selective disadvantage and in its turn be overgrown by cells of the original type, either survivors or back-mutants.

Bacteria are haploid organisms with no true nucleus or sexual reproductive cycle. The great expansion of knowledge of bacterial genetics followed the discovery (or discoveries) that genetic crosses of various kinds occur in bacteria. The organism about which most is known is *Escherichia coli* K12, whose single chromosome is a circular DNA molecule, circumference approximately 2 mm, i.e. about 1000 times longer than the *Esch. coli* cell (chapter 2). Other bacteria whose genetics have been studied also have single circular chromosomes—this is true of bacteria related to *Esch. coli* such as *Shigella* and *Salmonella* (Sanderson 1976) and also unrelated ones, *Pseudomonas aeruginosa* (Holloway *et al.* 1979) and *Bacillus subtilis* (Henner and Hoch 1980). Genes are mapped on these chromosomes and their linkages demonstrate the circular structure that, in the case of *Esch. coli*, has also been demonstrated physically (Cairns 1963). In many bacteria of medical importance, such as streptococci, staphylococci, clostridia, genetic studies are under way, but mapping many genes on chromosomes is not yet possible.

Bacterial chromosomes must, of course, replicate before cell division occurs and be distributed (*partitioned*) so that each daughter cell has its own copy. A complex organization exists to ensure that this happens. There may be other DNA molecules in a cell line, besides the chromosome, encoding heritable traits: these are plasmids. Such molecules are *replicons* (Jacob and Brenner 1963) and all bacterial replicons, as far as is known, are circular. The chromosome is the replicon in which the essential genes are encoded. A replicon cannot be made by circularizing any piece of DNA. It must include features of a natural replicon; a base sequence from which replication originates (*ori* sequence) and genes to promote and control replication.

Mutation and mutagenesis

Mutation can be defined as any permanent alteration in the sequence of bases of DNA even if this alteration does not have a detectable phenotypic effect. A given gene can exist in a variety of different forms as a result of mutational changes in its nucleotide sequence; the different mutational forms of a gene are called *alleles*. The form in which a given gene exists in a microorganism as it is first isolated from nature is defined as the *wild type* allele of that gene; altered forms resulting from mutations are called mutant alleles.

Mutations occur either spontaneously, or they are induced by mutagens which cause alterations in the structure of DNA. The spontaneous nature of mutation was demonstrated by an experiment devised in 1943 by Luria and Delbrück known as the **fluctuation test**. A small fluctuation in the number of phage-resistant colonies per plate was seen when approximately 10^9 *Esch. coli* cells were spread on plates seeded with bacteriophage. When small inocula (approximately 10^3 cells) of the same *Esch. coli* were grown up in a series of 50 separate broth cultures, plating of 10^9 cells from each showed a large fluctuation in the number of resistant colonies per plate. The results were explained by random spontaneous mutations occurring in the original culture, arising at different times during its growth. This experiment is depicted in Fig. 6.1.

A more direct way of demonstrating the spontaneous nature of mutations was devised by the Lederbergs in 1952. Of a phage-sensitive *Esch. coli* strain 10^8 cells were grown on a nutrient agar plate. These were replica-plated to a phage-coated nutrient agar plate on which only phage-resistant clones could grow and form colonies. Each colony came from an inoculum of a phage-resistant cell or cells present on the master plate. Approximately 10^5 cells from the location on the master plate corresponding to a colony on the replica-plate were transferred into a tube of broth

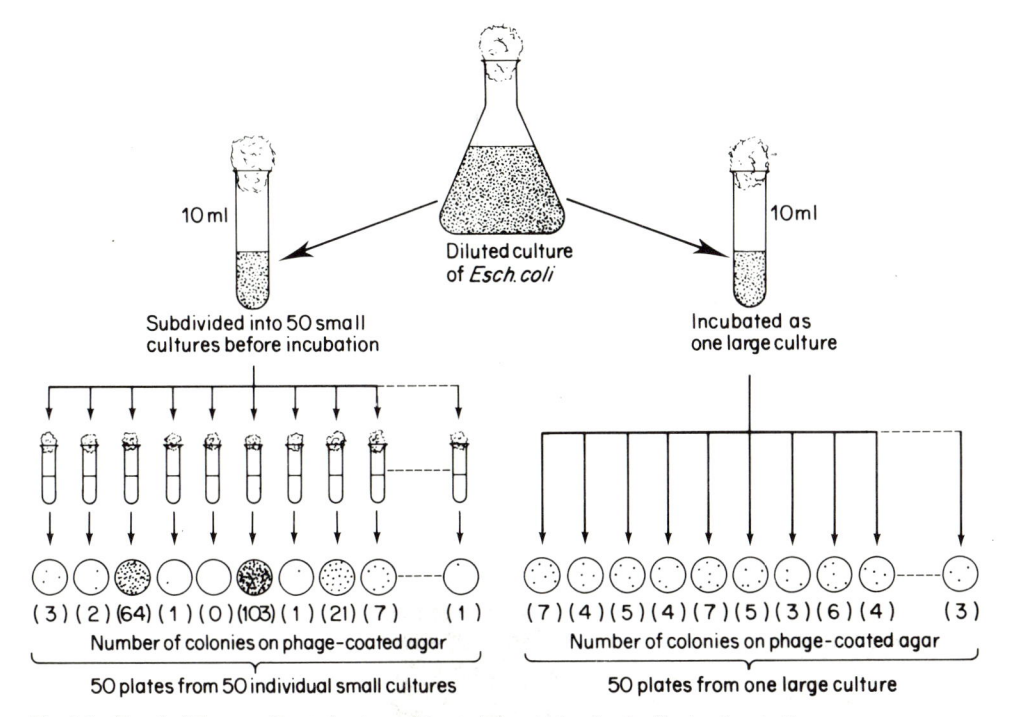

Fig. 6.1 Proof of the non-directed nature of bacterial mutation by the 'fluctuation test'.

and incubated for several hours. From this broth 10^5 cells were then plated on nutrient agar which was again replicated to phage-coated agar. By picking each time from a region on the *master plate* shown by replica plating to contain phage-resistant mutants it was possible eventually to obtain a pure culture of phage-resistant mutants. During the entire process the population used in the purification was never exposed to phage, demonstrating that resistance occurred in the complete absence of the selective agent.

Mutagenesis

Most of our understanding of the mechanism of mutation derives from experiments in which mutations are induced by chemical agents whose mode of action on DNA is known. Experiments on the chemical basis of mutation have been done mainly with *Esch. coli* and with a number of phages for which *Esch. coli* is the normal host. In experiments with bacteria and mutagens that are non-toxic, the mutagen is added to the growth medium for a number of generations after which the cells are plated on a medium selective for a particular class of mutant. Auxotrophic mutants requiring certain growth factors may be found by replica-plating to media lacking various growth factors, on which they will fail to form colonies. Their numbers can be increased to detectable levels by *penicillin screening*. A whole mutagen-treated culture is inoculated into liquid medium containing a penicillin but lacking a particular growth factor (amino acid, vitamin, purine or pyrimidine). The cells that do not require the factor start to grow and are killed by the penicillin. The relevant auxotrophs cannot grow, are protected from penicillin, and can be isolated upon removal of the latter.

Mutations can be divided into two groups—micro and macrolesions—depending on the extent of the alteration in the DNA base pairs. Microlesions are also known as point mutations.

Microlesions

Point mutations are of two classes, base pair substitutions and frame shift mutations. The first class affect one base pair at a time and can be subdivided into *transitions* and *transversions*. A transition occurs when a purine is replaced by a different purine, or a pyrimidine by a different pyrimidine. Transversions are those mutations in which a purine replaces a pyrimidine or vice versa.

All four of the bases in DNA can exist in different tautomeric forms which are related to each other by single proton shifts. The tautomeric forms arise when the keto (C=O) group is changed to the enol (C–OH) form or when the amino ($-NH_2$) group is changed to the imino form (=NH). The enol and imino groups can occur naturally but the keto and amino forms are the usual ones. The tautomeric forms can hydrogen-bond with a normally non-complementary base but pairing is still between purine and pyrimidine. These different forms are illustrated in Fig. 6.2. As well as occurring naturally, transitions may be formed by the incorporation of *base analogues* into the DNA. Among the most important analogues are the halogenated pyrimidines, particularly 5-bromouracil. This can replace thymine in DNA and pair with adenine. The halogen atoms are strongly electron negative and they pull electrons from the oxygen of the keto group resulting in an enol group. Bromouracil and similar analogues therefore tautomerize much more frequently than the natural base. Bromouracil in the keto form can pair with adenine in the template. If this undergoes a tautomeric shift to the enol form it can now base-pair with guanine resulting in an AT→GC transition. Similarly if it is incorporated in the enol form and base-paired with guanine, a tautomeric shift to the keto form results in a GC→AT transition.

Transitions can be formed by the action of nitrous acid. Nitrous acid deaminates amino groups converting them to hydroxyl groups. Deamination of the adenine of an AT pair will convert the adenine to hypoxanthine which can base-pair with cytosine.

Alkylating agents also cause transitions. Alkylating agents exert a variety of biological effects including mutagenesis and carcinogenesis. They all carry one, two or more alkyl groups in reactive form. An example is ethylmethane sulphonate (EMS), CH_3CH_2-O-SO_2-CH_3. The action of alkylating agents on DNA is complex. They are known to react with purine bases, particularly with guanine, and the bifunctional alkylating agents may thus bring about cross-linking between the opposing strands in the DNA molecule.

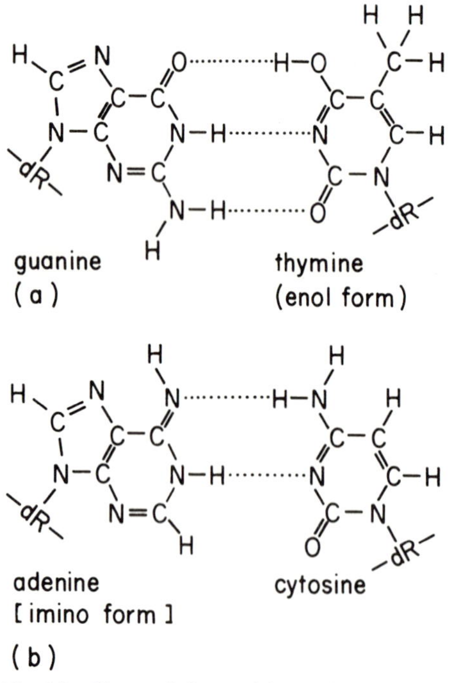

Transversions, substitution of a purine for a pyrimidine, or vice versa, may arise spontaneously and may also be induced by various mutagenic treatments. They nearly always appear along with transitions. Much less is understood about the mechanisms that generate transversions than about those generating transitions. Transversions may occur by a mechanism known as depurination, whereby purines are converted to pyrimidines. Depurination may result from two different treatments to DNA, heating at low pH or treatment with alkylating agents.

Acridine dyes induce a unique class of mutations known as *frame-shift mutations*. The properties of such mutants can be accounted for by assuming that acridines cause the insertion or deletion of one or a few base pairs in DNA. Since the message encoded in messenger RNA must be read as a series of triplets, insertion or deletion causes a shift in the reading frame with the result that all codons beyond the site of mutation are changed. A mutation caused by insertion can be reverted by a deletion at a site many base pairs away. A series of codons will be altered but beyond the deletion the mRNA will read normally. Although the resulting protein contains a few altered amino acids, unless they are at the active site of the protein, it may be functionally active. Although they revert to the wild type spontaneously and can be induced' to revert at a higher rate by acridine dyes, these mutations are never reverted by base analogues, nitrous acid or hydroxylamine. When reversion does occur, it always proves to be the consequence of a second-site, or suppressor mutation, within the same gene as the primary mutation. Furthermore, the suppressor mutation, when separated from the primary mutation by recombination, acts exactly like the primary mutation in every way.

One of the best known acridine dyes is proflavine. It is thought that proflavine intercalates between the stacked bases of DNA. It has a flat planar structure with a size similar to that of a base pair. It was suggested that proflavine exerted its mutagenic effect on DNA undergoing recombination. This was supported by experiments with partial diploids of *Esch. coli*. Partly diploid zygotes are formed in bacterial conjugation (p. 157) as the result of incomplete transfer of the chromosome of one bacterial cell to another. When zygotes of *Esch. coli* are treated with proflavine, frame-shift mutations are induced in the diploid region of the chromosome. The mechanism of recombination involves breakage and reunion of DNA

Fig. 6.2 Changes in base pairing as the result of tautomeric shifts in the enol form (a), thymine forms hydrogen bonds with guanine, instead of with adenine. In the imino form (b), adenine forms hydrogen bonds with cytosine, instead of with thymine. Similar shifts in guanine and cytosine will also cause changes in base pairing.

molecules (Low and Porter 1978). Proflavine appears to act at an early stage in this process, perhaps by binding to DNA in which a single-strand break has occurred thereby stabilizing a mispairing of strands. The duplex partly unwinds and the free end occasionally undergoes a displacement of its pairing sites, a displacement that may be stabilized by binding a proflavine molecule.

Transitions, transversions and frame-shift mutations lead to alterations in the sense of messenger RNA. A *missense* mutation gives rise to a triplet codon that encodes an amino acid different from that of the wild type. The effect of a missense mutation on a resulting polypeptide is variable depending on the position of the mutant amino acid. In some cases the polypeptide may show no alteration in function, in others all function is lost. Missense mutations can give rise to temperature-sensitive gene products in which the mutant protein can function at a lowered (*permissive*) temperature but not at higher (*lethal*) temperatures at which the wild type is still effective. These are known as *conditional-lethal* mutants. Many other examples of conditional-lethal mutants exist in which a normally lethal mutation can survive under an altered set of physiological conditions. Mutations may also give rise to the *nonsense* codons UAG, UAA or UGA in which case the product of the mutated gene is a polypeptide terminated prematurely, and incomplete.

The introduction of a nonsense codon into an early gene of an operon produces an effect called *polarity*. Translation ceases at the nonsense codon, and ribosomes are dissociated from messenger RNA. The discharged ribosomes may be able to reattach and re-initiate translation and the subsequent genes of the operon produce reduced amounts of protein. In some cases the polar effect is very strong and no protein is produced by the genes distal to the nonsense mutation. It is also thought that polarity may result from defective transcription when the amount of messenger RNA for genes subsequent to the mutation is reduced.

Suppressor mutations alter a mutated gene product so that it is functional again. These mutations are either intragenic, occurring within the same gene as the primary mutation, or are intergenic. Intragenic suppressors may act in two ways. One way is to insert or delete DNA and thereby compensate for a frame-shift mutation; the other way is to cause an amino acid substitution which cancels out the effect of the primary mutation. Intergenic suppressors act at the level of translation. They give rise to altered transfer RNA (tRNA) molecules that can read the nonsense codons UAA, UAG and UGA as an amino acid instead of terminating the translation.

Suppression must be highly inefficient if the cell is to survive. For example if a mutant enzyme is made in its suppressed form 5 per cent of the time, the other cell proteins will be made correctly for 95 per cent of the time. For many enzymes a level of 5 per cent of wild type activity is sufficient to permit growth of the cell.

Macrolesions

Alterations of the DNA molecule involving large numbers of base pairs fall into three groups: deletions, duplications, and inversions. Deletions are recognized by their inability to revert and by their failure to recombine with two or more point mutations. A number of different mechanisms can be postulated to generate deletions, involving errors in DNA replication, in genetic recombinations and in DNA repair. Errors in DNA replication (copy errors) are the most simple to imagine. A disruption in the binding between parental template, polymerase and daughter template could lead to either of two results outlined in Fig. 6.3. A jump ahead in copying produces a deletion in the

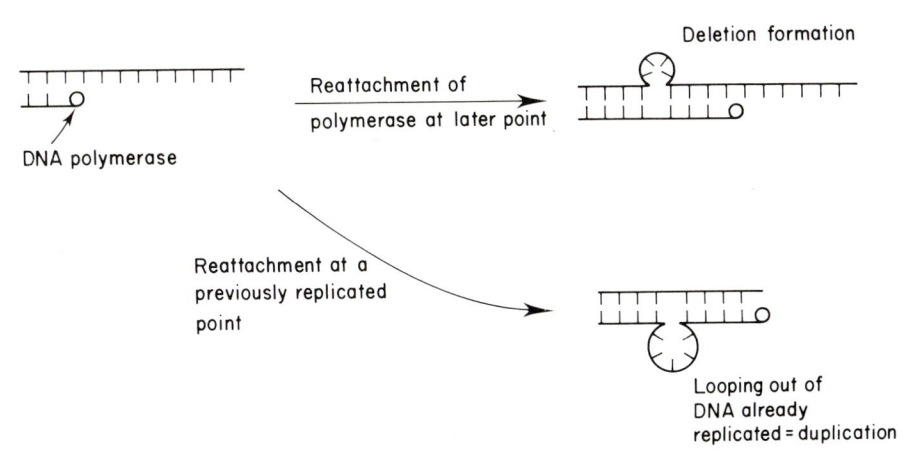

Fig. 6.3 Errors in DNA replication.

150 *Bacterial variation* Ch. 6

daughter strand or a jump behind produces a duplication.

Macrolesions may also be produced as a result of inter and intrachromosomal recombination as outlined in Fig. 6.4.

All three types of macrolesion may be produced by insertion sequences (p. 151).

UV-induced mutation Mutations may also be induced by ultra-violet light. DNA absorbs UV light strongly; the absorption maximum of DNA lies at a wavelength of 260 nm. Cells are rapidly killed by UV absorption and a high rate of mutation occurs among the survivors. Covalent bonds are formed between pyrimidine residues adjacent to each other on the same DNA strand, forming pyrimidine dimers. These dimers distort the shape of the DNA molecule and interfere with normal base pairing.

Treatments which lead to removal or cleavage of dimers also reverse most of the mutagenic effect of UV light. If UV-treated cells are immediately irradiated with visible light in the range of 300 to 400 nm, both mutation frequency and lethality are greatly reduced—this is called *photoreactivation* and results from the activation by light, at that particular wavelength, of an enzyme which hydrolyses pyrimidine dimers. The bacterial cell also has a set of enzymes for *dark repair* of UV light damaged DNA. Postirradiation conditions favouring the repair process whilst inhibiting DNA replication cause many potential UV light induced mutations to be lost.

Other types of mutation Spontaneous mutations may occur in the absence of known mutagenic treatment. There are probably many different mechanisms that can produce spontaneous mutations. Many products or intermediates of cell metabolism are demonstrably mutagenic; these include peroxides, nitrous acid, formaldehyde and purine analogues. Some spontaneous mutations may therefore in reality be induced by endogenous mutagens.

A mutation in one of the genes of *Esch. coli* and in a similar gene in *Salmonella typhimurium* has been shown to cause an increase in the spontaneous mutation rate for cell loci by a factor of 100 to 1000. These mutations, caused by the allele of the mutator locus, are transitions. The product of the bacterial mutator gene has not been identified, but there is a similar mutator gene in coliphage T4, whose product has been identified as DNA polymerase.

Flip-flop mechanisms Bacteriophage Mu contains a sequence of 3000 base pairs (bp) called the G segment which undergoes inversion. Inversion is mediated by a polypeptide encoded by a gene called *gin* that maps adjacent to the invertible region. The G segment is analogous to an insertion sequence and has inverted

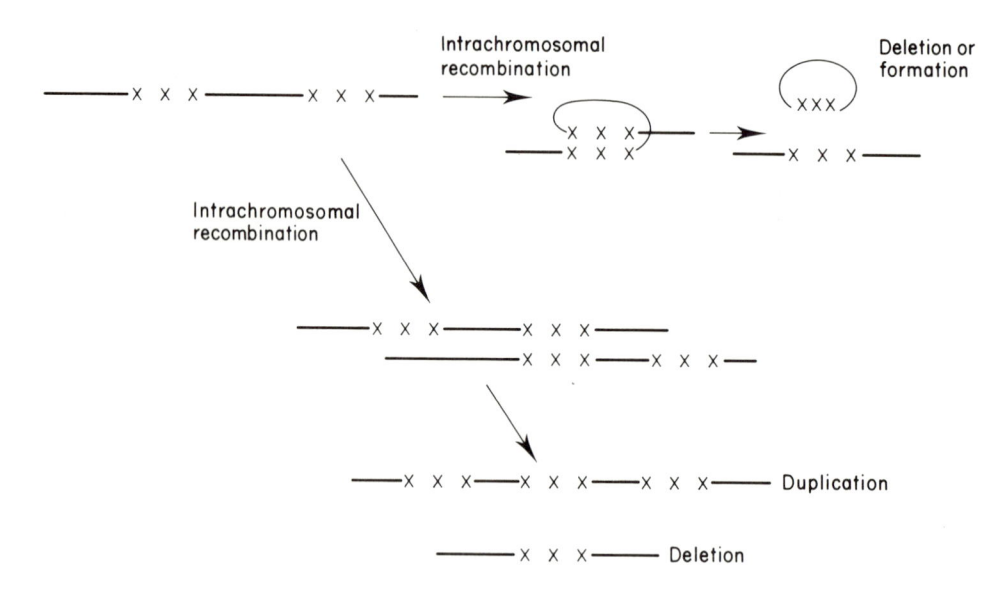

Fig. 6.4 Inter- and intra-chromosomal recombination.

repeat sequences at its end. The G segment controls host range, since its inversion changes the expression of genes which make proteins involved in specifying the host range of phage Mu. In one orientation (G+) the S and U proteins are made and the bacteriophage can infect *Esch. coli* K12. When G is in the opposite orientation, S' and U' proteins are made and the bacteriophage can infect strains of *Citrobacter* or *Shigella*. Other bacteriophages have analogous regions to G. A 3000-bp invertible loop with a great deal of homology to the G region of Mu has been found in bacteriophages P1 and P7.

In *Salmonella*, flagellar antigen phase variation involves a similar 'flip-flop' mechanism. Two genes, H1 and H2, code for different subunits of the flagellar protein, flagellin. The cell can alternate between expression of these two genes. A gene, *rh1*, codes for a repressor of H1 expression. When the cell is in phase II the H2 and *rh1* genes are transcribed. In phase I neither H2 nor *rh1* gene products are formed and the H1 gene is expressed. A 970-bp sequence adjacent to or overlapping the gene that encodes the H2 flagellar antigen is capable of inversion. This invertible region contains a promoter for the H2 gene and a gene called *hin* whose product mediates the inversion. The *hin* gene shows extensive homology with the transposase gene of Tn3 and with the *gin* of bacteriophage Mu.

Insertion sequences and transposons

During the past 20 years evidence has been growing for the existence of another type of mutation causing genetic variation. The first such to be recognized in bacteria was by Lederberg who in 1960 described an unusual class of Gal$^-$ mutants in *Esch. coli*. They reverted to Gal$^+$ at a frequency of 10^{-6} to 10^{-8} and this reversion frequency was not enhanced by mutagens, suggesting that the mutation was not a deletion or a point mutation. If the mutation was in the early genes of the galactose operon it could prevent the translation of any subsequent genes in the operon. It was unlike the polar mutations described earlier, which result from nonsense mutations in a gene, since these produce a reduction in the expression of subsequent genes in an operon and not complete cessation of expression. The new type of mutation was described as being a strong polar mutant. From measurements of the density of λgal transducing phages derived from the mutants it was shown that their *gal* operon contained extra DNA whilst revertants to Gal$^+$ had normal quantities of DNA. The mutation apparently resulted from the insertion of an extra piece of DNA now called an *insertion sequence* or IS (Jordan *et al*. 1968). Insertion of this extra DNA was not dependent on the *recA* gene product.

The five best characterized insertion sequences have been designated IS1 to IS5 (Table 6.1a). They have been found in several bacterial chromosomes and also in several plasmids. In 1974 a new class of insertion sequence encoding a recognizable gene product, the TEM-1 β-lactamase determining ampicillin resistance, was found by Hedges and Jacob. This new class was called a **transposon**. Since 1974 many other transposons carrying various genes have been identified. Because of their ability to insert into different chromosomes and plasmids they have been called 'jumping genes'. A few are described in Table 6.1b. Bacteriophage Mu resembles insertion sequences in that it can insert at many sites in the *Esch. coli* chromosome and mediate a variety of chromosomal rearrangements. Mu DNA, like the DNA of insertion sequences, does not replicate in a free form but only when inserted in host DNA. Insertion of Mu therefore amounts to transposition from one site to another (Chapter 7).

There are many similarities between insertion sequences and transposons. Heteroduplex analysis of plasmids containing insertion sequences or transposons reveals characteristic stem and loop structures visualized by the electron microscope (Fig. 6.5). This structure results from hybridization between bases at the ends of the IS or transposons which are arranged as inverted repeats i.e. the 5' to 3' sequence at one end is the same as the 5' to 3' sequence in the opposite DNA strand at the other end of the IS or transposon. The length of some of these inverted repeats is given in Table 6.1b. Several transposons are segments of DNA that are mobile because they are flanked by IS, for example Tn9 and Tn1681 are flanked by repeats of IS1.

It is becoming increasingly clear that insertion sequences and transposons are important in bringing about various changes in genetic information. They cause mutations by inserting into genes and cause all of the macrolesions previously described, deletions, duplications and inversions. By their very nature they can disseminate the genes that are encoded by them. In addition they can affect the expression of neighbouring genes by virtue of the transcriptional start and stop signals that they carry. IS1 contains nonsense codons in all three reading frames within the first 100 nucleotides. The strong polar effects are caused by these nonsense codons and the presence of a long untranslated stretch of messenger RNA (distal to these) will result in an interaction between a termination factor called *rho* with RNA polymerase to switch off transcription.

IS2 exerts polar effects when inserted in one of its two orientations. In orientation I, IS2 contains a *rho*

Table 6.1 Insertion sequences and transposons

(a) Insertion sequences

Insertion Sequence	Occurrence in *Esch. coli*	Length (b.prs)	Inverted Repeat b.prs	References
IS1	5–8 copies on chromosome	768	18/23	Fiant *et al.* 1972 Grindley 1978
IS2	5 on chromosome 1 on F	1327	32/41	Fiant *et al.* 1972 Saedler and Heiss 1973
IS3	5 on chromosome 2 on F	1400	32/38	Fiant *et al.* 1972 Hu *et al.* 1975
IS4	1 or 2 on chromosome	1400	16/18	Fiant *et al.* 1972 Habermann *et al.* 1979
IS5	Unknown	1250	Short	Blattner *et al.* 1974

(b) Transposons

Transposon	Marker	Length (b.prs)	Inverted Repeat	References
Tn1, 2 and 3	Ampicillin R	4957	38	Hedges and Jacob 1974 Cohen *et al.* 1979
Tn4	Ampicillin R Streptomycin R Sulphonamide R	20500	Short	Kopecko and Cohen 1975
Tn5	Kanamycin R	5400	1500	Berg *et al.* 1975
Tn7	Trimethoprim R Streptomycin R	14000	n.d.	Barth *et al.* 1976
Tn9	Chloramphenicol R	2638	18/23	Gottesman and Rosner 1975
Tn732	Gentamicin R Tobramycin R	11000	n.d.	Nugent *et al.* 1979
Tn951	*lac*	16000	Short	Cornelis *et al.* 1978
Tn1681	Heat-stable enterotoxin	2088	768(IS1)	So *et al.* 1979

b.prs = base pairs
n.d. = none described

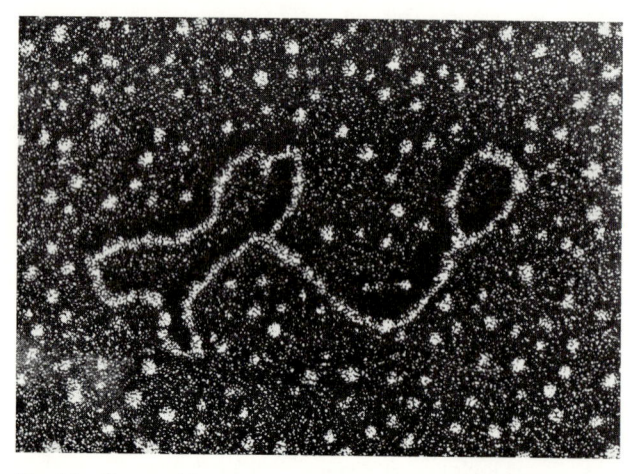

Fig. 6.5 Stem and loop structure characteristic of a transposon. The electron micrograph (magnification × 115 000) shows a heteroduplex molecule formed from one strand of DNA of a small plasmid with one strand from the same plasmid, carrying an inserted transposon. The stem represents the inverted repeat sequences that are part of the transposon. Electron micrograph by S. N. Cohen (from Cohen and Shapiro 1980, with permission).

dependent transcription termination site. In orientation II it contains a promoter sequence with the ability to switch on gene expression. Mutations in wild type IS2 which create promoters have also been described which result from duplications and rearrangement of pre-existing sequences.

Specificity and mechanism of insertion

When transposable elements are inserted into recipient DNA they produce duplications of part of the target site on either side of the inserted DNA. Duplications of five base pairs adjoining the transposed DNA are seen for IS2, Tn3 and $\gamma\delta$, nine base pairs are seen with IS1, Tn5, Tn9 and Tn10 and 11 base pairs with IS4. It is proposed that a staggered cut is made in the recipient DNA molecule into which the transposon inserts and the resulting single-stranded sequences are filled in at the end of transposition.

A wide range of insertion specificities has been found for different transposons. Some insert at many places on a plasmid with little recognizable specificity, others such as IS4 insert into one position in the *gal* region of the chromosome. Work with Tn10 suggests that it does not recognize specific target sites, but may interact with specific recognition sites before transposing into a target site. This is similar to the mechanism of action of some restriction enzymes. Recent work with IS1 and Tn9 also suggests that there is more frequent transposition into AT-rich regions. RNA polymerase binding sites are also characteristically AT-rich, and it is supposed that this is due to the need to denature DNA during initiation of transcription. An analogous step in transposition may be facilitated by denaturation of target site DNA.

The process of transposition is not dependent on the *recA* gene product, and a copy of the transposon remains at the original position when transposition occurs. Three models have been proposed for the 'mechanism' of transposition. Grindley and Sherratt (1979) propose that one strand of the transposon is attached to a staggered nick at the target site and this strand is copied followed by the return of the donor strand to its donor molecule. The new strand on the recipient molecule is used as a template to make the complete transposon.

A similar model has been proposed by Shapiro (1979) and is outlined in Fig. 6.6. During transposition a co-integrate structure is made in which donor and recipient replicons are fused, with a directly repeated copy of the transposon at each juncture point. Transposition is completed by recombination between the direct repeats of the transposon. These co-integrate structures have been seen in the transposition of IS1,

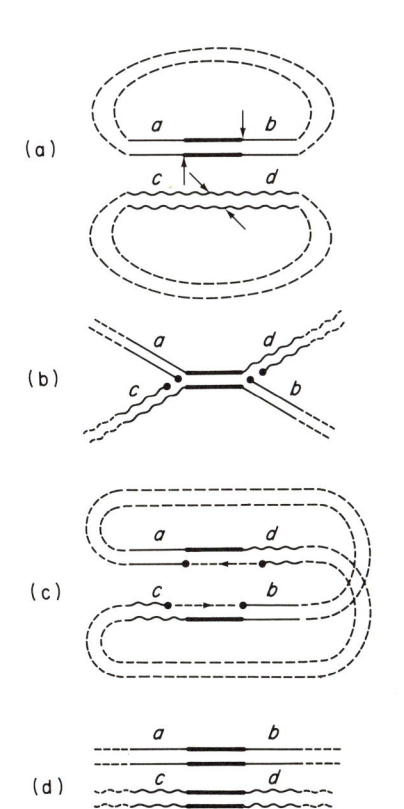

Fig. 6.6 A model for transposition (Shapiro 1979). (a) Single-stranded cuts are made at the ends of the transposable element, and a staggered cut of opposite polarity is made in the target DNA. (b) One end of the transposable element is attached by a single strand to each protruding end of the staggered cut. Two replication forks are thus created, and replication may proceed to copy the transposable element. (c) Semi-conservative replication has generated two new transposable elements and short direct repeat strands. If *ab* and *cd* were circles, *a*, *b*, *c* and *d* would now be covalently connected and transposable elements would form the joint regions of a fused replicon. (d) A site-specific cross-over between the transposable elements would resolve the cointegrate into the starting replicon *ab* and the target replicon *cd*, which now has a copy of the transposable element flanked by a short direct repeat.

154   *Bacterial variation*   Ch. 6

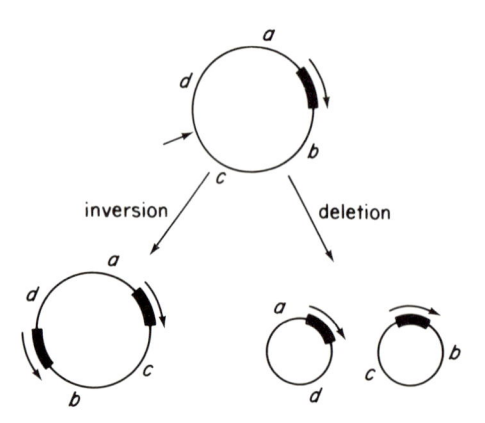

Fig. 6.7 Model for the generation of deletions and inversions (Shapiro 1979). If a transposable element residing between *a* and *b* inserts at a point between *c* and *d* in the same replicon, in one orientation a deletion will result, fusing the transposable element to *d* and generating a second copy of the transposable element fused to the deleted material, *b* and *c*. In the other orientation, an inversion of *b* and *c* will result, fused to *d* by a second copy of the transposable element in inverted orientation.

γδ, Mu, Tn3, Tn5 and Tn9. However, the *chi* structure (Fig. 6.6b) has not been seen in the electron microscope. This model differs from the early model of Grindley and Sherratt, in that replication to generate the new copy of the transposon is semi-conservative and presumably uses the host replication machinery.

A model in which the transposon forms a tandem repeat of itself which is active in transposition has also been proposed.

Deletions and insertions   Deletion formation was first seen with IS1. In the *Esch. coli* chromosome, the frequency of deletions near IS1 was 100 to 1000-fold more than that of spontaneous deletions elsewhere. Other IS and transposons have also been shown to mediate deletion formation. Several studies have demonstrated that the ends of the element remain intact after deletion formation with IS1, IS2, Tn3 and Tn9. It is suggested that deletions occur when a transposon inserts into a different position on the same replicon and recombination occurs between the two copies.

This is outlined in Fig. 6.7. If the orientation of the event is reversed, so that the opposite strand of the molecule receives the insertion sequence, an inversion results instead. Such inversions have been seen with IS1, Mu and Tn10.

Transposon-encoded proteins   Transposon 3 has been studied in detail by several research groups. As well as the TEM-1 β-lactamase, it encodes a transposase and a repressor which regulates transposition (Heffron *et al.* 1979). In addition to these genes which function in *trans*, Tn3 contains sites required in *cis* for normal transposition. These are the inverted repeat structures at the end of the transposon, which may be recognition sites for transposase, and a site near the centre of the transposon (Fig. 6.8). The latter is called the internal resolution site (IRS) and is required for resolution of co-integrates. The sequence of the IRS is similar to that of the terminally repeated sequences but the IRS cannot be substituted for a terminal repeat and still allow transposition. In addition to IRS there is a resolution function which may be supplied in *trans*.

Genetic and sequencing studies have also identified genes and gene products which mediate and regulate transposition in Tn5, Tn903 and bacteriophage Mu. However, no *in vitro* assay for transposition is available, nor has any precise enzymic activity been assigned to the transposases which are known.

Ubiquity of insertion sequences   The importance of insertion sequences and transposons in bringing about bacterial variation is becoming increasingly clear. However, they are not restricted to prokaryotes and may be responsible for a number of phenomena observed in eukaryotes. They can act as insertional mutagens in *Drosophila*. The control of yeast-mating type, antigenic variation in trypanosomes, and mammalian antibody diversity may all be due to DNA rearrangements analogous to those caused by insertion sequences. Several retroviruses (RNA tumour viruses) have recently been sequenced, and there are similarities between these and insertion sequences. These viruses integrate into chromosomal DNA of vertebrates and in this provirus state they all have inverted repeats of a few hundred base pairs at each end.

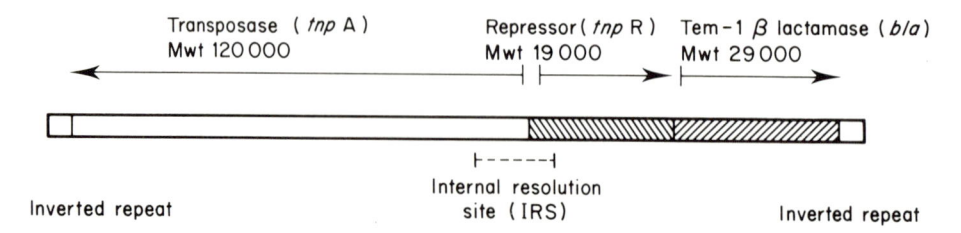

Fig. 6.8   Map of Tn3 showing the location of its three genes. The arrows indicate the direction of transcription. The internal resolution site (IRS) is necessary for resolution of cointegrates during transposition.

Acquisition of new genes

Genetic diversity in bacteria growing in isolation can occur only by mutations, as described in the previous section. Genetic recombination greatly enhances, by reassortment, the diversity created by mutations. In mixed populations of bacteria, diversity is provided by the transfer of DNA from one cell to another, with incorporation of new genes into the recipient cell, so that they are replicated and inherited by the descendants of that cell. The mechanisms of gene transfer, briefly described in this chapter, are transformation, transduction, conjugation and cell fusion.

Newly acquired genetic information, acquired by any of these four mechanisms, is of two kinds. The introduced DNA is either substituted for part of the recipient's own DNA or added to its complement. *Substitution* happens when chromosomal DNA from the same, or a closely related, species is introduced. The DNA is usually only a part of the chromosome, not in itself a replicon. By an efficient but completely mysterious means, it is brought into line with homologous DNA of the recipient, and recombination by a cross-over process occurs (Fig. 6.9). The recipient cell's copy of the exchanged DNA is lost. This is *general recombination* and requires a functional recombination (*recA*) gene. In many genetic experiments, the crossing-over involves wild-type genes and their mutant alleles. For example, a length of DNA that includes lactose-fermenting genes (*lac*+) from *Esch. coli* K12 is introduced into a K12 strain that is non-lactose fermenting because of a mutation in its *lac* operon. Recombination restores the *lac* genes to the wild type and the mutant sequence is eliminated. Genes in naturally occurring bacteria may also be allelic, for example those determining H antigens of *Salmonella*. When part of the chromosome of *Salm. paratyphi B* is transferred into *Salm. typhimurium*, recombination can insert the sequence that encodes antigen b into the chromosome, with elimination of the gene for antigen i. New salmonella serotypes can thus be created. Phylogenetic relationships among Enterobacteriaceae were reviewed by Sanderson (1976); and see Chapter 21.

Other newly acquired DNA is *added* to the genome of the recipient cell. Plasmids (p. 163) do not require recombination with a resident DNA molecule to survive; they are independent replicons. Insertion sequences (IS) or transposons (p. 151), not themselves independent replicons, when introduced into new host cells, may be inserted as extra sequences in a resident chromosome or plasmid without the necessity for homology or a functional *recA* gene.

If the newly introduced DNA fails to form a replicon or to be incorporated (rescued) by recombination with a replicon in the recipient, it will be lost as the cell grows and divides. Sometimes its genes may be fleetingly expressed before being diluted out as the host bacterium multiplies. When chromosomal genes for lactose fermentation are transferred into a *recA*⁻ host, they cannot be rescued by recombination with the chromosome, nor form an independent replicon, but survive for a matter of minutes and are transcribed and translated into protein that can be assayed (Bergmans *et al.* 1975). (This could be shown only when a high proportion of cells in a culture acquired the lac+ gene, as from an Hfr strain, p. 159.) Non-replicating DNA may survive and be expressed through several generations, as was shown in experiments by Stocker (1956) on *abortive transduction*. A gene determining motility, transduced (p. 162) into a non-motile salmonella, allowed the recipient cell to swim away from the inoculated area in a plate of soft agar. The motility gene could not replicate, so when the cell divided it produced one motile and one non-motile daughter. The non-motile one stayed where it was, divided, and formed a colony. The motile one swam off and, each time it divided, left behind a non-motile sister to form a colony. A trail of colonies resulted, along the track of the motility gene—an example of unilinear inheritance.

Modes of gene transfer and recombination in bacteria were reviewed by Low and Porter (1978). This is an area of fast expanding knowledge (see Sedgwick 1980, West *et al.* 1981).

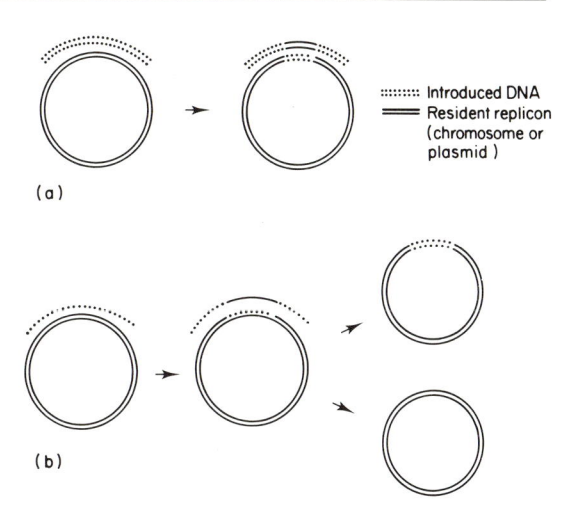

Fig. 6.9 General recombination. The figure shows two cross-overs in homologous DNA having occurred between (a) two duplex molecules and (b) a single-stranded and a duplex molecule. Multiple cross-overs (not illustrated) usually occur between homologous regions of DNA. The figure does not illustrate the mechanism of cross-over. The recA protein is required and some understanding of the mechanism is emerging (see text).

Transformation

The first observation of bacterial transformation was by Griffith in 1928 who was studying pneumococcal infection in mice. The virulence of pneumococci depends on the presence of capsular polysaccharide, which gives the colonies a smooth appearance (S). Colonies of non-virulent strains, without capsules, appear rough (R). Griffith observed that mice injected with either living, non-virulent R bacteria or heat-killed S bacteria were not harmed. However, injection of a mixture of the two preparations killed some animals and from these, virulent S bacteria were isolated whose capsular serotype was that of the killed strain. The dead S bacteria had liberated something that could change the R bacteria. This change was called **transformation**.

It was not until 1944 that Avery, Macleod and McCarty, in experiments *in vitro*, identified the transforming principle as deoxyribonucleic acid (DNA). Experiments using radio-actively labelled DNA demonstrated that the released DNA entered the R bacteria during transformation. Transforming DNA enters recipient cells most effectively when in the duplex state, but only one of the two donor strands is used for providing genetic information after uptake. It recombines with the host chromosome, after which it is replicated as part of the recipient chromosome.

The ability of certain bacteria to be transformed is called *competence*. When once obtained under suitable growth conditions it is not a permanent feature but a transient phase in the life of a population. The time at which competence is obtained and its duration is characteristic for each bacterial species. Commonly it is seen towards the end of logarithmic growth. Two theories have been put forward to explain competence. One is that during the competent phase the structure of the cell wall is such that it permits uptake of DNA. The other is that competence is also associated with synthesis of specific receptor sites at the cell surface; this is supported by studies of the kinetics of interaction of competent pneumococci with DNA which have been found to follow classical enzyme substrate reactions, and also by the finding that new proteins are synthesized when competence is acquired.

In certain species of bacteria the ability to be transformed is a specially evolved physiological state; in others competence has to be artificially induced. Species which have physiological competence include *Acinetobacter calcoaceticus, Azotobacter vinelandii, Bacillus subtilis, Haemophilus influenzae, Moraxella, Neisseria, Pasteurella novicida, Pseudomonas, Rhizobium, Streptococcus pneumoniae* and *Xanthomonas phaseoli*. Species that can be made competent artificially include *Escherichia coli, Pseudomonas putida, Salmonella typhimurium* and *Staphylococcus aureus*.

Physiological competence

The best characterized transformation systems in the physiological group include those of *Streptococcus pneumoniae, Bacillus subtilis* and *Haemophilus influenzae*. In all three, double-stranded donor DNA is cut on the cell surface, giving fragments of 15–30 kilobases (Kb) in *Bac. subtilis*, 8–9 Kb in *Str. pneumoniae* and approximately 18 Kb in *H. influenzae*. Transforming activity of chromosomal DNA increases with increasing length of the donor DNA and there is a minimum size requirement for the donor DNA below which no transforming activity is seen. In *Bac. subtilis* and *Str. pneumoniae* there is a period known as eclipse during which recovered donor DNA has no transforming activity until the fragments have been incorporated into the recipient chromosome.

Bacillus subtilis and *Str. pneumoniae* take up DNA from unrelated (heterospecific) bacteria as well as related (homospecific) DNA. However, *H. influenzae* only takes up homospecific DNA efficiently. An endonuclease is located on the surfaces of *Str. pneumoniae* and *Bac. subtilis* which degrades one strand of the double-stranded donor DNA, while the complementary strand is drawn into the cell. No such endonuclease has been found on the surface of *H. influenzae*, and donor DNA remains double-stranded for several minutes after uptake. Deich and Smith (1978) have found a membrane protein in *H. influenzae*, capable of specifically binding donor DNA, which is found only in competent cells. Sisco and Smith (1979) find that *H. influenzae* DNA contains a specific base sequence 8–12 base pairs long which binds to this membrane protein. There are approximately 600 copies of this specific sequence on the *H. influenzae* chromosome, but it occurs at a much lower frequency in heterologous DNA species. The fate of transformed donor DNA depends on its source. If the DNA is linear chromosomal DNA, it is incorporated into the chromosome of the recipient by recombination and is subsequently replicated with the recipient chromosome. This is not true for donor bacteriophage DNA where mature virus particles appear in the recipient (**transfection**), or with plasmid DNA where the plasmid retains its autonomous character.

Artificial competence

Methods have been devised to make certain strains of *Esch. coli, Ps. putida, Salm. typhimurium* and *Staph. aureus* artificially competent. The common feature of these methods is the treatment of bacteria with solutions of calcium chloride at 0°. This method was first used to enable *Esch. coli* to take up non-infective phage DNA and has since been used for uptake of chromosomal and plasmid DNA.

The transformation system of *Esch. coli* has been investigated in the greatest detail. It is impossible to transform *Esch. coli* with chromosomal DNA unless a recipient is used which lacks the gene for exonuclease V (recBC nuclease), since this nuclease destroys incoming linear transforming DNA. There is no such requirement for transformations when using plasmid DNA. However some plasmids, including ColE1 and related plasmids, fail to transform strains with the *sbcB* mutation which encodes exonuclease I. The exact reason for this is unknown, but may be related to plasmid maintenance.

It is not clear how the method for artificial competence works. Competence is not observed when the cells are treated with calcium ions at temperatures above 0°, or in the absence of calcium ions. The cells remain competent as long as the temperature is held at 0°. Evidence suggests that structural changes in the cell envelope are concerned. Calcium ions can bind to acidic phospholipids, cell envelope proteins, lipopoly-saccharides and peptidoglycans. This binding may lead to changes in conformation and consequent changes in membrane permeability. At physiological temperatures these interactions may be limited by an active cellular pump which pumps calcium ions out of the cell. These explanations are to some extent supported by the work of Sabelnikov and Domaradsky (1979) who induced competence in *Esch. coli* at room temperature in the presence of an uncoupler of oxidative phosphorylation.

Artificial competence in *Staph. aureus* requires helper phages, in addition to calcium chloride treatment at 0°, which facilitate the uptake of DNA by modifying the cell surface in an as yet undefined manner.

Conjugation

Conjugation in bacteria means the transfer of DNA directly from one cell to another without the intermediary of a phage particle, as in transduction, and without its being extracellular and therefore susceptible to nucleases, as in transformation.

Bacterial conjugation was discovered by Lederberg and Tatum (1946) in testing pairs of mutant strains of *Esch. coli* K12 for evidence of recombination. From a mixture of a strain that required phenylalanine, cystine and biotin (a triple *auxotroph*) with another that required threonine, leucine and thiamine, they obtained recombinant clones that were *prototrophs*, requiring none of these preformed nutrients, and others with various combinations of these requirements. Further studies, over several years, showed that recombination was not the result of sexual reproduction in the ordinary sense of equal participation of both parents in zygote formation. Rather it appeared that only a small amount of genetic information from the 'male' parent was incorporated into the recombinants, which otherwise showed all the characteristics of the other, 'female', parent. The capacity to donate genetic information was shown to depend upon carriage of the F (fertility) factor. This was recognized because strains were found that had become infertile, having lost the F factor, but they could regain fertility by contact with an F^+ strain (Hayes 1953a). At the time of these experiments the role of DNA and the nature of the F factor were unknown; F can now be described as a conjugative plasmid (p. 164).

In the first conjugation experiments, the transferred bacterial characters were determined by chromosomal genes, the F factor having mediated the transfer of chromosomal DNA from an F^+ to an F^- cell. Crossing over of the incoming donor DNA with homologous sequences in the recipient chromosome resulted in recombinant clones of bacteria with newly assorted characters.

Conjugation is not, as far as is known, a normal function of bacteria themselves, but is determined by a variety of plasmids. Transfer of chromosomal DNA by plasmids is a relatively infrequent event. It may be considered, like transduction of chromosomal genes by phages, as an accidental by-product of the main result of conjugation, which is the transfer to a new host of plasmid DNA. Conjugative plasmids convert their host cells into donors of DNA. In bringing about their own spread between bacterial cells they also *mobilize* other DNA molecules in the donor cells, including the chromosome.

Conjugation determined by F

The functions of F that determine conjugation have been analysed in detail and a brief description of them follows. The genes determining them are arranged as a continuous segment of the F molecule, called the transfer (*tra*) region, and consist of one operon (Willetts and Skurray 1980). Conjugative systems determined by some other plasmids are analogous; for others little information is so far available.

Conjugative pili

F^+ bacteria produce on their surface thin protein outgrowths or pili (Fig. 6.10). Cells without conjugative pili cannot transfer DNA, so pili must have an essential function, but its nature has not been proved. The current view is that they act by allowing an F^+ cell to make contact with an F^- one and then, by retracting into the cell, bring the two cells into direct contact. The protein, pilin, of which F pili are composed, exists as a pool of unpolymerized molecules in the outer membrane.

The pili can be recognized by electron microscopy, in which they can be distinguished morphologically, or by labelling with specific antibodies, and by the fact that they act as phage-receptors. F^+ bacteria are therefore susceptible to certain phages that cannot lyse F^- bacteria (pp. 164, 169, see Fig. 6.15).

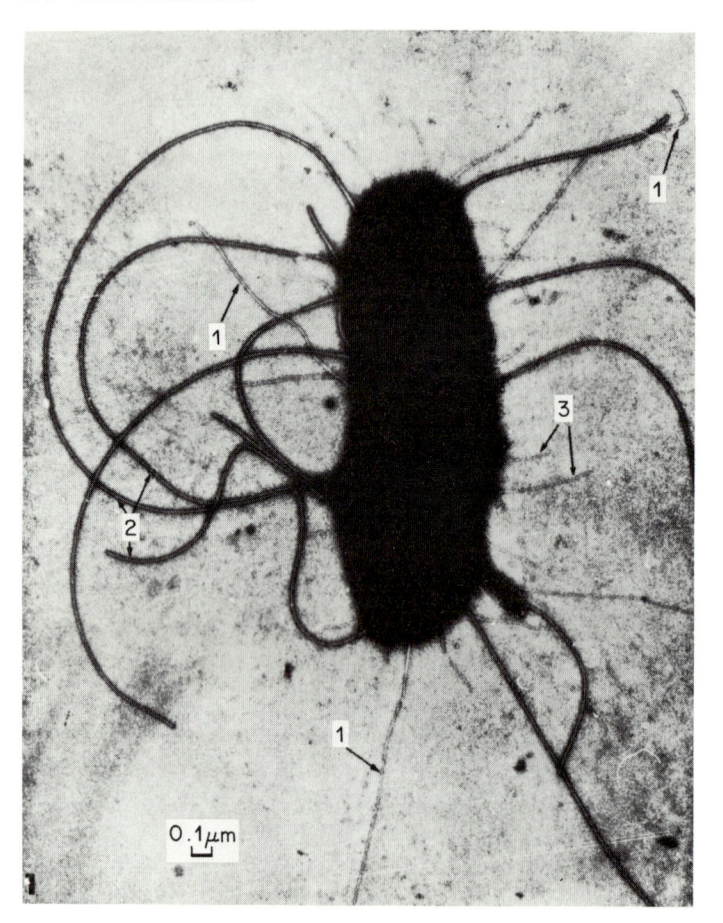

Fig. 6.10 *Escherichia coli* K12 F⁺. The electron micrograph shows three kinds of appendage to the cell (1) F pili, (2) flagella and (3) common pili or fimbriae. (From Meynell, Meynell and Datta, 1968).

Mating pair formation

Random collision of pairs of bacteria is the first stage in 'mating'; therefore none is detected unless the density of cultures is high. Clumping of cells occurs, and when a motile donor is mated with a non-motile, morphologically distinguishable recipient, paired cells can be seen moving together (Meynell 1972). Mating pairs or aggregates become stabilized, in the sense that they now need greater force than pipetting to separate them, but they can still be separated by whirlimixer or electric blender.

Surface exclusion

Mating is much less efficient between two F⁺ strains than between an F⁺ and an F⁻. In a cell, F determines *surface exclusion*, that prevents establishment of a stable mating pair with, and DNA transfer to, another F⁺ cell, though it does not affect conjugation determined by plasmids unrelated to F. The mechanism of surface exclusion is not fully understood. It is reduced in stationary phase cultures and the term F⁻ phenocopy refers to the finding that F⁺ cells, after long aeration, are good genetic recipients (Lederberg et al. 1952).

Transfer of DNA

Transfer of F plasmid DNA from donor to recipient cell proceeds from a specific place on its circular molecule, called *oriT* for origin of transfer. One strand of the DNA is cut (nicked) at *oriT*, and this strand goes across into the recipient cell. It is always the same strand that is transferred, from the 5' end. As the two strands separate, the retained strand is reconverted into double-stranded DNA by a replication process, and similarly, in the recipient, a new strand is synthesized complementary to the transferred one. Once in the new cell, the transferred F plasmid is recircularized.

Hfr and F' formation

The circular DNA molecule of F can enter into the circular chromosome to produce a single circle (Fig. 6.11). When this happens, the whole chromosome behaves as a conjugative plasmid, since it includes the origin of transfer, *oriT*, and all the conjugative functions of F. Because chromosomal genes are thus carried into all recipient cells, cultures with chromosomally integrated F are called Hfr, high frequency of recombination (Hayes 1953*b*). Transfer of F DNA, once a stable mating pair is formed, is very quick, but the chromosome is some 100 times longer than F, and its transfer by conjugation takes about 100 minutes. In fact, it is seldom transferred in its entirety. F can be inserted at numerous loci in the chromosome and in either orientation. For each Hfr strain, therefore, the chromosome is transferred to recipients from the *oriT* of F, always in one direction and with the genes nearest the transferring end of F going first (Fig. 6.12). These are found in greatest numbers in recombinants, and genes further away from the leading end of F are found in smaller and smaller numbers, because mating in any pair of cells seldom proceeds for the whole 100 minutes. A gradient of gene transfer can be plotted that shows the map positions of genes in the *Esch. coli* chromosome in relation to the site of F insertion. The same gradient, more directly related to the time of entry of each gene into recipient cells, can be plotted from the results of interrupted matings, in which samples of a mating mixture are removed at intervals, diluted so that no further pair formation occurs, sheared in a blender or mixer to separate stabilized pairs, and then plated on selective media to detect recombinants (Fig. 6.12). Because of these time-of-entry gradients, the map of the *Esch. coli* K12 chromosome is depicted as a circle with 100 minutes as reference points (Bachmann and Low 1980).

The integrated F of an Hfr strain can come out of the chromosome and resume autonomous replication. Sometimes the excision is not precise and some chromosomal DNA, adjacent on one side or both, is excised with F and becomes part of the plasmid: sometimes a remnant of F DNA is left in the chromosome (see Fig. 6.11). A plasmid in which chromosomal DNA is carried by an autonomous F molecule is an F' plasmid. F'*lac*, for example, is F incorporating the *lac* operon of *Esch. coli*. The operon is expressed, so F'*lac* in a non-lactose-fermenting culture, makes it *lac*⁺. *Esch. coli* K12 carrying an F' plasmid is diploid for genes that are present in both chromosome and plasmid. Crossing-over, or recombination between the

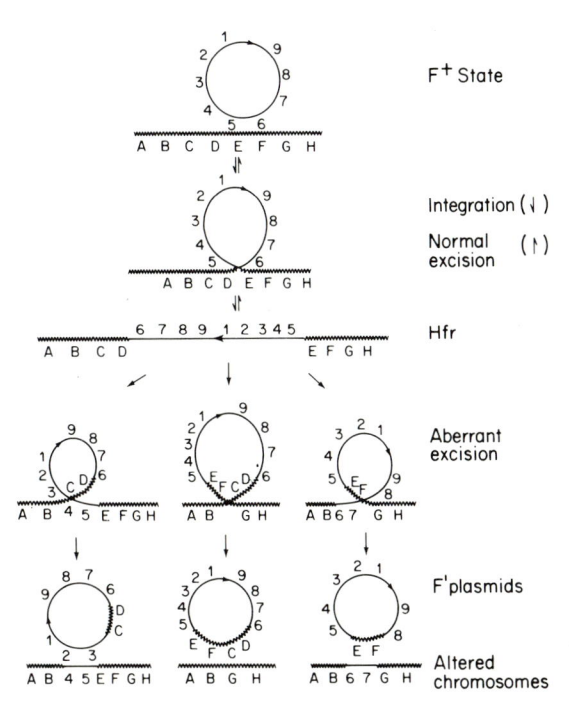

Fig. 6.11 The formation of (a) Hfr strains and (b) F' factors by reciprocal recombination between (a) two circles and (b) two points on the Hfr chromosome. (Reproduced with permission from *Plasmids* by Paul Broda 1979, W. H. Freeman and Co., Oxford and San Francisco).

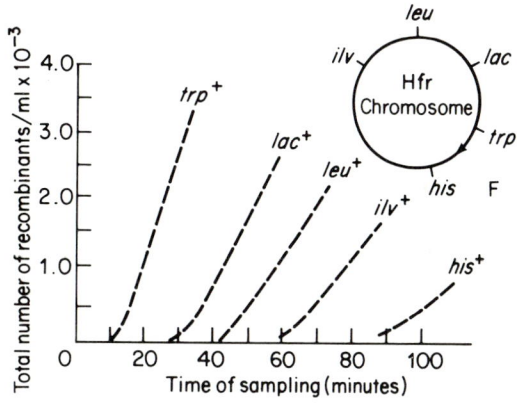

Fig. 6.12 Interrupted mating experiment. An Hfr strain of *Esch. coli*, with F inserted at the position shown, was mixed with an F⁻ strain that was *lac* (non-lactose fermenting) *trp*, *leu*, *ilv*, *his* (required tryptophan, leucine, isoleucine-valine and histidine for growth) *str* (streptomycin-resistant). Samples removed from the mixture at times shown were 'whirli-mixed' and plated on medium selective for the respective recombinants and containing streptomycin (to inhibit the Hfr donor).

homologous sequences, is a frequent event, and results in re-entry of F into the chromosome to give an Hfr strain with the same origin and direction of chromosome transfer as the Hfr from which the F' came. An F' culture transfers chromosomal DNA at frequencies between those of an ordinary F+ culture (with autonomous F) and those of an Hfr (with integrated F). It is a mixture of cells, some with F' out of the chromosome and some with it in.

Mobilization by autonomous F

Conjugative plasmids, including F, can bring about the transfer of other DNA molecules besides their own, a function known as *mobilization*. Autonomous F transfers chromosomal genes apparently at random, without a gradient of transfer such as is seen with Hfr or F' cultures. Mobilization happens in at least three ways: 1) Integration by recombination at a site of DNA homology to produce Hfr strains as described above. This integration requires a normally functioning recombination (*recA*) system. In an F+ culture, many of the chromosomal recombinants, such as were first observed by Lederberg and Tatum, result from newly formed Hfr clones. The homologous sequences that are sites of integration are IS sequences (p. 151) that occur in F and at numerous loci in the *Esch. coli* chromosome; 2) transfer of chromosomal genes occurs, but at very low frequency, in the absence of a functioning *recA* recombination system. Here too, it is believed to depend upon IS sequences, but a homology between chromosome and mobilizing plasmid is not involved. Transposition functions (p. 151) bring about the temporary co-integration of two replicons. ISs, in transposing from F to the chromosome or vice versa, produce co-integrates, in which F and the chromosome are covalently linked, and in this way chromosome transfer may occur as a by-product of transposition. The chromosome, as far as is known, is mobilized only by methods such as these, involving covalent linkage between it and the mobilizing plasmid. 3) When a conjugative plasmid such as F brings about the transfer of a small non-conjugative plasmid such as ColEl or R300B, covalent linkage is not required. This is because many plasmids (p. 164) that are strictly non-conjugative, i.e. do not determine bacterial conjugation, have an origin of transfer (*oriT*) and genetic information determining the transfer of their DNA to a recipient cell so long as a true conjugative plasmid is with them in the cell, providing conjugative pili and able to bring about stable mating pair formation.

Intergeneric F-mediated conjugation

F or F' plasmids can be transferred to other genera than *Esch. coli*, including such distantly related ones as *Yersinia* and *Vibrio*, and the *Esch. coli* genes are expressed. Incorporation of chromosomal DNA from *Esch. coli* into the chromosomes of other genera is not found unless there is a close phylogenetic relationship, as in *Shigella*. Even in *Salmonella*, chromosomal recombination with *Esch. coli* is rare (chapter 21). Insertion of F into *Shigella* or *Salmonella* chromosomes generates Hfr strains.

Conjugation other than F-mediated

Repression of transfer genes

Soon after antibiotic-resistance plasmids (p. 165) were discovered it was found that many of them, when transferred to an F+ or Hfr culture, inhibited F−-determined fertility (Nakaya *et al.* 1960, Watanabe 1963). They were called fi^+ (fertility inhibition). Many fi^+ plasmids determine pili that are antigenically related to those of F and that act as receptors for F-specific phages. Generally, however, F-like pili determined by other plasmids are present only on a minority ($<1\%$) of cells in the culture, even though the plasmid and pilus-determining genes are present in all the cells. This is because the transfer genes, including those determining pilus formation, are repressed. As a result, transfer of these plasmids occurs at low frequency compared with F transfer, and a culture of *Esch. coli* K12 carrying one of them, such as R1, is not visibly lysed by F-specific phages. Repression can be considered as a protective adaptation, since unrepressed (constitutive) pilus production leaves all the cells in a population susceptible to lysis by F-specific phages, which are common in sewage. The plasmid-determined repressor, which prevents full expression of conjugative ability by plasmids such as R1, acts also *in trans* upon the F conjugative functions; it represses pilus synthesis determined by F, and is responsible for the fi^+ phenotype. Mutants of R1, and other F-like plasmids, defective in repressor function, produce pili constitutively, allow plaque-formation by F-specific phages and mobilize chromosomal genes much as F itself does (Table 6.2) (Meynell and Datta 1967).

In a culture carrying a plasmid with repressed transfer properties, a few cells escape repression and these account for transfer at low frequency (Table 6.2). The transferred plasmid is temporarily freed from repression, because the repressor substance, absent in the new cell, is not synthesized as early as conjugative pili and enzymes required for transfer. The plasmid can be rapidly transferred to another cell, which becomes in its turn a donor, and so the plasmid can spread through the culture, if a sufficient concentration of recipient cells is available. A high frequency transfer (HFT) system results, phenotypically resembling a culture in which derepressed transfer is genetically determined (Datta *et al.* 1966, Bradley 1980). This provides a potent mechanism for change of genotype in a whole population of bacteria. The transfer genes of plasmids unrelated to F (p. 168 et seq) may or may not be repressed (Table 6.7). Derepressed mutants of some such plasmids have been obtained e.g., R144drd3 (Table 6.2).

induction of the resistance (Chapter 5). Quoting some previous examples, they suggest that induction of plasmid transfer by induction of a gene, such as a resistance gene, carried on the plasmid may be a usual

Table 6.2 Frequencies of transfer of plasmid and chromosomal genes

Plasmids in donor	F transfer	R transfer	Chromosome transfer	F phage sensitivity
F	1		10^{-5}	+
R1($fi+$)		10^{-3}	$<10^{-8}$	−
R1drd19		1	10^{-5}	+
F and R1	10^{-3}	10^{-3}	$<10^{-8}$	−
R144 ($fi-$)		10^{-3}	$<10^{-8}$	−
F and R144	1	10^{-3}	10^{-5}	+
R144drd3		1	10^{-5}	−

Figures show numbers of transfers per donor cell in 1-hour matings. R1 belongs to Incompatibility group FII and represses the transfer properties of F. R1drd19 is a mutant of R1, derepressed in its own transfer properties. R144 belongs to group I_1 and does not affect the transfer properties of F. R144drd3 is a derepressed mutant of R144. F phage sensitivity + means visible lysis by phage MS2, indicating the presence of F pili on all cells in the culture.

Conjugation in other genera

Conjugative plasmids have been identified in all the enterobacteria, and in *Pasteurella, Yersinia* and many species of non-fermenting gram-negative aerobes such as *Pseudomonas* and *Acinetobacter* and in *Haemophilus, Neisseria, Campylobacter, Bacteroides, Streptomyces, Streptococcus, Clostridium* (p. 168). Conjugative pili have not been identified in all these genera. Conjugative plasmids in *Pseudomonas* (Holloway *et al.* 1979), *Streptomyces* (Hopwood *et al.* 1973), *Acinetobacter* (Towner and Vivian 1976) bring about chromosome transfer in ways that seem to be analogous to those of F in the Enterobacteriaceae.

In some genera, conjugation and transfer of antibiotic resistance has been observed without any plasmid DNA being identifiable (Stuy 1980). Conjugative genes seem to be incorporated into bacterial chromosomes, but the behaviour of the strains is not analogous to that of Hfr strains of *Esch. coli*. In streptococci, for example, a transposon encoding tetracycline resistance, chromosomally integrated, appears to determine conjugative transfer of its own DNA, even between Rec$^-$ hosts and in the absence of any demonstrable plasmid (Franke and Clewell 1981).

Conjugative plasmids in other genera than those of the Enterobacteriaceae may hold their transfer genes under repressor control. In *Esch. coli*, the only mechanism known to lift repression, other than mutation, is transfer to a new host (p. 160), but in other genera induction of the conjugative function has been reported. Privitera and his colleagues (1981) describe induction of the conjugative function of a tetracycline-resistance plasmid in *Bacteroides fragilis* by low concentrations of tetracycline, in parallel with

phenomenon in bacteria. In *Streptococcus faecalis* a different and complex system of induction has been discovered, mediated by *sex pheromones* excreted by potential recipient cells. Each plasmid determines response to a specific pheromone (or clumping-inducing agent) of which plasmid-less strains produce a variety. Upon acquisition of a plasmid, the strain ceases to excrete the pheromone that induces the transfer genes of that plasmid (Dunny *et al.* 1978, 1979).

Phage-mediated conjugation

Changes in biotype, phage type and antibiotic resistance in *Staphylococcus aureus* may result from genetic exchange. Chromosomal genes and plasmids are transferable by transduction. In addition, gene transfer can take place between pairs of cells in mixed culture or on skin. The term *phage-mediated conjugation* (Lacey 1980) has been used to describe some genetic transfers in staphylococci. Plasmid transfer has not been observed between non-lysogenic strains of *Staph. aureus*, but often occurs when the donor strain is lysogenic. It may sometimes occur, and with high frequency, from a non-lysogenic recipient (Lacey 1980). In the same report, Lacey describes the carriage of penicillinase and cadmium resistance genes by factors with some of the characters of defective phages. Not only were these genetic factors transmissible between non-lysogenic staphylococci, but they could also bring about the transfer—in mixed culture but not in lysates—of unlinked resistance plasmids.

Transduction

Gene transfer between bacteria brought about by bacteriophages is called *transduction*. A normal phage particle consists of the phage genome, more often DNA than RNA, in a protein coat. Occasional, apparently accidental, particles are synthesized that contain, instead of phage nucleic acid, DNA of the host bacterium. Such a particle can inject its DNA into a new host cell. This is *generalized transduction* and can be brought about by many phages in many bacterial species. Of the order 10^{-5} to 10^{-6} particles in a phage lysate may be transducing particles. Random sequences of chromosomal DNA can thus be transferred to a new host where they can be rescued by recombination, if there is homologous DNA and a functional recombination system. Expression of transduced chromosomal genes is therefore possible only when donor and recipient cells are closely related. Generalized transduction can also bring about transfer of plasmids. The size of phages determines the amount of DNA that can be transduced. Large plasmids cannot be carried by phage coats too small to accommodate them, but when a phage particle transfers plasmid DNA that includes essential genes the plasmid may be able to establish itself in the new host as a smaller molecule than in the donor, having undergone *transductional shortening*. When only non-essential plasmid genes are transduced, they may be rescued by recombination if there is homologous DNA in the recipient in the form of a resident plasmid or chromosomal insert. Transposons may be rescued without this necessity.

Generalized transduction can be brought about by virulent or temperate phages (chapter 7), but is easier to detect with temperate ones. This is because the majority of phage particles in a lysate are ordinary non-transducing ones; recipient cells are lysed by virulent phage and transductant clones lost.

In generalized chromosomal transduction, the transducing particles contain no phage DNA, but each carries one continuous sequence of chromosomal DNA, randomly selected. When small plasmids are transduced, each having too little DNA to fill the phage head, there may be co-transduction of coexisting plasmids (Grubb and O'Reilly 1971).

Exposure of a transducing phage preparation to a small dose of UV light before it is added to potential recipient bacteria increases the numbers of recombinants. This applies to transduction of chromosomal genes, not to intact plasmids. Irradiation-damaged DNA, injected into the recipient, stimulates repair mechanisms that are also concerned in recombinational rescue of the transduced DNA. In the establishment of transduced plasmids, repair and recombination are not required.

Specialized transduction occurs, also as an exceptional event, with temperate phages such as λ that, as prophages, are inserted into a specific attachment site in the bacterial chromosome. Upon induction (chapter 7), excision of phage DNA from its chromosomal site is in most cells precise, only phage DNA being replicated and wrapped in coat protein. But in some cases (about 10^{-4}) excision is imprecise, and bacterial DNA from one side or other of the chromosomal attachment site is excised with the phage genome and replicated as phage. With phage λ the genes *gal* (galactose utilization) and *bio* (biotin synthesis) flank the chromosomal insertion site and are liable to be transduced. Chromosomal DNA, covalently linked to phage DNA, is thus incorporated into phage particles and can be injected into new cells. Because the total DNA content of the phage particle is limited, a variable length of phage DNA is missing. Phage λ particles that include the *gal* gene are called λdg, as they are *defective* (*d*) and transduce gal^+ (*g*). λdg phage cannot lyse λ-sensitive bacteria, though its DNA can be injected into them and inserted into their chromosome. Transduction of gal^+ is observable if the recipient strain has a mutation in its *gal* operon. The resulting gal^+ clone is a partial diploid, since the new *gal* gene is inserted as part of λ and the host's *gal* region is retained; the diploid state is unstable. The transductants are not *lysogenic* in that they do not liberate lytic phage into the growth medium, but they are immune to phage λ. Induction of λdg in such a gal^+ strain does lead to lysis and production of a high frequency transducing (HFT) preparation in which every particle carries, and can transduce, the bacterial gal^+ gene.

Phage conversion is not transduction but resembles specialized transduction. There are many examples of phages, not only defective ones, whose genomes include genes that determine a bacterial character. Every bacterium infected by such a phage acquires the character. The genes so introduced are not transduced from the genome of the previous host of the converting phage, but are part of the phage itself. The first known example (see Barksdale 1970) was toxigenicity in *Corynebacterium diphtheriae* (chapter 25), which depends upon the presence of a converting phage. Many other bacterial characters are determined by phages or prophages, including immunity to superinfection by related phages, changes in surface antigens, the synthesis of enzymes that modify and restrict DNA etc. (see chapter 7). An example of a converting phage that evolved during a laboratory experiment is P1CM (Kondo and Mitsuhashi 1964), in which the gene for chloramphenicol transacetylase is incorporated into phage P1. A naturally occurring counterpart is phage Øamp (Smith 1972), now renamed P7 (Scott et al. 1977), in which the TEM β-lactamase gene is incorporated into a P1-like phage. Evidently phage genomes can accept transposons without necessarily losing their normal phage characteristics.

Protoplast fusion

Genetic recombination can occur after fusion of bacterial protoplasts; fusion of membranes is followed by the formation of cytoplasmic bridges between the cells. Various methods have been used to induce fusion in a variety of bacterial genera. Fusions can be obtained between unrelated cells, even between members of different kingdoms. Gene transfer between fused cells has been called *genetic transfusion*.

As in transformation, conjugation or transduction, plasmid or virus DNA can be transferred interspecifically or intergenerically, but for chromosomal recombination close relationship between the participating cells is necessary. In this case two complete genomes are brought into contact to make an artificial diploid. Not all the genetic potential of the parents is expressed by the fused organism, yet all the potential is there since haploid recombinants can be recovered with many combinations of parental traits. Protoplast or spheroplast fusion is likely to prove a valuable method with the advance of biotechnology. Progress in this field was reviewed by Ferenczy (1981).

Mapping of chromosomal genes

A genetic map shows the order of genes on the chromosome and the distances between them. A map can be plotted if the chromosome is transferred to a recipient from a set origin i.e. an integrated conjugative plasmid. With *Esch. coli*, a series of Hfr strains in which F is inserted at different positions and in either

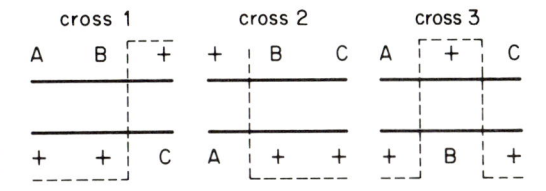

Fig. 6.13 To find the order of genes A, B and C, three crosses are made between three double mutants AB, BC and AC and three single mutants, C, A and B respectively, and prototrophic (wild type) recombinants are counted. + = wild type for each gene. The fact that far fewer prototrophs are recovered in cross 3 shows that gene B lies between A and C.

orientation makes approximate mapping easy. The order of three fairly close genes is determined by a three-factor cross (Fig. 6.13) in which the initial DNA transfer may be by transformation, conjugation or transduction. For very close mapping, within genes, complementation tests are used.

Plasmids

Plasmids are independent replicons that live in bacteria. The term *plasmid* was proposed by Lederberg (1952) for all extrachromosomal genetic structures that can replicate autonomously. The adjectives 'independent' and 'autonomous' here mean self-governing rather than capable of separate existence; plasmids are parasites or symbionts of bacteria. They differ from bacterial viruses (phages) in having no extracellular form.

The term *episome* is sometimes used as a synonym for plasmid. Episomes were defined (Jacob and Wollman 1958) as genetic elements that could exist and replicate in either of two modes, autonomously or incorporated into the host chromosome. This definition fits F and phage λ (Chapter 7) in *Esch. coli*, but it does not fit F in other hosts, e.g. *Proteus*, and the term is no longer useful for defining a class of replicon (Novick *et al.* 1976).

The practical importance of plasmid-determined genes was first recognized twenty years ago with the discovery of transferable drug resistance (Watanabe 1963). At present more and more characters of more and more kinds of bacteria are being reported as encoded by plasmids; although they represent a minor part of the genetic content of bacteria, they play a disproportionately large part in bacterial variation.

When F and the first R plasmids were discovered, their extrachromosomal nature was deduced from their genetic behaviour. One feature of this is the tendency for clones in a culture to lose the plasmid, the plasmid-determined characters thus being lost irreversibly. The stability of plasmids in growing cultures varies; some are lost at the rates of 10^{-2} to 10^{-3} per generation, much higher than mutation rates. Others are very stable; loss of a character at perhaps 10^{-7} per generation could result either from mutation or from plasmid loss. Some plasmids can be eliminated (*cured*) from a culture by certain treatments. The F plasmid is eliminated by treatment with acridine orange (Hirota 1960), and some plasmids of both staphylococci and enterobacteria by growth at 42° or 44° (May *et al.* 1964, Yokota *et al.* 1969). Many other *curing agents* (e.g. see Riva *et al.* 1972) have been described, but none can be relied upon as generally effective.

With the development of methods for isolating the DNA of plasmids, more direct evidence of their presence and genetic information is available. Plasmids can be rendered visible in bacterial lysates after gradient ultracentrifugation or gel electrophoresis. Their

DNA can be purified and reintroduced into plasmid-free cells by transformation. In this way, a bacterial character can be directly correlated with carriage of a particular species of plasmid DNA.

Plasmid-determined genes

Essential genes

The essential genes of a plasmid are those determining the controlled replication of the plasmid itself. Plasmids of all sizes, from about 2 kilobases (0.7 μm circumference) to over 300 kilobases (100 μm circumference, about 1/20 of the *Esch. coli* chromosome), occur naturally. The smallest, such as p15A (Chang and Cohen 1978), have only enough DNA to encode two or three proteins. Small, *cryptic*, plasmids that do not alter the phenotype of the host in any known way are commonly seen in bacterial lysates. Large plasmids can be whittled down in the laboratory by means of restriction enzymes: the resulting mini-plasmids, with minimum genetic information, are comparably small molecules (Danbara *et al.* 1980). Plasmids use the host's gene products to bring about the replication of their DNA, but they require a sequence from which replication can begin, an origin (*oriV*) of vegetative replication (not the same as *oriT*, the origin of transfer by conjugation p. 158). They also require mechanisms to ensure that their replication is kept synchronous with the growth of the bacterial host, and that their daughter molecules are inherited by host daughter cells. Too few rounds of plasmid replication would lead to loss of the plasmid from a proportion of progeny cells, and too many would fill the host cell with plasmid DNA and kill it. All plasmids depend upon their bacterial hosts for enzymes required in DNA replication as well as for the transcription and translation of plasmid genes. A few, notably ColE1, require DNA polymerase I—determined by host gene *polA*—but most known plasmids can replicate in *polA* mutant hosts.

F and many other plasmids maintain themselves in a growing culture with a low copy number, i.e. with 1–2 copies of the plasmid DNA per bacterial chromosome. They have been described as being under *stringent* replication control. Other plasmids, usually small ones, exist as multiple copies, of the order of 20 per chromosome, and theirs has been called *relaxed* replication control. For each plasmid, there is a characteristic copy number, although this can sometimes be disturbed. Some plasmids, in some bacterial hosts, continue to replicate after the host culture has entered the stationary phase, with accumulation of large numbers of plasmid molecules. In the case of the ColE1 plasmid, treatment of the culture with chloramphenicol has the same effect. The essential genes controlling replication include a repressive function which, as a secondary effect, determines plasmid *incompatibility*.

Two variants of the same plasmid are incompatible, i.e. they cannot be maintained stably together in a growing culture. The repression that each exerts over its own replication affects also that of its relative, so that there is a double dose of plasmid repressor activity in the bacterial cells. The result is that, as the bacteria grow and divide, they give rise to clones with one or other plasmid but not both (Uhlin and Nordström 1975). Unrelated plasmids have distinct mechanisms for control of replication that do not affect one another and so such plasmids are *compatible* (p. 168).

Genes determining conjugation

Conjugation (p. 157) determined by F or many other plasmids is a complex process controlled by about 20 genes, encoded in about 33 Kb of DNA (Willetts and Skurray 1980). Conjugative plasmids cannot be very small although some of them must have simpler transfer functions than F. The smallest conjugative plasmids known in enterobacteria are about 30 Kb. (For comparisons of conjugative pili of different plasmids see p. 169.)

Non-conjugative plasmids are mobilized for transfer to new hosts by conjugative ones (p. 160). Plasmids such as ColE1 (approx. 6.5 Kb) encode considerable transfer information, not only an origin of transfer (*oriT*), but genes for its own mobilization (*mob*) (Dougan and Sherratt 1977). A miniplasmid such as p15A does not possess its own mobilization genes but can use the *mob* products of ColE1 when both plasmids are in an Hfr culture. The various components of plasmid conjugative function show specificities, different non-conjugative plasmids being successfully mobilized by different conjugative ones (Willetts and Crowther 1981).

Other plasmid genes

Replication, copy-number control, maintenance and transfer may be considered inherently plasmid characteristics. Plasmids also carry a great variety of additional genes, determining many bacterial properties by means of which they provide advantages of various kinds to their bacterial hosts.

Antibiotic resistance Antibiotic resistance determined by plasmids was first discovered in Japan in 1959. Until then it had been assumed that antibiotic resistance in bacteria, which was already a clinical problem, resulted from the selection of mutants, multiple resistance resulting from the sequential accumulation of mutations. It was the epidemiology of drug-resistance in *Shigella flexneri* that gave the suggestion that resistance might be transmissible between bacteria. Bacillary dysentery was common in Japan after the second world war. By the mid-1950s, most Japanese *Sh. flexneri* were sulphonamide-resistant, and treatment with streptomycin, tetracycline or chloramphenicol was introduced. A few strains resistant to four unrelated drugs, sulphonamides (Su), streptomycin (Sm), chloramphenicol (Cm), and tetracycline (Tc) were then isolated, and such strains became progressively commoner (Table 6.3). Isolates

Table 6.3 Multiple drug resistance in *Sh. flexneri* isolated in Japan

Year	No. of isolates tested	% multi-resistant
1956	4399	0.02
1958	6563	2.9
1960	497	15
1962	6853	23
1964	5388	45
1966	4292	75
1968	1237	64
1970	562	74
1972	824	76

Multi-resistant strains were resistant to chloramphenicol, tetracycline, streptomycin and sulphonamide.
Figures from Mitsuhashi (1977).

resistant to two drugs, as would be expected if sulphonamide-resistant clones mutated to resistance to another drug, were rare. In outbreaks of infection with a single *Sh. flexneri* serotype, strains from some patients had the multiple-resistance pattern, while others did not. *Esch. coli* with the same pattern of resistance was found in faeces of dysentery patients. Given these observations, Akiba and, independently Ochiai, in 1959 (cited by Watanabe 1963) tested the hypothesis that multiple resistance might be carried by an infective agent analogous to the F factor of *Esch. coli*, and experiments proved that this was so. From mixtures of resistant *Sh. flexneri* with sensitive *Esch. coli*, clones of *Esch. coli* with the same four resistances as the shigella were isolated. The infective agents, *R factors*, are now known as R plasmids.

The increasing frequency of drug resistance in *Sh. flexneri* in Japan was not an isolated phenomenon. The emergence of resistance was seen wherever records were kept, such as in *Shigella sonnei* in the London area (Davies *et al.* 1968), in salmonella strains in the Netherlands (Manten *et al.* 1966), and in *Salm. typhimurium* in England (Anderson 1968). These examples differed from the Japanese experience in that there was a stepwise accumulation of resistances, first to sulphonamides and streptomycin, later to tetracycline, and only little to chloramphenicol. Within a decade, ampicillin resistance had become common in salmonella and shigella strains in Europe, though it remained uncommon in Japan. Much of the observed resistance was plasmid-determined.

At the time of the discovery of R factors in enterobacteria, multiply-resistant strains of *Staphylococcus aureus* were a cause of concern in hospital infection. Penicillinase producing strains of *Staph. aureus*, first recognized in the 1940s, had become common and had acquired resistance to each new antibiotic as it was introduced. In 1963 it was shown that penicillinase synthesis by *Staph. aureus* was a plasmid-determined character (Novick 1963), as are many other resistances in that species (Lacey 1975). Penicillinase plasmids in *Staph. aureus* have been extensively studied. They encode not only the enzyme, β-lactamase, but also its regulation. Its synthesis is ordinarily repressed but exposure of the staphylococcus to penicillin induces β-lactamase production, the whole operon being plasmid-borne. Penicillinase plasmids often determine resistance to mercury and/or cadmium salts and sometimes to erythromycin or fusidic acid. Resistance to tetracycline, chloramphenicol and neomycin is usually carried on other plasmids in staphylococci.

There has been confusion and controversy about the genetic determinant(s) for methicillin resistance in *Staph. aureus*. Recent work suggests that methicillin resistance, though chromosomally located, is carried by a DNA insert with no allele in methicillin-sensitive strains (Stewart and Rosenblum 1980).

Many new antibacterial drugs have been developed and used in medicine over the last 20 years, and plasmids conferring resistance to them have often appeared within a short time. Not only has the range of plasmid-determined resistance broadened, but also more and more bacterial genera are found carrying R plasmids (Tables 6.4, 6.5). The new appearance in the mid-1970s of plasmid-determined β-lactamase synthesis in *Haemophilus influenzae* and *Neisseria gonorrhoeae* was of importance in clinical medicine and also in the study of plasmid evolution. In both these species the β-lactamase was the TEM enzyme, genetic information for which is carried on a transposon (p. 151). Conjugative R plasmids in strains of *H. influen-*

Table 6.4 Antibacterial drugs to which plasmids determine resistance

Penicillins and cephalosporins
Erythromycin, lincomycin and streptogramin B

Streptomycin	Tetracyclines
Neomycin	Chloramphenicol
Kanamycin	Fusidic acid
Gentamicin	Sulphonamides
Tobramycin	Trimethoprim
Amikacin	

Table 6.5 Genera in which R plasmids have been found

Enterobacteriaceae i.e. *Escherichia, Salmonella, Shigella, Proteus, Providencia, Klebsiella, Serratia* etc.
Pseudomonas

Acinetobacter	*Staphylococcus*
Vibrio	*Streptococcus*
Yersinia	*Bacillus*
Pasteurella	*Clostridium*
Campylobacter	*Corynebacterium*
Haemophilus	
Neisseria	
Bacteroides	

zae isolated in widely different geographical areas are closely related as shown by DNA hybridization studies. This applies not only to those determining penicillin resistance, but also to chloramphenicol- and tetracycline-resistance plasmids; possibly a ubiquitous conjugative plasmid, resident in haemophilus strains, has acquired a variety of resistance transposons (Laufs and Kaulfers 1977). In *N. gonorrhoeae*, β-lactamase is determined by small (7 Kb or 5.2 Kb) non-conjugative plasmids. Many strains possess, in addition, a conjugative plasmid that can bring about the transfer (mobilization) of the R plasmid to other neisseriae and also, intergenerically, to *H. influenzae* or *Esch. coli* (Flett *et al.* 1981). The small R plasmids of *N. gonorrhoeae* are related to one another in their DNA sequence, and indistinguishable small β-lactamase plasmids are also found in some wild strains of *H. influenzae* (Laufs *et al.* 1979).

The mechanisms of antibiotic resistance determined by plasmids are discussed in Chapter 5.

Antibiotic synthesis

Plasmids determine the production of bacteriocins and microcins (chapter 7). They also play a part in the production of a variety of antibiotics including some of those used in medicine. The subject was reviewed by Hopwood (1978), who has shown that a plasmid, SCP1, in *Streptomyces coelicolor*, determines production of the antibiotic methylenomycin. The same plasmid confers methylenomycin resistance. Plasmids carry, in some cases at least, structural or control genes determining production by streptomyces of oxytetracycline, chloramphenicol and perhaps other antibiotics. The genetics of antibiotic-producers is of great potential importance in the pharmaceutical industry and further discoveries in this field can be expected.

Resistance to other agents

Plasmids confer resistance to a variety of metal ions and organometallic compounds. Plasmid-determined resistance to mercuric ions (Hg^{2+}) is very common in enterobacteria and staphylococci, frequently linked to antibiotic resistance. In different surveys, 25–60% of R plasmids determined Hg^{2+} resistance. In general, high frequencies of antibiotic resistance are found in environments where much antibiotic is present, hospitals being the obvious example. It is not clear what selective pressure has led to Hg^{2+} resistance being so frequently found in the same bacteria. Resistance to cadmium, lead, antimony, arsenic, tellurium and silver compounds are plasmid-determined in staphylococci and/or gram-negative bacteria (Summers and Silver 1978, Summers and Jacoby 1977). Resistance to silver salts was first reported after the use in burns units of silver-sulphonamide compounds (McHugh *et al.* 1975).

Plasmids may change the sensitivity of their bacterial hosts to UV light and other mutagens (Mortelmans and Stocker 1976) and to bacteriocins (see chapter 7).

The *degradation* of many different toxic organic compounds, such as camphor, naphthalene, octane, toluene, is determined by plasmids in soil bacteria. These, often *Pseudomonas* species, not only detoxify such compounds but also use them as nutrients. Plasmid-mediated detoxification processes may in the future be put to important use in overcoming environmental pollution (Salkinoja-Saloner and Sundman 1979).

Virulence factors

Enterotoxins in *Escherichia coli*, and surface antigens that allow colonization of the small intestine in specific mammals, including man, are plasmid-determined. These plasmids are associated with some kinds of *Esch. coli* enteropathogenicity (chapter 71). The role of plasmids in the determination of staphylococcal enterotoxin B has been a subject of controversy that is not finally resolved (Kahn and Novick 1982).

Strains of *Esch. coli* isolated from parenteral infections are more frequently haemolytic, and more frequently produce colicin V, than comparable faecal strains. Haemolysins and colicin V are plasmid-determined characters but, despite their association with bacterial invasion, neither can be classified as a virulence factor. Carriage of a ColV plasmid increases the virulence of *Esch. coli* strains for calves, chickens and mice (Smith and Huggins 1976); the effect is not from the colicin itself but from other determinants of the Col plasmid. The ColV plasmid facilitates iron uptake by its bacterial host (Williams 1979) and also increases serum-resistance (Binns *et al.* 1979). Other plasmids besides ColV confer serum resistance on *Esch. coli*, a resistance associated with an outer membrane protein absent from plasmid-free control cultures (Moll *et al.* 1980).

The S→R variation in *Shigella sonnei* is accompanied by loss of the form I (S) antigen and loss of virulence. It was recently shown (Kopecko *et al.* 1980) that the form I antigen is determined by a large plasmid in *Sh. sonnei* strains, loss of which brings about the S→R change. Transfer of this plasmid to *Shigella flexneri* 2a or *Salmonella typhi* gave cultures of *Sh. flexneri* and *Salm. typhi* that were agglutinable by their own specific antisera and also by antiserum specific to *Sh. sonnei* form I antigen. In *Sh. sonnei* the form I antigen, and hence the plasmid that carries it, is a necessary, but not the sole, requirement for virulence as demonstrated in the Serény test (chapter 36).

Strains of *Staphylococcus aureus* responsible for scalded-skin syndrome produce exfoliative toxins which, in some cases, are determined by plasmids (Wiley and Rogolsky 1977). Other examples of pathogenic properties determined by or associated with plasmids have been described and reviewed by Elwell and Shipley (1980). These authors also discuss R plasmids as possible virulence factors. Resistant bacteria, such as chloramphenicol-resistant *Salmonella typhi*, penicillin-resistant *Neisseria gonorrhoeae* or multiply-resistant *Staphylococcus aureus*, may sometimes appear to be of unusual virulence simply because the infections that they cause do not respond to the first line of antibacterial therapy. In addition, particular R plasmids may have specific effects upon virulence, either increasing or decreasing it, but generalizations without rigorous experimental backing are inappropriate.

No plasmid is known to produce tumours in animals, but one such exists in plants. *Agrobacterium tumifasciens* has long been recognized as the causative agent of crown gall, a tumorous disease that affects many plants and causes serious losses in fruit-growing. For an agrobacterium strain to have its effect, it must harbour a tumour-inducing (Ti) plasmid. In the plant tumours, plasmid DNA but no bacteria can be identified (Chilton *et al.* 1977).

Metabolic characters

Disease-producing bacteria may have plasmid-determined characters that make them difficult to identify in the diagnostic laboratory. Lactose-fermentation in strains of *Salmonella* (including *Salm. typhi*) is usually plasmid-determined. *Lac* plasmids have also been found in isolates of *Proteus* and *Yersinia enterocolitica* (Falkow *et al.* 1964, Cornelis *et al.* 1976). They are common in *Klebsiella* (Reeve and Braithwaite 1973), though in that genus, which has a chromosomal *lac* operon, they are not noticeable. *Lac* plasmids determine the regulation, as well as the structure, of the *lac* enzymes.

Genes for utilization of other sugars, especially sucrose (*suc*) and raffinose (*raf*) may be plasmid-borne. Plasmids with both *lac* and *suc* genes have been found in *Salmonella* (Johnson *et al.* 1976) and *raf* genes may be linked to those determining the K88 antigen, important in porcine *Esch. coli* enteritis (Shipley *et al.* 1978) (chapter 71). Plasmids determining H_2S production (Ørskov and Ørskov 1973), urease production (Farmer *et al.* 1977, Wachsmuth *et al.* 1979) or citrate utilization (Smith *et al.* 1978) occur at frequencies difficult to assess, since they are likely to go unrecognized. *Lac* plasmids in *Streptococcus lactis* cheese starter strains are important in industry (Farrow 1980).

Degradative plasmids mentioned under the heading 'resistance' can equally be classified as metabolic, and urease plasmids might come under 'virulence factors' if, as seems possible, they confer extra pathogenicity in the urinary tract.

Phage inhibition

A plasmid in a bacterial strain sometimes prevents its normal susceptibility to some phages. For example the F plasmid prevents *Esch. coli* being lysed by certain phages, such as ΦI, which have therefore been called *female-specific* (Meynell 1972). But this is just one example of a phenomenon of which there are many others. Plasmids of incompatibility groups I and H (see Plasmid Classification) prevent normal plaque formation by λ and other phages, and plasmids in *Salmonella typhi* or *Salm. typhimurium* alter phage types by preventing lysis. The mechanism of phage inhibition (phi) is in many such cases unknown. Some plasmids, however, prevent phage lysis by determining restriction enzymes (p. 173), endonucleases that attack the phage DNA as it enters the bacterial cell. *Eco*RI is an example of a plasmid-determined enzyme. These plasmids also determine *modification* of DNA (p. 173 and Chapter 7).

Another phenomenon that can be classified as phage inhibition is the repression of F pilus production by plasmids unrelated to F. The transfer genes of many plasmids, such as R100, R1 and R6, are alleles of F genes, but repressed. Repression affects the conjugative function of F and these plasmids are called fi^+ (see p. 160). Other plasmids, whose own pili are quite different from F-pili, are also sometimes fi^+ (Gasson and Willetts 1975). The evolutionary advantage of the fi^+ character may be its effect in protecting the host cell from lysis by ubiquitous phages that use F pili as their receptors.

Mapping of plasmids

Mapping of F and related plasmids has been achieved partly by genetic methods of recombination and complementation, and partly by physical methods of direct electron microscopic observation of heteroduplexes (Figs. 6.5, 6.14). Deletion-formation *in vitro*, the cloning of restriction fragments of plasmids into other

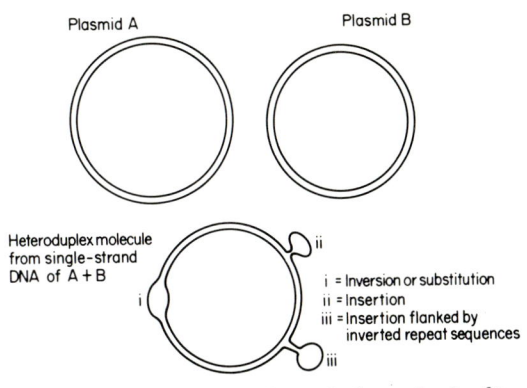

Fig. 6.14 Diagram to show heteroduplex molecule of two related plasmids.

replicons, and finally complete DNA sequencing of smaller plasmids, brings mapping to the ultimate point of accuracy.

The evolution and classification of plasmids

Plasmids are found in most, perhaps all, bacterial genera. Their origins are open to the same speculations as the origins of viruses. Each bacterial genus or group of genera has its set of plasmids, as each has its set of bacteriophages. The host specificities of plasmids have not been much studied but, as far as is known, plasmids from gram-positive bacteria cannot replicate in gram-negative ones or vice versa. Plasmids of staphylococci will replicate in *Bacillus subtilis* (Ehrlich 1977), and at least one streptococcus plasmid is transferable by conjugation to staphylococci (Engel et al. 1980) and lactobacilli (Gibson et al. 1979). Plasmids of Enterobacteriaceae can normally be transferred between all those genera (Harada et al. 1960) but not necessarily outside it. Some plasmids, transmissible between certain genera within the Enterobacteriaceae, cannot be transferred to proteus or providencia strains (Datta and Hedges 1972). Others have a wider host-range. Plasmids RP1 and RK2 are examples of a plasmid clone that was identified in the Birmingham Accident Hospital. These plasmids (of incompatibility group P, see below) were found in *Pseudomonas aeruginosa* to which they conferred high-level carbenicillin resistance, and also in many other bacterial species epidemiologically though not phylogenetically related (Roe et al. 1971). Such plasmids can be transferred by conjugation to a wide range of gram-negative bacteria (Olsen and Shipley 1973). In classifying plasmids, host-range should be an important feature, but in practice it is not a convenient one. When a plasmid is not transferred to a new host in a laboratory test it may be not because it is unable to replicate in that host. It may be that the conjugative (or other transfer) system is not appropriate, or restriction enzymes in the recipient cell may have destroyed the incoming plasmid DNA.

Plasmids can be classified by their incompatibility relations. Two variants of the same plasmid are incompatible because they have a common system of replication, under negative control, and each is subject to repression by the other. When two naturally occurring plasmids are incompatible, i.e. will not coexist stably in a growing culture, it is taken as evidence that they are closely related. When a collection of plasmids is tested, in pairs, each against all, they fall into *Incompatibility Groups*. This means of classification has been applied to plasmids in *Esch. coli* (Datta 1975), in *Pseudomonas* (Jacoby 1977) and in *Staph. aureus* (Novick et al. 1977). Plasmids that can replicate in both *Esch. coli* and *Pseudomonas* are included in both classifications (Hedges and Jacoby 1980).

Incompatibility is a good basis for grouping plasmids, since there is evidence that it reflects evolutionary relationships. Within a group, plasmids have much

Table 6.6 Incompatibility grouping of plasmids related to their molecular sizes and DNA homologies

Unlabelled DNA of: Plasmid	Inc gp	Size (Kb)	% hybridization with ^{3}H-labelled DNA of: R1	R144	R724	TP114	N3	RP4	S-a	R27	TP116
F	FI	166	43	—	—	—	—	—	—	—	—
R1	FII	104	100	6	—	—	6	9	14	5	1
ColIb-P9	I_1	104	—	—	—	2	—	—	—	—	—
R144	I_1	100	—	100	28	—	5	<1	2	0	3
R64	I_1	115	—	78	—	6	—	—	—	—	—
R621a	I_2	103	—	—	—	40	—	—	—	—	—
R724	B	93	—	—	100	—	—	—	—	—	—
TP113	B	91	—	—	96	3	—	—	—	2	4
TP114	I_2	66	—	—	—	100	—	—	—	0	1
N3	N	51	8	2	—	—	100	1	10	2	5
R15	N	64	—	—	—	75	—	—	—	—	—
RP4	P	54	—	3	—	—	4	100	3	—	—
R702	P	74	—	—	—	—	—	80	—	—	—
S-a	W	40	4	5	—	—	8	4	100	—	—
R388	W	34	—	—	—	—	—	—	79	—	—
R27	H	180	—	—	—	1	—	—	—	100	3
TP124	H	192	—	—	—	—	—	—	—	97	2
TP116	H	230	—	—	—	—	—	—	—	2	100

The figures are adapted from Grindley et al. 1973 and Falkow et al. 1974. A dash indicates experiment not done. In general, plasmids within a group have similar molecular weights. There is a high degree of homology between plasmids within a group (i.e. incompatible plasmids), a significant degree between compatible plasmids whose conjugative pili are antigenically related (see Table 6.7), and little between other pairs of plasmids. An exception is among plasmids of IncH. These are incompatible with one another, have antigenically related pili and high molecular weights but by DNA hybridization tests fall into two distinct groups, designated H1 and H2 (Smith et al. 1973).

DNA homology, i.e. a high proportion of the DNA of one, denatured to the single-stranded form, will hybridize with single-stranded DNA of another, creating heteroduplex molecules. Between groups, there is little such homology. Table 6.6 shows the proportions of hybridizable DNA in pairs of conjugative plasmids of the same or different incompatibility groups. Hybridization of DNA molecules can be demonstrated directly by electron microscopy (see Fig. 6.5). Fig. 6.14 demonstrates this diagrammatically. In the electron microscope, single-stranded DNA is distinguishable from double-stranded, and so it is possible to see not only how much hybridization there is between two plasmid molecules but also which segments of the two are homologous.

Another characteristic common to plasmids within an incompatibility group is the structure of their conjugative pili. These vary morphologically and serologically and act as receptors for different phages. F pili (p. 157) are thick (diameter about 9 nm) and flexible. They were described and analysed by Brinton (1965). Many conjugative plasmids of enterobacteria have pili morphologically similar to F pili: those whose transfer genes are alleles of F transfer genes have antigenically related pili. F pili were recognized as surface antigens and agglutinogens by Ørskov and Ørskov (1960) before their morphology had been determined (Brinton *et al.* 1964, Ishibashi 1967), but the antigenic relations of conjugative pili cannot usually be demonstrated by agglutination. Coating of pili by specific antibody can be seen by electron microscopy (Fig. 6.15) (Meynell *et al.* 1968, Bradley 1980a). Figure 6.15 also shows an F pilus with particles of the F-specific phage R17, adsorbed along its length—another method of discriminating between F-like and other pili. In electronmicroscopic studies of their morphology and serology, Bradley (1980a) has identified pili determined by conjugative plasmids of all known incompatibility groups of *Esch. coli* and several of *Ps. aeruginosa*. His work shows that, with very few exceptions, all plasmids within an incompatibility group determine the same type of pilus.

The converse, however, is not true. If one takes a collection of plasmids that all determine pili

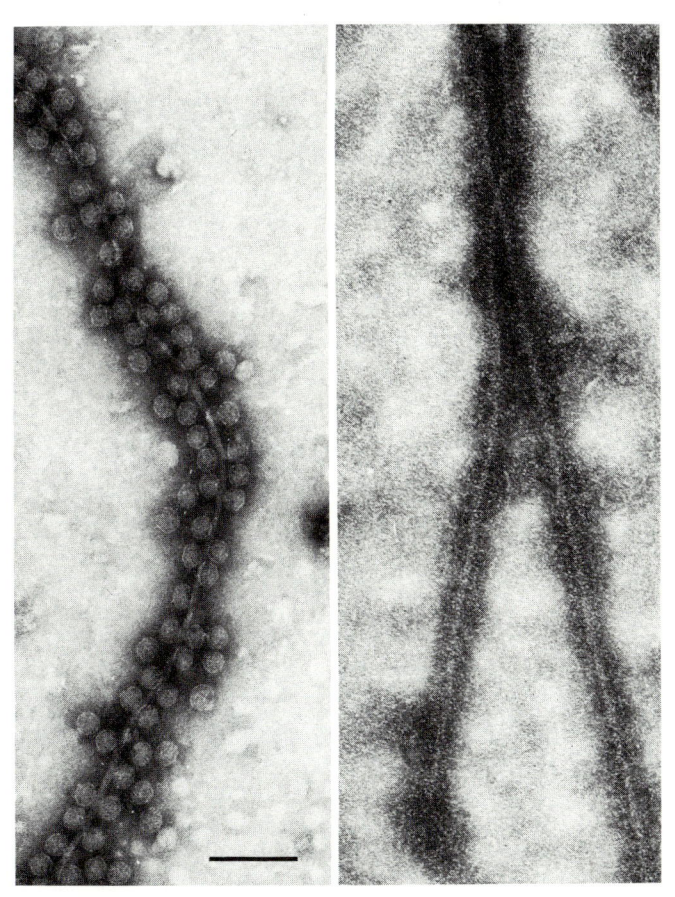

Fig. 6.15 Conjugative pili labelled with phage or antibody. (a) F pilus labelled with particles of phage R17 (from Bradley 1977, *J. gen. Microbiol.* **102,** 349, with permission). (b) D pilus labelled with specific antiserum (D. E. Bradley, unpublished).

serologically related to F pili, they are not all incompatible with one another. The fi^+ R plasmids that were first identified in Japan have F-like pili, yet are compatible with F. For historical reasons, the incompatibility group that includes plasmid F is called IncFI and many R plasmids, including R100, R1 and R6, which have been intensively studied in several laboratories, are classified as IncFII. Other plasmids with F pili have been grouped as FIII, FIV etc, and together constitute an incompatibility complex. In several instances the incompatibility groupings within the complex are not clear cut.

Other incompatibility groups are designated by capital letters (Table 6.7). A number of groups outside the F complex have thick flexible pili morphologically similar but serologically unrelated to those of F. Thin flexible pili (diameter 6 nm) (Fig. 6.16) are determined by plasmids of IncI groups designated I, because the first known example was a conjugative colicin I plasmid, Chapter 8. Plasmids determining pili which are morphologically and serologically I pili belong to several incompatibility groups. Some have been designated IncIα (synonym IncI$_1$), IncIδ etc; others for historical reasons are designated IncB and IncK.

Rigid pili of various serological types (Fig. 6.16, Table 6.7) are determined by plasmids of several groups. Plasmids with rigid pili are transferable very poorly, if at all, in liquid medium, but efficiently on the surface of agar media or membrane filters. Rigid pili are, in all examples studied, produced constitutively, i.e. present on all cells in plasmid-containing cultures. The production of flexible pili, whether thick or thin, is frequently repressed in cultures carrying wild-type plasmids (Table 6.7).

The incompatibility grouping of plasmids is associated, in general, with their properties of host range, DNA homology, approximate molecular size and, with conjugative plasmids, surface exclusion and pilus

Table 6.7 Incompatibility groups of plasmids in Enterobacteriaceae

Plasmid Incompatibility (Inc) group[a]	Conjugative pili Morphology	Serological specificity[b]	Receptors for phages[c] pilus sides	pilus tips	Expression repressed (R) or constitutive (C)[d]	Mating type[e]
B	Thin flexible	I$_1$	—	—	C or R	universal
C	Thick flexible	C	C-1	—	C or R	surface preferred
D	Thick flexible	D	—	fd	C	surface preferred
FI	Thick flexible	F	R17	fd	C or R	universal
FII	Thick flexible	F	R17	fd	R	universal
H	Thick flexible	H	—	—	R	universal
I$_1$	Thin flexible	I$_1$	—	Ifl	R	universal
I$_2$	Thin flexible	I$_1$	—	Ifl	R	universal
Iγ	Thin flexible	Iα	—	Ifl	R	universal
J	Thick flexible	C	C-1	—	R	universal
K	Thin flexible	Iα	—	—	R	universal
M	Rigid	M	—	X	C	surface obligatory
N	Rigid	N	—	Ike, PR4	C	surface obligatory
P	Rigid	P	PRR1	PR4, X	C	surface obligatory
T	Thick flexible	T	t	—	C	surface preferred
U	Rigid	U	—	X	C	surface obligatory
V	Thick flexible	nd	—	—	R	universal
W	Rigid	W	—	PR4, X	C	surface obligatory
X	Thick flexible	X	—	X	C	surface preferred

The table indicates the pilus type for plasmids within each incompatibility group and summarizes work by Bradley and his colleagues (Bradley 1980a, b; Bradley et al. 1980, 1981a, b). Although generally plasmids within an incompatibility group determine pili of uniform type, an exception has been reported, a plasmid of IncM determining F pili (Taylor et al. 1981).

(a) Incompatibility grouping (Novick et al. 1976). I$_1$ is synonymous with Iα and I$_2$ with Iδ. IncFIII, FIV, FVI are similar to FI and FII in their pilus type. Plasmid Folac (not listed) determines pili antigenically unrelated to F pili that act as receptors to phage fd. nd = none described.

(b) Serological specificity: antibody adsorbed to pili seen by electron microscopy. Antiserum of each specificity, prepared by immunizing rabbits with appropriate pili, showed no reaction with heterologous pili except where listed. The only unexpected cross-reaction (not shown) was with pili of plasmid R805a that reacted with anti I$_1$ and anti I$_2$ (Bradley 1980a).

(c) Examples are given for phages that attach to conjugative pilus e.g. numerous F-pilus specific phages have been described, but only two are listed here. It can be seen that phages that attach to the sides of pili reflect the serological type, but attachment to the tips tends to be less specific.

(d) Pilus synthesis listed as constitutive when large numbers of pili were seen in pure cultures by electron microscopy and repressed when none, or very few, were seen in pure cultures, but only in 'high-frequency transfer' mixtures of donor and recipient strains (see text).

(e) Mating type was designated universal when frequencies of transfer were similar in broth and on solid medium, surface preferred when frequencies were 50–500 times higher on surfaces and surface obligatory when they were >2000 times higher on surfaces. Rigid pili are associated with obligatory surface mating.
nd = not determined

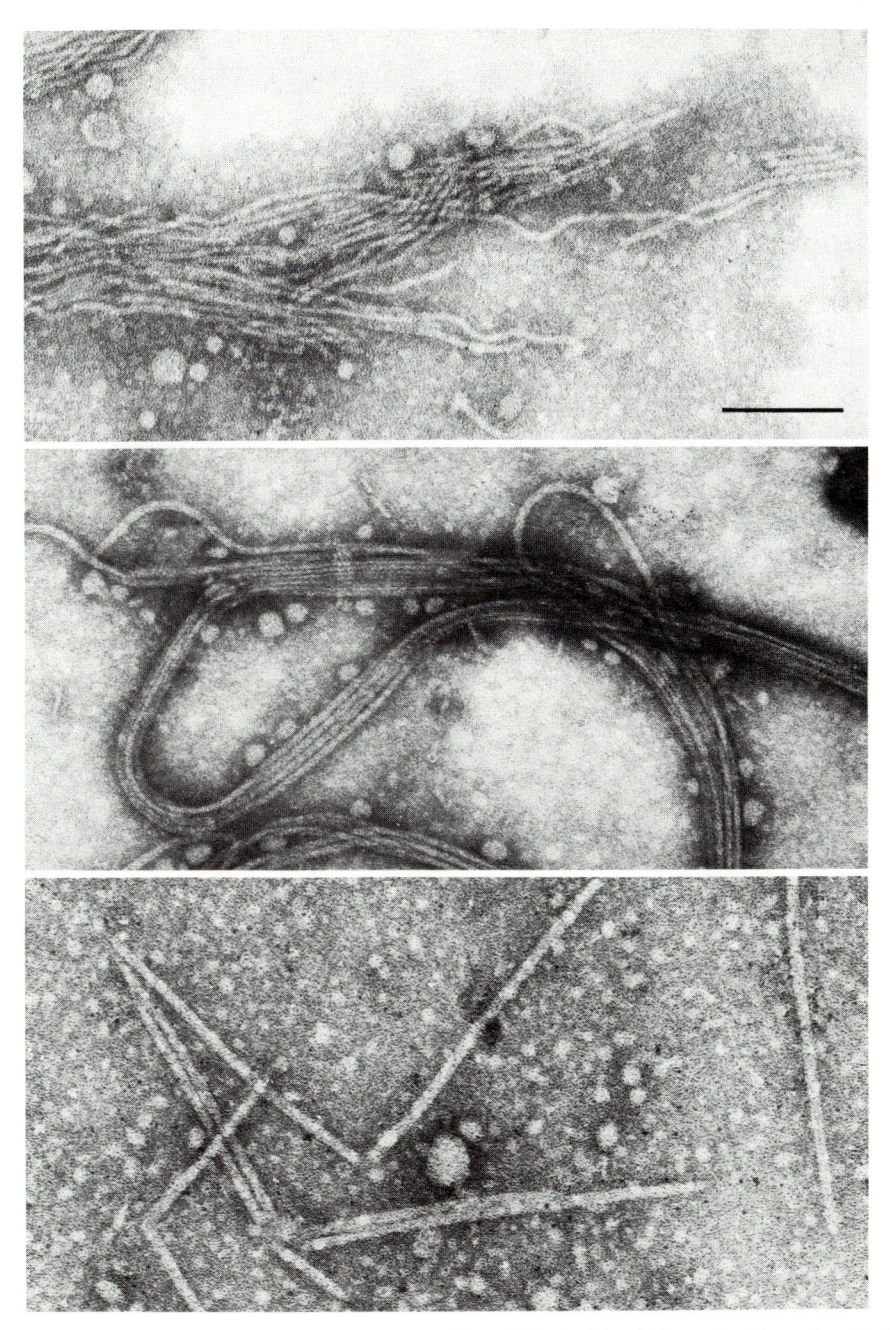

Fig. 6.16 Morphological types of conjugative pili. (a) Thin, flexible pili determined by a plasmid of Inc I$_1$. (b) Thick, flexible pili determined by a plasmid of IncH. (c) Rigid pili determined by a plasmid of IncN. (Electron micrographs by D. E. Bradley).

type. It is not related to other plasmid-determined characters, in particular antibiotic resistance genes. The same genes, identifiable either as DNA sequences or as protein products, can be found in a wide variety of otherwise unrelated plasmids. It seems evident that plasmids as they are identified today have evolved by the accumulation of inserted genes, often inserted transposons. An example is plasmids of IncFII whose

DNA molecules are hybridizable for much of their circumference but which determine different resistance patterns (Fig 6.17).

It is often stated that R plasmids consist of a *Resistance Transfer Factor* (RTF) or conjugative plasmid component linked to a *resistance determinant* (r-det) DNA segment. This is because members of one family of plasmids (Fig. 6.17) have a segment of their DNA, including some but not all of their resistance genes, with special properties. This, the resistance determinant segment, is flanked by copies of IS1, both in the same orientation; it can loop out of the entire plasmid to form a separate circular molecule, visible by electron microscopy (Clowes 1972) but not classifiable as a *replicon*, since there is no evidence that it can replicate autonomously in cells lacking the parent plasmid. The resistance determinant sequence has a higher G+C content and therefore higher density than the rest of the plasmid. In *Proteus mirabilis*, plasmid NR1 (synonyms R100 or 222) undergoes complex rearrangements that were first recognized by a gradual increase in the density of plasmid DNA in the cultures. The change, called *transition*, followed exposure of the R+ *Proteus* to chloramphenicol or streptomycin, resistance to which is encoded in the resistance determinant. The resistance determinant component of NR1 in *Proteus* forms monomers, dimers and polymers that can exist as separate molecules or can be inserted, as tandem repeat sequences, in the parent plasmid, producing larger and larger molecules (Perlman and Rownd 1975, Rownd *et al.* 1978). This is an example of gene *amplification* in response to a selective pressure. Transition with NR1 does not occur in *Esch. coli*, but there are other examples in *Esch. coli* (Mattes *et al.* 1979) and in *Streptococcus faecalis* (Yagi and Clewell 1976) of response to increased antibiotic concentration by the synthesis of a series of copies of DNA sequences that encode the relevant resistances, (Fig. 6.18). Amplification of a resistance-determinant component of R plasmids seems to be an exceptional, not a general, means of genetic adaptability. Plasmids of group IncW (Gorai *et al.* 1979) or IncP Villarroel *et al.* (1983), studied as heteroduplex molecules by electron microscopy, appear as a common, hybridizable molecule with a variety of heterologous inserts at varying positions in the various individual plasmids. There seem to be 'hot spots' for the inserts, but they are not limited to one defined resistance-determinant component. (For a review of bacterial plasmids, see Hardy 1981.)

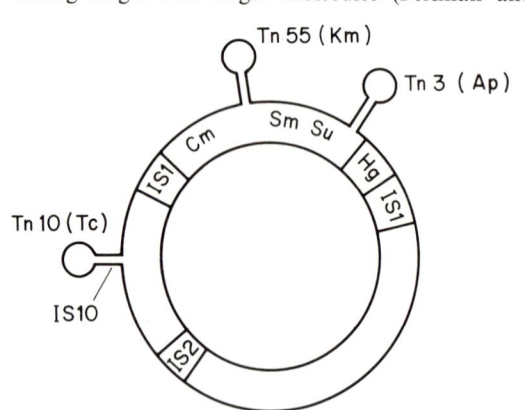

Fig. 6.17 Diagram to illustrate evolution of multiple-resistance plasmids. Composite heteroduplex diagram of three IncFII plasmids. R100, from *Shigella flexneri*, Japan, determines TcCmSmSuHg resistance, R1 from *Salmonella paratyphi* B, England, determines CmKmSmSuAp resistance and R6, from *S. typhimurium*, Germany, determines TcCmKmSmSu. Their DNA will hybridize one with another over most of their circumference, but transposons not present in all three show as loop structures in heteroduplex molecules (*see also* Figs. 6.5 and 6.14). Tn3 of R1 is within Tn4. (Adapted from Cohen and Shapiro 1980: Tc = tetracycline, Cm = chloramphenicol, Km = kanamycin, Sm = streptomycin, Su = sulphonamides, Ap = ampicillin, Hg = mercuric chloride.)

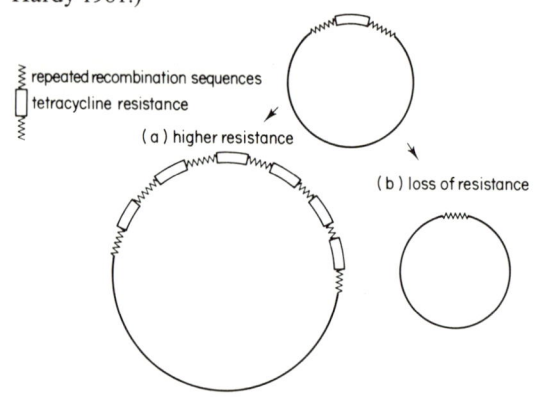

Fig. 6.18. Amplification or deletion of a resistance gene. A tetracycline-resistance plasmid in *Streptococcus faecalis* altered its structure when the bacteria were cultured on increasing concentrations of tetracycline. Analysis of the plasmid DNA showed molecules of various sizes, in which a 2.65 Md sequence that included the tetracycline resistance gene were present as tandem repeats (6 repeats shown (a) in the figure). In the absence of tetracycline, drug-sensitive variants of the host bacterium appeared in which the plasmid had lost the 2.65 Md sequence (b). The zig-zag segments represent direct repeats in which recombination, dependent upon the hosts *rec* genes, took place (Yagi and Clewell 1976, 1980).

Bacterial variation and epidemiology

In epidemiological studies, the aim is to identify a particular clone within a bacterial species in order to trace its spread. Characters used are biochemical and serological, phage-sensitivities and lysogeny, production of or sensitivity to bacteriocins, sensitivity or resistance to antibiotics and other antibacterial agents. These characters are usually sufficiently stable for them to be useful in tracing a bacterial strain, but none of them is immutable. In epidemiological studies, therefore, as many characters as possible of the strain under study should be used in its identification. The isolation of variants indicates either genetic change in an epidemic strain or the fortuitous finding of a different strain of the same species. It is not always possible to distinguish between these alternatives (Seal et al. 1981).

Genetic engineering

Bacterial variation may also result from *in vitro* manipulation of bacterial genes. This has been called genetic engineering and has recently emerged as an important new branch of biology. The techniques of genetic engineering make it possible to isolate one of the million odd genes of an animal cell, fuse that gene with part of a bacterial gene and insert the recombinant replicon into a bacterium. As these recombinant bacteria divide they make millions of copies of their own genes and of the animal genes inserted in them.

The bacterial chromosome contains some 3 million base pairs. In the human cell the DNA is packed into 46 chromosomes each one containing about 3×10^9 base pairs. The problem arises as to how a gene containing only a few thousand base pairs can be isolated from such a mass of DNA. Fortunately nature has devised a set of special enzymes called *restriction enzymes* that can specifically chop up DNA into smaller pieces. Restriction enzymes are produced by bacterial cells as part of their defence against foreign DNA. The host's DNA is protected from degradation by its own restriction enzymes, because it is *modified* by the addition of methyl groups to the DNA bases at the sites recognized by its restriction enzyme.

There are two types of restriction enzyme, I and II. Type I enzymes are complex multimeric proteins which require ATP, s-adenosyl methionine and magnesium ions as cofactors. Although they recognize specific sites in the DNA, they cleave randomly at a considerable distance from the recognition site. These are not of use in genetic engineering.

The type II enzymes have very specific target sites at which they cut double-stranded DNA. The target sites normally comprise 4 or 6 base pairs, although there are a few enzymes recognizing a 5-base-pair target site and one which has a 7-base-pair target site. These enzymes are one of the essential tools for genetic engineering.

There are two ways in which restriction enzymes cut double-stranded DNA. Some cut at positions on the DNA strands opposite to each other, giving blunt ends. Others give a staggered cut producing single-stranded 'sticky ends' (Fig. 6.19). The resulting fragments can be joined together by an enzyme called a *ligase* whose action requires magnesium ions and adenosine tri-phosphate. The most commonly used ligase commercially available is that from phage T4. Many restriction enzymes recognize palindromic sequences, such that the sequence read in the 5' to 3' direction in one strand is the same as that in the 5' to 3' sequence in the opposite strand (Fig. 6.19). The 'sticky ends' produced by these enzymes are the same, and therefore two fragments produced by the same restriction enzyme can be ligated together in either orientation (for a review see Smith 1979).

The technique for placing and maintaining a new gene in a bacterium is called *cloning*. There are two convenient types of genomes that can be used for cloning—plasmids and bacteriophages. The most commonly used such plasmids are small multicopy plasmids such as ColE1 and its derivatives, in particular pBR322. Foreign DNA is inserted into plasmid vectors by means of restriction enzymes and DNA ligase. pBR322 has approximately 4200 base pairs and single-cut sites for a number of restriction enzymes. The single *PstI* site is within the ampicillin resistance gene which encodes a β-lactamase. Cloning of DNA into this site results in a loss of ampicillin resistance. Similarly, cloning into the *Bam*HI, *Hind*III or *Sal*I sites results in a loss of tetracycline resistance.

Eukaryotic genes are not transcribed from their own promoters in bacteria, because bacterial RNA polymerase does not bind to eukaryotic DNA. This DNA has to be inserted into a bacterial gene and read from the bacterial promoter. The fundamental problem is to place only the desired DNA sequence into the bacterium. One approach, suitable for small proteins, is to synthesize a gene chemically and then insert it into a cloning vector. This was first used for the cloning of somatostatin—a pituitary hormone of 14 amino acids whose sequence is known. The genes for human growth hormone, thymosin and insulin have also been synthesized.

Another approach has to be used for larger peptides, this requires isolation of a particular messenger RNA and the synthesis of a complementary DNA fragment using reverse transcriptase. The appropriate messenger RNA may be isolated by a method involving immunoprecipitation of polysomes. Polysomes, consisting of messenger RNA, ribosomes, and nascent polypeptides, are treated with an antibody prepared against the desired protein. The polysome-antibody complex can be separated from other polysomes by fractionation on a sucrose gradient. The messenger RNA encoding the protein can be isolated from ribosomes, nascent polypeptides and antibody and used to make complementary DNA (cDNA).

A number of proteins are now being made by gene cloning in bacteria. These include somatostatin, growth hormone, insulin and interferon. Recently several groups have described the successful expression in *Esch. coli* of genes for human leukocyte (IFNα) and fibroblast (IFNβ) interferon. There have been reports that the IFNα from recombinant bacteria protects monkeys against myocarditis virus and may also inhibit growth of Burkitt lymphoma cells.

As well as producing large amounts of important

(a)

—GG CC—
—CC GG—   → Cutting produces blunt ends

Recognition sequence

(b)

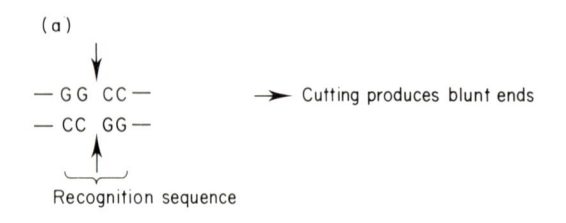

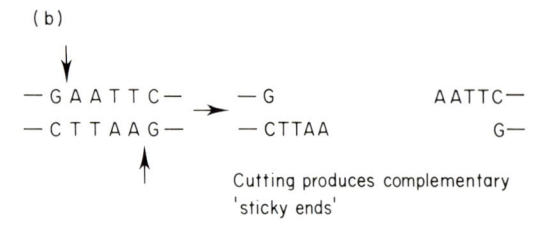

—GAATTC—   —G   AATTC—
—CTTAAG—   —CTTAA   G—

Cutting produces complementary 'sticky ends'

Fig. 6.19 Type II restriction enzymes. (a) *Hae*III from *Haemophilus aegyptius*. (b) *Eco*RI from *Escherichia coli* RY13.

proteins, genetic engineering is being used to add useful genes on to replicons. The methylotrophic bacteria produce protein from methanol and ammonia. The enzyme used to fix ammonia, glutamase synthase, is very energy-inefficient. A more efficient enzyme, glutamate dehydrogenase, has been cloned into methylotrophic bacteria resulting in a more efficient utilization of ammonia.

Genetic engineering in plants may soon be possible. The tumour-forming plant pathogen, *Agrobacterium tumifaciens*, contains plasmids called Ti plasmids, part of which, T DNA, can insert into the plant chromosome (p. 167). It may in the future be possible to introduce new genes into plants via the T DNA.

Genetic engineering may also be useful in the production of vaccines against viral or parasitic infections. Bacteria need make only the antigen against which antibody response is required, thus eliminating the danger of work with the intact pathogens. The techniques of molecular biology developed over the past ten years are being used to produce *in vitro* bacterial variation which promise to be of great use in the future.

References

Anderson, E. S. (1968) *Annu. Rev. Microbiol.* **22**, 131.
Bachmann, B. J. and Low, B. K. (1980) *Microb. Rev.* **44**, 1.
Barksdale, L. (1970) *Bact. Rev.* **34**, 378.
Barth, P., Datta, N., Hedges, R. W. and Grinter, N. (1976) *J. Bact.* **125**, 800.
Berg, D. E., Davies, J., Allen, B. and Rochaix, J-D. (1975) *Proc. nat. Acad. Sci. Wash.* **72**, 3628.
Bergmans, H. E. N., Hoekstra, W. P. M. and Zuidweg, E. M. (1975) *Molec. gen. Genet.* **137**, 1.

Binns, M. M., Davies, D. L. and Hardy, K. G. (1979) *Nature* **279**, 778.
Blattner, F. R., Fiant, M., Hass, K., Twose, P. and Szybalski, W. (1974) *Virology* **62**, 458.
Bradley, D. E. (1980a) *Plasmid* **4**, 155; (1980b) *J. Bact.* **141**, 828.
Bradley, D. E., Coetzee, J. N., Bothma, T. and Hedges, R. W. (1981a); (1981b) *J. gen. Microbiol.* **126**, 389, 397.
Bradley, D. E., Taylor, D. E. and Cohen, D. R. (1980) *J. Bact.* **143**, 1466.
Brinton, C. C. (1965) *Trans. N.Y. Acad. Sci.* **27**, 1003.
Brinton, C. C., Gemski, P. and Carnahan, J. (1964) *Proc. nat. Acad. Sci. Wash.* **52**, 776.
Cairns, J. (1963) *J. molec. Biol.* **6**, 208.
Chang, A. C. Y. and Cohen, S. N. (1978) *J. Bact.* **134**, 1141.
Chilton, M-D., Drummond, M. H., Merlo, D. J., Sciaky, D., Montoya, A. L., Gordon, M. P. and Nester, E. W. (1977) *Cell* **11**, 263.
Clowes R. C. (1972) *Bact. Rev.* **36**, 361.
Cohen, S. N., Casadaban, M., Chou, J. and Tu, C. (1979) *Cold Spring Harbor Symp. quant. Biol.* **43**, 1247.
Cohen, S. N. and Shapiro, J. A. (1980) *Scientific American* **242**, 36.
Cornelis, G., Bennett, P. M. and Grinsted, J. (1976) *J. Bact.* **127**, 1058.
Cornelis, G., Ghasal, D. and Saedler, H. (1978) *Molec. gen. Genet.* **160**, 215-224.
Danbara, H., Timmis, J. K., Lurz, R. and Timmis, K. N. (1980) *J. Bact.* **144**, 1126.
Datta, N. (1975) *Microbiology 1974*, p. 9. Ed. by D. Schlessinger. Amer. Soc. Microbiol., Washington D.C.
Datta, N. and Hedges, R. W. (1972) *J. gen. Microbiol.* **70**, 453.
Datta, N., Lawn, A. M. and Meynell, E. (1966) *J. gen. Microbiol.* **45**, 365.
Davies, J. R., Farrant, W. N. and Tomlinson, A. J. H. (1968) *J. Hyg., Camb.* **66**, 471.
Deich, K. and Smith, H. O. (1978) In: *Transformation*, p. 343. Ed. by Glover and Butler. Cotswold Press, Oxford, 1979.
Dougan, G. and Sherratt, D. S. (1977) *Molec. gen. Genet.* **151**, 151.
Dunny, G. M., Brown, B. L. and Clewell, D. B. (1978) *Proc. nat. Acad. Sci. Wash.* 75 3479.
Dunny, G. M., Craig, R. A., Carron, R. L. and Clewell, D. B. (1979) *Plasmid* **2**, 454.
Easterling, S. B., Johnson, E. M., Wohlhieter, J. A. and Baron, L. S. (1969) *J. Bact.* **100**, 35.
Ehrlich, S. D. (1977) *Proc. nat. Acad. Sci. Wash.* **74**, 1680.
Elwell, L. P. and Shipley, P. L. (1980) *Annu. Rev. Microbiol.* **34**, 465.
Engel, H., Soedirman, N., Rost, J., van Leeuwen, W. and van Embden, J. D. A. (1980) *J. Bact.* **142**, 407.
Falkow, S., Guerry, P., Hedges, R. W. and Datta, N. (1974) *J. gen. Microbiol.* **85**, 65.
Falkow, S., Wohlhieter, J. A., Citarella, R. V. and Baron, L. S. (1964) *J. Bact.* **88**, 1598.
Farmer, J. J., Hickman, F. W., Brenner, D. J., Shreiber, M. and Rickenbach, D. G. (1977) *J. clin. Microbiol.* **6**, 373.
Farrow, J. A. E. (1980) *J. appl. Bact.* **49**, 493.
Ferenczy, L. (1981) *Soc. gen. Microbiol. Symp.* **31**, 1.
Fiant, M., Szybalski, W. and Malamy, M. (1972) *Mol. gen. Genet.* **119**, 223.

Flett, F., Humphreys, G. O. and Saunders, J. R. (1981) *J. gen. Microbiol.* **125**, 123.
Franke, A. E. and Clewell, D. B. (1981) *J. Bact.* **145**, 494.
Gasson, M. J. and Willetts, N. S. (1975) *J. Bact.* **122**, 518.
Gibson, E. M., Chace, N. M., London, S. B. and London, J. (1979) *J. Bact.* **137**, 614.
Gorai, A. P., Heffron, F., Falkow, S., Hedges, R. W. and Datta, N. (1979) *Plasmid,* **2**, 485.
Gottesman, M. E. and Rosner, J. L. (1975) *Proc. nat. Acad. Sci. Wash,* **72**, 5041.
Grindley, N. D. F. (1978) *Cell* **13**, 419.
Grindley, N. D. F., Humphreys, G. O. and Anderson, E. S. (1973) *J. Bact.* **115**, 387.
Grindley, N. D. F. and Sherratt, D. (1979) *Cold Spring Harbor Symp. quant. Biol.* **43**, 1257.
Grubb, W. B. and O'Reilly, R. J. (1971) *Biochem. Biophys. Res. Comm.* **42**, 377.
Habermann, P., Klaer, R., Kühn, S. and Starlinger, P. (1979) *Molec. gen. Genet.* **175**, 369.
Harada, K., Suzuki, M., Kameda, M. and Mitsuhashi, S. (1960) *Jap. J. exp. Med.* **30**, 289.
Hardy, K. (1981) *Bacterial Plasmids.* Thomas Nelson & Sons, Sunbury-on-Thames.
Hayes, W. (1953a) *J. gen. Microbiol.* **8**, 72; (1953b) *Cold Spring Harbor Symp. quant. Biol.* 18, 75.
Hedges, R. W. and Jacob, A. E. (1974) *Molec. gen. Genet.* **132**, 31.
Hedges, R. W. and Jacoby, G. A. (1980) *Plasmid,* **3**, 1.
Heffron, F., McCarthy, B. J., Ohtsubo, H. and Ohtsubo, E. (1979) *Cell* **18**, 1153.
Henner, D. J. and Hoch, J. A. (1980) *Microb. Rev.* **44**, 57.
Hirota, Y. (1960) *Proc. nat. Acad. Sci. Wash.* **46**, 57.
Holloway, B. W., Krishnapillai, V. and Morgan, A. F. (1979) *Microb. Rev.* **43**, 73.
Hopwood, D. A. (1978) *Annu. Rev. Microbiol.* **32**, 373.
Hopwood, D. A., Chater, K. F., Dowding, J. E. and Vivian, A. (1973) *Bact. Rev.* **37**, 371.
Hu, S., Ptashne, K., Cohen, S. and Davidson, N. (1975) *J. Bact.* **123**, 687.
Ishibashi, M. (1967) *J. Bact.* **93**, 379.
Jacob, F. and Brenner, S. (1963) *C. R. Acad. Sci. Paris* **256**, 298.
Jacob, F. and Wollman, E. L. (1958) *C. R. Acad. Sci. Paris* **247**, 154.
Jacoby, G. A. (1977) In: *DNA Insertion Elements, Plasmids and Episomes,* p. 639. Ed. by A. I. Bukhari, J. A. Shapiro and S. L. Adhya. Cold Spring Harbor Laboratory.
Johnson, E. M., Wohlhieter, J. A., Placek, B. P., Sleet, R. B. and Baron, L. S. (1976) *J. Bact.* **125**, 385.
Jordan, E., Saedler, H. and Starlinger, P. (1968) *Molec. gen. Genet.* **102**, 353.
Kahn, S. A. and Novick, R. P. (1982) *J. Bact.* **149**, 642
Kondo, E. and Mitsuhashi, S. (1964) *J. Bact.* **88**, 1266.
Kopecko, D. J. and Cohen, S. N. (1965) *Proc. nat. Acad. Sci. Wash.,* **72**, 1373.
Kopecko, D. J., Washington, O. and Formal, S. B. (1980) *Infect. & Immun.* **29**, 207.
Lacey, R. W. (1975) *Bact. Rev.* **39**, 1; (1980) *J. gen. Microbiol.* **119**, 423.
Landman, O. E. and Halle, S. (1963) *J. molec. Biol.* **7**, 721.
Laufs, R. and Kaulfers, P.-M. (1977) *J. gen. Microbiol.* **103**, 277.
Laufs, R., Kaulfers, P.-M., Jahn, G. and Teschner, U. (1979) *J. gen. Microbiol.* **111**, 223.

Lederberg, J. (1952) *Physiological Rev.* **32**, 403.
Lederberg, J., Cavalli, L. L. and Lederberg, E. M. (1952) *Genetics* **37**, 720.
Lederberg, J. and Lederberg, E. M. (1952) *J. Bact.* **63**, 399.
Lederberg, J. and Tatum, E. L. (1946) *Nature* **158**, 558.
Low, K. B. and Porter, D. D. (1978) *Annu. Rev. Genet.* **12**, 249.
Luria, S. E. and Delbrück, M. (1943) *Genetics* **28**, 491.
McHugh, G. L., Moellering, R. C., Hopkins, C. C. and Swartz, M. N. (1975) *Lancet* **i**, 235.
Manten, A., Guinée, P. A. M. and Kampelmacher, E. H. (1966) *Zbl. Bakt. Parasit., Abt. I Orig.* **200**, 13.
Mattes, R., Burkhardt, H. J. and Schmitt, R. (1979) *Molec. gen. Genet.* **168**, 173.
May, J. M., Houghton, R. H. and Perret, C. J. (1964) *J. gen. Microbiol.* **37**, 157.
Meynell, E. and Datta, N. (1967) *Nature (London)* **214**, 885.
Meynell, E., Meynell, G. G. and Datta, N. (1968) *Bact. Rev.* **32**, 55.
Meynell, G. G. (1972) *Bacterial Plasmids.* Macmillan, London.
Mitsuhashi, S. (1977). In: *R Factor Drug Resistance Plasmid,* p. 3. Ed. by S. Mitsuhashi. University of Tokyo Press, Tokyo.
Moll, A., Manning, P. A. and Timmis, K. N. (1980) *Infect. & Immun.* **28**, 359.
Mortelmans, K. E. and Stocker, B. A. D. (1976) *J. Bact.* **128**, 271.
Nakaya, R., Nakamura, A. and Murato, Y. (1960) *Biochem. biophys. Res. Comm.* **3**, 654.
Novick, R. P. (1963) *J. gen. Microbiol.* **33**, 121–136.
Novick, R. P., Clowes, R. C., Cohen, S. N., Curtiss, R., Datta, N. and Falkow, S. (1976). *Bact. Rev.,* **40**, 168.
Novick, R. P., Cohen, S., Yamamoto, L. and Shapiro, J. A. (1977). In: *DNA Insertion Elements, Plasmids and Episomes,* p. 657. Ed. by A. I. Bukhari, J. A. Shapiro and S. L. Adhya. S. L. Cold Spring Harbor Laboratory.
Nugent, M. E., Bone, D. H. and Datta, N. (1979). *Nature,* **282**, 422.
Olsen, R. H. and Shipley, P. (1973). *J. Bact.,* **113**, 772.
Ørskov, F. and Ørskov, I. (1960). *Acta. path. microbiol. scand.,* **48**, 37.
Ørskov, I. and Ørskov, F. (1973). *J. gen. Microbiol.,* **77**, 487.
Perlman, D. and Rownd, R. H. (1975). *J. Bact.,* **123**, 1013.
Privitera, G., Sebald, M. and Fayolle, F. (1981). *Nature,* **278**, 657.
Reeve, E. C. R. and Braithwaite, J. A. (1973) *Genet. Res. Camb.* **22**, 329.
Riva, S., Fietta, A. M., Silvestri, L. G. and Romero, E. (1972) *Nature New Biol.,* **235**, 78.
Roe, E., Jones, R. J. and Lowbury, E. J. L. (1971) *Lancet* **i**, 149.
Rownd, R., Miki, T., Appelbaum, E. R., Miller, J. R., Finkelstein, M. and Barton, C. R. (1978). In: *Microbiology, 1978.* Ed. by D. Schlessinger. Amer. Soc. Microbiol. Washington D.C.
Sabelnikov, A. G. and Domaradsky, I. V. (1979) *Molec. gen. Genet.,* **172**, 313.
Saedler, H. and Heiss, B. (1973) *Molec. gen. Genet.,* **122**, 267.
Salkinoja-Salonen, M. S. and Sundman, V. (1979) In: *Lignin Biodegradation, Microbiology, Chemistry and Applications,* pp. 179. Ed. by T. K. Kirk. CRC Press, Cleveland, Ohio.
Sanderson, K. E. (1976) *Annu. Rev. Microbiol.* **30**, 327.

Scott, J. R., Kropf, M. and Mendelson, L. (1977) *Virology* **76,** 39.

Seal, D. V., McSwiggan, D. A., Datta, N. and Feltham, R. K. A. (1981) *J. med. Microbiol.* **14,** 295.

Sedgwick, S. (1980) *Nature* **287,** 676.

Shapiro, J. A. (1979) *Proc. nat. Acad. Sci. Wash.* **76,** 1933.

Shipley, P. L., Gyles, C. L. and Falkow, S. (1978) *Infect. & Immun.* **20,** 559.

Sisco, K. L. and Smith, H. O. (1979) *Proc. nat. Acad. Sci. Wash.* **76,** 972.

Smith, H. O. (1979) *Science* **205,** 455.

Smith, H. R., Grindley, N. D. F., Humphreys, G. O. and Anderson, E. S. (1973) *J. Bact.* **115,** 623.

Smith, H. W. (1972) *Nature, New Biol.* **238,** 205; (1974). *Lancet,* **ii,** 281.

Smith, H. W. and Huggins, M. B. (1976) *J. gen. Microbiol.* **92,** 335.

Smith, H. W., Parsell, Z. and Green, P. (1978) *J. gen. Microbiol.* **109,** 305.

So, M., Heffron, F. and McCarthy, B. (1979) *Nature* **277,** 453.

Stewart, G. C. and Rosenblum, E. D. (1980) *J. Bact.* **144,** 1200.

Stocker, B. A. D. (1956) *J. gen. Microbiol.* **15,** 575.

Stuy, J. H. (1980a). *J. Bact.*, 142, 925; (1980b) *J. Bact.* **144,** 999.

Summers, A. O. and Jacoby, G. A. (1977) *J. Bact.* **129,** 276.

Summers, A. O. and Silver, S. (1978) *Annu. Rev. Microbiol.* **32,** 637.

Taylor, D. E., Levine, J. G. and Bradley, D. E. (1981). *Plasmid.* **5,** 233.

Towner, K. J. and Vivian, A. (1976) *J. gen. Microbiol.* **93,** 355.

Uhlin, B. E. and Nordström, K. (1975) *J. Bact.* **124,** 641.

Villarroel, R., Maenhaut, R., Leemans, J., Engler, C. T., Schell, J., Montagu, M. van and Hedges, R. W. (1983) *Molec. gen. Genet.* (In press.)

Wachsmuth, I. K., Davis, B. R. and Allen, S. D. (1979) *J. clin. Microbiol.* **10,** 897.

Watanabe, T. (1963) *Bact. Rev.* **27,** 87.

West, S. C., Cassuto, E. and Howard-Flanders, P. (1981) *Nature* **290,** 29.

Wiley, B. B. and Rogolsky, M. (1977) *Infect. & Immun.* **18,** 487.

Willetts, N. S. and Crowther, C. C. (1981). *Genet. Res.* **37,** 371.

Willetts, N. and Skurray, R. (1980) *Annu. Rev. Gen.* **14,** 41.

Williams, P. H. (1979) *Infect. & Immun.* **26,** 925.

Yagi, Y. and Clewell, D. B. (1976) *J. molec. Biol.* **102,** 583; (1980) *J. Bact.,* **143,** 1070.

Yokota, T., Kanamaru, Y., Mori, R. and Akiba, T. (1969) *J. Bact.* **98,** 863.

7

Bacteriophages

D. A. Ritchie

Introductory	177	Replication of single-stranded DNA phages	198
General properties of phage	179	Replication of RNA phages	200
Bacterial lysis and the assay of phage	180	Lysis	202
The one-step growth experiment	182	Lysogeny	202
The Hershey-Chase experiment	183	The nature of prophage	203
Habitat and host specificity	184	Immunity of lysogenic strains	204
The phage particle	185	The regulation of lysogeny	205
Morphology	185	Induction of the lytic cycle	206
Structure and organization	186	The formation of transducing phages	206
Phage nucleic acids	187	The origin of specialized transducing phage	
Phage DNA	187	particles	206
Phage RNA	189	The origin of generalized transducing phage	
Lytic infection	189	particles	208
Adsorption	189	Genetics	208
Receptors of gram-positive bacteria	189	Mutation	209
Receptors of gram-negative bacteria	189	Genetic recombination	210
The dynamics of phage adsorption	190	Linkage analysis and genetic mapping	211
Penetration	191	Physical maps of phage genomes	211
Intracellular multiplication of phage	192	The mechanisms of genetic recombination	211
Replication of double-stranded DNA phages:		Host-controlled modification and restriction	213
Phage T4	192	Phage as vectors for genetic engineering	214
Phage T7	194	Cosmid vectors	216
Phage λ	196	Phage M13 as a vector	216
Phage Mu	198	Concluding remarks	216

Introductory

The study of bacteriophages is important for three reasons: (1) as simple experimental systems for the elucidation of basic principles in genetics and biochemistry, (2) as models for the study of virus-host interactions, and (3) as diagnostic reagents in clinical bacteriology. Of these the last is the least important. In fact the medical value of bacteriophage has been quite disappointing considering the early hopes that bacteriophages might provide a general purpose bactericidal agent suitable for the control and treatment of bacterial infections (d'Herelle 1917, 1918). Without doubt the contribution of bacteriophage research to the creation of molecular biology as a discipline and the resulting progress in our understanding of fundamental processes in genetics and biochemistry has been one of the outstanding successes in modern science.

The original observation of bacteriophage infection was made by the English bacteriologist Frederick Twort who described a curious degenerative change in colonies of a staphylococcus isolated from calf lymph (Twort 1915). These bacterial colonies appeared as contaminants on his culture plates during his attempts to cultivate virus in a cell-free system. Upon further incubation some of these colonies changed their appearance and underwent what Twort referred to as a 'glassy transformation'. Attempts to obtain viable bacteria from these colonies failed, indicating that the constituent bacteria had been killed; however, the ability to produce the glassy transformation could be readily transmitted to normal colonies of viable cells. Furthermore, highly diluted fluid from glassy colonies could still effect the transformation, as could extracts passed through filters capable of retaining bacterial cells. Twort suggested several possible explanations of his findings of which one was that the micrococcal disease was caused by a virus. The nature of filtrable viruses infecting plants and animals was quite well understood, having been established some years before Twort's discovery of the bacterial virus.

At about the same time a Canadian bacteriologist, Felix d'Herelle, had independently reached the same conclusion (d'Herelle 1917). On various occasions he had noticed circular clear spots in bacterial cultures growing on the surface of nutrient agar. The spots were devoid of bacteria. In a study arising from an outbreak of dysentery in a French cavalry regiment d'Herelle showed that the agent causing the clear spots on agar cultures of the dysentery bacillus could be isolated from the faeces of the infected soldiers. Moreover, the agent could produce the clear patches even after filtration through a porcelain Chamberland filter. In a remarkable series of experiments the course of the infection was followed by making filtered extracts of fecal samples taken daily from one infected patient. On each occasion a drop of the filtrate was added to a culture of the dysentery bacillus and a sample spread on nutrient agar. On the first three days the cultures were normal; however, after incubation with a sample taken on the fourth day d'Herelle noted that the liquid broth culture was completely clear and the agar plate was devoid of all bacterial growth. In recounting these experiences, d'Herelle later wrote, 'in a flash I had understood: what caused my clear spots was in fact an invisible microbe, a filtrable virus, but a virus parasitic on bacteria' (d'Herelle 1949). Reasoning that the events occurring in the test-tube had probably occurred also in the patient's intestine, he observed that the condition of the sick man had greatly improved during the night.

In d'Herelle's 1917 report proof is supplied that the lytic agent is living and can replicate. This was based on the observation that the agent could be passaged repeatedly and the demonstration that the clearings on an agar surface were produced by the agent in lysed cultures. As a result d'Herelle coined the word *bacteriophage* (eater of bacteria). Bacteriophage, *bacterial virus* and *phage* are used synonymously—the latter being most commonly used.

Twort's original papers described the essential features of the phenomenon (Twort 1915, 1922, 1925) but d'Herelle's observations, collected in three monographs (d'Herelle 1921, 1926, 1930), were far more detailed and extensive, and played a major part in the development of our present knowledge. By using an end-point dilution assay d'Herelle concluded that at the limiting dilution of a phage lysate those cultures which lysed had received only one phage particle. Furthermore, he noted that the clear spots, or *plaques* as they are generally known, represented the 'colonies' from single phage and thus could be used to assay a phage lysate. In broad terms the particulate or discontinuous nature of the phage was proposed, as were the rudiments of phage *adsorption* and its specificity and the *lysis* of a bacterial culture by repeated growth cycles from a small infecting phage inoculum. These events in the discovery of the bacteriophage were the subject of retrospective reviews by Twort and d'Herelle published in 1949 and have been reviewed recently in a critical and fascinating manner by Duckworth (1976).

D'Herelle's view of the bacteriophage as a filtrable self-reproducing virus parasitic on bacterial cells, in spite of being correct, met with considerable opposition in the early years. D'Herelle had not acknowledged Twort's published work and indeed he may not have known of it. When faced with the Twort phenomenon he asserted that this was not the same as his bacteriophage. This no doubt exacerbated the disagreement between d'Herelle and his critics particularly, as Gratia (1922) was soon to show that the 'glassy transformation' of *Staphylococcus* was identical in nature with d'Herelle's bacteriophage lysis. Undoubtedly, much of the early controversy regarding the nature of phage arose from a failure to distinguish between *lytic* and *lysogenic* phage multiplication. This failure to reconcile these two types of phage response was to divide phage research workers for many years. How far Twort would have anticipate d'Herelle's findings is difficult to say. Soon after his observations on the phage phenomenon he was called up for war service in Salonika and was therefore unable to pursue his researches.

The modern phase of phage research is usually recognized as beginning with the studies of Delbrück, notably with the paper by Ellis and Delbrück (1939) defining the quantitative characteristics of the lytic cycle (*the one-step growth experiment*) and showing how they could be measured with precision. Stent (1963) however, draws attention to the Australian microbiologist F. M. Burnet and the Hungarian chemist M. Schlesinger whose work during the early 1930's foreshadowed the rise of microbial genetics and molecular biology by bringing to bear on the subject an experimental precision and logic which much of the earlier work lacked. Delbrück's contribution stemmed from his training as a physicist, which provided the highly disciplined experimental approach and the powerful logic coupled with his interest in genetics and the nature of the gene. The phage was chosen as the experimental tool because of its simplicity. Of particular importance in the rapid development of phage research was that Delbrück and his collaborators confined their work to a very small number of phages infecting strain B of *Escherichia coli*. The contribution of Delbrück is summed up by Stent, who wrote in 1963 about his book *Molecular Biology of Bacterial Viruses* that 'most of the matter to be presented here derives from the work of Delbrück's school of phage workers, or persons who were working directly or indirectly under its influence'.

The nature of *temperate phage* was studied principally in France by Lwoff and his colleagues. In 1950 Lwoff and Gutman made a major contribution by showing conclusively that lysogenic bacteria carry a latent form of the bacteriophage which can be perpetuated indefinitely by being passed vertically to all descendants. The influence of Lwoff produced its own school of phage workers, a major aim of which was to understand the processes which control the transition of a phage between the lytic and lysogenic states. This line of work has been a dominating influence in the understanding of the molecular events in gene regulation.

General properties of phage

The principle characteristic of phages which, together with all other viruses, sets them aside from other self-replicating biological entities is the existence of the *phage particle* or *virion*. The phage particle is metabolically inert and has properties more characteristic of a complex organic macromolecule than of a stage in the life cycle of an organism. In fact it is customary to store phages as suspensions of virions in non-nutrient medium at 4°. However, the phage particle carries the potential for infectivity and under the appropriate conditions will be converted from the inert virion into the metabolically dynamic form known as *vegetative phage*. The usual conditions appropriate for conversion of virion to vegetative phage are the presence of sensitive bacteria growing logarithmically in a nutrient medium containing the correct balance of inorganic salts. Vegetative phage is the replicating stage of the life cycle and is parasitic causing the diversion of the biochemical capacity of the host cell from bacterial replication to phage replication. Phage replication proceeds by the synthesis of the various nucleic acid and protein components which form the phage particle followed by their assembly into infectious virions during the later stages of infection. The vegetative cycle ends with the release of the progeny phage particles, most usually by the sudden lytic disruption of the bacterial cell wall.

The alternation between the virion and vegetative phage forms is characteristic of all phages and is the means by which replication occurs. For one major group of phages, the *virulent phages*, this is the only phage-host interaction that occurs following infection. The other major group of phages, the *temperate phages*, can enter into a non-productive interaction with the host cell as an alternative to the productive vegetative phage. This non-productive phase is termed *prophage* and involves a non-lethal phage-host association in which the prophage replicates in unison with the host and is passed to each daughter cell at every cell division. This interaction is known as *lysogeny* and a cell carrying a prophage is said to be

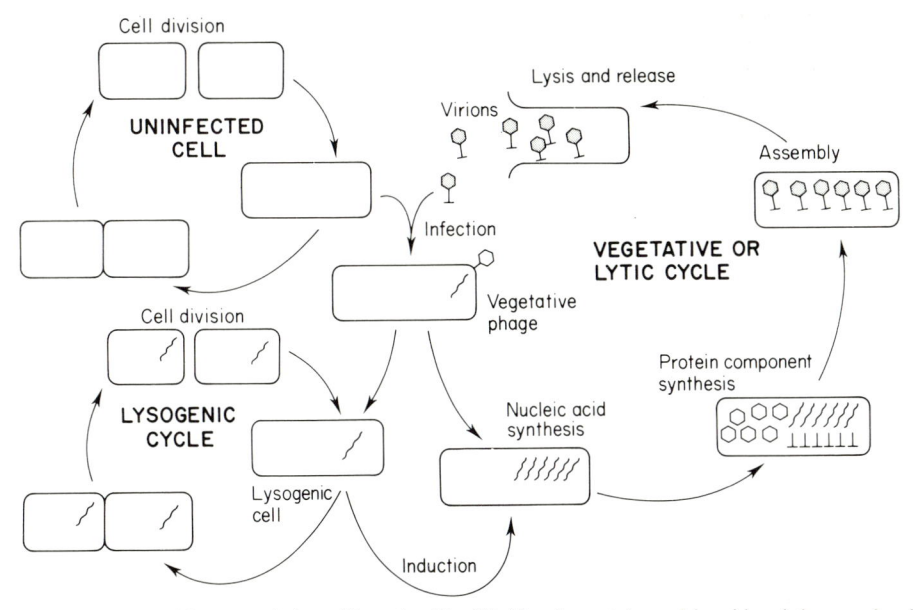

Fig. 7.1 General features of phage life cycles. The filled heads contain nucleic acid and the open heads are empty.

180 *Bacteriophages* Ch. 7

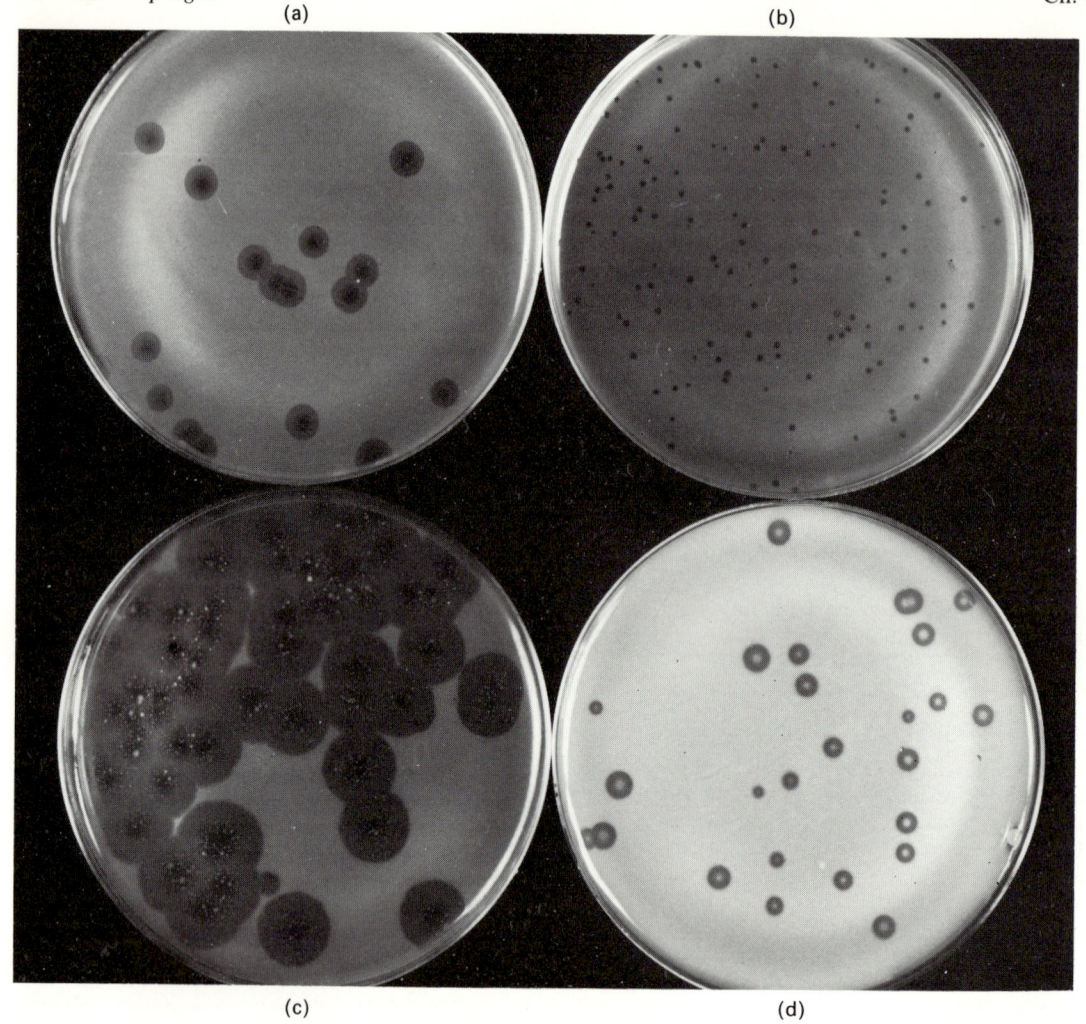

Fig. 7.2 Plaques of the virulent phages T1 (a), T4 (b), T7—the largest plaques of the four types shown (c), and the temperate phage P22—plaques with central white colony (d).

lysogenic. From time to time during the growth of a lysogenic culture this non-productive state passes into the vegetative phase with the ensuing replication of the phage and death of the host cell by lysis. The general features of these various forms of the phage and strategies of phage infection are shown in Fig. 7.1.

Bacterial lysis and the assay of phage

When nutrient agar is thickly inoculated with about 10^8 phage-sensitive bacteria and with measured samples of graded dilutions of a phage stock, the continuous sheet of bacterial growth produced upon incubation is interrupted by clearer circular areas which are confluent where the phage inoculum was concentrated and discrete where it was dilute (d'Herelle 1921). These circular clear areas are localized phage colonies or *plaques* (Löcher) produced by the lytic action of the bacteriophage (Fig. 7.2). After the original infecting phage particle has infected a cell and multiplied, the host cell is destroyed and the phage progeny released to infect new bacteria. A series of these lytic cycles of multiplication then ensues in the immediate area surrounding the original infection and progressing outwards in a radial manner. The plaque is therefore a colony of phage particles and the clear area results from the destruction of the bacteria within the localized growth zone.

Plaque formation, first observed by d'Herelle, remains the most generally used method for the titration of phage stocks. Figure 7.3 illustrates the result of an experiment in which the same volume of a series of doubling dilutions of a phage stock were assayed for plaques. The graph shows clearly that for each doubling in concentration of the phage stock plated there is a corresponding two-fold increase in the num-

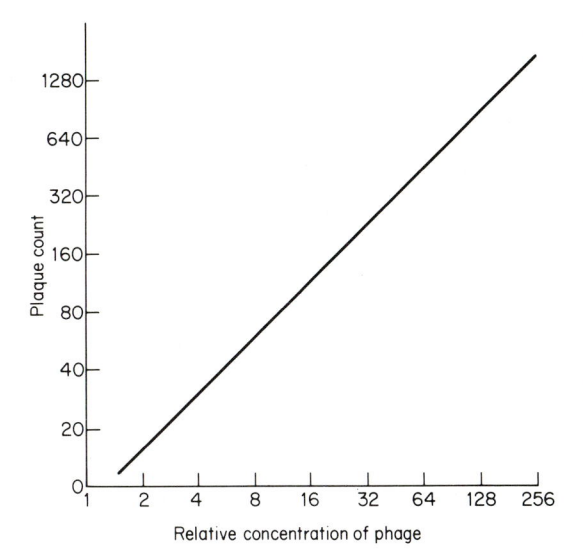

Fig. 7.3 The proportional relationship between plaque count and phage concentration (from the data of Ellis and Delbrück 1939).

ber of plaques appearing on the bacterial lawn. This is clear proof that each plaque is initiated by a single particle. (In genetic terms the phages present in a plaque originating from a single particle constitute a pure line or *clone* of individuals—a fact of major importance for the genetic analysis of bacteriophages.) Had this not been so, then the plot of plaque count versus concentration would have been non-linear; if, for example, more than one particle was necessary to form a plaque, the plaque count would have increased more rapidly than the phage concentration. If y plaques are produced by plating a volume v (in millilitres) of a dilution x, the titre n of *plaque forming units* (pfu) per ml is $n = y/vx$ (Luria *et al.* 1978). An average of 180 plaques per plate produced by plating 0.1 ml of a 10^{-6} dilution of a phage suspension gives a titre of $(180/0.1) \times 10^6 = 1.8 \times 10^9$ pfu/ml.

The direct relation between the number of plaques and the amount of material plated does not necessarily mean that every particle produces a plaque; a given phage stock when assayed under a variety of conditions may give widely differing plaque counts. Two of the more important factors affecting plaque-forming ability are medium and strain of indicator bacteria used. Variations in pH and the salt concentration of the medium will produce dramatic effects. For example, in the absence of sodium chloride or of tryptophan the titre of a phage T4 stock is reduced by several orders of magnitude; the presence of indole in the medium will equally reduce the titre of T2. Effects of the host cell may include the presence of a restriction system (see later) or simply phage resistance. Ellis and Delbrück (1939), having observed such variations,

introduced the concept of *efficiency of plating* (EOP) to define the ratio of the plaque titre under one set of conditions to that under a standard set of conditions. This standard is normally taken as the set of conditions producing the highest titre, which is then regarded as an EOP of 1. These comparisons provide an estimate of the relative efficiencies of plating but little of the absolute efficiency of plating, that is, the ratio of pfu to the total number of phage particles. The total phage particle count is usually estimated by direct observation with the electron microscope. This can be done by mixing a known volume of phage suspension with a known concentration of polystyrene latex particles and counting the ratio of phage particles to latex spheres (Luria *et al.* 1951). The absolute EOP determined for preparations of T2, T4 and T6 gave values of 0.4 to 1.4 and indicated an average value of greater than 0.5. Thus for most purposes all particles in a phage suspension are infectious and the plaque titre is a reasonably good estimate of the number of phage particles in a suspension. This is to be contrasted with animal virus preparations where it is not uncommon to find absolute EOPs of considerably less than 0.1.

The most commonly used method for making a plaque assay is a modification of that originally used by d'Herelle and is called the *agar layer method* or sometimes the *top layer method* (Adams 1959). The principle of this method is that the phage and bacteria are immobilized in a thin layer of agar on the surface of an agar plate. A 2.5-ml volume of molten 0.6 per cent agar, maintained as a liquid by incubation at 46°, is inoculated with a portion containing about 10^8 cells of a concentrated sensitive bacterial culture. An accurately dispensed volume of the phage suspension is then added and the mixture is poured over the surface of a nutrient agar plate. When the top layer has solidified, the plate is incubated. The large number of cells in the inoculum ensures that the lawn of indicator bacteria is even and continuous except where interrupted by a plaque. This method is a considerable improvement on the original in which the mixture was spread over an agar surface much in the way used to enumerate the number of colony-forming units in a bacterial suspension. In particular, the lawn is both dense and even, the plaques are very regular and uniform in size and show specific features more clearly, the EOP is often higher and volumes up to 1 ml can be plated on each petri dish.

The importance of this simple and reproducible assay technique cannot be overestimated. This was appreciated by Stent (1963) when he wrote, 'Probably more than any other single factor, it was the availability of the plaque assay that permitted the extraordinary development of bacterial virus research, and the later extension of this method to the assay of poliomyelitis virus by Dulbecco was to bestow similar benefits to the growth of quantitative animal virology'.

The one-step growth experiment

D'Herelle (1917) had observed that a bacterial culture infected with a highly diluted suspension of a phage lysate cleared suddenly after incubation for several hours. This lytic activity was serially transmissible over many generations. His interpretation of this phenomenon was that clearing resulted from a series of cyclic events involving infection, multiplication and release of progeny by lysis. Clearing was observed only after sufficient cycles had occurred to produce enough phage to lyse all the bacteria in the culture. The validity of this explanation became apparent with the description of the *one-step growth experiment* by Ellis and Delbrück (1939). This experiment defines in broad terms the events during one cycle of phage infection and serves to this day as the fundamental experiment on which all phage research is based. The basic growth characters of any phage are the *latent period* and the *burst size*. The latent period, or more precisely the minimum latent period, is the minimum time between the *adsorption* of phage to the host cell and *lysis* of the host cell with the release of progeny phage. The burst size is the average number of infectious progeny phage released per infected bacterium.

The principal features of the one-step growth experiment are described in detail by Adams (1959). A concentrated suspension of sensitive bacteria in the exponential phase of growth is infected with phage and incubated for a few minutes to allow the phages to adsorb. The infected culture is then diluted to such an extent that further chance collisions between phage and bacteria are effectively prevented. Usually this is between one-thousand and one-million-fold. This diluted culture of infected bacteria is incubated further and samples are withdrawn at intervals and assayed for plaque forming units. It is essential that the samples are plated immediately they are taken to ensure that each assay represents the plaque-forming titre at the time of sampling.

A typical one-step growth curve is shown in Fig. 7.4. The initial low plateau represents the latent period which ends when the plaque count begins to rise. After this rapid rise in plaque titre a final plateau is reached when no further change in plaque count occurs. The period during which the infectious titre is increasing is the *rise period*, which measures the time during which the infected cells in the culture are lysing, and the final plateau marks the completion of lysis of all infected bacteria. Since lysis of an individual bacterium happens very quickly, the rise period reflects the stagger between the time at which the first and last infected cells in the population are lysed. At least part of the variation in lysis time observed during the rise period comes from the variation in the time that each bacterial cell in the culture becomes infected. This will in turn produce an asynchrony in the time of lysis and the duration of the latent period will be correspondingly imprecise. This source of variation can often be eliminated by having a metabolic inhibitor present during phage adsorption; two commonly used inhibitors are potassium or sodium cyanide which inhibit respiration, and chloramphenicol which blocks protein synthesis. Alternatively the bacteria can be suspended in a non-nutrient medium. This block to metabolism does not usually prevent phage adsorption but does stop the infection proceeding beyond this stage. Once the block is removed, by dilution or addition of nutrient, the cycle of infection begins synchronously in all cells. Shorter adsorption periods also produce steeper rise periods.

Because the infected culture is diluted extensively before lysis there is little opportunity for the released progeny phages to infect any sensitive bacteria remaining in the culture; thus further incubation does not change the final phage titre. Had the infected culture not been diluted after adsorption then further cycles of infection might have occurred and the graph of infectious titres would have shown a series of stepwise rises, i.e. a multi-step growth curve.

As a rule the duration of the latent period is constant for a given phage-host combination provided that the growth conditions remain constant. Thus it is possible to state with confidence that at 37° the latent period for phage T4 infecting *Escherichia coli* strain B is 25 minutes. The latent period is sensitive to fluctuations in temperature and will increase as the temperature is reduced. No doubt this reflects the effect of temperature on the metabolic rate of the host. The host strain may also influence the time of lysis. Multiplicity of

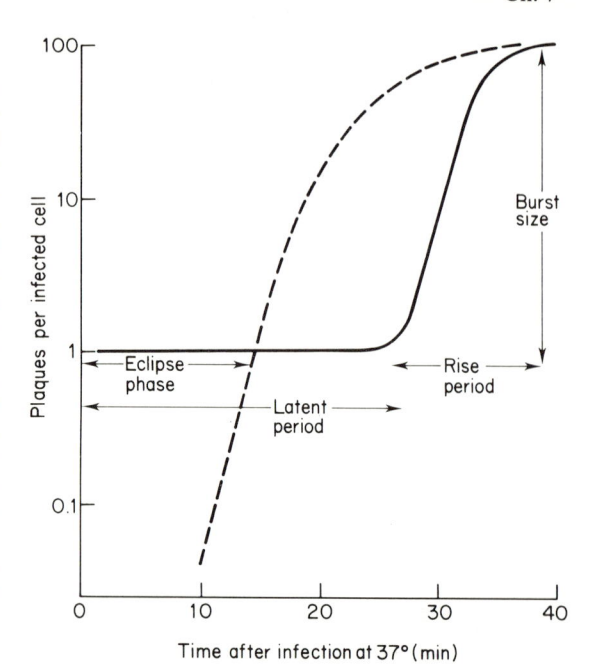

Fig. 7.4 One-step growth curve of phage T4. The solid line represents the one-step growth curve obtained by the spontaneous lysis of infected cells. The dashed line is the curve for prematurely lysed cells and shows the kinetics of intracellular phage production.

infection and type of growth medium, provided that it is not growth limiting, appear not to cause the latent period to vary.

Burst sizes are calculated as the ratio of the yield of progeny phages to the number of infected bacteria. To make accurate calculations of the burst size it is necessary to understand the origin of the plaques produced by samples taken from the infected culture during the latent period. At the moment of adding phage to bacteria the mixture will contain infectious phage particles and uninfected bacteria. As phages adsorb to bacteria an increasing proportion are converted from infectious phage particles to vegetative phage and so the proportion of free phage particles decreases until further adsorption is stopped by dilution. The growth tube will therefore contain a mixture of unadsorbed phage, infected bacteria and uninfected bacteria. The proportions of these three components will depend on factors such as the ratio of phage to bacteria at the beginning of adsorption, known as the *multiplicity* of infection, and the time allowed for adsorption to occur. The plaques that develop on a petri plate inoculated with a sample from the growth tube taken during the latent period will therefore derive from two sources, unadsorbed phage and phage-infected bacteria. Each infected cell, regardless of the number of progeny phage particles it contains, will produce a single plaque because the intact cell is immobilized in the top agar, and when it eventually lyses the large number of progeny particles released will be confined to a single site on the plate. Only after lysis of all infected bacteria, that is at the end of the rise period, will the plaques all arise from individual phage particles. Thus in calculating the burst size it is necessary to know what proportion of the plaques produced before lysis derive from infected cells. As an alternative, unadsorbed phage may be selectively removed by centrifugation at speeds which sediment bacteria but not phage, or inactivated by the addition of neutralizing antiphage serum to the mixture after phage adsorption but before dilution of the infected culture. Antiphage serum will not affect intracellular vegetative phage.

As with the latent period, the burst size is a characteristic of the phage, the host and the growth conditions. It may vary with the host strain and will be depressed by unfavourable growth conditions such as limiting nutrients and pH. However, the phage yield is not influenced by fluctuations in multiplicity of infection or temperature.

With concentrated suspensions of phage-infected bacteria the end of the latent period is marked by visible clearing of the turbid culture. The time at which the turbidity of the infected culture falls corresponds with the time at which the rise period begins (Doermann 1948). Both events in turn correlate with lysis as marked by the disappearance of bacterial cells when viewed with the microscope.

The one-step growth experiment is therefore the basic experimental procedure for all phage studies whether they are designed to analyse biochemical, genetical or physiological aspects of phage multiplication and development. Information gained from this experiment defines the basic characteristics of the particular phage-host system under a particular set of environmental conditions. However, little information is provided about the events which occur during the latent period, and a clear understanding of the process of phage multiplication requires a knowledge of events occurring within the phage-infected cell.

Early attempts to investigate the intracellular multiplication of phages involved breaking open phage-infected cells during the latent period and examination of the lysate for plaque-forming ability. This technique, known as *premature lysis*, was developed by Anderson and Doermann (1952) and Doermann (1948, 1952). Cells may be disrupted by sonication or chemically, although not all phages resist these treatments. The technique used by Doermann was that of 'lysis from without', a phenomenon described by Delbrück (1940) from the observation that bacteria infected by very high numbers of phage particles are disrupted rapidly. This effect does not involve phage multiplication and probably results from sudden and massive damage to the cell wall. In this classic experiment, bacteria infected with phage T4 under normal one-step growth experiment conditions were lysed from without by addition of an excess of phage T6 at intervals during infection. The cells were lysed in the presence of cyanide which prevented further metabolism during lysis and thus froze the infection at the time of sampling. The premature lysates were assayed on indicator bacteria which were sensitive to phage T4 but resistant to phage T6. This genetic trick, by eliminating T6 plaques from the assay plates, served to select only for the T4 phages in the sample. An observation of major importance emerged from this study. The intracellular events of the 25-minute latent period can be divided into two phases, an early phase lasting from the time of infection until about mid-latent period during which there is no detectable infectious phage inside the infected cell, and a late phase lasting until lysis, when the number of intracellular phages increases rapidly in a linear manner eventually reaching a level equal to that observed following spontaneous lysis (Fig. 7.4). The early period, known as the *eclipse phase*, demonstrated that the transition from infectious phage particle to vegetative virus was accompanied by a fundamental alteration in the nature of the phage, the most dramatic expression of which was the loss or eclipse of infectivity.

The Hershey-Chase experiment

From the results of Doermann's study of prematurely lysed infected cells it was clear that the multiplication of phage did not proceed by growth and division of the parent as it does for bacteria, otherwise it would be expected that infectivity could be detected at all times. Furthermore this mode of replication would suggest an exponential increase in progeny phages when in fact it is linear.

The search for an alternative explanation for phage multiplication was soon rewarded as a consequence of what has become known as the *Hershey-Chase Experiment* in recognition of the work of Hershey and Chase published in 1952. The design of this experiment depended on the knowledge that phage particles were composed only of protein and DNA (Schlesinger 1933, 1934, 1936). (Phage workers were to wait another nine years for publication of the discovery of an RNA-containing phage and even longer for lipid-containing phages.) Also it had been shown previously through the work of Anderson (1949, 1950) and Herriott (1951) that the phage nucleic acid and protein could be separated into free DNA and a protein shell or 'ghost', that phages adsorbed to bacteria by their tail and that by violently agitating a suspension of phage and bacteria in a Waring blender it was possible to stop infection.

The experimental basis of the Hershey-Chase experiment was to follow the fate of radioactivity incorporated into the protein and DNA of T2 phage particles during subsequent infection. Stocks of T2 were prepared labelled either in the DNA by growth in medium containing ^{32}phosphorus or labelled in the protein by growth in medium containing ^{35}sulphur. Since DNA contains phosphorus but not sulphur, and protein contains sulphur but no phosphorus, the ^{32}P and ^{35}S labels are specific for DNA and protein respectively. In parallel infections the two labelled phage stocks were mixed with non-radioactive bacteria and allowed to adsorb. After adsorption the infected cells were subjected to shearing in a Waring blender to remove the attached phage, and samples assayed for infectivity. After centrifuging the infected culture at low speed to sediment the bacteria, the pellet and supernatant fractions were assayed for radioactivity. The result can be summarized as follows: (1) blending did not significantly affect the viability of the infected cells, (2) blending removed most of the ^{35}S from the bacteria, and (3) most of the ^{32}P remained associated with the bacteria after blending. This result was interpreted as follows: the phage nucleic acid is the major particle component to enter the cell at infection, the phage coat remains at the cell surface and, since it can be removed by blending without affecting the subsequent infection, it is not required for virus replication but is the means for transmitting the DNA to the cell. There are several other important results and conclusions in this 1952 paper but the work is remembered principally as a significant landmark in the recognition of DNA as the genetic material of viruses (and by implication, all other organisms), and for showing that on bacteriophage infection only the DNA is required for phage replication, the protein coat from which it has been separated remaining at the cell surface taking no further part in phage multiplication. Although variations are recognized, there are no known exceptions to this rule. For animal and plant viruses, however, the picture is quite different.

Not only did this experiment throw light on the means whereby the cell becomes infected but also provided the reason for the eclipse of viral infectivity immediately following infection. Under the conditions normally used for phage assays isolated phage DNA is non-infectious and thus the eclipse phase represents that time during infection when the cell contains only free phage DNA; the end of the eclipse therefore marks the appearance of progeny phage particles of which at least the protein components have been synthesized *de novo*.

Habitat and host specificity

Phages are associated with most and probably all bacterial families and have been isolated wherever sought. A good indication of the range and distribution of phages among bacterial groups is given by Tikhonenko (1970). In addition phages have been described for yeasts (Wiebols and Wieringa 1936, Lindgren and Bang 1961), spirochaetes (Lewin 1960), blue-green algae (Safferman and Morris 1963, 1964) and mycoplasmas (Maniloff, Das and Christensen 1977).

Phages for a given species can usually be isolated from wherever that species occurs in nature; for instance, phages for intestinal bacteria are found in faeces, sewage and polluted water (d'Herelle 1921, Guélin 1952), phages for *Bacillus* species in sewage or soil (Cowles 1931, Lantos *et al.* 1960) and phages for actinomycetes in soil (Wellington and Williams 1981). Clearly, the most likely habitat for a phage is that of the bacterial host with which it is associated. Accordingly, phages are found in both terrestrial and aquatic environments and from fresh and salt water. In addition to phages being present in the environment as free phage particles they will also be present as prophage in lysogenic bacteria and these are frequently reported as the source of new temperate phages, e.g. Rountree (1949) and Boyd (1950).

A given phage type infects a limited range of susceptible bacteria and the host range usually follows the established taxonomic divisions among procaryotes. A coliphage, for instance, will not infect a *Staphylococcus* or a diphtheria bacillus, although it may infect *Shigella* of *Citrobacter*. Within these limits, however, some phages have a much wider host range than others. Some attack only one or a few closely related strains, whereas others may infect an entire bacterial species or several related species and occasionally genera (Wellington and Williams 1981). In this last example the bacteria usually belong to related genera;

T1 and T2 infect both *Escherichia* and *Shigella* strains, and certain *Pasteurella* phages also attack strains of *Shigella* and *Salmonella* (Lazarus and Gunnison 1947).

Where a phage adsorbs to a pilus rather than the cell wall, the susceptibility of the cell will be governed by the availability of pili. Some pili are known to be determined by plasmid-coded functions and the host range of the phage is effectively determined by the host range of the plasmid (Chapter 6). The extremely broad host range plasmid RP1 therefore provides the phage PRR1, which adsorbs specifically to the RP1 pilus, with an equally broad host range.

Generally the sensitivity of a bacterial strain to phage infection will depend on the ability of the phage to adsorb. Other factors influencing phage sensitivity and host specificity are prophage immunity and host-controlled restriction. Non-specific barriers to phage attachment, such as excessive mucopolysaccharide around the cell, will also affect phage susceptibility. Phage mutation and hybrid phage formation between related phages of different host specificities may also affect the host range. Phage mutation to extended host range is a common occurrence and has been fully documented for phage T2 where mutation to phage resistance by the host can be countered by mutation of the phage. Further mutation by the host to resistance to the host-range phage mutant can in turn be countered by further phage mutation (Hershey 1946a, Baylor et al. 1957).

Host specificity is therefore not the expression of a single phenomenon but the combination of the interactions of several unrelated phage-host functions.

The phage particle

Morphology

The morphology of phage particles has been the subject of widespread study. Most of this work is concerned with providing physical identification and classification of newly isolated phages but a few attempts have been made to provide detailed information on phage particle structure. Much of this detailed work has been with phages T2 and T4. As we are dealing with structures with dimensions of approximately 100 nm in diameter, these studies rely entirely on the use of high resolution electron microscopy.

In spite of the considerable variation in size and shape it is generally accepted that phages can be divided into six basic morphological types as defined by Bradley (1967). The most complex, type A, have a polyhedral head, usually hexagonal in outline, to which a long straight contractile tail is attached; the T-even phages are the best known examples but phages attacking a wide range of bacterial species belong to this type. Type B phages have a polyhedral head and long, flexible, non-contractile tails. This is the most common morphological type and includes phages T1, T5 and λ. Type C phages have polyhedral heads and short, stubby, non-contractile tails, e.g. phages T3, T7 and P22. Type D phages are tailless, icosahedral particles with large spikes attached to each apex, e.g. phages ΦX174 and S13. Type E phages are similar to type D but have no apical spikes and include the RNA-containing, male-specific phages MS2, R17 and Qβ. Type F phages have long, flexible, filamentous particles, e.g. the male-specific phages, containing single-stranded DNA, such as f1, fd and M13. Table 7.1 gives the dimensions of representative phages from each type; for comprehensive details the reader is referred to the monograph by Tikhonenko (1970).

Two principal techniques are used to prepare phages for electron microscopic visualization—*shadow casting* and *negative staining* (Tikhonenko 1970). Shadow casting reveals the three-dimensional shape of the particle. This is accomplished by vapourizing metals such as platinum, palladium and gold in a vacuum and at a low angle to the specimen. The stream of metallic vapour produces an oblique metal coating to the particles in which the thickness, and thus the electron opacity, of the metal varies with the shape of the object, much in the way that the intensity of light varies when falling at an angle on an object. The shadow thus cast reveals the shape and height of the object; however, a disadvantage is that the metal coating obscures surface detail. Negative staining is now the most widely used method, partly because of its simplicity and partly because it reveals much more detail than shadow casting. The technique consists in

Table 7.1 Dimensions of representative phage particles

Type	Phage	Head	Tail	Host genus
A	T2	100 × 70	110 × 25	Escherichia
	PST	126 × 95	115 × 25	Pasteurella
	PBS1	120	200 × 22	Bacillus
	P1	90	220 × 20	Shigella
B	T1	60	140 × 10	Escherichia
	λ	54	140 × 7	Escherichia
	B3	55	163 × 8	Pseudomonas
	MSP8	70 × 55	150 × 10	Streptomyces
C	T3	55	15 × 8	Escherichia
	P22	60	7 × 7	Salmonella
	Φ29	42 × 32	13 × 6	Bacillus
	AA-1	65	18 × 13	Acinetobacter
D	ΦX174	25	—	Escherichia
	S13	26	—	Escherichia
	M20	30	—	Escherichia
E	MS2	25	—	Escherichia
	μ2	24	—	Escherichia
	Φcb23	22	—	Caulobacter
	7s	25	—	Pseudomonas
F	M13	—	800 × 8	Escherichia
	fd	—	700 × 5	Escherichia
	f1	—	850 × 5	Escherichia

Dimensions are given in nm. Where the length and width of the head are the same only a single dimension is given.

mixing the virus suspension briefly with a weak solution, e.g. 2 per cent, of heavy metals such as phosphotungstic acid and uranyl acetate. A sample of the mixture is transferred to a carbon-coated microscope grid where it is air-dried and within a few minutes can be examined with the electron microscope. On drying, the electron-dense stain is deposited in any recess or hollow in the particle, which will then be opaque to the electron beam and contrast with the electron-transparent regions where the structural components of the particle are located and the stain excluded. Negative staining produces an extremely high level of detail, and most of our knowledge of the surface structure and architecture of virus particles comes from the use of this method. Because the stain can penetrate the virion head this method can readily distinguish full, that is nucleic acid-containing particles, from the empty particles which have lost their genome. This is useful when the electron microscope is used to determine the number of particles in a suspension and where some indication of viable particles is required.

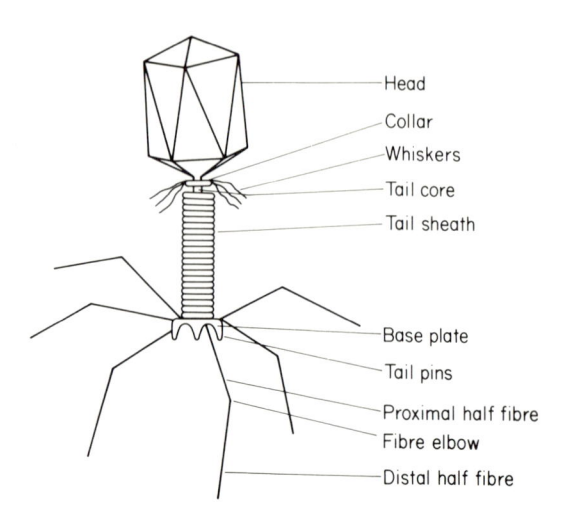

Fig. 7.5 The T4 phage particle. The particle is represented with the tail sheath in the uncontracted position.

Structure and organization

The shape and surface features of phage particles are determined by the protein moiety which surrounds and protects the internal nucleic acid genome. This protein shell or *capsid* is constructed from a large number of subunits often referred to as *capsomeres*. Capsomeres may be composed of individual subunits, or more often, groups of subunits. Two basic types of structure are known and this applies to all viruses whether they are animal, bacterial or plant viruses. One type of structure consists of capsid subunits arranged in a helical fashion to form a hollow tubular structure. The filamentous particles of type F phages and the tails of many of the phages in types A, B and C are composed of this type of structure. The other structural form is the hollow, near-spherical shell composed of capsid subunits arranged with *icosahedral* symmetry, that is have 20 faces and 12 apices. This type of capsid structure is found in the virions of phages from types D and E and the heads of type A, B and C phages. Classification in terms of helical and icosahedral symmetry is probably an oversimplification but nevertheless particles can be interpreted in these terms. In addition to protein and nucleic acid the particles of one group of phages also contain lipids (Palva and Bamford 1980).

As a general rule, the larger the information content of the phage genome, the more complex is the capsid and the larger the number of different types of protein forming the capsid.

The filamentous phages are relatively rare; the best known examples are the male-specific phages f1, fd and M13 infecting *Escherichia coli*. These phages attack only male bacteria, i.e. those harbouring the sex factor F (Chapter 6). This is because these phages adsorb specifically to the sex pili whose formation on the cell wall is determined by the F factor. The filamentous particles are not rigid but have a flexuous appearance in electron micrographs. Each particle is constructed from a single type of subunit.

Of the phages with icosahedral particles (types D and E) those of type E have the least complex construction. The RNA phage R17, for example, consists of 180 identical protein subunits forming the capsid together with one molecule of a second protein, the A or maturation protein, which is required for phage adsorption. Type D phages have spikes protruding from the 12 apices of the icosahedron. Three types of protein subunit are required to construct the capsid, a major coat protein forming the bulk of the particle and two spike proteins. These phages adsorb to bacteria by one of the apical capsomeres; there is evidence that one capsomere is unique and acts as the specific attachment site of the phage.

Among the more complex phages, those with a head and tail, there is a considerable variety of shape and structure. The shape of the head may range from distinctly icosahedral to oblong, and tails show an even greater variety of structure with those having contractile activity, the T-even phages, for example, being extremely complex.

The structure of the T4 phage particle, a type A phage, is probably known in greatest detail and will serve as an example of the structure of a complex virion (Fig. 7.5). Each particle is constructed from about 36 different proteins (Wood and Revel 1976). The head is an elongated icosahedron about 6 nm thick of a shape known as an elongated bipyramidal hexagonal prism of dimensions 100 nm long and 65 nm wide. The elongated shape is formed by an additional row of head capsomeres as an equatorial band. At the base of the head, the apex of one pyramid cap, is a collar with six whiskers to which the tail is attached. The tail is constructed from a hollow core of cylindrical composition measuring 80×7 nm, having a central hole 2.5 nm in diameter. Surrounding the core is a contractile sheath which is not fastened to the head and is composed of helically arranged capsomeres. In its uncontracted form the sheath has a length of 80 nm and a diameter of 16 nm. After contraction, the sheath is 35 nm long and 25 nm in width thereby maintaining its volume. Both core and sheath are constructed from a single type of protein subunit. At the distal end of the tail core is a hexagonal base plate to which six tail pins and six tail fibres are attached. The base plate is a complicated structure assembled from approximately 15 different protein subunits and, like the sheath, undergoes a structural rearrangement during phage adsorption. Each of the six tail fibres has a characteristically kinked structure and is constructed from two half fibres, each a different protein, and two minor distal half proteins which may serve as connectors (Wood and Revel 1976).

In the uncontracted state the tail fibres are attached to the tail plate and lie alongside the sheath with their distal tips attached to the collar. Under conditions favourable for adsorption the fibres break their connection with the collar and extend outwards from the base plate.

The long non-contractile tails characteristic of type B phages show cross-striations in negatively stained electron micrographs indicating the helical organization of the component capsomeres. A variety of termini have been described on the tails of these phages; some have small knobs with tail fibres attached, some have only tail fibres, and others show no structural differentiation at the tips of their tails.

Type C phages possess a small but nevertheless clearly defined tail. The heads are of the isometric polyhedral type or can be an elongated polyhedron as in the $\Phi 29$ of *Bac. subtilis*. The T7-like phages, of which there are many, have a short conical tail arising from one apex of the icosahedral head. The tail has no end plate and consists of a hollow core which may be terminated by two short projections, whether or not these projections function as tail fibres in adsorption is not known. The other group of C-type phages is the P22 group which has a more complex tail structure consisting of a short tail terminated by an end plate having six hexagonally arranged projections. The end plate is attached to the head by a short cylinder. The tail appears to consist of an outer sheath and an inner hollow rod-like structure.

Phage nucleic acids

The protein and nucleic acid molecules of phage particles are fairly easy to study, partly because they can be obtained in highly purified form uncontaminated by cellular components, and partly because of the relatively small number of different types of molecule within the particles of a given phage type. Furthermore, a great variety of high-resolution methods are now available for the analysis of proteins and nucleic acids, the ultimate of which permit the molecule to be sequenced. As a result much highly detailed and sophisticated information is available. However we shall confine ourselves to some fundamental generalizations.

We have already established the internal location of the nucleic acid, which may be DNA or RNA but not both.

The analysis of the structure, composition and function of phage nucleic acids requires the isolation of the nucleic acid in a purified and intact form. Three major hazards should be noted; (1) mechanical shear, (2) chemical degradation and, (3) enzymic degradation.

Mechanical shear breaks DNA and this is a particular problem with the DNA from the larger phages. Avoidance of vigorous shaking or forceful pipetting will minimize this problem (Hershey *et al.* 1962). Chemical alteration may affect both RNA and DNA below pH3 and above pH10 (Knight 1975). At acidic pH values depurination of adenine and guanine occurs. At alkaline pH values RNA is hydrolysed and duplex DNA is denatured. Extraction at near neutral pH is therefore recommended. Nucleases can be eliminated by using either nuclease inhibitors, such as diethylpyrocarbonate, and/or extraction methods which denature proteins and thereby inactivate nucleases.

A variety of procedures for phage nucleic acid extraction are available; a recent and detailed account is given by Knight (1975). These include hot salt, detergent, phenol, osmotic shock and heat release. Of the most universally used is the phenol method (Westphal *et al.* 1952). Phenol denatures protein and when added to an aqueous suspension of phage particles leads to release of the nucleic acid. The nucleic acid remains in solution in the aqueous phase, while the protein is extracted into the phenolic phase and to the interface between the two immiscible phases. The aqueous phase can then be removed and dialysed to remove residual phenol.

Phage DNA

All known DNA phages are haploid carrying only a single DNA genome. The DNA content varies over a 100-fold range from about 1.7×10^6 to 200×10^6 daltons, and the form in which this DNA exists shows considerable variety. Usually the DNA exists in a duplex form; this is true for all but the smallest phages where the DNA is single-stranded. These small phages include the type D icosahedral phages such as, $\Phi X174$ and S13 and the filamentous type F phages, e.g. M13, f1 and fd, all having genome molecular weights of the order of 2×10^6. Furthermore the DNA of these phages exists in the particle as a closed loop, generally referred to as circular (Fiers and Sinsheimer 1962). For the double-stranded genomes the usual conformation is linear although circular genomes such as the *Pseudomonas* phage PM2 (Espejo *et al.* 1971) and the mycoplasma phage MVL2 (Nowak and Maniloff 1978) have been reported.

Within these broad categories a fascinating variety of anatomical features have been reported (Thomas and MacHattie 1967) (Fig. 7.6). The linear duplex genomes of a number of phages have repeated DNA sequences at their molecular ends. The genome is thus said to be *terminally redundant* or *terminally repetitious*. Terminal redundancy was postulated by Streisinger, Edgar and Denhardt (1964) to explain the behaviour of a certain class of genetically heterozygous phage particle obtained in genetic crosses between mutants of phage T4. Physical evidence of terminal redundancy in the closely related phage T4 was provided by MacHattie, Ritchie, Thomas and Richardson (1967). They predicted that, if phage DNA was terminally redundant, treatment of complete phage DNA molecules with exonuclease III—an enzyme that removes nucleotides sequentially from the 3' ends of duplex DNA—would produce single-stranded ends which,

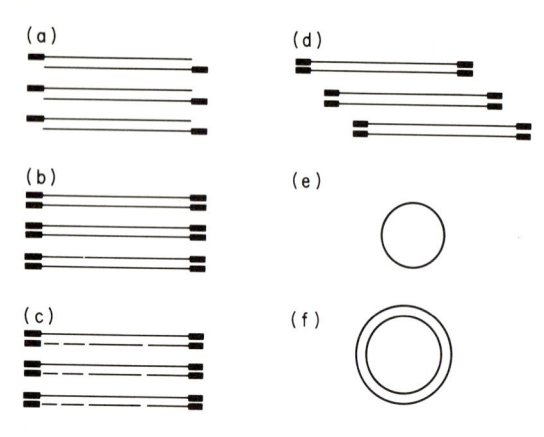

Fig. 7.6 The structural forms of phage DNA molecules: (a) unique sequence with cohesive ends, (b) unique sequence with terminally redundant ends, (c) unique sequence with terminally redundant ends and specifically-located single strand interruptions, (d) circularly permuted sequence with terminally redundant ends (this DNA form is also found with single strand interruptions), (e) single-stranded circle, and (f) double-stranded circle. The thickened regions indicate cohesive or terminally redundant sequences.

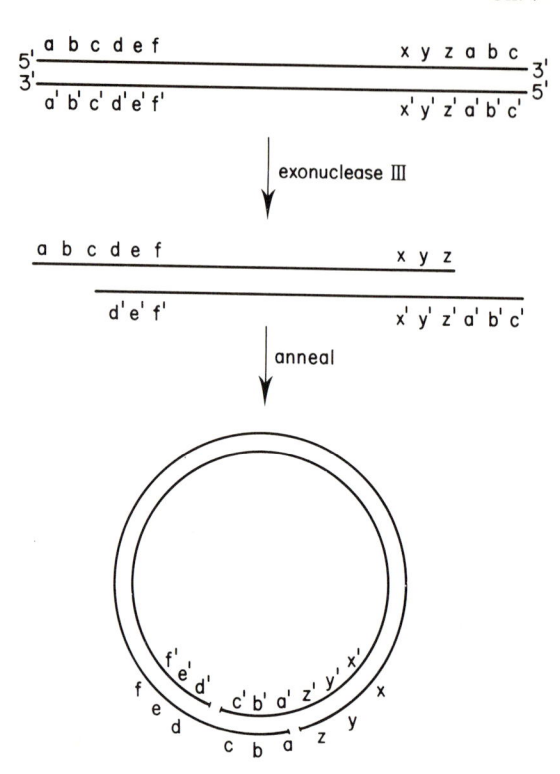

Fig. 7.7 Terminal redundancy in phage DNA. Treatment of terminally redundant DNA with exonuclease III exposes complementary single-stranded ends by removing nucleotides sequentially from the 3' termini. Annealing of the complementary ends converts the linear molecules to circles. Each letter represents a specific nucleotide sequence with the primed letters indicating the complementary sequence.

being complementary to each other should induce the formation of circles by the joining of the two ends of the linear molecule (Fig. 7.7). This prediction proved correct: controlled exonucleolytic digestion of T2 DNA molecules produced circles demonstrable by electron microscopy from intact genomes but not from broken genomes. The length of the redundant region was measured as 1–3 per cent of the total DNA length. These studies were extended to the DNA of phages T3 and T7 which showed similar results but in these the redundancy was shorter (Ritchie *et al.* 1967). For most phages the terminally redundant repeat sequences are direct repeats but for *Bacillus* phage Φ29 the repeat is inverted.

Another group of phages, of which λ is the type example, has complementary 5'-terminated single-stranded ends extending on opposite strands of the linear duplex DNA. These are known as *cohesive* or *'sticky'* ends because of their propensity for annealing together to form a stable duplex and thereby converting linear to circular λ genomes (Hershey *et al.* 1963).

A further major anatomical feature of phage DNA molecules is the relationship of genome sequences among molecules from the same phage type. Phages T7 and λ are representatives of those having *unique* DNA sequences, that is, all molecules in the population start with the same sequence (Hershey *et al.* 1963, Ritchie *et al.* 1967). Others, such as T4 and T1 begin with different sequences and are referred to as *circularly permuted* because the sequences of the linear molecules represent cyclic permutations of a common sequence (Thomas and MacHattie 1964, Gill and MacHattie 1976). The number of permutations varies from phage to phage with T2 having an apparently infinite number of sequence arrangements, whereas other phages have a very limited number, T1 for example, has 3 or 4 (MacHattie *et al.* 1967, Gill and MacHattie 1976, Ramsay and Ritchie 1980).

The DNA of T5, unlike that of most phages, has interruptions in one of the two strands of the genome. This can be seen following denaturation when the molecule falls into several single strand pieces. These fragments are of specific sizes showing that the breaks are at specific locations along the molecule (Bujard 1969). Other phages such as the *Bac. subtilis* phages SP50 and PBS1 also have single strand breaks in their DNA which in these cases appear to be random but may result from the circularly permuted arrangement of DNA sequences (Yamagishi 1968, Reznikoff and Thomas 1969).

Finally, it should be mentioned that most phage DNA molecules are constructed from the four standard DNA bases, adenine, thymine, guanine and cytosine. Several exceptions are known, however, in which this pattern is varied. The best known example of unusual bases comes from the T-even phages where 5-hydroxymethyl cytosine is present in place of cytosine. In addition, the 5-HMC residues are glucosylated in specific patterns which distinguish T2, T4 and T6. One function of the glucose residues is to protect the DNA from degradation by host restriction enzymes. The DNA of several phages active against strains of *Bac. subtilis* contain either hydroxymethyl uracil or uracil in place of thymine (Kornberg 1980).

Phage RNA

The RNA phages fall into two major classes in terms of the genome structure. Most have a linear, single-stranded chromosome with a molecular weight of about 1.2×10^6 (Paranchych 1975). This group includes the Group I phages such as f2, R17 and MS2 and the Group III phages such as Qβ. The other class, of which Φ6 is the type and only example, contains double-stranded RNA in three segments of molecular weights, 2.2, 2.5 and 4.5×10^6 (Semancik, Vidaver and van Etten 1973). Thus while the number of known RNA phages is small the variation in genome structure mimics that of their better known cousins infecting plant and animal cells. However, unlike the eukaryotic virus RNAs the phage RNA molecules are not polyadenylated at their 3′ ends nor do they carry the 5′ methylated 'cap'. The single stranded RNA genomes show properties which indicate that under normal conditions there is extensive secondary structure and it has been estimated that over 60 per cent of the molecule has a complementary helical configuration (Strauss and Sinsheimer 1963, Boedtker 1967). Moreover, the secondary structure is probably specific and sequencing analysis of part of the molecule indicate a flower arrangement of hairpin loops (Min Jou, *et al.* 1972). This secondary structure is important in the regulation of gene expression.

Lytic infection

The cycle of four stages of phage infection was defined by d'Herelle (1926); *adsorption* to the host bacterium, *penetration, intracellular multiplication* and finally *lysis* of the bacterium accompanied by release of new phage. These events are encompassed within the terms of the one-step cycle of growth.

Adsorption

The first step in phage infection is the encounter between phage and bacterial host. This reaction is controlled by diffusion and involves the interaction of the adsorption organelles of the phage with the receptor sites on the bacterial surface. Under normal experimental conditions for adsorption studies, i.e. the bacterial receptors are in sufficient excess of the phage particles not to be significantly depleted as the reaction proceeds, adsorption occurs exponentially. In other words, what is in essence a second order reaction is adequately described by the kinetics of a first-order-reaction.

Most phages adsorb to the bacterial cell wall. Well known examples of this common type are the T-phages and λ of *Esch. coli* and P22 of *Salmonella typhimurium*. Some phages, notably the filamentous and icosahedral male-specific phages f1, f2, MS2, M13 attach specifically to pili. These phages are specific for male bacteria because only these cells have the F-factor determined pilus structures on the cell surface. The filamentous phages adsorb to the end of F-pili whereas the icosahedral phages adsorb along the entire length of the pilus structure. Other phages will adsorb to other pili, e.g. the lipid-containing RNA phage Φ6 of *Pseudomonas* (Palva and Bamford 1980). Yet other phages are known to adsorb to flagella such as PBS1 of *Bacillus subtilis*, PBP1 of *Bac. pumilus* and χ which infects a broad range of motile bacteria (Archibald 1980, Meynell 1961).

Phages adsorb to specific receptor sites on the bacterial cell wall. In gram-negative bacteria the receptors have been identified as protein and lipopolysaccharide components of the outer membrane layer surrounding the peptidoglycan. For gram-positive bacteria the adsorption sites are the teichoic acid and polysaccharide components associated with the peptidoglycan layer. A particular phage or group of phages will adsorb to a specific site and different phages will adsorb to different sites. Thus on the surface of a given bacterial cell a variety of different receptors are present each type being present in a large number of copies. For example λ, T5, T6, BF23, T2 all infect *Esch. coli* but do so by adsorbing to different receptor sites. Phage receptors are often an antigenic or chemically or morphologically recognizable component of the cell surface. In many cases the receptor components have been isolated and purified thus allowing chemical analysis and *in vitro* studies on phage adsorption (Randall and Philipson 1980).

The receptors of gram-positive bacteria Phages for gram-positive bacteria adsorb to receptors associated with flagella and with the cell wall and plasma membrane. Cell wall and plasma membrane receptors include the surface protein layers, peptidoglycan and teichoic acid and peptidoglycan and polysaccharide (Archibald 1980). The regular array of protein on the outer cell wall surface of *Bacillus brevis* is a receptor for several phages including phage M. It has been shown that the isolated protein will inactivate the contractile-tailed phage M and that bacterial mutants resistant to M have altered surface subunits (Howard and Tipper 1973). The involvement of teichoic acid as phage receptor material has been documented for the group B typing phages of *Staphylococcus* and for *Bacillus subtilis* phages of Bradley's groups A (SP8, SP50), B (SPP1, SPO2) and C (Φ29) (Young 1967, Yasbin et al. 1976, Shaw and Chatterjee 1971). Polysaccharide components in the peptidoglycan layer have been implicated in the adsorption of phages of *Lactobacillus, Streptococcus* and *Micrococcus* (Archibald 1980).

The receptors of gram-negative bacteria Lipopolysaccharides are a major component of the outer membrane of gram-negative bacteria. They are constructed from a lipid portion and a polysaccharide portion and are located mainly in the outer leaflet of the outer membrane bilayer. The general structure is of a lipid, referred to as lipid A, an oligosaccharide unit, the R-core and a polysaccharide, the O-antigen. Typical R-core specific phages are the enteric phages P1, ΦX174, S13, T3 and T4. Phages binding to O-antigen receptors include the *Salmonella* phages E^{15}, g341 and P22, *Esch. coli* phages Ω8 and *Shigella* phage Sf6 (Wright *et al.* 1980). Because there is less structural variation in the R-core region of the lipopolysaccharide, phages specific for the R-core often exhibit broader host ranges than O-antigen receptor specific phages.

Protein components of the outer cell membrane also serve as phage receptors and it would appear that in several instances the protein in conjunction with lipopolysaccharide is the receptor rather than the pure protein (Schwartz 1980). Phages λ, T1, T5, Φ80 and T6 are examples of coliphages which adsorb to protein receptors. Protein receptors have also been documented for phages of *Salmonella typhimurium* and *Pseudomonas aeruginosa*. These receptors also can serve as adsorption sites for bacteriocines, for example, colicine E3 adsorbs to the same site as phage BF23, a T5-related phage.

In addition to protein components of the cell envelope serving as adsorption sites the protein appendages such as flagella and pili play a similar role. However, as noted above, flagellotropic phages are not restricted to gram-negative bacteria. In pilus-specific phages the adsorption sites are not identical; the isometric phages adsorb along the whole length of the pilus whereas the filamentous particles attach specifically to the pilus tips.

The dynamics of phage adsorption

The adsorption of phage to bacteria is classically considered to be a two-stage process, the first being reversible adsorption which is followed by irreversible adsorption. However, this view is not without its opponents (Weidel 1958). The generally accepted indication of a reversible stage in adsorption is the observation that phage particles will remain attached to their host bacteria during sedimentation, but upon dilution or chloroform treatment are released in a form that is still infectious. Dilution of irreversibly bound phages will not lead to their release and chloroform treatment prevents the formation of an infectious centre.

The T-even phages adsorb to *Escherichia coli* strain B in two stages. Reversible attachment occurs when three or more of the six tail fibres attach to the diglucosyl components of the lipopolysaccharide layer. This reaction evidently consists simply in the formation of electrostatic bonds between the tail fibre tips and the receptor molecules. It depends solely on the cationic composition of the medium and is not temperature dependent (Puck and Sagik 1953). Adsorption appears to occur at points on the cell surface where there is fusion between the outer and inner membranes (Bayer 1979). The baseplate, initially 100 nm from the cell surface, is then brought to within 10 nm of the surface with apparent contact between the six tail pins on the base plate and the cell surface. This is the irreversible stage and is temperature dependent. After attachment the tail sheath contracts and the tail tube tip approaches the inner cell membrane (Simon and Anderson 1967, Mathews 1977, Goldberg 1980) (Fig. 7.8). Stages of reversible and irreversible adsorption have been indicated for several other phages, e.g. T1, λ, T5, P22 ΦX174 (Goldberg 1980).

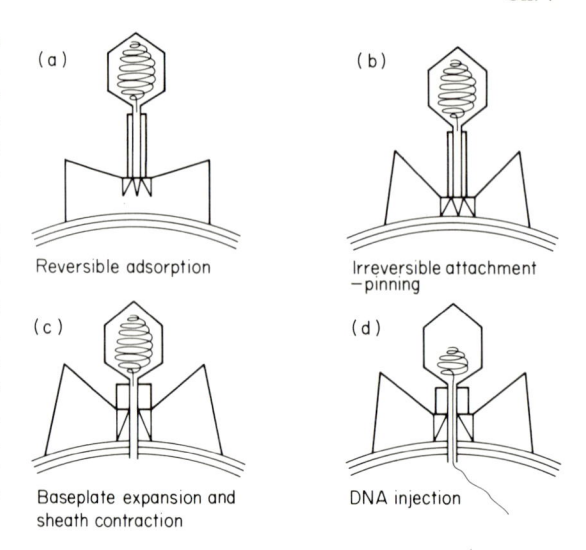

Fig. 7.8 Early stages of phage T4 infection.

(a) Reversible adsorption
(b) Irreversible attachment – pinning
(c) Baseplate expansion and sheath contraction
(d) DNA injection

When a very large number of particles of the T-even phages are adsorbed by a bacterium the cell undergoes what has been termed *lysis from without* (Delbrück 1940), that is it swells and lyses almost immediately and without multiplication of the phage. At the same time the phage loses infectivity. Lysis from without also occurs at a lower multiplicity of infection when the energy supply of the bacterium is curtailed; the effect does not require viable phage, since particles inactivated by irradiation or phage 'ghosts', particles which have lost their DNA (Duckworth 1970), can also cause lysis from without. This phenomenon has been also reported for phages of *Staphylococcus* (Ralston *et al.* 1957) and mycophages (Millman 1958).

Adsorption requires the appropriate ionic environment; for each phage there is an optimum concentration of cations which gives optimum rates of adsorption. The specificity between the phage tail and the bacterial cell surface receptor will be altered if either component changes its structure. These changes can arise by mutation. Changes to the bacterial cell surface receptor which leads to failure of the phage to adsorb results in bacterial *resistance* to phage infection. Mutations of the phage which permit the phage to adsorb to resistant bacteria are referred to as *host range* mutations.

Penetration

The experiments of Hershey and Chase (1952) established that only the phage nucleic acid and minor components such as the internal protein entered the cell at infection while the bulk of the protein remained outside. An exception to this rule has since been provided by the small filamentous phages.

Phages exhibit a variety of mechanisms for transferring their genome from the virion to its intracellular location at infection. Phages with complex contractile tails such as T4 eject their DNA by a different mechanism from the simple-tailed λ or the short-tailed P22. The non-tailed phages such as ΦX174 and the filamentous M13 display a different set of mechanisms.

The sequence of events at this stage of infection are: irreversible adsorption, ejection of nucleic acid (or uncoating for the small single-stranded phages) and *penetration*. It is not easy to separate these stages in time; they may not even be separate processes. For a complex phage with a head and tail structure, the process of ejection follows a sequence in which the DNA leaves the head and enters the tail, passes through the tail, emerges from the distal end of the tail and finally leaves the tail (Goldberg 1980). This process is understood in some detail for phage T4 (Fig. 7.8). After reversible adsorption by the tips of the tail fibres to the lipopolysaccharide receptors the baseplate attaches to the cell surface to form an irreversible attachment. This 'pinning' event leads to baseplate expansion and sheath contraction causing the tip of the tail tube to approach the inner cell membrane. Sheath contraction is not in itself sufficient to release the T4 DNA; the release signal appears to come from the cell. Ejection of T4 DNA is probably initiated by interaction of the tail tube tip with the cytoplasmic membrane surface. Goldberg (1980) has suggested that during particle morphogenesis the end of the DNA moves from the head into the tail shortly after head-tail joining. This would place one end of the DNA molecule at the end of the tail tube and so eliminate the need for a mechanism to transmit the signal to eject from the distal end of the tail to the head, a distance of 100 nm. The electron micrographs of Simon and Anderson (1967) suggest that the tail tip is in contact with the cell membrane and does not breach the membrane and thereby transmit the DNA directly into the cytoplasm. Probably the DNA emerges directly on to the membrane and then enters the membrane. At this stage the emerging DNA is resistant to digestion by intracellular exonuclease V because, like several other phages, the free end of the DNA molecule is protected by protein as it enters the cell.

The mechanism by which a T4 DNA molecule 50 times the length of the *Escherichia coli* cell passes from the phage head to the cell cytoplasm remains a mystery. Passage along the tail tube takes no more than a minute and probably involves rotation if it is packed in the head in coils (Hendrix 1978). Goldberg (1980) considers that a nicking and repairing mechanism involving a DNA gyrase type of protein must be required to avoid the problems of the torque produced by the DNA rotating in the cytoplasm. Another possibility would be to rotate the DNA spool in the phage head.

The fact that T4 particles can eject their DNA *in vitro* in the absence of an external energy source suggests that the energy must be contained in the phage head. Transport of the major bulk of the DNA into the cell probably does require cellular energy but details of this energy process are quite obscure, as also is information relating to the mechanism of DNA transport across the cell membrane. However, it has been suggested that restriction enzymes are required for DNA uptake (Sisco and Smith 1979).

Phage T5 is another interesting example, since *in vivo* the DNA is injected in two distinct stages. The first stage is the injection of the first 8 per cent of the molecule into the cell. This is followed five minutes later by injection of the remainder (Lanni 1968). After reversible and then irreversible binding of T5 by its tail fibres to its receptor (the *ton*A protein) the tip of the DNA molecule attaches to the cell membrane (Labedan 1976). This attachment prevents ejection of the rest of the DNA and occurs at 0°. At higher temperatures the initial 8 per cent of the DNA enters the cell by a process which does not require protein synthesis and is called first step transfer (FST). Expression of the genetic functions coded by the FST DNA then leads to the synthesis of proteins of which two are required for the penetration of the remainder of the genome or second step transfer (SST). What causes the arrest of DNA penetration at the end of FST is not known, but a signal in the form of a specific base sequence is thought to be a likely candidate. Cell wall fragments will elicit ejection of T5 FST DNA but SST ejection is arrested unless the cytoplasmic membrane is removed, suggesting a role of the cell membrane in this interrupted DNA penetration process. The energy for DNA transport comes mostly from the cell and after emergence of DNA from the phage tail the head plays no further role in DNA release.

The isometric single stranded DNA phages, such as ΦX174, exhibit a different response from the tailed phages. The tailed phages can use their tails to deliver the genome directly into the cell. The isometric phages have no such ability, and penetration and ejection give way to a process rather more like *uncoating*. For this reason the isometric phages exhibit a stage during infection when the DNA is sensitive to deoxyribonuclease, no doubt because uncoating requires alterations to the phage coat which lead to exposure of the DNA to nucleases in the medium. Irreversible adsorption involves the spike protein, which protrudes from each of the 12 vertices of the capsid, binding to a lipopolysaccharide component of the cell surface. The spike protein, the product of gene H, has been thought to have two functions; it signals uncoating and also binds to the phage DNA to direct its entry into the cell and initiate replication (Jazwinski et al. 1975).

The RNA phages of *Escherichia coli*, as noted earlier, adsorb to the side or base of the F-pili present on the surface of cells carrying the F plasmid. The isometric virion capsid consists of 180 molecules of the major capsid protein and one molecule of the A protein (gpA), which is essential for the interaction between phage particle and receptor. After primary adsorption the phage particle is transferred to the base of the pilus at its junction with the cell wall. It is still not clear whether this process occurs by phage migration along the pilus or by retraction of the pilus into the cell (Novotny and Fives-Taylor 1978). At the cell surface the RNA passes through an exposed, ribonuclease-sensitive phase as the RNA is taken up by the cell and uncoating requires the cleavage of gpA. The formerly held opinion that the RNA passed down the central hole in the cylindrical pilus is not now accepted because of the ability of the pilus to be retracted and extruded. This view also applies to DNA transfer during conjugation, the two processes apparently sharing common reactions in the uptake of nucleic acid.

Of the filamentous single-stranded DNA phages the best studied examples are the male-specific phages, such as M13, f1 and fd, which adsorb to the tips of the F pili of *Escherichia coli*. As with the RNA phages, the belief that the genome is ejected down the inside of the pilus into the cell is no longer held. This change of view is based on observations that M13 DNA penetration and uncoating is intimately linked to genome replication and that the coat protein is not discarded at the cell surface but enters the cell membrane (Marco et al. 1974). The particle proteins consist of the major capsid protein gp8, of which there are 2000 molecules per particle, and gp3 a pilot protein present in four molecules which are located at one end of the filament. This end is the end by which the particle adsorbs. A modification of gp3 leads to its cleavage and entry into the cell together with the genome and the rest of the coat protein. As the genome replicates to produce the duplex form it is uncoated and enters the cell. Inhibition of DNA replication does not affect adsorption but uncoating is blocked.

Intracellular multiplication of phage

The virion is a metabolically inert particle with the capacity for replication to produce progeny phage particles. This capacity is manifest following injection of the genome into the cell when the phage is converted to the metabolically dynamic vegetative phage. The phage genome is expressed largely through the metabolic machinery of the host cell to which is added the phage-coded gene products to direct the cellular processes towards phage production. Phages vary in their effect on their host and in the extent to which they depend on the host for replication. In general, the larger, more complex phages are less dependent on host functions than the smaller phages and at the same time cause greater disruption of the host's metabolism.

Virulent phages have only one course open to them after infection. This is the *lytic* or *productive* response leading to cell death and release of progeny following a brief latent period devoted to phage multiplication. Temperate phages also use the productive response in order to manufacture progeny phage. In addition, however, they may enter the *lysogenic* phase in which they are maintained in the host cell as prophage under conditions which permit continued function and reproduction of the host cell. The prophage is maintained at a level such that all daughter cells are lysogenic, thereby ensuring the vertical transmission of the prophage genome. The interplay between lysogenic and lytic development is controlled by signals at the level of gene expression.

Replication of double-stranded DNA phages

A vast amount of information is available about the large number of phages within this broad grouping. It is fair to say, however, that the principles of replication can be illustrated by reference to just a few phage-host systems, namely, the virulent T-phages and temperate phages λ and P22. We shall draw largely on these phages for the fundamental concepts of phage replication and refer to others when special cases arise.

Phage T4

The best studied of all virulent phages, T4 is perhaps not the most typical of this group. Firstly, its DNA contains the unusual base 5-hydroxymethyl cytosine in place of cytosine and this is reflected in some aspects of the genetic and biochemical functions of T4 (Mathews 1977). Secondly, being such a large phage, with a genome size of 130×10^6, sufficient information to code for 160–170 average-sized proteins, its life style is extremely complex. Nevertheless, the principles of phage biology were established with T4 and it has yielded more fundamental ideas than any other phage. About 140 genes have been identified genetically and to some extent characterized functionally (Wood and Revel 1976). Consequently there is a fairly detailed picture available of the processes of viral replication.

Infection of *Escherichia coli* by T4 occurs when the linear genome is injected into the cytoplasm of the host cell. About 3 per cent of the particle protein, known as internal proteins, enters the cell at infection. The product of gene 2, gp2, is probably attached to the leading end of the DNA molecule as a protective device to prevent degradation by exonuclease V, a host-coded DNA nuclease which degrades linear DNA

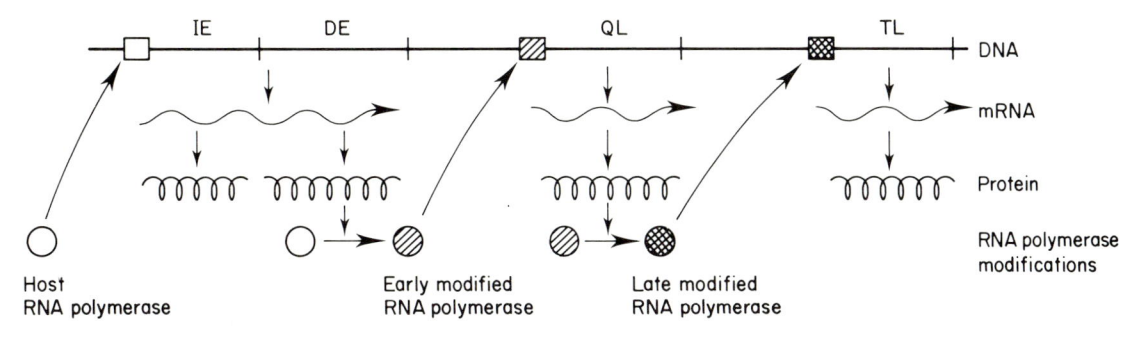

Fig. 7.9 The regulation of phage T4 gene expression. Immediate Early (IE) and Delayed Early (DE) genes are transcribed from IE promoters (□) by the host's RNA polymerase (○). Early gene products modify the host RNA polymerase (◐) to transcribe from Quasi Late (QL) promoters (▨). The polymerase is further modified (●) to recognize the True Late (TL) promoters (▩) and transcribe from the late genes.

(Silverstein and Goldberg 1976). Within a few minutes of infection bacterial DNA replication stops and the DNA is degraded to be re-used later as precursor material for T4 DNA. Host messenger RNA (mRNA), and stable RNA (transfer RNA and ribosomal RNA) synthesis is inhibited and host protein synthesis stops. Bacterial enzymes are, however, still functional and the multiplying phage uses the metabolism of the host for its own supply of small molecular weight substances.

The sequence of events following T4 infection has been investigated in considerable detail (Rabussay and Geiduschek 1977, Mathews 1977). Expression of the entering genome is strictly controlled and sequential, the order of expression being referred to as *immediate early, delayed early, quasi late* and *late*. These classes of T4 gene function are distinguished primarily by the time at which they are transcribed or the gene products synthesized. However, they reflect the activity of control systems which determine the times during the infectious cycle at which the different classes of genes are transcribed (Fig. 7.9). The immediate early genes are transcribed from about the first minute of infection by means of the host RNA polymerase as it exists in cells at the time of infection. This requires no new protein synthesis, as shown by its occurrence in the presence of the protein synthesis inhibitor chloramphenicol, and by the synthesis of immediate early mRNA *in vitro* using T4 DNA and *Escherichia coli* RNA polymerase. The delayed early genes are not transcribed in the absence of new protein synthesis, and *in vitro* this requires products of the immediate early genes. Delayed early expression occurs from the same *promoters* as immediate early and the delay in their appearance is because of their distal position to their promoters compared to immediate early genes. The function of the immediate early gene products in controlling delayed early transcription might be through an anti-terminator function, since the appearance of delayed early transcripts is prevented by the host termination factor *rho*. Transcription of the quasi-late mRNA requires the chemical modification of the *Escherichia coli* RNA polymerase by early T4 gene products. This modification permits recognition of the quasi-late (or middle) promoters and prevents recognition of host promoters, and so no further host transcription occurs. Quasi-late transcripts are detectable from about two minutes after infection at 37°. A further modification of the *Escherichia coli* RNA polymerase occurs about five minutes post-infection; this leads to a second alteration in promoter specificity which permits the recognition of the late gene promoters. At the same time transcription of early regions ceases. The transcription of the late genes differs in several ways from early and middle mRNA synthesis, in particular in using the other template strand of the T4 DNA molecule, and in its requirement for T4 DNA replication. This coupling of DNA replication and late transcription appears to be associated with the requirement of the presence of nicks in the template DNA which is a feature of replicating as opposed to non-replicating phage DNA.

The grouping of functions into temporally regulated transcriptional blocks agrees with their grouping according to function. Early functions are required for the shut-off of host cell protein synthesis and for the replication of phage DNA. Quasi-late (middle) functions are mostly concerned with DNA replication, phage tRNA synthesis and the regulation of late transcription. Late gene products are predominantly used for the structural components of phage particles and capsid assembly. Many of these related functions are clustered on the genetic map and constitute regions of coordinate transcription under the control of single promoters, i.e. they fit the description of *operons*.

T4 is therefore an example of regulation at the level of transcription. At any given time during infection only certain classes of mRNA can be synthesized; this control is exerted by changes in the ability of the host cell RNA polymerase to recognize promoters; thus at different times different promoter classes become available for transcription. The *Bacillus subtilis* phage SPO1 and coliphage N4 are further examples of the T4 pattern (Rabussay and Geiduschek 1977).

The genomes of the T-even phage particles are circularly permuted, terminally repetitious molecules which, during replication, give rise to a series of progeny particles whose genomes are circular permutations of the parental genome. Replication of the T4 genomes requires about 30 phage-induced proteins and is a complex process depending, in addition, on the activity of some host functions. Kornberg (1980) has classified the phage-induced replication proteins into three functional categories; nucleotide biosynthesis, DNA synthesis, and DNA modification.

The first group comprise functions which supply the building blocks for DNA synthesis. These enzymes supplement the supply of deoxynucleotides derived from the degradation of host DNA and thus enable the synthesis of DNA to proceed at a higher rate than in the uninfected cell. In addition to providing deoxynucleotides there is a requirement for functions to exclude cytosine from T-even phage DNA and to provide for hydroxymethyl cytosine. The second category includes functions concerned with RNA primer formation, DNA synthesis initiation, helix destabilization and DNA polymerization or chain elongation. DNA modification includes, in the case of T-even phages, the glucosylation of hydroxymethyl cytosine residues. This occurs after polymerization and follows a distinctive pattern. Additionally, a variety of nucleases both endo- and exonucleases are included in the general pattern of DNA replication.

T4 DNA synthesis is initiated at a specific *origin* near gene 42 and proceeds bidirectionally. It is possible that secondary origins of replication are also used. As is general for phage DNA synthesis, chain elongation in T4 is initiated with an RNA primer and proceeds in the $5' \rightarrow 3'$ direction by the ligation of short DNA fragments (*Okazaki fragments*). Re-initiation of DNA synthesis occurs repeatedly and before the previous round is complete to produce a highly complex structure of rather massive proportions and with multiple replication forks. Biochemical studies reveal these structures to have the properties of *concatemers*, i.e. structures containing several genome equivalents of DNA as one large molecule. A clearer view of concatemer formation and the role of these structures as precursors for progeny genome formation has come from studies with less complex phages such as λ, T7 and T1 and will be considered in detail later. The important point, however, is that concatemer formation is an essential feature of phage DNA replication and that these extended DNA forms are required for packaging into phage particles.

Packaging of T-even phage DNA proceeds hand in hand with particle morphogenesis by a mechanism known as *headful packaging*. This mechanism uses concatemeric DNA as a substrate and produces a collection of genomes which are circularly permuted and terminally redundant (Streisinger et al. 1964, Streisinger et al. 1967, Frankel 1968). Packaging starts at a specific site and encapsidates a complete genetic text plus the terminally redundant duplication. Having filled one head the DNA is cut and the filling of a second head begins. This second headful of necessity begins at a different site on the DNA from the first headful and will in turn be cut off when a complete genetic text plus redundancy is encapsidated. This headful cutting proceeds along the concatemer producing a set of packaged genomes each being a permutation of the other (Fig. 7.10). The number of permutations among a collection of T-even phage genomes seems to be very large, suggesting that processive cutting progresses uninterrupted along very long concatemers, although the possibility of several initiation sites must be considered.

The final stage in particle formation is the *assembly* of capsids or *maturation*. An extremely detailed account of this process is now available largely as a result of the work of Edgar and Wood and their colleagues (Edgar and Wood 1966, Edgar and Lielausis 1968, Muralido and Becker 1978). The essential feature of particle morphogenesis is that assembly occurs in a manner resembling a motor car production line. Components are assembled from their various constituent parts and the completed components (heads, tails, tail fibres) are brought together for assembly into complete particles. There are four main sub-assembly reactions: 1) neck, collar and whiskers, 2) DNA-filled head, 3) baseplate, tube and sheath, and 4) tail fibres. Some of the stages in the formation of active phage particles were established by the

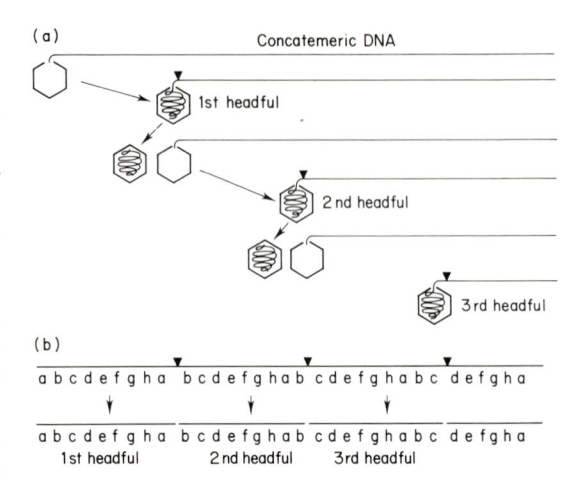

Fig. 7.10 DNA packaging by the headful mechanism. (a) Empty head structures (proheads) are filled with DNA from a concatemeric precursor. After completion of one headful the remaining DNA is released by a cutting reaction (▼) and proceeds to fill second and third headfuls. (b) A headful of DNA is equivalent to a complete copy of the genetic text (sequences *a* to *h*) plus a terminal redundancy—for the first headful this is sequence *a*. This mechanism produces a series of headfuls with DNA sequences arranged in circularly permuted fashion.

use of mutants blocked at different stages of phage assembly. These mutant infections accumulate components which are detectable in the electron microscope or by serological methods (Epstein et al. 1963). Some of the assembly steps occur spontaneously and can be studied *in vitro* by the use of extracts of mutant strains. For example, mutants in gene 23, which controls synthesis of the major head protein, yield extracts containing tails and tail fibres but no heads or complete particles; when these are mixed with extracts from mutants producing heads but no tails, active particles form spontaneously (Edgar and Wood 1966). The assembly of tail fibre components, the attachment of fibres to the baseplate, and the joining of completed heads and tails all occur spontaneously *in vitro* and presumably also *in vivo*. More recently it has been possible to demonstrate DNA packaging *in vitro*. This process however requires ATP and cofactors and is not a spontaneously occurring reaction.

There are 55 identified assembly gene functions of which the gene products of 36 are structural proteins of the T4 particle, and another 7 appear to be required to control or promote assembly. The role of the remaining 12 is not certain (Wood and Revel 1976).

Phage T7

This phage is an excellent example of a less complex virulent phage and is a member of a large group of closely-related coliphages which includes T3, H, W31 and ΦNII. It has been intensively investigated and its biology is known in remarkable detail.

The properties of T7 which make it a useful contrast to T4 are: 1) the DNA molecules are terminally redundant and have a unique sequence, 2) the genes are clustered into large functionally related blocks which are expressed in a coordinate manner, 3) the DNA composition is standard with the four usual DNA bases, adenine, thymine, guanine and cytosine, 4) the mechanism controlling gene expression differs fundamentally from the T4 type, and 5) DNA replication is a less complicated process and is therefore better understood.

Twenty five T7 genes have been identified, mostly by *conditional lethal* mutations, which account for over 80 per cent of the expected gene complement (Hausmann 1976, Mathews 1977). Polyacrylamide gel electrophoresis of T7-specified proteins shows that T7 codes for about 30 polypeptide gene products. Not all the known T7 genes are essential and under standard laboratory conditions 19 of the T7 genes are absolutely required for successful replication in wild type bacteria. This definition of an essential function is of course somewhat arbitrary and it is likely that under other conditions additional gene functions may become indispensable. This is certainly true for the gene 1.3 gene product, a DNA ligase, which is dispensable for infection of *Escherichia coli* containing a functional bacterial DNA ligase but is essential in a ligase-defective host. This arises because T7 can use either its own or the host-specified ligase function.

Most of the T7-coded polypeptides have been assigned to the genes which code for them and their functions are reasonably well characterized. A comparison of the genetic and functional maps reveals clearly the clustering of related functions (Fig. 7.11). Starting from the left end as the map is conventionally drawn is a group of early functions (genes 0.3 to 1.3). These are concerned with the primary events in replication such as, escape from the host's restriction system, inhibition of host function and regulation of the subsequent stages of T7 gene expression. These early or Class I functions are transcribed and translated between about 4 and 8 minutes postinfection at which time their expression is turned off. The second group, Class II functions, include genes 1.7 to 6 and control phage DNA replication and recombination. Class II functions are expressed between 6 and 15 minutes after infection and map in the centre of the genome. At the right hand side of the map are the class III, or late, functions required for the synthesis of capsid proteins, DNA packaging and assembly of progeny virions. This group, from genes 7 to 19, are expressed from 7 minutes after infection until lysis at about 25 minutes.

Like T4, the pattern of T7 gene expression is temporally regulated; unlike T4, however, the T7 genome is transcribed entirely from one strand. Transcription occurs from left to right on the conventional map and in the sequence of expression, that is, early, middle and late. Also different is the mechanism controlling gene expression. T4 infection is controlled by a series of modifications to the host cell RNA polymerase which alter promoter recognition. T7, on the other hand, switches from Class I to Class II expression by changing from one RNA polymerase to another. Class I promoters are recognized and transcribed by the bacterial RNA polymerase, whereas the Class II and III promoters are recognized by a T7-coded RNA polymerase. This phage enzyme is the product of gene 1, a Class I function, and therefore Class II and III genes cannot be expressed until the Class I products have been made (Chamberlin *et al.* 1970). Another Class I product, that of gene 0.7, is a protein kinase which inhibits the host RNA polymerase and is presumed to

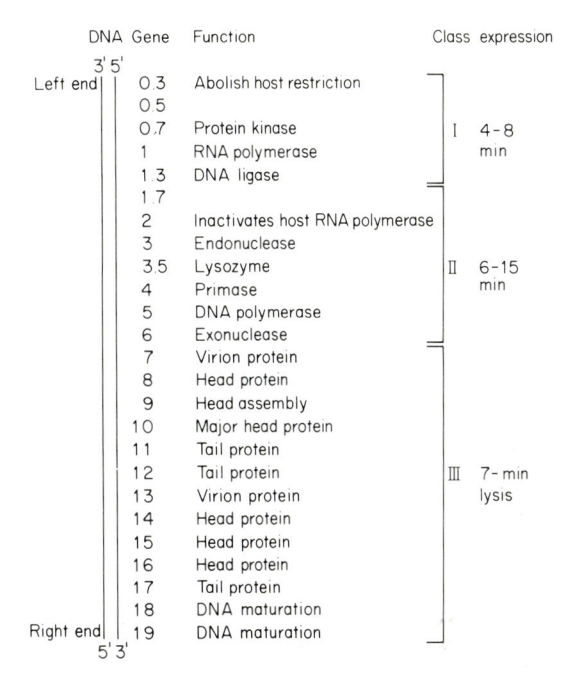

DNA	Gene	Function	Class	expression
Left end 3' 5'	0.3	Abolish host restriction		
	0.5			
	0.7	Protein kinase	I	4-8 min
	1	RNA polymerase		
	1.3	DNA ligase		
	1.7			
	2	Inactivates host RNA polymerase		
	3	Endonuclease		
	3.5	Lysozyme	II	6-15 min
	4	Primase		
	5	DNA polymerase		
	6	Exonuclease		
	7	Virion protein		
	8	Head protein		
	9	Head assembly		
	10	Major head protein		
	11	Tail protein		
	12	Tail protein	III	7-min lysis
	13	Virion protein		
	14	Head protein		
	15	Head protein		
	16	Head protein		
	17	Tail protein		
	18	DNA maturation		
Right end 5' 3'	19	DNA maturation		

Fig. 7.11 The genetic structure of phage T7. The DNA molecule is presented in the correct orientation to the genetic map showing the known genes (numbered 0.3 to 19) together with their functions. The genes are clustered into three classes according to their period of expression during infection at 37°.

be responsible for the early cessation of Class I transcription (Rahmsdorf *et al.* 1974). Transcription from the early promoters stops at the end of the early region by a termination signal which responds to the host termination factor *rho*. T7 expression is therefore regulated by both positively and negatively acting controlling elements (Rabussay and Geiduschek 1977).

Replication of T7 DNA also provides an example of general interest. Host cell DNA is degraded during infection by phage-coded enzymes to provide the major, if not only, substrate for phage DNA synthesis. Electron microscopic analysis of infecting phage genomes during replication shows a precise pattern of DNA synthesis starting at a fixed site, the origin of replication, located 17 per cent of a genome length from the left end. Replication is *bidirectional*, i.e. proceeds outwards in both directions (Fig. 7.12). This forms initially a 'bubble' structure at the site of replication but as DNA synthesis proceeds outwards the leftwards growing points reach their terminus before the rightwards moving replication form to give a Y-shaped molecule (Dressler *et al.* 1972). DNA synthesis is initiated on RNA primers under control of the gene 4 product and proceeds by the formation of short Okazaki fragments, at least on one strand, which are ligated into continuous deoxypolynucleotide chains.

Intracellular T7 DNA extracted shortly after the start of DNA synthesis and purified away from protein has the properties of concatemeric DNA. When examined by electron microscopy or zone sedimentation these molecules are seen to be several times unit length and to have single strand interruptions, often at sites corresponding to molecular ends

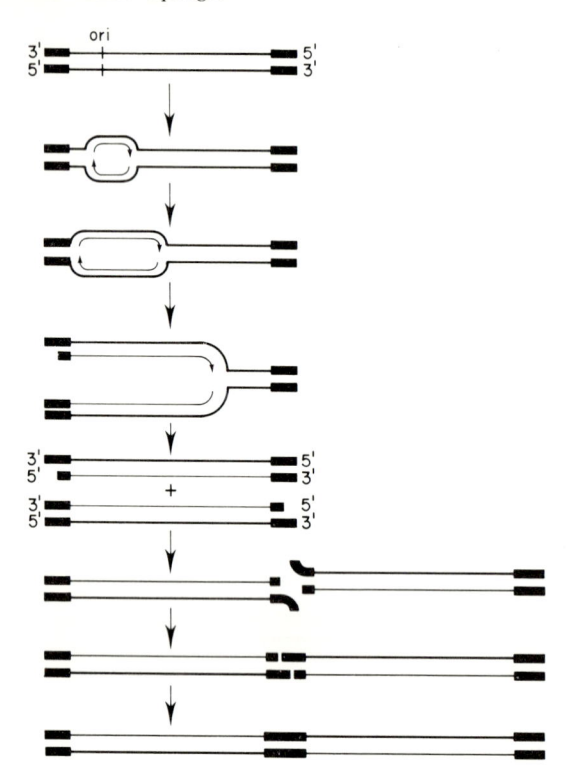

Fig. 7.12 T7 DNA replication. DNA synthesis starts at the unique origin (ori) and proceeds bidirectionally to produce successively a bubble form, a Y form and finally two daughter molecules in which the 5′ ends of the newly replicated strands are not fully replicated. The resulting 3′ single-stranded ends, because they are part of the terminal redundancy, are of complementary base sequence. The ends of two daughter molecules can therefore anneal by a recombination event which trims and then covalently seals the joint.

(Schlegel and Thomas 1972). The exact structure of T7 concatemers and more particularly, their role in replication, is still not clear. A commonly accepted view is that concatemers are required to complete the synthesis of progeny strands. This view, put forward by Watson (1972), is attractive in providing a solution to the dilemma that the need for RNA priming of DNA synthesis leads to the inability to complete the replication of linear DNA molecules. Formation of concatemers by the end-to-end joining of the partial replicas would allow completion of the synthesis of molecular ends and thereby ensure their preservation (Fig. 7.12). End-to-end joining is seen as a recombination event; this follows from observations with T7 and several other phages (T1, T4 and P22), that DNA replication and genetic recombination are intimately linked processes (Doermann 1974, Susskind and Botstein 1978, Ritchie and Joicey 1980). For example, neither the formation of concatemeric DNA nor recombinant molecules occur in the absence of the T7-coded exonuclease product of gene 6 (Kerr and Sadowski 1975).

In addition to the phage-coded replication functions, T7 DNA replication requires several host gene products. For instance, the T7 DNA polymerase depends on a host factor for its activation—this factor, identified as thioredoxin, complexes with the phage-coded product. However, T7 DNA synthesis is not dependent on the host DNA and recombination functions.

Assembly and DNA packaging has not been studied in T7 to the same extent as in T4, however, it is quite clear that the principles are the same and rely on the association of capsid components synthesized separately and later assembled in part at least by self-assembly reactions. Empty head precursors, proheads, synthesized on scaffolding proteins are filled with DNA and then completed before being attached to tails.

The packaging reaction which converts concatemeric precursor DNA into unit length genomes for introduction into proheads is not understood in detail. It must, however, be fundamentally different from the T4 headful packaging mechanism. This conclusion arises from the fact that T7 genomes have a unique base sequence and are not circularly permuted; this requires that the concatemer is cut to precise lengths at specific base sequences which mark the ends of virion DNA. Two possible mechanisms have been suggested.

In the first, two closely spaced single-strand interruptions (nicks) are introduced into opposite strands on either side of the terminally redundant region. (This is similar to the mechanism responsible for making the 'sticky ends' of phage λ DNA.) The single-stranded ends are then converted to double strands. Alternatively full-length genomes could be excised directly by double-strand cuts. Because of the requirement for each genome to have a terminal redundancy this mechanism would produce an incomplete genome lacking both ends for every complete one. Mutations in genes 8, 9, 10, 18 and 19 accumulate concatemeric DNA. Genes 8, 9 and 10 code for head proteins and failure to form a normal head structure leads to failure to package DNA. Genes 18 and 19 are considered to control DNA packaging and may function by introducing the endonucleolytic cleavages into concatemeric precursor DNA when it is introduced into proheads. Recently an *in vitro* system for packaging DNA into heads has been developed for T7 in which cell-free extracts will catalyse the formation of infectious phage particles from free DNA, and head and tail precursors. This system requires energy and mimics many of the intracellular events that occur *in vivo* (Kerr and Sadowski 1974).

Phage λ

Phage λ is typical of yet another class of duplex DNA phages—those which are temperate and so have the genetic capacity either to replicate productively in a lytic infection or to enter the lysogenic cycle. Lysogeny will be considered later; for the present λ is considered as an illustration of features of lytic replication not encountered with T4 or T7. The major differences are that λ DNA replication occurs in two modes—an early circular mode and a later concatemeric mode, the mechanism of DNA packaging, and some aspects of the regulation of gene expression.

After infection λ transcription is initiated at two promoter sites, pL transcribing leftwards and pR transcribing to the right (Fig. 7.19). These transcripts are synthesized from different DNA strands and are catalysed by the host RNA polymerase as indeed is all subsequent transcription. A similar series of events follows the induction of a lysogenic cell. This early phase of gene expression leads to the synthesis of gene products mostly with regulatory functions such as the gene *N* product which allows transcription to proceed into the next phase (delayed early transcription), and express genes controlling DNA replication and recombination which are necessary for lytic development. The transition from delayed early to late gene expression requires the activity of the product of early gene *Q*. Current belief is that the *Q* gene product, like that of gene *N*, has an antiterminator role and activates late gene expression by allowing transcription to proceed beyond a stop signal. Late transcription leads to expression of the structural and assembly functions and of the cell lysis genes *R* and *S* (Skalka 1977).

Replication of λ DNA is initiated on circular DNA templates formed very soon after infection by recombination between the complementary cohesive ends of the infecting genome. Once joined, the two ends are covalently closed by ligation. Replication is in two phases which differ in the mode of replication (Fig. 7.13). The early phase starts at a fixed origin (*ori*) situated in the DNA synthesis block of genes and to the left of gene *O* and requires the products of the λ genes *O* and *P* and in addition host proteins including RNA polymerase, *dna*B protein, primase and DNA polymerase III. The λ gene *N* protein is also required as a control element to ensure continued transcription of genes *O* and *P*. From the origin the covalently closed supercoiled circular molecules replicate bidirectionally in the classical theta form giving rise to two supercoiled circular daughter molecules (Skalka 1977). Replication of this early type continues for the first 10–15 minutes of the infectious cycle and this form of λ DNA can be transcribed, used as a template for further theta-type replication, used as a template for late λ replication, or integrated into the host chromosome as prophage. It cannot, however, be packaged into virions; this presumably is one reason why the mode of replication changes, since the substrate for DNA packaging is a concatemeric DNA structure produced by the late phase of replication. This requirement is consistent with the principle established with T4 and T7 and is observed commonly if not universally, among this group of phages. In the case of λ the concatemer molecules are the product principally of rolling circle replication rather than recombination. In this phase of replication nicked circles are used as templates with the 3'-OH of the nicked strand priming DNA synthesis and the uninterrupted strand serving as the template. This produces concatemeric DNA as the growing point passes repeatedly around the circular template. The molecular basis for the switch from early to late replication is not fully understood but may result from the activity of the *cro* gene product in depressing early transcription. This removes the transcriptional activation component and *O* and *P* production (Herskowitz and Hagen 1980). Another factor of undoubted importance is the appearance of the λ *gam* protein, a delayed early function. This protein binds to the host *rec*BC exonuclease thereby protecting the rolling circle structures which, unlike the circular DNA, have free ends and are therefore susceptible to *rec*BC degradation. In the absence of the *gam* protein circle replication continues and the transition to rolling circle replication does not occur

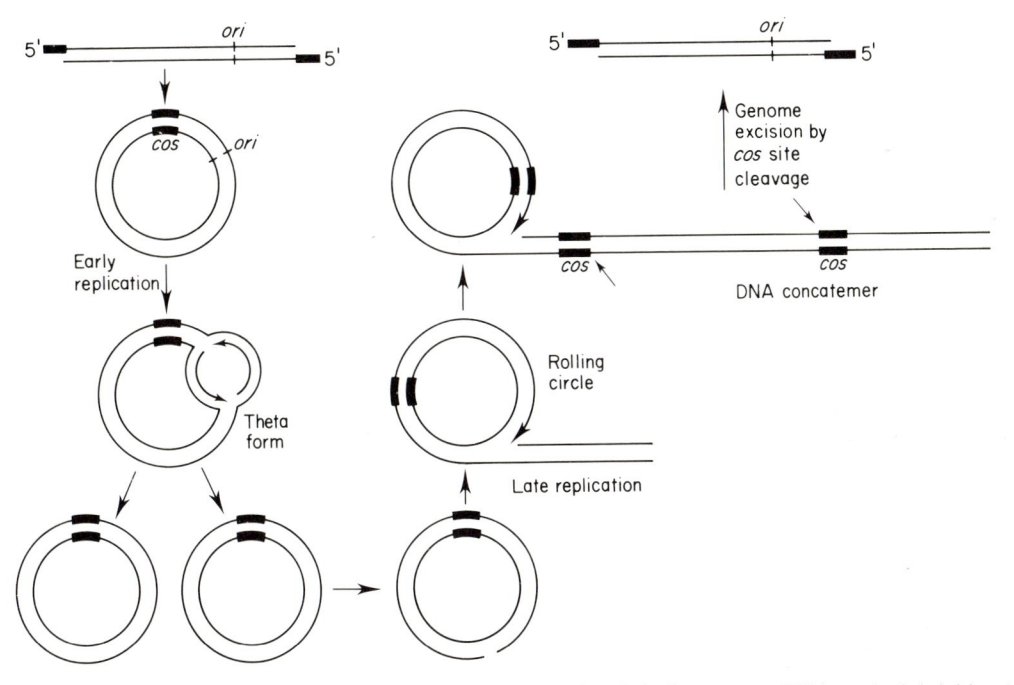

Fig. 7.13 Replication of phage λ DNA. After circularization of the infecting genome, DNA synthesis is initiated at a fixed origin (ori) and proceeds bidirectionally to produce daughter circles via a theta form replicating structure. This early phase of replication changes to the late rolling circle phase which produces concatemeric molecules. These are cut at specific *cos* sites which identify the cohesive ends of mature genomes. The excision of genomes from concatemeric DNA occurs during DNA packaging and particle morphogenesis.

(Skalka 1977). Concatemers may also be formed by recombination.

The details of λ particle assembly have been extensively analysed (Murialdo and Becker 1978). The pattern is similar to that described for T4. A prohead free of DNA is constructed from the major head protein, gpE, by a process involving the fusion and cleavage of several proteins. The prohead is filled with DNA and head formation is completed by addition of gpD the other major head protein into the lattice formed by gpE and in equivalent amounts, and by incorporation of gpF into the base vertex. At this point the tails, which have been assembled on a separate pathway, are joined to the heads.

DNA packaging, however, is different from the T4 and T7 patterns. This is because the processing of concatemeric DNA produces genomes which are unique and have cohesive ends. Concatemeric DNA is an essential substrate for packaging and it is clear from both *in vivo* and *in vitro* studies that monomeric DNA is not efficiently packaged. Maturation requires that the concatemeric DNA is nicked at staggered sites on either side of the 12 base cohesive end sites (*cos* sites) to produce the single-stranded cohesive ends (Fig. 7.13). This reaction is catalysed by the head gene product gpA, in conjunction with other λ and host functions.

Phage Mu

The notable and as yet unique features of this temperate phage are that it is able to integrate at any site on the host chromosome, and that the prophage state is an essential stage in vegetative replication (Howe and Bade 1975). For this reason phage Mu is considered an example of that category of genetic structures known as *transposable genetic elements* (see Chapter 6).

This fascinating feature of Mu arose from the initial discovery that it had mutator properties resulting from integration into a gene and thereby destroying the function of that gene (Taylor 1963). Mu replication apparently occurs not as with λ by prophage excision followed by replication of the vegetative DNA but by repeated transpositions to new sites on the host chromosome. Each transposition is accomplished by duplication of the transposed genome with one copy remaining at the original site of insertion and the other at the site to which transposition has occurred. After several rounds of DNA duplication and transposition the host chromosome contains a series of copies of Mu integrated at different sites. Maturation occurs with the phage DNA copies being excised from their chromosomal sites and packaged by the usual route. Excision includes short sequences of host DNA at either end of the Mu genome and, because each has originated from a different integration site, each particle carries different host sequences.

One further feature of Mu phage biology deserves mention. This is the G segment, a region of the genome comprising 8 per cent of the length and located in the right half of the map. The sequence of the G segment is invertible and can be found in either orientation with respect to the remainder of the DNA sequence. This invertible property provides an excellent example of *phase variation*, since with one orientation the phages are viable when infecting *Escherichia coli* and defective on *Citrobacter*, whereas the reverse is true for particles with the G segment in the other orientation. Phase variation appears in this case to affect tail fibre structure and therefore phage adsorption to one or the other host.

Replication of single-stranded DNA phages

This class of phages includes the isometric type D phages, e.g. ΦX174 and S13, and the male-specific filamentous type F phages, e.g. f1 and M13. Both types are characterized by having small circular genomes of $1.5-2 \times 10^6$ daltons molecular weight. The entire DNA sequences are known for several members of this group—ΦX174 contains 5386 nucleotides, G4 contains 5577, fd contains 6389 (Godson *et al.* 1978, Schaller *et al.* 1978). Furthermore the specific locations of each gene and the control signals are also known. Such observations make these phages the best understood biological systems and ideal models not only for viruses but for all genetic systems. Such is the power of DNA sequencing as a means of interpreting genetic structure!

ΦX174 codes for 9 known genes. However, at least 16 gene products have been identified by analysis of the proteins coded by ΦX174 during infection. In part this results from two genes, A and G, each coding for several polypeptides clearly distinguished by size. In gene A two major proteins are synthesized A and A* with A* being an internal fragment of A. In addition the A and A* proteins are each made in two discrete sizes differing slightly from each other. This gives four protein products from one gene (Tessman and Tessman 1978). It is not known whether A and A* have distinct functions. A further complication of the genetic structure of ΦX174 is that the same DNA sequence can be translated to give two quite different gene products by altering the reading frame (Fig. 7.14). Thus gene E is coded for by a fraction of the D gene translated in a second reading frame; similarly gene B is encoded within the A gene (Sanger *et al.* 1977). These remarkable findings mean that for this DNA region the base sequence codes for functional proteins in reading frames offset by 1 base. A current view of the signif-

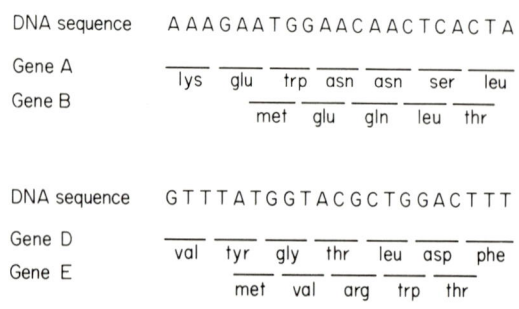

DNA sequence	A A A G A A T G G A A C A A C T C A C T A
Gene A	lys glu trp asn asn ser leu
Gene B	met glu gln leu thr

DNA sequence	G T T T A T G G T A C G C T G G A C T T T
Gene D	val tyr gly thr leu asp phe
Gene E	met val arg trp thr

Fig. 7.14 Overlapping reading frames of ΦX174 DNA. Each DNA sequence marks the site at which the triplet code is read in two overlapping frames and shows the corresponding amino acid sequences. The upper sequence corresponds to the site within the A gene at which the B gene starts. The lower sequence shows the site within gene D at which translation of the gene E product begins. The complete sequence of ΦX174 DNA and the translational products are described by Sanger *et al.* 1978.

icance of such phenomena is that they represent the products of selection for the most effective use of these limited genomes. Similar observations are being reported for the small animal papova viruses. Clearly our views on the arrangement and topology of genes will require some revision in the light of these revelations.

As discussed above the adsorption of members of these two groups is rather different, the isometric phages adsorbing to cell wall receptors, whereas the filamentous phage particles adsorb to the tips of the F or sex pili present on male cells and coded for by F-factor genes. Following these early events the subsequent stages of replication are very broadly the same for the two groups. Radical differences again become apparent during virus release from the infected cells.

Infection by ΦX174 is initiated when the single strand of infecting DNA is transported into the cell in association with the gene H or capsid spike protein (Jazwinski et al. 1975). The H protein (pilot protein) not only facilitates adsorption but is required for DNA entry and for the initiation of DNA replication. The DNA strand that infects the cell is the 'plus' strand, i.e. has the same informational sense as the mRNA. Thus to begin transcription it is first necessary to synthesize a 'minus' strand on the 'plus' strand template. This first stage of DNA replication produces double-stranded circular supercoiled replicative form (RF) molecules and, because no phage transcription has occurred at this stage, relies entirely on host functions (Sinsheimer 1968).

This reaction has been studied in considerable detail particularly by Kornberg and his colleagues and has been analysed both in vivo and in vitro. In fact, ΦX174 DNA was the first to be synthesized entirely in vitro and shown to be infectious (Goulian et al. 1967). The reaction is primed by RNA and requires at least 20 proteins; several of the stages have been reproduced in vitro (Kornberg 1980). Synthesis of the first RF molecule and its transcription takes place during the first minute of infection.

Thereafter follows stage II which continues until about 20 min post-infection and is a period of RF replication in which about 60 progeny RF molecules per cell are synthesized. The currently accepted view of this process is that RF replication is by rolling-circle replication (Baas and Jansz 1978). However, unlike the late phase of λ DNA replication in which rolling-circle replication forms concatemeric DNA, the ΦX174 process is terminated after every duplication event. This produces two covalently closed, duplex daughter RF molecules at each replication cycle. The positive strand is made by continuous DNA synthesis by 3'-OH extension of a primer and the negative strand is synthesized discontinuously. RF Stage II replication requires the function of ΦX174 proteins. Specifically the products of gene A are essential and function in several ways including the introduction of a nick into the 'plus' (viral) strand. This serves to initiate RF synthesis which also requires the activity of the host rep protein; this functions in a complex with the A product to promote replication fork movement and so allow the growing point to progress along the template.

The last phase of ΦX174 replication, Stage III, is the formation of viral single-stranded DNA—the product to be packaged into progeny phage particles. This phase occurs between about 25 and 30 min after infection and leads to the synthesis of approximately 500 viral strands per infected cell. An interesting aspect of DNA replication is revealed at this stage, for there is no difference in the type of DNA synthesis occurring during Stages II and III. What does differ is that Stage III is associated with capsid assembly such that as the 'plus' (viral) strand is being synthesized by rolling-circle replication on the RF template it is being simultaneously encapsidated. In this way it cannot be used as a template for minus strand synthesis and RF formation is therefore suppressed. Stage III replication requires seven phage-specified proteins. These are, of course, the capsid proteins in addition to the gene A product. The requirement for the rep protein

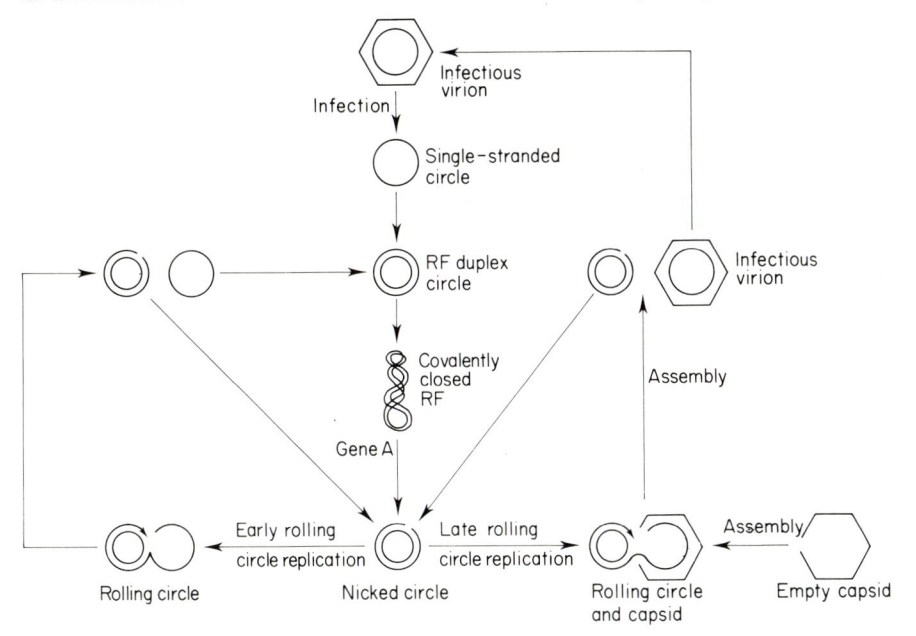

Fig. 7.15 The stages of phage ΦX174 DNA replication. Details are given in the text.

200 Bacteriophages   Ch. 7

also continues during this stage. The morphogenetic pathway occurs in a series of steps which forms a prohead constructed from 12 subunits each containing 5 molecules of proteins F and G attached to a gene H protein. This structure is catalysed by the gene B protein and is thought to be supported by a scaffold of gene D protein (Hayashi 1978). The prohead is filled by a DNA viral single strand as it is synthesized on the RF template (Fig. 7.15). Continuation of rolling circle single-strand synthesis is stopped by the single strand cutting activity of the gene A product. The introduction of the DNA into the prohead is facilitated by the gene C protein and possibly protein J. The loss of the gene D scaffold protein precedes the final maturation of an infectious particle (Hayashi 1978).

Expression of the ΦX174 genes requires the RF duplex form of DNA molecule, since it is the 'minus' DNA strand and not the viral 'plus' strand that is transcribed. Transcription is exclusively from the 'minus' strand; the strand switching observed for T4 and λ does not occur for ΦX174. Characterization of mRNA transcribed *in vivo* from ΦX174 DNA indicates a range of molecules from less than 10 per cent of the genome in length to considerably greater than one genome. The sizes of the various mRNA species indicate that most are *polycistronic*, i.e. larger than a single gene, and that a given phage gene is encoded in more than one species of mRNA (Fujimura and Hayashi 1978). The very long molecules arise by transcription continuing beyond one complete round of transcription. Mapping the transcripts to the DNA molecule shows three regions at which transcription begins but that for mRNA initiated at any one site there are several termination points. This pattern would produce the overlapping transcriptional pattern that is observed (Fig. 7.16).

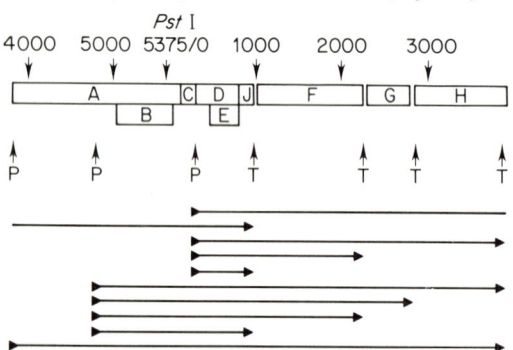

Fig. 7.16 Transcription of ΦX174 DNA. The genetic map indicates the relative locations of each gene including the overlapping genes A/B and D/E. The arrows marked P and T identify the sites of promoters and terminators respectively. The lower part shows the mRNA transcripts made during infection with the arrows indicating their length and direction of synthesis. The numbers refer to the number of nucleotides starting from the unique *Pst* I site and the circular genome is presented in linear form starting with gene A. (Based on data of Fujimura and Hayashi 1978.)

There is no evidence for the temporal control of gene expression and no division into early and late functions. The most likely controls of gene expression are differences in promoter and ribosome binding sites for transcription and translation and variations in the rate at which different polypeptides are synthesized. This is postulated in the absence of any known control elements and the observation that different gene products are made in different amounts. For example, the products of genes D and F are the most abundant.

As the archetype of the isometric phages, ΦX174 reveals the extreme efficiency with which the genetic information is used together with a much simpler form of regulation of gene expression.

We have referred to the broadly similar replication patterns of the isometric and filamentous single-stranded DNA phages. However there are a few points of major distinction that must be mentioned.

The first is that cells infected by filamentous phages continue to function more or less normally and they will even divide (Marvin and Hohn 1969). On the other hand ΦX174, while depending extensively on host functions, ultimately inhibits host DNA synthesis and kills the cell. This particular property of the filamentous phages is related to their non-lytic mechanism of release. A second major difference, also related to phage release, is the mechanism of particle morphogenesis. The viral DNA molecule is associated with a pilot protein which leads the DNA to the inner cell plasma membrane where the capsid protein accumulates. This protein is hydrophobic and forms an α-helical tube around the DNA in the form of a left-handed helix (Marvin and Wachtel 1975). As the DNA leaves the cell the protein is assembled around it and the DNA-binding pilot protein is released. The virion therefore emerges through the cell membrane and cell wall by a budding process. M13 DNA of any length can be packaged, whether it is shorter or longer than normal. The packaging of shorter fragments produces defective mini M13 particles. The packaging of longer fragments, M13 DNA linked by *in vitro* recombinant techniques to non-M13 DNA, makes M13 and its relatives useful as a gene cloning vector and it has been used for one of the major methods for DNA sequencing.

Replication of RNA phages

The discovery by Loeb and Zinder (1961) that several male-specific phages contained RNA as their genetic material raised important and interesting questions about genome replication and the control of gene expression which had previously been the preserve of DNA. Although three groups of RNA phages are distinguishable by serological, chemical and physical criteria, most of the experimental work has concentrated on the group I phages, MS2, f2 and R17, and the group III phages Qβ (Zinder 1975, Eoyang and August 1974).

The linear single-stranded genome codes for 3 genes in the group I phages and 4 for Qβ. Two genes code for coat proteins: one the major subunit present as 180 molecules per particle and referred to as the coat protein, and a second known as the A or maturation protein which contributes one molecule per virion. The third gene product is a component of the RNA replicase enzyme which catalyses genome replication. This complex consists of four subunits of which the phage replicase is one; the other three are the host protein synthesis factors SI (ribosomal protein) and the protein synthesis elongation factors EF-TU and EF-TS. The fourth gene product of Qβ is a further coat protein subunit resulting from the translation of a region at the end of the coat protein gene.

Because RNA phages do not undergo genetic recombination it is not possible to derive gene orders by classical genetic mapping. This order, derived from analysis of the protein products of specific regions of the genome, is; A gene, coat protein gene, RNA replicase gene, when reading from the 5' end. With this minimal information content the RNA phages are the simplest of the known phages and it is not surprising that they depend considerably on host functions for their replication. While RNA phage replication is independent of DNA synthesis, nevertheless, damage to the host DNA by ultraviolet light, or inhibition of host DNA transcription with Actinomycin D, does reduce the ability of the cell to support phage replication. Doi and Spiegelman (1962) showed that phage-infected cells did not contain any DNA hybridizable with phage RNA. This combined information ruled out the possibility that viral RNA was first copied into DNA which could then be transcribed by the normal DNA-dependent RNA polymerase of the host and suggested that replication and transcription occurred entirely through RNA. Confirmation of this mode of replication came with the isolation of enzymes with the properties of RNA-dependent RNA polymerases from cells infected with MS2, f2 and Qβ. A clue to the replication process came with the identification of intracellular replicative forms which were double stranded and contained both virus 'plus' strands and the complementary 'minus' strands. It is now clear that replication occurs by synthesis of 'minus' strands on the 'plus' strand template. The 'minus' strand can then in turn serve as template for daughter 'plus' strands used as mRNA for translation into proteins and as progeny strands for encapsidation into virions (Fig. 7.17). All RNA synthesis proceeds in the 5'→3' direction starting at the 3' end of the template and catalysed by the RNA replicase complex. Interestingly the replicase recognizes and binds initially at an internal site, the coat protein translation initiation site before being transferred to the 3' end to begin replication.

Translation of the 'plus' strands is affected by the secondary structure of the RNA molecule which involves about 60 per cent of the molecule in the formation of hydrogen-bonded complementary regions. This normally leads to masking of the ribosome binding sites for the A and replicase genes, leaving only the coat protein initiation site available for protein synthesis. As translation of the coat protein gene gets under way the duplex regions are opened up and in doing so expose the replicase initiation site for ribosome binding. The A gene initiation site is uncovered during replication as the growing daughter strand passes along the parent strand. In addition to their structural role the coat protein molecules have a regulatory function which is exerted by their capacity to bind to the RNA molecule such that the replicase initiation site is blocked. This inhibits replicase production and presumably also A gene product synthesis since RNA replication is catalysed by the replicase. Thus it is *translational repression* by the coat protein which leads to cessation of replicase and A protein synthesis late in infection.

The double-stranded RNA phage Φ6 has not received much attention as yet. To date the notable features of its intracellular development are that there is an RNA polymerase activity associated with the virion used in the replication/transcription of the segmented genomes and that each of the three segments codes for several proteins. The virus codes for between 9 and 11 different polypeptides (Cuppels *et al.* 1980).

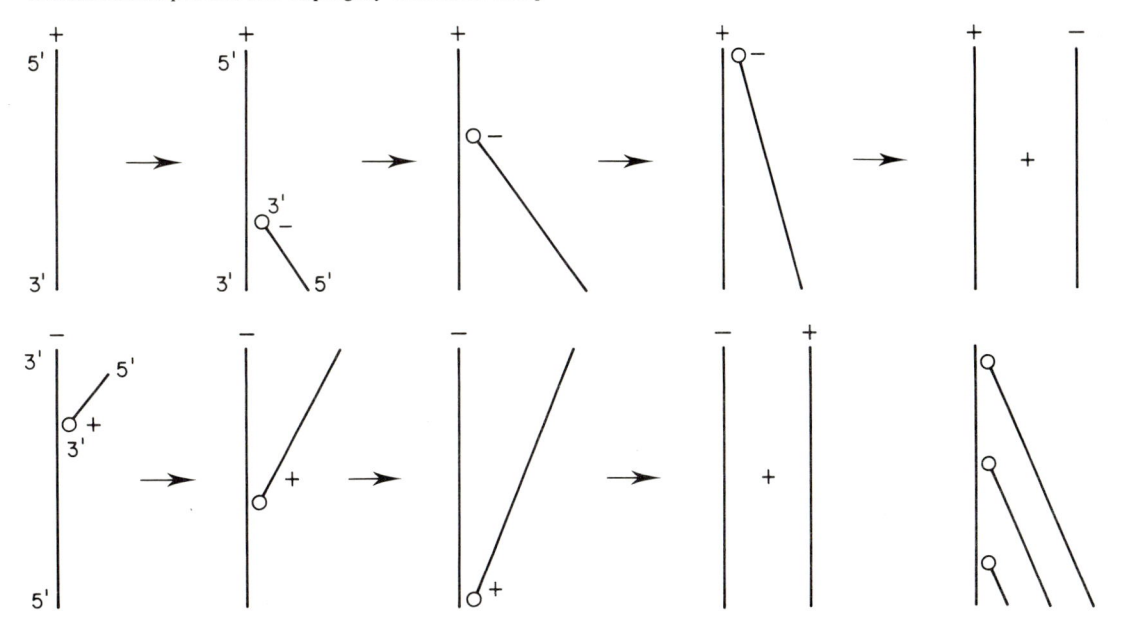

Fig. 7.17 Replication of phage RNA. The infecting viral plus strand (+) is used as a template for the synthesis of the complementary minus (−) strand in a reaction catalysed by the RNA replicase enzyme (o). Plus strands are then synthesized on minus strands. All new strands are synthesized in the 5'→3' direction. The last diagram shows how several progeny strands may be replicated simultaneously.

Notwithstanding their diminutive size and relatively simple lifestyle, the RNA phages have made important and fascinating contributions to our understanding of the principles of RNA replication and of the ways in which the expression of RNA genomes can be regulated.

Lysis

Lysis of phage-infected bacteria is the disruptive process which terminates the latent period by rupturing the cell and liberating the accumulated crop of progeny phage particles. Microscopic observation of lysis is marked by the sudden disappearance of the bacterial cell (Lwoff and Gutman 1950); the phage particles released on lysis appear as a granular cloud (Anderson *et al.* 1959). Lysis is observed in liquid suspensions of infected cells by clearing of the turbid culture or by loss of absorbance when detected spectrophotometrically (Doermann 1948). Synchronously infected cultures clear within very few minutes and the release of phages by individual cells, determined by Hutchison and Sinsheimer (1963) with ΦX174, took less than 30 seconds. The filamentous DNA phages, which are released by budding through the bacterial cell wall, are the only exception to the rule that phages are released from infected cells by lysis.

Bacterial lysis occurs by the action of the group of enzymes known as *lysozymes*, which degrade the cell wall by hydrolysing the peptidoglycan layer. Many phages code for a lysis function without which lysis of the infected cell is inhibited and in several (T4, T3, T7, T5) the gene product has been identified as a lysozyme (Streisinger *et al.* 1961, Inouye *et al.* 1973, De Martini *et al.* 1975). The small DNA phages code for a lysis function but none has been identified for the RNA phages.

Although lytic enzymes are well known as phage-specified gene products and their chemical activity has been analysed, there is a distinct lack of understanding about the timing of lysis. The mere presence of lysozyme within the infected cell is not sufficient to lyse the cell. This conclusion follows from the knowledge that lysozyme is present within infected cells for some time prior to lysis and this must mean that its activity is being controlled. Two other lines of evidence suggest that, firstly the intracellular lysozyme cannot hydrolyse the cell wall until it penetrates the cell membrane, and secondly that lysis is somehow associated with a cessation of cellular metabolism. For example, cells infected with T4 mutants defective in phage lysozyme production fail to lyse, nevertheless at the time that lysis would normally have taken place cell metabolism, measured as O_2 uptake, ceases. It has long been known that cells infected by wild type T4 fail to lyse when exposed to secondary infection (usually referred to as superinfection) a few minutes later. Under such conditions of superinfection lysis inhibition the cessation of O_2 uptake at the usual lysis time does not occur. Additional studies of this phenomenon have led to the 'equilibrium theory' of phage lysis (Mukai *et al.* 1967). This postulates that lysis involves a combination of the presence of lysozyme and unrelated events which lead to the termination of cell metabolism. Before this metabolic death the action of lysozyme on the cell wall is counteracted by cell wall repair. This equilibrium is disrupted when cell wall repair ceases and lysis ensues. In a similar manner Hausmann (1976) has reported that the addition of cyanide to T7-infected cells as early as 6 minutes after infection—but not earlier—induced immediate lysis. T7 lysozyme synthesis starts at about this time after infection and cyanide acts by stopping cellular metabolism. Additional evidence to support this theory comes from the identification of the *t* gene of T4 (Joslin 1971). The *t* gene product damages the cytoplasmic membrane and arrests energy coupling in the membrane. This allows the lysozyme to come into contact with the peptidoglycan layer of the cell wall. Lysis inhibition resulting from superinfection is due to the membrane acquiring resistance to the *t* protein. Phage λ has two similar gene functions; gene S codes for a protein which, like the T4 *t* protein, causes membrane damage and gene R codes for an endolysin with a lysozyme activity.

Lysogeny

Lysogeny defines the relationship that exists between a bacterial host cell and the *prophage* form of a temperate phage. As mentioned previously, temperate phages have two possible ways of infection. One is the lytic cycle, which is the same fundamental process for temperate and virulent phages and results in cell death with the concomitant synthesis and release of a crop of progeny phage particles. The other is lysogeny in which the phage DNA persists indefinitely in the host cell in a non-replicating form. On occasion a lysogenic cell may liberate infectious progeny into the surrounding medium.

Phage-carrying cultures are of two kinds. The first consist of strains which produce heterogeneous cultures of phage-sensitive and phage-resistant bacteria, and are merely contaminated with phage. The resistant bacteria, which are unaffected by the phage, produce some sensitive daughter cells at each generation which enable the phage to continue to multiply (Wahl and Blum-Emerique 1952). Strains of this type are called *carrier strains* and are distinguished from true lysogenic strains, in which the capacity to produce phage persists without the intervention of exogenous phage (Lwoff and Gutman 1950).

Phage cannot be eradicated from a true lysogenic strain by such means as washing, serial single colony isolation (Bordet 1925), or growth in the presence of anti-phage serum (Burnet 1934)—means which eliminate free phage particles and rid a carrier strain of its contaminating phage. In a true lysogenic strain the phage is present in a non-infectious form and no infectious phage is liberated when the bacteria are artificially disrupted (Burnet and McKie 1929, Wollman and Wollman 1936) and no phage antigen can be detected.

It was at first not clear whether the free infective particles in lysogenic strains were produced by continuous secretion of small numbers of phage from all cells in the culture or of larger amounts by a few bacteria as the result of lysis. Lwoff and Gutman (1950) proved that the second hypothesis was correct. Single bacteria of a lysogenic strain of *Bacillus megaterium* were grown in micro-drops of growth medium. These were observed microscopically and at division the daughter cells were separated and the culture medium titrated for infectious phage. A given bacterium went through as many as 19 generations without producing phage, though it retained the ability to produce a lysogenic colony. On the rare occasions when free phage particles were produced the bacterium was lysed and the surrounding medium contained phage in the numbers obtained by lytic growth. Thus the phage precursor in a lysogenic strain, to which Lwoff gave the name *prophage*, is usually transmitted at each bacterial generation without the production of free phage. Occasionally, however, the prophage enters the vegetative state, multiplies autonomously and produces phage particles that are released by bacterial lysis, in the same manner as with the lytic infection of a sensitive bacterium. Phage is therefore produced at random by a small fraction of cells in a lysogenic population, with a probability determined by the phage concerned, the bacterium and the prevailing conditions (Bertani 1951, Six 1959). It is for this reason that cultures of lysogenic bacteria contain a low proportion of infectious free phage particles identified by their sensitivity to phage neutralizing antibody and their presence in the supernatant of a lysogenic culture from which the bacterial cells have been removed by centrifugation.

Lysogenic strains can be recognized by plating on a strain of sensitive bacteria. This can be done either by cross-streaking on agar medium or by plating a sample of the lysogen on a lawn of sensitive cells. By the first method a zone of lysis will occur where the two streaks cross and by the second plaques will appear. Both tests rely on the presence of phage particles in the lysogenic culture. They also rely on the availability of a host strain sensitive to the phage under study. The plaques produced by temperate phage are turbid, that is they have a central growth of bacteria. This feature of temperate phage plaques arises because not all of the infected bacteria lyse—some are *immune* to infection. Immunity occurs when a lysogenic cell is infected by the same phage. Under these conditions the *superinfecting* phage fails to replicate and lyse the cell. The cell therefore survives and continues to grow within the zone of the plaque. Immunity is quite distinct from *resistance*, which results from failure to adsorb to the cell surface receptor sites, and is an intracellular event caused by repression of the replication of the superinfecting genome.

Our knowledge of the lysogenic state comes largely from studies with the phage λ-*Escherichia coli* and the phage P22-*Salmonella typhimurium* systems. These two systems provide general models relevant to most known temperate phages and have established the principles of lysogeny.

The nature of prophage

Lysogeny is usually a stable bacterial character and, in general, every cell in a lysogenic culture transmits to its daughter cells the ability to produce phage and immunity to superinfection by homologous phage. The efficiency with which this vertical transmission occurs indicates that the prophage must be represented many times in the cell or be associated with the bacterial genome so that its replication and segregation are coordinated with cell division. In either case this would ensure transmission of the prophage to all daughter cells. Early studies with λ showed that the number of copies of the prophage per cell was small and of the same order as the number of bacterial genomes (Jacob and Wollman 1953). Moreover, conjugation studies provided clear proof that the λ prophage behaved like a bacterial genetic marker in its pattern of transmission from donor to recipient bacteria with a site which was mapped between *gal* and *bio* (Lederberg and Lederberg 1953).

Studies by Jacob and Wollman (1957) on a series of λ-related (lambdoid) phages showed that each also had a fixed prophage location but that the genetic site was different for each phage.

The lambdoid phage pattern is not, however, universal. Phage P2 can occupy several sites although there is a pecking order for preference so that integration will occur primarily at one site, but should that site be occupied then other integrations will occur at less preferred sites (Bertani and Bertani 1974). Phage Mu integrates at a large number of sites and as far as is known can integrate anywhere in the bacterial genome or for that matter into any other DNA (e.g. plasmid) that is present in the cell (Howe and Bade 1975). Other phages are known in which the prophage state is not integrated but resembles a plasmid, being free in the cell and having autonomous replication. Phage P1 is the best known example of this plasmid-type prophage and exists as a low number of copies per cell. Some λ mutants, which are defective and unable to integrate, are also known to exist as plasmids or independent replicons.

The mechanism by which the phage genome becomes associated with the host chromosome is known as *integration*. Elucidation of the features of prophage integration have come mostly from genetic studies with phage λ although parallel studies with P22 have shown a pattern similar in every respect (Susskind and Botstein 1978).

Early investigations of the prophage state established two important facts. The first was that the phage genome was inserted linearly into the bacterial genome—for example, the genetic linkage between bacterial genes at the integration site becomes looser indicating that they are farther apart and deletions affecting adjacent bacterial genes can extend into the prophage. The second observation was that the order of genetic markers on the λ prophage map was a permutation of that on the vegetative map.

In 1962 Campbell proposed a mechanism for the insertion of the λ genome into the *Esch. coli* chromosome which took account of these facts and provided a mechanism which is now generally accepted for insertion and excision of temperate phages. The linear genome is converted to a covalently closed circle which becomes inserted into the larger circle of the host chromosome by a single reciprocal recombination between the two structures (Fig. 7.18).

The first requirement, circularization of the linear phage genome, occurs during infection with λ and a number of other temperate phages of *Escherichia coli*. In λ and its close relatives circularization occurs by annealing of the cohesive ends followed by ligation to join the strands in covalent fashion. For P22, which does not have cohesive ends, circularization is by recombination within the terminal repetition (Weaver and Levine 1977). The conversion of linear to circular λ DNA occurs both *in vitro* and *in vivo* following infection of both λ-sensitive and λ-lysogenic strains (Skalka 1977).

It was originally considered that insertion depended on homology between the specific host attachment site and the specific phage attachment site and required the normal recombination pathway. However recent studies show clearly that the attachment site is complex and consists of a very short region of genetic homology, the O region, only 15 bases long surrounded by longer non-homologous regions serving a recognition function. The phage attachment site is designated as POP′ where P and P′ are phage specific recognition sequences. The bacterial attachment site is designated BOB′, where B and B′ denote bacterial specific recognition sequences. The recombination event between the two sites can therefore be written:

$$\text{POP}' \times \text{BOB}' = \text{POB}' \times \text{P}'\text{OB}$$
$$(\text{attP}) \quad (\text{attB}) \quad (\text{attR}) \quad (\text{attL})$$

where attR and attL refer to the right and left junctions of the prophage (Herskowitz and Hagen 1980).

Neither does integration use the normal pathway for general recombination. This is most clearly seen by genetic studies which show that mutations defective in general recombination have no effect on the integration reaction. In addition mutants unable to integrate but able to undergo normal recombination have been isolated and characterized. Known as *int* mutants they code for an integrase function which only catalyses the site-specific recombination event which leads to prophage integration (Miller *et al.* 1979).

Prophage excision occurs by the reverse of the process of integration producing thereby the excised circular phage genome, which goes on to replicate vegetatively, and the bacterial chromosome. Excision also requires the *int* gene function but in addition needs the activity of a further gene known as *xis* for excision. *Xis* mutants integrate normally but are unable to excise the prophage. Integration and excision also require the activity of several host functions. Interestingly, *int*, *xis* and *red* (the general λ recombination system) occupy contiguous positions on the genetic map.

Immunity of lysogenic strains

Lysogenic strains are immune to the phage they produce (Burnet and Lush 1936). This immunity is an intracellular phenomenon and not due to failure of adsorption or, as superinfection tests show, failure to enter the cell. When a lysogenic strain is superinfected with the homologous phage, which must be genetically distinguishable from the prophage, the superinfecting phage survives in the cell but cannot multiply as long as the cell contains the prophage as such. Consequently, at cell division the superinfecting

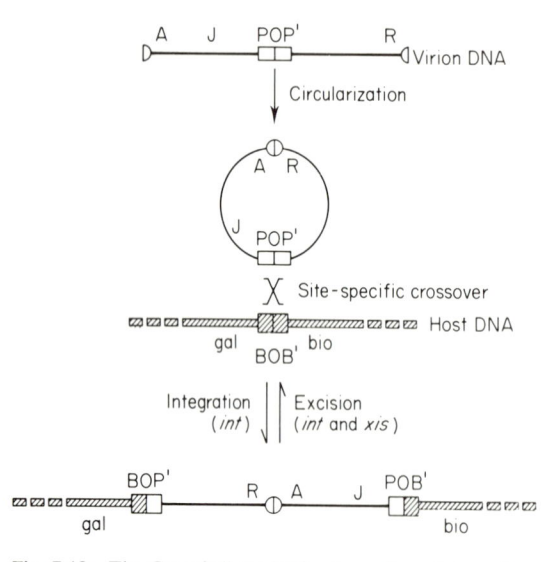

Fig. 7.18 The Campbell model for λ prophage integration and excision. Integration of the circular genome requires a site-specific reciprocal cross-over between the phage attachment site (POP′) and the bacterial integration site (BOB′). λ genes A, J and R are given as reference points and the cohesive ends are indicated by a semi-circle. Only a short region of the circular bacterial DNA molecule is shown.

passes to only one of the two daughter cells (Bertani 1954). Occasionally it takes the place of the existing prophage (prophage substitution), or joins in at the same place in the genome to give a double lysogenic strain in which genetic recombination between the prophages can occur (Arber 1960).

Superinfection immunity is highly specific and operates only against homologous phages. Even closely related phages, such as λ and 434, which can form intertypic recombinants, have different immunities. λ can therefore replicate in a 434 lysogen and *vice versa*. Moreover a hybrid recombinant with the genome of λ but the immunity region of 434 will behave like 434 in immunity tests.

The regulation of lysogeny

The phenomenon of superinfection immunity is a facet of the lysogenic state and is the responsibility of a diffusible cytoplasmic repressor or immunity substance whose production is controlled by the prophage and whose role is to repress vegetative phage growth (Weisberg *et al.* 1977, Herskowitz and Hagen 1980). In the lysogenic strain the repressor both maintains the phage in the prophage state (at its chromosomal location with integrated prophages) and prevents replication of an homologous superinfecting phage. The control of the lysogenic state depends on the expression and interaction of two regulatory repressor proteins. The first, the *cI* protein, represses vegetative phage growth and promotes lysogeny. The second, the *cro* protein, represses lysogeny and promotes vegetative growth. The balance between lysis and lysogeny and the decision to proceed along one of the two pathways is determined by the antagonism of these two repressors.

After infection the phage can enter either the lytic or the lysogenic cycles. The response chosen is influenced by the environment. The fraction of cells entering lysogeny increases with the multiplicity of infection, the magnesium concentration and the concentration of cyclic AMP (Grodzicker *et al.* 1972). Herskowitz and Hagen (1980) have suggested that after infection the λ genes are expressed in a sequential manner having three phases: *uncommitted growth*, *commitment* and *execution*.

During the early phase of uncommitted growth the infection proceeds along a single pathway (Fig. 7.19). This entails transcription from the promoter *pL* to produce *N* gene product and from *pR* to produce *cro* gene product. The availability of *N* allows transcription from *pL* and *pR* to extend beyond the *rho* dependent terminators *tL1* and *tR1* respectively and this leads to expression of *cIII* from *pL* and *cII* from *pR*. *cII* and *cIII* are required to establish lysogeny by expression of the *cI* and *int* genes. *N*-mediated transcription also leads to expression of genes *O, P* and *Q*, required for vegetative growth.

The *cII* and *cIII* products activate transcription of the *immunity operon*—the sole region of the λ genome to be expressed in lysogenic cells. This transcription is initiated at the site *cre* (controller for repressor establishment) and specifies synthesis of the *cI* repressor from the promoter *pRE*. The exact role of *cII* is unclear; models in which the *cII*

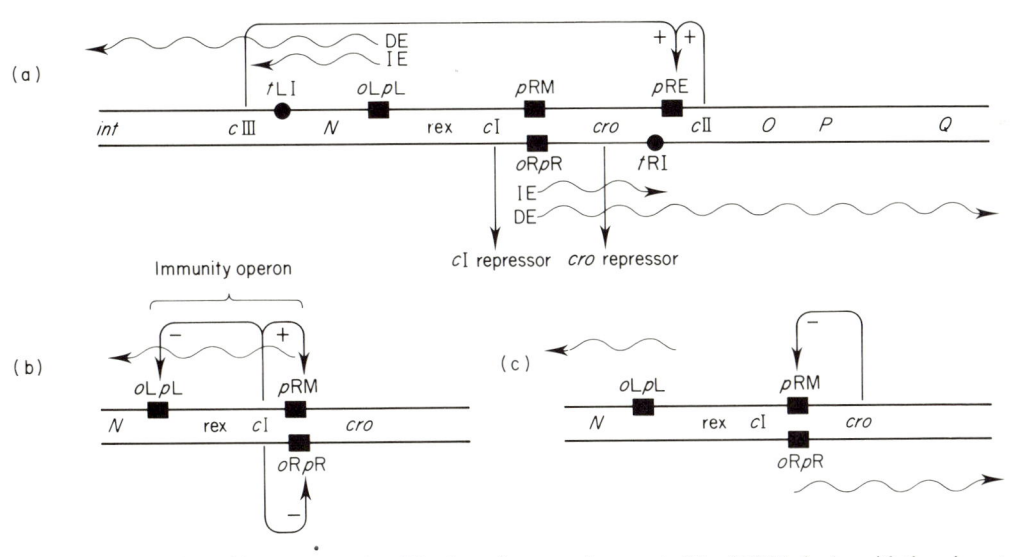

Fig. 7.19 Regulation of λ gene expression. The three diagrams show part of the λ DNA duplex with the relevant genes and control signals. The boxes and circles denote promoter and termination sites respectively, and the wavy lines refer to RNA transcripts with the leftward transcripts synthesized from the top strand and the rightward transcripts synthesized from the bottom strand of the genome. The + signs denote that the gene product stimulates transcription (positive regulation) and the − signs indicate inhibition of transcription (negative regulation). (a) Pattern of gene expression during the early uncommitted phase of regulation, (b) pattern of expression under conditions favouring lysogeny (excess of *cI* repressor) and (c) pattern of expression under conditions favouring lytic development (excess of *cro* repressor).

product facilitates transcription and prevents transcription termination have been suggested. *c*III appears to have an indirect action in stabilizing *c*II by inhibiting a host cell protease which destroys the *c*II protein.

Therefore during this early phase both *c*I and *cro* repressors accumulate. As their concentration increases a point is reached where early λ transcription is inhibited since both repressors will bind to and inactivate *o*L and *o*R.

The phase of commitment revolves around the concentration of the *c*II product. *c*II promotes the lysogenic response by stimulating *c*I and *int* synthesis and also by inhibiting the lytic response. If the *c*II level is high then lysogeny ensues, if *c*II is low the cell proceeds to the lytic cycle. The *c*III product also promotes the lysogenic response, in this case by inhibiting the *c*II inhibitor. Furthermore the *c*II inhibitor level is controlled by the metabolic state of the cell and *c*III concentration is influenced by the multiplicity of infection. Thus the *c*II–*c*III involvement in determining lysogeny is a function of their concentrations which in turn depend on multiplicity of infection and metabolic state. This interaction is transmitted to the level of *c*I repressor synthesis.

During the stage of execution there is an interaction between the two repressors *c*I and *cro*. This depends on their relative concentrations and their effect on another controlling site *p*RM—promoter for *c*I repressor maintenance. This promoter transcribes mRNA for *c*I repressor product, and *p*RM transcription is stimulated by *c*I and inhibited by *cro*. Therefore in an intracellular environment with excess *c*I the continued synthesis of *c*I occurs from *p*RM, this leads to *c*I binding at *o*L and *o*R to prevent further synthesis of *N*, *O* and *P* gene products. Late gene products, dependent on *N* and DNA synthesis, dependent on *O* and *P* fail to be made. In addition *int* synthesis is stimulated. This all results in inhibition of lytic functions, stimulation of prophage integration and maintenance of the lysogenic state by continued expression of *c*I.

If however, the *cro* repressor predominates (low *c*II/*c*III activity) there is no expression from *p*RM and no synthesis of *c*I. This enables continued transcription from *p*L and *p*R, operation of the *N–Q* switch and expression of late genes leading to lytic development terminated by lysis. At a critical concentration, *cro* binds to *o*L and *o*R to turn off early functions at a time when late gene expression can occur.

λ therefore exemplifies a complex and interconnected system of control circuits which have provided a model for gene regulation.

It now becomes clear why immunity to superinfection occurs. On entering a lysogen which is expressing only the immunity operon the incoming genome is immediately subjected to *c*I repression and fails to express any genes other than those controlled by *p*RM.

Induction of the lytic cycle

A lysogenic bacterium can be induced to enter the lytic cycle and liberate phages by agents such as ultraviolet light (Lwoff *et al.* 1950), x-rays (Marcovich 1956), a variety of chemical substances mostly having carcinogenic and mutagenic activity (see Lwoff 1953), certain antibiotics (Hall-Asheshov and Asheshov 1956), and agents inhibiting DNA synthesis (Melechen and Skaar 1962). This process is called *induction*. When a large proportion of the bacteria in a culture is induced the culture undergoes visible lysis to release progeny phage after a latent period. The development of phage originating from prophage in a lysogenic cell follows the same course as development from an infecting particle of the same kind.

Induction arises from destruction of the *c*I repressor molecules. This is illustrated well by the induction of a λ mutant with a temperature-sensitive repressor. At low temperatures the lysogens are stable; however, raising the temperature destroys the thermolabile repressor and induces the prophage. The inducing agents which affect DNA, such as UV light, lead to induction because they stimulate the degradation of *c*I repressor by proteolytic cleavage. The proteolytic activity in this case is the *Rec*A protein which in another role promotes general recombination in the bacterial host and to some extent in λ DNA. It is believed that the damage to DNA activates the *Rec*A protease activity (Roberts *et al.* 1978).

Phage P22 codes for an *anti-repressor* which inhibits binding of the repressor to the early operator sites, thereby promoting early expression and induction.

Zygotic induction, which occurs when the prophage is transferred by conjugation into a non-lysogenic, i.e. non-immune, cell can be seen as a direct result of the prophage being moved into an intracellular environment from which the *c*I repressor is absent.

The formation of transducing phages

Transduction is the process whereby bacterial genes are carried from one host cell to another by infection with a phage particle. Two categories of transduction are known: *specialized transduction*, in which phage prepared from a lysogen carry specific regions of the host chromosome, usually close to the prophage integration site (Morse 1954), and *generalized transduction* in which all segments of the host chromosome may be transferred (Zinder and Lederberg 1952). Both forms of transduction have been used as methods for the genetic analysis of bacterial chromosomes (Chapter 6). In this section, however, the formation of transducing phage particles will be considered from the point of view of their contribution to the understanding of phage replication.

Both mechanisms are the result of failure of the normal activities of phage, and the two mechanisms are quite unrelated. Specialized transducing phage arise primarily by the faulty excision of an integrated prophage, whereas generalized transducing phage arise during the lytic cycle by faulty DNA packaging at maturation.

The origin of specialized transducing phage particles

Specialized transduction is restricted to temperate phages, λ being the classical example. Prophage exci-

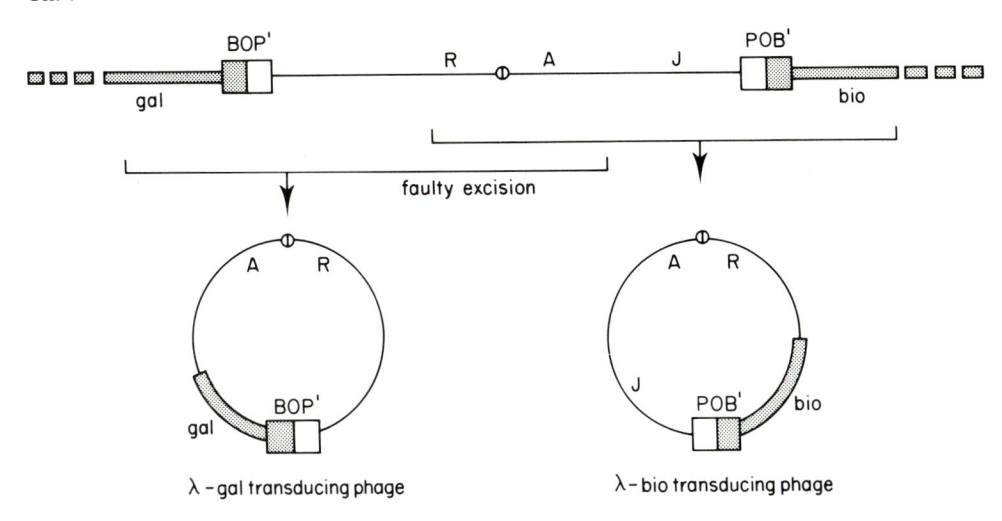

Fig. 7.20 Formation of specialized λ transducing phage. Faulty excision of the prophage leads to the incorporation of host DNA into the phage particle at the expense of some phage DNA. Details of the symbols are given in the legend to Fig. 7.18.

sion is normally precise and by a reversal of the integration process produces a complete λ genome capable of replication with the production of infectious progeny. Rarely this excision is imprecise and produces molecules that are partly viral and partly host DNA (Fig. 7.20). This occurs with a frequency of about $1/10^5$ for λ. The bacterial DNA excised is adjacent to and chemically associated with the phage section as it existed in the integrated state. Thus bacterial genes to one side or the other of the prophage can be excised. In the case of λ these regions include genes for galactose utilization (*gal*) and biotin synthesis (*bio*) located on either side of the prophage. Because the product of imprecise excision is part phage and part host it may be defective and able only to replicate and be packaged provided the cell also carries wild type phage to supply the missing phage functions. λ-*bio* phages on the other hand are frequently able to form plaques. This difference is a consequence of the λ genes lost; for λ-*gal* it is the essential capsid functions without which particles cannot be made, whereas for λ-*bio* phages it is the integration and recombination functions, which are not essential for plaque production, that are replaced.

λ-*gal* defective particles can be replicated in co-infection with a normal, helper phage. Under these circumstances large yields of up to 50 per cent of λ-*gal* particles are produced. These lysates are referred to as High-Frequency Transduction (HFT) lysates in contrast to lysates produced by induction of normal lysogens in which the frequency of transducing particles to total particles is about $1/10^5$ (Weigle 1957). These lysates are LFT—*Low-Frequency Transduction*. The composition of transducing phages in LFT lysates is heterogeneous, since many independent events have occurred and each will be different. For HFT lysates the phages produced from a transduced cell, a *heterogenote*, will be homogeneous since they are derived by replication following induction of a clone. Different HFT lysates, however, like LFT, will contain different phages.

Transduction of a recipient cell by a specialized transducing phage leads to lysogenization which incorporates the section of bacterial DNA at the prophage attachment site. This results in gene addition and the cell is now diploid for the transduced bacterial material, and this situation is stable while the prophage state is maintained. However, rarely it is also possible for the transduced genes to recombine with the homologous chromosomal genes of the recipient cell to give a permanently stable haploid transductant.

Two additional mechanisms for specialized transducing phage formation have been reported. The first is the packaging of prophage DNA directly from the chromosome of a lysogen. The second is the insertion into the phage genome of translocatable elements. This is the mechanism proposed to account for the tetracycline transducing P22 phages (Susskind and Botstein 1978), as the tetracycline genes were originally sited on an R factor carried by the host cell.

Occasionally λ will integrate at other chromosomal sites and phages transducing host genes adjacent to these alternative attachment sites can be derived from such lysogens. Other well documented specialized transduction systems include: Φ80, a λ related phage, which transduces the tryptophan genes, phage T1 resistance, and tyrosine tRNA, phage P22, which transduces for proline A and B genes in *Salmonella typhimurium* and P1 for the *Escherichia coli* proline and lactose genes. P1 and P22 are better known as generalized transducing phages but, because of their ability to integrate into the bacterial chromosome and the occasional faulty excision of the prophage, they can also produce specialized transducing phage particles.

The origin of generalized transducing phage particles

Generalized transducing phage particles arise by the encapsidation of cellular DNA into a phage coat. This is a rare event; it has been estimated that about 1 per cent of phages in transducing lysates contain cellular instead of phage DNA and are therefore capable of transducing a recipient bacterium. For any particular bacterial marker the frequency of transducing particles in a lysate is about $1/10^5$. Unlike specialized transducing phage, the generalized transducing particles contain only bacterial DNA (Ebel-Tsipis et al. 1972).

The formation of generalized transducing phage particles occurs during lytic growth and is quite unrelated to the lysogenic stage. For example, strongly virulent phages such as T1 and T4 and virulent mutants of temperate phages P1 and P22 unable to lysogenize all form transducing particles. This is because the transducing phages arise as a consequence of DNA packaging which in all phages able to produce generalized transducing phage is by the headful mode. A major feature of this mechanism is that once the signal for the commencement of packing is recognized, the headful machine progresses along the concatemeric molecule excising headfuls by cutting at sites determined by length and not by sequence. Should, by chance, the phage start to package cellular DNA in error it will progress along the bacterial genome excising headfuls of host DNA. Because the initiation of packaging for phage DNA begins at a specific site (the *pac* site) and presumably is determined by the recognition of a particular DNA sequence by a protein, it seems likely that the packaging of host DNA requires a similar protein-nucleic acid interaction (Susskind and Botstein 1978). The lower efficiency of host DNA packaging argues that the host carries *pac*-like sites, recognized at a lower efficiency presumably because they resemble but are not identical with phage *pac* sites. The observation that different bacterial genes are transduced with different efficiency can be taken to mean that different host sequences are packaged with different efficiencies. This would be an expected consequence of packaging sites on the host chromosome, which vary in their initiation efficiency, and of the distance of bacterial genes from such a site.

Phage mutants with altered efficiencies of transduction have been studied in both T1 and P22. In the former case one class of high frequency transduction mutants are located in a gene affecting the degradation of host DNA. T1 being a virulent phage normally degrades host DNA and utilizes the products as precursors for T1 DNA. The high frequency transducing T1 are produced by mutants which are defective in host DNA degradation (Roberts and Drexler 1981). The effect of this is to preserve the host DNA in a form more suitable for DNA packaging. P22 mutants with high frequency transducing ability (HT mutants) have been isolated which exhibit an altered specificity of packaging (Schmeiger 1972). This conclusion is derived from the observation that not only is host DNA packaging altered but so is that of phage DNA, which exhibits random rather than limited permutation. This result argues strongly that phage and host DNA are packaged by the same mechanism and that the specificity is reduced in HT mutants thereby increasing the number of packaging initiation sites.

After infection of recipient bacteria by generalized transducing phage particles, the majority of the transduced DNA retains its original size, and does not replicate or become integrated into the host chromosome. This produces the phenomenon known as '*abortive*' *transduction* in which the phenotypic change is passed to only one daughter cell at each division (Stocker et al. 1953, Ozeki 1959). The minority fraction, 10 per cent or less, produce '*complete*' *transduction* in which the transduced DNA is integrated stably into the host chromosome to produce a genetically inherited change that is passed to all cell progeny. This requires the recombination of the fragment into the host chromosome—a process which results in the integration of large pieces of conserved duplex DNA.

A further example of the separation of generalized transducing phage production from the integration of phage into host DNA is phage P1. This phage is present in an autonomous state during lysogeny and is only rarely integrated. The opportunity for faulty prophage excision is therefore very low. Nevertheless transducing phages are produced at the same rate as for P22.

Genetics

It is now widely recognized that genetic studies with phages have laid the foundation of modern genetics by providing the basis for molecular studies of genetic mechanisms. The initiative for this development came from the desire to understand the structure and function of genes; this led to the choice of phage since, as a simple organism, it offered the best possibility of success. Much of the early work in the 1940s and 1950s by Delbrück, Hershey and their colleagues, helped to lay these foundations upon which today's sophisticated genetical research is based. One of the most striking success stories of modern science is the understanding of the two major genetic processes in molecular terms. The first is *mutation*, the source of genetic variation in organisms. The second is *recombination*, the mechanism by which this diversity is distributed among a population of organisma. Over the last few years it has become apparent that genetic recombination occurs in all organisms by much the same process and thus the contribution of phages to genetical research is a major one.

Mutation

Phages, like all other organisms, transmit copies of their genetic information to progeny after replication of the genetic material. Nucleic acid replication is a high-fidelity process ensuring that from generation to generation the genetic information of the phage remains largely unaltered and variation is infrequent. Nevertheless, alterations do occur and, if sufficiently large numbers of phage from a given population are examined, it is possible to identify rare *mutant* individuals with properties different from the normal or *wild type* individual. Once a mutation has arisen it will be copied with the same precision as the wild type. That is to say, phages are genetically stable but not absolutely so. Genetic variation is the basis of phage genetics and the study of mutants is vital for genetic analysis.

Early genetic studies relied mainly on two types of mutation, *plaque morphology* and *host range*. Plaque morphology mutants are easy to observe because they alter the major visible phenotypic character of a phage—the appearance of the plaque. One example is the $r^+ \to r$ mutation of the T-even phages (Hershey 1946a, 1946b), and another is the *clear plaque* mutants of temperate phages (Levine 1957, Kaiser 1957). The *r* mutations (*r* is the symbol for rapid lysis) do not elicit the lysis-inhibiting response that occurs when wild type infected cells are superinfected. As a result, all infected cells within a zone of infection by *r* phages are lysed giving the characteristic clear sharp-edged *r* plaque in contrast to the fuzzy-edged r^+ plaque with its peripheral zone of partial lysis. The failure of a temperate phage to establish lysogeny will lead to clear plaques since, in the absence of immunity to superinfection, the colony of lysogenic cells that develops in the centre of a temperate phage plaque will not be produced. Other plaque variants have *minute* or *turbid* plaques. The range of distinguishable plaque types may be extended by the use of agar containing various indicator dyes (Bresch and Trautner 1955). Other variations in plaque morphology must arise from differences in the host-phage response.

Host range mutants infect a wider range of host bacteria than wild type. These mutants are usually isolated by plating high concentrations of the phage on bacteria resistant to the parent phage. Under these selective conditions the rare host range mutants will be detected by their ability to form plaques (Sertic 1929, Hershey 1946a, Bresch 1953, Appleyard 1956, Baylor et al. 1957).

The main advantage of these two classes of mutation is the ease with which they can be identified in a mixed population. This makes them useful as unselected markers in genetic analysis. Their disadvantage is that they leave much of the chromosome unmapped, because they affect specific and limited functions and often do not identify functions of major biochemical interest, i.e. their function is not essential for phage replication.

The 1960s saw the identification of a new class of phage mutants known as *conditional lethal mutations* (Campbell 1962, Epstein et al. 1963). The isolation of conditional lethal mutations has greatly increased the number of loci available for genetical and biochemical analysis, facilitating the identification of genes otherwise not amenable to analysis but whose function is essential for phage production. These mutants, which can be isolated from most genes and from all viruses, have allowed the construction of virtually complete genetic maps for several phages. Because they affect essential functions, their biochemical analysis has provided important information about the major events in phage replication.

Conditional lethal mutations are so called because they are unable to replicate to produce viable progeny phage under one set of conditions but can replicate more or less normally under another set of conditions. The two major classes of conditional lethal mutations are *nonsense mutations* and *temperature-restricted* mutations. Nonsense mutations arise by base changes which convert a sense codon (one that codes for an amino acid) to a nonsense codon (one that specifies the termination of polypeptide synthesis). Three types are known which correspond to the three chain terminating codons, these are UAG (*amber mutants*), UAA (*ochre mutants*) and UGA mutants. These mutants can replicate only in bacterial strains carrying suppressor mutations that specify mutant *t*RNA species able to translate a nonsense codon as an amino acid; in this way the lethal effect of the mutation is eliminated. The non-suppressing and suppressing strains are referred to as *non-permissive* and *permissive* strains respectively.

Temperature-restricted mutants will replicate within a more restricted range of temperature than wild type phage. The most commonly used phenotype is the *temperature-sensitive* (*ts*) mutation which cannot replicate at the upper range of temperatures. For phages of the Enterobacteriaceae the preferred permissive and non-permissive temperatures are 30 and 42° respectively. *Cold-sensitive* mutants with the reverse phenotype are also known. Temperature-sensitive mutations are base changes leading to *missense mutations* in which the mutant gene product is unable to function or unable to be synthesized at the non-permissive temperature.

The *r*II mutants of T4 which have contributed so extensively to our knowledge of gene structure and the genetic code are also conditionally lethal since they are unable to replicate in strains of *Escherichia coli* lysogenic for phage λ. Non-lysogenic hosts are permissive for *r*II mutants (Benzer 1957).

Phages are sensitive to a variety of mutagenic treatments which enhance the normally low spontaneous frequency of mutations. Mutagens can be used either on phage particles or on phage-infected cells (Drake 1970). Chemical mutagens such as nitrous acid, hydroxylamine and ethane methane sulphonate will penetrate the protein shell of the phage particle and react chemically with the nucleic acid. The chemical reactions lead to some inactivation of infectivity but the survivors will produce progeny containing an enhanced frequency of mutations. The particular value of these *in vitro* mutagenic treatments is that each mutant plaque picked from platings of the survivors will have been induced by an independent event. Therefore there can be no doubt that all mutants are separate and distinct from each other and are not clonally related. Mutagens which are effective during phage replication (*in vivo*) include 5-bromodeoxyuridine and 2-amino-purine (for DNA phages) and 5-fluorouracil (for RNA phages). These chemical mutagens are incorporated in the nucleic acid during replication and cause replication errors which may become fixed as a mutation. The chemical agents referred to above induce base pair changes which lead to *point mutations*.

Physical agents such as high temperature and ultraviolet light irradiation also increase the mutation rate. Mutations with more drastic DNA sequence alterations can be induced with intercalating compounds such as the acridines or with transposable genetic elements such as insertion sequences and transposons.

Genetic recombination

Delbrück and Bailey (1946) and Hershey (1946a) were the first to demonstrate genetic recombination between viruses. Their first observation of importance was that more than one phage could infect and replicate within the same bacterial cell. From this it was a short step to infect cells with mixtures of phages differing by two genetic characters and show that the progeny exhibited a reassortment of characters derived from different parents. In this manner Delbrück and Bailey showed that the progeny from a mixed infection by phages T2r and T4r^+ contained not only the *parental genotypes* but also the recombinant genotypes, T2r^+ and T4r. Hershey used h and r mutants of phage T2 with similar results, i.e. the progeny from the cross $h^+r^+ \times hr$ contained four genotypes: the parental genotypes h^+r^+ and hr and the recombinant genotypes h^+r and hr^+.

A genetic cross with phage is made by infecting logarithmically-growing bacteria with equal proportions of each parental phage type in sufficient numbers that each cell receives several phages of each parent (a multiplicity of infection of 5 per parent is usually sufficient). The infected cells are incubated until lysis, when the progeny phage are examined for the proportions of the various possible genotypes (Adams 1959). This can be done by plating on selective and non-selective indicator bacteria, by direct observation of plaque type, or by a combination of both depending on the particular markers being used. In some cases the only available assay may require biochemical tests on the phage isolated from single progeny plaques.

In a typical cross the progeny phages under examination will have originated from a very large number of infected cells. It is clear, therefore, that the end result does not arise from a single mating event but represents the yield of bacterial cells in each of which many replications of the genetic material, and perhaps many mating events, have taken place. Therefore the analysis of a phage cross, and even the events within a single cell, must be considered in terms of population genetics. From the analysis of the progeny of genetic interactions with T-even phages, Visconti and Delbrück (1953) formulated a general description of phage recombination. The major points of the *Visconti-Delbrück theory* are:

(1) The infecting viral genomes multiply as vegetative genomes to form a pool of viral precursor genomes.

(2) From the end of the eclipse phase precursor genomes are withdrawn at random from this pool for maturation into infectious progeny phage.

(3) The vegetative pool remains at a constant size and maintains a steady state by the rate of replication, which increases the pool, and the rate of maturation, which depletes the pool, being equal.

(4) Matured intracellular particles do not replicate or take part in genetic interactions.

(5) Within the vegetative pool the genomes undergo repeated pairwise matings leading to the exchange of genetic material.

(6) These matings occur at random with respect to both partner and time.

(7) Having completed a mating event the genomes can undergo further random matings until the mating experience is terminated by maturation.

(8) At the time of lysis not all genomes in the pool will have been matured and so will not be represented among the progeny.

The following aspects of phage recombination are consistent with the expectations of the Visconti-Delbrück theory. (1) The proportion of recombinants increases the longer phage multiplication continues; this is a prediction of a population undergoing repeated matings. (2) Among the progeny from a cross involving three genetically distinct parents some have genetic markers derived from all three parents. This also is a result of repeated matings. (3) From infections with two parental phages differing at three linked loci more double recombinants are isolated than would be expected if recombination between each pair of loci was independent; this is known as *negative interference*.

The observed recombination frequency is therefore a function both of the probability of recombination occurring between the given markers during each mating and of the average number of mating events, or *rounds of matings*, which each particle undergoes. For T2 the average number of rounds of mating is about 5, while for λ it is about 0.5, and for T1 about 1.0 (Wollman and Jacob 1954, Bresch and Trautner 1955).

Linkage analysis and genetic mapping

The first detailed study of phage recombination was made by Hershey and Rotman (1949) using different *r* mutants of T2 crossed against T2*h*. Their analysis of the results of two-factor crosses revealed that the frequency of recombinants was constant for a given *r* mutant but varied for different *r* mutants. Moreover, the frequencies of the two recombinant classes were equal for a particular cross. These results indicated the existence of genetic linkage, and showed that the relative locations of the different mutant loci on the chromosome were reflected in the frequency with which they produced recombinants. This information, as with higher organisms and bacteria, can be used to construct a linkage map which establishes the order of loci on the genetic map.

Genetic maps have been constructed for many phages by the use of both 2-factor and 3-factor crosses. In all cases there is only one linkage group and the map is unbranched, as would be expected for an organism with a single chromosome. At least one phage with a segmented genome is known. This is the RNA phage $\Phi 6$. It would be expected that each RNA molecule would behave as a single linkage group; however, as recombination has never been observed for RNA phages this may not be subject to testing by linkage analysis. Genetic maps may be linear as for λ, T7, T1, or circular as for T2, T4 and P22.

Two facts should be borne in mind when making genetic maps with phage. Firstly, there is a lack of additivity between a series of map distances, the sum of a series of short distances being greater than the map distances between markers at either end of the series. This results from an apparent excess of double cross-overs or negative interference and is a consequence of the ability of genomes to undergo repeated matings. The outcome is that an individual which has mated once and recombined in one marked genetic interval may mate again and have an opportunity to recombine in another genetic interval and so produce a doubly recombinant genome. The second fact is that, for phages such as T2 and T4 which recombine frequently, the maximum frequency of recombinants will be produced between markers which are relatively close together. That is to say, only over short genetic distances can linkage values be used to order genetic markers. These aspects of phage recombination are dealt with at length by Stent (1963).

Physical maps of phage genomes

Modern techniques have made it possible to construct maps which relate a series of DNA fragments produced by restriction endonucleases to the order in which they occur on the DNA molecule. These *physical maps* identify the locations of restriction enzyme cleavage sites and so do not require the use of mutants nor do they rely on the formation of recombinants in genetic crosses. This is an advantage in that it obviates the extensive effort needed to establish a collection of mutants and is not subject to vagaries of recombination, such as non-additivity of map distances. These maps are made by using either combinations of enzymes to identify which fragments cut by one enzyme are cut by the second, or by using partial digestion to identify adjacent fragments.

A correspondence between genetic and physical maps can be obtained by the identification of which gene functions are carried by each DNA fragment (see for example, Weisbeek and van Arkel 1978). Several approaches are available to this end. For example, bacteria can be infected with mutant phages together with specific purified DNA fragments from wild type. The progeny can be examined for the formation of wild type recombinants, the presence of which are proof that the particular fragment carried the wild type allele of the mutant being tested. Another approach makes use of complementation rather than recombination. This can be used for conditional lethal mutants where the ability of wild type DNA fragments to complement non-permissive cells infected by mutant phage can be assessed (Edgell *et al.* 1972).

The mechanism of genetic recombination

Recombination in phage, as in bacteria and, as far as is known, all organisms, is by a process in which the two parental genomes taking part in the recombination event are physically associated in the recombinant molecule. This was first demonstrated in phage λ by density labelling two genetically marked phages and showing that recombinants could be detected which carried label from both parents (Meselson and Weigle 1961, Kellenberger *et al.* 1961a, Meselson 1964). A similar process was later confirmed in phage T4 (Tomizawa and Anraku 1965).

Studies with a number of phages have shown that the progeny from mixed infections contains a small proportion of phages which are heterozygous for one or more pairs of alleles in the cross. The most detailed analysis has come from studies of the *r* and *r*$^+$ alleles of the T-even phages where the heterozygous particles are readily observed by the characteristic 'mottled' plaques they produce as a result of subsequent segregation of the *r* and *r*$^+$ alleles (Hershey and Chase 1951, Levinthal 1954, Streisinger *et al.* 1967). Heterozygous phage arise from the maturation of phage

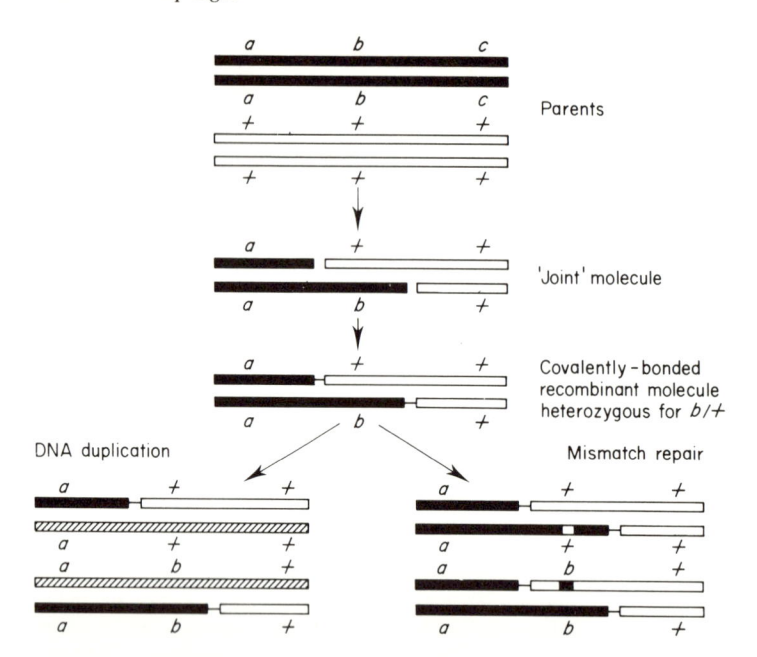

Fig. 7.21 Heteroduplex formation during T4 recombination. A recombination event between the *abc* and + + + parental molecules leads to a 'joint' molecule in which the two parental structures are linked by hydrogen bonds. This molecule is heterozygous at the *b* locus and recombinant for the *a* and *c* loci. Repair of the interruptions forms a covalently bonded recombinant molecule. The heterozygous region is made homozygous by either DNA duplication to produce *a*+ + and *ab*+ recombinants or by mismatch repair producing either *a*+ + by correction of the *b* site or *ab*+ by correction of the + allele at the *b* locus.

genomes which have engaged in a recombination event but have not completed the process. The region of heterozygosity consists of a short heteroduplex segment in which the two strands are derived from different parents. For the r/r^+ heterozygotes one strand carries the *r* allele and the other is r^+. Levinthal (1954) observed that genetic markers on either side of the heterozygous region were frequently recombinant. Completion of the recombination event requires conversion of the heterozygous region to one which is homozygous and is effected either by DNA duplication or by a correction mechanism (*mismatch repair*) which chemically corrects the non-complementary base pair (Fig. 7.21).

Confirmation that phage recombination proceeds by the formation of short heteroduplex overlap structures at the site of crossing-over came from the biochemical studies of Tomizawa and Anraku (1964). *Escherichia coli* was infected with a mixture of T4 phages labelled with the density label 5-bromouracil (5-BU) and phages labelled with the radioactive label ^{32}P; in the presence of KCN to inhibit replication, 'joint' molecules were isolated composed partly of 5-BU DNA and in part by ^{32}P DNA. The contributions from the two parents were not covalently bonded but held together by hydrogen bonding through an overlap region. When DNA synthesis was permitted the 'joint' molecules were converted to recombinant molecules in which the strands of different parental origin were joined by covalent bonds. This conversion required the combined activity of the T4 DNA polymerase and DNA ligase (Fig. 7.21).

Single-stranded regions, located internally (gaps) or at the ends of duplex DNA, promote recombinant formation. For gapped DNA this leads to the formation of characteristic X- and H-shaped molecules as intermediate structures; for single-stranded regions at the ends of DNA the product is the concatemeric molecule (Fig. 7.22). Current models of recombination also require the opening of regions of duplex DNA to permit the introduction of a complementary single strand.

Enzymically, recombination is catalysed by a variety of DNA enzymes, such as endonucleases, exonucleases, DNA binding proteins, DNA polymerases and ligases (see Kornberg 1980). The nucleases are probably responsible for the formation of the tails and gaps, the polymerase and ligase reactions being required for gap-filling and sealing reactions which restore the covalent structure of the exconjugant duplex DNA (Fig. 7.22). Mutants defective in the ability to recombine have been isolated from several phages. A common biochemical defect of these recombination-defective mutants is the loss of an exonuclease activity. For phages λ and T1 this is the only known type of recombination defective mutant. Mutants of phage T7

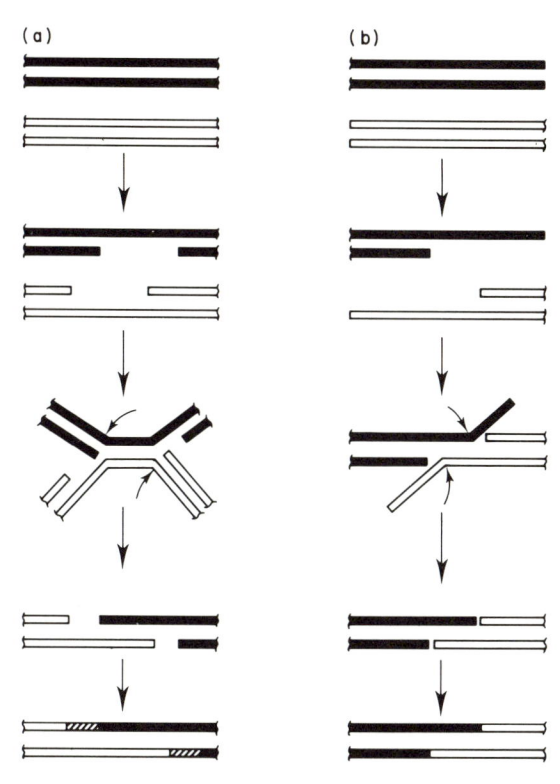

Fig. 7.22 Recombination of phage DNA. (a) Recombination between homologous DNA regions at an internally located gap leads to the formation of H joint molecules. After removal of terminal arms the gaps are filled and covalently sealed. (b) Recombination between homologous single-stranded sequences at the ends of DNA molecules is followed by the removal of unpaired single strands and the formation of covalent bonds. The single-stranded regions are produced by the action of exonucleases which in the case of internal gaps act at the site of nicks in the DNA molecules. The structure of the recombinant molecules formed depends on which strands are removed at the third stage in each diagram.

defective in endonuclease and exonuclease activities and for DNA polymerase all exhibit a phenotype defective in recombination; a similar pattern, which also includes a ligase gene function, applies to phage T4.

Genetic recombination has not been observed for RNA phages. However, with the discovery of RNA phages with segmented genomes such as $\Phi 6$, it seems likely that reassortment of chromosomes as is found for influenza virus and reovirus will occur during maturation. This is formally equivalent to the independent assortment of chromosomes that occurs during eukaryotic cell division.

Host-controlled modification and restriction

It has long been known that viruses can be adapted to give improved efficiency of replication on a given host strain by recycling or passaging the virus through this host. One important aspect of this adaptive process in phage is known variously as *host-controlled modification*, *host-induced modification*, or *host-controlled variation* (Luria and Human 1952, Arber 1965, 1974). This is not a process of mutation and selection of more adapted viruses; rather it is a non-inherited modification of the phage by the host cell. Being non-heritable, the particular modification or adaptation persists only so long as the phage is grown in the particular host. Host-controlled modification is a widespread phenomenon and is a common property of the interaction of phage and bacteria. At the practical level it forms an important part of the typing systems used to identify new isolates of bacteria; the restriction enzymes used in genetic manipulation studies are part of this phenomenon.

The phenomenon is best illustrated by reference to the example of phage λ growing in *Escherichia coli* strains K and B. When λ grown on strain B is used to infect strain K only about 10^{-4} of the infected cells yield active phage and give rise to a plaque when compared with a parallel infection of strain B. Phage λ grown on strain B is therefore said to be *restricted* on strain K. The rare phages that do succeed in escaping the restriction barrier to replicate produce progeny which now plate with high efficiency, i.e. 100 per cent, on strain K. These phages are said to be modified and now carry the K modification. Succeeding generations of λ continue to plate with high efficiency on strain K provided that they are always maintained on this strain. Similarly, λ phages grown on strain K plate with low efficiency when plated on strain B; the few progeny that do arise, however, now have an efficiency of plating (eop) of 1 on strain B and will continue to plate with this high efficiency while stocks are maintained in B (Fig. 7.23). When discussing host-controlled modification the convention is to indicate the last host strain on which the phage was grown; therefore $\lambda.B$ refers to phage λ whose last cycle of replication was on *Escherichia coli* strain B. We can now restate the position in terms of $\lambda.K$ having an eop of 1 on K and an eop of 10^{-4} on B, and $\lambda.B$ with eops of 1 and 10^{-4} on strains B and K respectively. The implication of this terminology is that the last host modifies the phage in such a way that it evades the restriction barrier to which non-modified phages fall

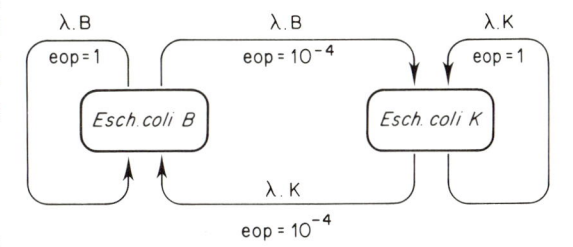

Fig. 7.23 The host-controlled restriction and modification of phage λ growing in *Escherichia coli* strains B and K.

victim. Furthermore the modification is impermanent; one cycle of growth of λ.K in strain B will produce λ.B phages which will be restricted by strain K. This is why the phenomenon is not inherited and why the modification properties of the last host are important in being solely responsible for the type of modification carried by a phage.

Some naturally occurring *Escherichia coli* strains, strain C for example, appear not to express any restriction-modification system and are therefore a universal acceptor of phages grown on any host. Restriction-modification systems can be coded for by phages as well as bacteria; phage P1 for example. *Escherichia coli* strain K lysogenic for phage P1 therefore expresses two restriction-modification systems, that of the host cell, the K system, and that of the phage, the P1 system. λ.K and λ.K(P1), since they both carry the K modification, have an eop of 1 on K, whereas the eops of λ.K and λ.K(P1) are 10^{-4} and 1 respectively on strain K(P1). λ.B will be subject to both the K and P1 restriction systems and so will plate with an eop of 10^{-4} on K but this will be even lower, e.g. 10^{-7}, on K(P1).

Restriction has nothing to do with phage adsorption; it occurs after injection of the phage DNA into the cell. λ.K and λ.B phages adsorb with equal efficiency to strains K and B. Furthermore, the small numbers of phage that do replicate in a restricting host and become modified are derived from infections of exceptional non-restricting bacteria in the population and not usually from infection by mutant phages capable of withstanding the restriction barrier (such phages do however, exist). The rare phages that do replicate produce progeny which show the same properties as the entire phage population. In addition, the extent of the restriction is affected by the physiological state of the host bacteria. For example, UV irradiation of host bacteria before infection, incubation at high temperature, and entry into stationary phase all increase the eop of non-modified phage.

The details of the mechanisms underlying restriction and modification have come from studies of the biochemistry of the process. Infection of bacteria by non-modified phage-containing DNA labelled with ^{32}P has been used to follow the fate of the infecting genomes. Soon after infection the phage DNA is broken down to low molecular weight components (Dussoix and Arber 1962, Hattman *et al.* 1966). This breakdown appears to be the work of several nucleases. The initial reaction in this degradation sequence is catalysed by a class of endonucleases generally referred to as *restriction enzymes* or *restriction endonucleases*. These act by cutting duplex DNA to produce fragments (Chapter 6). After this initial fragmentation the DNA is degraded to nucleotides by exonuclease digestion.

Two types of restriction-modification systems are known. One is based on the glucosylation of DNA and is known only for the T-even phages; the much more common system is based on DNA methylation. The hydroxyl groups of the 5-hydroxymethyl cytosine residues present in T-even phage DNA may either be free or glucosylated and the pattern of glucosylation differs between T2, T4 and T6 with each phage having its own highly specific pattern. Following infection these phages induce the synthesis of two glucosyl transferases, one catalysing α-glucosyl linkages and the other β-glucosyl linkages, which catalyse the transfer of glucose from uridine diphosphoglucose (UDPG) to 5-hydroxymethyl cytosine (HMC). Infection of strains defective in the transfer of glucose from UDPG leads to the formation of progeny phages with non-glucosylated DNA. This progeny, designated T*2, T*4 and T*6, is restricted in *Escherichia coli* B or K12 but can grow in *Shigella dysenteriae* (Hattman and Fukasawa 1963, Shedlovsky and Brenner 1963). Mutations in either of the phage glucosyl transferase genes has the same effect. The failure of non-glucosylated phage to grow on restrictive hosts is due to the rapid degradation of the non-glucosylated DNA.

The methylation-based modification reaction is brought about by the transfer of methyl groups to DNA bases at specific sites and is catalysed by methylase enzymes. One group, involved with type I restriction-modification systems, uses S-adenosylmethionine as the methyl donor and requires ATP and Mg^{++} and normally produces a pair of 6-methylaminopurine residues at the modification site (Vovis and Zinder 1975). A second group, the type II modification enzymes, are those associated with the restriction endonucleases commonly employed in genetic engineering. These methylases react specifically at adenine and cytosine residues (Rubin and Modrich 1977).

Our discussion has concentrated on the modification and restriction of phage DNA but the same principles apply to any DNA entering a bacterial cell be it from phage, bacteria, plasmids or eukaryotic cells. It appears that restriction is a means whereby the genetic interactions occurring between endogenous and exogenous DNA may be limited to those of closely related types. In this sense genetic interactions would be confined to DNA molecules which in their recent past had sojourned within cells of the same restriction-modification class.

Phage as vectors for genetic engineering

With the development of techniques for cloning DNA sequences using *recombinant DNA technology* (*genetic engineering*) has come the need for *vector* DNA molecules for the isolation of clones of foreign DNA and for their maintenance and propagation. Good vectors should be easily maintained and propagated, give high yields, be easy to purify, have means whereby the presence of inserted DNA can be readily identified and contain single sites (or pairs of sites) for commonly used restriction endonucleases. Phages fulfil many of these requirements and consequently have been developed as cloning vectors. By the clever application of genetic and biochemical techniques phage λ has been developed as a cloning vector as useful and important as plasmid vectors. The single-stranded DNA phage M13 is an excellent vector for cloning DNA for base sequence studies, and hybrid molecules constructed from parts of λ fused to plasmid DNA, known as *cosmids*, have been developed. These properties of phages properly come under the heading of applications and not principles; however, the topic is far too important to be left out on these grounds. Full and clear discussions of the principles and techniques for genetic engineering are given in the books by Old and Primrose (1981) and Glover (1980).

The suitability of a phage as a cloning vector depends on the ability to attach lengths of foreign DNA to the phage DNA such that the hybrid DNA molecule can be packaged and replicated much as a normal phage. The phage therefore provides a vehicle for propagating the foreign DNA away from its normal environment. This is achieved with λ by the construction of λ derivatives which are deleted for large sections of their genome and replacement of the deleted phage DNA by foreign DNA. The chemical association of vector and foreign DNA is brought about by cutting both DNA types with the same restriction enzyme, thereby producing complementary end sequences which can hybridize.

It has long been known that considerable regions of the λ DNA molecule can be deleted without seriously affecting the ability of the phage to reproduce vegetatively (Kellenberger *et al.* 1961*b*). The deleted region occupies a central position on the DNA molecule and codes for functions particularly concerned with the integration and excision of the prophage (*att*, *int*, and *xis* genes) and with the general recombination functions (*red*). Such phages, while able to replicate lytically, are unable to lysogenize. The general recombination functions are not essential, and in any case the recombination functions of the host cell provide a suitable substitute. In this way mutants of phage λ with deletions of about 20 per cent of the DNA can be obtained. DNA can also be deleted from a region containing one of the termination control sites giving the *nin* deletion (Fig. 7.24). Further loss of DNA occurs if the λ immunity region is removed and replaced by the same region from the related phage *Φ*80. These deletions can remove up to 25 per cent of λ DNA without too much impairment of reproductive ability and allow an equivalent amount of foreign DNA to be inserted (Murray *et al.* 1977, Blattner *et al.* 1977).

An additional factor of importance concerns the packaging of λ DNA into phage particles. Wild type λ genomes have a molecular weight of 31 megadaltons and are formed by the excision of specific genome length molecules from a concatemeric precursor. Though the excision process specifically recognizes the *cos* sites which mark the ends of genomic DNA, the amount of DNA between the *cos* sites may vary between 79 per cent and 109 per cent of the normal λ genome and still be packaged normally. This flexibility is useful in allowing foreign DNA of a variety of lengths to be inserted in the λ vector DNA.

These factors thus provide λ phage vectors capable of carrying foreign inserts equivalent to 25 per cent of the λ genome, that is, about 8 megadaltons or 12 kilobase pairs of DNA in size (Fig. 7.24).

The two most commonly used restriction sites for inserting foreign DNA into a λ vector are those produced by the restriction enzymes *Eco*RI and *Hind*III. Wild type λ contains 5 *Eco*RI and 6 *Hind*III sites. Derivatives of λ have been constructed to contain either single sites or a pair of sites defining a fragment that can be removed without loss of replication functions. For the single site vector the foreign DNA is inserted at the site whereas for double site vectors the phage DNA between the two sites is removed and replaced by foreign DNA.

λ vectors have been subsequently improved by

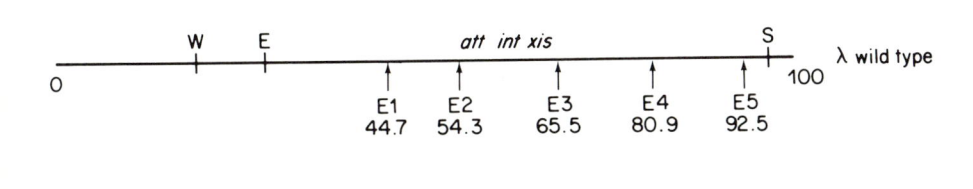

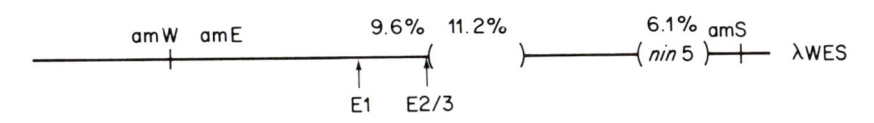

Fig. 7.24 The λWES recombinant DNA cloning vector. The wild type λ DNA shows the positions of genes W, E and S together with the five cleavage sites for the *Eco*RI restriction endonuclease (E1–E5) and their locations as a percentage of the λ DNA molecule. λWES is a cloning vector in which the 11.2 per cent region between sites E2 and E3 has been removed and the ends rejoined to give a single site E2/3. Sites E4 and E5 have also been eliminated and there is a 6.1 per cent deletion referred to as *nin* 5. Foreign DNA can be inserted in place of the λ DNA between the two remaining *Eco*RI sites, E1 and E2/3. These deletions account for 26.9 per cent of the λ DNA molecule. The vector also carries amber mutations in genes W, E and S as a biological safety measure.

modifications affecting their safety, in terms of biological containment (when cloning fragments which may be hazardous), the ease of identifying λ genomes containing inserted DNA, and the expression of the inserted DNA to form active protein products. Improvements to safety rely on the introduction of amber mutations in essential genes. These mutations are lethal to phage replication when infecting most laboratory strains of *Escherichia coli* and can only be propagated on bacteria carrying suppressor mutations, which are very uncommon in nature. The amber mutations currently used affect the synthesis of structural proteins and lytic functions. Identification of hybrid phages from those without inserted foreign DNA is rendered possible by using vectors into which the β-galactosidase gene of *Escherichia coli* has been incorporated. This particular phage carries only a single *Eco*RI site which is within the β-galactosidase gene. When infecting *lac*⁻ bacteria these phages will use the substrate 'X-gal' present in the plating medium to give blue plaques. Insertion of foreign DNA into this *Eco*RI site eliminates the β-galactosidase function and under the same plating conditions will produce colourless plaques. Screening plates for these colourless plaques provides a direct means for identifying potential hybrid phages. A second technique depends on the insertion of foreign DNA into a site in the phage immunity region. This destroys the phage repressor making the recombinant phage unable to lysogenize. Such a phage will produce a clear plaque which is readily distinguished from the turbid plaques of phages which do not carry recombinational inserts and retain a functional repressor.

λ DNA with or without a foreign DNA insert can be introduced into a bacterial cell and will replicate to produce progeny phages. This can be done by *transfection* with pure DNA or by using an *in vitro* system to package the DNA into phage particles which are then used to infect cells in the normal manner. Transfection is possible in *Escherichia coli* induced to be competent by treatment with $CaCl_2$ before and during DNA uptake (Mandel and Higa 1970). Under these conditions the phage DNA enters the cell and proceeds to replicate. This $CaCl_2$—shock competence technique is a vital component of genetic engineering technology and is used for the uptake of all vector DNA molecules into *Escherichia coli* whether plasmid or phage. *In vitro* packaging is a reaction in which a λ-infected cell extract is used to provide the structural and enzymic components necessary to package DNA provided from an exogenous source (Hohn and Murray 1977). The exogenous λ DNA, which in the present discussion will contain foreign material inserted by *in vitro* gene manipulation techniques, will be packaged in this test-tube reaction to produce infectious λ recombinant phage particles. The infectivity of *in vitro* packaged DNA is about 100-fold greater than the same amount of λ DNA used for transfection—a factor of importance when the material is available in small amounts.

Cosmid vectors Cosmids are hybrid DNA structures composed of plasmid DNA and λ phage DNA (Collins and Hohn 1979). The plasmid contributes the replication functions, the origin of replication, and selectable drug resistance genes; the λ phage contributes the *cos* sites required to package DNA into λ phage particles. Of particular value is their capacity for cloning large pieces of foreign DNA. This is achieved because λ heads will package DNA from a concatemer only if the length of DNA between adjacent *cos* sites falls within certain limits. A cosmid vector of 5 megadaltons is too small and will be packaged only if a further 18–27 megadaltons of DNA is inserted into the vector. This amount is considerably greater than the 8 megadaltons of DNA that can be accommodated within a λ phage vector. After insertion of the foreign DNA into the cosmid, the DNA is packaged *in vitro* into λ particles which are then used to infect bacteria. The injected DNA circularizes by joining at the cohesive ends and the circular molecule thereafter replicates as a plasmid. Maintenance is ensured by drug selection.

Phage M13 as a Vector M13 is a single-stranded DNA phage with a circular genome. This phage can be used to prepare cloned fragments of DNA in single-stranded form—a property which is particularly useful when DNA sequencing of the inserted fragment is the primary requirement (Messing and Gronenborn 1978, Ray and Kook 1978). Insertion of the foreign DNA is accomplished by standard techniques using the double-stranded intracellular replicative form of the DNA as the vector. Infection by the hybrid DNA leads to its replication and amplification to about 300 copies per cell.

Concluding remarks

This chapter has concentrated on one aspect of bacteriophage study—their contribution to the understanding of basic principles in genetics and biochemistry. In this respect they continue to provide new insight into the mechanisms by which biological molecules interact. It should not be forgotten, however, that as models for the study of virus-host interactions bacteriophages have provided the pattern for the analysis of animal and plant virus replication. From this basis the study of eukaryotic viruses has now advanced to the point where it is a separate discipline, able to contribute fundamental information about the genetic structure and function of higher organisms. Lastly, reference should be made to the medical importance of bacteriophages. The early hopes that bacteriophages could be used as bactericidal agents for the control of bacterial infections has remained unfulfilled, their place having been taken by antibiotics and vaccines. However, a new and more important role in medicine has emerged with the development of phages as cloning vectors for genetic engineering. In this new role they will contribute to medical problems of a much more far reaching nature than those concerned with the prevention of bacterial disease.

References

Adams, M. H. (1959) *The Bacteriophages*, p. 450. Interscience, New York.
Anderson, T. F. (1949) *Botan. Rev.* **15,** 464.
Anderson, T. F. (1950) *J. appl. Physiol.* **21,** 70.
Anderson, E. S., Armstrong, J. A. and Niven, J. S. F. (1959) *Symp. Soc. gen. Microbiol.* **9,** 224.
Anderson, T. F. and Doermann, A. H. (1952) *J. gen. Physiol.* **35,** 657.
Appleyard, R. K. (1956) *J. gen. Microbiol.* **14,** 573.
Arber, W. (1960) *Virology* **11,** 250.
Arber, W. (1965) *Ann. Rev. Microbiol.* **19,** 365.
Arber, W. (1974) *Prog. Nucleic Acid Res.* **14,** 1.
Archibald, A. R. (1980) *Virus Receptors, Part I, Bacterial Viruses*, p. 5. Ed. by L. L. Randall and L. Philipson. Chapman and Hall, London.
Baas, P. D. and Jansz, H. S. (1978) *The Single Stranded DNA Phages*, p. 215. Ed. by D. T. Denhardt, D. Dressler and D. S. Ray. Cold Spring Harbor Laboratory.
Bayer, M. (1979) *Bacterial Outer Membranes*. Ed. by M. Inouye. Wiley Interscience, New York.
Baylor, M. B., Hurst, D. D., Allen, S. L. and Bertani, E. T. (1957) *Genetics* **42,** 104.
Benzer, S. (1957) *The Chemical Basis of Heredity*, p. 70. Ed. by W. D. McElroy and B. Glass. Johns Hopkins Press, Baltimore.
Bertani, G. (1951) *J. Bact.*, **62,** 293; (1954) *Ibid.* **67,** 696.
Bertani, L. E. and Bertani, G. (1974) *Advanc. Genetics* **16,** 199.
Blattner, F. R. *et al.* (1977) *Science* **196,** 161.
Boedtker, H. (1967) *Biochem.*, **6,** 2718.
Bordet, J. (1925) *Ann. Inst. Pasteur* **39,** 717.
Boyd, J. S. K. (1950) *J. Path. Bact.* **62,** 501.
Bradley, E. D. (1967) *Bact. Rev.* **31,** 230.
Bresch, C. (1953) *Ann. Inst. Pasteur* **84,** 157.
Bresch, C. and Trautner, T. (1955) *Z. Naturf.* **10,** 436.
Bujard, H. (1969) *Proc. nat. Acad. Sci., Wash.* **62,** 1167.
Burnet, F. M. (1934) *J. Path. Bact.* **38,** 285.
Burnet, F. M. and Lush, D. (1936) *Aust. J. exp. Biol. med. Sci.* **14,** 27.
Burnet, F. M. and McKie, M. (1929) *Aust. J. exp. Biol. med. Sci.* **6,** 277.
Campbell, A. (1961) *Virology* **14,** 22; (1962) *Advanc. Genetics* **11,** 101.
Chamberlin, M., McGrath, J. and Waskell, L. (1970) *Nature, Lond.* **228,** 227.
Collins, J. and Hohn, B. (1979) *Proc. nat. Acad. Sci., Wash.* **75,** 4242.
Cowles, P. B. (1931) *J. Bact.* **21,** 161.
Cuppels, D. A., Van Etten, J. L., Burbank, D. E., Lane, L. C. and Vidaver, A. K. (1980) *J. Virol.* **35,** 249.
Delbrück, M. (1940) *J. gen. Physiol.* **23,** 643.
Delbrück, M. and Bailey, W. T. (1946) *Cold Spring Harb. Symp. quant. Biol.* **11,** 33.
De Martini, M., Halegoua, S. and Inouye, M. (1975) *J. Virol.* **16,** 459.

Doermann, A. H. (1948) *J. Bact.*, **55,** 257; (1952) *J. gen. Physiol.*, **35,** 645; (1974) *Annu. Rev. Genet.*, **7,** 325.
Doi, R. H. and Spiegelman, S. (1962) *Science*, **138,** 1270.
Drake, J. (1970) 'The Molecular Basis of Mutation', Holden-Day Inc., San Francisco.
Dressler, D., Wolfson, J. and Magazin, M. (1972) *Proc. nat. Acad. Sci., Wash.*, **69,** 998.
Duckworth, D. H. (1970) *Bact. Rev.*, **34,** 344; (1976) *Ibid.* **40,** 793.
Dussoix, D. and Arber, W. (1962) *J. molec. Biol.* **5,** 37.
Ebel-Tsipis, J., Fox, M. S. and Botstein, D. (1972) *J. molec. Biol.* **71,** 449.
Edgar, R. S. and Lielausis, A. (1968) *J. molec. Biol.* **32,** 263.
Edgar, R. S. and Wood, W. B. (1966) *Proc. nat. Acad. Sci., Wash.* **55,** 490.
Edgell, M. H., Hutchison, C. A. and Sclair, M. (1972) *J. Virol.* **8,** 574.
Ellis, E. L. and Delbrück, M. (1939) *J. gen. Physiol.* **22,** 365.
Eoyang, L. and August, J. T. (1974) *Comprehensive Virology*, Vol. 2, p. 1. Ed. by H. Fraenkel-Conrat and R. R. Wagner. Plenum Press, New York.
Epstein, R. H. *et al.* (1963) *Cold Spring Harb. Symp. quant. Biol.* **28,** 375.
Espejo, R. T., Espejo-Canelo, E. S. and Sinsheimer, R. L. (1971) *J. molec. Biol.* **56,** 597.
Fiers, W. and Sinsheimer, R. L. (1962) *J. molec. Biol.* **5,** 408.
Frankel, F. R. (1968) *Cold Spring Harb. Symp. quant. Biol.* **33,** 485.
Fujimura, F. K. and Hayashi, M. (1978) *The Single-Stranded DNA Phages*, p. 485. Ed. by D. T. Denhardt, D. Dressler and D. S. Ray. Cold Spring Harbor Laboratory.
Gill, G. S. and MacHattie, L. A. (1976) *J. molec. Biol.* **104,** 505.
Glover, D. M. (1980) *Genetic Engineering-Cloning DNA*, Chapman and Hall, London.
Godson, G. N., Fiddes, J. C., Barrell, B. G. and Sanger, F. (1978) *The Single-Stranded DNA Phages*, p. 51. Ed. by D. T. Denhardt, D. Dressler and D. A. Ray. Cold Spring Harbor Laboratory.
Goldberg, E. (1980) *Virus Receptors, Part I, Bacterial Viruses*, p. 117. Ed. by L. L. Randall and L. Philipson, Chapman and Hall, London.
Goulian, M., Kornberg, A. and Sinsheimer, R. L. (1967) *Proc. nat. Acad. Sci., Wash.* **58,** 2321.
Gratia, A. (1922) *Brit. med. J.* **ii,** 296.
Grodzicker, T., Arditti, R. R. and Eisen, H. (1972) *Proc. nat. Acad. Sci., Wash.* **69,** 366.
Guélin, A. (1952) *Ann. Inst. Pasteur* **82,** 78.
Hall-Asheshov, E. and Asheshov, I. N. (1956) *J. gen. Microbiol.* **14,** 174.
Hattman, S. and Fukasawa, T. (1963) *Proc. nat. Acad. Sci., Wash.* **50,** 297.
Hattman, S., Revel, H. R. and Luria, S. E. (1966) *Virology* **30,** 427.
Hausmann, R. (1976) *Curr. Top. Microbiol. Immunol.* **75,** 77.
Hayashi, M. (1978) *The Single-Stranded DNA Phages*, p. 531. Ed. by D. T. Denhardt, D. Dressler and D. S. Ray, Cold Spring Harbor Laboratory.
Hendrix, R. W. (1978) *Proc. nat. Acad. Sci., Wash.* **75,** 4779.
D'Herelle, F. (1917) *C. R. Acad. Sci., Paris* **165,** 373; (1918) *Ibid.* **167,** 970; (1921) *Le Bactériophage: son rôle dans l'immunité*, Masson, Paris. Engl. transl. Williams & Wilkins, Baltimore and London; (1926) *Le Bactériophage: et son comportment*, Masson, Paris. Engl. transl. Williams &

Wilkins, Baltimore and London; (1930) *Bacteriophage and its Clinical Applications*, Charles C Thomas, Springfield, Illinois; (1949) *Science News* No. **14**, p. 49, Penguin, Harmondsworth.
Herriott, R. M. (1951) *J. Bact.* **61**, 752.
Hershey, A. D. (1946a) *Cold Spring Harb. Symp. quant. Biol.* **11**, 67; (1946b) *Genetics* **31**, 620.
Hershey, A. D., Burgi, E. and Ingraham, L. (1962) *Biophys. J.* **2**, 423; (1963) *Proc. nat. Acad. Sci., Wash.* **49**, 748.
Hershey, A. D. and Chase, M. (1951) *Cold Spring Harb. Symp. quant. Biol.* **16**, 471; (1952) *J. gen. Physiol.* **36**, 39.
Hershey, A. D. and Rotman, R. (1949) *Genetics* **34**, 44.
Herskowitz, I. and Hagen, D. (1980) *Annu. Rev. Genet.* **14**, 399.
Hohn, B. and Murray, K. (1977) *Proc. nat. Acad. Sci., Wash.* **74**, 3259.
Howard, L. and Tipper, D. J. (1973) *J. Bact.* **113**, 1491.
Howe, M. M. and Bade, E. G. (1975) *Science* **190**, 624.
Hutchison, C. A. III and Sinsheimer, R. L. (1963) *J. molec. Biol.* **7**, 206.
Inouye, M., Arnheim, N. and Sternglanz, R. (1973) *J. biol. Chem.* **248**, 7247.
Jacob, F. and Wollman, E. L. (1953) *Cold Spring Harb. Symp. quant. Biol.* **18**, 101.
Jacob, F. and Wollman, E. L (1957) *The Chemical Basis of Heredity*, p. 468. Ed. by W. D. McElroy and B. Glass. Johns Hopkins Press, Baltimore.
Jazwinski, S. M., Marco, R. and Kornberg, A. (1975) *Virology* **66**, 293.
Joslin, R. (1971) *Virology* **44**, 101.
Kaiser, A. D. (1957) *Virology* **3**, 42.
Kellenberger, G., Zichichi, M. L. and Weigle, J. (1961a) *Proc. nat. Acad. Sci., Wash.* **47**, 869; (1961b) *J. molec. Biol.* **3**, 339.
Kerr, C. and Sadowksi, P. D. (1974) *Proc. nat. Acad. Sci., Wash.* **71**, 3545; (1975) *Virology* **65**, 281.
Knight, C. A. (1975) *Chemistry of Viruses*, p. 80. Springer-Verlag, New York.
Kornberg, A. (1980) *DNA Replication*, p. 555. Freeman, San Francisco.
Labedan, B. (1976) *Virology* **75**, 368.
Lanni, Y. T. (1968) *Bact. Rev.* **32**, 227.
Lantos, J., Varga, I. and Ivanovics, G. (1960) *Acta microbiol. Acad. Sci. hung.* **7**, 31.
Lazarus, A. S. and Gunnison, J. B. (1947) *J. Bact.* **53**, 75.
Lederberg, E. M. and Lederberg, J. (1953) *Genetics* **38**, 51.
Levine, M. (1957) *Virology* **3**, 22.
Levinthal, C. (1954) *Genetics* **39**, 169.
Lewin, R. A. (1960) *Nature, Lond.* **186**, 901.
Lindgren, C. C. and Bang, Y. N. (1961) *Leeuwenhoek ned. Tijdschr.* **27**, 1.
Loeb, T. and Zinder, N. D. (1961) *Proc. nat. Acad. Sci., Wash.* **47**, 282.
Luria, S. E., Darnell, J. E., Baltimore, D. and Campbell, A. (1978) *General Virology*, 3rd Edn. p. 25. Wiley, New York.
Luria, S. E. and Human, M. L. (1952) *J. Bact.* **64**, 557.
Luria, S. E., Williams, R. C. and Backus, R. C. 1951) *J. Bact.* **61**, 179.
Lwoff, A. (1953) *Bact. Rev.* **17**, 269.
Lwoff, A. and Gutman, A. (1950) *Ann. Inst. Pasteur* **78**, 711.
Lwoff, A., Siminovitch, L. and Kjeldgaard, N. (1950) *Ann. Inst. Pasteur* **79**, 815.
MacHattie, L. A., Ritchie, D. A., Thomas, Jr., C. A. and Richardson, C. C. (1967) *J. molec. Biol.* **23**, 355.

Mandel, M. and Higa, A. (1970) *J. molec. Biol.* **53**, 159.
Maniloff, J., Das, J. and Christensen, J. R. (1977) *Advanc. Virus Res.* **21**, 343.
Marco, R., Jazwinski, S. M. and Kornberg, A. (1974) *Virology* **62**, 209.
Marcovich, H. (1956) *Ann. Inst. Pasteur* **90**, 303.
Marvin, D. A. and Hohn, B. (1969) *Bact. Rev.* **33**, 172.
Marvin, D. A. and Wachtel, E. J. (1975) *Nature, Lond.* **253**, 19.
Mathews, C. K. (1977) *Comprehensive Virology* Vol. 7, p. 179. Ed. by H. Fraenkel-Conrat and R. R. Wagner. Plenum Press, New York.
Melechen, N. E. and Skaar, P. D. (1962) *Virology* **16**, 21.
Meselson, M. (1964) *J. molec. Biol.* **9**, 734.
Meselson, M. and Weigle, J. (1961) *Proc. nat. Acad. Sci., Wash.* **47**, 857.
Messing, J. and Gronenborn, B. (1978) *The Single-Stranded DNA Phages*, p. 449. Ed. by D. T. Denhardt, D. Dressler and D. S. Ray. Cold Spring Harbor Laboratory.
Meynell, E. W. (1961) *J. gen. Microbiol.* **25**, 253.
Miller, H. I., Kikuchi, A., Nash, H. A., Weisberg, R. A. and Friedman, D. I. (1979) *Cold Spring Harb. Symp. quant. Biol.* **43**, 1121.
Millman, I. (1958) *Proc. Soc. exp. Biol., N.Y.* **99**, 216.
Min Jou, W., Haegeman, G., Ysebaert, M. and Fiers, W. (1972) *Nature, Lond.* **237**, 82.
Morse, M. L. (1954) *Genetics* **39**, 984.
Mukai, F., Streisinger, G. and Miller, B. (1967) *Virology* **33**, 398.
Murialdo, H. and Becker, A. (1978) *Microbiol. Rev.* **42**, 529.
Murray, N. E., Brammar, W. J. and Murray, K. (1977) *Molec. gen. Genet.* **150**, 53.
Novotny, C. P. and Fives-Taylor, P. (1978) *J. Bact.* **133**, 459.
Nowak, J. A. and Maniloff, J. (1978) *J. Virol.* **29**, 374.
Old, R. W. and Primrose, S. B. (1981) *Principles of Gene Manipulation*, 2nd Edn. Blackwell, Oxford.
Ozeki, H. (1959) *Genetics* **44**, 457.
Palva, T. and Bamford, D. (1980) *Virus Receptors, Part I, Bacterial Viruses*, p. 95. Ed. by L. L. Randall and L. Phillipson. Chapman and Hall, London.
Paranchych, W. (1975) *RNA Phages*, p. 85. Ed. by N. E. Zinder. Cold Spring Harbor Laboratory.
Puck, T. T. and Sagik, B. (1953) *J. exp. Med.* **97**, 807.
Rabussay, D. and Geiduschek, E. P. (1977) *Comprehensive Virology* Vol. 8, p. 1. Ed. by H. Fraenkel-Conrat and R. R. Wagner. Plenum Press, New York.
Rahmsdorf, H. J. et al. (1974) *Proc. nat. Acad. Sci., Wash.* **71**, 586.
Ralston, D. J., Leiberman, M., Baer, B. and Krueger, A. P. (1957) *J. gen. Physiol.* **40**, 791.
Ramsay, N. and Ritchie, D. A. (1980) *Molec. gen. Genet.* **179**, 669.
Randall, L. L. and Philipson, L. (1980) *Virus Receptors, Part I, Bacterial Viruses*, Chapman and Hall, London.
Ray, D. S. and Kook, K. (1978) *The Single-Stranded DNA Phages*, p. 455. Ed. by D. T. Denhardt, D. Dressler and D. S. Ray, Cold Spring Harbor Laboratory.
Reznikoff, W. S. and Thomas, Jr., C. A. (1969) *Virology* **37**, 309.
Ritchie, D. A. and Joicey, D. H. (1980) *Virology* **103**, 191.
Ritchie, D. A., Thomas, Jr., C. A., MacHattie, L. A. and Wensink, P. C. (1967) *J. molec. Biol.* **23**, 365.
Roberts, M. D. and Drexler, H. (1981) *Virology* **112**, 670.

Roberts, J. W., Roberts, C. W. and Craig, N. (1978) *Proc. nat. Acad. Sci., Wash.* **75**, 4714.

Rountree, P. M. (1949) *J. gen. Microbiol.* **3**, 153.

Rubin, R. A. and Modrich, P. (1977) *J. biol. Chem.* **252**, 7265.

Safferman, R. S. and Morris, M. (1964) *J. Bact.* **88**, 771; (1963) *Science*, **140**, 679.

Sanger, F. et al. (1977) *Nature, Lond.* **265**, 687.

Sanger, F. et al. (1978) *The Single-Stranded DNA Phages*, p. 659. Ed. by D. T. Denhardt, D. Dressler and D. S. Ray. Cold Spring Harbor Laboratory.

Schaller, H., Beck, E. and Takanami, M. (1978) *The Single-Stranded DNA Phages*, p. 139. Ed. by D. T. Denhardt, D. Dressler and D. S. Ray. Cold Spring Harbor Laboratory.

Schlegel, R. A. and Thomas, Jr., C. A. (1972) *J. molec. Biol.* **68**, 319.

Schlesinger, M. (1933) *Biochem. Z.* **264**, 6; (1934) **273**, 306; (1936) *Nature* **138**, 508.

Schmeiger, H. (1972) *Molec. gen. Genet.* **119**, 75.

Schwartz, M. (1980) *Virus Receptors, Part I, Bacteriol Viruses*, p. 74. Ed. by L. L. Randall and L. Philipson. Chapman and Hall, London.

Semancik, J. S., Vidaver, A. K. and Van Etten, J. L. (1973) *J. molec. Biol.* **78**, 617.

Sertic, V. (1929) *Compt. rend. Soc. Biol.* **100**, 612.

Shaw, D. R. D. and Chatterjee, A. N. (1971) *J. Bact.* **108**, 584.

Shedlovsky, A. and Brenner, S. (1963) *Proc. nat. Acad. Sci., Wash.* **50**, 300.

Silverstein, J. and Goldberg, E. (1976) *Virology* **72**, 195.

Simon, L. D. and Anderson, T. F. (1967) *Virology* **32**, 279.

Sinsheimer, R. L. (1968) *Progr. nucl. acid Res. Molec. Biol.* **8**, 115.

Sisco, K. L. and Smith, H. O. (1979) *Proc. nat. Acad. Sci., Wash.* **76**, 972.

Six, E. (1959) *Virology* **7**, 328.

Skalka, A. M. (1977) *Curr. Top. Microbiol. Immun.* **78**, 201.

Stent, G. S. (1963) *Molecular Biology of Bacterial Viruses*. Freeman, San Francisco.

Stocker, B. A. D., Zinder, N. D. and Lederberg, J. (1953) *J. gen. Microbiol.* **9**, 410.

Strauss, J. H. and Sinsheimer, R. L. (1963) *J. molec. Biol.*, **7**, 43.

Streisinger, G., Edgar, R. S. and Denhardt, G. H. (1964) *Proc. nat. Acad. Sci., Wash.* **51**, 775.

Streisinger, G., Emrich, J. and Stahl, M. M. (1967) *Proc. nat. Acad. Sci., Wash.* **57**, 292.

Streisinger, G., Mukai, F., Dreyer, W. R., Miller, B. and Horiuchi, S. (1961) *Cold Spring Harb. Symp. quant. Biol.* **26**, 25.

Susskind, M. M. and Botstein, D. (1978) *Microbiol. Rev.* **42**, 385.

Taylor, A. L. (1963) *Proc. nat. Acad. Sci., Wash.* **50**, 1043.

Tessman, E. S. and Tessman, I. (1978) *The Single-Stranded DNA Phages*, p. 9. Ed. by D. T. Denhardt, D. Dressler and D. S. Ray. Cold Spring Harbor Laboratory.

Thomas, Jr., C. A. and MacHattie, L. A. (1964) *Proc. nat. Acad. Sci., Wash.* **52**, 1297; (1967) *Annu. Rev. Biochem.* **36**, 485.

Tikhonenko, A. S. (1970) *Ultrastructure of Bacteriol Viruses*, Plenum Press, New York.

Tomizawa, J. and Anraku, N. (1965) *J. molec. Biol.* **11**, 509.

Twort, F. W. (1915) *Lancet* **ii**, 1241; (1922) *Brit. med. J.*, **ii**, 293; (1925) *Lancet* **ii**, 642; (1949) *Science News* No. **14**, p. 33, Penguin, Harmondsworth.

Visconti, N. and Delbrück, M. (1953) *Genetics* **38**, 5.

Vovis, G. F. and Zinder, N. D. (1975) *J. molec. Biol.* **95**, 557.

Wahl, R. and Blum-Emerique, L. (1952) *Ann. Inst. Pasteur* **82**, 29.

Watson, J. D. (1972) *Nature, Lond.* **239**, 197.

Weaver, S. and Levine, M. (1977) *Virology* **76**, 29.

Weidel, W. (1958) *Annu. Rev. Microbiol.* **12**, 27.

Weigle, J. (1957) *Virology* **4**, 14.

Weisbeek, P. J. and van Arkel, A. (1978) *The Single-Stranded DNA Phages*, p. 31. Ed. by D. T. Denhardt, D. Dressler and D. S. Ray. Cold Spring Harbor Laboratory.

Weisberg, R. A., Gottesman, S. and Gottesman, M. E. (1977) *Comprehensive Virology* Vol. 8, p. 197. Ed. by H. Fraenkel-Conrat and R. R. Wagner. Plenum Press, New York.

Wellington, E. M. and Williams, S. T. (1981) *Zbl. Bakt.*, *I. Abt.* Suppl. **11**, 93.

Westphal, O., Luderitz, O. and Bister, F. (1952) *Z. Naturforsch.* **7b**, 148.

Wiebols, G. L. W. and Wieringa, K. T. (1936) *Fonds Landbouw Export Bureau, 1916–1918*, No. 16.

Wollman, E. L. and Jacob, F. (1954) *Ann. Inst. Pasteur* **87**, 674.

Wollman, E. and Wollman, E. (1936) *Ann. Inst. Pasteur* **56**, 137.

Wood, W. B. and Revel, H. R. (1976) *Bact. Rev.* **40**, 847.

Wright, A., McConnell, M. and Kanegasaki, S. (1980) *Virus Receptors, Part I, Bacterial Viruses*, p. 93. Ed. by L. L. Randall and L. Philipson, Chapman and Hall, London.

Yamagishi, H. (1968) *J. molec. Biol.* **35**, 623.

Yasbin, R. E., Maino, V. C. and Young, F. E. (1976) *J. Bact.* **125**, 1120.

Young, F. E. (1967) *Proc. nat. Acad. Sci., Wash.* **58**, 2377.

Zinder, N. (1975) *RNA Phages*. Cold Spring Harbor Laboratory, New York.

Zinder, N. D. and Lederberg, J. (1952) *J. Bact.* **64**, 679.

8

Bacterial ecology, normal flora and bacteriocines
Graham Wilson

Bacterial ecology

Bacterial Ecology: Introductory	220	Microbial ecology of water	224
Microbial ecology of soil	221	Cycle of elements in nature	225
The soil population	222	Other ecological associations	227
The soil environment	222	Pathogenicity of soil and water organisms	228
Decomposition by soil microbes	223		
Synthesis by soil microbes	223	**Normal flora**	**230**
Nitrogen transformations	223	**Bacteriocines**	**247**
Sulphur transformations	224		
Other elements	224		

Introductory

In this book we are interested chiefly in the application of the principles of bacteriology and immunity to the problems of human and animal diseases. To the general microbiologist, these problems concern but a small part of the vast field of microbial ecology—the study of the interactions between micro-organisms and their environment. The ecological systems of the medical bacteriologist are the tissues of the higher vertebrates and the microbial parasites that may establish therein either a manifest infection or a carrier state. Disease of this kind is but one small aspect of microbial parasitism—a condition widespread among living organisms of all kinds, and one which in turn is but a small part of microbial activities in the world as a whole, upon which the maintenance of life on the planet depends.

Ecology is a study of the relations between living things and their environment. It is not a property of the living things themselves. It cannot be studied apart from their environment. It is something extrinsic to the living thing, not intrinsic.

In its earlier days, medical bacteriology was largely orientated towards the prevention and cure of microbial disease. It is still very properly so orientated. Nevertheless, as medical bacteriologists and virologists, we have in more recent years acquired a greater and more sophisticated insight into our problems, and a greater versatility in our ways of attacking them. From the standpoint of general biology, they range widely, from associations between vertebrate hosts and free-living organisms, the pathological manifestations of microbial parasitism of intercellular tissues and of the cells themselves, and the phenomena of the carrier state and the latent infection, to the opposite extreme, of a symbiotic state like that of bacterium and lysogenic phage, where even the hereditary apparatus of the two organisms is intimately and stably linked.

Whether we recognize it or not, all the major problems posed by living matter—either by individual organisms or by the interactions of different organisms—arise in our comparatively restricted field of inquiry; and we shall solve them the better the more we can relate them to wider aspects of biology in general.

The ecological task of the medical bacteriologist is in one respect often simple, in that he has to consider the relation of the host tissues to one microbe only; though it is often difficult to reconcile the properties of the microbe as studied in isolation with its behaviour as an infecting agent. It is relatively simple, too, when the microbe is a strict parasite, as with, for example, a virus that infects only a given type of animal cell, because, by and large, the ecological system to be studied may profitably be limited to virus and cell. When, however, a complex microbial flora is in question, as in ecological systems like the intestinal contents, the soil, or the sea, a knowledge of the numbers and the behaviour in isolation of the different microbial constituents of the system may be positively misleading, so profoundly may the complex natural environment of a given microbe affect its behaviour.

It is our purpose in this chapter not to review bacterial ecology in any detail, but broadly to outline the place of bacteria in the world in general; and thereby provide a background against which our detailed exploration of medical bacteriology and immunity may be seen in perspective. For excellent short accounts of general microbiology in relation to microbial ecology we refer the reader to the text-books by Hawker and Linton (1972) and Doetsch and Cook (1973), the second of which pays special attention to the ecological implications of bacterial metabolism.

We have already seen (Chapter 6) the enormous capacity for variation shown by micro-organisms. These fall into two main groups—variations in genetic constitution (see Shepherd 1957) and variations in the response of micro-organisms with a stable genetic constitution to changes in environment (see also Symposium 1961). The simplest situation is one in which a genetically stable strain of micro-organism multiplies in the laboratory in a defined reproducible medium in constant conditions of pH, temperature, aeration and so forth. Even in the laboratory, however, these conditions are seldom realized, though we are constantly striving to get as near to them as possible. But in nature we have to reckon with the complicated interplay of genetic variation and environmental changes and of competition between bacteria, viruses, protozoa, fungi, algae and other organisms. The bacteria concerned are far more varied, in morphology, modes of growth and metabolic specialization than those—mainly rods and cocci—which parasitize warm-blooded animals; and they flourish in physical environments that differ widely in pH, oxidation-reduction potential, salinity, temperature, pressure and so forth. Bacteria in fact prove to have been the most adaptable of all living organisms in their modes of obtaining energy for growth, ranging from the chemo-autotrophic organisms that oxidize, e.g. H_2, CO, CH_4 and S, reduce sulphates, fix nitrogen, through the photosynthetic species, to heterotrophs requiring complex nutrients.

It will be useful to consider first the main sources of micro-organisms in nature. Inevitably we shall be concerned more with bacteria than with viruses, since viruses, though they may survive outside the cell, are incapable of multiplying except in the presence of living cells. Bacteria, on the other hand, are ubiquitous in nature, being plentiful not only in close association with animals and plants but in the land, fresh water, sea-water, and the air. The air, as we relate in Chapter 9, is to be regarded not as a primary source of microbes but as a vehicle for their spread. The two main primary sources are land and water. Since the microbial flora of flowing inland waters is largely derived from and contributes to the flora of the soil, we shall confine our discussion to the soil, fresh water lakes and the sea.

Microbial ecology of soil

About the middle of the last century, little was known of the chemical processes in soil, except that the production of humus—the dark-coloured sticky material in farmyard manure or earth—was due to decomposition of vegetable matter, and that the production of nitrates by soil was demonstrable. Liebig had contended that plant nutrition was dependent on inorganic matter in soil, produced from organic matter by slow oxidation; and Boussingault in France that it depended on nitrification. Schloesing and Müntz (Schloesing 1889, Schloesing and Müntz 1877, 1878, 1879) showed that the production of nitrates from ammonia by preparations of soil *in vitro* could be inhibited by treating the preparations with chloroform vapour, or by boiling, and restored by adding a little fresh vegetable soil—indicating the activity of living organisms. Warington at the Rothamsted Experimental Station distinguished two stages in the conversion of ammonia to nitrate—the oxidation of ammonia to nitrite and nitrite to nitrate; and Winogradsky (1890, 1891) isolated two autotrophic organisms, now known as *Nitrosomonas europoea* and *Nitrobacter winogradskyi*, capable respectively of catalysing these processes *in vitro*. The assimilation of nitrate by plants, and its reconversion to ammonia when the plants decay, and the further nitrification for the use of new plants plausibly explained the cycle of nitrogen usage (Fig. 8.2) in the soil; but did not allow for loss of nitrates washed out of the soil by rain and so forth. Berthelot suggested that nitrates were replenished as a result of the biological fixation of atmospheric nitrogen. Microbial fixation of nitrogen proved to depend on two kinds of activity: one, as Hellriegel and Wilfarth found, exhibited by leguminous plants possessing root nodules, which can grow in nitrate-free soil, and the other, as demonstrated by Winogradsky (1890, 1891, 1925, 1950) and Beijerinck (1888), exhibited by free-living bacteria. The root nodules proved to contain bacteria of the genus *Rhizobium*, which stimulate the cells of the host plant to fix nitrogen. (For accounts of these and other developments of agricultural bacteriology, see Russell 1923, 1957, Winogradsky 1950.)

The soil population In neutral soils, bacteria and actinomycetes are the most abundant of the microflora.

The following figures exemplify the content of one gram for a typical fertile soil: bacteria 2.5×10^9; actinomycetes, 7×10^5; fungi and yeasts, 4×10^5; algae, 5×10^4; protozoa (mainly colourless flagellates and amoebae), 3×10^4. Small numbers of other types of microbes, like the myxomycetes, are also present. The commoner genera include *Achromobacter, Actinomyces, Arthrobacter, Bacillus, Clostridium, Flavobacterium, Micrococcus* and *Pseudomonas*. Heterotrophic species, which include the actinomycetes, outnumber the autotrophic, and are responsible for the decomposition of all types of organic matter in the soil. The soil fungi are present both as spores and as actively growing mycelium; one gram of fertile soil may contain 100 metres of mycelium. The larger size of the fungal cells, in comparison with bacteria, means that in spite of their small numbers, their total mass is of the same order of size as that of the bacteria; one estimate of Rothamsted soil indicated 1500–3500 lb bacteria, 1500 lb fungi but only 150 lb protozoa per acre. The fungi are active decomposers of organic matter, more especially in acid soils. The algae include heterotrophic species, and in surface soils, photosynthetic species.

In the average compact fertile soil, the protozoa and algae are confined to the upper 4–6 in and the fungi to the upper 14 in; the bacteria are most abundant in the upper 2–9 in, but both bacteria and actinomycetes are found as deep as 30 in, the bacteria becoming progressively less abundant with increasing depth.

The figures just given apply to one kind of soil, and represent the average for a relatively large sample. The population varies widely in both numbers and type of microbe according to the nature of the soil. In poor or virgin soils the number may be as low as 10^5/g, and as high as 10^8/g in cultivated soils. The number is dependent on the amount of nutrient available, but single factors like, e.g., acidity or lack of phosphates, may be limiting for some groups. Anaerobic conditions tend to predominate in excess of water; and in acid soils there may be deleteriously high concentrations of inorganic salts. But aside from considerations such as depth, moisture, aeration, temperature, pH, O–R potential and nutrients, there is considerable variation, even within small samples, because of the different micro-environments that obtain in soil.

The soil environment

The grains of soil consist of mineral particles coated with a film of colloid, partly clay, in which most of the microbes lie. These are covered by the soil solution, containing mixtures of salts and dissolved air. Intermixed with these are particles of plant and other remains, each with a rich microbial flora. The whole is bound together in a 'crumb' structure by the soil colloids, including humus, and to some extent by polysaccharide gums of microbial origin. The soil water and gases occupy about half the soil volume. The soil atmosphere between the particles differs from air chiefly in its higher humidity and in containing more CO_2 and less oxygen. To these varieties of environment must be added the regions on and around the rootlets and root-hairs of plants—the *rhizosphere*— which differ from the surrounding soil not only because of the metabolic activities of the roots in taking up nutrients, but also of the metabolic products and secretions of the plants that pass into the soil. As a result, the rhizosphere is a region of high microbial activity by microbes differing both quantitatively—up to 30 times the number in soil proper—and qualitatively from those in the surrounding soil. As an example of the direct influence of the plant on the flora of the rhizosphere, we may note that the amino acids secreted by the plant induce a high proportion of bacteria requiring amino acids for their nutrition. The selection may be even more specific; thus, two species of flax were observed to have strikingly different fungal flora in the rhizosphere—a difference which appeared to be determined by the different antifungal substances secreted by the two plants (Timonin 1941).

The relation between any given organism and its immediate physico-chemical and biotic environment, both internal and external, is frequently referred to as its *ecosphere* or *micro-environment*.

As Thornton (1956) points out, it is against the background of these varied micro-environments that the microbial activity must be considered. The microbial interactions may be mutually favourable or antagonistic. Some of the favourable relations we discuss below in connection with the various 'cycles' whereby nutrients in different stages of oxidation become available for successive utilization by different microbes. The synthesis by one microbe of accessory food substances required by a second is another example. Among the antagonisms, we may note first those in which one organism preys directly on another, as in the feeding of soil amoebae on bacteria and the growth of actinomycetes in fungal mycelium. Second, one organism may secrete substances that either kill or inhibit the growth of others. Some of these substances are lytic; others, loosely referred to as 'antibiotics', and including the bacteriocines (p. 247), kill without lysis. The wide variety of antibiotics active against bacteria and fungi produced *in vitro* by pure cultures of soil organisms, particularly the actinomycetes, suggests that antibiotics may play a part in maintaining the equilibrium between different microbial species in soil. Direct evidence of antibiotic action in natural soils, however, is scanty. Antibiotics have been produced by cultivation in sterile soil of the organism producing them, and deformations of fungal hyphae characteristic of an antibiotic attack have been observed in soils containing actinomycetes known to produce the antibiotic *in vitro*. Effective antibiosis *in situ* would, however, depend on factors such as the optimal supply of nutrients for antibiotic production, and absence of substances or microbes that would otherwise neutralize, adsorb or decompose the antibiotic; and the operation of such factors is largely unknown.

Last, we may note antagonisms due to competition for limiting nutrients; thus, of two organisms requiring the nutrient, the faster growing species is likely to pre-empt it, to the exclusion of the slower growing species.

It is, of course, the aim of certain agricultural procedures to alter the ecological status of a given soil, as in manurial treatments. The addition to soil of selective nutrients can profoundly alter the microbial population. Indeed, on an experimental scale, this technique was devised by Beijerinck for the isolation of soil bacteria. Thus, the addition of cellulose to a soil culture enriches the cellulose decomposers, and the addition of sulphur the sulphur-oxidizing bacteria. The deliberate introduction into soil of useful microbes also has agricultural applications. The outstanding example is the nodule organism *Rhizobium* (see below), which is added to soil to induce nitrogen fixation in leguminous crops; the bacteria find their appropriate ecological niche in the rhizosphere of the legumes.

Decomposition by soil microbes

There appears to be no class of natural organic compounds which is insusceptible to degradation by one or other of the microbes found in soils. Synthetic compounds, such as the weed-killers and insecticides now extensively used in agriculture, often with structures not known to occur in nature, are also attacked. Some are attacked only slowly, but ultimately disappear—a fortunate circumstance that obviates perhaps disastrous accumulations of these substances in productive soils (see Alexander 1965a, b, 1971). The burden put upon the degradative capacity of the soil flora may, nevertheless, be very great. It has recently been estimated, for example, that the foliage and soil in the United States annually receive some 10^9 lb of synthetic pesticides (see Wright 1971, Cripps 1971). Herbicides, especially in amounts that produce extensive defoliation, may irreversibly alter the vegetable organic matter upon whose continuing supply the particular nature of the soil depends. But whether the alterations are reversible or irreversible, herbicides, pesticides and gross pollution with toxic wastes may profoundly disturb the dynamic equilibrium in the flora and fauna of a given soil.

Carbohydrates—The cellulose $(C_6H_{10}O_5)_n$ of plants is probably the most abundant form of organic carbon, and is broken down largely by aerobic and anaerobic bacteria in neutral, and by fungi in acid soils. The first product appears to be the disaccharide cellobiose, which is converted by cellobiase-producing bacteria to glucose. A variety of fungi and anaerobic bacteria then oxidize the glucose, either completely to CO_2 and H_2O, or less completely, e.g., to CH_4, H_2, ethanol and short-chain fatty acids. The soil bacteria also attack the hemicelluloses, the pectins and starches of plants, and the chitins of fungi and of insects.

Proteins—Proteins are destroyed by proteolytic bacteria, heterotrophs like *Clostridium* and *Pseudomonas* spp., and the resulting amino acids and oligopeptides either used by soil flora, or converted to ammonia ('ammonification') by numerous species, among which spore-bearing aerobes, *Pseudomonas*, *Chromobacterium*, *Klebsiella* and certain fungi have been implicated. Ammonia is also formed by the bacterial decomposition of urea and hippuric acid in farmyard manures.

Fats—Oils, fats and waxes are oxidized to CO_2 and water, via fatty acids and glycerols.

Hydrocarbons—In aerobic conditions, virtually all kinds and classes of hydrocarbons appear to be attacked by bacteria and fungi, so that, except in peculiar environments, like those that give rise to accumulation of petroleum, little hydrocarbon persists in soils. The compounds attacked include natural gases, petrols, paraffins, tars, asphalts and rubber (see Zobell 1950).

Lignins—The lignin of wood is one of the many aromatic compounds decomposed by fungi, and to some extent by bacteria. Others include the phenols and cresols, formed in manured soils during the decomposition of proteins.

Synthesis by soil microbes

Nitrogen transformations We noted above the conversion of ammonia to nitrite by the deaminating enzymes of *Nitrosomonas* spp. and of nitrite to nitrate by *Nitrobacter* spp., thus making available nitrate for plant nutrition. Soils also contain a variety of microbes which, especially in poorly aerated, waterlogged soil, convert nitrates into N_2O or N_2; this loss of nitrogen in gaseous form is referred to as denitrification. (For nitrifying organisms, see Wallace and Nicholas 1969.)

The reduction of dinitrogen to ammonia is carried out by an oxygen-sensitive binary enzyme referred to as nitrogenase. Chemically, such a conversion is most unusual requiring, as it does in the Haber-Bosch process, red-hot magnesium or a catalyst at elevated pressure and temperature. Nitrogenase, however, can perform this operation at ordinary temperatures and pressures, even in water under an atmosphere of oxygen—both of which would interfere drastically with reagents in the Haber-Bosch system (Postgate 1978). Nitrogenases extracted from different nitrogen-fixing organisms appear to be very much alike.

Aside from the nitrogen derived, through ammonia, from the breakdown of organic nitrogenous compounds, the soil nutrients are enriched by the fixation of atmospheric nitrogen. This is due to (*a*) free-living and (*b*) symbiotic microbes. Next to the assimilation of atmospheric CO_2 by the higher plants and (see below) by marine algae (Fig. 8.1), the fixation of atmospheric nitrogen is one of the most fundamental biochemical processes in nature. It is estimated that some 10^8 tons of nitrogen are fixed annually, mainly by the action of the root-nodule organisms (Postgate 1974).

(a) Among the free-living microbes, we may note the anaerobe *Cl.. butyricum*, shown by Winogradsky to fix atmospheric nitrogen in pure culture, the *Azotobacter* species of Beijerinck, the blue-green alga *Nostoc*, *Klebs. aerogenes* and certain photosynthetic bacteria. Feeble fixation is demonstrable in a wide variety of other bacteria, including fungi, actinomycetes and pseudomonads. Both the *Clostridium* and the *Azotobacter* species convert the nitrogen, via ammonia, to nitrates. The process is endothermic, the required energy being obtained by the oxidation of carbohydrates and other readily decomposable nutrients, so that fixation does not occur in poor soils. Fixation is also limited when nitrate is readily available in the soil for the use of the nitrogen-fixing bacteria. Which of these many microbes are efficacious in soil is not yet clearly determined. Those that are feebly active *in vitro* may be highly important in nature, either by reason of their numbers, or because they may be very active in association with other soil organisms (see Wilson 1958, Nicholas 1963).

(For further information on nitrite formation, see Cutler and Mukerji 1931; for a short historical review, see Barritt 1933; for a general review of the biology of nitrogen fixation, see Quispel 1974, Postgate 1978; for soil sampling, see Hodgson 1978; and for a taxonomic study of the aerobic nitrogen-fixing bacteria, see Thompson and Skerman 1979.)

(b) The best known example of symbiotic nitrogen fixation is that induced in leguminous plants by *Rhizobium* species. Beijerinck (1888) first isolated *R. leguminosarum* and demonstrated 'infection' of the plants by the organism, and the establishment of a symbiotic association, which later proved to be responsible for fixing nitrogen and supplying the plant with an abundance of nitrates. *Rhizobium* itself cannot fix nitrogen except under special conditions.

The infective cycle is complex. It appears that root products escaping into the rhizosphere attract the minute, free-living mobile rhizobia, and these in turn induce curling of the root hairs by the indole acetic acid they produce from the tryptophan excreted by the plant. Interaction of host and rhizobium results in secretion by the root of a polygalacturonase, which may assist penetration of the root hairs by the bacteria. After penetration the rhizobia grow down the root hair to reach and infect the cells of the cortex—where cell proliferation is induced, with the formation of the nodule. In infected cortical cells the rhizobium multiples, and changes to the bacteroid form—vacuolated, highly pleomorphic and sometimes filamentous rods. The mature nitrogen-fixing system consists of a cluster of bacteroids surrounded by an envelope derived from the host; the active nodule contains haemoglobin. The process of infection is complex. Nutman (1963), for example, discusses 8 recognizable stages in pre-infection, 5 in the invasive phase, and 6 in the intracellular phase. The actual fixation of nitrogen appears to take place in the membranes of the host cells.

Though for many years it was held that *Rhizobium* could not fix nitrogen away from the host plant, this is now known to be untrue. Under special conditions, including careful control of the dissolved oxygen concentration and the presence of some source of fixed nitrogen, it can be grown in liquid medium and shown to fix nitrogen. (See Vincent *et al.* 1980.)

Sulphur transformations Reduced inorganic sulphur compounds in soil are oxidized largely by chemosynthetic, autotrophic sulphur bacteria, of the genus *Thiobacillus*; and sulphates from rotting organic matter are reduced, ultimately to H_2S, by anaerobic *Desulphovibrio* species. By contrast, in the sea, though *Desulphovibrio* species are concerned in sulphate reduction, the photosynthetic green and purple sulphur bacteria are active oxidizers of sulphides (see Fig. 8.3 and below).

Other elements Phosphates are released from organic matter by heterotrophic bacteria, actinomycetes and fungi—a process that is very effective in anaerobic conditions—in the presence of acids formed during carbohydrate breakdown; and from insoluble salts like calcium phosphate by the action of acids formed during the metabolism of soil organisms. They are used directly by other microbes and by plants, and thus made available in organic form for animals.

For further information on the microbial ecology of soil, we refer the reader to the monograph of Russell (1957), the paper by Thornton (1956) and the textbook of Cray and Williams (1971). Four symposia, on microbial ecology (1957), on symbiotic associations (1963) and on soil ecology (1967a, b) contain reviews of many important aspects of the subject. (See also Fenchel and Blackburn 1979.)

Microbial ecology of water

We shall discuss (Chapter 9) in some detail the microbial content of different waters—mainly as regards species that can be grown on the media commonly used by medical bacteriologists—and its fluctuations with physical and biological alterations of the environ-

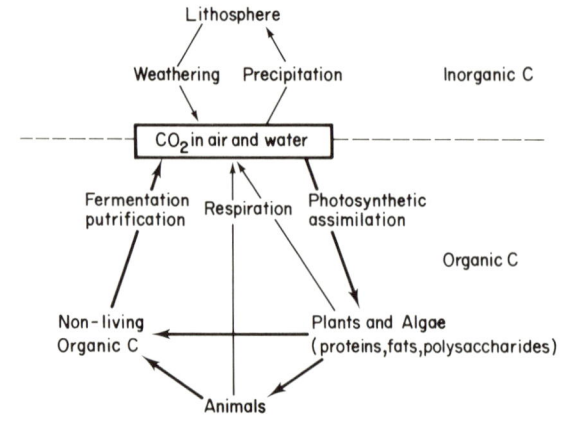

Fig. 8.1 Scheme for a biological carbon cycle (modified from Doetsch and Cook 1973).

ment. Here we may indicate the relation of water bacteria to the biological activities of water as an ecological system, exemplified in large bodies of fresh water and in the sea. There are many points of general resemblance in the economy of microbial communities on land and in water. The most striking difference lies in the type of organism that takes advantage of solar energy. The chief photosynthetic process, whereby CO_2 is transformed by an endothermic reaction into stores of energy such as sugars, fats and starches, and oxygen is released, is carried out on land largely by the higher plants, and to a very small extent by algae. In water, the situation is reversed. The photosynthetic contribution of, for example, seaweeds is small, whereas that of the microscopic marine algae is predominant; not only in the sea but, in terms of activity, in the world as a whole; the CO_2 assimilated in the oceans is estimated to be 8 times greater than that fixed by all land plants—the process being probably the largest single chemical process carried out on earth. The primary production of food in fresh and sea-water depends on the phytoplankton of the upper layers, which consists of floating, motile microbes. These, with the zooplankton (protozoa and small animals), constitute one important ecological system, the *plankton*. The phytoplankton is in fact the basis of ocean life. In well populated regions, the plant life under a given area of sea is estimated to be greater than that on the same area of tropical forest; and in terms of capacity to support stock, in the agriculturist's sense of the word, the 'productivity' of the sea is probably the higher of the two. Phytoplanktonic organisms, by reason of their small size, and therefore large surface: volume ratio, are well adapted to take full advantage of the relatively dilute nutrients and the moderate intensities of light that obtain in aquatic habitats.

The second ecological system is the *benthos*, communities in and near the surface of the beds of rivers, lakes and oceans; here, by reason of its depth and therefore the absence of light, there is little or no photosynthetic activity.

Provided that there was no major drain on the biological energy it contained, a large water community like the ocean could maintain itself for a considerable period, through the cycles of synthesis from inorganic matter by 'plant' forms, maintenance of an animal population, microbial degradation of dead organisms of all kinds, and re-synthesis from the degradation products. But land communities, especially man, remove fish at an estimated rate of the order of 5×10^7 metric tons annually, to produce which some 5×10^{10} metric tons of phytoplankton are needed. In fresh-water and estuarine communities this loss is in part compensated for by organic matter of terrestrial origin, but in the oceans, without the energy accumulated by photosynthesis, the loss would grossly deplete ocean communities. It is a moot point today whether the annual loss by fishing exceeds the annual gain by photosynthesis; it is indeed possible that ocean communities are remaining at their present abundance because they are living on reserves of photosynthetic energy accumulated in the remoter past (Wood 1958).

The photosynthetic algae—diatoms, blue-green algae and flagellates—are found at considerable depths in the sea, but are concentrated mainly in the surface layers. In high latitudes photosynthesis does not occur at depths greater than 100 m or below 300–400 m in low latitudes. This 'photic' zone may, of course, be much shallower, depending on the amount of light reaching the water, and the presence of light-absorbing matter, inert or living, in the water. In the plankton, the bacteria, like the fungi and protozoa, are in greatest numbers where organic matter is richest; that is, their distribution tends to coincide with with that of the algae; and is subject therefore to the fluctuations of temperature, light, nutrients—seasonal or other—that determine the algal populations. The bacteria include autotrophic and heterotrophic species, the latter abundant in water rich in organic matter. Aside from intestinal bacteria, fresh-water plankton contains *Pseudomonas, Flavobacterium, Chromobacterium, Achromobacter* and *Micrococcus* species, and actinomycetes, spore-bearing bacilli and nitrifying bacteria from soil. The species in the sea include *Pseudomonas, Flavobacterium, Achromobacter, Vibrio*, sulphur bacteria and actinomycetes. The bacteria, fungi and protozoa of the sea have the same kind of inter-relations and play much the same part in the transformations and conversions of nitrogen, sulphur, phosphorus and carbon compounds as they do in soil.

As in soil, there is likewise some evidence that competition between the different planktonic organisms may in part be determined by extracellular 'antibiotic' substances.

In benthic communities, algae are absent, and anaerobic bacteria and fungi predominate, being specially abundant in and just above fresh-water and sea muds, and in shallow waters round vegetation.

Fresh waters are more subject to environmental fluctuations than the oceans, and their flora is consequently the more qualitatively variable. The species present are determined by ionic balance and by pH; acid and alkaline waters have strikingly different florae. Marine waters, on the other hand, vary little in salinity and pH, and the flora is more uniformly salt-tolerant. The diatoms and bacteria of deep-sea benthic communities are adapted to life at high pressures; and cannot live at lower pressures. (For further reading, see Harvey 1955, Hardy 1956, Wood 1958, Hawker and Linton 1972, Doetsch and Cook 1973.)

The cycle of the elements in nature

It will be evident from the foregoing brief sketches of the microbial ecology of soils and waters that

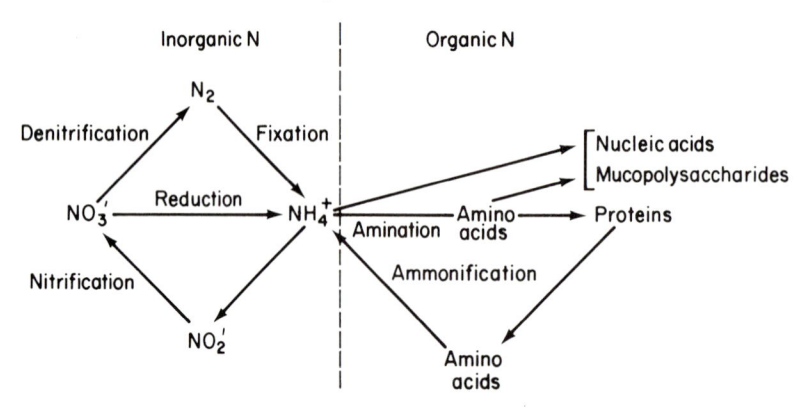

Fig. 8.2 Scheme for a biological nitrogen cycle (modified from Doetsch and Cook 1973).

autotrophic organisms constitute the starting point of the 'food-chains' that ultimately reach the higher plants and animals; and that microbes are responsible for the degradation of dead organic matter to the point where the products of degradation are available for use again by living organisms. It is customary, and useful, to represent the passage of the chief elements in this process as cycles. Thus, carbon in its fully oxidized form is reduced by photosynthetic organisms, and converted to oxidizable forms like cellulose, starch and sugars, and, in combination with compounds of nitrogen, phosphorus, sulphur and so forth, used for the synthesis of cellular matter. The carbon of living matter is oxidized to CO_2 during the metabolism of the living organism, but the greater part is produced by saprophytic microbes after death of the organism. The diagrams of some of these cyclic (Figs. 8.1, 8.2, 8.3) processes broadly indicate the interrelations of the biological changes mediated by living organisms in whose metabolism the organic and inorganic forms of the elements are implicated, but they are misleadingly over-simplified in a number of respects.

The ecological systems of regions like soil or the ocean are complexes of highly interdependent organisms, many of them capable of a wide variety of biochemical activities; so that widely different species may each be concerned in a given stage of a cycle. Moreover, each stage of a cycle for a certain class of compound is ultimately bound up with the metabolism of other classes of compounds. To take a simple example, the formation of H_2S in sea muds by desulphovibrios makes phosphate available for the benthic community by releasing it from the ferric phosphates present there. Again, the cycle as depicted suggests a steady succession of activities by the organisms concerned. It is truer to picture an ecological system as simultaneously containing all the components necessary for the transformations in question, the component needed for a given stage becoming dominant as dictated by the complete environmental conditions of the system, which do not necessarily change so as always to induce completion of the cycle. Lastly, the cycles illustrate the movement of relatively abundant elements. Although the evidence is scantier for other elements, analogous cycles are postulated for phosphorus, iron, calcium, potassium, magnesium, zinc, copper, molybdenum, cobalt and boron; and equally good cases can be made out for hydrogen and oxygen. Indeed, any molecular transformation regularly involving more than one organism can be accorded the dignity of cycle; and the elucidation of the cycle even of trace elements essential to life—especially if they are contained in limiting nutrient concentrations—may be of biological importance.

The inorganic part of the biological carbon cycle is chiefly a matter of the release of carbon from carbonates like limestone and shell marls, and its return to mineral form by deposition—as occurs, for example, in marine sediments and in corals. The main lines of the organic part concern photosynthetic assimilation by plants and marine algae, and the degradation of both plant and animal carbon compounds to CO_2; chiefly by bacterial action. To the cycle as depicted in Fig. 8.1 might be added a relatively small assimilation of CO_2 by chemo- and photosynthetic bacteria; and a segregation of organic carbon in the form of coals and oils, whose formation appears to depend on the deposition of organic matter in conditions that preclude extensive microbial degradation, and its return to the atmosphere as CO_2 on combustion.

It will be noted that the carbon cycle implies an oxygen cycle, since oxidation of organic C to CO_2 necessitates the reduction of O_2 to water, and the photosynthetic reduction of CO_2 the release to the atmosphere of free O_2.

Since the blue-green algae and the green plants also produce large amounts of protein and other organic nitrogen compounds, photosynthesis may be regarded as a means of converting N_2 as well as C into organic forms. The photosynthetic organisms use nitrates and ammonium compounds in amination and other reactions to form proteins, nucleic acids and muco-

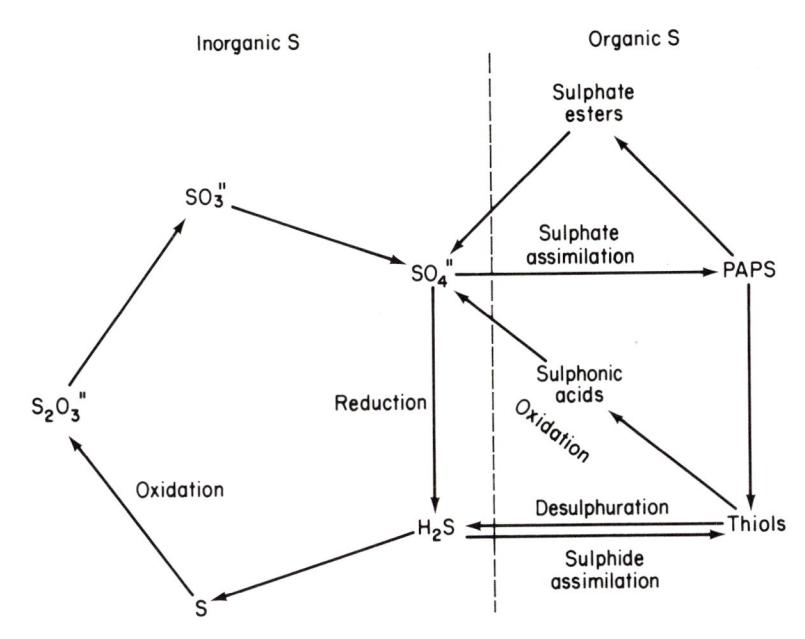

Fig. 8.3 Scheme for a biological sulphur cycle (modified from Doetsch and Cook 1973).

polysaccharides which, in plant or animal form, are in turn microbially degraded to ammonia (Fig. 8.2). In the sulphur cycle (Fig. 8.3), sulphate may be regarded as the starting point of organic assimilation—though H_2S is used for organic syntheses by some organisms—resulting in the formation of phosphoadenosine phosphosulphate (PAPS), from which the sulphur is ultimately reduced to the stage at which it is incorporated in thiol compounds—the sulphur-containing amino acids—and compounds like biotin and thiamine. In the putrefactive part of the cycle the plant and animal organic sulphur is mineralized via H_2S, which then undergoes microbial oxidation to S, thiosulphate, sulphite and finally sulphate. (For the sulphate-reducing bacteria, see Postgate 1979.)

Other ecological associations

The microscopical flora and fauna of habitats like the soil and the sea clearly form ecological associations in which each of a number of the species constitutes a significant part of the environment of organisms of other species. That is to say, there is a mutual selective action between at least some of the living components of the association. The association is loose, in that the mutual selectivity is variably expressed, the balance between components being grossly altered by changes in the environment. The ecological niche provided by soil or the sea is obviously variable, and imposes both quantitative and qualitative variability on its contained populations. There is nevertheless a real sense in which these associations can be called symbiotic, in that at least some of the species live together in close spatial and physiological relationship, and in doing so may constitute a unit which, though variable in its individual components, is stable enough to have characteristic structure and biochemical activity. In a less inherently variable environment, like the interior of the abomasum of a ruminant animal (see p. 237), the structure and activity of the microbial association, though varying in detail, are much more stable and the microbes are more strictly adapted to life in that particular ecological niche. The various species not only display symbiotic relations between themselves, but the population as a whole is in evident symbiosis with the ruminant whose gut it inhabits. The animal takes in the vegetable food and provides a stable environment for its rumen population; and the population breaks down the food to provide the animal with its essential nutrients. This is an example of mutualism in symbiosis, a reciprocal dependence of microbes and animal (see Hungate 1966). The population of microbes may indeed be regarded as a ruminant organ—or 'quasi-organ'—for the digestion of cellulose. The relation between microbes and ruminants is in some ways analogous to that between plants and the microbes of the rhizophere.

With closer physical association, as occurs when the microbes live in the tissues or inside the cells of the plant, the environment of the microbes is more restricted, and the association is the more specific—that is to say, the mutual selection imposed by the partners is the more strict. The association of legumes and *Rhizobium* spp. discussed in connection with nitrogen fixation is one good example of a specific form of mutualism. Another is that between plants and mycorhizal fungi; these associations range from that between mycorhizae and the roots of forest trees, where there is but little penetration of the plant tissue by the fungus, to the intimate symbiosis of certain fungi in the intercellular spaces and in the cortical cells of certain orchids which, being without chlorophyll, are incapable of photosynthesis. In all these instances there is evidence, either direct or indirect, of mutual benefit. The photosynthetic algae are widespread intracellular symbionts with protozoa and coelenterates and in some instances the alga appears to be necessary to the host. In others the relation is variously interpreted as parasitism, one partner benefiting at the expense of the other; or as commensalism, one partner benefiting without damaging the other.

The distinction between these various kinds of symbiotic association is by no means clear cut; nor is it in the interests of biological description that it should be so. The adaptation of two organisms to some form of common life is highly unlikely to have occurred as the result of purely favourable mutual selection. The most apparently harmonious partnership probably represents the end-result of a long process of trial and error in which variants of each member were destroyed to suit the requirements of the other. Moreover, the antagonisms, as well as the common benefits, demonstrable in the looser, less specific associations of free-living forms, are probably operative, though not readily demonstrable, in the more specific instances of mutualism. In those associations of a fungus and an alga that have resulted in the wide variety of lichens that populate the world—the classical example of symbiotic mutualism—there are features which suggest that one partner is parasitic on the other.

The elucidation of the interactions in a given symbiotic relationship, determining which of the activities of the partners are essential to the stability and specificity of the association and which are accidental, is an exceedingly complex task. We have in this chapter touched on only a few aspects of a universal phenomenon; of which associations between one microbe and another, or between microbes and the more highly organized plants and animal, are but a small part. Microbial associations are the concern of the agricultural, aquatic and industrial microbiologists, of those interested in the biological and evolutionary significance of parasitism in general, and of the plant and animal pathologists (cf. Symposium 1963, Hawker and Linton 1972). There are numerous instructive examples of the phenomenon to be found in all these microbiological disciplines; but few instances of the complete elucidation of any one of them.

Pathogenicity of soil and water bacteria

Most organisms giving rise to disease are believed to have come originally from the soil and gradually adapted themselves to growth in the human or animal body. Some of these organisms have remained harmless commensals or symbionts; others have gone further and adopted a pathogenic role. There are, however, species that still inhabit soil and water which, when given the opportunity of gaining access to the body, can cause disease, often severe and fatal. For example, *Clostridium tetani* and the gas-gangrene group of anaerobic spore-bearing bacteria in the soil may contaminate wounds; so also may the halophilic vibrios in sea-water. *Clostridium botulinum* and the halophilic vibrios may be ingested with food. Inhalation of humidifier water may enable *Legionella pneumophila* to invade the lungs. In hospitals, soil and water bacteria may multiply in medicaments, catheters, urinals and other articles, and then be injected, infused, or inhaled. In sterile tissues they meet with no competition; in others they may overgrow the normal flora, especially in patients on antibiotic treatment, or whose immunological state is defective. These organisms are passed from man to man only in exceptional circumstances, and generally in populations heavily exposed to antibiotics (Chapters 58 and 61).

Our concern in this book is for the special case of parasitism—either deleterious or lethal—of man and some of the higher vertebrates by bacteria and viruses. Only incidentally do we consider the wider aspects of microbial ecology, and then mainly in order to understand the epidemiology of parasitic events in the individual animal. In its earlier days, medical bacteriology was largely orientated towards the prevention and cure of microbial disease. It is still very properly so orientated. Nevertheless, as medical bacteriologists and virologists, we have in more recent years acquired a greater and more sophisticated insight into our problems, and a greater versatility in our ways of attacking them. From the standpoint of general biology, they range widely, from associations between vertebrate hosts and free-living organisms, the pathological manifestations of microbial parasitism of intercellular tissues and of the cells themselves, and the phenomena of the carrier state and the latent infection, to the opposite extreme, of a symbiotic state like that of bacterium and lysogenic phage, where even the hereditary apparatus of the two organisms is intimately and stably linked.

Whether we recognize it or not, all the major problems posed by living matter—either by individual organisms or by the interactions of different organisms—arise in our comparatively restricted field of inquiry; and we shall solve them the better the more we can relate them to wider aspects of biology in general. (For some basic principles of human ecology, see Stapledon, G. 1971; for a general review of microbial ecology, see Lynch and Poole 1979, Wolin 1979; for a contemporary review, see Ellwood *et al.* 1980; for advances in microbial ecology, see Alexander 1979; and for the methods of experimental ecology, see Aaronson 1970.

References

Aaronson, S. (1970) *Experimental Microbial Ecology.* Academic Press, London and New York.
Alexander, M. (1965a) *Advanc. appl. Microbiol.* **7**, 35; (1965b) *Proc. Soc. Soil Sci.* **7**, 29; (1971) *Microbial Ecology.* Wiley, New York.
Alexander, M. (Ed.) (1979) *Advances in Microbial Ecology,* Vol 3. Plenum Press, New York.
Barritt, N. W. (1933) *Ann. appl. Biol.* **20**, 165.
Beijerinck, M. W. (1888) *Bot. Ztg.* **46**, 724, 740, 756, 780, 796.
Cray, T. R. T. and Williams, S. T. (1971) *Soil Micro-organisms.* Oliver and Boyd, Edinburgh.
Cripps, R. E. (1971) In: *Microbial Aspects of Pollution.* p. 255. Ed. by G. Sykes and F. A. Skinner, Academic Press, London.
Cutler, D. W. and Mukerji, B. K. (1931) *Proc. roy. Soc. B* **108**, 384.
Doetsch, R. N. and Cook, T. M. (1973) *Introduction to Bacteria and their Ecobiology.* Medical and Technical Publishing Co., Lancaster.
Ellwood, D. C., Hedger, J. N., Latham, M. J., Lynch, J. M. and Slater, J. H. (1980) *Contemporary Microbial Ecology.* Academic Press, London.
Fenchel, T. and Blackburn, T. H. (1979) *Bacteria and Mineral Cycling.* Academic Press, London.
Hardy, A. C. (1956) *The Open Sea.* Collins, London.
Harvey, H. W. (1955) *The Chemistry and Fertility of the Sea Water.* Cambridge University Press, Cambridge.
Hawker, L. E. and Linton, A. H. (1972) *Micro-organisms: Function, Form and Environment.* Arnold, London.
Hodgson, J. M. (1978) *Soil Sampling and Soil Descriptions.* Clarendon, Oxford University Press.
Hungate, R. E. (1966) *The Rumen and its Microbes.* Academic Press, New York; (1966) *The Rumen and its Microbes.* Academic Press, New York; (1975) *Annu. Rev. ecol. Syst.* **6**, 39.
Lynch, J. M. and Poole, N. J. (1979) *Microbial Ecology. A Conceptual Approach.* Blackwell Scientific Publications.

Nicholas, D. J. D. (1963) *Symp. Soc. gen. Microbiol.* **13**, 93.
Nutman, P. S. (1963) *Symp. Soc. gen. Microbiol,* **13**, 51.
Postgate, J. R. (1974) *J. appl. Bact.* **37**, 185; (1978) *Nitrogen Fixation.* Edward Arnold, London; (1979) *The Sulphate-reducing Bacteria.* Cambridge University Press, Cambridge.
Quispel, A. (Ed.) (1974) The Biology of Nitrogen Fixation. In: *Frontiers of Biology,* Vol. 33, p. 746. North-Holland Publishing Co., Amsterdam.
Russell, E. J. (1923) *The Micro-organisms of the Soil.* Longmans, Green, London; (1957) *The World of Soil.* Collins, London.
Schloesing, T. (1889) *C. R. Acad. Sci.* **109**, 210, 423, 883.
Schloesing, T. and Müntz, A. (1877) *C. R. Acad. Sci.* **84**, 301; *Ibid.* **85**, 1018; (1878) *Ibid.* **86**, 892; (1879) *Ibid.* **89**, 891, 1074.
Shepherd, C. J. (1957) In: *Microbial Ecology,* p. 1. Cambridge University Press.
Stapledon, G. (1971) *Human Ecology,* 2nd edn. Chas. Knight & Co. Ltd, London.
Symposium. (1957) *Symp. Soc. gen. Microbiol.* **7**, 1–388; (1961) *Ibid.* **11**, 1–416; (1963) *Ibid.* **13**, 1–343; (1967a) *Soil Biology.* Ed. by A. Burges and F. Raw, Academic Press, New York; (1967b) *The Ecology of Soil Bacteria.* Ed. by T. R. C. Gray and D. Parkinson, Liverpool University Press.
Thompson, J. P. and Skerman, V. B. D. (1979) *The Taxonomy and Ecology of the Aerobic Nitrogen-fixing Bacteria.* Academic Press, London.
Thornton, H. G. (1956) *Proc. roy. Soc. B* **145**, 364.
Timonin, M. I. (1941) *Soil Sci.* **52**, 395.
Vincent, J. M., Nutman, P. S. and Skinner, F. A. (1980) *Identification Methods for Microbiologists.* Academic Press, London.
Wallace, W. and Nicholas, D. J. D. (1969) *Biol. Rev.* **44**, 359.
Wilson, P. W. (1958) *Handbuch der Pflanzenphysiologie.* **8**, 9. Springer, Berlin.
Winogradsky, S. (1890) *Ann. Inst. Pasteur* **4**, 213; (1891) *Ibid.* **5**, 92, 577; (1925) *Ibid.* **39**, 299; (1950) *Nature, Lond.* **165**, 826.
Wolin, M. J. (1979) In: *Advances in Microbial Ecology.* Vol. 3, p. 49. Ed. by M. Alexander. Plenum Press, New York and London.
Wood, E. J. F. (1958) *Bact. Rev.* **22**, 1.
Wright, S. J. L. (1971) In: *Microbial Aspects of Pollution.* p. 233. Ed. by G. Sykes and F. A. Skinner, Academic Press, London.
Zobell, C. E. (1950) *Advanc. Enzymol.* **10**, 443.

The normal bacterial flora of the body

Normal bacterial flora of the body	230	Intestinal microbes and resistance	
Introductory	230	to infection	236
Normal flora of the alimentary tract	231	Intestinal microbes and nutrition	237
Mouth	231	Normal flora of the respiratory tract	237
Intestine	232	Normal flora of the vagina	239
The importance of micro-organisms in the intestine	234	Normal flora of the urethra	240
		Normal flora of the skin	240
Intestinal toxaemia	235	Bacteria in the blood and	
The effect of antibiotics	235	internal organs	242

Introductory

From considering microbial ecology in general we pass on now to bacterial ecology in particular. Qualitative, and still more quantitative, assessment of the flora in any given part of the body is beset with difficulties. In the first place the variety of metabolic requirements—among which we must include nutrients, optimum temperatures and suitable atmospheres—of the different species in a mixed flora means that no one medium can be expected to detect all the viable microbes present. Moreover, less abundant species cannot be detected at all unless selective media are used to inhibit the growth of the predominant microbes. Thus for a broad survey of the distribution of the main types of bacteria in the intestine, in which identification is carried no further than genera or groups of related genera, Haenel (1960) used no fewer than eleven quantitative cultural tests for each specimen.

In the second place, whether the cultural methods are extensive or restricted, the definition of 'normal' flora arrived at depends on the adequacy of each sample examined as fully representative of the flora, and the extent to which the individual animals examined represent the average 'normal' animal living in a given community under certain environmental conditions. A 'normal' flora in one community may be abnormal in another consuming a different diet, or exposed to a different climate.

There are certain bacterial species which, so far as our information goes, are constantly present in particular situations. Thus, *Esch. coli* is a normal inhabitant of the intestine in all parts of the globe; α-haemolytic streptococci are always present in the nasopharynx, and *Staph. epidermidis* on the skin. There are, however, species whose range is limited in space and time. Many of these are potential pathogens, such as the pneumococcus, the meningococcus, the influenza bacillus, and haemolytic streptococci among the nasopharyngeal flora; but it is probable that many non-pathogenic species have a similarly localized distribution. If our knowledge was more extensive than it is, we should probably be able to describe a basal flora, characteristic of mankind under all conditions; a supplementary flora, varying in frequency, but ranging widely as normal parasites of man; and various species showing a restricted range or a temporary prevalence, conditioned by local or transient environmental factors. This aspect of bacterial ecology has, however, received but little attention, except with regard to the incidence of a few important pathogenic species. The intestinal flora of man, for example, to some extent varies with diet, gastric acidity, degree of peristalsis, external temperature and so on; in the upper part of the small intestine it is more copious in persons after gastrectomy than in normal persons. Again, the nasopharyngeal flora is constantly changing both in the same individual and in the same community. At one time, pneumococci and *H. influenzae* may be present in large numbers in apparently healthy persons, whereas at another time they may be comparatively uncommon. It must be realized, therefore, that it is not at present possible to give a complete description of the microbes that may be found in different parts of the body, or to indicate more than roughly their frequency and relative proportions.

The species found in the regions of the body which are normally colonized by bacteria include those generally accepted as pathogens, those generally accepted as saprophytes, and many that are intermediate. We know very little of the nature of the association between the bacterium and the healthy host tissue in any of these three groups. We may regard all typical members of a pathogenic species as virulent, and postulate a high level of immunity, perhaps specific, in order to explain the prevalence of the organism in the normal flora; or we may postulate a wide variation of virulence in the species, and assume a lesser virulence for the strains found in healthy tissue. And by analogy with a large number of instances of biological associations variously described as parasitism, saprophytism, commensalism, symbiosis or mutualism, we may expect that the associations of saprophytes and healthy tissue cells are equally dynamic. (For a discussion of the multiple factors affecting bacterial virulence, see Wilson 1957.)

Moreover, equilibria exist not only between host and 'normal' flora, but between the different species within the flora; and the association of various bacterial species on a given surface of the body is relevant to any discussion of infection. It cannot be too strongly emphasized that, except perhaps in the immediately newborn animal, none of those regions of the body to which bacteria can gain entrance by the normal portals provides a virgin soil in which any newcomer may flourish. It is just as impossible to ensure the proliferation of a particular bacterial species by introducing it into the mouth, as it is to ensure a crop of a particular plant by scattering seeds in a field already occupied by a pre-existing plant association. The newcomer will have little chance of survival, unless it is adapted to occupy some definite place in its new environment. It is fairly certain that those pathogenic bacteria which spread readily from host to host owe their capacities in this direction to their ability to colonize on skin or mucous surfaces, or to escape from the superficial environment to the underlying tissues. This aspect of the problem, in relation to the upper part of the respiratory and alimentary tracts, was investigated in some detail by Bloomfield and his colleagues (Bloomfield 1919, 1921, 1922a, b, c, d, e, Bloomfield and Felty 1923a, b, c, Felty and Bloomfield 1923; see Chapter 90 of 6th edition; and Savage 1972).

Many workers have studied the development of the natural flora of the intestine in rats and mice, either in ordinary animals or in germ-free animals in which various organisms have been implanted. The results indicate that different organisms find the environment that is most congenial to them, and establish themselves so firmly that it is difficult or impossible to replace them except by drastic measures (see Gibbons et al. 1964, Schaedler et al. 1965, Ducluzeau and Raibaud 1968, Savage et al. 1968, Syed et al. 1970, Mushin et al. 1970, Gibbons and van Houte 1971, Savage 1972, Skinner and Carr 1974).

Normal flora of the alimentary tract

The bacteria present in the **mouth** are subject to great variation both in number and in kind. Even the fairly clean and healthy mouth contains a considerable amount of detritus and other organic matter derived from particles of food, desquamated epithelium, pharyngeal mucus, and other sources; these provide nutriment for a diverse flora which is of necessity undergoing frequent change. Moreover, inflammatory processes of the teeth and gums are often present, favouring a predominance of certain gram-positive cocci, neisseriae, fusiform bacilli and H_2S-producing organisms (see, e.g., Richardson and Jones 1958). What adds to the complexity is the change in the micro-environment within the mouth. Thus the conditions differ between those on (a) the lips, cheek and palate; (b) the tongue; (c) and the tooth surface, whether buccal and lingual, gingival, approximal, or occlusal with its pits and fissures. These conditions vary not only in stages from aerobic to anaerobic, but also in the milieu provided for the metabolic activity of the different micro-organisms. Interactions between the numerous species of bacteria are constantly present, some of them favourable, others unfavourable.

The **saliva** as initially secreted appears to be sterile. It contains substances that promote the growth of some species, and inhibit that of others. They include immunoglobulin, lysozyme, lactoferrin, lactoperoxidase, and the tetrapeptide sialin, which leads to the formation of alkali (see Umemoto 1961, Bowden et al. 1979). The saliva is rapidly colonized by organisms that belong, according to Russell and Melville (1978), to the following genera: *Staphylococcus*, *Micrococcus*, *Lactobacillus*, *Corynebacterium*, *Actinomyces*, *Clostridium*, *Propionibacterium*, *Escherichia*, *Proteus*, *Pseudomonas*, *Klebsiella*, *Rothia*, *Bacterionema*, *Bacteroides*, *Pasteurella*, *Campylobacter*, *Bacillus*, *Fusobacterium*, *Selenomonas*, *Spirillum*, *Treponema*, *Borrelia*, *Leptotrichia*, and *Mycoplasma*.

On the **teeth** streptococci are prominently represented, and are concerned in the formation of dental plaque. This consists of bacteria embedded in an organic matrix derived partly from salivary glycoprotein and partly from microbial extracellular polymers, such as dextran. Certain streptococci, notably *Str. mutans* and *Str. sanguis*, form an insoluble dextran from sucrose, and become more firmly attached to the tooth surface than do most other organisms. Plaque varies in its microbial content according to age; and is strongly associated with dental caries and periodontal disease (see Ch. 57). The earliest colonizers—streptococci and neisseriae—are soon outnumbered by gram-positive rods and filaments, including actinomycetes. Later, gram-negative rods and spirochaetes appear. Anaerobic bacteria and spirochaetes flourish in the gingival crevice region. In the presence of periodontal disease *Bacteroides* and *Fusobacterium* abound (Duerden 1980a; see Ch. 26). Lactobacilli are more frequent in the presence of caries than in its absence. Streptococci belong mainly to the α-haemolytic and non-haemolytic groups; but β-haemolytic streptococci are generally found in 5-10 per cent of swabs from healthy throats. Yeasts, chiefly *Candida* *albicans*, may be present in the healthy mouth, and Vincent's spirillum, usually accompanied by fusiform bacilli (see Brooke 1938). Other organisms include *Bifidobacterium* and *Nocardia*. The **tongue** has its own flora, the most prominent members of which are *Str. salivarius* and *Veillonella*. With the enormous variety of constantly changing micro-organisms, it is not surprising that Russell and Melville (1978) refer to the oral cavity as a 'complex continuous culture vessel'. Nevertheless, the implantation of foreign organisms is very difficult, as it is likewise in the gastro-intestinal tract lower down. The rarity of infections of the salivary glands is probably due, in part, to the flushing action of the saliva as it is secreted, and (Bloomfield 1922c) the constant movement, dependent apparently on suction currents, of mouth bacteria towards the oesophagus. (See Chapter 90 of 6th edition.)

Most organisms normally present in the mouth are sensitive to penicillin and many other antibiotics, and are virtually eliminated by their use. They are replaced by pre-existent streptococci (Phillips *et al.* 1976), and often by coliform bacilli, pseudomonads, and yeasts (Long 1947).

Repeated sampling shows that the proportion of species in saliva remains remarkably constant for a given individual (Kraus and Gaston 1956, Richardson and Jones 1958). As regards numbers, the following estimated mean counts per ml saliva from ten persons are recorded: cultivable, on optimal medium, 6.3×10^7; lactobacilli, 2.5×10^7; 'proteolytic' organisms, 1.2×10^7; *Veillonella* and *Fusobacterium* spp. each 4×10^6; 'staphylococci,' 6.3×10^3; 'proteus,' 30 (Haenel *et al.* 1958a).

The *infant's* mouth becomes colonized shortly after birth, staphylococci, streptococci, lactobacilli and coliform bacteria being readily detectable; within a few days, the flora is largely that of the adult, except for streptococci, which appear only after dentition. The maternal vagina, and then the upper respiratory tract, appear to be the source of the colonizing organisms (see Witkowski 1935).

Eruption of the teeth in early life heralds the establishment of *Streptococcus sanguis*. With increasing age *Bacteroides melaninogenicus* and spirochaetes appear. (For papers and reviews of the normal flora of the mouth, see Bisset and Davis 1960, Socransky and Manganiello 1971, Socransky *et al.* 1977, Skinner and Carr 1974, Hardie and Bowden 1974, Russell and Melville 1978, Sanyal and Russell 1978, Medical Research Council Review of Dental Research 1977, Bowden *et al.* 1979).

The **intestine** at birth contains at most a few bacteria; it is colonized rapidly, *per os* and to some extent *per anum*. Schild (1895) demonstrated organisms in the meconium sometimes within 4 hours of birth, and usually within 10–17 hours; disinfection and protection of the anus delayed colonization until after 20 hours, when food bacteria appeared in the stools. Others have cultivated bacteria, from a proportion of the stools examined, a few hours after birth (e.g. Snyder 1936; see also Haenel 1961).

The intestinal flora of the *breast-fed infant* consists largely of anaerobic bacilli of the *Bifidobacterium* group. They may constitute 99 per cent of the total viable organisms in the faeces (Cruickshank 1925). The predominance of *B. bifidum* has been confirmed by many workers (see, e.g., Snyder 1940, Dehnert 1957a, b, Haenel *et al.* 1970). Seeliger and Werner (1962) record counts of 10^{10} per g of stool, as compared with 10^9 in adults, and a predominance of 3- to 300-fold over the remainder of the flora. Coliforms, enterococci, staphylococci and aerobic lactobacilli are also present. *Staph. aureus* may occur in small numbers (Crowley *et al.* 1941, Martyn 1949, Buttiaux and Pierret 1949; see also Williams 1963), probably derived from the infant's nose. Among the anaerobic flora, *Bacteroides fragilis* is usually present in numbers almost equal to those of *Bifidobacterium* (Rotimini and Duerden 1981). During the first six weeks of the infant's life the pH of the faeces is given by Bullen and his colleagues (1977) as 5.1–5.4. (For fuller information on the anaerobic flora of the mouth, see Ch. 26.)

With weaning, or in *bottle-fed infants*, the flora tends to resemble that of the adult. The number of anaerobic lactobacilli declines slightly, bacilli of the *Fusobacterium* group and 'proteolytic' bacilli (capable, e.g., of digesting and blackening an egg-white liver broth) appear in large numbers (see Haenel 1960, 1961); and small numbers of aerobic and anaerobic spore-bearing organisms. The *faeces* are more alkaline than those of breast-fed infants ranging from pH 5.9 to 8.2 (Bullen *et al.* 1977).

In the *adult*, numerous workers have found that the empty **stomach** is generally sterile. Immediately after a meal it contains numerous organisms which have been ingested with food; but these, with the exception of acid-resistant vegetative bacilli and sporing bacteria, appear to be killed off rapidly. If, however, the motility of the stomach is excessive, or the acidity is below normal, this sterilizing effect of the gastric juice is probably incomplete. Thus in cases of gastric disease—particularly of carcinoma—sarcinae, saprophytic bacilli, and other organisms may multiply in the stomach (Goodsir 1842, Oppler 1895). In the healthy adult the **jejunum and upper ileum** are practically sterile (Cregan and Hayward 1953). The number of organisms—mainly facultatively anaerobic viridans streptococci, staphylococci, lactobacilli, and fungi—increases from the stomach to the ileocaecal valve, beyond which the flora becomes much more abundant and qualitatively different. The **duodenum** may contain 100–1000 organisms per ml, the jejunum 1000–10 000, the upper ileum about 100 000, and the lower ileum 1 000 000; but there is much variation from sample to sample (Gorbach *et al.* 1967b). Coliforms appear in the jejunum in certain diseases, such as infantile gastro-enteritis (Thomson 1955), hepatic cirrhosis

(Martini *et al.* 1957), after gastrectomy, and in chronic debilitated states (Haenel and Müller-Beuthow 1958, Haenel 1961). The flora of the lower part of the ileum is relatively scanty, but includes most of the organisms found in abundance in the large intestine (see Niszle 1928, Cregan and Hayward 1953, Haenel and Müller-Beuthow 1958). The flora of the **large intestine and faeces** comprises a variety of organisms, such as coliform bacilli, enterococci, staphylococci, aerobic and anaerobic spore-bearing bacilli, bifidobacteria, aciduric bacteria, species of *Bacteroides*, *Pseudomonas*, and *Proteus*, spirochaetes and yeasts. (For the anaerobic flora of the gastro-intestinal tract, see Ch. 26.) The extensive observations of Haenel and his collaborators (1956, 1957, 1958*a*, *b*) and of Seeliger and Werner (1962, 1963) on the faeces are especially noteworthy, but are open to the criticism that, owing to technical deficiencies, they underestimated the number of obligatory anaerobic bacteria (Braun 1959, Drasar 1967). Gram-negative non-sporing anaerobic bacilli are sensitive to oxygen (see Spears and Freter 1967) and unless strict anaerobic methods such as those of Hungate (1960, 1963), either in their original or their modified form (Moore 1966, Drasar 1967, Mitsuoka *et al.* 1969), are used, together with suitable selective media, members of the *Bacteroides* and *Bifidobacterium* groups, in particular, are liable to be underestimated. By improved methods it can be shown that anaerobic bacteria make up about 99 per cent of the faecal flora. Most of these organisms belong to the gram-positive bifidobacteria and the gram-negative bacteroides (see Duerden 1980*b*). Their absolute numbers vary, but both of them are commonly present in the order of 10^9 to 10^{10} per gram of faeces (Mitsuoka *et al.* 1965, Drasar 1967, Gorbach *et al.* 1967*a*). Lactobacilli, clostridia and fusobacteria average each about 10^3 to 10^5 per gram, enterobacteria 10^6 and enterococci 10^5 per gram. Other organisms, such as *Veillonella*, staphylococci and yeasts seldom exceed 10^3 per gram (Drasar 1967, Levison and Kaye 1969). Less frequent are *Proteus* spp., *Ps. aeruginosa*, aerobic spore-bearers and spirochaetes (see Bendig *et al.* 1968*a*). *Clostridium perfringens* (*Cl. welchi*), though a fairly regular inhabitant of the large intestine, is but a minor component of the faeces in healthy persons (Willis 1969, Collee 1974); but in certain circumstances it is associated with such illnesses as food poisoning, pig-bel, and enteritis necroticans (see Chapter 72). Like *Esch. coli*, its presence in water and soil is generally indicative of faecal contamination. In vegetarians, the gram-positive anaerobic organism known as *Sarcina ventriculi* may be present in the faeces; Crowther (1971) found it in about 75 per cent of such subjects in numbers up to 10^8/gram.

Viruses are often demonstrable in the gut, but our knowledge of their nature and distribution is not yet sufficient to say to what extent they can be regarded as part of the normal flora. (For the effect of antimicrobial drugs on the faecal flora, see Sutter and Finegold 1974; and for the transfer of antibiotic resistance, see Burton *et al.* 1974.)

Among the enterobacteria—or more strictly the coliform bacilli—*Esch. coli* is far the commonest. Organisms of the *Klebsiella* and *Enterobacter* groups are found in only a proportion of healthy persons and as a rule in only small numbers (Bendig *et al.* 1968*b*). The antigenic types of *Esch. coli* in any given person vary in number and persistence (see Kauffmann and Perch 1943, Sears and Brownlee 1952) and cannot easily be displaced by new ones given by the mouth (Seeliger and Werner 1963, Ozawa and Freter 1964).

It must be pointed out that, with a few exceptions, most of the investigators in this field have failed to distinguish, among the gram-positive non-sporing anaerobic bacilli, between lactobacilli and bifidobacteria. The term lactobacillus as used by them has generally included organisms of both groups and in this respect is misleading. In fact, as more recent work has shown, bifidobacteria greatly outnumber the lactobacilli, most of which are facultative rather than obligatory anaerobes. (See van der Wiel-Korstanje 1973.)

It is evident, that, as emphasized in the past by Eggerth and Gagnon (1933), Weiss and Rettger (1937), and Misra (1938) among others, obligatory anaerobic non-sporing bacteria are predominant in the faeces, being in a ratio to aerobic organisms of something like 100 to 1 or 200 to 1 (Ellis-Pegler *et al.* 1975). Altogether bacteria are estimated to compose $\frac{1}{4}$ to $\frac{1}{3}$ by weight of the faeces (see Stephen and Cummings 1980). The greater part of them seem to be dead.

Provisionally we may conclude that in healthy persons the majority of the organisms ingested in the food and swallowed with the saliva are destroyed in the stomach by the gastric juice. The few surviving bacteria that reach the duodenum and upper jejunum are passed on too rapidly for any multiplication to take effect even during a temporary colonization of the lumen. Conditions for their growth become favourable as they reach the lower part of the small intestine. The caecum and colon allow the establishment of the relatively stable microbial complex that characterizes the flora of the large intestine and faeces. (For a description of anaerobic techniques in the study of intestinal organisms, and of interactions between different organisms in the gut, see numerous papers in Report 1972.)

The variations of intestinal flora within **mammalian species** may be very great. For example, the normal flora of the guinea-pig intestine was found by Crecelius and Rettger (1943) to be fairly simple; a lactobacillus resembling *Lacto. acidophilus* constituted approximately 80 per cent of the total cultivable flora, the remainder being accounted for chiefly by soil and air bacteria, yeasts, and an anaerobic sarcina. Coliform organisms and enterococci were seldom present. Again, according to Porter and Rettger (1940), the flora of the white rat on a normal mixed diet consists mainly of lactobacilli and coliform organisms; very few cocci are found. Lactobacilli were common in the stomach, where little else but yeasts were demonstrable. Coliforms and lactobacilli were present in the lower parts of the small intestine, and abundant in the large intestine, where spore-bearing anaerobes also appeared. Except in the lower gut, it may be noted, there was little correlation between the pH of the gut contents and the predominating flora. In the faeces of the cow *Esch. coli, Str. faecium*, and micrococci are always present; *Bacillus* spp. are found in about 50 per cent of specimens, but *Cl. perfringens* and *Str. faecalis* can practically never be demonstrated (Wilssens and Buttiaux 1958). In a detailed analysis of small numbers of faecal specimens from various animals, Haenel and Müller-Beuthow (1956) found a similar ratio of anaerobic lactobacilli, aerobes, coliforms, staphylococci and enterococci in omnivorous animals—man, rat, hen and dog—the anaerobic organisms being greatly predominant. In the herbivorous horse, guinea-pig and rabbit, the anaerobe numbers were lower and the coliform and enterococcal counts very low. The rabbit had relatively large numbers of 'staphylococci,' the dog none. The proportion of coliforms and spore-bearers was high in the cow. For further information on the intestinal flora of different species of animals reference may be made to Smith and Jones (1963) and van der Heyde and Henderickx (1964) for the pig, Raibaud *et al.* (1966) for the rat, Mushin and Dubos (1966) for the mouse, Eichel (1964) for the mink, Ochi *et al.* (1964) for the hen, Soucek and Mushin (1970) for certain antarctic birds and mammals, and Mushin and Ashburner (1964) for various other animals and birds. For streptococci, see Ch. 30.

The individual variations in the estimates, both in man and animals, of bacterial numbers, are great; and it is evident that, quantitatively speaking, the intestinal flora is 'constant' only within broad limits. Consequently, only gross and continued alterations in the number, proportions or distribution of the various species can be used as indicators of a disturbance or an abnormality of the intestinal flora (Seeliger 1957, Haenel 1961).

Geographical differences occur in the intestinal flora, notably between the omnivorous peoples of the West and the peoples of Africa and Asia living mainly on a vegetable diet (Drasar 1974). Moreover, the mucosa of the small intestine of normal persons, both indigenous and expatriate, living in the Tropics, differs in structure and function from that of normal persons living in temperate regions (Lindenbaum 1968).

Among the factors that influence the intestinal flora are the gastric and intestinal secretions, such as hydrochloric acid, bile salts, and lysozyme; diet; immune globulins principally IgA (see Savage 1977); antibacterial drugs; bacterial interactions; and mobility, i.e. the rate at which the intestinal contents are passed through the alimentary canal (Drasar and Hill 1974).

Of these factors, Clarke and Bauchop (1977) would place most stress on acidity and mobility. Areas in which large numbers of micro-organisms are present usually have a pH between 5.0 and 7.5.

The high acidity of the stomach is unfavourable for bacterial multiplication. The rate at which the food material is passed through the ileum may be as much as 10 times that in the colon, where complete stasis occurs (Luckey 1924). It is therefore in the colon that microbes are found in the greatest numbers and where fermentation is at its maximum. The conditions here are anaerobic, and it is not surprising that anaerobic organisms constitute over 99 per cent of the faecal flora (Werner 1966, Peach *et al.* 1974, Clarke and Bauchop 1977). *Esch. coli* which, up till the introduction by Hungate (1960, 1963) of special anaerobic methods was regarded as one of the main intestinal bacteria, constitutes not more than about 0.01 per cent of the total population in the lower gut of man (Clarke and Bauchop 1977). (For a description of the micro-organisms, including protozoa, present in the gut of higher animals, see Smith 1965, Clarke and Bauchop 1977; and for the gnotobiotic animal in the study of gut microbiology, see Clarke and Bauchop 1977.) According to Drasar and Hill (1974), *Bacteroides* is 'probably the most important genus of intestinal bacteria'.

The importance of micro-organisms in the intestine

It was at one time widely believed that the presence of bacteria in the intestine was essential for the life of the host, that they assisted in the digestive processes, and that without them much of the food taken in would be passed out of the body in an unassimilable condition. This belief arose as a very natural deduction from Pasteur's work on the microbial nature of fermentation. It was largely owing to Metchnikoff, who came to regard the intestinal flora with suspicion, that the possibility of maintaining life with a sterile intestinal tract was seriously considered.

Numerous workers had already devoted themselves to the realization of this possibility; notable among them was Schottelius (1899, 1902, 1908), who spent several years endeavouring to rear chicks under sterile conditions, and failed. Nuttall and Thierfelder (1895–96) met with some success; they removed an embryo guinea-pig by Caesarean section, and maintained it uncontaminated for 8 days. Cohendy (1912) succeeded in showing that prolonged life was possible under such conditions. He reared chicks under completely sterile conditions for from 12 to 40 days, and found that they developed as well as control, non-sterile chicks. The only difference noted was that the food of the sterile chicks was less perfectly digested than that of the control chicks; but the sterile chicks made up for this by eating more. (See also Küster 1915.)

Germ-free animals Since the 1930s, as a result of the work of Reyniers in the United States and of Glimstedt in Sweden, the technique of rearing animals in microbiologically sterile conditions has been perfected to the point where not only small laboratory vertebrates, but even larger animals, are available for study in the 'germ-free' state (see Reyniers 1959, Glimstedt 1959). There is no doubt that on a suitable diet the smaller animals both grow and can in some cases be bred for several generations in this state; but they differ from animals reared in conventional environments in a number of respects. Thus, germ-free chickens, rats and to a smaller extent mice are less hydrated and have less connective and reticulo-endothelial tissue than the corresponding conventional animals. The germ-free guinea-pig has little lymphoid tissue. The relative absence of lymphoid-macrophage tissue and of γ globulin in the plasma appears to result from lack of stimulus by environmental microbial antigens. The striking feature of germ-free mammals like the guinea-pig, rat and rabbit is enlargement of the caecum, sometimes with a thinning of the gut wall. The small intestine may be smaller than in conventional animals. The caecal effect appears to be directly due to absence of organisms. Thus in the rat, colonization by faecal flora, and to a lesser extent by pure cultures of either a coliform or a *Proteus* sp., restored the size and structure of the caecum to normality (Gustafsson 1959*a*; see also Gordon and Bruckner-Kardoss 1961, Loesche 1969).

The germ-free animal provides systems for exploring a number of problems concerning the contribution of the normal intestinal flora to the state of the animal—as a source of nutrients or of toxic substances, and resistance to intestinal pathogens. These we shall consider along with evidence from conventional animals. For discussions of the detailed implications of the germ-free state, we refer the reader to two Symposia (1959*a*, *b*).

Intestinal toxaemia It was for long held that an imbalance of the intestinal flora, particularly an increase in 'proteolytic' (or 'putrefactive') microbes at the expense of the lactobacillary flora, led to production, and absorption by the host, of deleterious toxic products. There was little experimental evidence to support this notion of 'intestinal toxaemia' (see Discussion 1913, Dudgeon 1926). It has, however, been found that germ-free rats in which intestinal strangulation and obstruction have been produced live five times or more longer than conventional rats treated similarly; and that the peritoneal fluid that collects in these circumstances is not lethal to mice injected intraperitoneally when taken from germ-free animals but causes death within 24 hours when taken from conventional rats (Amundsen and Gustafsson 1963, Trippestad and Midtvedt 1970). There is also evidence that in rats autoantibodies to colonic tissue, though formed in the presence of certain clostridia, are not formed in germ-free animals (Hammarström *et al.* 1969). Attempts to replace the putrefactive flora—such as *Cl. sporogenes*, *Bac. mesentericus*, *Proteus* and *Pseudomonas* spp.—by direct feeding of cultures of organisms like *Lacto. bulgaricus*, failed completely. Some success was claimed for changing the flora by alterations of diet intended to alter the ecological balance in favour of non-putrefactive, 'saccharolytic' organisms. Thus a diet with a high content of lactose or dextrin—carbohydrates which are slowly absorbed and so pass into the large intestine—was reported greatly to increase the aciduric bacilli in the rat intestine (Rettger and Cheplin 1921, Cannon and McNease 1923, Cruickshank 1928); and the addition of 12 per cent lactose to cows' milk, to convert the flora of infants to that characteristic of breast-fed infants (Gerstley *et al.* 1932). Haenel and his colleagues (1958*b*), however, found that lactose in the diet had no effect on the aciduric flora. In infants only mothers' milk suppressed the putrefactive organisms in the intestine (Feldheim *et al.* 1960).

Qualitative and quantitative changes in the faecal flora can be brought about by extreme changes in the diet. Both in man (Hoffmann 1964) and in rats (Schulze *et al.* 1970) a high carbohydrate diet favoured the growth of the obligatory anaerobic bacteria, particularly the bifidobacteria, and led to an increase in the total germ content. A high fat diet (Hoffmann 1964) repressed the enterococci and the bifidobacteria but encouraged proliferation of the bacteroides. On a high-protein diet the flora did not differ significantly from that on a normal diet. On a diet rich in cellulose the total germ content of the human intestine diminished, as did likewise the frequency of the proteus, staphylococcal, clostridial and halophilic aerobic spore-bearing groups, and the yeasts (Haenel *et al.* 1964).

In this chapter we do not consider the various physiological processes that are carried out by micro-organisms in the gut, but it may be mentioned that there is some reason to believe that the production of nitrosamines, and of certain metabolites of the biliary steroids, may be associated with the development of cancer affecting the colon and other parts of the body (see Drasar and Hill 1974).

The therapy of intestinal disorders by measures intended to change the intestinal flora is discussed in the monograph of Rettger and his colleagues (1935; see also Weinstein *et al.* 1938). It seems difficult in adults to produce more than minor changes by ordinary means (Torrey and Montu 1931, Haenel *et al.* 1957), and little progress has so far been made along these lines. (See also Orla-Jensen *et al.* 1945.)

The effect of antibiotics

Under the influence of antibiotics, sensitive organisms in the normal gut flora are replaced by resistant ones. For this reason, when antibiotic treatment is required for prophylactic purposes in surgery it should not be begun long before operation and should necessarily therefore be of short duration.

Bacterial overgrowth following the administration of antibiotics may lead to diarrhoea. An example of this is afforded by the cholera-like disease due to enterotoxin in the gut resulting from the overgrowth by resistant staphylococci (see Ch. 60).

A more destructive action leading to necrosis of the intestinal mucosa is seen in the so-called post-antibiotic enterocolitis. Clindamycin is the antibiotic most frequently concerned; and the disease itself is attributed to a cytotoxic agent produced by *Clostridium difficile* (see Ch. 63). However, not all cases of post-antibiotic diarrhoea are of staphylococcal or clostridial origin, and not all cases of enterocolitis in hospital are attributable to antibiotics.

The inter-relation of gut flora and health is strikingly evident in the observation of Stokstad and Jukes (1950, 1951) that the addition of chlortetracycline to the diet stimulated the growth of chicks, pigs and turkeys. This effect was subsequently observed by many workers in a variety of laboratory and farm animals, using a variety of chemotherapeutic agents and antibiotics (see the reviews of Braude et al. 1953, Stokstad 1954). Not only do the animals grow faster, but, in terms of weight gained per unit weight of food, foodstuffs are used more efficiently. The effect has been variously attributed to more complete absorption of the nutrients; increased production of microbial vitamins or other growth factors needed by the animal; inhibition of bacteria that otherwise compete with the animal for nutrients in the gut; and inhibition of bacteria that have deleterious effects on the animal. We consider the nutritional aspects in the next section. But whether the antibiotics ultimately affect nutrition or not, there is a large body of circumstantial evidence that their nutritional effect depends primarily on the inhibition of certain microbes of the intestinal flora.

The effect is unlikely to be due to a specific growth stimulation by the antibiotic, since a wide variety of structurally different antibacterial agents is effective. Moreover, though there is some evidence of a direct action of antibiotics on the animal itself (see Luckey 1959), in general antibiotics do not appear to promote the growth of germ-free animals (Jukes 1955, Lev and Forbes 1959). That antibiotics act by decreasing or eliminating deleterious flora in the gut is strongly suggested by their inefficacy in animals kept under good hygienic conditions. Thus chickens on a good diet in clean premises grow as well as antibiotic-fed chickens in old premises, and do not respond to antibiotics with further improvement of growth; but the growth rate of 'clean' chickens is depressed to that of untreated chickens from old premises when the two kinds of flock are kept in contact (Coates et al. 1951, 1953, 1955). There is little to suggest that the growth depression relieved by the antibiotic is due to changes in the gross numbers or proportions of the intestinal flora in general (see also Eyssen and de Somer 1963). A decrease in intestinal clostridia, however, was noted by Elam and his colleagues (1953) in treated chickens, and later Lev and his colleagues (1957) found an association between the presence of toxigenic Cl. perfringens type A in the gut contents and suceptibility to growth promotion by penicillin. In both germ-free chicks and in germ-free chicks whose gut was contaminated with a mixture of lactobacilli, a streptococcus and Esch. coli—a flora that was not itself growth-depressant—contamination with Cl. perfringens type A depressed growth; and the growth depression was largely relieved by penicillin (Lev and Forbes 1959). In this connection it may be noted that in germ-free chicks contaminated with Cl. perfringens, the gut walls, characteristically thin in the germ-free state, approximate in thickness to those of animals with a normal intestinal flora (Wostman et al. 1959-60).

Continued small antibiotic supplementation of the diet has no ill effect on the farm animals receiving it; but it has proved to be associated with the appearance and persistence in the gut flora of antibiotic-resistant strains, for example, of salmonellae (Huey and Edwards 1958, Garside et al. 1960, Hobbs et al. 1960), faecal streptococci (Barnes 1958), Esch. coli (Smith and Crabbe 1956, 1957) and Cl. perfringens (Smith 1959). There is thus a risk of spreading infection by antibiotic-resistant pathogens to other animals and man. Since the likelihood of the emergence of populations of resistant pathogens increases with the duration of the use of antibiotic supplements, and since supplementation tends to be used as a substitute for efficient farm husbandry, it is advisable to limit it to stock for meat production, maintaining the longer-lived breeding stock without its aid. Moreover, the antibiotics included in the diet of animals should be different from those used in human therapeutics. There is now ample evidence to show that gastrointestinal disease may follow the introduction into the gut of resistant strains of Salmonella sp. derived from animals receiving supplementary antibiotic feeding.

No positive effect on growth was observed in under-nourished, adult human subjects treated with oral chlortetracycline (Gabuzda et al. 1958).

Intestinal microbes and resistance to infection

The diminution of the intestinal flora observed in antibiotic-treated, and its absence in germ-free, animals may profoundly affect their resistance to certain pathogens. It is evident from work with germ-free animals that absence of intestinal bacteria, among other environmental microbes, is associated with poor development of antibody-forming tissue (Chapter 11) and apparently also with diminished leucocytic mobilization (Abrams and Bishop 1965); and that alteration in the flora of the gut by antibiotics increases the susceptibility of experimental animals to infections with the cholera vibrio and certain enterobacteria.

The contribution of the intestinal flora to resistance to bacterial infections is still in an early stage of analysis. It must be emphasized that besides ecological relations in the intestinal contents themselves, and the status of the walls of the intestine as barriers to infective microbes dependent on the presence of the flora, systemic factors may be concerned. We have noted the lymphoid immaturity of the germ-free animal, indicating that the immunological competence of normal animals depends on stimulation of environmental antigens; and we shall see in Chapter 12 that 'natural' antibodies may owe their origin to intestinal microbes. Endotoxins may well be absorbed from the gut and, according to circumstance, either stimulate or intoxicate the defences.

Lastly, the intestinal flora may affect resistance by altering the nutritional status of the animal.

Intestinal microbes and nutrition

The successful rearing of fowl and laboratory animals like the rat and guinea-pig in a germ-free state (see, e.g., Symposium 1959b) indicates that intestinal microbes are not necessary for life; but there is both suggestive and direct evidence that, provided the diet is adequate, the bacteria contribute to the well-being of the host.

The special case of ruminants is reviewed by Hungate (1960, 1963). The complex flora—both microbial and protozoal—of the *rumen* is adapted for the anaerobic digestion of celluloses, starch and protein, and for the conversion of the products of carbohydrate and protein breakdown into metabolites suitable for the nutrition of the animal; the ruminant is in effect a feeder on the cells and waste products of a complex plankton that it cultivates in its rumen. The bacteria concerned belong to such groups as *Bacteroides, Ruminococcus, Succinomonas, Lactobacillus, Selenomonas, Streptococcus, Veillonella,* and *Methanobacterium*. Their numbers range from 1×10^{10} to 5×10^{10} per gram of rumen contents. There is not only a relation between the host—cow or sheep—and the microbial population, but also between the different microbes themselves. The products of one group of organisms serve as a source of energy for other groups. Moreover the microbial community provides all the B vitamins required by the animal. The protozoa are almost as numerous, though less important than the bacteria. They are unable to synthesize amino acids, and therefore depend on the bacteria for their nitrogen (Wolin 1979).

In the non-ruminant, the intestinal bacteria have been investigated chiefly as a source of vitamins. *In vitro*, the intestinal bacteria of man—for instance *Esch. coli* and *Kl. aerogenes*—are reported as synthesizing biotin, riboflavin, pantothenate and pyridoxin (see Burkholder and McVeigh 1942, Peter 1960) and vitamin K (Dam *et al.* 1941).

Active microbial synthesis *in vivo* is deduced in various ways: by noting the continued excretion of a vitamin in an animal upon a vitamin-free diet—depletion of the body reserves being excluded as a possible source; by induction of avitaminosis by partial sterilization of the intestinal contents with chemotherapeutic agents, including antibiotics, followed by the cure of the avitaminosis during drug therapy by giving adequate doses of the appropriate vitamin (Gant *et al.* 1943); and by studies in germ-free animals.

The results of applying the first two methods strongly suggest that in the alimentary canal of the rat, vitamin K (Black *et al.* 1942), biotin, folic acid (Nielsen and Elvehjem 1942), the B-complex, vitamin E (Daft and Sebrell 1942), and possibly certain essential amino acids (Martin 1944) are synthesized by bacteria and used by the host. In man, a similar dependence upon intestinally synthesized thiamin (Najjar and Holt 1943); but see Alexander and Landwehr 1945), riboflavin (Najjar *et al.* 1944), nicotinamide (Najjar *et al.* 1946), and nicotine acid (Ellinger *et al.* 1944) has been postulated (see also Johansson and Sarles 1949).

Germ-free animals, moreover, may fail to survive on deficient diets that are reasonably adequate for conventional animals (see, e.g., Horton and Hickey 1961); and the rat appears normally to depend on its gut flora for biotin (Luckey *et al.* 1955), and for vitamin K (Gustafson 1959b). The vitamin K deficiency of the germ-free rat can be removed by contaminating the gut, with, e.g., *Esch. coli*. The flora in the rat also appears capable of transforming the dietary precursors of vitamin K into the form active in the mammal (Gustafson *et al.* 1962). It is noteworthy, in respect of the presence of vitamin in the intestinal contents as evidence for its synthesis there, that large quantities of vitamins may be found concentrated in the caecum of germ-free chicks (Reyniers *et al.* 1950).

Partial sterilization of the gut contents by antibiotics may result in nutritional advantages. Several workers report a vitamin-sparing effect of oral antibiotics. Thiamin appears to be the B vitamin most readily spared in this way (see Waibel *et al.* 1953). The effect may be due to the more ready absorption of the vitamin by the thinner-walled gut characteristic of antibiotic-treated animals. Eyssen and de Somer (1963) observed that the improved growth of chicks under the influence of dietary virginiamycin was characterized by a greater absorption of fats and carbohydrates; and suggest, by analogy with steatorrhoea in man, that the relative malabsorption in untreated chicks may be associated with poor absorption of amino acids, vitamins and minerals.

Intestinal bacteria may destroy essential foodstuffs. Thus, in the absence of a more readily fermented carbohydrate, *Esch. coli* and a number of other intestinal species will decompose vitamin C in media containing organic nitrogen. It is not known to what extent dietary vitamin C is destroyed in the alimentary canal, but the hypothesis of its decomposition by bacteria is in accord with the observation that certain scorbutic patients respond to injected, but not to oral, vitamin C (see Young and Rettger 1943). Again, nicotinic acid may be used as the sole source of nitrogen by organisms of the *Pseudomonas* group and by certain chromobacteria, though in the absence of nicotinic acid these organisms synthesize the vitamin (Koser and Baird 1944, Benesch 1945). The production of a vitamin by the intestinal flora will clearly be determined by a multiplicity of factors, of which the presence of vitamin-producing and vitamin-destroying bacteria is only one.

Numerous observations have been made in *gnotobiotic animals* on the effect of nutrition and metabolism on the normal flora of the mouth and gastro-intestinal tract (see Heneghan 1973).

(For general reviews of the intestinal flora, see Drasar and Hill 1974, Skinner and Carr 1974, Clarke and Bauchop (1977).)

The normal flora of the respiratory tract

For our present purpose we may divide the respiratory tract into three sections: an upper part, including the anterior and posterior nares and the nasopharynx; an intermediate zone, common to the respiratory and alimentary tracts, including the oropharynx and the tonsils; and a lower part, including the larynx, trachea, bronchi and lungs.

The bacterial flora of the **nasal passages** differs in several respects from that of the nasopharynx. As determined by direct plating of swabs, it is less copious. Moreover, diphtheroid bacilli and staphylococci—both *Staph. aureus* and *Staph. epidermidis*—are far more frequent in the nose than in the nasopharynx, whereas α-haemolytic and non-haemolytic streptococci, and gram-negative cocci of the *N.*

subflava type, are far less frequent. These streptococci and gram-negative cocci appear to constitute the basal normal flora of the **nasopharynx** in most communities, so far as this is revealed by the usual methods of cultivation; and the flora of the oropharynx seems to be composed mainly of the same species (Neumann 1902, Shibley *et al.* 1926, Noble *et al.* 1928, personal observations). Cultivation that catered for a greater variety of bacterial species might yield a very different picture. For example, in the nasal secretions of 10 persons there were, per ml, an average of 6.3×10^6 anaerobic 'lactobacilli,' 1.6×10^6 'staphylococci' and 4×10^4 proteolytic organisms (Haenel *et al.* 1958a).

It would appear (Calamida and Bertarelli 1902) that the accessory nasal sinuses are normally sterile. According to Thompson and Hewlett (1896), the **trachea and bronchi** contain few, if any, bacteria. The culture of fluid aspirated from the trachea of patients subsequently shown to be free from infection of the lower respiratory tract amply confirms this (see Schreiner and Digranes 1979). A certain number of organisms—mainly contained in particles of about 1 μm in diameter (see Ch. 9)—are drawn through the larynx with the inspired air, but the highly effective filter provided by the tortuous nasal passages traps most of the bacteria entering the nose. Those penetrating beyond the vocal cords are removed: in part the organisms are propelled upwards in the mucous sheath on the epithelial surface by the action of the cilia and are then swallowed or expectorated; and part by removal through the lymphatics.

The diphtheroid bacilli and staphylococci usually present in the nose, and the non-haemolytic streptococci and gram-negative cocci of the nasopharynx and tonsils, are seldom the only species isolated from these situations. Various other species, including many potential pathogens, are often found. The relative frequency of three such species—*Str. pneumoniae*, haemolytic streptococci and *H. influenzae*—compared with the frequency of gram-negative cocci, in the nose, nasopharynx, tonsils, and oral cavity in general, is shown in Table 8.1.

It will be noted that all four species were far less frequent in the nose than in the nasopharynx. This difference is particularly striking for the gram-negative cocci; but it is clear that there is less tendency for the pathogenic species to vegetate in the nose than in the nasopharynx. Mouth-washing, in this series of cases at least, appears to be a less effective method of isolating the pneumococcus and the influenza bacillus, but equally effective for the haemolytic streptococci and the gram-negative cocci (see also Gundel and Okura 1933, Gundel 1933).

It must not be supposed that these are constant figures. They fluctuate widely both from time to time and from place to place. Thus the nasopharyngeal carrier rate of haemolytic streptococci may vary from zero to 20 per cent in a sample of the normal adult population subjected to repeated swabbing. The fluctuations in pneumococci and influenza bacilli are usually smaller (see also Burky and Smillie 1929).

Table 8.2 summarizes the results of a large survey which included a group of adults sampled at intervals

Table 8.2 Showing the carrier rates for certain bacterial species in the nasopharynx and nose for a group of adults in an urban population (Straker *et al.* 1939).

Organism	Nasopharyngeal carrier rate %	Nasal carrier rate %
Str. pneumoniae	20–40	5–15
Haemolytic streptococci	5–15	<1
H. influenzae	40–80	5–10
N. meningitidis	5–20	0–4
Gram-negative cocci other than *N. meningitidis*	90–100	0–15

over a period of seven years. In children, the pneumococcal carrier rates tended to be higher than those recorded in the table, but otherwise the rates were not significantly different.

The rates varied to some extent with the season. In the nasopharynx, pneumococci and to a slight extent *H. influenzae* were found more frequently in cold, damp seasons than in the dry hot periods; in the nose a similar, but more definite seasonal variation occurred. In the summer months the organisms were confined largely to the nasopharynx, but in late winter and early spring they tended to colonize downwards into the trachea and forward into the nasal cavity. (See also Masters *et al.* 1958, Box *et al.* 1961.) The carrier rates of haemolytic streptococci in other countries are discussed in Chapter 59.

Such studies often show a striking tendency for any given individual to maintain a characteristic nasopharyngeal flora over a considerable period of time.

Table 8.1 Showing the frequency of certain bacterial species in the nasopharynx, nose, tonsils and oral cavity among a sample of the general population (Report 1930).

Group	No. of swabs	Sites sampled	Pneumo-coccus %	*H. influ-enzae* %	Haemolytic streptococci %	Gram-negative cocci %
I and II	412	Nasopharynx	40.98	59.01	7.77	97.09
		Nose	16.79	11.68	0.05	12.11
III	137	Nasopharynx	42.34	59.54	5.84	97.08
		Tonsil	44.53	54.89	5.11	98.54
IV	252	Nasopharynx	49.80	51.01	11.90	95.24
		Mouth-wash	25.70	25.10	11.95	94.82

For instance, a person may carry one or more given types of pneumococcus in the nasopharynx over a period of months and years. Moreover, when pneumococci are isolated from a carrier after a long period during which samples contained none, the newly appearing pneumococcus may be of the same type as that originally isolated. In other carriers, there may be replacement of one type by another (Gundel and Schwarz 1932, Straker *et al* 1939, Smillie *et al.* 1943; see also Chapter 67).

Since the demonstration of high *Staph. aureus* carrier rates in hospital patients (Hallman 1937) and healthy adults (McFarlan 1938), it has become clear that potentially pathogenic staphylococci are part of the normal flora of the anterior nares. The subject is fully reviewed by Williams (1963). Carrier rates of 20 to 50 per cent are recorded for healthy adults, and of up to 80 per cent for healthy nurses working in hospitals or institutions. As a rule, large numbers of *Staph. aureus* are present. Repeated sampling over long periods indicates that some 10–20 per cent of persons in a community carry persistently, in many cases probably the same type or types of staphylococcus, and some 10–20 per cent never do so. The remainder are intermittent carriers, often of different phage types of staphylococci in successive periods of carriage, suggesting successive transient colonization by environmental *Staph. aureus*. It is noteworthy that colonization by one type appears to prevent colonization of the nares by other types (see Rountree and Barbour 1951). In general, the nose appears to be the primary source of staphylococci found in the throat in which carrier rates of 4–7 per cent are recorded in Great Britain and the United States, 8 per cent in the Netherlands (Noble *et al.* 1967) and 45 per cent or more in Scandinavia. The nose is likewise the primary source of both skin (see below) and intestinal staphylococci. The organisms are carried on the squamous epithelium of the nasal vestibule. Other organisms that may be present on the skin are discussed later (see p. 240). (See also Chapter 60.)

The characteristic nasopharyngeal flora, with its basic constituents such as streptococci of the viridans type and *N. subflava*, together with the fluctuating population of potentially more pathogenic bacteria, is established soon after birth; and there is no doubt that the infant derives these organisms from the adults or older children in its immediate entourage.

The nasopharynx is virtually sterile *at birth*. Gundel and Schwarz (1932) record colonization in the first 2 days of life by viridans streptococci, gram-negative cocci of the *subflava* type, and occasional diphtheroid and coliform bacilli. Pneumococci and haemophilus bacilli appeared by the 2nd or 3rd day, with a frequency dependent on their distribution in the mothers and to a lesser extent in the nurses in attendance on the children (see also Kneeland 1930). Pneumococci, and to a lesser degree haemophilus bacilli, appear to establish themselves more frequently than streptococci in the infant's upper respiratory tract (see Laurell *et al.* 1958, Masters *et al.* 1958, Box *et al.* 1961).

Torrey and Reese (1945) found that the nose, nasopharynx, and throat were usually free from aerobic organisms till after the infants had come into close contact with their mothers for feeding—as a rule at about 12 hours after birth. *Staph. albus* and streptococci of the viridans and non-haemolytic groups were the first to appear; *Staph. aureus* came later. Cunliffe (1949) observed a rise in the incidence of *Staph. aureus* in the nose of infants from 8.6 per cent on the 1st day to 76 per cent on the 4th day. By the end of a fortnight practically every infant delivered in hospital was carrying this organism, but the carrier rate fell steeply to a minimum of about 20 per cent at 6–24 months. Subsequent work has in general confirmed these findings; carrier rates approaching 100 per cent are established by the 10th day, and the decline to 20 per cent at 2 years is followed by a rise to adult carrier rates by the 5th year (see Williams 1963).

Various workers have described the presence of small gram-negative, filter-passing, anaerobic bacilli, in suspensions obtained by washing out the nares, nasopharynx and oropharynx with sterile broth (Olitsky and Gates 1921*a, b*, Mills *et al.* 1928, Burky and Freese 1931). They are now considered to belong to the genus *Bacteroides*. The organism *Dialister pneumosintes* (see Ch. 26), at one time regarded as a possible cause of epidemic influenza, apparently belongs to this group. The information at present available on the relative frequency of these bacilli in normal persons, and in those suffering from colds, does not indicate that they play any significant role in such infections. Garrod (1928) described the isolation, by the same technique, of minute, gram-negative, filter-passing, anaerobic cocci.

The normal flora of the vagina

An important distinction must be drawn between the flora of the vulva and vestibulum on the one hand and that of the vagina proper on the other. Except immediately after parturition, and during the first few days of the puerperium, the vaginal flora is quite distinct from that of the vulval flora.

The *vulva* of the newly born child is sterile; organisms make their appearance in about 7 to 8 hours. The normal flora of the vulva is rich and varied, and depends largely on the organisms present in its immediate environment. According to Wegelius (1909) it consists of: (1) Obligatory aerobes and aerophilic bacteria, of which the chief types are pseudo-diphtheria bacilli and *Micro tetragenus*. (2) Coliform bacilli. (3) Facultative anaerobes, usually more or less suceptible to acid. (4) Bacilli derived from the vagina, including yeasts. (5) Obligatory anaerobes. On aerobic plates the commonest organisms to form colonies are staphylococci, diphtheroids, enterococci, sarcinae, and coliform bacilli. Besides these organisms, yeasts—*Candida* and *Saccharomyces*—are very common; and the smegma bacillus is not infrequently demonstrable in smear preparations stained with Ziehl-Neelsen. Under anaerobic conditions of cultivation numerous colonies appear, consisting of organisms, many of which have been only imperfectly studied. Pathogenic bacteria are uncommon.

The normal flora of the *vagina* seems to depend largely on the glycogen content of the vaginal epithelium which, in its turn, is dependent on ovarian activity (Miura 1928, Cruickshank and Sharman 1934). The vagina of the newly born child is sterile; organisms make their appearance in 12–24 hours. At first they consist of staphylococci, enterococci, and diphtheroids, but these are often replaced in 2 or 3 days by a practically pure culture of lactobacilli (Döderlein's bacillus) (see Chapter 29). At this time glycogen is demonstrable in the vaginal epithelium, and the vaginal secretion itself is acid. The occurrence of glycogen appears to be due to the presence of oestrin derived from the maternal circulation. Soon this is excreted in the urine, glycogen is no longer demonstrable in the epithelium, Döderlein's bacillus disappears, and the vaginal secretion reverts towards alkalinity. From now on till puberty the vaginal secretion remains alkaline, and there is a varied flora, the commonest organisms being *Staph. epidermidis* and diphtheroid bacilli, followed by α-haemolytic streptococci and lactobacilli, and in smaller numbers still by non-haemolytic streptococci and *Esch. coli* (Hammerschlag *et al*. 1978). At puberty glycogen is again deposited in the vaginal wall, the secretion becomes acid, and Döderlein's bacillus establishes itself as the predominant anaerobic organism. The flora, however, still remains mixed and Döderlein's bacillus is accompanied by many of the organisms that we have just mentioned, especially streptococci, diphtheroids, fungi and yeasts. Haenel records the following average numbers per ml in the vaginal secretion of 8 women: anaerobic lactobacilli, 5×10^7; proteolytic bacteria, 3×10^7 'Proteus', 3×10^3; no enterococci, and occasional coliforms, staphylococci and fungi (Haenel *et al*. 1958*a*, Ohm and Galask 1975). Though lactobacilli, of which there are many species (Bartlett *et al*. 1977), account for most of the acid production in the vagina (Ohm and Galask 1975), other organisms, such as streptococci and corynebacteria, probably contribute towards it (Corbishley 1977).

The streptococci are varied. Those belonging to Lancefield's group D are common; groups F and G much less so. Group B streptococci were isolated from 18 per cent of college women (Baker *et al*. 1977), and are said to be most numerous in the middle phase of the menstrual cycle (Galdiero *et al*. 1977). In addition to streptococci, *Bacteroides* (Duerden 1980*a*) (Ch. 26 and 62), *Candida albicans* and *Mycoplasma* spp. may be present in the vagina, and sometimes an anaerobic vibrio requiring a raised concentration of CO_2 for growth (Moore 1954). *Clostridium difficile* was cultured in 17 per cent of young women, and *Gardnerella vaginalis* in 6 per cent, the latter being associated with an offensive discharge (Bramley *et al*. 1979). *Trichomonas vaginalis* is occasionally present. Taking all groups of bacteria together, the plate counts are said to range from 100×10^6 to $10\,000 \times 10^6$ per ml (Lindner *et al*. 1978), and to be lower in infertile than in pregnant women (Moberg *et al*. 1978). At the menopause, when ovarian activity ceases, the conditions revert to those met with before puberty. After parturition, and during the first few days of the puerperium, the vaginal flora resembles that of the vulva. At the time of parturition organisms ascend from the vulva to the uterus, where they may be demonstrated for several days (Wegelius 1909).

The vaginal secretion has a moderate bactericidal action (see Küster 1929). To a large extent this may be ascribed to the presence of lactic acid, though other factors are probably concerned. The lactic acid itself is derived from the glycogen, partly perhaps as the result of natural enzymes, but mainly on account of the fermentative activity of Döderlein's bacillus.

The normal flora of the urethra

In a study of the anterior urethral flora of apparently normal men Bowie and his colleagues (1977) cultured aerobic and anaerobic bacteria from a high proportion, including such organisms as lactobacilli, *Gard. vaginalis*, α-haemolytic streptococci and *Bacteroides* species. *Chlamydia trachomatis* was found in 3 per cent, and *Ureaplasma urealyticum* in 59 per cent. A short gram-negative diplobacillus, not to be mistaken for the gonococcus, was noted by Kutscher (1909); it was probably a member of the *Acinetobacter* group. The female urethra is either sterile or contains a few non-pathogenic cocci. A saprophytic acid-fast bacillus, *Mycobacterium smegmatis*, is present in the preputial secretion of both sexes; this organism may occasionally find its way into the urine, and be mistaken for the tubercle bacillus.

Normal flora of the skin

The skin is the largest organ in the body, weighing 5 kg and having an area of about 1.75 sq. metres (Noble and Somerville 1974). It contains two types of glands: the *apocrine*, present mainly in the axillae, the ano-genital region, and around the nipples, losing part of their cells when functioning; and the *eccrine* sweat glands, which do not lose their protoplasm when functioning, and which secrete fluid containing salt, urea and lactate. Modified apocrine glands exist in the ear (ceruminous glands), the eyelids (Moll's glands), and the breast (mammary glands). The apocrine glands are not mature till puberty is well advanced; they function in response to both mental and physical stimulation.

Price (1938) distinguishes the 'transient' flora, consisting of multifarious organisms dependent on personal hygiene and environmental conditions, from the 'resident' flora, consisting of organisms that are more or less constantly present. These comprise mainly staphylococci and micrococci, aerobic and anaerobic diphtheroids; in addition, gram-negative bacilli of the genus *Acinetobacter* (see Ch. 32), enterobacteria, aerobic spore-bearing bacilli, and *Streptomyces* may be found (Kligman 1965, M. J. Marples 1965, Noble and Somerville 1974, Kloos and Musselthwaite 1975). In moist areas, such as the axillae, gram-positive cocci and diphtheroids constitute nrly 90 per cent of the aerobic flora (M. J. Marples 1974). According to Pitcher (1977), the aerobic diphtheroids, using this term in the sense defined in Chapter 25, are more common than the pseudodiphtheroids ('diphtheroids') but both vary in the skin sites that they colonize (Noble and Somerville 1974). Together they constitute the major part of the aerobic skin flora. The so-called anaerobic diphtheroids are not true members of the *Corynebacterium* genus; they are often classified in the genus *Propionibacterium*. Among them is the well known lipolytic organism *P. acnes*. The anaerobic flora outnumbers the aerobic, usually in a ratio of between 10:1 and 100:1 (Evans *et al.* 1950, Haenel *et al.* 1958a), and flourishes most in the sebaceous glands where anaerobic conditions prevail. The potentially pathogenic *Staph. aureus* is present as a rule in small numbers. Coliform and proteus bacilli are not uncommon (Horwood and Minch 1951). *Acinetobacter* is found particularly in the axillae and groin; its numbers increase in hot weather.

At birth the skin flora consists of *Staph. albus* and diphtheroid bacilli, with a few coliform and proteus bacilli, and occasional streptococci—all derived from the birth canal (Sarkany and Gaylarde 1968). The skin flora of children is more diverse than that of adults. As in the alimentary canal, so on the skin, the resident flora interferes with the implantation of fresh organisms. This may perhaps depend on the production of bacteriocines or penicillinase, or on the liberation of free fatty acids from sebum (M. J. Marples 1969). The healthy skin, moreover, appears to have some natural *self-disinfecting mechanism* that is responsible for the rapid disappearance of living organisms implanted on it (see Colebrook and Maxted 1933, Arnold and Bart 1934). Bathing has little effect on the resident skin flora. Sweating brings about a temporary increase.

Partly because most of the microbes on the skin occur in the form of large colonies, and partly because of the difficulty of sampling the resident flora in the glands and deeper layers of the epidermis, estimates of the number of organisms present can never be more than approximate, and from a medical point of view are of less interest than the species to which they belong. (For information on the effect of age on the skin flora, see Somerville 1969, Kloos and Musselthwaite 1975, Carr and Kloos 1977.)

Quantitatively the results depend largely on the method of sampling, the aim of which must be to remove all the bacteria from the surface of a given area of skin, and to prevent their aggregation before being cultured. Williamson (1965), who used the glass cylinder method of Pachtman, Vicher and Brunner (1954), with Triton X-100^R in 0.075 M phosphate buffer as a diluent, obtained wide variations not only between different persons but also between areas of skin on different parts of the body. The subjects he examined could be divided into those supporting large, medium, or small populations of aerobic bacteria. On the forearm the counts ranged from 400 to 19 000, and in the axilla from 6300 to 16.7×10^6 per 3.8 cm^2 area of skin. This technique, of course, did not sample the organisms situated deep in the hair follicles or the anaerobic bacteria in the sebaceous glands. Mustakallio and his colleagues (1967) describe a suction technique that raises a blister separating the epidermis, which can then be examined. Another technique (Holland *et al.* 1974) extracts the organisms from the pilo-sebaceous ducts by treatment of the skin with a drop of cyanoacrylate within the confines of a Teflon ring, followed by pressing on the skin with a glass sampler for 20 seconds.

Staph. aureus is found, in relatively small numbers, in some 8–20 per cent of samples taken from various parts of the skin, and in about 40 per cent of samples from the hand (see Williams 1963). Skin carriage is usually associated with nasal carriage, and the serological or phage types of cocci are usually the same (Gillespie *et al.* 1939, Miles *et al.* 1944, Williams 1946, 1963). To some extent, therefore, it appears that the skin is continually being infected with *Staph. aureus* from the nose. This conclusion was supported by the finding that reduction of nasal carriage by topical treatment with penicillin leads to a significant reduction in the proportion of skin carriers (Moss *et al.* 1948). Though *Staph. aureus* is usually transient in Price's sense, some persons carry this organism so deeply in the recesses of the skin that prolonged sweating is required to bring it to the surface (Devenish and Miles 1939). In such persons it may be regarded as part of the resident flora (see also Gillespie *et al.* 1939, Vierthaler 1940). It appears also to be part of the resident flora of the perineum (Ridley 1959).

In general it may be said that the density of the resident flora depends on the humidity of the skin, on the presence of inhibitory or stimulating free fatty acids formed by the metabolism of sebum triglycerides (Woodroffe and Shaw 1974), and on co-actions between different species of organisms (Selwyn and Ellis 1972). How much the flora of the skin is determined by microbial antagonisms is difficult to say. The subject is considered by Selwyn, Marsh and Sethna (1976) and Wright and Terry (1981).

Hair is of special interest in that it frequently harbours *Staph. aureus* and forms a reservoir for cross infection (see Summers *et al.* 1965, Noble 1966). (For a general review of the skin flora, see M. J. Marples 1965.)

In the skin lining the *external auditory meatus*, as well as the usual staphylococci, diphtheroid bacilli are not uncommon, and sometimes saprophytic acid-fast bacilli, derived from the cerumen, are met with. In the conjunctival sac, organisms are comparatively scanty, and consist chiefly of diphtheroids, such as *C. xerosis*. Anaerobic lactobacilli are reported to predominate in both cerumen and conjunctival fluid (Haenel *et al.* 1958a). The conjunctiva may owe its comparative freedom from bacteria, in part, to the highly potent lytic enzyme, *lysozyme*. Fleming and Allison (1927) found that tears had a high bactericidal power, capable of dissolving certain saprophytic cocci in a dilution of 1/40 000. With the exception of sweat and urine, this enzyme is distributed to a greater or less extent in practically all human secretions (see also Fleming 1929). For further information on the flora of the skin, see the series of articles in the book by Maibach and Hildick-Smith (1965).

Human milk may contain large numbers of bacteria, plate counts of over 500 000 per ml being not uncommon (Scherer 1951).

The contamination comes mainly from the skin, *Staphylococcus epidermidis* being much the commonest organism. When strict aseptic precautions are practised, the plate count at 37° should be under 2500 per ml (Wright 1947). Most of the organisms are expelled with the initial flow. If the first 10 ml are rejected, manually expressed milk may be expected to have a plate count of not more than 5000 per ml (West et al. 1979). Freshly collected milk has some antibacterial activity (see Evans et al. 1978, Welsh and May 1979). To preserve this, milk intended for a milk bank should be kept in a deep freezer at a temperature of −18°. Milk stored at about 4° in the domestic refrigerator allows the growth of psychrotrophic bacteria, including some coliform organisms, that lead to spoilage. The effect of pasteurization on the antibacterial action of human milk is not fully known.

Bacteria in the blood and internal organs

We know little about the occurrence of bacteria in the blood and viscera of healthy persons. The technical limitations of the usual method of performing blood cultures are such that small numbers of organisms may quite easily escape detection. Even when a sufficiently large sample of blood is withdrawn, it is difficult to ensure that the few organisms contained in it will grow; they may be destroyed by the bactericidal power of the blood or ingested by leucocytes. Moreover the blood that is examined is usually drawn from a peripheral vein, coming from one limited portion of the body, and not therefore so likely to contain bacteria as blood from, let us say, the pulmonary artery through which the blood from all parts of the body has to pass. It has been stated that bacteria frequently gain access to the blood stream under normal conditions, but are filtered off in such organs as the liver, lungs, and spleen, where they are destroyed by cellular and bactericidal action. We do not know how frequently this occurs, what the channels of transit are, or what bacterial species may be concerned. It is also stated that in the newborn infant the epithelial lining of the intestinal mucosa is defective in places, and that passage of bacteria into the lymph stream and blood is very much more frequent during the first few days of life than it is subsequently. Whether the normal intestinal mucosa of the adult presents an effective barrier to the passage of bacteria is not definitely known; nor is it known why some organisms are able to pass through it very much more easily than others. There is good evidence that organisms, particularly streptococci, pass into the blood stream of patients suffering from focal infections in different parts of the body (Chapters 57 and 62).

Examinations of normal mesenteric lymph glands have been few. It seems probable that bacteria of different sorts are frequently absorbed through the intestinal mucosa. The experimental evidence of the mode of infection in enteric diseases (Chapter 68) and of the destruction by x-irradiation of the resistance of the animal to infection by its own intestinal flora strongly suggests that the mesenteric lymph nodes constitute the main barrier to further invasion by intestinal bacteria that pass through the mucosa; a few may perhaps make their way to the blood stream, from which they will be removed by phagocytic cells. In support of this possibility is the observation that the muscles and viscera of apparently healthy cattle and pigs not infrequently contain bacteria (Zwick and Weichel 1911, Reith 1926). Different animals, however, seem to vary in this respect. Schweinburg and Sylvester (1953), for example, found that the organs of rats, guinea-pigs, and golden hamsters were sterile, but that 80 per cent of samples of tissues from dogs and 40 per cent from rabbits yielded clostridia—mainly *Cl. perfringens*. (For the place of Bacteroidaceae in the normal flora see Ch. 26.)

References

Abrams, G. D. and Bishop, J. E. (1965) *Arch. Path.* **79**, 213.
Alexander, B. and Landwehr, G. (1945) *Science* **101**, 229.
Amundsen, E. and Gustafsson, B. E. (1963) *J. exp. Med.* **117**, 823.
Arnold, L. and Bart, A. (1934) *Amer. J. Hyg.* **19**, 217.
Baker, C. J. et al. (1977) *J. infect. Dis.* **135**, 392.
Barnes, E. M. (1958) *Brit. vet. J.* **114**, 333.
Bartlett, J. G., Onderdonk, A. B., Drude, E., Goldstein, C., Anderka, M., Alpert, S. and McCormack, W. M. (1977) *J. infect. Dis.* **136**, 271.
Bendig, J., Haenel, H. and Braun, I. (1968a) *Zbl. Bakt.* **209**, 81; (1968b) *Ibid.* **209**, 90.
Benesch, R. (1945) *Lancet* **i**, 718.
Bierman, H. R. and Jawetz, E. (1951) *J. Lab. clin. Med.* **37**, 394.
Bisset, K. A. and Davis, G. H. G. (1960) *The Microbial Flora of the Mouth.* Heywood, London.
Black, S., Overman, R. S., Elvehjem, C. A. and Link, K. P. (1942) *J. biol. Chem.* **145**, 137.
Bloomfield, A. L. (1919) *Amer. Rev. Tuberc.* **3**, 553; (1921) *Johns Hopk. Hosp. Bull.* **32**, 33; (1922a) *Amer. Rev. Tuberc.* **5**, 903; (1922b) *Johns Hopk. Hosp. Bull.* **33**, 61; (1922c) *Ibid.* **33**, 145; (1922d) *Ibid.* **33**, 252; (1922e) *Amer. J. med. Sci.* **164**, 854.
Bloomfield, A. L. and Felty, A. R. (1932a) *Arch. intern. Med.* **32**, 386; (1923b) *Johns Hopk. Hosp. Bull.* **34**, 393; (1923c) *Ibid.* **34**, 414.
Bowden, G. H. W., Ellwood, D. C. and Hamilton, I. R. (1979) In: *Advances in Microbial Ecology.* Vol. 3, p. 135. Ed. by M. Alexander. Plenum Press, New York and London.
Bowie, W. R. et al. (1977) *J. clin. Microbiol.* **6**, 482.
Box, Q. T., Cleveland, R. T. and Willard, C. Y. (1961) *Amer. J. Dis. Child.* **102**, 293.
Bramley, H. M., Dixon, R. A. and Jones, B. M. (1979) *Brit. med. J.* **ii**, 442.
Braude, R., Kon, S. K. and Porter, J. W. G. (1953) *Nutr. Abstr. Rev.* **23**, 473.
Braun, O. H. (1959) *Zbl. Bakt.* **174**, 390.
Brooke, J. W. (1938) *J. dental Res.* **17**, 57.
Bullen, C. L., Tearle, P. V. and Stewart, M. G. (1977) *J. med. Microbiol.* **10**, 403.
Burkholder, P. R. and McVeigh, I. (1942) *Proc. nat. Acad. Sci., Wash.* **28**, 285.
Burky, E. L. and Freese, H. L. (1931) *J. Bact.* **22**, 309.
Burky, E. L. and Smillie, W. G. (1929) *J. exp. Med.* **50**, 643.
Burton, G. C., Hirsh, D. C., Blenden, D. C. and Zeigler, J. L. (1974) In: *The Normal Microbial Flora of Man.* Symp. Series Soc. appl. Bact. No. 3. Ed. by F. A. Skinner and J. G. Carr, Academic Press, London.
Buttiaux, R. and Pierret, J. (1949) *Ann. Inst. Pasteur* **76**, 480.
Calamida, U. and Bertarelli, E. (1902) *Zbl. Bakt.* **32**, 428.
Cannon, P. R. and McNease, B. W. (1923) *J. infect. Dis.* **32**, 175.
Carr, D. L. and Kloos, W. E. (1977) *Appl. envir. Microbiol* **34**, 673.

Clarke, R. T. J. and Bauchop, T. (1977) *Microbial ecology of the gut*. Academic Press, London.
Coates, M. E. *et al.* (1951) *Nature, Lond.* **168**, 332; (1953) *J. Sci. Fd. Agric.* **3**, 43; (1955) *Ibid.* **6**, 419.
Cohendy, M. (1912) *Ann. Inst. Pasteur* **26**, 106.
Colebrook, L. and Maxted, W. R. (1933) *J. Obstet. Gynaec.* **40**, 966.
Collee, J. G. (1974) In: *The Normal Microbial Flora of Man*. Symp. Series Soc. appl. Bact. No. 3. Ed. by F. A. Skinner and J. G. Carr, Academic Press, London.
Corbishley, C. M. (1977) *J. clin. Path.* **30**, 745.
Crecelius, H. G. and Rettger, L. F. (1943) *J. Bact.* **46**, 1.
Cregan, J., Dunlop, E. E. and Hayward, N. J. (1953) *Brit. med. J.* **ii**, 1248.
Cregan, J. and Hayward, N. J. (1953) *Brit. med. J.* **i**, 1248.
Crowley, N., Downie, A. W., Fulton, F. and Wilson, G. S. (1941) *Lancet* **ii**, 590.
Crowther, J. S. (1971) *J. med. Microbiol.* **4**, 343.
Cruickshank, R. (1925) *J. Hyg., Camb.* **24**, 241; (1928) *Brit. J. exp. Path.* **9**, 318; (1934) *J. Path. Bact.* **39**, 213.
Cruickshank, R. and Sharman, A. (1934) *J. Obstet. Gynaec.* **41**, 190, 369.
Cunliffe, A. C. (1949) *Lancet* **ii**, 411.
Daft, F. S. and Sebrell, W. H. (1942) *Publ. Hlth Rep., Wash.* **58**, 1542.
Dam, H., Glavind, J., Orla-Jensen, S. and Orla-Jensen, A. (1941) *Naturwiss.* **29**, 287.
Dehnert, J. (1957a) *Zbl. Bakt., Ref.* **163**, 481; (1957b) *Zbl. Bakt.* **169**, 66.
Devenish, E. A. and Miles, A. A. (1939) *Lancet* **i**, 1088.
Discussion (1913) *Proc. R. Soc. Med.* **6**, Part i, Gen. Rep. p. 1.
Döderlein, A. (1892) *Das Scheidensekret und seine Bedeutung für das Puerperalfieber*. Leipzig.
Drasar, B. S. (1967) *J. Path. Bact.* **94**, 417.
Drasar, B. S. (1974) In: *The Normal Microbial Flora of Man*. Symp. Series Soc. appl. Bact. No. 3. Ed. by F. A. Skinner and J. G. Carr, Academic Press, London.
Drasar, B. S. and Hill, M. J. (1974) *Human intestinal flora*. Academic Press, London.
Ducluzeau, R. and Raibaud, P. (1968) *Ann. Inst. Pasteur* **115**, 941.
Dudgeon, L. S. (1926) *J. Hyg., Camb.* **25**, 119.
Duerden, B. I. (1980a) *J. med. Microbiol.* **13**, 89; (1980b) *J. med. Microbiol.* **13**, 69; (1980c) *Ibid.* **13**, 79.
Eggerth, A. H. (1935) *J. Bact.* **30**, 277.
Eggerth, A. H. and Gagnon, B. H. (1933) *J. Bact.* **25**, 389.
Eichel, H. (1964) *Zbl. Bakt.* **192**, 324.
Elam, J. F. *et al.* (1953) *J. Nutrit.* **49**, 307.
Ellinger, P., Coulson, R. A. and Benesch, R. (1944) *Nature, Lond.* **154**, 270.
Ellis-Pegler, R. B., Crabtree, C. and Lambert, H. P. (1975) *J. Hyg., Camb.* **75**, 135.
Evans, C. A., Smith, W. M., Johnston, E. A. and Giblett, E. R. (1950) *J. invest. Derm.* **15**, 305.
Evans, T. J. *et al.* (1978) *Arch. Dis. Childh.,* **53**, 239.
Eyssen, H. and Somer, P. de (1963) *J. exp. Med.* **117**, 127.
Feldheim, G., Schmidt, E. F. and Haenel, H. (1960) *Zbl. Bakt.* **177**, 62.
Felty, A. R. and Bloomfield, A. L. (1923) *Johns Hopk. Hosp. Bull.* **34**, 379.
Finland, M., Brown, J. W. and Barnes, M. W. (1940) *Amer. J. Hyg.* **32**, B, 24.
Fleming, A. (1929) *Lancet* **i**, 217.

Fleming, A. and Allison, V. D. (1927) *Brit. J. exp. Path.* **8**, 214.
Gabuzda, G. J. *et al.* (1958) *Arch. intern. Med.* **101**, 476.
Galdiero, F., Tufano, M. A., Rossano, F. and Romano, C. (1977) *Bol. 1st. sieroter, milan.* **56**, 429.
Gant, O. K., Ransone, B., McCoy, E. and Elvehjem, C. A. (1943) *Proc. Soc. exp. Biol., N.Y.* **52**, 276.
Garrod, L. P. (1928) *Brit. J. exp. Path.* **9**, 155.
Garside, J. S., Gordon, R. F. and Tucker, J. F. (1960) *Res. vet. Sci.* **1**, 184.
Gerstely, J. R., Howell, K. M. and Nagel, B. R. (1932) *Amer. J. Dis. Child.* **43**, 555.
Gibbons, R. J. and Houte, J. van (1971) *Infection & Immunity* **3**, 567.
Gibbons, R. J., Socransky, S. S. and Kapsimalis, B. (1964) *J. Bact.* **88**, 1316.
Gillespie, E. H., Devenish, E. A. and Cowan, S. T. (1939) *Lancet* **ii**, 870.
Glimstedt, G. (1959) *Ann. N.Y. Acad. Sci.* **78**, 281.
Goodsir, G. (1842) *Edinb. med. surg. J.* **57**, 430.
Gorbach, S. L., Nahas, L., Lerner, P. I. and Weinstein, L. (1967a) *Gastroenterology* **53**, 845.
Gorbach, S. L., Plaut, A. G., Nahas, L., Weinstein, L., Spanknebel, G. and Levitan, R. (1967b) *Gastroenterology* **53**, 856.
Gordon, H. A. and Bruckner-Kardoss, E. (1961) *Amer. J. Physiol* **201**, 175.
Gundel, M. (1933) *Z. Hyg. InfektKr.* **114**, 659.
Gundel, M. and Okura, G. (1933) *Z. Hyg. InfektKr.* **114**, 678.
Gundel, M. and Schwarz, F. K. T. (1932) *Z. Hyg. InfektKr.* **113**, 411.
Gustafsson, B. E. (1959a) *Recent Progress in Microbiology*. Almqvist and Wiksell, p. 327. Stockholm; (1959b) *Ann. N.Y. Acad. Sci.,* **78**, 166.
Gustafsson, B. E. *et al.* (1962) *J. Nutrit.* **78**, 461.
Haenel, H. (1960) *Zbl. Bakt.* **176**, 305; (1961) *J. appl. Bact.* **24**, 242.
Haenel, H., Feldheim, G. and Müller-Beuthow, W. (1958a) *Zbl. Bakt.* **172**, 73.
Haenel, H., Gaszmann, B., Grütte, F. K. and Müller-Beuthow, W. (1964) *Zbl. Bakt.* **192**, 491.
Haenel, H. and Müller-Beuthow, W. (1956) *Zbl. Bakt.* **167**, 123; (1958) *Ibid.* **172**, 93.
Haenel, H., Müller-Beuthow, W. and Grütte, F. K. (1970) *Zbl. Bakt.* **215**, 333.
Haenel, H., Müller-Beuthow, W. and Scheunert, A. (1957) *Zbl. Bakt.* **168**, 37.
Haenel, H. *et al.* (1958b) *Zbl. Bakt.* **173**, 76.
Hallman, F. A. (1937) *Proc. Soc. exp. Biol., N.Y.* **36**, 789.
Hammarström, S., Perlmann, P., Gustafsson, B. E. and Lagercrantz, R. (1969) *J. exp. Med.* **129**, 747.
Hammerschlag *et al.* (1978) *Pediatrics* **62**, 57.
Hardie, J. M. and Bowden, G. H. (1974) In: *The Normal Microbial Flora of Man*. Symp. Series Soc. appl. Bact. No. 3. Ed. by F. A. Skinner and J. G. Carr, Academic Press, London.
Heneghan, J. B. (Ed) (1973) *Germ-free Research. Biological Effect of Gnotobiotic Environment*. Academic Press, London.
Heyde, H. van der and Henderickx, H. (1964) *Zbl. Bakt.* **195**, 215.
Hobbs, B. C. *et al.* (1960) *Mon. Bull. Minist. Hlth. Lab. Serv.* **19**, 178.

Hoffmann, K. (1964) *Zbl. Bakt.* **192,** 500.
Holland, K. T., Roberts, C. D., Cunliffe, W. J. and Williams, M. (1974) *J. appl. Bact.* **37,** 289.
Holt, R. J. (1971) *J. med. Microbiol.* **4,** 319.
Horton, R. E. and Hickey, J. L. S. (1961) *Proc. Animal Care Panel* **11,** 93.
Horwood, M. P. and Minch, V. A. (1951) *Food Res.* **16,** 133.
Huey, C. R. and Edwards, P. R. (1958) *Proc. Soc. exp. Biol., N.Y.* **97,** 550.
Hungate, R. E. (1960) *Bact. Rev.,* **24,** 353; (1963) *Symp. Soc. gen. Microbiol.* **13,** 266.
Johansson, K. R. and Sarles, W. B. (1949) *Bact. Rev.* **13,** 25.
Jukes, T. H. (1955) *Antibiotics in Nutrition.* Antibiotic Monograph No. 4. Medical Encyclopedia, New York.
Kauffmann, F. and Perch, B. (1943) *Acta path. microbiol. scand.* **20,** 201.
Kligman, A. (1965) In: *Skin Bacteria and their Role in Infection,* p. 13. Ed. by H. I. Maibach and G. Hildick-Smith. McGraw-Hill Book Co., New York.
Kloos, W. C. and Musselthwaite, M. S. (1975) *Appl. Microbiol.* **30,** 381.
Kneeland, Y. (1930) *J. exp. Med.* **51,** 617.
Koser, S. A. and Baird, G. R. (1944) *J. infect. Dis.* **75,** 250.
Kraus, F. W. and Gaston, C. (1956) *J. Bcet.,* **71,** 703.
Krunwiede, E. (1949) *Pediatrics* **4,** 634.
Küster, E. (1915) *Arb. Kaiserl Gesundh.* **48,** 1; (1929) See *Kolle and Wassermann's Hdb. path. Mikroorg.* 3te Aufl. **6,** 372.
Kutscher, K. (1909) *Berl. Klin. Wschr.* **66,** 2059.
Laurell, G., Tunevall, G. and Wallmark, G. (1958) *Acta paediat.* **47,** 34.
Lev., M., Briggs, C. A. E. and Coates, M. E. (1957) *Brit. J. Nutrit.* **11,** 364.
Lev. M. and Forbes, M. (1959) *Brit. J. Nutrit.* **13,** 78.
Levison, M. E. and Kaye, D. (1969) *J. infect. Dis.* **119,** 591.
Lindenbaum, J. (1968) *Amer. J. clin. Nutr.* **21,** 1023.
Lindner, J. G. E. M., Plantema, F. H. F. and Hoogkamp-Korstanje, J. A. A. (1978) *J. med. Micriobiol.* **11,** 233.
Loesche, W. J. (1969) *J. Bact.* **99, 520.**
Long, D. A. (1947) *Brit. med. J.* **ii,** 819.
Lowbury, E. J. L. (1969) *Brit. J. Derm.* **81** (Suppl. 1), 55.
Luckey, T. D. (1924) *Amer. J. Nutr.* **27,** 1266; (1959) *Ann. N.Y. Acad. Sci.* **78,** 127.
Luckey, T. D. *et al.* (1955) *J. Nutrit.* **57,** 169.
McFarlan, A. M. (1938) *Brit. med. J.* **ii,** 939.
Maibach, H. I. and Hildick-Smith, G. (1965) *Skin Bacteria and their Role in Infection.* McGraw-Hill Book Co., New York.
Marples, Mary J. (1965) *The Ecology of the Human Skin.* Charles C. Thomas, Springfield, Ill.; (1969) *Brit. J. Derm.* **81** (Suppl. 1), 2; (1974) In: *The Normal Microbial Flora of Man.* Symp. Series Soc. appl. Bact. No. 3. Ed. by F. A. Skinner and J. G. Carr, Academic Press, London.
Marples, R. R. (1969) *Brit. J. Derm.* **81** (Suppl. 1), 47.
Martin, G. J. (1944) *Proc. Soc. exp. Biol., N. Y.* **55,** 182.
Martini, G. A. *et al.* (1957) *Clin. Sci.* **16,** 35.
Martyn, G. (1949) *Brit. med. J.* **i,** 710.
Masters, P. L. *et al.* (1958) *Brit. med. J.* **i,** 1200.
Mieth, H. (1960) *Zbl. Bakt.* **183,** 68.
Miles, A. A., Williams, R. E. O. and Clayton-Cooper, B. (1944) *J. Path. Bact.* **56,** 513.
Mills, K. C., Shibley, G. S. and Dochez, A. R. (1928) *J. exp. Med.* **47,** 193.
Misra, S. S. (1938) *J. Path. Bact.* **46,** 204.
Mitsuoka, T., Morishita, Y., Terada, A. and Yamamoto, S. (1969) *Jap. J. Microbiol.* **13,** 383.
Mitsuoka, T., Sega, T. and Yamamoto, S. (1965) *Zbl. Bakt.* **195,** 455.
Miura, H. (1928) *Mitt. med. Akad. Kioto* **2,** 1.
Moberg, P., Eneroth, P., Harlin, J., Ljung-Wadström, A. and Nord, C. E. (1978) *Med. Microbiol. Immunol.* **165,** 139.
Moore, B. (1954) *J. Path. Bact.* **67,** 461.
Moore, W. E. C. (1966) *Int. J. syst. Bact.* **16,** 173.
Moss, B., Squire, J. R., Topley, E. and Johnston, C. M. (1948) *Lancet* **i,** 320.
Mushin, R. and Ashburner, F. M. (1964) *J. appl. Bact.* **27,** 392.
Mushin, R. and Dubos, R. (1966) *J. exp. Med.* **123,** 657.
Mushin, R., Ford, F. C. P. and Hughes, J. C. (1970) *J. med. Microbiol.* **3,** 573.
Mustakallio, K. K., Salo, O. P., Kiistala, R. and Kiistala, U. (1967) *Acta path. microbiol. scand.* **69,** 477.
Najjar, V. A. and Holt, L. E. (1943) *J. Amer. med. Ass.* **123,** 683.
Najjar, V. A., Holt, L. E., Johns, G. A., Medairy, G. C. and Fleischmann, G. (1946) *Proc. Soc. exp. Biol., N.Y.* **61,** 371.
Najjar, V. A., Johns, G. A., Medairy, G. C., Fleischmann, G. and Holt, L. E. (1944) *J. Amer. med. Ass.* **126,** 357.
Neumann, R. O. (1902) *Z. Hyg. InfektKr.* **40,** 33.
Nielsen, E. and Elvehjem, C. A. (1942) *J. biol. Chem.* **145,** 713.
Niszle, A. (1928) See *Kolle and Wassermann's Hdb. path. Mikroorg.* 3rd edn. 1928–29, **6,** 391.
Noble, W. C. (1966) *J. clin. Path.* **19,** 570; (1971a) *Brit. J. Derm.* **81** (Suppl. 1), 27; (1971b) *J. clin. Path.* **22,** 249; (1981) *Microbiology of human skin.* Lloyd-Luke, London.
Noble, W. C., Fisher, E. A. and Brainard, D. H. (1928) *J. prev. Med., Baltimore* **2,** 105.
Noble, W. C. and Somerville, J. A. (1974) *Microbiology of human skin.* Lloyd-Luke, London.
Noble, W. C., Valkenburg, H. A. and Wolters, C. H. L. (1967) *J. Hyg., Camb.* **65,** 567.
Nuttall, G. H. and Thierfelder, H. (1895–96) *Z. physiol. Chem.* **21,** 109.
Ochi, Y., Mitsuoka, T. and Sega, T. (1964) *Jap. J. vet. Sci.* **26,** 80.
Ohm, M. J. and Galask, R. P. (1975) *Amer. J. Obst. Gynec.* **122,** 683.
Olitsky, P. K. and Gates, F. L. (1921a) *J. exp. Med.* **33,** 713; (1921b) *Ibid.* **34,** 1.
Oppler, B. (1895) *Dtsch. med. Wschr.* **21,** 73.
Orla-Jensen, S., Olsen, E. and Geill, T. (1945) *Senility and Intestinal Flora.* Ejnar Munksgaard, Copenhagen.
Ozawa, A. and Freter, R. (1964) *J. infect. Dis.* **114,** 235.
Pachtman, E. A., Vicher, E. E. and Brunner, M. J. (1954) *J. invest. Derm.* **22,** 389.
Peach, S., Fernandez, F., Johnson, K. and Drasar, B. S. (1974) *J. med. Microbiol.* **7,** 213.
Peter, A. (1960) *Zbl. Bakt.* **177,** 26.
Phillips, I., Warren, C., Harrison, J. M., Sharples, P., Ball, L. C. and Parker, M. T. (1976) *J. med. Microbiol* **9,** 393.
Pitcher, D. G. (1977) *J. med. Microbiol.* **10,** 439.
Porter, J. R. and Rettger, L. F. (1940) *J. infect. Dis.* **66,** 104.
Postgate, J. R. (1974) *J. appl. Bact.* **37,** 185.
Price, P. B. (1938) *J. infect. Dis.* **63,** 301.
Raibaud, P. *et al.* (1966) *Ann. Inst. Pasteur* **110,** 568, 861.
Reith, A. F. (1926) *J. Bact.* **12,** 367.
Report. (1930) *Rep. publ. Hlth med. Subj. Minist. Hlth, Lond.* No. 58; (1972) *Amer. J. clin. Nutr.* **25,** 1292–1494.

Rettger, L. F. and Cheplin, H. A. (1921) *A Treatise on the Transformation of the Intestinal Flora with Special Reference to the Implantation of Acidophilus*. New Haven, Connecticut.
Rettger, L. F., Levy, M. N., Weinstein, L. and Weiss, J. E. (1935) *L. acidophilus: Its therapeutic application*. Yale University Press, New Haven, Connecticut.
Reyniers, J. A. (1959) *Ann. N.Y. Acad. Sci.* **78**, 3, 47.
Reyniers, J. A. et al. (1950) *J. Nutrit.* **41**, 31.
Richardson, R. L. and Jones, M. (1958) *J. dent. Res.* **37**, 697.
Ridley, M. (1959) *Brit. med. J.* **i**, 270.
Rotimi, V. C. and Duerden, B. I. (1981) *J. med. Microbiol.* **14**, 51.
Rountree, P. M. and Barbour, R. G. H. (1951) *J. Path. Bact.* **63**, 313.
Russell, C. and Melville, T. H. (1978) *J. appl. Bact.* **44**, 163.
Sanyal, B. and Russell, C. (1978) *Appl. environm. Microbiol.* **35**, 670.
Sarkany, I. and Gaylarde, C. C. (1968) *J. Path. Bact.* **95**, 115.
Savage, D. C. (1972) In: *Microbial Pathogenicity in Man and Animals* p. 25 Symposium Soc. gen. Microbiol. 1972. Cambridge University Press, Cambridge; (1977) *Annu. Rev. Microbiol.* **31**, 107.
Savage, D. C., Dubos, R. and Schaedler, R. W. (1968) *J. exp. Med.* **127**, 67.
Schaedler, R. W., Dubos, R. and Costello, R. (1965) *J. exp. Med.* **122**, 59, 77.
Scherer, P. (1951) *Z. Hyg. InfektKr.* **132**, 217.
Schild, W. (1895) *Z. Hyg. InfektKr.* **19**, 113.
Schottelius, M. (1899) *Arch. Hyg.* **34**, 210; (1902) *Ibid.* **42**, 48; (1908) *Ibid.* **47**, 177.
Schreiner, A. and Digranes, A. (1979) *J. Infect.* **1** (Suppl. 2), 23.
Schulze, J., Müller-Beuthow, W. and Grutte, F. K. (1970) *Zbl. Bakt.* **215**, 77.
Schweinburg, F. B. and Sylvester, E. M. (1953) *Proc. Soc. exp. Biol., N.Y.* **82**, 527.
Sears, H. J. and Brownlee, I. (1952) *J. Bact.* **63**, 47.
Seeliger, H. P. R. (1957) *Zbl. Bakt.* **170**, 288.
Seeliger, H. P. R. and Werner, H. (1962) *Z. Hyg. InfektKr.* **148**, 27; (1963) *Ann. Inst. Pasteur* **105**, 911.
Selwyn, S. and Ellis, H. (1972) *Brit. med. J.* **i**, 136.
Selwyn, S., Marsh, P. D. and Sethna, T. N. (1976) In: *Chemotherapy*, Vol. 5, p. 391. Ed. by J. D. Williams and A. M. Geddes. Plenum Press, London.
Shibley, G. S., Hanger, F. M. and Dochez, A. R. (1926) *J. exp. Med.* **43**, 415.
Skinner, F. A. and Carr, J. G. (Eds) (1974) *The Normal Microbial Flora of Man*. Symp. Series Soc. appl. Bact. No. 3. Academic Press, London.
Smillie, W. G., Calderone, F. A. and Onslow, J. M. (1943) *Amer. J. Hyg.* **37**, 156.
Smith, H. W. (1959) *J. Path. Bact.* **77**, 79; (1965) *J. Path. Bact.* **89**, 95.
Smith, H. W. and Crabbe, W. E. (1956) *Vet. Rec.* **68**, 274; (1957) *Ibid.* **69**, 24.
Smith, H. W. and Jones, J. E. T. (1963) *J. Path. Bact.* **86**, 387.
Snyder, M. L. (1936) *J. Pediat.* **9**, 624; (1937) *J. infect. Dis.* **60**, 223; (1940) *Ibid.* **66**, 1.
Socransky, S. S. and Manganiello, S. D. (1971) *J. Periodontol.* **42**, 485.
Socransky, S. S. et al. (1977) *J. periodont. Res.* **12**, 90.
Somerville, D. A. (1969) *Brit. J. Derm.* **81** (Suppl. 1), 14.
Somerville, D. A. and Noble, W. C. (1973) *J. med. Microbiol.* **6**, 323.
Soucek, Z. and Mushin, R. (1970) *Appl. Microbiol.* **20**, 561.
Spears, R. W. and Freter, R. (1967) *Proc. Soc. exp. Biol., N.Y.* **124**, 903.
Sprunt, K. Leidy, G. and Redman, W. (1970) *Pediatrics* **46**, 84.
Stephen, A. M. and Cummings, J. H. (1980) *J. med. Microbiol.* **13**, 45.
Stokstad, E. L. R. (1954) *Physiol. Rev.* **34**, 25.
Stokstad, E. L. R. and Jukes, T. H. (1950) *Proc. Soc. exp. Biol., N.Y.* **73**, 523; (1951) *Ibid.* **76**, 73.
Straker, E., Hill, A. B. and Lovell, R. (1939) *Rep. publ. Hlth med. Subj. Minist. Hlth, Lond.* No. 90.
Sukchotiratama, M., Linton, A. H. and Fletcher, J. P. (1975) *J. appl. Bact.* **38**, 277.
Summers, M. M., Lynch, P. F. and Black, T. (1965) *J. clin. Path.* **18**, 13.
Sutter, V. L. and Finegold, S. M. (1974) In: *The Normal Microbial Flora of Man*. Symp. Series Soc. appl. Bact. No. 3. Ed. by F. A. Skinner and J. G. Carr, Academic Press, London.
Syed, S. A., Abrams, G. D. and Freter, R. (1970) *Infection & Immunity* **2**, 376.
Symposium. (1959a) *Recent Progress in Microbiology*. p. 259. Almqvist and Wiksell, Stockholm; (1959b) *Ann. N.Y. Acad. Sci.* **78**, 1.
Thompson, St. C. and Hewlett, R. T. (1896) *Lancet* **i**, 86.
Thomson, S. (1955) *J. Hyg., Camb.* **53**, 357.
Torrey, J. C. and Montu, E. (1931) *J. infect. Dis.* **49**, 141.
Torrey, J. C. and Reese, M. K. (1945) *Amer. J. Dis. Child.* **69**, 208.
Trippestad, A. and Miltvedt, T. (1970) *Acta path. microbiol. scand., B* **78**, 219.
Umemoto, Y. (1961) *Germ-free Human Saliva*. Osaka Dental College, Japan.
Vierthaler, R. W. (1940) *Z. Hyg. InfektKr.* **123**, 126.
Waibel, P. E., Cravens, W. W. and Baumann, C. A. (1953) *J. Nutrit.* **50**, 441.
Wegelius, W. (1909) *Arch. Gynaek.* **88**, 249.
Weinstein, L., Weiss, J. E. and Gillespie, R. W. H. (1938) *J. Bact.* **35**, 515.
Weiss, J. E. and Rettger, L. F. (1937) *J. Bact.* **33**, 423.
Welsh, J. K. and May J. T. (1979) *J. Pediat.* **94**, 1.
Werner, H. (1966) *J. appl. Bact.* **29**, 138.
West, P. A., Hewitt, J. H. and Murphy, O. M. (1979) *J. appl. Bact.* **46**, 269.
Wiel-Korstanje, J. A. A. van der (1973) *Bifidobacteriën en Enterococcen in de Darmflora van de Mens*. Drukkerij Elinlwijk, Utrecht.
Williams, R. E. O. (1946) *J. Path. Bact.* **58**, 259; (1963) *Bact. Rev.* **27**, 56.
Williamson, P. (1965) In: *Skin Bacteria and their role in Infection*. p. 3. Ed. by H. I. Maibach and G. Hildick-Smith, McGraw-Hill Book Co., New York.
Willis, A. T. (1969) *Clostridia of wound infection*. Butterworth, London.
Wilson, G. S. (1957) *Sympos Soc. gen. Microbiol.* No. 7, p. 338.
Wilssens, A. and Buttiaux, R. (1958) *Ann. Inst. Pasteur* **94**, 332.
Witkowski, R. (1935) *Zbl. Bakt.* **133**, 331.
Wolin, M. J. (1979) In: *Advances in Microbial Ecology*, Vol.

3, p. 49. Ed. by M. Alexander, Plenum Press, New York and London.

Woodroffe, R. C. S. and Shaw, D. A. (1974) In: *The Normal Microbiol Flora of Man*. Symp. Series Soc. appl. Bact. No. 3. Ed. by F. A. Skinner and J. G. Carr, Academic Press, London.

Wostman, B. S., Wagner, M. and Gordon, H. A. (1959-60) *Antibiot. Annu. New York*.

Wright, J. (1947) *Lancet* **ii,** 121.

Wright, P. and Terry, C. S. (1981) *J. med. Microbiol.* **14,** 271.

Young, R. M. and Rettger, L. F. (1943) *J. Bact.* **46,** 351.

Zwick and Weichel. (1911) *Arb. ReichsgesundhAmt.* **38,** 327.

Bacterial antagonism: bacteriocines

Bacterial antagonism: bacteriocines	247	vibriocines	249
Introductory	247	*Serratia* and *Yersinia* bacteriocines	249
Colicines	248	Bacteriocines of gram-positive bacteria	249
Bacteriocines of other gram-negative organisms	249	The biological significance of bacteriocines	249
Pyocines	249	Bacteriocine typing	250

Introductory

The subject of bacteriocines is one that does not fit easily into any of the chapters in this book. We shall discuss it here on the supposition that strains forming them may be better able to establish themselves or maintain their position in the flora of their natural environment than they otherwise would be able to do. Many different strains of micro-organisms can be shown by means of in-vitro tests to form substances that inhibit the growth of other strains but to which they are themselves insusceptible. These substances include the antibiotics (Chapter 5)—low-molecular-weight compounds that kill or inhibit the growth of micro-organisms, and are often uniform in their action on all members of a species or genus. Bacteriocines, on the other hand, are substances of high molecular weight—generally proteins—that are formed by bacteria and have strain-specific lethal activity on other bacteria, nearly always including some members of the same species as the producer. Bacteriocine production has been detected in most genera in which it has been sought. When collections of strains from a particular site are examined, e.g. from the throat, the skin or the faeces, they are generally found to include some that form bacteriocines active on other strains from the same source. This suggests that, despite the frequency of bacteriocinogeny, it is not a decisive factor in controlling the composition of the bacterial flora (p. 248).

The first examples of bacteriocine activity to attract attention were those between various gram-positive bacteria, e.g. the inhibition of diphtheria bacilli by staphylococci, which were investigated early in the twentieth century as possible means of eliminating pathogens from the human body (see Abraham and Florey 1949). When Gratia (1925, 1932) demonstrated antagonism between different strains of *Esch. coli* he noted the narrow range of specificity of this activity and concluded that it was due to the production of substances with properties analogous to those of bacteriophages; these were later called 'colicines' (Gratia and Fredericq 1946). Jacob and his colleagues (1953) introduced the term 'bacteriocine' to describe colicines and similar agents formed by other sorts of bacteria.

Many terms intended to designate the species or genus of origin of a particular series of inhibitory substances have since been coined: 'pyocine' (*Ps. aeruginosa*); 'pesticine' (*Yersinia pestis*); 'vibriocine' (*Vibrio cholerae* and related species); 'megacine' (*Bac. megaterium*); 'staphylococcine' (*Staphylococcus*); and so on; many writers omit without justification the final e from these names, e.g. 'colicin'.

Bacteria form many inhibitory substances that are clearly not bacteriocines. These include simple chemicals such as H_2O_2, lytic toxins, including the 'haemolysins' of staphylococci and streptococci and the phospholipases of clostridia, cell-wall-destroying enzymes like the active principle of lysostaphin (Chapter 30), and a number of 'classical' antibiotics, notably the peptide antibiotics of *Bacillus* (Chapter 5). Bacteriocines resemble phages in that a specific receptor on the sensitive cell is necessary for their adsorption (Fredericq 1946); indeed agents of the two sorts may share the same receptor. However, bacteriocines differ from phages in that when they destroy bacterial cells they do not undergo multiplication. Colicines are further distinguished from phages in that they are non-particulate, though other agents usually classified as bacteriocines, for example, certain of the pyocines, can be shown electronmicroscopically to be rod-shaped bodies, some of them resembling phage tails.

The bacteriocines of gram-negative bacteria, like phages, tend to have a narrow range of activity confined to the species of the producer strain and closely related species, but there are exceptions to this. The agents produced by gram-positive bacteria usually have a wider range of activity.

Jacob and his colleagues (1953) defined bacteriocines as proteins the biosynthesis of which leads to the death of the producer cell, that are active at the intraspecific level, and that require a specific receptor for adsorption. Other workers would add further characters to the definition, for example, that their action is bactericidal, that the producer culture is immune to its own bacteriocine, or that the genetic determinant is a plasmid. A clear definition applicable to the bacteriocines of gram-negative and gram-positive bacteria is in fact difficult to formulate, and distinctions between them and 'classical' antibiotics, bacterilytic enzymes or defective phages are at times hard to make. Tagg and his colleagues (1976) point out that essential information about many of the agents that have been described as bacteriocines is still lacking and suggest that until this becomes available the term 'bacteriocine-like' should be used. They also give practical advice on how to distinguish bacteriocines from other inhibitory agents.

The colicines

These form the most thoroughly investigated group of bacteriocines. They are produced by and act upon strains within the family Enterobacteriaceae, but their spectrum of activity tends to centre around the species of the producer strain. They are usually formed spontaneously in liquid and on solid media, but increased production is induced by ultraviolet light or mitomycin C.

Colicines were initially classified into a series of groups (Fredericq 1948, 1957, Hamon and Péron 1963) according to the indicator strains to which they were adsorbed, i.e. their specificity for colicine receptors. Members of these groups may differ in other ways; for example, colicines E1, E2 and E3, though sharing the same receptor, differ in their mode of action (see below). Even within these subdivisions, the colicines E1 and E2 produced by wild strains do not form homogeneous groups but can be distinguished by differences in the immunity of indicator strains to them and in their serological specificity (Lewis and Stocker 1965). In describing a particular colicine, therefore, the identity of the producer strain should always be given.

The colicines examined chemically by Goebel and his colleagues (Goebel et al. 1955, Goebel and Barry 1958, Hutton and Goebel 1963, Barry et al. 1963) were lipopolysaccharide-protein complexes with an antigenic specificity identical with that of the O antigen of the producer strain. However, their antibiotic activity was destroyed by trypsin. When K-colicinogeny had been transferred from *Esch. coli* strain K235 to *Sh. sonnei*, the colicine was isolated as a lipopoly-saccharide-protein complex with the specificity of the *Sh. sonnei* O antigen. Despite differing antigenicity, preparations from the escherichia and shigella strains were identical in their antibacterial spectrum and in eliciting K-neutralizing antibody (Hinsdill and Goebel 1966). These studies were made with spontaneously released colicines, but induced cultures yielded active preparations of protein free from lipopolysaccharide (Herschman and Helsinki 1967, Jesaitis 1970). The colicines are now considered to be proteins of molecular weight 50 000-90 000 (see Hardy 1975).

In all wild strains that have been adequately studied the genetic determinants for colicinogeny are on plasmids; these are referred to as Col factors or Col plasmids. They can be divided into two groups (see Hardy 1975, Hughes et al. 1978): group I comprises plasmids of molecular weight $ca\ 5 \times 10^6$, specifying, for example, colicines E1, E2, E3 and K; and group II, larger plasmids ($6-9 \times 10^7$), present in fewer copies and specifying, for example, colicines I, B and V. Group II plasmids are self-transmissible, and many belong to the F or I incompatibility groups (Chapter 6). For the transmission of the group I plasmids the presence of another self-transmissible plasmid is necessary. Chromosomally integrated Col factors, though formed in the laboratory, have not been identified in wild strains.

Colicines kill but do not lyse sensitive strains. Colicines A, E1 and K act as uncouplers of oxidative phosphorylation (Fields and Luria 1969, Nagel de Zwaig 1969). Colicine E2 inhibits DNA synthesis and degrades DNA (Nomura 1963) and is a DNA endonuclease (Schaller and Nomura 1976). Colicine E3 inhibits protein synthesis by attacking the 30S ribosomal subunit (Konisky and Nomura 1967, Senior et al. 1970). Colicines Ia and Ib inhibit all macromolecular synthesis (Levisohn et al. 1968).

The kinetics of cellular killing by colicines suggested with a certain probability that the process resulted from contact with a single bacterial cell ('quantal killing'; Luria 1970). The lethal effect was reversible by trypsin for a few minutes after the adsorption of the colicine to the sensitive cell (Nomura 1963), and radioactively labelled colicine appeared to remain on the surface of the cells (Maeda and Nomura 1966). This led to the concept of a two-stage process in which reversible attachment to the cell envelope led to irreversible changes which were transferred to and amplified in the cell membrane, causing widespread biochemical disturbances (see Nomura 1967). However, it soon became apparent that adsorption to the receptor site was not an essential prelude to the lethal action of the colicine. In the absence of the cell envelope, for example in L-forms, this was enhanced rather than diminished (Šmarda and Taubeneck 1968), and when the cell membrane was rendered accessible to the colicine direct damage to it could be demonstrated (Bhattacharyya et al. 1970). Moreover, some colicines, e.g. E2 and E3, were shown to act directly on specific biochemical targets, respectively in the genome and the ribosome, in cell-free preparations.

Bacteria that possess suitable receptors on the cell envelope may nevertheless be insusceptible to the colicine in question. Such strains form a small-molecular-weight protein ('immunity protein') that specifically neutralizes the colicine (Sidikaro and Nomura 1974); the gene for its production is present on the corresponding Col plasmid, e.g. the E3 immunity protein is coded for on the Col E3 plasmid (Sidikaro and Nomura 1975). Colicine preparations from which the immunity protein has been removed have a remarkably broad range of activity; thus, purified colicine E2 degrades cell-free DNA from bacteria, phages and simian virus 40 (Schaller and Nomura 1976). (For the action of purified bacteriocines on eukaryotic cells, see Farkas-Himsley 1980.)

Bacteriocines of other gram-negative organisms

Pyocines (Jacob 1954) are widespread in *Ps. aeruginosa* (Hamon *et al.* 1961); most of them act only on members of this species but a few also attack strains of *Ps. fluorescens*. Enterobacteria in the smooth state are resistant but rough mutants are said to be sensitive (Hamon 1956). Some of the pyocines (S type) are non-particulate and appear to be proteins resembling the colicines (Ito *et al.* 1970), but others are particulate (Kageyama 1964, Ishii *et al.* 1965, Takeya *et al.* 1967). The particulate pyocines are either (1) contractile sheathed rods (R type) resembling headless phages (Kageyama 1964) or (2) longer, thinner, non-contractile rods (F type; Govan 1974). (For illustrations of the particulate pyocines, see Bradley 1967 and Govan 1974.) Some pyocines of the R type act on gonococci and serologically ungroupable meningococci (see Chapter 31).

Vibriocines act not only on members of the genus *Vibrio* but on some members of a variety of enterobacterial species (Datta and Prescott 1969). They seem to be associated with the presence of structures resembling contractile phage tails (Jayawardene and Farkas-Himsley 1968, Lang *et al.* 1968). One vibriocine was found to inhibit DNA synthesis and to degrade DNA (Jayawardene and Farkas-Himsley 1970).

Members of the genus *Serratia* form two groups of inhibitory agents, one resembling colicines in their range of inhibitory action on enterobacteria and the other attacking only strains belonging to the genus of the producers (Chapter 34). Most strains of *Yersinia pestis* release inhibitory substances that kill *Y. pseudotuberculosis*, a character which Beesley and Surgalla (1970) suggest may be useful in the identification of plague bacilli.

Bacteriocines of gram-positive bacteria

Inhibitory substances resembling bacteriocines have been reported in *Bacillus, Clostridium, Corynebacterium, Lactobacillus, Listeria, Mycobacterium, Staphylococcus* and *Streptococcus*. Many of them have not been sufficiently well characterized to establish their true identity, but a number have several properties in common with the colicines. Most of them differ from colicines in not being inducible by ultraviolet light or mitomycin C and in having a wider range of activity; and host-cell immunity to a number of them is only partial. We shall not attempt to detail the known properties of the agents, but refer readers to the monograph of Tagg and his colleagues (1976) and to relevant chapters of Volume 2 of this book.

The production of bacteriocines and similar agents by gram-positive organisms is often dependent on the conditions of growth, such as the presence of particular nutrients and sometimes of oxygen; a few are formed only on solid media. Most of the well characterized agents have a protein component; the molecular weights quoted for these vary over a wide range, but some are considerably lower (*ca* 10 000) than those of the colicines. These proteins differ in their susceptibility to proteolytic enzymes, some being resistant to trypsin. In others the protein has not been separated from lipid and carbohydrate components, and a few have not been demonstrated in cell-free preparations.

With certain of the agents, notably in those of *Bacillus, Clostridium, Listeria* and *Mycobacterium*, activity is associated with the presence of phage-like particles, sometimes after induction. The ability to form an inhibitory agent is often irreversibly lost in culture, and in some the rate of loss is accelerated by treatment with acridines or growth at a higher temperature, suggesting that bacteriocinogeny is plasmid-determined. A few of the relevant plasmids have been characterized, e.g. the one that is responsible for bacteriocine production and haemolysis in *Str. faecalis* (Chapter 29) and the one found in *Staph. aureus* strains of phage-type 71 (Chapter 30). The inhibitory substances formed by gram-positive organisms are generally specific for the producer strain, and usually also attack some members of other genera. Thus, agents from staphylococci often act on a number of streptococci and corynebacteria, from *Bacillus* spp. on micrococci, and from clostridia on strains of *Bacillus*; agents produced by streptococci, staphylococci and corynebacteria are said to inhibit various gram-negative bacilli.

The biological significance of bacteriocines

Convincing evidence that bacteriocinogeny promotes or maintains colonization is difficult to find (see Hardy 1975, Tagg *et al.* 1976). Effective action by trypsin-sensitive agents such as colicines in the gut is difficult to envisage. Evidence of an effect on the flora of the skin is equivocal (see Woodroffe and Shaw 1974); moreover a strain of *Staph. aureus* (502A) that has undoubted powers of interfering with the growth of other strains of this species on human skin and mucous membranes (Chapter 60) has not been shown to form an inhibitory agent *in vitro*. The inhibitory activities of the resident streptococcal flora of the throat might perhaps prevent colonization by group A streptococci for which bacteriocines were responsible (Crowe *et al.* 1973, Dajani *et al.* 1976), but the evidence is far from conclusive. Streptococcal bacteriocines tend to be inactivated by proteases in the saliva (Kelstrup and Gibbons 1969), and, in any case, strains of *Str. mutans* can coexist with bacteriocines to which they are susceptible in dental plaque, probably being protected from their action by the extracellular polysaccharide (Rogers 1974).

Plasmids coding for bacteriocinogeny sometimes

also specify other important biological characters: a virulence factor in *Esch. coli* (Col V; see Chapter 34), the production of one of the epidermolytic toxins in *Staph. aureus* (Chapter 30), and haemolysis in *Str. faecalis* (Chapter 29), but in none of these is there clear evidence that the bacteriocine is itself responsible for the other effect.

Bacteriocine typing. Strains of bacteria can be characterized by the range of activity of their bacteriocines on a set of indicator strains, by their sensitivity to a set of bacteriocine-producing strains, or by both. The use of bacteriocine typing is discussed in Chapter 20, and many typing systems are mentioned in appropriate chapters in volumes 2 and 3 of this book.

(For general reviews of bacteriocines, see Fredericq 1963, Reeves 1965, Nomura 1967, Mayr-Harting *et al.* 1972, Hager 1973, Hardy 1975, Tagg *et al.* 1976, and Konisky 1978.)

References

Abraham, E. P. and Florey, H. W. (1949) In: *The Antibiotics*, Vol 1, Chapter 11. Ed. by H. W. Florey *et al.* Oxford University Press, London.
Barry, G. T., Everhardt, D. L. and Graham, M. (1963) *Nature, Lond.* **198**, 211.
Beesley, E. D. and Surgalla, M. J. (1970) *Appl. Microbiol.* **19**, 915.
Bhattacharyya, P., Wendt, L., Whitney, E. and Silver, S. (1970) *Science* **168**, 998.
Bradley, D. E. (1967) *Bact. Rev.* **31**, 230.
Crowe, C. C., Sanders, W. E. and Longley, S. (1973) *J. infect. Dis.* **128**, 527.
Dajani, A. S., Tom, M. C. and Law, D. J. (1976) *Antimicrob. Agents Chemother.* **9**, 81.
Datta, A. and Prescott, L. M. (1969) *J. Bact.* **98**, 849.
Farkas-Himsley, H. (1980) *J. antimicrob. Chemother.* **6**, 424.
Fields, K. L. and Luria, S. E. (1969) *J. Bact.* **97**, 57.
Fredericq, P. (1946) *C.R. Soc. Biol., Paris* **140**, 1189; (1948) *Rev. belge Path.* Suppl. 4; (1957) *Annu. Rev. Microbiol.* **11**, 7; (1963) *Ergebn. ImmunForsch.* **37**, 114.
Goebel, W. F. and Barry G. T. (1958) *J. exp. Med.* **107**, 185.
Goebel, W. F., Barry, G. T., Jesaitis, M. and Miller, E. M. (1955) *Nature, Lond.* **176**, 700.
Govan, J. R. W. (1974) *J. gen. Microbiol.* **80**, 1, 17.
Gratia, A. (1925) *C. R. Soc. Biol., Paris* **93**, 1040; (1932) *Ann. Inst. Pasteur* **48**, 413.
Gratia, A. and Fredericq, P. (1946) *C.R. Soc. Biol., Paris* **140**, 1032.
Hager, L. P. (Ed.) (1973) *Chemistry and Functions of Colicines*. Academic Press, London and New York.
Hamon, Y. (1956) *Ann. Inst. Pasteur* **91**, 82.
Hamon, Y. and Péron, Y. (1963) *C.R. Acad. Sci., Paris* **257**, 309.
Hamon, Y., Véron, M. and Péron, Y. (1961) *Ann. Inst. Pasteur* **101**, 738.
Hardy, K. G. (1975) *Bact. Rev.* **39**, 464.
Herschman, H. R. and Helsinki, D. R. (1967) *J. biol. Chem.* **242**, 5360.
Hinsdill, R. D., and Goebel, W. F. (1966). *J. exp. Med.* **123**, 881.
Hughes, V., Le Grice, S., Hughes, C. and Meynell, G. G. (1978) *Molec. gen. Genet.* **159**, 219.
Hutton, J. J. and Goebel, W. F. (1963) *J. gen. Physiol.* **45**, 125.
Ishii, S., Nishi, Y. and Egami, F. (1965) *J. molec. Biol.* **13**, 428.
Ito, S., Kageyama, M. and Egami, F. (1970) *J. gen. appl. Microbiol, Tokyo* **16**, 205.
Jacob, F. (1954) *Ann. Inst. Pasteur* **86**, 149.
Jacob, F., Lwoff, A., Siminovitch, L. and Wollman, E. (1953) *Ann. Inst. Pasteur* **84**, 222.
Jayawardene, A. and Farkas-Himsley, H. (1968) *Nature, Lond.* **219**, 79; (1970) *J. Bact.* **102**, 382.
Jesaitis, M. A. (1970) *J. exp. Med.* **131**, 1016.
Kageyama, M. (1964) *J. Biochem., Tokyo* **55**, 49.
Kelstrup, J. and Gibbons, R. J. (1969) *J. Bact.* **99**, 888.
Konisky, J. (1978) In: *Bacteria: a Treatise on Structure and Function*, Vol 6, p. 71. Ed. by L. C. Gunsalus and L. Ornston. Academic Press, New York.
Konisky, J. and Nomura, M. (1967) *J. molec. Biol.* **26**, 181.
Lang, D., McDonald, T. O. and Gardner, F. W. (1968) *J. Bact.* **95**, 708.
Levisohn, R., Konisky, J. and Nomura, M. (1968) *J. Bact.* **96**, 811.
Lewis, M. J. and Stocker, B. A. D. (1965) *Zbl. Bakt.* **196**, 173.
Luria, S. E. (1970) *Science* **168**, 1166.
Maeda, A. and Nomura, M. (1966) *J. Bact.* **91**, 685.
Mayr-Harting, A., Hedges, A. J. and Berkeley, R. C. W. (1972) In: *Methods in Microbiology*, Vol. 7A, p. 315. Ed. by J. R. Norris and D. W. Ribbons. Academic Press, London.
Nagel de Zwaig, R. (1969) *J. Bact.* **99**, 913.
Nomura, M. (1963) *Cold Spr. Harb. Symp. quant. Biol.* **28**, 315; (1967) *Annu. Rev. Microbiol.* **21**, 257.
Reeves, P. (1965) *Bact. Rev.* **29**, 24.
Rogers, A. H. (1974) *Antimicrob. Agents Chemother.* **6**, 547.
Schaller, K. and Nomura, M. (1976) *Proc. nat. Acad. Sci., Wash.* **73**, 3989.
Senior, B. W., Kwasniak, J. and Holland, I. B. (1970) *J. molec. Biol.* **53**, 205.
Sidikaro, J. and Nomura, M. (1974) *J. biol. Chem.* **249**, 445; (1975) *Ibid.* **250**, 1123.
Šmarda, J. and Taubeneck, U. (1968) *J. gen. Microbiol.* **52**, 161.
Tagg, J. R., Dajani, A. S. and Wannamaker, L. W. (1976) *Bact. Rev.* **40**, 722.
Takeya, K., Minamishima, Y., Amako, K. and Ohnishi, Y. (1967) *Virology* **31**, 166.
Woodroffe, R. C. S. and Shaw, D. A. (1974) In: *The Normal Microbiol. Flora of Man*. Symp. Series Soc. appl. Bact., No. 3. Ed. by F. A. Skinner and J. G. Carr. Academic Press, London.

9

The bacteriology of air, water and milk
Graham Wilson

Air

Introductory	251	Estimating human pollution of the air	255
Outdoor air	251	Measures for controlling airborne infection	256
Indoor air	252	Ventilation	256
Dust	252	Safety cabinets	256
Droplets	252	Filtration	256
Droplet nuclei	253	Infection caused by droplet nuclei	257
Measurement and degree of air contamination	254	Ultraviolet irradiation	257
Sedimentation	254	Disinfectant sprays and vapours	257
Impaction	254	The control of respiratory infection	257
Impingement	254		
Filtration	255	**Water**	**260**
Animal method	255	**Milk**	**279**
Particle-size distribution	255		

Introductory

In chapter 60 we discuss the role of the staphylococcal disperser in the spread of pyogenic infection, and in Chapter 51 the apparently conflicting views of Cornet and of Flügge on the respective parts played by dust and by droplets in the spread of pulmonary tuberculosis. In the present chapter we shall enter a little more fully into the mechanics of airborne infection; but for a further study of this subject we must refer our reader to the Symposia on Aerobiology held at Chicago in 1942 (Report 1942), in London in 1967 (Report 1967), at Brighton in 1969 (Report 1970), and in Zürich (Report 1973*a*) and Enschede (Report 1973*b*); to the special volumes on airborne infection and experimental aerobiology by Jennison (1942), Rosebury (1947), Magill, Holden and Ackley (1956), Riley and O'Grady (1961), and Dimmick and Ackers (1969); to reports by the Committee on Research and Standards (Report 1947), and the Medical Research Council (Report 1948); to papers by Williams (1948, 1949*a*, 1951, 1967); to various authors in the review on Air Hygiene (Report 1961); and for a description of air-sampling instruments to Report (1966).

Outdoor air

The degree of bacterial contamination of the outside air depends on a number of factors, such as the density of the human and animal population, the amount of vegetation, the nature of the ground or soil, the temperature and humidity of the atmosphere, the direction and speed of the wind, solar radiation, and rainfall in different proportions (see Rüden *et al.* 1978). According to Errington and Powell (1969), the number of bacteria is very small—one viable organism in some tens of litres. Spores and fragments of moulds are about 100 times as frequent. Aeroplane surveys (see Proctor 1934, 1935, Proctor and Parker 1938) show that bacteria in the upper air consist largely of aerobic spore-bearing bacilli, and to a much less extent of organisms belonging to the *Achromobacter*, *Sar-*

cina, and *Micrococcus* groups, suggesting that they are derived mainly from soil and surface dust. Their numbers are much the same over land and sea (Pady and Kelly 1953), those over the sea being contained in small droplets liberated by bursting air bubbles to which organisms in the water are adsorbed (Baylor *et al.* 1977). These organisms can be carried vertically for several miles into the air, and for long distances horizontally (Hirst and Hurst 1967). Though there is evidence to suggest that some resistant organisms, such as the foot-and-mouth virus, may be borne by wind currents from one area or country to another and, on reaching the ground, set up disease in susceptible animals (see Chapter 89), most of our observations indicate that infective material is seldom carried for more than short distances by air, or that, when it is, it is too diluted or damaged by atmospheric agencies to cause recognizable disease. Autotrophic bacteria may undergo some multiplication in cloud areas, where moisture and traces of ammonia and carbon dioxide are present, but it is very improbable that pathogenic organisms can grow in air. For the medical bacteriologist the infective capacity of the flora of the outside air has still to be demonstrated.

Indoor air

Air-borne bacteria may be distributed in three different forms: (*a*) attached to dust particles; (*b*) contained in gross droplets expelled from the nose and mouth; (*c*) present in droplet nuclei, which result from evaporation of the smaller droplets expelled from the nose and mouth.

Dust. Dust consists of varying sized particles of animal, vegetable, or mineral origin. The medical bacteriologist is concerned particularly with dust derived from human beings, who are the ultimate source of most of the common pathogenic organisms. Secretion from the nose gets carried on to the alae nasi and the upper lip, and thence by the hands on to the skin, clothing, and bedding, from which it may be liberated in the form of dust (Rubbo and Benjamin 1953, Hare and Thomas 1956). Organisms, such as staphylococci, are often detached directly from the healthy or diseased skin of different parts of the body, including the perineum (Hare and Ridley 1958), as well as from septic wounds. Intestinal organisms may be disseminated in dried particles of faeces, which are abundant in the napkin stage of infancy, in young children, and in incontinent persons. However they are liberated, the heavier particles fall rapidly to the ground, but can be stirred up by sweeping and draughts; the lighter particles, which may be 1 μm or less in diameter, remain more or less permanently suspended in the atmosphere. It was shown by White (1936), Cruickshank (1938), Hare (1941), Thomas (1941) and others that haemolytic streptococci might be found in the dust of wards harbouring cases or carriers of these organisms. Tubercle bacilli have often been demonstrated in the dust of sanatoria (see Pressman 1937). Wright, Shone and Tucker (1941); and Crosbie and Wright (1941) demonstrated diphtheria bacilli in large numbers in floor dust collected near diphtheria patients (see also Herrmann and Pütz 1942). Staphylococci are frequent in the dust of hospital wards. The survival of these organisms is affected by numerous factors, such as the temperature and humidity of the atmosphere and the amount of light to which they are exposed (see Engley 1955). Under favourable conditions they may remain alive for many weeks (but see Hinton *et al.* 1960). Quantitatively, bedclothes are one of the most prolific sources of dust. So also are personal clothes, which provide most of the bacteria present in the air of occupied rooms (Duguid and Wallace 1948, Williams 1949*b*). Skin fragments containing squamous epithelial cells, when airborne, have a mean diameter of about 8 μm: 50 per cent of them fall between 4 and 20 μm (Noble *et al.* 1963, Mackintosh *et al.* 1978). They appear to be the chief constituent of household dust (Clark 1974), and can be identified by their content of squalene (Clark and Shirley 1973). Physical activity of the clothed or naked body increases the liberation of these particles (Chapter 60). Hand-washing has a similar effect (Mears and Yeo 1978).

The interesting observations of Lewis and his colleagues (1969) on the aerodynamics of the human microenvironment may be quoted here. By the use of *Schlieren* photography they found that the human body was closely enveloped in a stream of air, 1–2 cm thick, which, starting at the feet, passed upwards at increasing velocity till by the time it reached the head it was travelling at the rate of half a metre per second. Bacteria and other particles from the skin and from the atmosphere were entrained by this stream of air, some of which was inevitably inhaled when it reached the nose. The part played by the boundary layer in auto-infection, and in contamination of the ambient air as it leaves the body, is still a subject for inquiry.

Droplets. Since Flügge (1899) first drew attention to the role of minute droplets expelled from the mouth and nose of phthisical patients in conveying tuberculosis, the concept of droplet-borne infection has been widely applied to the transmission of respiratory disease in general. Renewed interest in its mechanism was stimulated by the studies of Wells and his colleagues in the United States (Wells 1934, Wells and Wells 1936, 1938). The observations of these workers, and of others such as Weyrauch and Rzymkowski (1938) and Jennison (1942), amplified the earlier findings of Lange and Keschischian (1925) and Strauss (1926), and threw fresh light on the size and fate of droplets emitted from the mouth during sneezing, coughing, whistling, singing, and talking. A brief abstract of the earlier reports is given in the chapter on tuberculosis (Chapter 51), but it will be well here to expand this description in the light of more recent work.

During a sneeze large numbers of droplets—up to a million or more—are expelled, mainly through the closed teeth, at an initial velocity of about 150 feet per second (Jennison, 1942, Hatch 1942, Duguid 1945). Smaller numbers of droplets are also projected during coughing, though less rapidly. The still smaller numbers produced during talking, especially in the pronunciation of sharp consonants such as p, b, t, f, k, and s, are of larger size and lower velocity; few travel more than a foot.

The number and size of the particles generated by sneezing and coughing are compared by Gerone and his colleagues (1966) in Table 9.1, from which it is seen that over 90 per cent are under 5 μm in diameter.

Table 9.1 Size of droplets produced by sneezing and coughing (Gerone et al. 1966)

Particle diameter in μm	No. of particles	
	Sneeze	Cough
<1–1	800 000	66 000
1–2	686 000	21 300
2–4	101 000	2 800
4–8	16 000	700
8–15	1 600	38
Total	1 604 600	90 838

The fate of these droplets depends on their size and initial velocity. Those larger than 1 mm in diameter, when projected during a sneeze from the mouth of a man of average height, fall in a curved trajectory and reach the ground in about 4.5 metres (15 feet). On the other hand, those less than 0·1 mm possess too little kinetic energy to enable them to travel more than a very short distance and are so little affected by gravitation and possess such a large surface in relation to their volume that they evaporate before they have fallen any appreciable distance. The evaporation time in seconds of droplets of different sizes is given in Table 9.2. After evaporation, their size is of course much smaller. (For photographs of droplet expulsion in a sneeze, see Weyrauch and Rzymkowski 1938, Bourdillon and Lidwell 1941, Jennison 1942.)

Table 9.2 Evaporation time of water droplets in unsaturated still air at 22°. (Taken from Jennison 1942, modified from Wells 1934.)

Diameter of droplet in μm	Evaporation time in seconds
2000	515.0
1000	129.0
500	32.0
200	5.2
100	1.3
50	0.31
25	0.08
12	0.02

Droplet nuclei. On evaporation, the smaller droplets become converted into what Wells calls *droplet nuclei*. These behave very much like minute particles of smoke. In absolutely still air they fall at the rate of probably 1 to 3 feet per hour, but in the ordinary atmosphere they remain suspended almost indefinitely, being dependent for their transportation on air currents. It has been shown experimentally that micro-organisms attached to droplet nuclei are rapidly dispersed throughout rooms and even buildings, and are inhaled by anyone breathing air containing them (Wells and Wells 1936).

Trillat (1938) described an interesting experiment in which he atomized one litre of a fluid suspension of *Serratia marcescens* in a central court of the Pasteur Institute in Paris. A hundred agar plates were exposed for more than 100 metres around. Every single plate became infected, even those to windward of the court.

What proportion of droplets and droplet nuclei contain bacteria, and of these what proportion are infective, probably varies greatly from patient to patient. Duguid (1946a), who held plates 3 inches in front of the mouth of patients during a series of six voluntarily produced coughs, found that 39 out of 87 patients suffering from scarlet fever or carrying haemolytic streptococci in the throat expelled infected droplets. The number of these varied with individual patients from 0 to 400. Of all droplets expelled by the 87 patients, only 10 per cent contained haemolytic streptococci. Similar observations were made on patients affected with diphtheria and pulmonary tuberculosis.

The survival of bacteria and viruses within airborne particles is determined by a multiplicity of factors that debars the drawing of any trustworthy generalization. Information on the subject will be found in studies by de Jong and Winkler (1964), Cox (1966), Anderson and Cox (1967), Hatch and Wolochow (1969), Akers (1969), and Fedorak and Westlake (1978).

Apart from the usual factors that tend to the destruction of bacteria, Druett and May (1968) observed one, which they refer to as the *open air factor OAF*, that was responsible for an unexpectedly high death rate among cells of *Esch coli* exposed to the night air at Porton. This substance was stable only in fresh air; it decomposed on contact with a surface. It appeared to be an olefin-ozone complex, formed as the result of the combination of olefins contained in the exhaust fumes of internal combustion engines with ozone in the atmosphere. Its uses as a germicidal agent for indoor air is explored by Druett and Packmann (1972). In the Netherlands its concentration was found to vary greatly from day to day and from one part of the country to another (de Mik and de Groot 1977).

Droplet nuclei are, of course, much smaller than droplets and hence less often contain bacteria; on the other hand there are far more of them.

From experiments carried out on patients during sneezing and coughing, Duguid (1946b) found that droplets and droplet nuclei ranged from 0.5 to 2,000 μm in diameter, 95 per cent being between 2 and 100 μm. Of the droplet nuclei proper 97 per cent were between 0·5 and 12 μm, the commonest being 1–2 μm. Assuming that saliva, which provides most of the material for droplet nuclei, contains 30 million organisms per ml, Duguid calculated the proportion of droplets containing one or more bacteria (Table 9.3):

Table 9.3 Proportion of small droplets containing one or more bacteria (Duguid 1946b).

Diameter of droplet in μm	Percentage infected
1–	0.0059
2–	0.047
4–	0.38
8–	3
16–	12
24–	30
32–	51
40–	76
50–	98
75–	100

Since about 90–97 per cent of the droplets expelled during sneezing are less than 50 μm in diameter, and therefore evaporate to droplet nuclei within a fraction of a second, it follows that 10 000, 100 000, or 1 000 000 droplet nuclei may result from a single sneeze. Of these probably 50 per cent or so contain one or more bacteria, though most of these will be non-pathogenic even when the patient is suffering from respiratory disease. In the examination of air taken from hospital wards, operating theatres and offices Noble, Lidwell and Kingston (1963) found that most organisms associated with human beings were attached to fairly large particles in the range of 4–20 μm equivalent diameter. This suggests that they were derived mainly from dried saliva and nasal mucus, skin scales and clothes.

The proportion of dust particles and droplet nuclei that reach the lung when inhaled depends mainly upon their size. Experiments on animals and to a less extent on man indicate that practically all particles over 5 μm in diameter are retained in the nose; below 5 μm an increasing proportion pass the nasal filter; most particles of 1 μm reach the lung and are retained in the alveoli, but below 1 μm the proportion retained in the lung diminishes. Moreover the slower the rate of breathing, the greater is the proportion of particles that can pass the nasal filter (see Wells and Lurie 1941, Boyland et al. 1947, Brown et al. 1950, Ostrom et al. 1958, Druett 1967).

The part played by air-borne droplet nuclei in the spread of respiratory infection was studied by Wells and Wells (1942), and is discussed in Chapter 51. An experiment, however, may be quoted to prove the reality of airborne infection. Couch and his colleagues (1966) exposed 12 volunteers to contact with subjects infected by Coxsackie A21 virus but separated from them by a wire barrier. Contamination of the air resulted mainly from coughing. Every one of the volunteers contracted infection.

Infective or potentially infective droplets may be liberated in the form of aerosols not only from the human respiratory tract but by various laboratory procedures, such as the opening of screw-capped culture bottles (Tomlinson 1957), the removal of a cotton plug from a shaken culture flask, the streaking of an agar plate with a loopful of broth culture, and the intraperitoneal injection of a guinea-pig (see Wedum 1964). Aerosols are also produced in various dental manipulations, particularly during the use of the air-turbine drill (Mazzarella and Flynn 1969, Hausler 1970); and in the flushing of water-closets (Darlow and Bale 1959).

Measurement and degree of air contamination

Several methods have been devised for measuring the bacterial contamination of the atmosphere. The results are often widely at variance, mainly because of the differences in methods of sampling. We shall draw attention here to some of the various methods that have been described, but for a discussion of the physical principles underlying them we would refer our readers to papers by May (1967) and Akers and Won (1969), and for a description of air-sampling instruments to Report (1966).

Most samples are taken of moving air, whether indoors or outdoors. Ideally the air should be allowed to flow into the collecting orifice without change in direction or velocity. This is known as *isokinetic sampling*. When the velocity of sampling is greater than that of the moving air small particles will predominate: when it is slower large particles will predominate. Owing to what May (1967) refers to as 'small-scale turbulence,' laminar air flow seldom occurs under natural conditions. Departure from isokinetic sampling will therefore result in a disproportion of the different-sized particles collected from the atmosphere.

Briefly, the methods used are sedimentation, impaction, impingement, filtration, precipitation, and animal inoculation.

Exposure in still air of agar plates in different parts of a room allows the larger particles and dust to settle, but unless it is continued for several hours, few droplet nuclei will be deposited. Unrepresentative as the results may be, **sedimentation** nevertheless does afford a measure of the number of organisms that are likely to contaminate an exposed surface (Blowers and Wallace 1960).

By **impaction** is meant the collection of particles on to a solid or semi-solid surface. Instruments such as the air centrifuges of Wells, Phelps and Winslow (1937) and of Hollaender and Dalla Valle (1939) have been devised for this purpose, but one of the most effective is Bourdillon's slit sampler (Bourdillon et al. 1948). As Blowers and Wallace (1960) point out, this instrument has the merit of relating the time of appearance of contaminated particles in the air to events taking place in the room.

Impingement is the conventional term for the collection of particles in a liquid medium. Various instruments for this purpose have been described by May and Harper (1957), Akers and Won (1969), and Errington and Powell (1969). In all impinger types of apparatus the fluid in which the organisms are collected should be such as to preserve their viability without permitting their multiplication.

Unless a membrane **filter** is used (Albrecht 1957), or alginate wool which can be dissolved in phosphate buffer solution (Hammond 1958), filtration suffers from the difficulty of freeing the organisms from the material in which they are enmeshed.

Particles can be **precipitated** from the air by thermal or electrostatic means (see Akers and Won 1969, Luckiesh et al. 1946, Morris et al. 1961).

The **animal method** is limited to the detection, or occasionally the counting, of pathogenic organisms in the air. Viruses or other organisms may be trapped in the respiratory tract without giving rise to disease, but their presence may be determined by cultural and other methods.

Measurement of the **particle-size distribution** of viable airborne bacteria may be made by the cascade impactor of May (1945) and the multistage stacked plate impactor of Andersen (1958). A multistage liquid impinger devised by May (1966) separates the organisms into three sizes — over 6 μm, 3–6 μm, and less than 3 μm.

The culturing of any sample is a subject reviewed by Noble (1967). Suitable media must be used for different species of bacterium. Viruses are best collected in a fluid medium and inoculated into tissue cultures or fertile eggs..

For intermittent sampling an apparatus may be used in which airborne bacteria are distributed in a spiral pattern on to an agar plate 150 mm q at a flow rate of 1 to 7 litres a minute. The spiral track is 2·5 metres (100 inches) long and is traversed at intervals of 15 minutes to 6 hours according to choice. The best results are obtained when the number of colony-forming bacteria is of the order of 0·1–10 per litre (Carlberg et al. 1965).

An ingenious and effective means for capturing atmospheric particles is by use of the very fine silk threads made by small spiders (Dessens 1949). May and Druett (1968), who adapted this method for airborne microbes, using spider threads 0·5 μm in diameter, found that sampling was independent of wind direction, turbulence, or other sources of error.

Estimating human pollution of the air

In the environment of human beings air is contaminated from various parts of the body, particularly the respiratory tract and the skin.

For estimating the degree of nasopharyngeal pollution of the air, Gordon (1902–3) long ago, in his study of the ventilation of the House of Commons, suggested that '*Str. viridans*' should be used, in much the same way as *Esch. coli* is used for measuring the excretal pollution of water. This suggestion was revived by American workers (Buchbinder et al. 1938, Wells, Wells and Mudd 1939). Williams and Hirch (1950) found that the best method was to collect the bacteria in a slit sampler on 5 per cent serum agar containing 5 per cent sucrose, 1/100 000 crystal violet and 1/400 000 potassium tellurite. The growth of most staphylococci and micrococci was inhibited on this medium, *Str. salivarius* formed distinctive mucoid colonies, and other streptococci could be identified by suitable sampling of colonies. 'Viridans and non-haemolytic streptococci are useful for testing the efficacy of bactericidal agents designed for the destruction of potentially pathogenic organisms in the air. They can be sprayed into a closed room, and the disinfection rate estimated by plate counts on air collected at measured intervals. The sampling of dust for haemolytic streptococci is described by Williams (1949b).

There are no accepted standards of bacterial pollution of the air for occupied rooms nor are we yet in a position to formulate them. Our present technique of air examination is rather like that of water examination at the end of the last century before Houston introduced the coli test. For the sake of interest, however, a few more or less representative counts are given in Table 9.4.

In their observations on the air of occupied schoolrooms in a London suburb Williams, Lidwell and Hirch (1956) found a bacterial content ranging from 20 to 250 colonies per cu ft, with a mean of about 70. Over 80 per cent of the colonies consisted of undefined micrococci. The remainder were made up chiefly of diptheroid, coliform, and spore-bearing bacilli, streptococci, aerococci, and *Staph. aureus*.

Table 9.4 Colonial counts observed in different types of occupied premises in England (modified from Williams 1949a, Williams et al. 1950, Hirch 1951).

	Colonies per cubic foot		
	General flora	*Str. salivarius*	All streptococci
Infant and junior schools	80	0.2	1.25
Clerical offices	30	0.05	0.9
Boot and shoe factories	30	0.1	—
Shops	70	0.07	0.86
Underground trains	40	—	0.28

Most of the organisms seemed to come from clothing and floor dust, but *Str. salivarius* was associated with the amount of talking. During broom-sweeping the air of hospital wards may contain as many as 600 bacterial particles per cu ft (Blowers 1960).

For hospital operating theatres Bourdillon and Colebrook (1946) suggest that, when the room is quiet, the colonies developing on 5 per cent blood agar in 24 hours at 37° should not exceed the following numbers per cubic foot of air:

Minor operations and dressing small wounds	20
Major operations on healthy tissues	10
Burns, and operations on central nervous system	0.1–2

Further observations have confirmed the general truth of these conclusions. When the count per cubic foot exceeds 35 colonies there is a risk of infection from the air. Below 5 colonies it is slight (Report 1972). Much lower counts than these are obtainable by modern methods of ventilation, but whether they reduce the risk of infection is uncertain. (See Ch. 58.)

The demonstration of respiratory pathogens in barrack rooms, school rooms, or hospital wards in which infection is current demands the examination of large volumes of air. Artenstein and Miller (1966), for example, could find only one viable meningococcus particle in every 100 cubic feet of air, and one adenovirus particle in a quantity of air ranging from 300 to 3000 cubic feet.

Measures for controlling airborne infection

The various measures necessary for the control of infection in hospitals are described in Chapter 58. In the present chapter we shall confine ourselves largely to true airborne infection, that is to say infection borne by droplet nuclei, omitting that borne by dust and droplets.

Ventilation There is a wealth of information on the principles and practice of both natural and forced ventilation (see Riley and O'Grady 1961, Daws 1967, Lidwell and Towers 1969, 1970, Report 1972). Special attention has been paid to the ventilation of operating theatres (Blowers and Wallace 1960, Lidwell 1962, Report 1962), in which, however, problems other than those due solely to airborne infection are concerned. For the removal of droplet nuclei from an occupied room Riley and O'Grady (1961) recommend at least one change of air a minute. It is difficult to attain this rate by air conditioning without an intolerable noise and draught. Consequently other methods may have to be used aimed at the destruction of organisms in the droplet nuclei. Nevertheless a great reduction in the airborne microbial flora may be brought about by taking advantage of the laminar airflow system developed by Whitfield (1962). Lidwell and Towers (1969), for example, found that a horizontal flow of air at about 40 feet per minute in a room in which spores of *Bac. subtilis* had been liberated led to a diminution in their numbers 100–1000 times greater than that to be expected in a room where turbulent ventilation was employed. Particles did not go against the air current, and the movement of persons in the room did not transport the particles far transversely. A great deal has still to be learned about the mode and rate of ventilation under different structural conditions. The ultimate criterion of its effectiveness must be the diminution of cross infection, but the measurement of this under controlled conditions presents a major problem in itself.

Safety cabinets are of three classes. Class 1 (exhaust protection cabinets) protect the operator by maintaining a rapidly moving stream of air, which entrains infectious particles and deposits them on a high-efficiency filter. The air is then discharged outside the building (Wedum 1953, Williams and Lidwell 1957). They have an opening in front through which manipulations can be performed. Class II (laminar flow cabinets) are open-fronted, and are designed to protect the material that is being handled from extraneous contamination (McDade *et al.* 1968, Favero and Berquist 1968, van der Waaij and Andreas 1971). They do not protect the operator, though they can be modified to do so (Newsom 1979). Class III cabinets, for the handling of very dangerous material, are completely enclosed. The material is handled through gas-tight glove ports. Air enters through a non-return valve and is extracted through a high-efficiency filter (Evans *et al.* 1972). The ventilation of safety cabinets is dicussed briefly by Thomas (1970), and very fully by Chatigny and Clinger (1969).

Filtration In closed halls and offices in which forced ventilation is used, the air may be filtered to remove bacteria-carrying particles before being admitted to the rooms. Several types of mechanical filters are in use, varying in efficacy (see Report 1948). In the operating theatre of the Birmingham Accident Hospital Bourdillon and Colebrook (1946) used a combination of filters that succeeded in removing about 99.5 per cent of the bacteria in the atmosphere. Cotton-wool filters are effective provided they are not too tightly packed and are kept dry. When used for filtration on a large scale they are less satisfactory, owing principally to the appearance of channels in the packed mass. Glass wool or slag wool is to be preferred. Other mechanical methods of filtration include precipitation of the organisms in the air by steam or by *electrostatic* means (Lidwell 1951). Even with ultra-high filters removal from the air of more than 99.99 per cent of particles 1 to 5 μm in diameter cannot be guaranteed (Decker *et al.* 1963). Filtration of the outgoing air is mainly relied upon for the removal of living organisms from the type of safety cabinets used in the laboratory.

Most filters consist of randomly oriented fibres arranged so that the interstices are, in general, much larger than the particles to be removed. The filtering action depends on the particles coming into contact with and adhering to the fibres.

Infection carried by droplet nuclei When droplet nuclei cannot be removed by ventilation, an attempt may be made to destroy the micro-organisms that are attached to them. Hitherto, the main lines of attack have been by ultraviolet irradiation and by disinfectant sprays and vapours.

1. Ultraviolet irradiation The chief advocates of ultraviolet irradiation have been Wells and his colleagues in the United States (Wells and Wells 1936, 1938, Wells 1940, 1942, Wells, Wells and Wilder 1942). The whole room may be exposed to ultraviolet rays; or ultraviolet light barriers may be used in hospitals to prevent the air of cubicles from becoming contaminated with air from the general ward. This method, in conjunction with air conditioning, is said to have given satisfactory results in the Cradle at Evanston (Rosenstern 1942), but in an infants' hospital at Boston it failed to diminish the incidence of cross infection (Brooks *et al.* 1942). Lurie (1944) used it successfully to prevent cross infection of normal rabbits in one set of cages adjacent to another set containing rabbits infected with tubercle bacilli (see also Riley and O'Grady 1961, Riley *et al.* 1971). On the whole, however, the promise of ultraviolet irradiation has proved delusory, and little use is now made of this method in practice (see Du Buy *et al.* 1945, Langmuir *et al.* 1948, Downes 1950, Report 1954).

2. Disinfectant sprays and vapours In the early days of antiseptic surgery Lister introduced a carbolic spray to sterilize the air over the wound. How far it was effective is doubtful; it was later abandoned for other reasons. Sulphurous acid and particularly formalin vapours were, and are occasionally still, used for terminal disinfection in isolation hospitals and sick rooms, but are too irritating for disinfecting the air of occupied rooms. What appeared to be a great advance in this respect was the finding of Douglas, Hill and Smith (1928) that it was possible to destroy the cells of *Esch. coli* in the atmosphere by means of sodium hypochlorite solution sprayed into the air in very high dilution. In spite of this impressive demonstration no general interest in the possibilities of aerial disinfection was aroused till the publication of Wells's paper on droplet nuclei and Trillat's (1938) paper on bacterial aerosols. Masterman (1938, 1941) then toook up the subject, followed by Andrews (1940), Bourdillon, Lidwell and Lovelock (1942), Challinor (1943), Pulvertaft (1944) and others, all of whom reported favourable results under experimental conditions from the spraying of a 1 per cent sodium hypochlorite solution in a quantity to give a concentration in the air of 1–5 per million parts or so.

Drawbacks to the use of hypochlorite led to the trial of several other vapours, such as those of resorcinol, propylene glycol, triethylene glycol, various hydroxy-acids, and ozone. It is not proposed to discuss these here (see 5th ed., pp. 2493–95), because after many trials under practical conditions none of them has proved its worth in the prevention of cross infection, and their use has been generally abandoned.

The mode of action of vaporized disinfectants has been the subject of much controversy. Trillat (1938) and several others maintained that they acted in the form of minute droplets or aerosols, whereas Masterman (1941) and others believed that they acted as simple gaseous disinfectants. The principles underlying aerial disinfection are discussed by Puck (1947), Nash (1951), and Report (1948), to which the reader may be referred.

It may be mentioned that to both ultraviolet light and disinfectant vapours gram-negative bacilli are more susceptible than gram-positive; that organisms are harder to kill when dried on to particles of dust than when in the form of droplet nuclei; that a varying proportion of organisms in dust and in droplet nuclei die naturally in the absence of any active interference; and that some organisms, such as staphylococci, tend to lose their virulence when dried, before actually dying (Hinton *et al.* 1960).

For multiple papers on the provision of clean indoor air, see Report (1973a) of the 1972 symposium at Zürich.

The control of respiratory infection

It is as yet too early to assess the value of the different methods just discussed for the control of respiratory disease. One of the great difficulties is to ensure the continuous absence of pathogenic organisms from the atmosphere during the 24 hours of each day. In the prevention of enteric infections reliable measures are available for the destruction of the causative organisms in water and in milk; and provided due sanitary care is exercised, food should seldom be contaminated except by accident. In this way the entire population of a city can be more or less effectively protected. But with respiratory disease, the problem is far more complex. Even if good ventilation is provided in the factory, measures are taken to allay dust, overcrowding is avoided, and pathogenic organisms in the air are destroyed by ultraviolet irradiation or by disinfectant gases, the operatives are still exposed to the risk of infection in the buses, trains, and tubes that take them to and from their work, in the shops where they buy their food and other necessaries, in the places of amusement they frequent, and probably for 10 or 12 hours of every day in the houses where they live and to which their children return bringing infection with them from school. Whether it will ever prove practible to afford continuous protection to the worker in all these varied situations, it is impossible to say. To do so will require an intensive and prolonged education of the general public, greatly improved ventilation of houses and other buildings, and a willingness on the part of both private persons and of public authorities to incur the necessary expenditure to combat the spread of aerial infection.

For the present it will probably be wise to confine our attention to the prophylaxis of air-borne infection in such situations as operating theatres, hospital wards, out-patient departments, and certain institutions, where reliable bacteriological and epidemiological records on the type and incidence of respiratory disease can be collected. By this means we should find out in course of time how effective different measures are, alone or in combination, and be in a better position to decide whether their extension to other fields is desirable.

In conclusion, it would probably be fair to say that the revival of interest during the thirties and forties of this century in the importance of air in the spread of respiratory disease has not been maintained, and that the pendulum is swinging back to the views previously held on the necessity for close contact with the source of infection (see Tyrrell 1967).

References

Akers, T. G. (1969) In: *An introduction to Experimental Aerobiology*, p. 296. Ed. by R. L. Dimmick and A. B. Akers, Wiley, New York

Akers, A. B. and Won, W. D. (1969) In: *An Introduction to Experimental Aerobiology*, p. 59. Ed. by R. L. Dimmick and A. B. Akers. Wiley, New York.

Albrecht, J. (1957) *Arch. Hyg., Berl.* **141**, 210.

Andersen, A. A. (1958) *J. Bact.* **76**, 471.

Anderson, J. D. and Cox, C. S. (1967) *Airborne Microbes*, p. 203. Cambridge University Press.

Artenstein, M. S. and Miller, W. S. (1966) *Bact. Rev.* **30**, 571.

Baylor, E. R. Baylor, M. B., Blanshard, D. C., Syzdek, L. D. and Appol, C. (1977) *Science* **198**, 575.

Blowers, A. R. (1960) *Mon. Bull. Minist. Hlth. Lab. Serv.*, **19**, 207.

Blowers, R. and Wallace, K. R. (1955) *Lancet* **i**, 1250; (1960) *Amer. J. publ. Hlth.* **50**, 484.

Bourdillon, R. B. and Colebrook, L. (1946) *Lancet* **i**, 561, 601.

Bourdillon, R. B. and Lidwell, O. M. (1941) *Lancet* **ii**, 365.

Bourdillon, R. B., Lidwell, O. M. and Lovelock, J. E. (1942) *Brit. med. J.* **i**, 42.

Bourdillon, R. B., Lidwell, O. M. and Schuster, E. (1948) *Spec. Rep. Ser., Med. Res. Coun., Lond.* No. 262, p. 12.

Boyland, E., Gaddum, J. H. and McDonald, F. F. (1947) *J. Hyg., Camb.* **45**, 290

Brooks, G. L., Wilson, U. and Blackfan, K. D. (1942) *Aerobiology*, p. 228. Publ. Amer. Ass. Advanc. Sci., Wash., No. 17.

Brown, J. H., Cook, K. M., Ney, F. G. and Hatch, T. (1950) *Amer. J. publ. Hlth* **40**, 450.

Buchbinder, L., Solowey, M. and Solotorovsky, M. (1938) *Amer. J. publ. Hlth* **28**, 61.

Carlberg, D. M., Bell, R. T., Scheir, R., Finkelstein, H. and Burns, D. H. (1965) *Bact. Proc., Abstr.*, p. 12.

Challinor, S. W. (1943) *J. Hyg., Camb.* **43**, 16.

Chatigny, M. A. and Clinger, D. I. (1969) In: *An Introduction to Experimental Aerobiology*, p. 194. Ed. by R. L. Dimmick and A. B. Akers, Wiley, New York.

Clark, R. P. (1974) *J. Hyg., Camb.* **72**, 47.

Clark, R. P. and Shirley, S. G. (1973) *Nature, Lond.* **246**, 39

Couch, R. B., Cate, T. R., Douglas, R. G., Gerone, P. J. and Knight, V. (1966) *Bact. Rev.* **30**, 517.

Cox, C. S. (1966) *J. gen. Microbiol.* **43**, 383.

Crosbie, W. E. and Wright, H. D. (1941) *Lancet* **i**, 656.

Cruikshank, R. (1938) *Lancet* **i**, 841.

Darlow, H. M. and Bale, W. R. (1959) *Lancet* **i**, 1196.

Daws, L. F. (1967) In: *Airborne Microbes*, p. 31. Cambridge University Press.

Decker, H. M., Buchanan, L. M., Hall, L. B. and Goddard, K. R. (1963) *Amer. J. publ. Hlth* **53**, 1982.

Dessens, H. (1949) *Quart. J. R. met. Soc.*, **75**, 23.

Dimmick, R. L. and Akers, A. B. (1969) *An Introduction to Experimental Aerobiology*. Wiley, New York.

Douglas, S. R., Hill, L. and Smith, W. (1928) *J. industr. Hyg* **10**, 219.

Downes, J. (1950) *Amer. J. publ. Hlth* **40**, 1512.

Druett, H. A. (1967) *Airborne Microbes*, p. 165. Cambridge University Press.

Druett, H. A. and May, K. R. (1968) *Nature, London.* **220**, 395.

Druett, H. A. and Packman, L. P. (1972) *J. appl. Bact.* **35**, 323.

DuBuy, H. G., Dunn, J. E., Brackett, F. S., Dreessen, W. C., Neal, P. A. and Posner, I. (1948) *Amer. J. Hyg.* **48**, 207.

DuBuy, H. G., Hollaender, A. and Lackey, M. D. (1945) *Publ. Hlth Rep., Wash.*, Suppl. No. 184.

Duguid, J. P. (1945) *Edin. med. J.* **52**, 385; (1946a) *Brit. med. J.* **i**, 265; (1946b) *J. Hyg. Camb.*, **44**, 471.

Duguid, J. P. and Wallace, A. T. (1948) *Lancet*, **ii**, 845.

Ende, M. Van Den and Thomas, J. C. (1941) *Lancet*, **ii**, 755.

Engley, F. B. (1955) *Texas Rep. Biol. Med.* **13**, 72.

Errington, F. P. and Powell, E. O. (1969) *J. Hyg., Camb.* **67**, 387.

Evans, C. G. T., Smith, R. H. and Stratton, J. E. D. (1972) In: *Safety in Microbiology*, p. 21. Ed. by D. A. Shapton and R. G. Board. Academic Press, London.

Favero, M. S. and Berquist, K. R. (1968) *Appl. Microbiol.* **16**, 182.

Fedorak P. M. and Westlake, D. W. S. (1978) *Canad. J. Microbiol.* **24**, 618.

Flügge, C. (1899) *Z. Hyg. InfektKr.* **30**, 107.

Gerone, P. J., Couch, R. B., Keefer, G. V., Douglas, R. G., Derrenbacher, E. B. and Knight, V. (1966) *Bact. Rev.* **30**, 576.

Gordon, M. H. (1902-03) *Suppl. 32nd ann. Rep., M.O.H., loc. Govt. Bd.*, p. 421.

Hammond, E. C. (1958) *J. gen. Microbiol.* **19**, 267.

Hare, R. (1941) *Lancet* **i**, 85.

Hare, R. and Ridley, M. (1958) *Brit. med. J.* **i**, 69.

Hare, R. and Thomas, C. G. A. (1956) *Brit. med. J.* **ii**, 840.

Hatch, T. F. (1942) *Aerobiology*. p. 102. Publ. Amer. Ass. Advance. Sci., Wash., No. 17.

Hatch, M. T. and Wolochow, H. (1969) In: *An Introduction to Experimental Aerobiology*, p. 267. Ed. by R. L. Dimmick and A. B. Akers. Wiley, New York.

Hausler, W. J. (1970) In: *Silver's Aerobiology*, p. 76. Academic Press, London.

Herrmann, W. and Pütz, T. (1942) *Dtsch. med. Wschr.* **68**, 1101.

Hinton, N. A., Maltman, J. R. and Orr. J. H. (1960) *Amer. J. Hyg.* **72**, 343.

Hirch, A. (1951) *Brit. J. industr. Med.* **8**, 8.
Hirst, J. M. and Hurst, G. W. (1967) *Airborne Microbes*, p. 307. Cambridge University Press.
Hollaender, A. and Dalla Valle, J. M. (1939) *Publ. Hlth. Rep., Wash.*, **54**, 574.
Jarrett, E. T., Zelle, M. R. and Hollaender, A. (1948) *Amer. J. Hyg.*, **48**, 233.
Jennison, M. W. (1942) *Aerobiology. Publ. Amer. Ass. Advanc. Sci., Wash.* No. 17, p. 106.
Jong, J. G. de and Winkler, K. C. (1964) *Nature, Lond.* **201**, 1054.
Lange, B. and Keschischian, K. H. (1925) *Z. Hyg. InfektKr.* **104**, 256.
Langmuir, A. D., Jarrett, E. T. and Hollaender, A. (1948) *Amer. J. Hyg.* **48**, 241.
Lewis, H. E., Foster, A. R., Mullan, B. J., Cox, R. N. and Clark, R. P. (1969) *Lancet* **i**, 1273.
Lidwell, O. M. (1951) *J. Instn. heat. vent. Engrs.*, **19**, 139; (1962) *J. int. Coll. Surg.* **38**, 200.
Lidwell, O. M. and Towers, A. G. (1969) *J. Hyg., Camb.* **67**, 95; (1970) In: *Silver's Aerobiology*, p. 109. Academuc Press, London.
Luckiesh, M., Taylor, A. H. and Holladay, L. L. (1946) *J. Bact.* **52**, 55.
Lurie, M. B. (1944) *J. exp. Med.* **79**, 559.
McDade, J. J., Sabel, F. L., Akers, R. L. and Walker, R. J. (1968) *Appl. Microbiol.* **16**, 1086.
Mackintosh, C. M., Lidwell, O. M., Towers, A. G. and Marples, R. R. (1978) *J. Hyg., Camb.* **81**, 471.
Magill, P. L., Holden, F. R. and Ackley, C. (1956) *Air Pollution Handbook*, McGraw-Hill, New York.
Masterman, A. T. (1938) *J. industr. Hyg.*, **20**, 278; (1941) *J. Hyg., Camb.* **41**, 44.
May, K. R. (1945) *J. sci. Instrum.*, **22**, 187; (1966) *Bact. Rev.*, **30**, 559; (1967) In: *Airborne Microbes*, p. 60. Cambridge University Press.
May, K. R. and Druett, H. A. (1968) *J. gen. Microbiol.* **51**, 353.
May, K. R. and Harper, G. J. (1957) *Brit. J. industr. Med.* **14**, 287.
Mazzarella, M. A., and Flynn, D. D. (1969) In: *An Introduction to Experimental Aerobiology*, p. 437. Ed. by R. L. Dimmick and A. B. Akers. Wiley, New York.
Meers, P. D. and Yeo, G. A. (1978) *J. Hyg., Camb.* **81**, 99.
Mik, G. de and Groot, I. de (1977) *J. Hyg., Camb.* **78**, 175.
Morris, E. J., Darlow, H. M., Peel, J. F. H. and Wright, W. C. (1961) *J. Hyg., Camb.* **59**, 487.
Newsom, S. W. B. (1979) *J. clin. Path.*, **32**, 505.
Noble, W. C. (1967) In: *Airborne Microbes*. p. 81. Cambridge University Press.
Noble, W. C., Lidwell, O. M. and Kingston, D. (1963) *J. Hyg., Camb.* **61**, 385.
Ostrom, C. A., Wolochow, H. and James, H. A. (1958) *J. infect. Dis.* **102**, 251.
Pady, S. M. and Kelly, C. D. (1953) *Science* **117**, 607.
Pressman, R. (1937) *Amer. Rev. Tuberc.* **35**, 815.
Proctor, B. E. (1934) *Proc. Amer. Acad. Arts Sci.* **69**, 315; (1935) *J. Bact.* **30**, 363.
Proctor, B. E. and Parker, B. W. (1938) *J. Bact.* **36**, 175.
Pulvertaft, R. J. V. (1944) *J. Hyg. Camb.*, **43**, 352.
Report. (1942) *Aerobiology. Publ. Amer. Ass. Advance. Sci.*, Washington, D.C., No. 17; (1947) *Amer. J. Publ. Hlth* **37**, 13; (1948) *Spec. Rep. Ser. med. Res. Coun., Lond.* No. 262; (1954) *Spec. Rep. Ser. med. Res. Coun., Lond.* No. 283; (1961) *Bact. Rev.* **25**, No. 3; (1962) *Lancet* **ii**, 945; (1966) *Amer. Conf. governm. industr. Hygienists.* 3rd ed., Cincinnati, Ohio; (1967) *Airborne Microbes. 17th Symposium Soc. Gen. Microbiology.* Cambridge University Press; (1970) *Silver's Aerobiology.* Proc. 3rd int. Symposium, Brighton 1969. Academic Press, London; (1973a) *Reinraumtechnik I. Ber. int. Symp., Zürich* 1972; (1973b) *Airborne Transmission and Airborne Infection.* 4th int. Symp. on Aerobiology at Enschede, The Netherlands. Oosthoek Publishing Co., Utrecht; (1972) *Ventilation in Operation Suites. Med. Res. Coun. & Dept. Hlth Social Security,* London.
Riley, R. L. and O'Grady, F. (1961) *Airborne Infection, Transmission and Control.* Macmillan New York.
Riley, R. L., Permutt, S. and Kaufman, J. E. (1971) *Arch. environm. Hlth* **23**, 35.
Rosebury, T. (1947) *Experimental Air-borne Infection.* Williams & Wilkins, Baltimore.
Rosenstern, I. (1942) *Aerobiology*, p. 242. *Publ. Amer. Ass. Advanc. Sci., Wash.*, No. 17.
Rubbo, S. D. and Benjamin, M. (1953) *J. Hyg., Camb.* **51**, 278.
Rüden, H., Fischer, P. and Thofern, E. (1978) *Zbl. Bakt. Ite Abt. Orig. B.*, **166**, 132.
Strauss, W. (1926) *Z. Hyg. InfektKr.* **105**, 416.
Thomas, G. (1970) In: *Silver's Aerobiology*, p. 98. Academic Press, London.
Thomas, J. C. (1941) *Lancet* **i**, 433.
Tomlinson, A. J. H. (1957) *Brit. med. J.* **ii**, 15.
Trillat, M. A. (1938) *Ann. Hyg. publ., Paris* **16**, 49.
Tyrrell, D. A. J. (1967) In: *Airborne Microbes*, p. 286. Cambridge University Press.
Waaij, D. van der and Andreas, A. H. (1971) *J. Hyg., Camb.* **69**, 83.
Wedum, A. G. (1953) *Amer. J. publ. Hlth* **43**, 1428; (1964) *Ibid.* **54**, 1669.
Wells, W. F. (1934) *Amer. J. Hyg.* **20**, 611; (1940) *J. Franklin Inst.* **229**, 347 (1942) *Arch. phys. Therap.* **23**, 143.
Wells, W. F. and Lurie, M. B. (1941) *Amer. J. Hyg.* **34**, 21.
Wells, W. F., Phelps, E. B. and Winslow, C.-E. A. (1937) *Amer. J. publ. Hlth., Suppl. Yearbook* 1936-37, **27**, 97.
Wells, W. F. and Wells, M. W. (1936) *J. Amer. med. Ass.*, **107**, 1698, 1805; (1938) *Amer. J. publ. Hlth.*, **28**, 343; (1942) *Aerobiology*, p. 99. *Publ. Amer. Ass. Advanc. Sci., Wash.* No. 17.
Wells, W. F., Wells, M. W. and Mudd, S. (1939) *Amer. J. publ. Hlth.* **29**, 863.
Weyrauch, F. and Rzymkowski, J. (1938) *Z. Hyg. InfektKr.* **120**, 444.
White, E. (1936) *Lancet* **i**, 941.
Whitfield, W. J. (1962) *A New Approach to Clean Room Design.* Sandia Corp., Albuquerque, N. M., Tech. Rep. No. SC-4673 RR. Quoted by McDade *et al.* 1968.
Williams, R. E. O. (1948) *J. R. sanit. Inst.*, **68**, 583; (1949a) *J. Instn. heat. vent. Engrs.* **16**, 404; (1949b) *J. Hyg., Camb.* **47**, 416; (1951) *Brit. med. Bull.* **7**, 171; (1967) In: *Airborne Microbes*, p. 268. Cambridge University Press.
Williams, R. E. O. and Hirch, A. (1950) *J. Hyg., Camb.* **48**, 504.
Williams, R. E. O., Hirch, A. and Lidwell, O. M. (1950) *Lancet* **i**, 128.
Williams R. E. O. and Lidwell, O. M. (1957) *J. clin. Path.* **10**, 400.
Williams, R. E. O., Lidwell, O. M. and Hirch, A. (1956) *J. Hyg., Camb.* **54**, 512.
Wright, H. D., Shone, H. R. and Tucker, J. R. (1941) *J. Path. Bact.* **52**, 111.

Water

Bacterial flora in water	260	Protozoal content	263
Natural water bacteria	260	Rainfall	263
Soil bacteria	260	Season	263
Sewage bacteria	260	Storage	264
Factors determining the kinds and numbers of bacteria in water	261	Filtration	264
		Self-purification of rivers	264
Surface waters	261	Bacteriological analysis	264
Rain	261	Coliform count	266
Snow	261	Confirmatory and differential coliform tests	267
Hail	261	Faecal streptococci	267
Ice	261	*Clostridium perfringens*	268
Shallow wells	261	Pathogenic organisms	268
Upland surface waters	261	Plate count	268
Rivers	261	Sea-water	268
Lakes	261	Interpretation of bacteriological analysis	268
Sea-water	262	Swimming-bath waters	270
Mineral springs	262	Water-borne disease	270
Deep waters	262	Bacterial	270
Nutrition	262	Viral	272
Temperature	262	Danger of bathing	272
Light	263	Watercress	272
Acidity	263	Mode of contamination of water supplies	273
Salinity	263	Shell-fish	273
Dissolved oxygen	263	Sewage	274

Bacterial flora in water

It is convenient to divide the bacteria found in water into three groups (see Marshall 1921).

(A) Natural water bacteria In this group are included those organisms that are commonly found in waters free from gross pollution. They consist mainly of saprophytic organisms belonging to the genera *Micrococcus, Pseudomonas, Serratia, Flavobacterium, Chromobacterium,* and *Achromobacter.*

(B) Soil bacteria These organisms, though not normally inhabitants of water, are frequently washed into it during heavy rains. Most of them belong to the group of aerobic spore-bearing bacilli, such as *Bac. subtilis, Bac. megaterium,* and *Bac. mycoides.* Others, such as *Klebs. aerogenes* and *Enterobacter cloacae,* which may be found on grain, plants and decaying vegetation, and which may conveniently be treated as soil organisms, are aerobic non-sporing bacilli. By the use of special media other organisms, such as the nitrifying bacteria, may be isolated. Growth of some types of soil micro-organisms, such as *Actinomyces, Streptomyces,* and certain fungi, may impart an earthy odour and taste to the water, particularly if it is allowed to stagnate at temperatures suitable for their multiplication (see Mackenzie 1938, Ferramola 1949).

Polythene pipe lines favour the growth of bacteria, some of which produce copious quantities of slime often leading to blockage (Böing 1957, Collins 1964).

(C) Sewage bacteria Many of the organisms in this group are normal inhabitants of the intestine of man and animals. Others live chiefly on decomposing organic matter of either animal or vegetable origin. Occasionally pathogenic organisms are included.

(1) *Intestinal Bacteria.*
   Escherichia coli group.
   Streptococcus faecalis.
   Cl. perfringens.
   Pathogenic organisms, such as *Salm. typhi* and *V. cholerae.*
(2) *Sewage Bacteria proper.*
   Proteus: *Pr. vulgaris.*
   Anaerobic spore-bearing bacilli: *Cl. sporogenes.*

Studying about 800 strains of bacteria isolated from lakes and streams in the English Lake District, all but one of which were subject to both human and animal pollution, C. B. Taylor (1942) found that about 95 per cent were gram-negative rods, about 4 per cent gram-positive spore-forming rods, and less than 1 per cent cocci. A large proportion of the organisms were chromogenic; growth on ordinary media was often very poor, and biochemical activity was slight. Taylor points out that these characters differ from those commonly found in soil bacteria; he maintains, in effect, that there is a natural water flora distinct from the organisms washed in with soil.

Numerous viruses have been found in water, belonging to the entero, ECHO, adeno, parvo, and reo groups. Like the bacteria just mentioned, they are derived from sewage, in which many of them can survive for a long time (Schäfer 1970).

Factors determining the kinds and numbers of bacteria in water

Waters may be divided according to their source into (1) surface and (2) deep waters. The former comprise all those that are found on or near the surface of the earth, and that have not been filtered through any considerable thickness of soil; the latter comprise those which, in order to reach the underground stratum that they occupy, have percolated, often for several hundred feet, through porous layers of soil. Since surface waters are frequently exposed to contamination from dust, soil, sewage, factory wastes, and other decomposing organic matter, they may contain large numbers of bacteria, many of which are of intestinal origin. Deep waters, on the other hand, are generally pure, having had most of their surface contaminants filtered off on their downward passage through the soil; and, though it is not unusual to find bacteria in larger numbers than one would expect, these are generally of a harmless type.

(1) Surface waters

(a) Rain In falling to the earth, the raindrops come into contact with particles of suspended dust, and carry these down with them. The more dust there is in the atmosphere, the greater is the bacterial count. In the open country the number of organisms may not exceed 10 or 20 per litre, but in the town there may be 5000 to 10 000 or so.

(b) Snow This tends to be less pure than rain, probably because the snowflakes have a greater surface on which to collect suspended particles in the atmosphere; and also because their low temperature conduces to the survival of bacteria. In snow situated on the tops of high mountains, where it will be remembered that Pasteur found the air to be practically sterile, there are hardly any organisms.

(c) Hail Curiously enough, hail contains more bacteria than either rain or snow. Belli (1902) examined hail that fell in Padua during July, 1901, and found no fewer than 140 000 organisms per litre. Examination of the bacteria showed that they belonged to nine different types. During the formation of hail, it seems probable that rapidly ascending currents of air carry the raindrops up into a region of the atmosphere where they are solidified; falling down they are melted, and again swept upwards and frozen. After they have been frozen and thawed a number of times the hailstones are thrown out on the periphery of the storm centre and finally come to earth (Mason 1902). It is suggested that the air currents carry up to the cloud region quantities of dust, which is thus incorporated in the hail. It is difficult to explain otherwise the presence in it of vegetable cells, and of fluorescent and soil bacteria.

(d) Ice The number of organisms in ice depends on the nature of the water from which the ice is formed. With the exception of the ice of glaciers, it is generally impure. Its low temperature is favourable to the survival of most bacteria; hence the self-purification that occurs in waters on storage occurs hardly at all, or very slowly, in ice. For a short review on the bacteriology of ice, see Jensen (1943).

(e) Shallow wells When protected from contamination in the immediate vicinity by brick sides, and provided with a pump, shallow wells may contain only a few bacteria; but the water of an open draw-well, subject to the influx of dust and of surface washings, is generally very impure. In an examination of over 50 shallow wells, Savage (1906) found that the gelatin plate count at 20° ranged from 100 to 20 000 or more per ml, and the agar count at 37° from less than 10 to over 100 per ml.

(f) Upland surface waters When derived from open moorland that is protected from human and animal excretions, these waters are fairly pure. Most of the organisms they contain belong to the soil group of bacteria. The agar count is generally low, not more than 10 or 20 per ml. The gelatin count is fairly high, up to 1000 per ml; after heavy rainfall it may rise to several thousands.

(g) Rivers In most countries rivers are heavily contaminated, and contain not only the natural and the soil bacteria but large numbers of organisms derived from sewage. In the raw Thames water taken at Hampton during 1906–11 Houston (1913) found the average number of colonies per ml on gelatin to be 4310, and on agar 368; typical *Esch. coli* was found in 49.3 per cent of samples of 0.1 ml and in 99.2 per cent of samples of 100 ml. In 1954 the average number of *Esch. coli* in the Thames at Walton was 4405 per 100 ml. From a high proportion of rivers in England and Wales organisms of the *Salmonella* group, including *Salm. paratyphi B*, can be isolated.

(h) Lake waters Owing to the natural storage of water in lakes, there is a continuous process of self-purification occurring; hence lake water is purer than the streams that feed it. During long rainless periods the bacterial content tends to fall. After rain it rises once more, owing partly to washed-in bacteria and partly to multiplication of existing organisms in the organic matter carried in by the rain (Collins 1957). In

English lakes bacterial numbers are highest in winter (Collins 1960). Water taken from the middle of the lake contains fewer organisms than that taken near the shore. Lake Geneva contained as many as 150 000 bacteria per ml near the shore, but further out as few as 38 (Marshall 1921). In lake Windermere Taylor (1940) found that within 200 metres of the entry of the main river the plate count was 200 times less than at the mouth of the river itself.

(i) **Sea-water** The number of bacteria in sea-water is generally less than in fresh. Russell (1892) found that the water taken near the coast in the Gulf of Naples contained 70 000 organisms per ml, but 4 kilometres out there were only 57. In the sea off heavily polluted bathing beaches the coliform content may rise to 200 000 or more per 100 ml and salmonellae are not infrequently present (Report 1959a, b). In deep sea the bacteria seem to be distributed evenly, there being almost as many near the bottom as on the surface. The ooze on the bed of the ocean is very rich in bacteria; Russell found 20 000 organisms per ml. Fresh sea-water is said to destroy sewage bacteria fairly rapidly (ZoBell 1936); coliform and salmonella organisms usually perish within 2 to 3 days (for references, see Moore 1954b); so likewise do viruses. In estuaries, however, containing a large amount of organic matter sewage bacteria may actually increase (Steiniger 1951, Meyer 1961). The inactivating property of sea-water is greater in summer than in winter; it is destroyed by heating at 45° for 1 hour, by chlorination, and by Seitz filtration (Magnusson et al. 1966). How far the saline content is responsible is doubtful. The destruction of enteric organisms is ascribed partly to the action of unidentified heat-labile antibiotic substances elaborated by marine bacteria (Vaccaro et al. 1950), partly to predatory protozoa, copepods, and other lower animal forms (Waksman and Hotchkiss 1937), and partly to competition for a limited food supply (see Nabbut and Kurayiyyah 1972, and reviews by Moore 1954b, Greenberg 1956).

(j) **Mineral springs** These are usually pure, and most of the organisms found in water derived from them come from imperfectly sterilized bottles (Duhot and Hutin 1933). The reaction of these waters varies considerably. Bance and Caillon (1929) give the following figures: Badoit pH 6·1, Vals pH 6.05–6.4, Vichy pH 6.4–6.9, Contrexéville pH 6.9–7.0, Vittel pH 7.1, Rubinat pH 7.3–7.6. (For the microbiology of mineral waters, see Report 1960.)

(2) Deep waters

Some of the purest waters that we know come from deep wells and springs. Fifteen driven wells in the neighbourhood of Boston, Mass., contained an average of only 18 colonies per ml (Prescott and Winslow 1913). In a period of 7 consecutive years, out of 1565 samples of waters from the deep wells in Kent, Houston (1913) found *Esch. coli* in only 5.7 per cent, even when 100 ml were examined. The purity of springs depends mainly on their source and surroundings. In waters which have percolated through thick strata, the flora differs from that in surface waters; the organisms develop slowly at room temperature, there are few liquefying colonies, and chromogenic organisms are relatively numerous (Prescott and Winslow 1913).

Nutrition The amount of available food supply is probably the most important factor of all in determining the number of bacteria in a given water. When organic matter is plentiful, organisms abound; when it is scarce they are few, and tend to die out.

Temperature The effect of temperature varies with the amount of organic matter present. A rise of temperature in a water containing an ample food supply for the bacteria causes them to multiply rapidly: but when the organic matter is small in quantity, a rise in temperature has the reverse effect; this is probably due to early exhaustion of the food supply, and the consequent diminution in rate of multiplication of the bacteria.

A low temperature, independent of the amount of organic matter present, favours the survival, though not the multiplication, of bacteria. Houston (1913) added typhoid bacilli to raw Thames water, and maintained the samples at temperatures varying from 0° to 37°. The initial number added was 103 000 ca per ml of water. Table 9.5 shows how much more rapid was the death of the organisms at 37° than at freezing-point. Hamilton (1935), who made observations on the Whangpoo river, found that in the short run from

Table 9.5 Influence of temperature on the survival of typhoid bacilli in river water. (Houston 1913.)

| Degrees | No. of bacilli per ml surviving after weeks |||||||||
	1	2	3	4	5	6	7	8	9
0°	47 766	980	65	34	3	3	2	1	0.0
5°	14 894	26	6	3	0.3	0.1	0.0	—	—
10°	69	14	3	0.3	0.0	—	—	—	—
18°	39	3	0.4	0.0	—	—	—	—	—
27°	19	0.1	0.0	—	—	—	—	—	—
37°	5	0.0	—	—	—	—	—	—	—

Shanghai to Woosung there was a diminution of 16 per cent in the colon bacteria during the winter months, and of no less than 97–99 per cent during the summer months.

Light Under natural conditions the actinic rays of the sun probably exert little or no bactericidal effect in water. This results partly from the opacity, and partly from the movement of the water, which cuts down the exposure time of any given organism. Even in very clear water ultraviolet rays do not penetrate for more than about 1.5 metres below the surface. When they are used for artificial purification, the water must be clear and exposed in a shallow layer for a sufficient length of time.

Acidity Winslow and Lochridge (1906) showed that *Esch. coli* in tap water was destroyed by 0.0123 normal hydrochloric acid in 40 minutes. This corresponded to 12.8 parts per million of dissociated hydrogen. Organic acids were likewise bactericidal, but usually in a higher concentration; their effect appeared to be due not only to the dissociated hydrogen, but also to the undissociated molecule or to the anion (see Chapter 4).

Many natural waters have an acid reaction. In the moorland streams acidity is due chiefly to the presence of peaty acids—the so-called humic and ulmic acids. Though it is difficult to estimate the effect of acidity on the destruction of bacteria and on the inhibition of their growth, there is little doubt that it does play a considerable part in purifying some waters.

Salinity The antiseptic and disinfectant action of salts has already been described in the chapter on Disinfection. It is not necessary here to do more than remind the reader of the inimical effects on bacterial growth and survival that certain salts may exercise. This factor may account for some of the difference in the numbers and species of bacteria found in sea-water (see p. 268).

Dissolved oxygen Whipple and Mayer (1906) found that *Salm. typhi* and *Esch. coli* remained viable in sterile water containing dissolved oxygen much longer than in water kept under anaerobic conditions. Thus, in one experiment, *Salm. typhi* survived in filtered tap water exposed to the air at room temperature for nearly 2 months, but died in 4 days in an atmosphere of hydrogen. They suggest that this may partly explain why this organism dies more rapidly in polluted than in pure water, and why it survives for a shorter time in summer than in winter. The importance of oxygen in favouring the growth and survival of aerobic and facultative anaerobic organisms has also been stressed by Müller (1912). Taylor (1940), on the other hand, was unable to demonstrate any close relationship between the amount of dissolved oxygen in lake water and the bacterial content; but other factors may possibly have obscured the effect.

Protozoal content Huntemüller (1905) showed that flagellates contributed notably to the extermination of bacteria in water. River water naturally polluted by bacteria, or suspensions of *Salm. typhi*, could be cleared in 4 days if flagellates—*Bodo saltans* or *Bodo ovatus*—were added (see also Kyriasides 1931). Stokvis and Swellengrebel (1911) demonstrated a similar action by infusoria—*Colpoda cucullus*. The bacterial destruction was preceded by a rise in the number of protozoa in the water, and was probably due to active ingestion, though this was not demonstrated conclusively. Aerobic conditions and a temperature between 10° and 30° were essential. König (quoted from Thresh and Beale 1925) found that in 1 ml of water from a well at the Hygienic Institute of Munich 21 million added typhoid bacilli perished in 24 hours. This he attributed to the action of protozoa. In pure water the death-rate in this time was trifling. Increasing attention has been paid to the action of predatory plankton in the self-purification of naturally polluted waters (Butterfield *et al.* 1931, Hoskins 1935). By keeping the bacterial population below the saturation point, it is suggested that the plankton favours the continuous multiplication of bacteria in the water, which results in its turn in a progressive oxidation of organic matter.

Rainfall The effect of rainfall on the bacterial content of water is complex. Rain falling after a drought washes large numbers of soil organisms into the water, and hence increases the numbers of bacteria. If the rain continues for some days, fewer organisms may be carried in during the later period of rainfall so that the stream is diluted with water purer than its own; its bacterial content therefore decreases. A pure stream may be contaminated by rain: an impure stream may be benefited by dilution. As a rule, rivers and upland surface waters contain their greatest numbers of bacteria after heavy rainfall. In Lake Windermere Taylor (1940) observed a close association between the fluctuations in bacterial content and the amount of rain that had fallen in the drainage area during the previous week. The effect is due to the influx of organisms, to the addition of nutritive substances, and to the increased oxygenation resulting from the beating effect of the rain upon the surface of the water. (See also Collins 1960.)

Season The monthly variation in the bacterial content of waters depends chiefly on the temperature and the rainfall. In Great Britain, the highest counts are generally found in the winter months, when the temperature is low and the rainfall greatest (see Collins (1960). Rivers show more variation than upland surface waters. The Thames in winter is swollen by heavy rain, and the quality deteriorates in consequence of the scouring effect over the whole drainage area. In summer many sources of pollution have dried up, and much of the water in the river is virtually stored or filtered water, derived from underground sources of supply; hence the bacterial content falls (Houston 1917). In upland surface waters, a rise in the bacterial content is not infrequent during July and August,

Table 9.6 Effect of storage. 1907–8. Average results. (Houston 1913.) Bacteria per ml.

	Gelatin, 20°–22°	Agar, 37°	No. of samples with *Esch. coli* in 0.01 ml
River Thames before storage	4465	280	10.1%
River Thames after 15 days' storage at Chelsea	208	44	1.1%
Reduction	95.3%	84.3%	89.1%

probably owing to dust and soil washings carried in by the summer rains.

In countries in which the water supply is augmented by melting snows the bacterial content rises considerably in spring time. Oslo obtains its water supply from a lake about 160 metres above sea level. During most of the year, Schmelck (1888) found that the number of organisms per ml was 10 to 60, rising up to 200 after heavy rain. But in the spring, when the snow was melting, the organisms increased to as many as 2500 per ml. This increase was probably due to the large amount of earth and detritus which was brought down by the glacial streams.

Storage The simple storage of water in a reservoir suffices to decrease its bacterial content enormously. Houston (1913) found that after only 15 hours storage the New River water showed a reduction in the agar count of 40 per cent. He maintained that storage acted in three ways: (1) Sedimentation: the organisms adherent to particles of suspended matter, and those in clumps or zoogloeae, sink to the bottom, leaving the supernatant water purer. (2) Equalization: this factor comes into play only when several waters of different qualities are collected into one reservoir. It ensures a thorough mixing of the different waters, and prevents the excess distribution of a bad supply on any one day. Even in a river, the water may not be completely homogeneous; samples taken from one side may be different in their bacterial content from those taken near the other side. In a reservoir, whether natural or artificial, a greater degree of homogeneity is attained. (3) Devitalization: the organisms die in large numbers, probably from lack of food supply, and ingestion by protozoa. Houston added cholera vibrios to raw river water, and found that after 1 week's storage their numbers had been reduced by 99.9 per cent; after 3 weeks they could not be isolated even from 100 ml of water. Table 9.6 shows the effect of storage on the London water.

The reduction occurs not only in pathogenic organisms and *Esch. coli*, but in organisms of all sorts, though not always equally. Houston states that even 1 week's storage would be more efficacious, in reducing the initial numbers of typhoid bacilli or cholera vibrios in a water, than sand filtration. Though it is true that pathogenic organisms usually survive longer in pure than in impure water, nevertheless pollution of purified water in reservoirs and water mains should be guarded against very carefully (E. W. Taylor 1958).

It is probably owing to storage that lake waters are so much purer than the streams that feed them. Some rivers with a very low gradient may offer conditions suitable for sedimentation.

Filtration Natural filtration occurs on a large scale, resulting in the accumulation of the underground deposits of water that are tapped by deep wells and main springs. Its efficacy in the removal of bacteria depends on the nature of the soil, and the depth of the strata penetrated. In loose, porous soils a greater depth must be traversed to ensure the same degree of purification that is attained by filtration through a more compact soil. In chalk and limestone rifts are not uncommon, and water may sometimes pass through these to a great depth without undergoing the purifying process of filtration. Evidence suggests that, in a soil of moderate density, the greater part of the bacteria are removed in the first 3 or 4.5 metres. This is the reason why deep well water is so pure.

Artificially, sand filtration is used to remove bacteria from water in order to render it potable. Houston (1913) finds that this process, which in the London water follows storage for 30 days, removes 98 per cent of the residual bacteria.

Self-purification of rivers

Observations by numerous workers, such as those of Frank (1888) on the Spree below Berlin, of Jordan (1900) on the Illinois, of Brezina (1906) on the Danube below Vienna, and of Hamilton (1935) on the Whangpoo, have shown that, in the absence of fresh contamination, polluted rivers tend to purify themselves during their natural flow. Organic matter is broken down, and a varying proportion of the micro-organisms perish, owing to the action of some of the factors we have just been considering, particularly sedimentation, destruction by protozoa, and exhaustion of nutritive material.

Bacteriological analysis

We do not propose to give a detailed description of the bacteriological analysis of water, since the Ministry of Health (Report 1969*a*) in Great Britain, the American Public Health Association (1971*c*) in the United States, and the World Health Organization (Report 1970) have each described standard methods for its performance, to which reference should be made by those desirous of further knowledge. Great attention has to be paid to the sampling of the water (for details see Report 1969*a*), and to the technical performance of the various procedures involved (for error of these see Wilson *et al.* 1935). For reference to the methods used in France see Buttiaux (1951), in the Argentine see Ferramola (1947), and for general information on the bacteriology of water and water supplies see E. W. Taylor (1958) and Holden (1970). Changes in the coliform flora of samples stored at different temperatures are considered in Reports (1952, 1953*c*).

Though in the early days of bacteriology the plate count was used in an attempt to determine the quality of a water supply, it was soon found that it bore little relation to safety for drinking purposes. It was gradually replaced by search for an organism specifically indicating the presence of excretal pollution. Largely owing to the work of Alexander Houston round about 1900 (see Houston 1913), *Escherichia coli* came to be regarded as the most suitable organism for this purpose. It is constantly present, usually in large numbers, in human faeces; and its presence in water indicates that excretal pollution has occurred. Unfortunately it is not specifically confined to the human intestine; it is present in the faeces of many domestic animals and birds. Since, however, the *Esch. coli* in sewage, with which water is liable to be contaminated, is mainly of human origin, this drawback is not as great as might appear. Again, there are numerous coliform organisms that closely resemble *Esch. coli* but have their chief habitat in soil, vegetation, and decaying organic matter (see Wilson *et al.* 1935, Randall 1956, Geldreich *et al.* 1962, Schubert and Mann 1968, and Table 9.7).

Many of these gain access adventitiously to the human body with food. They are, however, not natural inhabitants of the intestine; they persist in it for only a short time, and are excreted in the faeces, as a rule in much smaller numbers than *Esch. coli*. Their presence in water is therefore of less significance in indicating the occurrence of excretal pollution. They may have come from faeces or from some other more usual source, but their presence in water cannot be neglected. Outside the body they survive longer than *Esch. coli*, and they may therefore be found in water from which this organism has died out, indicating perhaps remote rather than recent pollution. Their presence, however, in the absence of *Esch. coli*, may be due to contamination with non-polluted dust or soil, or with old sacking, leather washers, jute packing, or decaying leaves in which they may be growing.

Other inhabitants of the intestine that may be used to indicate the presence of excretal pollution are faecal streptococci and *Clostridium perfringens*. Whether faecal streptococci survive longer outside the body than *Esch. coli* or not is doubtful; there is evidence both ways, but for practical purposes both organisms may be regarded as having much the same significance, namely as evidence of recent excretal pollution (Leiguarda *et al.* 1947, Mallmann and Litsky 1951, Cuthbert *et al.* 1955). *Cl. perfringens*, on the other hand, is a sporing organism. Its spores may survive for many weeks or months, and when found in water in the absence of *Esch. coli* indicate the occurrence of remote pollution.

There are times when a search for pathogenic organisms is required. These organisms are usually present in only very small numbers; large amounts of water have to be examined; and their isolation and identification take time. For routine control of the quality of the water they are quite unsuitable. Moreover, since

Table 9.7 Classification of coliform strains met with in Great Britain (Wilson *et al.* 1935)

Type	MR	VP	Growth in citrate	Indole	Gas in MacConkey at 44°	Gelatin liquefaction, 7 days	Probable habitat
Esch. coli, type I, faecal	+	−	−	+	+	−	Human and animal intestine.
Esch. coli, type II	+	−	−	−	−	−	Doubtful; possibly partly intestinal.
Citrobacter, type I	+	−	+	−	−	−	Mainly soil.
Citrobacter, type II	+	−	+	+	−	−	Mainly soil.
Klebs. aerogenes, type I	−	+	+	−	−	−	Mainly vegetation.
Klebs. aerogenes, type II	−	+	+	+	−	−	Mainly vegetation.
Enterobacter cloacae	−	+	+	−	−	+	Mainly vegetation.
Irregular, type I	+	−	−	+	−	−	Human and animal intestine.
Irregular, type II	+	−	−	−	+	−	Doubtful.
Irregular, type VI	−	+	+	−	+	−	Jute and hemp yarn.
Irregular, other types	Reactions variable						Doubtful.

266 The bacteriology of air, water and milk

their survival outside the body is short, they may easily have died out by the time suspicion of the water is aroused.

Plate counts reflect the amount of organic matter in the water available for bacterial nutrition. They are of value in the food manufacturing industry, partly for the purpose of excluding waters with large numbers of bacteria, and partly to detect undesirable organisms that are likely to cause deterioration of the product. They are likewise of use to the water undertaker in the control of sand filtration and chlorination. With slow sand filters the count on the filtered water should show a 95–98 per cent reduction on that of the raw water. A rise in the colony count is the signal of some defect in the filter beds, demanding instant attention.

Coliform count

Two methods are available for estimating the number of coliform bacilli in water—the one known as the liquid dilution or multiple tube method, the other membrane filtration.

(1) **Multiple tube method** This method, though not so rapid as membrane filtration, has the advantage of rendering evident the production of gas, which distinguishes coliform from most non-coliform bacteria. The medium used in Britain is either MacConkey broth or a glutamate medium based on Folpmers's (1948) original description. MacConkey broth has proved its value since the beginning of the century, but suffers from variation in the inhibitory power of the bile salts it contains. These may be replaced by 1 per cent Teepol 610 (Jameson and Emberley 1956, Report 1968). The resulting product gives as high a coliform count as the original MacConkey broth, and is particularly useful when the coliform bacilli are of lowered vitality, as in sea-water. Folpmers's medium was improved by Gray (1959, 1964), and further slightly modified (Report 1969b). The resulting lactose formate glutamate medium is superior to MacConkey broth, not only in yielding a higher count of coliform bacilli in both raw and chlorinated water with fewer false positive reactions, but in the constancy of its composition and the cheapness of its constituents (Report 1968, 1969b).

Many other media are used for the detection of coliform bacilli. In the United States, for example, plain lactose broth or lauryl tryptose broth are recommended by the American Public Health Association (Report 1971c). Lactose broth does not distinguish between coliform and other lactose-fermenting organisms. The lauryl tryptose broth, however, is an excellent medium, but, according to the Water Committee of the Public Health Laboratory Service, is slightly inferior to the modified lactose formate glutamate medium (Report 1980a).

In British practice a series of MacConkey, Teepol, or glutamate broths are inoculated with falling quantities of water, namely one 50 ml quantity, five 10 ml, five 1 ml, and, when necessary, five 0.1 ml quantities, incubated at 37° for 2 days, and the number of tubes showing acid and gas counted. The results are reported in terms of the probable number of coliform bacilli per 100 ml, given in tables slightly modified from the original ones of McCrady (1918); see Report 1969b).

Since acid and gas formation may occasionally be due to certain aerobic spore-bearers (see Greer 1928, Porter et al. 1935), to anaerobic spore-bearers (see Mackenzie et al. 1948), or to the synergic action of two different species of bacteria, one of which, such as *Str. faecalis*, breaks down the lactose, and the other of which, such as *Morganella morgani*, produces acid and gas from the monosaccharide so formed (see Atkinson and Wood 1938b), the result is best referred to as the '*presumptive coliform count*'. In raw water false positive results from these causes are uncommon (Bardsley 1938a), but in chlorinated water they occur in about 5 per cent of samples, chiefly from the action of anaerobic spore-bearers. Where, as in the United States, lactose broth free from bile salts is used, they are much more common.

(2) **Membrane filtration** The principle of this method is to filter a measured quantity of water through a membrane that retains bacteria, to transfer the membrane to a petri dish where it is laid face upwards on an absorbent pad soaked in a suitable liquid medium, and to count the acid-forming colonies after incubation for a given time. The practical application of the method is described fully in Report (1969b, 1972).

The most satisfactory medium is the Membrane Enriched Teepol Broth, described in Report (1969a). Since, however, Teepol, which is a proprietary product, became unavailable, the Public Health Laboratory Service Committee, after numerous comparative trials, recommended its replacement by 0.1 per cent sodium lauryl sulphate—a substance having the advantage of being chemically defined, cheap, and readily obtainable (Report 1980c).

The temperature and length of incubation may be varied according to the nature of the sample and the rapidity with which the result is desired. With raw waters incubation for 4 hours at 30° followed by 14 hours at 35° is said to give as high a count as 48 hours' incubation at 37° by the multiple tube method. With chlorinated waters 6 hours at 25° followed by 18 hours at 35° are recommended (see Report 1969b). When there is no urgency, then incubation at a single temperature for a longer time may be employed (see Taylor and Burman 1964, Burman 1967).

The colony count, though not subject to quite such a big error as that of the multiple tube method, nevertheless varies considerably, as shown by replicate counts on the same water sample. When for example, a colony count, C, of over 20 is observed, the 95 per cent confidence limits for the true number may be calculated as follows:

$$\text{Upper limit} = C + 2 \times (2 + \sqrt{C})$$

$$\text{Lower limit} = C - 2 - \times (1 + \sqrt{C})$$

Thus, with an observed count of 100 colonies, the true count will lie 19 times out of 20 between 78 and 124.

The membrane filtration method has many advantages. It gives a rapid result; it is economical in glassware and labour; with care, membranes may be re-used, thus effecting a saving in cost; and false positive results due to aerobic or anaerobic spore-bearing organisms are avoided. On the other hand, it is unsuitable for use with turbid waters having a low coliform count, since the membrane may become blocked before sufficient water has passed through. It is likewise unsuitable for waters having a low coliform but a high non-coliform count, and for waters containing numerous lactose-fermenting but non-gas-producing organisms, because of the high proportion of false positive results that will be registered (Leiguarda and Muratorio 1958). The absence of visible gas formation by colonies is a serious disadvantage in the examination of waters in the distribution system, rendering it necessary for all yellow colonies to be subcultured for purposes of confirmation; this delay partly nullifies the rapidity of the result that is one of the chief merits of the membrane method. An investigation in which several laboratories took part showed that the membrane method gave rather higher counts of coliform organisms in both chlorinated and unchlorinated waters than the multiple tube method, but lower counts for *Esch. coli* in chlorinated waters. With unchlorinated waters the results were equivocal (Report 1972).

It may be added that Danielsson (1965) and Danielsson and Laurell (1965) described a method for the direct demonstration of *Esch coli* on the membrane by use of the fluorescent antibody technique; that Levin and Bang (1968, see Evans *et al.* 1978) devised a *Limulus* lysate assay; and that Rao and Labzoffsky (1969) adapted the membrane technique for the demonstration of viruses in the water.

Confirmatory and differential coliform tests

To confirm the presence of coliform organisms it usually suffices to subculture one or more of the fermented tubes in the multiple tube method, or one or more colonies in the membrane method, into brilliant green bile broth (Mackenzie *et al.* 1948), or, better, 1 per cent lactose ricinoleate broth (Report 1969*a*) and incubate for 48 hours at 37°. Acid and gas production may be taken as adequate confirmation. At the same time it is advisable to plate out the original fermented tubes on to MacConkey medium, and examine the resulting colonies macroscopically for the typical coliform appearance and microscopically for the presence of gram-negative non-sporing rods. For greater assurance, the colonies may be submitted to the Kovács oxidase test to exclude the possibility of organisms of the *Aeromonas* group, which give a positive reaction in this test and are of no sanitary significance.

To distinguish between the different coliform organisms and, in particular, to test for the presence of *Esch. coli*, the simplest method is to subculture each of the fermented tubes into tubes of lactose ricinoleate broth and peptone water, incubate both at 44° in a carefully controlled water-bath, and examine for gas production in the first tube and test for indole production in the second. Positive results in these two tubes are confirmatory evidence of the presence of *Esch. coli*. The reason for testing all the fermented tubes in this way is that *Esch. coli* may have been present in the water in only minimal numbers and been inoculated into only one tube, or that it may have been overgrown by other coliform organisms; though it should be added that in the selective media used, incubated at 37°, *Esch. coli* is more likely to overgrow other coliform organisms than vice versa (see Habs and Langeloh 1960). To cut down the labour of subculturing, tests for the simultaneous production of gas and of indole at 44° may be combined in a single medium (Schubert 1956, Fennell 1972). Better than Schubert's medium is lauryl tryptose mannitol broth (Report 1980*b*, 1981).

As shown by Wilson and his colleagues (1935), and confirmed by numerous other workers (see p. 269), the production of gas in MacConkey broth at 44° is practically specific for *Esch coli*. Occasionally other coliform bacilli are met with that do this, but in Britain they are very few. The organism listed as Irregular vi (Table 9.7) is abundant in jute and hemp, and may multiply under suitable conditions in yarn used for packing the joints of water pipes (Taylor and Whiskin 1945); it differs from *Esch. coli* in being MR negative, VP positive and failing to produce indole. *Cl. perfringens* likewise may form gas in MacConkey broth at 44°. It is liable to cause confusion through being found in polluted water that has been chlorinated; its presence may be suspected by its failure to produce gas at 44° in either brilliant green bile broth or lactose ricinoleate broth. Should any doubt persist of the nature of coliform organisms in the water, colonies on MacConkey's medium should be put through the usual tests, namely gas production at 44°, MR, VP, growth in citrate, indole production, and gelatin liquefaction (Table 9.7). Confirmatory tests for *Cl. perfringens* include the blackening of a medium containing sulphite and iron and the production of stormy fermentation in milk when cultures are incubated anaerobically.

Examination for faecal streptococci

The colloquial term 'faecal streptococci' embraces a number of species of streptococci found in water polluted with human or animal faeces (see Chapter 29). These organisms are distinguished by their ability to grow at 45° in the presence of 40 per cent bile, and in concentrations of sodium azide or potassium tellurite that are inhibitory to most coliform organisms. For their demonstration in water the multiple tube method may be used, as with coliform bacilli, but with the substitution of either glucose azide (Hannay and Norton 1947) or ethyl violet azide broth (see Dutka and Kwan 1978) for MacConkey or glutamate broth. The tubes are incubated at 37° for 3 days, and those showing acid production are heavily subcultured into further tubes of the same medium and incubated at 45° for 48 hours. Alternatively, the membrane filtration method may be used, the membranes being cultured on well dried plates of Slanetz and Bartley's (1957) glucose azide agar. These should be incubated for 4 hours at 37° and then for 44 hours at 44–45°. All red or maroon colonies are regarded as faecal streptococci. Microscopical and other confirmatory tests may be necessary (Report 1969a).

Examination for *Clostridium perfringens*

For the demonstration of this organism, the water should be heated at 75° for 10 minutes to destroy non-sporing organisms. One 50 ml and five 10 ml quantities should be inoculated into the glucose sulphite iron medium of W. J. Wilson (1928) as modified by Gibbs and Freame (1965), and incubated anaerobically at 37° for 48 hours. Clostridial growth is evidenced by the blackening of the medium due to reduction of the sulphite and precipitation of ferrous sulphide. Each positive tube should be subcultured into a freshly boiled tube of litmus milk. The production of stormy fermentation within 48 hours at 37° is indicative of the presence of *Cl. perfringens*.

Search is sometimes required for *Pseudomonas aeruginosa*. This organism may suppress the growth of *Esch. coli* (Schiavone and Passerini 1957a, b, Reitler and Seligmann 1957); and in warm countries is found in 10 per cent of normal faeces. Its presence points to the possible occurrence of excretal contamination. Cultures of the water in plain lactose broth are plated out after 24 hours at 37° on to a variety of selective and indicator media, such as MacConkey's, Endo's, Levine's, or a medium containing tetrazolium. Suspicious colonies are picked off and tested for pigment production and other properties (see Chapter 31).

Pathogenic organisms

The demonstration of *typhoid bacilli* in water is difficult and demands bacteriological skill and experience. In general, a large volume of water—100 to 1000 ml—should be filtered through a special candle (see Report 1969a), and the organisms cultured in a selective medium such as selenite broth. After 24 hours' incubation at 37° tubes should be plated on to Wilson and Blair's (1931) bismuth sulphite agar. Colonies of *S. typhi* have a characteristic black appearance surrounded by a metallic sheen. For other salmonellae incubation in selenite broth at 40° or even 43° (Harvey and Thompson 1953) is advisable, followed by plating on to brilliant green MacConkey agar (Harvey 1956).

In the demonstration of *V. cholerae* advantage may be taken of its rapid growth in alkaline peptone water under aerobic conditions, and of the numerous differential media available (see Chapters 27 and 70).

Various methods have been used for the isolation of **viruses**. The methods proposed for concentrating and isolating **viruses** are reviewed by Sobsey (1976).

So far, the two methods of choice are: (1) ultrafiltration through soluble alginate filters with tangential flow across the membrane to prevent clogging (Gärtner 1966/7); and (2) flow through filter absorption-elution systems depending on the adsorption of viruses to cellulose membranes at low pH with added polyvalent cations, and subsequent elution with alkaline buffer containing glycerol or protein (see Wallis *et al.* 1972). A combination of these two methods is described by Slade (1977), who found between 12 and 49 plaque-forming units per litre in the river Thames.

Plate count

This consists in a plate count of the colonies formed on a standard nutrient agar medium incubated aerobically at 22° for 3 days or 37° for 1–2 days. Since not all the organisms in water are viable; since many viable organisms, such as those of the anaerobic and nitrifying groups, do not develop under these conditions, and since some of the organisms occur in groups that give rise to only a single colony, the colony count corresponds only to the number of bacterial units capable of multiplying under the nutritional, respiratory, and temperature conditions provided. For this reason it should be reported not as the number of organisms but as the 'number of colonies developing per ml,' or more simply as the 22° or 37° 'plate count per ml.'

Examination of sea-water The examination of sea-water follows much the same lines as those for fresh water. The high salt concentration of sea-water interferes with lactose fermentation by coliform organisms. For this reason the ratio of sea-water to medium should not be less than 1:10 (Papadakis 1975); and MacConkey's medium should be prepared with bile salt having only a weakly inhibitory action, and with brom-cresol purple rather than with neutral red (Report 1959a). Alternatively Teepol agar (Jameson and Emberley 1956) may be used, or plain lactose broth with subsequent confirmation of the presumed coliform organisms (Johannesson and Martin 1957, Christovão *et al.* 1958). The *roll-tube method* described under Shell-fish (p. 273) is likewise applicable, but again care must be taken to avoid the use of a medium that is too inhibitory (Report 1959a; see also Pretorius 1961).

Interpretation of the bacteriological analysis

Before attempting to give an opinion on the results of a bacteriological analysis, it is essential to gather particulars of the nature of the water, the method by which it was collected, the time of collection, and the amount of recent rainfall. It is sound practice, though not always possible, for the bacteriologist to make a topographical survey of the gathering ground, so as to ascertain the extent and the kind of pollution to which it is subject. If he is unable to do this personally, he should consult a map on which the source of the water and the immediate environment are indicated. Particular care must be given to the mode of collection of the sample; otherwise contamination, especially with coliform organisms, may disturb the interpretation of the results (for instructions, see Report 1969*a*).

Chlorinated water should be taken into bottles to which sufficient sodium thiosulphate is added before sterilization to give a fixed concentration of 18 mg per litre in the water sample, e.g., 0.1 ml of a 3 per cent solution for a 170 ml bottle. Bottles should be delivered to the laboratory packed in ice, or, failing that, in insulated containers. As with raw water, it should be examined within 6 hours of collection; beyond this time a significant fall in the numbers of *Esch. coli* may occur that will falsify the results of the analysis.

Where it is apparent from simple inspection that, as with a shallow well, the water is subject to human excretal pollution, there is no point in undertaking a bacteriological examination. Indeed it is unwise to do so. The contamination may be intermittent, and failure to find coliform organisms in a given sample of water may lead to a false sense of security.

The effect of rainfall should be noted. In general, the less the water is influenced by this factor, the better. Rain may carry in large numbers of undesirable organisms from the soil. A rise in the coliform count after heavy rain should be regarded as a danger signal.

The frequency of examination must depend on the size of the population served. The larger the population, the greater is the risk of catastrophe should pollution occur, and the more stringent must be the precautions be to safeguard the purity of the supply. Piped water for a population of 100 000 or more should be examined at least once a day, and for a population of less than 20 000 at least once a month (Report 1969*a*).

As already made clear, the whole aim of the bacteriological analysis of water is to detect evidence of excretal pollution. For this purpose reliance is placed mainly on the demonstration of coliform bacilli and, more specifically, of *Esch. coli*. In Great Britain *Esch. coli* is of particular significance, partly because it is usually derived from sewage; partly because coliform organisms in sewage are mainly of human origin; and partly because it does not multiply in water, so that the probable count, though not necessarily indicating the whole extent, does indicate at any rate the minimal degree of pollution that has occurred. British practice favours the virtual absence of this organism from piped water supplies intended for a large population, and the complete absence not only of *Esch. coli* but of other coliform organisms as well from chlorinated supplies.

Owing to the large sampling errors of both the multiple tube and the membrane method, an exact count of coliform organisms and of *Esch. coli* cannot be made with certainty. It is therefore necessary to allow a certain latitude on the reported counts. Throughout the year 95 per cent of samples of unchlorinated water taken from the distribution system in towns should be free from coliform organisms in 100 ml; and no sample should ever contain more than 10 coliform organisms or 2 *Esch. coli* in 100 ml (Report 1969*a*). The consumption of unchlorinated water, however, is never devoid of risk, and it is strongly recommended that all piped supplies should be chlorinated before distribution.

Waters intended to be taken into use for the first time should receive a full bacteriological examination supported by chemical analysis. In a small proportion of excretally contaminated waters faecal streptococci are said to be present in the absence of coliform bacilli (Leiguarda *et al.* 1956, Buttiaux 1958); such waters should be regarded with suspicion. When coliform bacilli of doubtful origin are found, the accompanying presence of streptococci is suggestive of an excretal source. As already mentioned, the spores of *Cl. perfringens* survive much longer than *Esch. coli*, and their presence in the absence of coliform organisms is indicative of remote contamination. This conclusion is often supported by chemical examination for organic matter of animal origin (see Report 1970).

Incidentally it may be remarked that chemical examination is mainly of value in the production stage of a water supply. It measures hardness; it detects the presence of heavy metals and of certain other toxic substances; it is essential for the proper control of chlorination; and it affords a guide to the suitability of a water for manufacturing purposes. It is unable, however, to reveal minimal degrees of excretal contamination; for this, bacteriological analysis is required, and that is why during the distribution stage bacteriological and not chemical examination is of prime importance. Standards for chemical pollutants, both those of the World Health Organization and the United States, are listed by Kenny (1978).

The interpretation of the results of bacteriological analysis varies to some extent in different countries according to local conditions. In the *tropics*, for example, a fairly high proportion of coliform organisms belong to the *Klebsiella*, *Enterobacter* and *Citrobacter* groups. Since careful sanitary surveys have shown that such waters may be free from exposure to excretal contamination, it is clear that reliance on the presumptive coliform count will result in the unnecessary condemnation of a number of unpolluted waters. Differentiation will generally be necessary. In this process it should be borne in mind that the yarn organism (Irregular vi in Table 9.7) is very common in some regions and, by producing gas in MacConkey's medium at 44°, may cause confusion with *Esch. coli*. (For further information see Pawan 1925, 1926, 1931, Raghavachari and Iyer 1939*a, b*, Burke-Gaffney 1932, 1933.) Deterioration in the quality of piped water supplies is often due to the use of plumbing materials

Swimming-bath waters

The main organisms found in swimming-bath water are aerobic spore-bearers, *Pseudomonas, Serratia*, coliform bacilli, *Proteus*, staphylococci of skin and salivary origin, streptococci, *Neisseria*, and *Cl. perfringens*. According to Bardsley (1938*b*) most of the coliform bacilli are of excretal origin and consist chiefly of organisms of the coli I type; but other coliform organisms are also present, and in chlorinated waters may actually exceed coli I in number. A little reflection will enable the student to understand why the bacteriological control of swimming-bath waters is not altogether satisfactory. The results vary from one part of the bath to another, and from minute to minute in the same part depending on the numbers and personal cleanliness of the bathers. Moreover infection, of either nasopharyngeal or excretal origin, may pass from one person to another with great rapidity so that no bacteriological examination of the water can afford an accurate assessment of the risks incurred. Fortunately the danger of disease transmitted by swimming-bath water can be greatly diminished by proper chlorination (Report 1929). If chlorine is added in sufficient concentration to give a residual figure of 0.2–0.5 part per million of *free* chlorine, pathogenic organisms of the enteric group are rapidly destroyed. To ensure that chlorine is present in the free state, the water must contain no ammonia; and to ensure that the chlorinated water is non-irritant to the eyes, the water must be alkaline—preferably pH 7.5–8.0. If ammonia is present, chloramines are formed which have a much lower disinfectant action than free chlorine.

For the bacteriological control of swimming-bath water numerous organisms have been proposed as indicators of respiratory or excretal pollution (see Report 1953*b*). For free chlorine the redox potential test is best (Victorin 1974).

Ferramola and Durieux (1951) regarded the gram-positive cocci—particularly enterococci—as superior to coliform bacilli as indicators of human pollution. In 1195 samples of water from swimming-baths they isolated coliform bacilli from 10 ml quantities in 19.3 per cent and gram-positive cocci growing in azide broth in 60.4 per cent. The cocci consisted of streptococci and staphylococci. Of 987 strains of streptococci studied, 931 were enterococci and 56 of viridans type. Of 1241 strains of staphylococci studied, 95 were coagulase-positive. The coliform bacilli appeared to be less resistant to chlorine than the gram-positive cocci and were never found in their absence. Tapley and Jennison (1941), on the other hand, favoured *Branhamella catarrhalis*, which in their experience was present in greater numbers than coliform bacilli and fluctuated more in accordance with the number of bathers in the pool.

The careful observations made by a special committee of the Public Health Laboratory Service set up to study the problem showed that with free chlorine in a strength of 0.2–0.5 ppm (*a*) the plate count at 37° did not exceed 10 colonies per ml in 76 per cent of samples or 100 colonies in 96 per cent; (*b*) coliform organisms and neisseriae were generally absent; (*c*) α-haemolytic streptococci and staphylococci were often present and were too resistant to serve as indicator organisms; (*d*) *Cl. perfringens* was not found at all; (*e*) the results were very little affected by the number of bathers using the bath; and (*f*) there was virtually no difference in the bacterial quality of the water at the inlet and outlet of the bath (Report 1953*b*). In practice the committee recommended that, in swimming-bath water containing 0.2–0.5 ppm of free chlorine, no sample should contain coliform organisms in 100 ml; and that in 75 per cent of the samples the plate count at 37° in 1 ml of water should not exceed 10 colonies and in the remainder it should not exceed 100 colonies. The American Public Health Association exacts a similar but slightly lower standard (Report 1964; see also Report 1967*b*). To maintain a residual concentration of 0.2–0.5 ppm of free chlorine in school baths marginal chlorination cannot be relied upon; the break-point method should be used (Report 1965*a*).

According to Amies (1956) secretions from the nose and mouth of the bathers, together with fatty substances from the skin, form a film on the surface of the water that is rich in bacteria. As these organisms are partly protected from the action of the chlorine in the water below, Amies suggests that a separate examination should be made of the surface film.

In natural waters used for bathing, such as rivers, lakes and sea, it is impracticable to lay down any bacteriological standard of purity; but where there is reason to believe that sewage is gaining access to the water a count of *Esch. coli* I exceeding 1000 per 100 ml (Moore 1954*a, b*), or of coliform organisms exceeding 1000 (Scott 1957) or 2500 per 100 ml (Report 1957), may be regarded as indicative of potential risk to the bathers (see Moore 1954*b*). (For otitis externa due to *Ps. aeruginusa*, see Ch. 61.)

Water-borne disease

Water is one of the chief vehicles of gastro-intestinal disease; and the provision of a safe water supply is one of the first tasks to be undertaken in the introduction of environmental sanitation. The task is not an easy one, as is shown by the fact that in the developing countries in 1970 only 10 per cent of the rural and 50 per cent of the urban population were supplied with safe water (Report 1971*a*).

The **bacterial** diseases carried by water are enteric fever, dysentery, cholera, infectious hepatitis, and gastro-enteritis. For a description of these reference should be made to Chapters 68, 69, 70, 72 and 101. Occasional outbreaks of Weil's disease and tularaemia are reported. John Snow (1855), in his study of cholera (see Chapter 70), was the first to bring conclusive evidence of the water-borne carriage of disease; but it was a long time before the full part played by water in this respect was realized. Indeed it is only recently that infectious hepatitis was added to the list of water-borne infections. In Great Britain, where over 95 per cent of the population has a public supply of piped drinking water—almost all of it chlorinated—outbreaks of water-borne disease are rare, and occur only as the result of accident. Formerly, however, as on some parts of the Continent and of the United States, they were not uncommon. For example, between 1911 and 1937 in England and Wales there were 21 outbreaks of disease conveyed by public water supplies, resulting in 1237 cases of enteric fever, 2800 of bacillary dysentery, and 7439 cases of gastro-enteritis (Report 1939). Since the Croydon outbreak of 1937, however, it is doubtful whether a single case of typhoid fever in England and Wales has been caused by a properly treated water supply. Between 1936 and 1975 there was a ten-fold decrease in the incidence of typhoid fever; the cases that still occur are nearly all traceable to infection contracted abroad (Report 1978). Viral hepatitis, which has so frequently occurred in epidemic form in the United States, India, and elsewhere, has been notably absent.

In the United States Gorman and Wolman (1939) collected records of 399 outbreaks between 1920 and 1936, affecting 115 645 persons. Among the cases of disease were 12 585 of typhoid fever and 101 603 of diarrhoea. During the period 1938 to 1945, 327 outbreaks were recorded, affecting 111 320 persons (Eliassen and Cummings 1948); during 1946 to 1960 there were 228 known outbreaks, affecting 25 984 persons (Weibel *et al.* 1964; Table 9.8); during 1961 to 1970 there were 121 known outbreaks, affecting 46 325 persons (Taylor *et al.* 1972), and during the eight years 1971 to 1978 there were 224 outbreaks (Craun 1981) affecting 79 115 persons (see Haley *et al.* 1980). (For later figures see Report 1982.)

According to Craun, from whose paper Table 9.8 was mainly constructed, the major causes of outbreaks associated with municipal and community supplies were contamination of the distribution system, primarily as a result of cross-connections or back-siphonage, and deficiencies in treatment processes. Those associated with semi-public or non-community supplies were due mainly to the use of untreated contaminated ground-water and deficiencies in treatment.

In South Wales Harvey and Price (1981) recorded numerous cases of infection as the result of drinking or bathing in contaminated fresh water—mainly rivers and brooks. Among the diseases met with were cercarial dermatitis, leptospirosis, typhoid fever, paratyphoid fever, and anthrax.

These figures all underestimate the true incidence. As with milk, so with water, small outbreaks are often not reported, or the mode of carriage not even recog-

Table 9.8 Reported outbreaks of water-borne disease in the United States 1946–70, and outbreaks only 1971–8

	1946–1970		1971–1978
	Outbreaks	Cases	Outbreaks
Gastro-enteritis	311	69 694	132
Typhoid fever	57	832	4
Infectious hepatitis	78	2 248	16
Dysentery	51	12 492	20
Salmonellosis*	20	24 133	7
Amoebiasis	5	75	0
Giardiasis	24	11 958	22
Other	37†	5 383	23
Total	583	126 815	224

* Numbers mainly attributable to one large outbreak
† Mainly chemical

nized. The fact that 36 per cent of the known outbreaks in Weibel's series were reported in New York State alone leaves no doubt about the inadequacy of the reporting system in many other States of the Union.

As is apparent from these records, *gastro-enteritis* figures prominently in water-borne disease. Its nature and causation are far from clear. No pathogenic organisms can be isolated. This applies even to large outbreaks, such as the one occurring among visitors to the Pennsylvania State Park in 1966 in which several thousand persons were affected (Lobel *et al.* 1969). The facts that the average incubation period is often around 30 hours and that family contacts may be affected subsequently suggest the presence of an infectious agent. Possibly pathogenic organisms of the coliform group may be responsible, as in an outbreak that occurred at a conference centre near Washington where *Esch. coli* 0 111:B4 was isolated both from the well water and from the patients' stools (Schroeder *et al.* 1968); but these organisms have not usually been sought for. There seems little doubt that the access of sewage in considerable amount to a water supply may be followed by gastro-enteritis or diarrhoea, though how it does so is a problem still unsolved (see Kathe and Königshaus 1932, Report 1936, 1937, Pharris *et al.* 1938, Eliassen and Cummings 1948, Ross and Gillespie 1952). Lead poisoning must, of course, be borne in mind, as well as the less common forms of chemical poisoning, such as that described by Campbell (1940) of naphthalene poisoning after the painting of a storage tank. The length of the incubation period and the usual absence of vomiting in most cases of water-borne gastro-enteritis render a chemical cause unlikely.

Dysentery is not often attributable to polluted water, but occasional quite large outbreaks are recorded, such as those south of Albany, N.Y. (Freitag 1960), and at Montrose, Col. (Green *et al.* 1968), both of which were caused by *Shigella sonnei*, and were apparently due to failure of the chlorination system. *Paratyphoid fever* is even less common, probably because the organisms are usually too few to cause clinical illness; but Franklin and Halliday (1937) reported one outbreak in Canada; and two or three small outbreaks are on record in Great Britain (see Report 1942, Page 1942). *Salm. typhimurium* may at times cause trouble, as in the big outbreak in the City of Riverside, California, in 1965, where over 16 000 persons drinking unchlorinated well water are reputed to have been affected within the space of 5 weeks (Report 1965*b*, 1971*b*). Of the three largest outbreaks recorded between 1971 and 1978, one of giardiasis in Rome, N.Y., affected 4800 persons; another, also of giardiasis, in Pico Rivers, Calif., affected 3500 persons (Craun *et al*. 1976); and one of *Campylobacter* infection in Vermont affected 3000 persons (Haley *et al*. 1980). In these years the greatest number of water-borne outbreaks occurred in June to September, and most were due to untreated ground-water. In only just over a half was an aetiological agent detected.

The possible role of **viruses** in the causation of water-borne disease has been studied intensively during the past few years. Apart from infectious hepatitis and gastro-enteritis following the consumption of contaminated water, there is little evidence of viral origin in any of the water-borne diseases. Though chlorination of purified water cannot be relied upon completely, no proof has been forthcoming, certainly in Great Britain, that any outbreak has occurred as the result of drinking properly treated water (Gamble 1979). With one or two doubtful exceptions, the same appears to have applied in the United States. No evidence of their presence could be found in the outbreak of gastro-enteritis in New York State investigated by Borden, Harris and Mosher (1970); and though Berg (1966–67) regards any amount of virus in drinking or recreational water that is detectable in appropriate cell cultures as a hazard, confirmation of this opinion in practice is still awaited. For many years poliomyelitis was thought to be water-borne, but the evidence was never conclusive. The first viral disease in which water was shown to be the occasional vehicle was *infectious hepatitis*. Mosley (1959, 1966) compiled a list of 28 outbreaks occurring between 1916 and 1957 attributed to the drinking of contaminated water; in the largest one, at Delhi in 1955–6, a total of 28 745 icteric cases was reported. Craun and McCabe (1973) reported 53 water-borne outbreaks of infectious hepatitis in the United States between 1946 and 1970; and no fewer than 178 outbreaks of viral gastro-enteritis affecting 45 255 persons. Most of these outbreaks followed the drinking of raw surface water. Analysis of infectious hepatitis rates in US cities using surface water in 1961 showed that the incidence was inversely proportional to the amount of residual chlorine (Taylor *et al*. 1966).

One of the parvoviruses is strongly suspected of causing outbreaks of *acute infectious non-bacterial gastro-enteritis*, but the incriminating evidence is so recent that, as Fox (1976) says, no extensive studies of this disease have yet been reported. And here again the outbreaks so far studied have followed the consumption of polluted unpurified water. In biology, a negative is impossible to prove; so that while retaining an open mind and being sceptical of the importance attached to it by the Americans, we prefer to regard viral disease attributable to water treated up to conventional standards of purity as an uncommon, if not indeed a rare, event. In this connection, it is not inappropriate to remember Hilaire Belloc's ironic couplet:

'Oh let us never, never doubt
What nobody is sure about'

The whole subject is reviewed in the American Public Health Association's publication (Viruses in Water. Ed. by C. Berg *et al*. 1976).

The danger of **bathing** in polluted water is difficult to assess. In *fresh water* cases of typhoid and paratyphoid fever are invariably sporadic (see Martin 1947), and other sources cannot always be excluded. Stevenson (1953) found a higher incidence of illness among bathers than non-bathers, and obtained some evidence suggesting that nose-and-throat and gastro-intestinal diseases were more frequent among swimmers in water containing 2500 or more coliform organisms per 100 ml than among those in less contaminated water. Both Baron and his colleagues (1982) and Koopman and his colleagues (1982) describe an outbreak of gastro-enteritis in persons who had bathed in the same pool in a recreational park in July 1979. The Norwalk virus was isolated from the stools, but was not seen by electronmicroscopy. The outbreak was not confined to the swimmers, though infection was commoner in these than in non-swimmers. The authors regard the circumstantial evidence as pointing to an association with bathing. In *sea-water* the risk is minimal unless gross pollution is present (Report 1959*a*, *b*). The contamination of beaches with human sewage, with grease from meat, wool, and other trade wastes (see Flynn and Thistlethwayte 1965), or with the discharge of slaughterhouses (see Meyer 1961), is objectionable on the grounds of amenity, but does not appear to be dangerous to health. This is a hard saying but none the less true (see Moore 1960). Sporadic cases of *Weil's disease* are not uncommon among bathers in rat-infested streams and canals. Infection probably occurs through the conjunctival and nasal mucosae and through cuts and scratches on the skin.

Watercress This is a potential source of enteric disease. It should be cultivated in beds supplied with pure water. Even so some growth of coliform organisms occurs in the beds, particularly in the summer (Baker and Billing 1958), thus diminishing the value of the coliform test as an index of excretal pollution. In a survey undertaken by the Public Health Laboratory Service (Report 1965*c*) coliform organisms were found in over 90 per cent of samples of watercress, irrespective of the hygienic condition of the beds; and even under the best conditions *Esch. coli* I was found in 48 per cent of samples taken at the beds themselves. These organisms come from the soil of the beds, and are present even when the inflowing water is of potable standard; they do not indicate excretal pollution. The coliform counts ranged so widely that it was impossible to suggest any standard for control purposes. Watercress may be treated with 50 ppm of free chlorine for 2 hours to remove possible danger. So far as England and Wales are concerned the only trouble traced to the consumption of watercress has been the occasional case of fascioliasis.

Mode of contamination and purification of water supplies

Contamination of a water supply may occur at the source, during storage, or during distribution. Among the commonest causes are pollution of shallow wells and surface waters, seepage of surface water into gravity conduits, cross connections between a purified and a polluted supply, and suction of contaminated soil water into fractured or imperfectly jointed water mains when only an intermittent supply is provided. Disease is likely to follow when polluted water is drunk raw, or when filtration or chlorination is insufficient or is interrupted.

An account of the measures necessary to provide a safe water supply is beyond the scope of this book (see Report 1967*a*, Holden 1970). Scrupulous care and supervision of every part of the system are demanded, including the protection of gathering grounds (Report 1948) and the health of the operatives associated with the water itself (see Report 1939). Supplies of pure deep well water are every year becoming fewer, and recourse has to be made more and more to surface water, some of which, such as river water, is often grossly polluted. Water of this sort has to be treated by storage, chemical precipitation, sand filtration, and chlorination. The resulting product, which is in fact purified sewage, can be drunk with impunity so long as examination shows that virtually all vegetative organisms of intestinal origin have perished or been destroyed. There is still some doubt about the resistance of the virus of infectious hepatitis to chlorine; but the freedom of cities in Great Britain from water-borne outbreaks of this disease suggests that the normal practice of chlorination is sufficient to ensure the safety of the water. In regard to chlorination, it must be understood that this process can be relied upon only in water freed from organic matter and ammonia. Otherwise, chloramine is formed, which has far less disinfecting activity than free chlorine. Normally a concentration of free chlorine of 0.5 ppm should be sufficient to destroy coliform organisms and viruses (Report 1970). Even free chlorine may fail to destroy all vegetative enterobacteria if crustaceans, such as *Cyclops* and *Daphnia*, are present in the water. In them, the organisms are protected from the chlorine, but may be liberated by the force of the water passing through a spigot. This sometimes accounts for the presence of coliform bacilli in the distribution system (Tracy *et al.* 1966). For observations on the resistance of viruses to chlorine and other disinfectants in water the reader is referred to a paper by Mahnel (1977).

Shell-fish

The chief importance of shell-fish from a public health point of view is their liability to give rise to enteric fever. In France alone it was estimated that within a period of 15 years more than 100 000 cases of typhoid fever due to the consumption of shell-fish occurred, of which 25 000 ended fatally (Bélin 1934). In Great Britain enteric fever attributable to this cause is very uncommon, mainly because of the much greater care that is now paid to the purification and control of shell-fish. The risk of typhoid fever arises chiefly from oysters, clams, and mussels, which are often eaten raw. Cockles, periwinkles, whelks, and escallops, which are cooked by boiling or steaming, are far less likely to be responsible. Occasional outbreaks of infectious hepatitis are traced to shell-fish, such as the Swedish outbreak in December 1955 due to oysters, in which there were over 600 primary cases (Roos 1956), the one in Mississippi and Alabama in 1961 with 80 cases, likewise due to oysters (Mason and McLean 1962), one mainly of gastro-enteritis with some cases of hepatitis in New Jersey in 1966 due to raw clams (Dismukes *et al.* 1969), and another in Connecticut during 1963–64 due to raw clams (Ruddy *et al.* 1969). Viruses may be found in shell-fish, but apart from the virus of infectious hepatitis it is doubtful whether they play any part in causing disease. Appleton (1981) records the finding of small round viruses in the faeces of patients in 11 outbreaks of food poisoning associated with the consumption of shell-fish though none could be isolated in culture.

Mussels and clams that have fed on marine dinoflagellates, such as *Gonyaulax catenella* or *G. tamarensis*, may give rise to a fatal form of paralytic food poisoning due to the action of a curare-like substance (see Schantz 1969). The dinoflagellates flourish best in enclosed water at a fairly high temperature, sometimes causing a red tide (Clark 1968); but they may grow in open water in cool weather such as that which preceded the outbreak on the north-east coast of England at the end of May, 1968 (McCollum *et al.* 1968). The death of sand-eels and of certain sea-birds, such as shags and Arctic terns, that have fed on them, may give a warning of the toxicity of the water before human cases occur (Clark 1968). (For further references see Sommer and Meyer 1937, Sommer *et al.* 1937, Sherwood 1952, Martin 1958, Evans 1965, Grindley and Sapeika 1969.)

Shell-fish become infected by being laid down in polluted water. According to Eyre (1924), about 2 litres of water enter and leave the shell of an oyster every hour. Large volumes of water also pass through mussels every 24 hours. The bacterial flora of these shell-fish is largely determined by the nature of the water in which they are immersed. This is often subject to rapid changes. It is quite possible, for example, for mussels to be heavily polluted on an ebb tide and to clean themselves on a flood tide. The only satisfactory way of ensuring their purity is to lay them down for a day or two in non-polluted water so as to give them time to clean themselves (see Dodgson 1936, Allen *et al.* 1950), though even this method cannot be relied upon for all pathogenic organisms (Janssen 1974). The water must be kept at a temperature high enough to enable active metabolism to proceed and should not contain free chlorine. Water cleansed by ultraviolet light may be used for the same purpose (Wood 1961). Enteroviruses, as well as bacteria, are taken up during feeding, and in cold weather may survive for some weeks (Metcalf and Stiles 1966); but they are usually excreted rapidly when the shell-fish are transferred to warm clean water (Hoff and Becker 1969, Hamblet *et al.* 1969). Barrow and Miller (1969) reported the finding in purified oysters of vibrios resembling *V. parahaemolyticus*, but their significance is doubtful. According to Cann and his colleagues (1981) raw shrimps should not contain more than 10^4/g of this organism.

Fish, as well as mollusca, caught in heavily polluted water and imperfectly cooked or cured, may give rise to enteric fever when eaten (see Steiniger 1951).

An attempt is often made to control the suitability of shell-fish for human consumption by bacteriological methods. The general principles of such an analysis are similar to those we have just discussed in the examination of water, particular attention being paid to the numbers of coliform bacilli. The difficulties, however, of obtaining an exact bacterial enumeration are considerably greater with shell-fish than with water. The chief reason for this is the extraordinary variation in numbers from one oyster or mussel to another, and the gross irregularity in distribution of organisms within the individual shell-fish themselves. Numerous methods have been devised for the bacteriological examination of shell-fish (see Eyre 1924, 1933, Dodgson 1928, 1936, 1938, Perry 1928, 1929, Bigger 1934, Beard and Meadowcroft 1935, Report 1947). In Great Britain no standard method of analysis or of interpretation is laid down, but the method devised by Clegg and Sherwood (1947) is widely used. This consists essentially in making a roll-tube count of *Esch. coli* in a modified MacConkey agar incubated at 44°. The method avoids the danger of using the presumptive coliform test. As Dodgson (1938) showed, the 37° test may be grossly misleading, because organisms of the aerogenes-cloacae type may under favourable conditions multiply enormously in mussels, barnacles, raw sea-water, and sterilized tank water. By the Clegg and Sherwood method the following standard of interpretation may be tentatively accepted (Sherwood and Thomson 1953):

Grade I   not more than 5 faecal coli per ml of shell-fish tissue
Grade II 6-15 faecal coli per ml.
Grade III Over 15 faecal coli per ml.

Grade I may be regarded as satisfactory, grade II as suspicious, and grade III as unsatisfactory.

These categories correspond roughly to those adopted by the Fishmongers' Company based on the examination of 0.2 ml quantities of fluid from 10 individual shell-fish, namely 80-100 per cent clean, 70 per cent, and 60 per cent or less (see Knott 1951). They are not applicable to viruses. Indeed there seems to be little relation between the coliform count and the virus count (Vaughn *et al.* 1980).

Just as with water, so in the control of shell-fish a careful topographical survey to exclude possible sources of pollution is desirable. Bacteriological examination must remain as an ancillary method to check the results of the sanitary investigation. Valuable information on the sanitary control of the shell-fish industry will be found in a report (1946) of the United States Public Health Service, and a general review of shell-fish in a paper by Martin (1958).

Sewage

In Great Britain crude sewage contains about 10 million organisms per ml capable of developing on gelatin at 20°, and from 1 to 5 million per ml on agar at 37°. In America the numbers appear to be lower. There is often a considerable rise in the summer months. The organisms making up these numbers are of many different kinds, and vary from one sewage to another. Prominent among them are the *Proteus* group, the coliform bacilli, streptococci, anaerobic spore-bearing bacilli, natural water bacteria, and the denitrifying bacilli. *Escherichia coli* may number 100 000 per ml, *Str. faecalis* 1000 to 10 000 per ml, and *Cl. perfringens* 100 to 1000 per ml. Pathogenic organisms of the typhoid, paratyphoid, and food-poisoning groups can often be found in crude sewage by the use of appropriate selective media. So also may tubercle bacilli (Pramer *et al.* 1950, Müller 1959). Poliomyelitis, Coxsackie, and ECHO viruses may be present in sewage. In Michigan enteroviruses were found by tissue-culture techniques in 118 out of 1403 (8.4 per cent) samples of sewage taken over a period of two years (Mack *et al.* 1958). Neither pathogenic bacteria nor viruses are entirely removed by treatment, though the numbers found in the effluent may be expected to be somewhat less than in the crude sewage. Sewage is an excellent source of a large variety of bacteriophages.

The process of purification is accompanied not so much by a diminution in the numbers of organisms,

Table 9.9 Boston sewage. (Modified from Winslow 1905.)

	Microscopic count	Gelatin	Agar	Microscopic ÷ gelatin	Agar × 100 ÷ gelatin
Crude sewage	29 000 000	1 690 000	1 400 000	17	83
Septic tanks	30 000 000	787 000	504 000	38	64
Contact filters	24 000 000	521 000	432 000	46	83
Trickling filters	17 000 000	451 000	284 000	39	63
Sand filters	65 000	9 160	10 800	7	120

as by a change in the distribution of different organisms. Thus in the septic tank, the anaerobic liquefying bacteria are prominent; on the contact beds the aerobic liquefying and the denitrifying bacteria gain the upper hand. It is evident that enormous numbers of bacteria must perish in the process; by a comparison of the microscopic and the gelatin counts, Winslow (1905) found that the ratio of the total to viable organisms was about 20 to 1 in crude sewage, 40 to 1 in the septic tanks and filter beds, and 70 to 1 in the sand filter effluents. Table 9.9 illustrates the numerical changes occurring during biological purification.

It will be seen that no striking fall occurs in the number of living organisms till the sand filter stage is reached.

A method of sewage disposal that is now widely used is the Activated Sludge Process. Briefly this consists in treating the sewage with about 15 per cent of bacterially active liquid sludge, in the presence of an ample supply of atmospheric oxygen. As a result of this treatment, a large proportion of the colloidal material undergoes coagulation, and subsequent sedimentation. The activated sludge itself results from the aeration of successive portions of sewage (see Martin 1927, Buswell 1928). As Wooldridge and Standfast (1936) showed, the purification is brought about by a series of catalysed oxidation-reduction reactions determined by bacterial enzymes present in the living or dead bacterial cells or liberated by them into the fluid of the reacting system. A reduction of as much as 98 per cent in the number of living bacteria may occur during the process (Miller 1936, Keller 1959). The activated sludge method of purification is said to be more effective in this respect than the trickling filter process (Theios *et al.* 1967; see also Askew *et al.* 1965). Viruses likewise suffer great diminution in the activated sludge process (Clarke *et al.* 1961).

The survival of typhoid bacilli in sewage has been studied by a number of workers (see Green and Beard 1938). The rate at which they die out depends on a variety of factors. As a rule, the great majority are dead within a week and all within a month. Their death is probably accelerated by the activated sludge process of treatment, though observations on *Salm. typhimurium* have shown that this organism may survive in drying sewage sludge for several months (Stokes *et al.* 1945).

Attempts to sterilize sewage by chlorination are beset with difficulties. Injurious chemical compounds may be formed, such as cyanogen which is toxic to fish; nitrification of the effluent may be inhibited leading to delay in self-purification; and though a high proportion of bacteria may be killed by a suitable dose of chlorine, multiplication of the survivors soon occurs leading to as high a bacterial population as before (see Allen and Brooks 1949). Viruses are greatly diminished in number by biological treatment but, like bacteria, are not completely destroyed by chlorination (Kelly and Sanderson 1959). In Vancouver, where the aerobic effluent treatment of sewage is practised all the year round, the effluent is said to be invariably virucidal; at no time have enteroviruses been isolated from it (Bursewicz 1975). In Great Britain the concentration of viruses in sewage effluent is reduced from the initial figure of 10 per litre (Geldreich and Clarke 1971) to about 1 in 10 000 litres. Gerba (1981) estimates the reduction of viruses by biological purification to be of the order of 90 to 99 per cent. To remove all viruses, the sludge, after anaerobic digestion, should be subjected to heat, preferably by composting it under aerobic conditions. By this process the temperature should rise to 60° or more. When this is maintained for a week or so, the resulting product is found to be free from viruses and to have the texture and odour of rich soil (Ward 1981).

Since the introduction by Moore (1948, 1950) of the sewer swab technique for the isolation of pathogenic organisms, the examination of sewage now plays an important part in the epidemiological investigation of enteric disease. The swabs may be inserted in main sewers, branch sewers, house drains, drains from food premises and abattoirs, or even water-closets, and the results used to follow back the organisms to their source in the human or animal carrier, or inanimate material such as processed egg or meat (see Harvey 1957). For a good general description of the production and purification of water and the disposal of sewage, see Tebbutt (1973), Curds and Hawkes (1975) and Sidwick and Murray (1976); and for a discussion on the disposal of sewage from coastal towns, see Stanfield (1982).

References

Allen, L. A. and Brooks, E. (1949) *J. Hyg., Camb.* **47**, 320.
Allen, L. A., Thomas, G., Thomas, M. C. C., Wheatland,

A. B., Thomas, H. N., Jones, E. E., Hudson, J. and Sherwood, H. P. (1950) *J. Hyg., Camb.* **48,** 431.
Amies, C. R. (1956) *Canad. J. publ. Hlth* **47,** 93.
Appleton, H. (1981) In: *Viruses and Wastewater Treatment.* Ed. by M. Goddard and M. Butler. Pergamon Press, Oxford.
Askew, J. B., Bott, R. F., Leach, R. E. and England, B. L. (1965) *Amer. J. publ. Hlth* **55,** 453.
Atkinson, N. and Wood, E. J. F. (1938*a*) *Aust. J. exp. Biol. med. Sci.* **16,** 103; (1938*b*) *Ibid.* **16,** 111.
Baker, L. A. E. and Billing, E. (1958) *J. appl. Bact.* **21,** 4.
Bance, J. and Caillon, L. (1929) *Arch. Inst. Pasteur Tunis* **18,** 199.
Bardsley, D. A. (1934) *J. Hyg., Camb.* **34,** 38; (1938*a*) *Ibid.* **38,** 309; (1938*b*) *Ibid.* **38,** 721.
Baron, R. C. et al. (1982) *Amer. J. Epidem.* **115,** 163.
Barrow, G. I. and Miller, D. C. (1969) *Lancet* **ii,** 421.
Beard, P. J. and Meadowcroft, N. F. (1935) *Amer. J. publ. Hlth* **25,** 1023.
Bélin, V. M. (1934) *Coquillages et Fièvres Typhoides. Un Point d'Histoire Contemporaine.* Les Presses universitaires de France, Paris.
Belli, C. M. (1902) *Zbl. Bakt.*, IIte Abt. **8,** 445.
Berg, G. (Ed.) (1966-67) *Transmission of Viruses by the Water Route.* Interscience Publishers, New York.
Berg, G., Bodily, H. L., Lennelte, E. H., Melnick, J. L. and Metcalf, T. G. (1976) *Viruses in Water,* Amer. publ. Hlth Ass., Inc., Washington D.C.
Bigger, J. W. (1934) *J. Hyg., Camb.* **34,** 172.
Böing, J. (1957) *Zbl. Bakt.* **168,** 324.
Borden, H. H., Harris, R. W. and Mosher, W. E. (1970) *Amer. J. publ. Hlth* **60,** 283.
Brezina, E. (1906) *Z. Hyg. InfektKr.* **53,** 369.
Burke-Gaffney, H. J. O'D. (1932) *J. Hyg., Camb.* **32,** 85; (1933) *Ibid.* **33,** 510.
Burman, N. P. (1967) *Proc. Soc. Wat. Treatm. Exam.* **16,** 40.
Burman, N. P. and Colbourne, J. S. (1977) *J. appl. Bact.* **43,** 137.
Bursewicz, A. M. (1975) *Lancet* **i,** 639.
Buswell, A. M. (1928) *The Chemistry of Water and Sewage Treatment.* Chemical Catalog Co. Inc., New York.
Butterfield, C. T., Purdy, W. C. and Theriault, E. J. (1931) *Publ. Hlth Rep., Wash.* **46,** 393.
Buttiaux, R. (1951) *L'Analyse Bactériologique des Eaux de Consommation.* Flammarion, Paris; (1958) *Ann. Inst. Pasteur* **95,** 142.
Campbell, M. S. (1940) *J. Amer. Wat. Works Ass.* **32,** 1928.
Cann, D. C., Taylor, L. Y. and Merican, Z. (1981) *J. Hyg., Camb.* **87,** 485.
Christovão, D. de A., Netto, J. M. de A., Jezler, H. and Brandão, H. (1958) *Rev. Dept. Aguas Esgótos, São Paulo* No. 30.
Clark, R. B. (1968) *Lancet* **ii,** 770.
Clarke, N. A., Stevenson, R. E., Chang, S. L. and Kabler, P. W. (1961) *Amer. J. publ. Hlth* **51,** 1118.
Clegg, L. F. L. and Sherwood, H. P. (1947) *J. Hyg., Camb.* **45,** 504.
Collins, V. G. (1957) *J. gen. Microbiol.*, **16,** 268; (1960) *J. appl. Bact.* **23,** 510; (1964) *J. appl. Bact.* **27,** 143.
Craun, G. F. (1981) *Amer. Water Works Ass. J.* **73,** 360.
Craun, G. F. and McCabe, L. J. (1973) *J. Amer. Water-Wks Ass.* **65,** 74.
Craun, G. F., McCabe, L. J. and Hughes, J. M. (1976) *J. Amer. Water-Wks Ass.* **68,** 420.

Curds, C. R. and Hawkes, H. A. Ed. (1975) *Ecological Aspects of Used-Water Treatment* Vol. 2. Academic Press, London.
Cuthbert, W. A., Panes, J. J. and Hill, E. C. (1955) *J. appl. Bact.* **18,** 408.
Danielsson, D. (1965) *Acta path. microbiol. scand.* **63,** 597.
Danielsson, D. and Laurell, G. (1965) *Acta path. microbiol. scand.* **63,** 604.
Dismukes, W. E., Bisno, A. L., Katz, S. and Johnson, R. F. (1969) *Amer. J. Epidem.* **89,** 555.
Dodgson, R. W. (1928) *Min. Agric. Fish., Lond., Fish. Invest.*, Ser. 2, **10,** No. 1; (1936) *Brit. med. J.* **ii,** 169; (1938) *Proc. R. Soc. Med.* **31,** 925.
Duhot, E. and Hutin, A. (1933) *C. R. Soc. Biol.* **112,** 195.
Dutka, B. J. and Kwan, K. K. (1978) *J. appl. Bact.* **45,** 333.
Eliassen, R. and Hutin, A. (1933) *C. R. Soc. Biol.* **112,** 195.
Evans, M. H. (1965) *Brit. J. exp. Path.* **46,** 245.
Evans, T. M., Schillinger, J. E. and Stuart, D. G. (1978) *Appl. environm. Microbiol* **35,** 376.
Eyre, J. (1924) *Publ. Hlth, Lond.,* **38,** 6; (1933) *J. Hyg., Camb.* **33,** 1.
Fennell, H. (1972) *Wat. Treatm. Exam.* **21,** 13.
Ferramola, R. (1947) *Examen Bacteriologico de Anguas.* El Ateneo, Buenos Aires; (1949) *Organo ofic. Asoc. interamer. Ingen, sanit.* **2,** 358.
Ferramola, R. and Durieux, J. E. (1951) *Rev. Obras sanit. Nacion, B. Aires* **15,** 173.
Flynn, M. J. and Thistlethwayte, D. K. B. (1965) *Int. J. Air Wat. Pollut.* **9,** 641.
Folpmers, T. (1948) *Leeuwenhoek ned. Tijdschr.* **14,** 58.
Fox, J. P. (1976) In: *Viruses in Water* p. 39. Ed. by G. Berg et al. Amer. publ. Hlth Ass., Inc Washington, D.C.
Frank, G. (1888) *Z. Hyg. InfektKr.* **3,** 355.
Franklin, J. P. and Halliday, C. H. (1937) *Canad. publ. Hlth J.* **28,** 82.
Freitag, J. (1960) *Hlth, News, N.Y. St. Dept. Hlth* **37,** April, p. 4.
Gamble, D. R. (1979) *Lancet* **i,** 425.
Gärtner, H. (1966) *See* Berg, p. 121.
Geldreich, E. E. and Clarke, N. A. (1971) In: *Proceedings of the 13th water quality conference* p. 103. Ed. by V. Snoeyink and V. Griffin, Urbana-Champaign University, Illinois.
Geldreich, E. E., Huff, C. B., Bordner, R. H., Kabler, P. W. and Clark, H. F. (1962) *J. appl. Bact.* **25,** 87.
Gerba, C. P. (1981) In: *Viruses and Wastewater Treatment* p. 39. Ed. by M. Goddard and M. Butler. Pergamon Press, Oxford.
Gibbs, B. M. and Freame, B. (1965) *J. appl. Bact.* **28,** 95.
Gorman, A. E. and Wolman, A. (1939) *J. Amer. Wat. Works Ass.* **31,** 225.
Gray, R. D. (1959) *J. Hyg., Camb.* **57,** 249; (1964) *J. Hyg., Camb.* **62,** 495.
Green, C. E. and Beard, P. J. (1938) *Amer. J. publ. Hlth* **28,** 762.
Green, D. M., Scott, S. S. Mowat, D. A. E., Shearer, E. J. M. and Thomson, J. M. (1968) *J. Hyg., Camb.* **66,** 383.
Greenberg, A. E. (1956) *Publ. Hlth Rep., Wash.* **71,** 77.
Greer, F. E. (1928) *J. infect. Dis.* **42,** 501.
Grindley, J. R. and Sapeika, N. (1969) *S. Afr. med. J.* **43,** 275.
Habs, H. and Langeloh, U. (1960) *Arch. Hyg., Berl.* **144,** 277.
Haley, C. E. et al. (1980) *J. infect. Dis.,* **141,** 794.
Hamblet, F. E., Hill, W. F., Akin, E. W. and Benton, W. H. (1969) *Amer. J. Epidem.* **89,** 562.

Hamilton, W. (1935) *A Study of the Pollution of the River Whangpoo as affecting its Use as a Source of Water Supply.* The Mercantile Printing Co., Ltd., Shanghai.
Hannay, C. L. and Norton, I. L. (1947) *Proc. Soc. appl. Bact.* No. 1, p. 39.
Harvey, R. W. S. (1956) *Mon. Bull. Minist. Hlth*, **15**, 118; (1957) *Brit. J. clin. Pract.* **11**, 751.
Harvey, R. W. S. and Thomson, S. (1953) *Mon. Bull. Minist. Hlth* **12**, 149.
Harvey, R. W. S. and Price, T. H. (1981) *J. appl. Bact.* **51**, 369.
Hoff, J. C. and Becker, R. C. (1969) *Amer. J. Epidem.* **90**, 53.
Holden, W. S. (1970) *Water Treatment and Examination.* Churchill, London.
Hoskins, J. K. (1934) *Publ. Hlth Rep., Wash.* **49**, 393; (1935) *Ibid.* **50**, 385.
Houston, A. C. (1913) *Studies in Water Supply*, London; (1917) *Rivers as Sources of Water Supply*, London.
Hughes, J. M., Merson, M. H., Craun, G. F. and McCabe, L. J. (1975) *J. infect. Dis.* **132**, 336.
Huntemüller, O. (1905) *Arch. Hyg., Berl.* **54**, 89.
Jameson, J. E. and Emberley, N. W. (1956) *J. gen. Microbiol.* **15**, 198.
Janssen, W. A. (1974) *Hlth Lab. Sci.*, **12**, 20.
Jensen, L. B. (1943) *Food Res.* **8**, 265.
Johannesson, J. K. and Martin, R. E. (1957) *J. appl. Bact.* **20**, 151.
Jordan, E. O. (1900) *J. exp. Med.* **5**, 271; (1903) *J. Hyg., Camb.* **3**, 1.
Kathe and Königshaus. (1932) *Arch. Hyg., Berl.* **109**, 1.
Keller, P. (1959) *J. Hyg., Camb.* **57**, 410.
Kelly, S. and Sanderson, W. W. (1959) *Sewage industr. Wastes, Wash.* **31**, 683.
Kenny, A. W. (1978) *Roy. Soc. Hlth J.* **98**, 116.
Knott, F. A. (1951) *Memorandum on the Principles and Standards employed by the Worshipful Company of Fishmongers in the Bacteriological Control of Shellfish in the London Markets.* Fishmongers' Hall, London Bridge, E.C.
Koopman, J. S. *et al.* (1982) *Amer. J. Epidem.* **115**, 173.
Kyriasides, K. (1931) *Z. Hyg., InfektKr.* **112**, 350.
Leiguarda, R. H. and Muratorio, A. J. (1958) *Rev. Obr. sanit. Nac., B. Aires* No. 177, p. 214.
Leiguarda, R. H., Peso, O. A. and Kempny, J. C. (1947) *Rev. Obr. sanit. Nac., B. Aires* **11**, 133.
Leiguarda, R. H., Peso, Mignone, Polcetti and Muratorio. (1956) *Rev. Obr. sanit. Nac., B. Aires* **20**, 166.
Lobel, H. O., Bisno, A. L., Goldfield, M. and Prier, J. E. (1969) *Amer. J. Epidem.* **89**, 384.
McCollum, J. P. K., Pearson, R. C. M., Ingham, H. R., Wood, P. C. and Dewar, H. A. (1968) *Lancet* **ii**, 767.
McCrady, M. H. (1918) *Publ. Hlth J. Toronto (Canad. publ. Hlth J.)* **9**, 201.
Mack, W. N., Mallmann, W. L., Bloom, H. H. and Kreuger, B. J. (1958) *Sewage industr. Wastes* **30**, 957.
Mackenzie, E. F. W. (1938) *33rd ann. Rep. met. Water Bd. London.* p. 25.
Mackenzie, E. F. W., Taylor, E. W. and Gilbert, W. E. (1948) *J. gen. Microbiol.* **2**, 197.
Magnusson, S., Hedström, C-E. and Lycke, E. (1966) *Acta path. microbiol., scand.* **66**, 551.
Mahnel, H. (1977) *Zbl. Bakt., IB* **165**, 527.
Mallmann, W. L. and Litsky, W. (1951) *Amer. J. publ. Hlth* **41**, 38.
Marshall, C. E. (1921) *Microbiology*, 3rd edn. London.

Martin, A. J. (1927) *The Activated Sludge Process.* London.
Martin, C. D. (1958) *Rev. Sanid. Hig. publ., Madr.* **32**, 187-229.
Martin, P. H. (1947) *Mon. Bull. Minist. Hlth. Lab. Serv.* **6**, 148.
Mason, J. O. and McLean, W. R. (1962) *Amer. J. Hyg.* **75**, 90.
Mason, W. P. (1902) *Water Supply*, 3rd edn. New York.
Merson, M. H. Barker, W. H., Craun, G. F. and McCabe, L. J. (1974) *J. Infect. Dis.* **129**, 614.
Metcalf, T. G. and Stiles, W. C. (1966) *See* Berg, p. 439.
Meyer, R. (1961) *Arch. Hyg., Berl.* **145**, 504.
Miller, R. E. (1936) *A quantitative and qualitative study of the bacterial flora of the sewage passing through activated sludge sewage treatment at Collingswood, N.J.* Dissertation, Univ. Pennsylvania.
Moore, B. (1948) *Mon. Bull. Minist. Hlth Lab. Serv* **7**, 241; (1950) *Ibid.* **9**, 72; (1954a) *J. Hyg., Camb.*, **52**, 71; (1954b) *Bull. Hyg., Lond.* **29**, 689; (1960) *R. Soc. Hlth. J.* **80**, 183.
Mosley, J. W. (1959) *New Engl. J. Med.*, **261**, 703, 748; (1966) *Transmission of Viruses by the Water Route*, p. 5. Ed. by G. Berg. Interscience Publishers, New York.
Müller, A. (1912) *Arb. ReichsgesundhAmt.* **38**, 294.
Müller, G. (1959) *Städtehyg.* **10**, 96.
Nabbut, N. H. and Kurayiyyah, F. (1972) *J. Hyg., Camb.* **70**, 223.
Page, G. F. B. (1942) *Med. Offr.* **67**, 80.
Papadakis, J. A. (1975) *J. appl. Bact.* **39**, 295.
Pawan, J. L. (1925) *Ann. trop. Med. Parasit.* **19**, 319; (1926) *Ibid.* **20**, 303; (1931) *J. trop Med. Hyg.* **34**, 229, 267, 288, 310, 317, 345, 360, 380, 391, 413.
Perry, C. A. (1928) *Amer. J. Hyg.* **8**, 694; (1929) *Ibid.* **10**, 580.
Pharris, C., Kittrell, F. W. and Williams, W. C. (1938) *Amer. J. publ. Hlth.* **28**, 736.
Plotkin, S. A. and Katz, M. (1966) *See* Berg, p. 151.
Porter, R., McCleskey, C. S. and Levine, M. (1935) *Proc. Soc. exp. Biol., N.Y.* **32**, 1032.
Pramer, D., Heukelekian, H. and Ragotzkie, R. A. (1950) *Publ. Hlth Rep., Wash.* **65**, 851.
Prescott, S. C. and Winslow, C. E. A. (1913) *Elements of Water Bacteriology*, 3rd edn. New York.
Pretorius, W. A. (1961) *J. appl. Bact.* **24**, 212.
Raghavachari, T. N. S. and Iyer, P. V. S. (1936) *Indian J. med. Res.* **23**, 619; (1939a) *Ibid.* **26**, 867; (1939b) *Ibid.* **26**, 877.
Randall, J. S. (1956) *J. Hyg., Camb.* **54**, 365.
Rao, N. U. and Labzoffsky, N. A. (1969) *Canad. J. Microbiol.* **15**, 399.
Reitler, R. and Seligmann, R. (1957) *J. appl. Bact.* **20**, 145.
Report. (1929) *The Purification of the Water of Swimming Baths.* Min. Hlth., H.M.S.O., London; (1936) *Ann. Rep. C.M.O., Min. Hlth., Lond.*, p. 192; (1937) *Ann. Rep. C.M.O. Min. Hlth., Lond.*, p. 162; (1939) *Memo., Min. Hlth., Lond.*, No. 221; (1942) *Mon. Bull. Emerg. publ. Hlth Lab. Serv.*, Feb., p. 1; (1946) *Publ. Hlth Bull., Wash.*, No. 295; (1947) *Amer. J. publ. Hlth*, **37**, 1121; (1948) *Gathering Grounds*, Min. Hlth, Lond.; (1952) *J. Hyg., Camb.* **50**, 107; (1953a) *Ibid.* **51**, 268; (1953b) *Mon. Bull. Minist. Hlth Lab. Serv.* **12**, 254; (1953c) *J. Hyg., Camb.* **51**, 559; (1957) *Recommended Practice for Design, Equipment and Operation of Swimming Pools and Other Public Bathing Places.* Amer. publ. Hlth Ass., N.Y.; (1958) *J. Hyg., Camb.* **56**, 377; (1959a) *J. Hyg., Camb.* **57**, 435; (1959b) *Sewage Contamination of Bathing Beaches in England and Wales.* Med. Res.

Coun., Lond., Memo. No. 37, H.M.S.O.; (1960) 3me Symposium int. Microbiol. alimentaire, *Ann. Inst. Pasteur, Lille*, **11**, 3–231; (1964) *Suggested Ordinances and Regulations covering Public Swimming Pools. Amer. publ. Hlth Ass.* New York; (1965a) *Mon. Bull. Minist. Hlth.* **24**, 116; (1965b) *The Times* **28** June; (1965c) *Mon. Bull. Min. Hlth Lab. Serv.* **25**, 146; (1967a) *Safeguards to be adopted in the Operation and Management of Waterworks*. Minist. Hous. Loc. Govern., H.M.S.O., London; (1967b) *Swimming Pools and Natural Bathing Places; an Annotated Bibliography*, 1957–66. *U.S. Publ. Hlth Serv.*, Publ. No. 1586. Washington D.C.; (1968) *J. Hyg., Camb.* **66**, 67; (1969a) *The Bacteriological Examination of Water Supplies. Rep. publ Hlth med. Subj.* No. 71, 4th edn., Minist Hlth, Lond; (1969b) *J. Hyg., Camb.* **67**, 637; (1970) *European Standards for Drinking-Water*. 2nd edn. World Hlth Org. Geneva; (1971a) *WHO Chron.* **25**, 70; (1971b) *Amer. J. Epidem.*, **93**, 33, 49; (1971c) *Standard Methods for the Examination of Water and Wastewater*. 13th edn. Amer. publ. Hlth Ass., New York; (1972) *J. Hyg., Camb.* **70**, 691. (1972) *J. Hyg., Camb.* **70**, 691; (1978) *Ibid.* **81**, 139; (1980a) *Ibid.* **85**, 35; (1980b) *Ibid.* **85**, 51; (1980c) *Ibid.* **85**, 181; (1981) *Ibid.* **87**, 369. Report (1982) Water-related disease outbreaks. CDC, Atlanta..

Roos, B. (1956) *Svenska Läkartidn.* **53**, 989.

Ross, A. I. and Gillespie, E. H. (1952) *Mon. Bull. Minist. Hlth Lab. Serv.* **11**, 36.

Ruddy, S. J., Johnson, R. F., Mosley, J. W., Atwater, J. B., Rossetti, M. A. and Hart, J. C. (1969) *J. Amer. med. Ass.* **208**, 649.

Russell, H. L. (1892) *Z. Hyg. InfektKr.* **11**, 165.

Savage, W. G. (1906) *The Bacteriological Examination of Water Supplies*. London.

Schäfer, E. (1970) *Arch. Hyg., Berl.* **154**, 299.

Schantz, E. J. (1969) *Agric. Fd. Chem.* **17**, 413.

Schiavone, E. L. and Passerini, L. M. D. (1959a) *Rev. Asoc. bioquim. Argent.* **22**, 86; (1957b) *Sem. méd., B. Aires*, **111**, 1151.

Schmelck, L. (1888) *Zbl. Bakt.* **4**, 195.

Schroeder, S. A., Caldwell, J. R., Vernon, T. M., White, P. C., Granger, S. E. and Bennett, J. V. (1968) *Lancet* **i**, 737.

Schubert, R. (1956) *Z. Hyg. InfektKr.* **142**, 476.

Schubert, R. H. W. and Mann, S. W. (1968) *Zbl. Bakt.* **208**, 498.

Scott, W. J. (1957) *Connecticut Hlth Bull.* **71**, 111.

Sherwood, H. P. (1952) *J. R. sanit. Inst.* **72**, 671.

Sherwood, H. P. and Thomson, S. (1953) *Mon. Bull. Minist. Hlth Lab. Serv.* **12**, 103.

Sidwick, J. M. and Murray, J. B. (1976) 'A brief history of sewage treatment'. *Effluent Water Treatment J.* **16**, 65 et seq.

Slade, J. S. (1977) *Water Engineering Science*, **31**, 3.

Slanetz, L. W. and Bartley, C. H. (1957) *J. Bact.* **74**, 591.

Snow, J. (1855) *On the Mode of Communication of Cholera*. 2nd edn. John Churchill, London.

Sobsey H. D. (1976) In: *Viruses in water*, p. 89. Ed. by G. Berg et al. Amer. publ. Hlth Assoc., Washington.

Sommer, H. and Meyer, K. F. (1937) *Arch. Path.* **24**, 560.

Sommer, H., Whedon, W. F., Kofroid, C. A. and Stohler, R. (1937) *Arch. Path.* **24**, 537.

Stanfield, G. (1982) *Roy. Soc. Hlth J.* **102**, 53.

Steiniger, F. (1951) *Z. Hyg. InfektKr.* **132**, 228.

Stevenson, A. H. (1953) *Amer. J. publ. Hlth* **43**, 529.

Stokes, E. J., Jones, E. E., Mohun, A. F. and Miles, A. A. (1945) *J. Proc. Inst. Sewage Purification* **1**, 1.

Stokvis, C. S. and Swellengrebel, N. H. (1911) *J. Hyg., Camb.* **11**, 481.

Tapley, G. O. and Jennison, M. W. (1941) *A Symposium on Hydro-biology*, p. 355. Univ. Wisconsin Press, Madison, Wis.

Taylor, A., Craun, G. F., Faich, G. A., McCabe, L. J. and Gangarosa, E. J. (1972) *J. infect. Dis.* **125**, 329.

Taylor, C. B. (1940) *J. Hyg., Camb.* **40**, 616; (1942) *Ibid.* **42**, 284.

Taylor, E. W. (1958) In: Thresh, Beale and Suckling's *The Examination of Waters and Water Supplies*. 7th edn. Churchill, London.

Taylor, E. W. and Burman, N. P. (1964) *J. appl. Bact.* **27**, 294.

Taylor, E. W. and Whiskin, L. C. (1945) *Trans. Inst. Water Engrs.* **50**, 219.

Taylor, F. B., Eagen, J. H., Smith, H. F. D. and Coene, R. F. (1966) *Amer. J. publ. Hlth*, **56**, 2093.

Tebbutt, T. H. Y. (1973) *Water Science and Technology*. John Murray, London.

Theios, E. P., Morris, J. G., Rosenbaum, M. J. and Baker, A. G. (1967) *Amer. J. publ. Hlth* **57**, 295.

Thresh, J. C. and Beale, J. F. (1910) *Lancet* **ii**, 1849; (1925) *The Examination of Waters and Water Supplies*, 3rd edn. London.

Tracy, H. W., Camarena, V. M. and Wing, F. (1966) *J. Amer. Wat. Wks. Ass.* **58**, 1151.

Vaccaro, R. F., Briggs, M. P., Carey, C. L. and Ketchum, B. H. (1950) *Amer. J. publ. Hlth* **40**, 1257.

Vaughn, J. M., Landry, E. F., Vicale, T. J. and Dahl, M. C. (1980) *J. Food Protection*, **43**, 95.

Victorin, K. (1974) *J. Hyg., Camb.* **72**, 101.

Wallis, C., Henderson, M. and Melnick, J. L. (1972) *Appl. Microbiol*, **23**, 476.

Waksman, S. A. and Hotchkiss, M. (1937) *J. Bact.* **33**, 389.

Ward, R. L. (1981) In: *Viruses and Wastewater Treatment*, p. 65. Ed. by M. Goddard and M. Butler, Pergamon Press Oxford.

Weibel, S. R., Dixon, F. R., Weidner, R. B. and McCabe, L. J. (1964) *J. Amer. Wat. Wks. Ass.* **56**, 947.

Whipple, G. C. and Mayer, A. (1906) *J. Infect. Dis.*, Suppl. 2, p. 76.

Wilson, G. S., Twigg, R. S., Wright, R. C., Hendry, C. B., Cowell, M. P. and Maier, I. (1935) *Spec. Rep. Ser. med. Res. Coun., Lond.* No. 206.

Wilson, W. J. (1928) *Final Rep. publ. Hlth. Congr. and Exhibition*, p. 203.

Wilson, W. J. and Blair, E. M. McV. (1931) *J. Hyg., Camb.* **21**, 138.

Winslow, C.-E. A. (1905) *J. infect. Dist.* Suppl. 1, p. 209.

Winslow, C.-E. A. and Lochridge, E. E. (1906) *J. infect. Dis.* **3**, 547.

Wood, P. C. (1961) *The Principles of Water Sterilization by Ultra-violet Light, and their Application in the Purification of Oysters*. Fishery Invest. Ser. 11, **23**, No. 6, Minist. Agric., Fish, Food. H.M.S.O., London.

Wooldridge, W. R. and Standfast, A. F. B. (1936) *Biochem. J.* **30**, 1542.

ZoBell, C. E. (1936) *Proc. Soc. exp. Biol., N.Y.* **34**, 113.

Milk

Sources of bacteria in milk	279	Antibacterial properties of human milk	283
Keeping quality	280	Bacteriological grading of milk	283
Production of clean milk	280	According to cleanliness	283
Types of bacteria in milk	280	Sampling	283
Acid-formers	280	Sediment test	284
Gas-formers	280	Cellular content	284
Proteolytic bacteria	280	Breed smear	284
Alkali-forming bacteria	280	Keeping quality	284
Inert bacteria	280	Plate count	284
Human milk	280	Coliform test	285
Pathogenic bacteria in milk	281	Modified methylene blue reduction test	286
Tubercle bacilli	281	Resazurin test	286
Brucella abortus	281	Laboratory pasteurization test	287
Streptococci	281	According to safety	287
Staphylococcus aureus	281	Milk designations	287
Corynebacterium diphtheriae	281	Statutory tests for milk designations	287
Typhoid, paratyphoid, food-poisoning, and dysentery bacilli	281	Diseases borne by milk	288
		Milk sickness and allergy	290
Other organisms	281	Methods for increasing safety	290
Quality in milk	282	Control of animal disease	290
Cleanliness	282	Control of human personnel	290
Keeping quality	282	Heat treatment; pasteurization, boiling, and sterilization	290
Pasteurizability	282		
Safety	282	Phosphatase test for pasteurized milk	290
Antibacterial properties of milk	283	Refrigeration of human milk	292
Antibacterial system	283	Churns and bottles	292
Antibiotics	283	Cream	292
Bacteriophage	283	Dried milk	292

Sources of bacteria in milk

Even when drawn with aseptic precautions milk is never sterile. The micro-organisms it contains are derived from the *cow's udder*. They vary in frequency from quarter to quarter and from cow to cow, and are highest in the fore-milk and lowest in the strippings. Several estimates have been made of their numbers in raw milk from apparently normal cows, but the figures are so discrepant that it is not worth quoting them. Suffice it to say that milk drawn aseptically from the healthy udder should not usually have a plate count of over 100 per ml (von Darányi 1941).

When aseptic precautions are not employed, the milk is liable to be contaminated with bacteria from the outside of the udder, the interior of milk vessels and utensils, and dust in the atmosphere of the milking shed. Of these various sources, far and away the most important is **unsterilized milking equipment**. The total number of organisms gaining access from the air and from dust (see Benham and Edgell 1970) is almost negligible compared with that derived from the surfaces of unsterilized pails, coolers, cans, strainers, clarifiers, pipe lines, and bottle fillers. Unless these utensils are actually sterilized—preferably by steam—their surfaces become coated with bacteria of various types, which may contribute enormous numbers of micro-organisms to the milk.

An additional source of contamination is the **human personnel**. Unless machine milking is employed, organisms get washed into the milk from the hands. Though few, they may belong to pathogenic species, particularly if the milker happens to be an intestinal carrier of typhoid, paratyphoid, dysentery, or food-poisoning bacilli. The milker constitutes a further source of danger if he is carrying haemolytic streptococci or diphtheria bacilli in his throat or nose, since these organisms may gain access to the milk via the cough spray.

Apart from initial contamination of the milk, **imperfect or delayed cooling** is often responsible for the presence of large numbers of bacteria in any given sample. However carefully milk is produced, sooner or later, provided it is kept at a suitable temperature, it will go sour or putrid as the result of bacterial multiplication. If it is to remain sweet for more than a few hours, all milk should be cooled immediately after production to a temperature of 10° or below. The lower the temperature at which it is stored, the less is the deterioration. Even at 4°, however, multiplication of psychrotrophic bacteria, such as certain members of the *Pseudomonas*, *Alkaligenes*, *Flavobacterium*, and *Enterobacter* groups, goes on, so that ordinary raw milk cannot be kept sweet indefinitely (see Panes and Thomas 1968, Shehata and Collins 1971).

It may be noted that the **keeping quality** of the milk is determined partly by the degree of initial contamination and partly by the temperature at which it is kept. Milk produced under really clean conditions has a considerable bacteristatic power, and shows little bacterial multiplication for several hours, even when incubated at a favourable temperature. On the other hand, the bacteristatic effect of milk produced under dirty conditions is very slight, and bacterial multiplication sets in rapidly. For this reason it is much easier to distinguish between a milk produced under sanitary, and one produced under insanitary, conditions if the examination is delayed till the milks have stood at a temperature of 15° or so for 12–18 hours. After this time a milk produced under clean conditions will still have a low bacterial count, whereas a milk produced under dirty conditions will have a high one.

The **production of clean milk** is largely a matter of technique, not of structural equipment or refinement. Provided all utensils are sterilized by steam, and the surface of the udder and the milker's hands are cleansed, it is possible to produce milk with a very low bacterial content even under unfavourable conditions. It is, however, not easy to maintain a satisfactory technique day in and day out unless suitable conditions and appliances are provided for the workers.

A full description of the principles and practice of milk hygiene on the farm and in the dairy is given in two reports of the World Health Organization (Report 1962, 1970a; see also Schulz 1979).

Types of bacteria in milk

The dairy bacteriologist is naturally interested in knowing what types of bacteria are met with in milk. The public health bacteriologist is concerned more with their action than with their species. Briefly, the organisms may be classified into the following groups:

(1) **Acid-formers** These organisms ferment the lactose in the milk with the production mainly of lactic acid, which combines with the calcium caseinogenate, liberating free caseinogen; this is precipitated in the form of a smooth gelatinous curd, which shows little tendency to contract, and which can be redissolved by the addition of alkali. The chief members of this group—lactic streptococci—are responsible for souring of the milk.

(2) **Gas-formers** These organisms ferment lactose with the production of acid and gas. The acid is largely acetic, which, besides clotting the milk, imparts an unpleasant flavour to it. Among these organisms the commonest are the coliform bacilli; less frequent are anaerobic spore-bearers, such as *Cl. perfringens* and *Cl. butyricum*. The capsulated members of the coliform group are responsible for ropiness in milk (see Parry and Tee 1958, Chalmers 1962).

(3) **Proteolytic bacteria** These organisms secrete one or other, or both, of two ferments—rennet and casease. Rennet acts in two stages: in the first, the caseinogen is converted into soluble casein; in the second, this is precipitated as calcium caseinate by the calcium salts in the milk. The resulting clot, which cannot be redissolved by alkali, contracts and squeezes out a more or less clear fluid known as whey, containing the sugar, salts, and the proteins lactalbumin and lactoglobulin. Casease is a proteolytic enzyme, which breaks down the proteins to proteoses, peptones, and amino acids. When the rennet predominates, a firm white curd is produced, which is slowly digested by the casease. When the casease is in excess, either no coagulation occurs, or a soft flocculent clot is formed that is rapidly peptonized. Organisms in this group comprise aerobic spore-bearers, such as *Bac. subtilis* and *Bac. cereus*, the latter of which is responsible for 'bitty' cream (see Donovan 1959). To these may be added *Proteus vulgaris*, staphylococci and micrococci. In ordinary raw milk proteolytic organisms are overgrown by the acid-forming group; only when these are few, as in milk produced under the best hygienic conditions, does their action become apparent.

Alkali-forming bacteria These organisms digest the protein rendering the milk alkaline. Some also secrete lipase, which saponifies the fat and converts the milk into a yellow translucent unsavoury whey-like fluid. Like the proteolytic bacteria, they are usually overgrown by the acid-formers, so that their action is not seen. Organisms included in this group are members of the *Alkaligenes* and *Achromobacter* species, and aerobic spore-bearers (see Abd-el-Malek and Gibson 1952).

Inert bacteria Many organisms produce no visible change in milk, even in pure culture. This group is notable in that it contains most of the pathogenic species.

According to Gavin and Ostovar (1977), **human milk** taken before the infant is fed contains the following organisms in small numbers: *Staph. epidermidis* in every sample, *Str. mitis* in 69 per cent of samples, '*Gaffkya tetragenus*' in 19 per cent, *Staph. aureus* in 13 per cent, and a few other species in 3–9 per cent. After feeding, the same organisms are present, but in large numbers, most of them apparently derived from the infant's mouth and the maternal skin. (See also p. 288)

Pathogenic bacteria in milk

Apart from the organisms that we have described as being normally present in milk, there are others which are sometimes found in pathological conditions of the cow's udder or intestine, or which gain access to the milk from an infected water supply or from some person handling the milk.

Tubercle bacilli The presence of these organisms in milk depends on the frequency of bovine tuberculosis. In most western European countries and in North America this disease has been virtually eliminated, so that tubercle bacilli rarely occur in milk unless some local breakdown in the Attested Herds system has occurred. Tubercle bacilli may be excreted in the absence of any detectable disease of the udder (Report 1909), but some degree of mastitis can generally be demonstrated in cows passing infected milk. The numbers of tubercle bacilli excreted by a single cow vary from day to day and from animal to animal; they may reach several hundred or even a thousand per ml (see Pullinger 1934). Other acid-fast bacilli may also be found (Hosty and McDurmont 1975). Milk must become infected from time to time with tubercle bacilli of human type coughed out by tuberculous farm hands or dairy workers, but their demonstration can practically never be accomplished because of the very small number in which they are present and because of their failure to multiply in milk.

Bruc. abortus This organism is responsible for contagious abortion of cattle. The udder is frequently infected, even in animals that have not aborted (see Chapter 56), though no lesions can be detected by clinical examination. The bacilli are excreted regularly or intermittently in the milk. According to Stockmayer (1936) their numbers are greatest at the beginning of lactation, when they may reach as many as 200 000 per ml, but in the later stages they diminish and seldom exceed 2000 per ml. Experience, however, of other workers has shown that they often increase towards the end of lactation. The udder may remain infected for years. Other organisms of the *Brucella* group, chiefly *Bruc. melitensis*, may be excreted by infected cows. (For review of brucellosis in dairy cattle, see Morgan 1970.)

Streptococci Mastitis streptococci are of various types, the most common being *Str. agalactiae*. They are excreted in variable numbers in the milk of cows suffering from mastitis. Most streptococci causing mastitis are probably non-pathogenic for man, but occasionally *Str. pyogenes* of human origin invades the udder, and is responsible for an outbreak of scarlet fever or septic sore throat in persons consuming the milk (see Bendixen and Minett 1938, Dublin *et al.* 1943). The milk may also be contaminated directly with these organisms from persons who are either suffering from streptococcal throat lesions, or who are carrying these organisms in their throat or nasopharynx (see Henningsen and Ernst 1939). Most outbreaks of milk-borne streptococcal infection appear to result from the former method of infection. Though pathogenic streptococci must often gain access to milk from human cases and carriers, their rate of multiplication in raw milk at ordinary temperatures is usually too slow for them to render the milk sufficiently infective to cause disease (see Pullinger and Kemp 1937).

Staph. aureus This organism is a not infrequent cause of mastitis. It is often found in the apparently healthy udder, but in cows suffering from mastitis it may be present in large numbers in the milk. Its public health importance lies in the fact that, under favourable conditions, it may multiply in the milk and give rise to toxin capable of producing gastro-enteritis in human beings (see Chapter 72). Staphylococcal food poisoning caused by liquid milk is uncommon, mainly because the growth of *Staph. aureus* in milk is very slow under 25° and is restrained by the presence of more rapidly growing organisms such as lactic and faecal streptococci (Smith 1957, Jones *et al.* 1957).

C. diphtheriae This organism occasionally finds its way into the milk from the throat or nasopharynx of a human carrier or case of diphtheria. Very occasionally it becomes implanted on ulcers on the cow's teats. Such an occurrence is peculiarly dangerous, since the milk is uniformly infected (for references, see Goldie and Maddock 1943).

Typhoid, paratyphoid, food-poisoning, and dysentery bacilli These important pathogenic organisms, which may occasionally contaminate the milk, are usually derived from human or other extraneous sources (see Chapters 69–72), though some members of the *Salmonella* group, particularly *Salm. typhimurium*, *Salm. enteritidis*, *Salm. dublin*, and *Salm. newport*, may get into the milk from the faeces of an infected animal or be actually excreted in the milk itself should the udder become infected.

Other organisms Of other pathogenic organisms contaminating milk *Coxiella burneti* can often be demonstrated in areas in which infection of cattle with this organism is common (see Chapter 77). *Listeria monocytogenes* may be excreted in the milk, often for a long time (Hyslop and Osborne 1959). So also may leptospirae, but the risk to man is low because whole raw milk contains a lethal antileptospiral factor (Kirschner and Maguire 1955, Steele 1960). Milk may occasionally be contaminated with *Streptobacillus moniliformis* from the nasal secretion of rats, and with *Campylobacter fetus* subsp. *jejuni* from the animals' faeces. *Yersinia enterocolitica* is not uncommonly found in milk (Schiemann and Toma 1978); when present in large numbers, it may give rise to gastro-intestinal disease (see Chapters 55 and 71). Of the viruses, foot-and-mouth virus, vaccinia, parainfluenza, enteroviruses, and tick-borne encephalitis virus may infect the udder; infectious hepatitis virus may gain access to the milk from contaminated water or from man, and poliovirus probably from human faeces (Sharp and Bramley 1977). In the USA the **bovine leukaemia virus** is said to be excreted in the milk of a high proportion of cows in commercial dairy herds (Ferrer *et al.* 1981). In goats the virus of Russian spring-summer encephalitis may be excreted in the milk and give rise to disease in man (see Moritsch and Krausler 1957). In human milk *Toxoplasma gondi* is sometimes found (Zardi *et al.* 1968).

Quality in milk. Cleanliness and safety

Quality is a composite, not a single, attribute of milk. There are, for example, (1) the nutritive quality, (2) the cleanliness, (3) the keeping quality, (4) the pasteurizability, and (5) the safety of milk, to mention only five of its most important attributes. Leaving aside the nutritive quality, which can be determined only by chemical analysis and animal feeding experiments, we may consider for a moment what we mean by the other four properties.

By **cleanliness** is generally understood the freedom of the milk from extraneous matter, from blood, and from an undue number of leucocytes and bacteria. It is an unsatisfactory term, but by general usage it has come to bear this connotation.

The **keeping quality** of the milk refers, of course, to the length of time it will remain sweet, and free from odours and tastes that render it unpalatable. Though cooling prolongs the life of milk, it must not be thought that it is a panacea against all spoilage problems. Murray and Stewart (1978), for example, estimate that about 10 per cent of specimens of milk from refrigerated farm tanks may have undesirable flavours, mainly owing to growth of psychrotrophic organisms such as are found in pipe-line milking plants and bulk tanks; about 90 per cent of these organisms are lipolytic or proteolytic. Moreover, the feeding of certain kinds of forage may impart undesirable flavours to the milk, or even the poisonous aflatoxin formed by the ubiquitous *Aspergillus flavus*.

The **pasteurizability** of the milk is a term devised to indicate the suitability of the milk for heat treatment. It is a mistake to believe that any milk can be pasteurized. A milk that is too acid will clot when the temperature is raised, and the resultant product will have to be discarded. Again, as Anderson and Meanwell (1933) showed, some milks contain large numbers of heat-resistant bacteria, which are not destroyed by pasteurization, and which may therefore prevent the pasteurized product from conforming to the legal standard.

The **safety** of milk is a term denoting its freedom from bacteria capable of giving rise to disease in man or animals. It is by some writers confused with cleanliness, but there is no necessary or constant relationship between these two properties. Dirty milks, if free from pathogenic bacteria, may be quite safe, and very clean milks are not infrequently dangerous.

It has already been pointed out that pathogenic bacteria in milk may come from (1) the udder, such as the tubercle bacillus, *Bruc. abortus*, and some streptococci and staphylococci; (2) the infected nasopharynx of human beings handling the milk, such as haemolytic streptococci and diphtheria bacilli; (3) excretal material gaining access either from the hands of human beings, or indirectly from water contaminated with urine or faeces—usually of human origin; in one or other of these ways typhoid, paratyphoid, dysentery, and food-poisoning bacilli may be carried into the milk. These are the organisms responsible for *dangerous* milk. Those responsible for *dirty* milk, on the other hand, come from various sources, particularly unsterilized milk utensils, caked mud and manure on the cow's udder, dirt on the milker's hands, and from the fore-milk in the teat canal. Provided the cow is healthy, provided that none of the human beings handling the milk is a carrier of pathogenic bacteria, and provided that the water supply is pure, dirt and bacteria may gain access to the milk in considerable quantities without endangering the health of those consuming it. On the other hand, if pathogenic bacteria find their way into the milk from one of the sources quoted above, no matter in how cleanly a manner the milk is produced, it is a potentially dangerous and unsafe milk.

These conclusions are borne out by epidemiological experience, which has shown that several outbreaks of disease have followed the consumption of milk of the highest standard of cleanliness (see Wilson 1942). Without discussing this subject further, we shall probably be wise to regard cleanliness and safety as two entirely separate attributes of milk.

For human consumption it is desirable, of course, that milk should be both clean and safe. Clean milk is more aesthetically desirable, it has a better flavour, and it keeps longer. Moreover, it is not likely to contain any of those toxic substances resulting from undue bacterial proliferation, which have an irritating effect on the gastro-intestinal tract—particularly of infants (see Park and Holt 1903). The fewer organisms there are in milk, and the more bacterial proliferation is checked, the less liable is the milk to give rise to digestive disturbance of this type.

From this it will be clear that *no raw milk can ever be regarded as completely safe for human consumption.* The frequency of disease in cattle, the risk of contamination from human and other sources, and the suitability of the milk itself as a medium for bacterial multiplication, combine to render the consumption of raw milk potentially dangerous. The only satisfactory way of eliminating this danger is by pasteurization or some other form of heat treatment.

Antibacterial properties of milk

Numerous observers have shown that the growth rate of organisms is much slower in clean than in dirty milk. In clean milk there is a lag phase of variable duration before multiplication begins, whereas in dirty milk kept at a suitable temperature the organisms start multiplying almost at once. Jones and Simms (1903) brought evidence to show that clean milk contained an inhibitory substance to which they gave the name *lactenin*.

Lactenin is present in the whey fraction of the milk. It will not pass through a dialysing membrane. It is inactivated by heating to 80°, but not by pasteurization. It will remain stable for some weeks at 6°, and is not digested by trypsin. It is irreversibly inactivated by the exclusion of atmospheric oxygen and by sulphur-containing reducing agents such as cysteine, glutathione, and thioglycollic acid. Further observations have shown that the antibacterial properties of milk are due to the presence of antibodies, and of other components such as lysozyme, lactoferrin, and the lactoperoxidase (LP) system in which lactoperoxidase, thiocyanate, H_2O_2, and catalase are all concerned. This system not only inhibits the growth of group A streptococci and of many coliform organisms, but actually kills them. The disinfectant action occurs in two stages: the first depends on the oxidation of thiocyanate by lactoperoxidase and H_2O_2, and can be reversed by a reducing agent; the second depends on the presence of accumulated H_2O_2 formed by catalase-negative organisms, such as the lactic acid bacteria, and is reversed by catalase. The bactericidal activity is greatest at pH 5.0 or below, and is related to the oxidation of bacterial sulphydrils to sulphonic acid and sulphydril thiocyanate derivatives (Björck *et al.* 1975, Reiter *et al.* 1976, Thomas and Aune 1978). The whole subject is reviewed by Reiter (1976). (See also Mickelson 1979.) Against leptospirae in milk is an inhibitor that is said to be heat-stable and resistant to glutathione (Kirschner *et al.* 1957). Hydrogen peroxide, formalin, boric acid, and benzoates added for preservative purposes may be present in milk, as may also hypochlorites, chloramine, and quaternary ammonium compounds used for the cleansing of churns and bottles (Lück 1956, McKenzie and Booker 1955, Report 1970*a*). Other contaminants include traces of pesticides, radio-nuclides, plant poisons, mycotoxins, and heavy metals (Report 1962, 1970*a*).

Besides intrinsic antibacterial agents, milk may contain **antibiotics** used in the treatment of mastitis. Aqueous solutions of penicillin injected into the udder are excreted in the milk for 24 hours, oil-water emulsions for 48 hours, and suspensions in vegetable or mineral oil for 72 hours or even longer (Edwards 1964, Rollins *et al.* 1970). For detection of antibiotic residues in milk, see Macaulay and Packard (1981).

Bacteriophage in milk, acting particularly on *Str. cremoris*, may delay clotting during the manufacture of cheese (see Hiscox and Briggs 1953).

Antibodies, such as those against diphtheria, measles, and poliomyelitis, are often present in the milk of lactating women (see, *e.g.*, John and Devaragar 1973).

Human milk likewise exerts a bacteristatic effect. It inhibits the growth of many species of bacteria, including enteropathogenic strains of *Esch. coli*. This property is inactivated by heat—partly even by pasteurization—and by iron-binding proteins in the milk (Ford *et al.* 1977, Dolby *et al.* 1977, Evans *et al.* 1978, Welsh and May 1979). Human milk is unusually rich in lactoferrin. How far this is responsible for the bacteristatic effect is doubtful. Brock (1980) thinks that lactoferrin is concerned mainly with the control of iron absorption, and possibly with facilitating the excretion of iron in the immediate postnatal period.

Bacteriological grading of milk

Grading according to cleanliness

Owing to the lack of a suitable test for safety, milk is graded almost entirely according to cleanliness (Wilson 1936). Numerous methods are available for this purpose. Some used in the past have now been generally abandoned. Of the remainder some are used in one country, others in another. Some are preferred by public health workers, others by the agriculturalists. Here we give an account of them at some length, not because they are all equally useful, but because many of them exemplify principles of general interest to the bacteriologist.

For a detailed description of the methods used and a study of their experimental errors, and for a critical discussion on the interpretation of their results, the reader is referred to the monograph by Wilson and his colleagues (1935).

Sampling Since the answer given by any test is determined so largely (*a*) by multiple factors involved in the production and subsequent handling of the milk; (*b*) by the care with which the milk is sampled, and (*c*) by the time-temperature conditions under which the sample is taken and kept prior to analysis, it follows that it is unwise to pay too much attention to the results of any single examination. The most satisfactory procedure is to make frequent and regular examinations of the milk of any given producer throughout the year, and insist that a given proportion, for example 75 per cent, of the samples should come up to a given standard. In this way some allowance will be made for factors over which the producer is unable to exercise complete control, while at the same time ensuring that the conditions of production as a whole are kept at a reasonably high level.

Every endeavour should be made in the sampling of the milk to obtain a homogeneous distribution of organisms and fat. For the subsequent treatment of the sample, two alternative procedures are available. If the intention is to test the cleanliness of production of raw milk or the keeping quality of pasteurized milk, it is desirable to expose the sample after collection to an agreed temperature for a given length of time in order that the latent contamination may have time to manifest itself. When, on the other hand, the effect of heat treatment or bottling is to be studied, the sample may be iced at once so as to prevent bacterial multiplication during transmission to the laboratory.

The sample may be examined by the following tests:

(1) **The sediment test** The amount of extraneous matter in the milk is determined either by filtration through a cotton pad, or by centrifugation. A simple sediment test of this sort is of value in controlling the gross dirt in milk and in education of the unhygienic farmer. Since, however, filtration through muslin gauze on the farm will remove dirt of this type, the absence of an obvious sediment must not be taken to mean that the conditions of production were satisfactory. Most of the bacteria in milk come not from gross dirt but from unsterilized utensils. Indeed, actual manure can be added to the milk without producing any serious increase in the bacterial content; for this reason a sediment test is useful for the detection of gross particulate matter, the presence of which is not likely to be detected by purely bacteriological tests.

(2) **The cellular content** Normal milk contains various types of cells, the number of which is estimated by different workers at between about 200 000 and 1 000 000 per ml. Blackburn (1968), who examined 50–60 cows at each of seven lactations, found that in milk from healthy udders the average cell count rose from 0.19 million per ml during the first lactation to 0.67 million during the seventh. Udders infected with *Staph. aureus* yielded milk with an average cell count of 3.6 million per ml, and with streptococci 2.41. An abnormally high cell count may occur in the absence of infection (Heidrich *et al.* 1964); it is often due to retention of milk in the udder, which leads to desquamation of the epithelial cells in the alveoli and ducts. In the presence of infection most of the cells are polymorphonuclear leucocytes. A mere count of the cells affords little information about the quality of the milk, and it is unwise to condemn a milk solely because of an increased cell content.

(3) **The Breed smear method** The general technique of this method is to spread 0.01 ml of the milk over 1 cm² on a glass slide, to fix and de-fat, to stain with methylene blue, to count the number of individual organisms—the Breed 'individual' count—or preferably the number of groups of organisms, individual organisms each being regarded as a group—the Breed 'clump' count—in a given number of fields, and finally to calculate the number of organisms in 1 ml of the original milk. The result obtained affords an index of the total *stainable* organisms present, not necessarily of the total organisms. Though it enables a rapid opinion to be formed of the cleanliness of individual churn milks, its use has been generally discontinued, partly because of the apparatus and skilled workers it demands, and partly because of the growing replacement of churn milk by refrigerated tank milk. It is of value, however, to the farm inspector in indicating the probable nature of the defect in unsatisfactory milk. Thus the presence of large masses and clumps of bacteria suggests unsterilized milk utensils; the presence of numerous organisms arranged in pairs or small units suggests inadequate cooling; the presence of long-chained streptococci associated with a high leucocyte count indicates mastitis, and so on (see Breed 1929).

(4) **The keeping quality test** The milk is kept in a special bottle at a given temperature, such as 15° or 21°, and examined every 8 or 12 hours until it becomes sour or putrid. The fact that the end-point is largely subjective necessarily exposes the result to a big experimental error. Some degree of objectivity can be introduced by means of the *alcohol-precipitation* or the *clot-on-boiling test*.

If 1 ml of a 68 per cent solution of ethanol is added to 1 ml of the milk and the tube is inverted, precipitation of the casein occurs in milks that are near the souring point. The clot-on-boiling test is carried out by heating 5 ml of milk in a clean test-tube in boiling water for about 5 minutes; samples showing no sign of precipitation are regarded as passing the test.

Though the alcohol-precipitation method is slightly more delicate, both these tests give similar results and are in general agreement with the grading of the samples by taste (see Anderson and Wilson 1945, Rowlands *et al.* 1950). The subjective keeping quality test, though useful for special purposes, is manifestly unsuitable for routine grading, but the clot-on-boiling test has a good deal to recommend it to the dairy worker (see Rowlands and Hosking 1951). It is really an indirect test for the measurement of acidity, though a small proportion of milks may clot on boiling in the absence of acidity if enzymes of the rennet type are formed in sufficient concentration—'sweet curdling' (see Taylor and Clegg 1958). The modified methylene blue reduction test carried out at 25° or 30° may be used as a quality test of the milk arriving at the creamery (Thomas and Druce 1971), or at 15° as an indirect measure of keeping quality (Provan *et al.* 1936).

(5) **The plate count** This is one of the most widely used methods for the bacteriological grading of milk according to cleanliness. It is, however, open to criticism on many grounds. Ostensibly the plate count measures the numbers of bacteria in the milk, but in fact it does not. The bacteria in raw milk are of many different kinds and come from varied sources. Often a large proportion are dead, and those that are alive differ in their nutritional, respiratory, and temperature requirements, so that no one medium incubated for a given length of time at a given temperature under aerobic conditions can possibly afford a true estimate even of the living organisms present. More important still, however, is the fact that many of the bacteria are distributed not individually but in chains and groups

of varying sizes. The colonies developing on plates are derived, therefore, not solely from individual organisms but also from aggregates containing varying numbers of bacteria. In consequence the plate count merely registers the number of bacterial particles capable of forming colonies under the particular conditions selected. Since the average number of bacteria per clump is variable from one milk to another, and not constant from time to time even in milk of the same origin, it follows that the figures yielded by the plate count are arbitrary, are not strictly comparable, and have no absolute significance.

When it is further pointed out that these clumps of bacteria may disintegrate to a variable and quite uncontrollable extent during the process of dilution, leading in extreme instances to errors of 1000 per cent (Ward 1926), and that many of the individual steps in the technique of the count are very difficult to standardize, and even when standardized as nearly as possible are attended by a large experimental error, it will be realized that the figures yielded by the plate count are not merely arbitrary, but are also only approximate. (For a description and numerical assessment of the various errors of dilution, incubation, counting, personal bias and personal variability, sampling, overcrowding, and so on, see Wilson *et al.* 1935.)

Because the results are expressed quantitatively in figures extending over a wide range, they afford a fictitious appearance of accuracy which leads, not only in laymen but even in public health officials, to a wholly unjustifiable feeling of confidence in their value. It is not denied that the plate count can be used as a yardstick for the grading of milk, but it is denied that it can be used as a millimetre rule. In the grading of bulk-collected milk, dairy bacteriologists find that a high count of thermoduric organisms indicates neglected cleansing of the milking plant, and that a high count of psychrotrophic organisms leads to the deterioration of refrigerated milk (Thomas and Thomas 1978). A rapid method for counting bacteria in milk is described by Pettipher and Rodrigues (1981).

(6) **The coliform test** Like the plate count, this method has been extensively used in the past for the grading of milk without any real appreciation of the fundamental assumptions on which it rests. We have seen that in the analysis of *water* the coliform test affords us the most delicate index we have of excretal pollution. The main reasons for this are that (1) in Great Britain something like 90 per cent of waters giving a positive presumptive coliform reaction in MacConkey broth are proved on further examination to contain typical *Esch. coli* of excretal origin (Bardsley 1934); (2) organisms of the *Esch. coli* type in water in Great Britain are very frequently derived from human excretal material, and since human faeces and urine may contain pathogenic bacteria, the presence of any excretal pollution, as indicated by the coliform test, must be regarded as potentially dangerous; and (3) as a rule, coliform organisms do not seem to multiply in water under natural conditions; on the contrary, they tend to die out rather rapidly (Houston 1913). In Great Britain, therefore, the coliform count in water indicates the minimal amount of excretal pollution that has occurred.

When we turn to *milk* we find that (1) a considerable proportion—something like 50–70 per cent—of raw milks in Great Britain in which coliform bacilli are found contain not the true *Esch. coli* but organisms of the intermediate or aerogenes-cloacae types. These organisms come mainly not from faeces and urine but from soil and vegetation, and their presence therefore in milk cannot be regarded as an index of excretal pollution; (2) the true *Esch. coli* organisms that are found in milk appear to come either directly from cow-dung and manure, or indirectly from unsterilized milk utensils in which bacterial multiplication has occurred. If they are derived from the latter source, they clearly afford no index of *direct* excretal pollution. If they are derived from the first source, their presence may be considered objectionable on the ground that organisms pathogenic for man are sometimes present in the intestinal canal of the cow. The presence, however, in milk of excretal material of bovine origin must be regarded as very much less dangerous for man than that of human origin. (3) Unlike water, milk affords an admirable medium for the growth of coliform bacilli. If it is kept at a temperature of 50° F or over, a great increase in their numbers may take place, a rise of several thousandfold often occurring at 60° F within 24 hours (see Ayers and Clemmer 1918, Finkelstein 1919, Sherman and Wing 1933). Unless, therefore, the milk has been kept at 40° F or less, an estimation of the coliform bacilli affords only a very imperfect, and often entirely misleading, index of the extent of the original contamination. With a milk produced and kept under ordinary conditions, it is impossible to tell how many of the coliform bacilli gained access at the time of production, and how many are due to the multiplication of those originally present.

It will thus be seen that none of the three premises on which the scientific application of the coliform test to water in this country is based holds true for milk. Even a differentiation of the coliform group into its constituent types is not likely to help us, since (*a*) it is impossible to tell how many of the different types were original contaminants and how many have resulted from growth, and (*b*) organisms of the faecal coli group are not necessarily derived directly from excretal material, but may come from unsterilized milk utensils on which they have been multiplying perhaps for several generations.

The presence of coliform bacilli in small numbers in water or milk is not in itself objectionable; it is merely as an index of the possible accompanying presence of pathogenic organisms that it is of importance. In milk, the coliform test cannot be used as an index of direct excretal pollution, for though it is true that if manure gains access to milk, coliform organisms will probably be found, the reverse conclusion does not hold true. There are so many other sources on the farm for these organisms that their presence in milk cannot be held to justify the conclusion that the milk has necessarily been contaminated with excretal material. The coliform test on milk fails to provide us with that specific qualitative information which it supplies in the case of water, and as a general method of grading, therefore, it seems to be unsuitable.

(7) **The modified methylene blue reduction test** The general technique of this test is to add 1 ml of a standard methylene blue solution to 10 ml of milk in a test-tube, to insert a sterile rubber cork, to invert the tube once or twice so as to mix the dye with the milk, to incubate in a constant-temperature water-bath at 37–38° in complete darkness, to invert the tube every half-hour, and to observe the time at which the dye is decolorized. If reduction has not occurred within a given time, then no further observations need be made (see Wilson *et al.* 1935).

Cows' milk contains (*a*) a natural reducing system active under anaerobic conditions dependent to some extent on xanthine oxidase and largely destroyed by pasteurization; (*b*) the Schardinger enzyme active in the presence of formol and destroyed by heating to 80°; (*c*) a reducing substance, probably an aldehyde derived from lactose, present in autoclaved milk; (*d*) a system active in both raw and boiled milk that reduces methylene blue in the presence of light; and (*e*) reducing substances derived from bacteria which are active in both raw and pasteurized milk. (For references see Wilson *et al.* 1935, Nilsson 1959.)

In the methylene blue test commonly used for the grading of milk, reduction depends partly on the absorption of oxygen and the production of reducing substances by the organisms, and partly on the presence of a natural reducing system in the milk itself (see Hobbs 1939). Since some of the reducing enzymes and some of the organisms are adsorbed on the fat, and since the fat globules themselves play an important mechanical part in increasing the surface available for enzyme action and in affecting the visual depth of colour, the necessity of preventing an accumulation of cream at the surface will be realized; hence the necessity of half-hourly inversion.

The rate at which methylene blue is reduced in milk depends partly on the number of organisms and partly on their metabolic activity. For this reason it affords a very good index not only of the bacterial cleanliness of the milk but also of its keeping quality. The test is unsuited for the milk of individual cows, and for freshly pasteurized milk, but affords quite a good index to the keeping quality of pasteurized milk as delivered to the consumer (Ciani 1939). It is likewise unsuitable for milk collected on the farms in refrigerated tanks, unless it is pre-incubated for some time in the laboratory, since the flora is composed largely of psychrotrophic organisms having little reducing ability (see Tolle 1973, Murray and Stewart 1978). Its chief value lies in affording a fairly rapid means of assessing the general hygienic quality of raw mixed milk. Its technical simplicity, its cheapness, and its extremely small experimental error render it admirably suited for the routine control of this type of milk. The reduction time to be laid down must depend on the degree of cleanliness desired. A milk of high quality, which has been kept for about 12 hours at atmospheric temperature in this country, should not reduce methylene blue in less than about 8 hours in the winter or 7 hours in the summer, and one of fairly good quality should not reduce the dye in less than about 6 and 5 hours respectively.

Besides the advantages already noted, the reduction test has certain other advantages over the plate count. The first is that the result does not appear to be affected to any considerable extent by the degree of aggregation of the bacteria, which is one of the most disturbing features of the plate count on milk. The second is that the test affords a much more delicate index of bacterial growth than the plate count. It will be remembered (Chapter 3) that during the lag phase bacteria increase in size for some time before they commence to divide. The plate count therefore remains approximately constant during this phase. The reduction time, on the other hand, falls rapidly, owing to the active metabolism of the organisms. It serves, in fact, as a remarkably sensitive index of growth, and is hence peculiarly fitted to gauge the keeping quality of the milk. The plate count yields a *static* picture of the bacterial population: the methylene blue reduction test affords a *dynamic* picture. (For comparison between the two tests see Wilson *et al.* 1935, Nichols and Edwards 1936, Asdrubali 1939, Barkworth, Irwin and Mattick'1941, Pullinger 1945, 1946, Provan and Rowlands 1949, Galesloot 1949, Rowlands *et al.* 1950, Johns 1952.)

(8) **The resazurin test** The principle of this test, which owes its practical introduction mainly to the work of Ramsdell, Johnson and Evans (1935), is the same as that of the methylene blue test.

The test consists in adding to the milk a solution of resazurin to give a final concentration of 1/200 000, incubating at a given temperature, and observing the time taken to bring about a change in the colour of the dye, or alternatively noting the colour reached after a given time. The colour passes from blue, through various shades of purple and mauve, to pink, and ultimately to complete decolorization. The change from blue to pink is due to the irreversible reduction of resazurin to resorufin: the change from pink to colourless is due to the reversible reduction of resorufin to hydroresorufin. The colour can be assessed by means of a suitable comparator.

The advantages claimed for the resazurin over the methylene blue test are mainly two. Firstly, resazurin changes in colour rather more rapidly, and though complete reduction takes as long as or even longer than methylene blue, useful information can be obtained by noting a change in colour from blue to lilac or pink. Secondly, resazurin is more sensitive to the cell content of milk, and is therefore regarded by some workers as of value in the detection of mastitis. The practical importance of these two advantages is dubious. Admittedly a 10-minute test is useful as a screen for grossly contaminated milk; but if the resazurin test is to be used as an index of keeping quality and the reduction of the dye observed to the pink stage, then there is practically no saving of time over the methylene blue test (Anderson and Wilson 1945). Watts and Stirling (1944), in a very careful investigation, found that the resazurin test, even when carried out on individual quarter samples, was of limited value for the routine diagnosis of mastitis, and practically valueless for the detection of milk from infected cows when mixed with normal milk; in fact, as much as 73 per cent of the herd milk could be derived from cows suffering from mastitis without significantly altering the result of the resazurin test.

The disadvantages of the resazurin test are that the dye is not easy to standardize and is very sensitive to sunlight, that the colour change has to be followed by means of a comparator, and that the lilac and pink stages are not easily appreciated by persons whose vision is insensitive to red colours. In general terms it may be said that the resazurin test is more influenced by abnormal milk such as colostrum, late-lactation milk, milk from cows with mastitis, and by milk containing organisms with low reducing powers, but that the methylene blue test affords the better index of bacterial contamination of the milk.

(9) **The laboratory pasteurization test** This consists in making a plate count before and after pasteurization in the laboratory, and noting the numerical reduction that has occurred (Anderson and Meanwell 1933). Its main purpose is to detect the presence of heat-resistant and thermophilic bacteria in the milk, which may be present in such numbers as to prevent the pasteurized product from conforming to the official standards laid down. It is essentially a test for use in processing depots.

It will be realized that numerous tests are available for assessing the general cleanliness of milk, and that each of these tests affords an answer to a different question. Complete correspondence of the results of the different tests cannot therefore be expected. Experience has shown, however, that the Breed clump count, the plate count, the clot-on-boiling test, and the modified methylene blue reduction test are all fairly highly correlated with each other. On individual milks, of course, the reduction test and the plate count may give widely different results, owing mainly to the extraordinarily high experimental error of the plate count, but on an average the results of these tests are very similar. For practical purposes, therefore, there seems to be no advantage in using the complex, expensive, and highly inaccurate plate count in preference to the simple, inexpensive, and accurate reduction test. From a public health point of view, probably only two divisions need be made on the basis of *cleanliness*, namely (*a*) into milk that is suitable and (*b*) into milk that is not suitable for human consumption in the liquid state. We would again emphasize that a *clean* milk is not necessarily a *safe* milk.

Grading according to safety

The pathogenic organisms for which search is commonly made in milk are the tubercle bacillus, *Bruc. abortus*, and haemolytic streptococci. The methods employed in searching for the first two organisms are described in Chapters 51 and 56 respectively. The presence of mastitis streptococci in milk may be demonstrated microscopically or by culture, but according to Minett (1935) the microscopic method detects only about half of the cases that are positive culturally. The cultural method is, therefore, the method of choice. Individual quarter samples of the fore-milk should be plated on to a suitable blood agar medium and a study made of the colonies that develop. The separation of *Str. agalactiae* from other closely related types is discussed in Chapter 29.

Milk designations

The grades of milk in England and Wales are defined in the Milk (Special Designation) Regulations, 1963 (Report 1963*b*). Three main grades of milk are laid down: Untreated, Pasteurized, and Sterilized (Table 9.10). All milk, whether raw (untreated) or heat-treated, for consumption in the liquid state must come from TT (Tuberculin Tested) herds.

Statutory tests for milk designations For a full description of the tests, reference must be made to Report (1963*b*, 1965*a*). It may be noted that (*a*) both Untreated and Pasteurized milk are submitted to a ½-hour methylene blue test for cleanliness. (*b*) Pasteurized milk is submitted in addition to a phosphatase test, carried out by the Aschaffenburg-Mullen method, for adequate heat treatment. (*c*) Samples of Untreated and Pasteurized milk for the methylene blue test are transported to the laboratory in an insulated container and are then kept at atmospheric shade temperature during the period 1 May to 31 October, and at $65 \pm 2°$ F during the period 1 November to 30 April, till 9.30 the following morning when they are submitted to the test. (*d*) Samples of Pasteurized milk for the phosphatase test are examined on arrival at the laboratory or kept at a temperature of 3–5° until they are examined. (*e*) Milk may be pasteurized by the holder process at 145–150° F for not less than 30 minutes, or by the High-Temperature Short-Time process at 161° F for not less than 15 seconds, and must be

Table 9.10 Milk designations in England and Wales

Designation	General conditions	Bacteriological requirements	Remarks
Untreated	Must come from an Attested herd	Sample must be kept overnight at shade temperature during May to Oct., and at 65°F during Nov. to April, and satisfy the ½-hour methylene blue test carried out at 9.30 a.m. the following day.	(i) Must not be heat-treated. (ii) Must be labelled Untreated Milk. (iii) If bottled on farm, may be labelled Farm Bottled milk.
Pasteurized	Must come from an Attested herd. Milk must be kept (i) at a temperature of 145–150°F for at least 30 min, or (ii) of 161°F for at least 15 sec, or (iii) at some other approved time-temperature combination, and be immediately cooled to 50°F or below.	Must satisfy the ½-hour methylene blue test as above, and in addition the phosphatase test (see p. 286).	Pasteurizing plant must be approved by the licensing authority, and must be equipped with indicating and recording thermometers. Temperature records must be preserved for at least a month. HTST plant must be provided with an automatic flow-diversion valve. Must be labelled Pasteurized milk or Tuberculin Tested (Pasteurized) milk.
Sterilized	Must come from an Attested herd. Milk must be kept at a temperature of not less that 212°F for a time sufficient to ensure compliance with the turbidity test.	Must satisfy the turbidity test (see p. 288).	Sterilizing plant must be approved by the licensing authority, and must be provided with the specified thermometers and pressure gauges. Milk must be heated in bottles, and the bottles must be sealed afterwards with an airtight seal. Must be labelled Sterilized milk.
Ultra Heat Treated (UHT)	Must come from an Attested herd. Milk must be exposed to a temperature of not less than 270°F for not less than 1 second, and filled into sterile airtight containers.	The unopened sample is pre-incubated at 30–37°F for 24 hours. The 48-hour 30–37° count in yeastrel agar seeded with a loopful of milk shall be less than 10. The loop is of 4 mm internal diameter holding about 0.01 ml.	Heating plant must be approved by the licensing authority and provided with the specified indicating and recording thermometers and automatic diversion device. Milk must be labelled Ultra Heat Treated or UHT Milk.

immediately cooled to a temperature of 50°F or below. (*f*) Pasteurized milk for delivery to the consumer must be bottled or put into other sealed containers on the same premises as those at which it is pasteurized. (*g*) Sterilized milk is filtered or clarified, homogenized, and heated in bottles that can be hermetically sealed to a temperature of not less than 212°F for a time long enough to ensure its compliance with the turbidity test. (*h*) Besides pasteurization and sterilization, milk may be submitted to the ultra heat treatment process, consisting of exposure to a temperature of not less than 270°F for not less than 1 second, followed by filling into sterilized airtight containers.

It should be pointed out that Untreated milk should be boiled or pasteurized in the home in order to ensure the destruction of any pathogenic organisms it may contain.

Diseases borne by milk

From the description that has already been given of the pathogenic organisms that may be found in milk, it will be realized that raw milk serves as an important vehicle in the spread of disease to man. The frequency of tuberculosis of bovine origin and of undulant fever due to milk-borne organisms is considered in Chapters 51 and 56 respectively. In these diseases infection of the milk occurs in the udder of the cow itself. The same is true also of Q fever caused by *Coxiella burneti*. In scarlet fever, septic sore throat, and diphtheria, infection occurs from the udder of the cow or from the nasopharynx of some person handling the milk, whereas in typhoid fever, paratyphoid fever, and dysentery the causative organisms generally gain access to the milk from the infected hands of some human attendant, or probably less often from a contaminated water supply. Salmonella food poisoning may be due to contamination of the milk from infected herds or from the faeces of the cow, or to excretion of the organism in the milk of a cow whose udder is infected. Food poisoning caused by staphylococci is usually due to milk from a cow affected with staphylococcal mastitis, though a nasal human carrier is occasionally responsible. As already mentioned, *Camp. fetus* subsp. *jejuni* probably reaches the milk from the animal's faeces, or possibly the udder itself, and the viruses of poliomyelitis and infective hepatitis from human excreta or sewage.

Occasional outbreaks are on record of *rat-bite fever* caused by contamination of the milk with *Actinobacillus moniliformis* (Parker and Hudson 1926, Place and Sutton 1934). In the acute stage of *anthrax* a small amount of altered milk may be secreted containing anthrax bacilli. The organisms do not sporulate at a temperature less than 15°, so that they are readily destroyed by pasteurization. Rarely, in animals that recover, anthrax bacilli are excreted for weeks or months on end from an inflamed udder (Weidlich 1934). In countries in which the disease is indigenous in cattle, spores may reach the milk from the environment and, provided the temperature is over 25° (Minett and Dhanda 1941), may germinate and give rise to large numbers of bacilli. *Vaccinia virus* in milk is liable to infect man by contact rather than by ingestion. Disease, sporadic or epidemic, may be caused by goats' milk containing the virus of Russian spring-summer encephalitis (Report 1970a).

The true prevalence of milk-borne disease is extremely difficult to estimate, partly because the disease may be insidious in its onset and suspicion of the milk supply is not aroused, and partly because isolated cases of milk-borne infections, even if they are recognized as such, are not usually recorded (see Bendixen *et al.* 1937). In Great Britain Savage (1912) attempted to collect the figures referring to milk-borne epidemics prior to 1912. Those from 1912 to 1937 were collected by Wilson (1942) from various sources, and those from 1938 to 1978 almost entirely from the records of the Public Health Laboratory Service. In the last two columns of Table 9.11, 1961–70 and 1971–8, none but outbreaks of gastro-enteritis were notified. Apart from three *Campylobacter* and two staphylococcal outbreaks, all were due to salmonellae. The largest outbreak of milk-borne infection with *Campylobacter jejuni* occurred in Bedfordshire in the spring of 1979; no fewer than 2500 children who drank raw milk at school were affected (Jones *et al.* 1981; see also Porter and Reid 1980, Report 1981). In Scotland during the 10-year period 1970–79 there were 29 outbreaks of salmonellosis affecting at least 2428 persons who had consumed raw milk (Sharp, Peterson and Forbes 1980).

These outbreaks represent merely those of which records are readily available, and almost certainly underestimate the real number. They pay no attention to sporadic cases of milk-borne infection, nor do they include the 6000 or more cases of tuberculosis of bovine origin or the 400–500 cases of undulant fever that used to occur annually in Great Britain up till the end of the 2nd world war or the large outbreaks of summer diarrhoea that occurred up to 1921 for which milk was partly responsible (see 6th edn, p. 2058). Likewise cases and outbreaks of Q fever spread by milk and those of poliomyelitis that may have been spread by milk are excluded. Figures for the more recent outbreaks are given in Table 9.11. It should be noted that these records are not strictly comparable with those before 1937, since the laboratory services after 1939 were greatly expanded and led to much better reporting.

We have no space to discuss the epidemiology of milk-borne disease. Apart from the explosive nature of the outbreak, it is interesting to note that in milk-borne scarlet fever and diphtheria the average age incidence if often very much higher than in non-milk-borne epidemics. The majority of the cases may occur in persons of 15 years and over (see Godfrey 1929, Clarke 1936, Hill and Mitra 1936). It may be mentioned that the first recorded example of milk-borne disease in Great Britain was one of typhoid fever at Terling in Essex in 1867, reported by Lord Rayleigh (Rayleigh 1924).

Table 9.11 Recorded outbreaks of milk-borne disease, excluding tuberculosis, undulant fever, Q fever, and outbreaks due to milk products, such as ice-cream, cheese, milk powder and cream buns. The 1912–60 figures refer to Great Britain, the 1961–78 figures refer to England and Wales only.

Disease	1912–60 Outbreaks	1912–60 Persons ill	1961–70 Outbreaks	1961–70 Persons ill	1971–78 Outbreaks	1971–78 Persons ill
Scarlet fever and Septic sore throat	54	6 206	0	0	0	0
Diphtheria	26	810	0	0	0	0
Enteric fever	39	3 229	0	0	0	0
Typhoid fever	8	57	0	0	0	0
Paratyphoid fever	15	334	0	0	0	0
Dysentery	21	2 630	0	0	0	0
Gastro-enteritis	97	8 916	46	1 383	47	833
Infectious hepatitis	2	152	0	0	0	0
Campylobacter enteritis	0	0	0	0	3	177
Total	262	22 334	46	1 383	50	1 010

Note. In 1979 an outbreak of campylobacter enteritis affected over 2 000 children in schools in Hertfordshire.

(For the isolation of pathogenic organisms from milk, see the relevant chapters in Volume 3, and for reviews of milk-borne disease McCoy 1966, Parry 1966.)

Milk sickness, milk allergy, and related disorders Not all illness caused by milk is of bacterial origin. For diagnostic purposes it should be remembered that milk sickness may result from the ingestion of milk containing toxic substances derived from poisonous plants eaten by the cow (see Wilson 1942, Schoental 1959), Abraham Lincoln's mother is said to have died of it; that milk allergy is by no means uncommon in infants and children who have become sensitized to proteins in the milk (see Williams 1936, Gunther *et al.* 1960, Parish *et al.* 1960); and that substances such as penicillin injected into the udder for therapeutic purposes, and the chlorinated organic pesticides used in agriculture (Clifford 1957), may be excreted in milk in toxic concentrations.

Methods for increasing the safety of the milk supply

These fall into three main categories.

A. Control of animal disease Reference is made in Chapters 51 and 56 to the methods available for diminishing the frequency of tuberculosis and contagious abortion in cattle, and in Chapter 57 to the elimination of mastitis. The success attendant on these measures will depend very largely on the vigour with which they are prosecuted. The eradication of disease, however, from the animal population is essentially an agricultural and veterinary problem to be undertaken in the economic interests of the farmer and affords no more than a partial solution to the public health problem of providing a safe milk for the human population.

B. Control of the human personnel handling the milk However healthy the animals producing the milk may be, there is always danger of its becoming infected directly or indirectly from the nasopharynx or faeces of human carriers of *Str. pyogenes*, *C. diphtheriae*, *Salm. typhi*, *Salm. paratyphi* A and B, dysentery bacilli, and bacilli of the food-poisoning group. In the United States some attempt has been made to control the health of those engaged in handling the milk. The most frequent examinations made comprise a physical inspection, a careful inquiry into the history of infectious disease, a series of faecal and urinary examinations for organisms of the enteric and dysentery groups, and nose and throat swabs for streptococci and diphtheria bacilli (see Borman *et al.* 1935). There is no question that close supervision of milk operatives, particularly of those working in pasteurizing depots, is highly desirable. The procedure is, however, costly and is not free from administrative difficulties. Moreover, since the carrier state is often transient or intermittent, it is impossible without almost daily supervision to detect all dangerous persons.

C. Heat treatment of the milk: pasteurization, boiling and sterilization The dangers of infection of milk from cattle, from the human personnel, and from water are so great that we shall not be overstating the case if we assert that no raw milk can be regarded as perfectly safe for human consumption. Clearly we must interpret that axiom with discretion, remembering always that the stringency of our precautions must be commensurate with the size of the population at risk. It is very difficult to assess in any particular instance the real danger run by persons consuming a given milk supply, and our safest course is therefore to insist as far as possible that all pathogenic organisms shall be destroyed by some form of heat treatment.

Pasteurization is the most satisfactory means for this purpose. When carried out in properly designed plant supervised by intelligent and conscientious operatives it can be relied upon to destroy all pathogenic bacteria in the milk (Scott and Wright 1935, Dalrymple-Champneys 1935, Report 1956). It allows, however, some viruses such as the foot-and-mouth virus (Blackwell and Hyde 1976), the infectious hepatitis virus, the bovine leukaemia virus (Baumgartner *et al.* 1976), and probably some others to survive. *Coxiella burneti*, too, needs a slightly higher temperature (162° F) for complete destruction (Enright *et al.* 1957). Two main methods are used in England and Wales—the Holder method and the HTST method. The HTST method is more rapid and requires less space, and has largely replaced the Holder method in all big concerns. Other methods, such as *Uperization* in which the milk is heated to 150° F for less than a second (see Tentoni 1955); the *Vacreator* method in which it is heated under reduced pressure (see Tracy *et al.* 1950); and the Ultra High Temperature method in which the milk is exposed to a temperature of not less than 270° F for not less than a second have been described, but only the last method producing UHT milk is officially recognized in England and Wales. (For information on this method, see Burton 1965, Report 1965a, Ridgway 1970; and on the planning, operating, and testing of pasteurizing plants, see Report 1953, 1966a.)

The adequacy of heat treatment may be controlled by the **phosphatase test** (Kay and Graham 1935, Kay, Aschaffenburg and Neave 1939, Memo. 1943, Report 1944), which depends on the fact that the enzyme phosphatase, which is normally present in milk, is destroyed by a temperature of 145° F within 30 minutes or of 162° F within 15 seconds.

The test is carried out by incubating the milk in the presence of a suitable substrate. Under the influence of the phosphatase, the substrate is broken down with the liberation of a radicle that is either coloured itself or gives a colour with a test reagent. The intensity of colour is in proportion to the amount of phosphatase present. In the original Kay and Graham (1935) test the substrate used is disodium phenyl phosphate. Phenol is split off, Folin and Ciocalteu's reagent is added, and the resulting blue colour is estimated by a Lovibond tintometer or comparator. Raw milk gives figures of about 50 blue units, but in properly pasteurized milk this figure is reduced to 2.3 or less. The results of this test are liable to be disturbed by traces of phenolic substances in the air, the reagents, and the apparatus; partly for this reason and partly for the shorter time it takes, the modified test devised by Aschaffenburg and Mullen (1949) is now recognized officially in England and Wales. The substrate used is disodium *p*-nitrophenyl phosphate. The *p*-nitrophenol split off by the phosphatase has a yellow colour the intensity of which can be measured by a comparator. The phosphatase test is most valuable and can be used in practice to ensure that the processing of the milk is carried out satisfactorily.

Phosphatase is slightly more resistant to heat than the tubercle bacillus, so that a negative phosphatase result is strong presumptive evidence in favour of the safety of the milk.

The objections that have been raised to pasteurization, though often made on apparently scientific grounds, are determined almost entirely by economic or political motives. The farmer, especially the producer-retailer, may not obtain such a high return for his milk if he has to pasteurize it himself or send it to a pasteurizing depot as if he sells it directly to the customer in its original raw and possibly dangerous state. Producer-retailers, moreover, are afraid that pasteurization will lead to their being gradually absorbed by the big milk combines, which can of course afford to pasteurize and bottle their milk under very much more favourable conditions than is possible for a small producer. Though appreciating the force of these motives, we must not allow the public health of the nation to be sacrificed to the financial interests of a small section of the community.

One of the chief objections that has been raised is that pasteurization lowers the nutritive value of the milk. The evidence for and against this contention has been reviewed by several workers (see Stirling and Blackwood 1933, Report 1934, 1936, 1937–39, Bendixen *et al.* 1937, Wilson 1942). The main effects of pasteurization by the Holder method at 145–150° F for half an hour followed by immediate cooling to 50° F or below are as follows: (1) The cream line is reduced by about 10–30 per cent. (2) About 5 per cent of the lactalbumin is coagulated. (3) There is a diminution of about 5 per cent in the soluble calcium and phosphorus. (4) There is some destruction, amounting usually to about 10 per cent, and at the most to 25 per cent, of vitamin B_1. (5) There is a variable diminution in the vitamin C content, dependent on the degree of previous exposure of the milk to light, the presence of copper in the pasteurizing plant, and the amount of dissolved oxygen; at present, the average reduction is about 20 per cent. (6) Some of the iodine—almost certainly less than 20 per cent—may be driven off as the result of volatilization. (7) A high proportion of the bacteria in the milk are destroyed, leading, as a rule, to a considerable increase in the keeping quality.

Against these positive changes may be set a number of negative findings. Thus it is found from feeding experiments on rats and calves that the biological value and digestibility of the proteins are not decreased, that there is no detectable change in the availability of calcium or phosphorus, that there is no loss of vitamin A or D, and that the energy value of the milk remains unaltered. So far as human beings are concerned, it may be pointed out that milk is a poor source of vitamins C and D, and that infants fed on milk, whether raw or pasteurized, should therefore always be given a supplement of fruit or vegetable juice and cod-liver oil. The differences between raw and pasteurized milk are far less than those between cows' milk and human milk, or even between summer and winter samples of raw milk (see Kay 1939). Extensive feeding experiments on rats, mice and calves, and comparative investigations on children have failed to show that pasteurization has any significant effect in lowering the total nutritive value of the milk for the growing animal, and no fear need therefore be entertained that a policy of pasteurization will prejudice the nutritional development of the child population. It may be added that HTST pasteurization destroys about 20 per cent of any penicillin there may be in the milk (Garrod 1964).

The UHT process renders the milk sterile or practically so without imparting to it the caramelized flavour of sterilized milk. When kept cold, the milk may remain sweet for several months, and is therefore of particular value in places where fresh milk is not obtainable. Little immediate effect is noticeable on the proteins or vitamins of the milk, though during subsequent storage the proteins undergo changes leading ultimately to gelation (Samel *et al.* 1971), and there is a loss in some of the water-soluble vitamins, especially ascorbic acid and folic acid (Report 1968–69, Law 1979).

There are manifestly strong reasons why all milk intended for consumption in the liquid state by a community of any size should be submitted to pasteurization, the process being, of course, adequately controlled by the appropriate supervising authorities. There is reason to believe that if pasteurization was rendered compulsory and universal, milk-borne disease would practically cease to exist (see Wilson 1942).

When pasteurization is impracticable, the milk should be *boiled* in the individual household. The most satisfactory way of doing this is to bring it to the boil in a closed vessel, preferably a double saucepan, and cool it immediately to as low a temperature as possible.

Sterilization is the term used for a process which is applied commercially, and which is intended to destroy all but the most resistant spore-bearing organisms. In some plants the temperature is not raised above boiling-point, but more usually the milk is heated under a slight pressure. It is distributed in hermetically sealed bottles. Bacteriologically the milk is perfectly safe, but the partial caramelization that it undergoes gives it a distinctive flavour. A phenolic flavour may occasionally be imparted to the milk by the growth of *Bac. circulans* (Chalmers 1962). In milk that is properly sterilized the soluble proteins are completely denatured so that when the caseinogen is precipitated with ammonium sulphate a clear filtrate is obtained. This forms the basis of the **turbidity test** devised by Aschaffenburg (1947).

20 ml of sterilized milk are run into a 50-ml conical flask containing 4 g of ammonium sulphate. The flask is shaken to dissolve the salt, and not less than 5 minutes later the contents of the flask are filtered through a folded paper (Whatman No. 12) into a test-tube. When 5 ml of the filtrate have been collected, the tube is placed in a beaker containing boiling water and kept there for 5 minutes. After cooling, the contents of the tube should show no sign of turbidity when examined by indirect light.

Human milk As already mentioned, human milk is liable to contamination with a variety of organisms derived mainly from the skin. Of these, the commonest is *Staph. epidermidis*. The numbers depend on the hygienic precautions taken during collection. As in cows, contamination is greatest in the fore-milk. If the first 10 ml or so are discarded, the bacterial content of the remainder is found to be much lower (West *et al.* 1979). Mothers who are contributing to a milk bank should be advised to adopt this procedure. After withdrawal, the milk should be refrigerated as quickly as possible. To prevent the growth of psychrotrophic bacteria, storage of up to 3 days may be permitted in the freezing compartment of a domestic refrigerator; but if longer storage is required, the milk should be transferred to a deep-freeze.

Churns and bottles

However thoroughly the processing of the milk is carried out, the safety of the final product may be jeopardized if it is filled into contaminated churns or bottles. These utensils must be treated either by heat, by alkaline detergents, or by a combination of the two in such a way as to ensure the destruction of all non-sporing organisms. The bottles should be subsequently capped with an overlapping seal, preferably by automatic machinery and not by hand.

The cleanliness of churns and bottles can be assessed by means of a plate count carried out on suitable rinsings. The method of examining bottles is described in full by Hobbs and Wilson (1943) in their report on the comparative efficacy of different methods of washing and sterilizing bottles. Briefly it consists in taking 6 or 12 representative bottles, adding to each 20 or 30 ml of ¼ strength Ringer's solution according to the size of the bottle, rolling the bottle to ensure that the whole of the internal surface is thoroughly wetted, allowing the bottle to stand for 15–30 minutes, again rolling it, and then preparing duplicate plates each with 5 ml of the rinsings. The plates are prepared with Yeastrel milk agar (Memo 1937), and are incubated one at 22° and the other at 37° for 3 days, after which the colonies are counted. The following standards are widely used:

Less than 600 colonies per bottle	Satisfactory
600 to 2000 ,, ,, ,,	Fairly satisfactory
Over 2000 ,, ,, ,,	Unsatisfactory

but there is a tendency to raise them by substituting the figures: not more than 200; over 200 to 1000; and over 1000, for those just quoted.

The examination of churn rinsings is carried out in an essentially similar way.

Cream

In accordance with the Food Standards Committee on Cream (Report 1967), the Cream Regulations (Report 1970*b*) define six grades based upon the fat content.

Clotted cream must contain not less than 55 per cent fat, double cream 48, whipping cream 35, sterilized cream 23, single cream (i.e. coffee cream) 18, and half cream 12 per cent. Like milk, cream is potentially dangerous in the raw state, and should always be pasteurized for human consumption. The time-temperature conditions required for the destruction of pathogenic organisms in cream are similar to those in milk (Aschaffenburg *et al.* 1956). No official method for pasteurizing cream has yet been formulated in England and Wales, or any standard laid down for the resulting product. Provided that the Holder or HTST methods are used, the phosphatase test can be relied on as a guide to the efficacy of the heat treatment. The cream should be kept cold after processing; otherwise some reactivation of the phosphatase may occur (Wright and Tramer 1954, Fram 1957). As a screening test for gauging the keeping quality of the product as delivered to the consumer, the 4-hour modified methylene blue reduction test serves a useful purpose (Report 1958). Occasional anomalous results occur (Jenkins and Henderson 1969, Cox 1970), such as when a sample with a high plate count fails to reduce the dye (Report 1971). 'Bitty' or 'broken' cream is usually due to the lecithinase of *Bac. cereus* or *Bac. mycoides* which, by breaking down the lecithin on the surface of the fat globules, allows the globules to aggregate into larger particles (Chalmers 1962).

Dried milk Milk may be dried either by (*a*) the roller process, in which it is fed on to hot rotating drums and scraped off mechanically in the form of a thin sheet, or (*b*) the spray process, in which after preliminary heat treatment it is sprayed into a hot vacuum chamber and collected as a powder. In the roller process the milk is heated to a higher temperature than in the spray process, and is bacteriologically more satisfactory, though less acceptable in other ways. In the spray process the heat treatment is barely sufficient to destroy all pathogenic organisms; in the United States, for example, salmonellae have been isolated from both the dried product and the plant (Report 1966*b*). Under certain conditions, not perfectly defined, staphylococci may grow in the milk during its passage through the plant and form enterotoxin. Outbreaks of acute food poisoning have been traced to this source (Anderson and Stone 1955, Hobbs 1955, Armijo *et al.* 1957; see also Hawley and Benjamin 1955, Crossley and Campling 1957), and also to toxins of apparently non-staphylococcal origin (Dickie and Thatcher 1967). The bacterial quality of dried milk is generally gauged by the plate count method. It should be pointed out, however, that a low plate count does not necessarily guarantee the safety of the dried product; though the staphylococci may be destroyed, the enterotoxin they have produced may persist.

References

Abd-el-Malek, Y. and Gibson, T. (1952) *J. Dairy Res.* **19**, 294.
Anderson, E. B. and Meanwell, L. J. (1933) *J. Dairy Res.* **4**, 213.
Anderson, E. B. and Wilson, G. S. (1945) *J. Dairy Res.* **14**, 21.
Anderson, P. H. R. and Stone, D. M. (1955) *J. Hyg., Camb.* **53**, 387.
Armijo, R., Henderson, D. A., Timothée, R. and Robinson, H. B. (1957). (1957) *Amer. J. publ. Hlth* **47**, 1093.
Aschaffenburg, R. (1947) *Mon. Bull. Minist. Hlth Lab. Serv.* **6**, 159.
Aschaffenburg, R., Briggs, C. A. E., Crossley, E. L. and Rothwell, J. (1956) *J. Dairy Res.* **23**, 24.
Aschaffenburg, R. and Mullen, J. E. C. (1949) *J. Dairy Res.* **16**, 58.
Asdrubali, M. (1939) *Ann. Igiene (sper.)* **49**, 218.
Ayers, S. H. and Clemmer, P. W. (1918) *U.S. Dep. Agric. Bull.* No. 739.
Bardsley, D. A. (1934) *J. Hyg., Camb.* **34**, 38.
Barkworth, H., Irwin, J. O. and Mattick A. T. R. (1941) *J. Dairy Res.* **12**, 265.
Baumgartner, L., Olson, C. and Onuma, M. (1976) *J. Amer. vet. med. Ass.* **169**, 1189.
Bendixen, H. C., Blink, G. J., Drummond, J. C., Leroy, A. M. and Wilson, G. S. (1937) *Bull. Hlth Org., L.o.N.* **6**, 371.
Bendixen, H. C. and Minett, F. C. (1938) *J. Hyg., Camb.* **38**, 374.
Benham, C. L. and Egdell, J. W. (1970) *J. Soc. Dairy Technol.* **23**, No. 2.
Björck, L., Rosén, C. G., Marshall, V. and Reiter, B. (1975) *Appl. Bact.* **30**, 199.
Blackburn, P. S. (1968) *J. Dairy Res.* **35**, 59.
Blackwell, J. H. and Hyde, J. L. (1976) *J. Hyg., Camb.* **77**, 77.
Borman, E. K., West, D. E. and Mickle, F. L. (1935) *Amer. J. publ. Hlth* **25**, 557.
Breed, R. S. (1929) *N.Y. St. agric. exp. Sta. Bull.* No. 566.
Brock, J. H. (1980) *Arch. Dis. Childh.* **55**, 417.
Burton, H. (1965) *Dairy Ind.* **30**, 792.

Chalmers, C. H. (1962) *Bacteria in Relation to the Milk Supply*. 4th ed., Edward Arnold, London.
Ciani, G. (1939) *Ann. Igiene (sper.)* **49**, 290.
Clarke, J. H. (1936) *Med. Offr.* **55**, 40.
Clifford, P. A. (1957) *Publ. Hlth Rep., Wash.* **72**, 729.
Cox, W. A. (1970) *J. Dairy Technol.* **23**, 195.
Crossley, E. L. and Campling, M. (1957) *J. appl. Bact.* **20**, 65.
Dalrymple-Champneys, W. (1935) *The Supervision of Milk Pasteurising Plants. Min. Hlth, Lond., Rep. publ. Hlth med. Subj.* No. 77.
Darányi, J. von. (1941) *Arch. Hyg., Berl.* **127**, 1.
Dickie, N. and Thatcher, F. S. (1967) *Canad. J. publ. Hlth* **58**, 25.
Dolby, J. M., Stephen, S. and Honour, P. (1977) *J. Hyg., Camb.* **78**, 235.
Donovan, K. O. (1959) *J. appl. Bact.* **22**, 131.
Dublin, T. D., Rogers, E. F. H., Perkins, J. E. and Graves, F. W. (1943) *Amer. J. publ. Hlth* **33**, 157.
Edwards, S. J. (1964) *Vet. Rec.* **76**, 545.
Enright, J. B., Sadler, W. W. and Thomas, R. C. (1957) *Publ. Hlth Monograph*, No. 47 Washington, D.C.
Evans, T. J. *et al.* (1978) *Arch. Dis. Childh.* **53**, 239.
Ferrer, J. F., Kenyon, S. J. and Gupta, P. (1981) *Science* **213**, 1014.
Finkelstein, R. (1919) *J. Dairy Sci.* **2**, 460.
Ford, J. E., Law, B. A., Marshall, V. M. E. and Reiter, B. (1977) *J. Pediat.* **90**, 29.
Fram, H. (1957) *J. Dairy Sci.* **40**, 19, 1649.
Galesloot, T. E. (1949) *Ned. Melk-Zuiveltijdschr.* **3**, 205.
Garrod, L. P. (1964) Pers. Comm.
Gavin, A. and Ostovar, I. (1977) *J. Food Protect.* **40**, 614.
Godfrey, E. S. (1929) *Amer. J. publ. Hlth* **19**, 257.
Goldie, W. and Maddock, E. C. G. (1943) *Lancet* **i**, 285.
Gunther, M., Aschaffenburg, R., Matthews, R. H., Parish, W. E. and Coombs, R. R. A. (1960) *Immunology* **3**, 296.
Hawley, H. B. and Benjamin, M. I. W. (1955) *J. appl. Bact.* **18**, 493.
Heidrich, H. J., Grossklaus, D. and Mülling, M. (1964) *Berl. Münch tierärztl. Wschr.* **77**, 83, 85.
Henningsen, E. J. and Ernst, J. (1939) *J. Hyg., Camb.* **39**, 51.
Hill, A. B. and Mitra, K. (1936) *Lancet* **ii**, 589.
Hiscox, E. R. and Briggs, C. A. E. (1953) *J. Dairy Res.* **20**, 381.
Hobbs, B. C. (1939) *J. Dairy Res.*, **10**, 35; (1955) *J. appl. Bact.* **18**, 484.
Hobbs, B. C. and Wilson, G. S. (1943) *J. Hyg., Camb.* **43**, 96.
Hosty, T. S. and McDurmont, C. I. (1975) *Hlth. Lab. Sci.* **12**, 16.
Houston, A. C. (1913) *Studies in Water Supply*. Macmillan's Science Monographs, London.
Hyslop, N. St. G. and Osborne, A. D. (1959) *Vet. Rec.* **71**, 1082.
Jenkins, H. R. and Henderson, R. J. (1969) *J. Hyg., Camb.* **67**, 401.
John, T. J. and Devaragar, L. V. (1973) *Indian J. med. Res.* **61**, 1009.
Johns, C. K. (1952) *J. Milk Tech.* **15**, 8.
Jones, A. C., King, G. J. G., Fennell, H. and Stone, D. (1957) *Mon. Bull. Minist. Hlth Lab. Serv.* **16**, 109.
Jones, F. S. and Simms, H. S. (1930) *J. exp. Med.* **51**, 327.
Jones, P. H., Willis, A. T., Robinson, D. A., Skirrow, M. B. and Josephs, D. S. (1981) *J. Hyg., Camb.* **87**, 155.

Kay, H. D. (1939) *Nutrit. Abstr. Rev.* **9**, 1.
Kay, H. D., Aschaffenburg, R. and Neave, F. K. (1939) *Tec. Commun.* No. 1. Imp. Bur. Dairy Sci., Shinfield.
Kay, H. D. and Graham, W. R. (1935) *J. Dairy Res.* **6**, 191.
Kirschner, L., Maguire, T. and Bertaud, W. S. (1957) *Brit. J. exp. Path.* **28**, 357.
Law, B. A. (1979) *J. Dairy Res.* **46**, 573.
Lück, H. (1956) *Dairy Sci. Abstr.* **18**, 363.
Macaulay, D. M. and Packard, V. S. (1981) *J. Food Protection* **44**, 696.
McCoy, J. H. (1966) *J. Dairy Res.* **33**, 103.
McKenzie, D. A. and Booker, E. M. K. (1955) *J. appl. Bact.* **18**, 401.
Memorandum. (1937) *Bacteriological Tests for Graded Milk.* Memo 139/Foods. Min. Hlth, London; (1943) *The Phosphatase Test for Heat Treated Milk.* Addendum to Memo 139/Foods, Min. Hlth, London.
Mickelson, M. N. (1979) *Appl. environm. Microbiol.* **38**, 821.
Minett, F. C. (1932) *Off. int. Epizoöties* **6**, 124; (1935) *Proc. 12th int. vet. Congr.*, 511.
Minett, F. C. and Dhanda, M. R. (1941) *Indian J. vet. Sci.* **11**, 308.
Morgan, W. J. B. (1970) *J. Dairy Res.* **37**, 303.
Moritsch, H. and Krausler, J. (1957) *Wien klin. Wschr.* **69**, 921, 952, 965.
Murray, J. G. and Stewart, D. B. (1978) *J. Soc. Dairy Tech.* **31**, 28.
Nichols, A. A. and Edwards, S. J. (1936) *J. Dairy Res.* **7**, 258.
Nilsson, G. (1959) *Bact. Rev.* **23**, 41.
Parish, W. E., Barrett, A. M., Coombs, R. R. A., Gunther, M. and Camps, F. E. (1960) *Lancet* **ii**, 1106.
Park, W. H. and Holt, L. E. (1903) *Med. News*, N.Y. **83**, 1066.
Parker, F. and Hudson, N. P. (1926) *Amer. J. Path.* **2**, 357.
Parry, A. H. and Tee, G. H. (1958) *Mon. Bull. Minist. Hlth Lab. Serv.* **17**, 19.
Parry, W. H. (1966) *Lancet* **ii**, 216.
Pettipher, G. L. and and Rodrigues, U. M. (1981) *J. appl. Bact.* **50**, 157.
Place, E. H. and Sutton, L. E. (1934) *Arch. intern. Med.* **54**, 659.
Porter, I. A. and Reid, T. M. S. (1980) *J. Hyg.*, Camb. **84**, 415.
Provan, A. L., Dudley, F. J. and Thomas, S. B. (1936) *Welsh J. Agric.* **12**, 130.
Provan, A. L. and Rowlands, A. R. (1949) *12th int. Dairy Congr.* Stockholm, p. 586.
Pullinger, E. J. (1934) *Lancet* **i**, 967; (1935) *J. Dairy Res.* **6**, 369; (1945) *S. Afr. vet. med. Ass.* **16**, 110; (1946) *Ibid.* **17**, 15.
Pullinger, E. J. and Kemp, A. E. (1937) *J. Hyg.*, Camb. **37**, 527.
Ramsdell, G. A., Johnson, W. T. and Evans, F. R. (1935) *J. Dairy Sci.* **18**, 705.
Rayleigh (4th Baron) (1924) *John William Strutt: Third Baron Rayleigh.* Edward Arnold, London.
Reiter, B. (1976) In: *Inhibition and Inactivation of vegetative microbes* p. 32. Ed. by F. A. Skinner and W. B. Hugo, Academic Press.
Reiter, B., Marshall, V. M. E., Björck, L. and Rosén, C. F. (1976) *Infect. & Immun.* **13**, 800.
Report. (1909) *3rd interim Rep., roy. Comm. Tuberc.* H.M.S.O., London; (1934) *Econ. advis. Coun., Comm. Cattle Dis.*, H.M.S.O. London; (1936) *The Nutritive Value of Milk. Min. Hlth & Dept. Hlth*, Scotland, H.M.S.O., London; (1937-39) Milk and Nutrition. Parts i-iv. Poynder & Son, Reading; (1944) *Statutory Rules and Orders*, No. 349. H.M.S.O., London; (1953) *Milk Pasteurization.* World Hlth Org., Monograph Ser. No. 14; (1956) *Mon. Bull. Minist. Hlth Lab. Serv.* **15**, 232; (1958) *Ibid.* **17**, 77; (1962) *Milk Hygiene. Monogr. Ser. World Hlth Org.*, No. 48; (1963a) *Antibiotics in Milk in Great Britain.* Min. Agric., Fish, Food, H.M.S.O., London; (1963b) *The Milk (Special Designation) Regulations 1963. Stat. Instrum.* No. 1571. H.M.S.O., London; (1965a) *Ibid.* 1965, No. 1555; (1965b) *Mon. Bull. Minist. Hlth Lab. Serv.* **24**, 34; (1966a) U.S. Dept. Hlth, Educ. & Welfare, *Publ. Hlth Serv. Publcn*, No. 731; (1966b) *C.D.C. Salmonella Surveil.* No. **53**; (1967) Minist. Agric., Fish., Food, H.M.S.O., London; (1968-69) *Annu. Rep. Agric. Res. Coun.*, p. 76; (1970a) 3rd Rep. Joint FAO/WHO Committee on Milk Hygiene. *Wld Hlth Org. techn. Rep. Ser.*, No. 453; (1970b) *Stat. Instrum.* No. 752. H.M.S.O., London; (1971) *J. Hyg.*, Camb. **69**, 155; (1981) *CDR* 81/39, p. 3.
Ridgway, J. D. (1970) *R. Soc. Hlth J.* **90**, 33.
Rollins, L. D., Mercer, H. D., Carter, G. G. and Kramer, J. (1970) *J. Dairy Sci.* **53**, 1407.
Rowlands, A., Barkworth, H., Hosking, Z. and Kempthorne, O. (1950) *J. Dairy Res.* **17**, 159.
Rowlands, A. and Hosking, Z. (1951) *J. Dairy Res.* **18**, 268.
Samel, R., Weaver, R. W. V. and Gammack, D. B. (1971) *J. Dairy Res.* **38**, 323.
Savage, W. G. (1912) *Milk and the Public Health.* Macmillan, London.
Schiemann, D. A. and Toma, S. (1978) *Appl. environm. Microbiol.* **35**, 54.
Schoental, R. (1959) *J. Path. Bact.* **77**, 485.
Schulz, T. (1979) In: *Lehrbuch der Tierhygiene*, p. 615. Ed. by G. Mehlhorn. Gustav Fischer, Jena.
Scott, A. W. and Wright, N. C. (1935) *An Inquiry into the Design, Operation and Efficiency of Pasteurising Plants.* Hannah Dairy Res. Bull., No. 6.
Sharp, J. C. M., Paterson, G. M. and Forbes, G. I. (1980) *J. Infection* **2**, 333.
Sharpe, M. E. and Bramley, A. J. (1977) *Dairy Industr. int.* **42**, 24.
Shehata, T. E. and Collins, E. B. (1971) *Appl. Microbiol.* **21**, 466.
Sherman, J. M. and Wing, H. U. (1933) *J. Dairy Sci.* **16**, 165.
Smith, H. W. (1957) *Mon. Bull. Minist. Hlth Lab. Serv.* **16**, 39.
Steele, J. H. (1960) *J. Amer. vet. med. Ass.* **136**, 247.
Stirling, J. D. and Blackwood, J. H. (1933) *The Nutritive Properties of Milk in relation to Pasteurisation.* Hannah Dairy Res. Bull. No. 5.
Stockmayer, W. (1936) *Z. InfektKr. Haustiere* **49**, 46.
Taylor, M. M. (1967) *Dairy Industries*, April.
Taylor, P. B. and Clegg, L. F. L. (1958) *J. Dairy Res.* **25**, 32.
Tentoni, R. (1955) *G. Batt. Immunol.* **47**, 264.
Thomas, E. L. and Aune, T. M. (1978) *Infect. and Immun.* **20**, 456.
Thomas, S. B. and Druce, R. G. (1971) *Dairy Sci. Abstr.* **33**, 339.
Thomas, S. B., Druce, R. G. and Jones, M. (1971) *J. appl. Bact.* **34**, 659.
Thomas, S. B. and Thomas B. F. (1978) *Dairy Industr. int.* **40**, 338.
Tolle, A. (1973) *Arch. LebensmittHyg.* **24**, 149.

Tracy, P. H., Pedrick, R. and Lingle, H. C. (1950) *J. Dairy Sci.* **33**, 820.
Ward, A. R. (1926) *Dairy Prod. Merchandising* **6**, No. 5. March.
Watts, P. S. and Stirling, A. C. (1944) *Vet. Rec.* **56**, 83.
Weidlich, N. (1934) *Wien. tierärtl. Mschr.* **21**, 289.
Welsh, J. K. and May, J. T. (1979) *J. Pediat.* **94**, 1.
West, P. A., Hewitt, J. H. and Murphy, O. M. (1979) *J. appl. Bact.* **46**, 269.

Williams, D. A. (1936) *Brit. med. J.* **ii**, 1081.
Wilson, G. S. (1936) *Vet. Rec.* **xvi**, No. 16, 494; (1942) *The Pasteurization of Milk*. Edward Arnold, London.
Wilson, G. S., Twigg, R. S., Wright, R. C., Hendry, C. B., Cowell, M. P. and Maier, I. (1935) *Spec. Rep. Ser. med. Res. Coun., Lond.* No. 206.
Wright, R. C. and Tramer, J. (1954) *J. Dairy Res.* **21**, 37.
Zardi, O., Giorgi, G., Carenza, L., Camilli, A. and Deragna, S. (1968) *Boll. Ist. sieroter. Milano* **47**, 547.

10

The normal immune system
Heather M. Dick, P. Wilkinson, S. Powis

Immunity	296	Specific immunological responses	305
Specific and non-specific immunity	297	Lymphocytes and other mononuclear cells	305
Antibody versus cell-mediated responses	297	Lymphocyte recirculation	305
Phagocytic cells	297	The origin of T and B lymphocytes	307
Types of cell	297	B cells (lymphocytes)	308
The neutrophil leucocyte	297	Differentiation of B lymphocytes	308
Formation and fate	297	T cells (lymphocytes)	309
Neutrophils in inflammation	298	Natural killer cells	310
Exit from blood vessels	298	Recognition of antigen	311
Neutrophil locomotion and chemotaxis	298	Surface receptors on macrophages	311
Chemotactic factors	299	T-dependent and T-independent antigens	311
Phagocytosis	300	Cell cooperation: cell-mediated immunity	311
Microbicidal activity	301	T-cell suppression	312
Eosinophils	302	The major histocompatibility complex (MHC)	312
Mononuclear phagocytes: monocytes, macrophages and tissue histiocytes	302	MHC and the recognition of antigen	314
		MHC Restriction	314
Macrophage functions in clearance and inflammation	303	The Thymus	315
		Soluble factors and T-cell activation	315
Accessory or 'antigen-presenting' cells	304	Lymph nodes and spleen	316

Immunity

The importance of a specific response to infection was recognized by early observers in the immunity conferred on individuals who recovered from an attack of infectious disease. The study of the physiological processes by which such protection is brought about forms the basis of the subject of modern immunology. By definition, 'immunity' refers to freedom from infection (literally, exempt or secure from taxation, contagion or poison-OED). In the study of the means by which such a state is developed, modern immunologists have tended to make experimental use of animals other than man (mice, rabbits and chickens, for example), both for reasons of convenience and ethics; many of the substances used to stimulate the immunological response are far removed from the agents of infectious disease, but the results of the carefully designed experiments are relevant to our understanding of human responses and have increased our knowledge of both normal and abnormal processes in disease.

Metchnikoff (1899) recognized the potentially protective effects which might be the end result of the activities of the mobile or wandering cells which he observed in simple protozoans and aquatic species. Subsequently, certain basic observations were made, including the phenomenon of phagocytosis or engulfment of foreign particulate matter, including bacteria, and the destruction or survival and multiplication of the engulfed material. The characteristics of phagocytic cells and the distribution in the tissues of many species were described.

Soluble factors present in fresh blood or serum were also discovered—'antibodies' which could neutralize the effects of noxious substances or microbes (see Ehrlich 1900). The importance of antibodies to the survival from infection was separated in the early work from the cellular events which had been noted. Thus, understanding of the processes by which immunity was developed was clouded by the controversy which developed over the relative importance of cellular (phagocytic) and humoral (antibody-mediated) events. Lengthy and often bitter arguments ensued between the proponents of each mechanism. In retrospect we may deplore the futility of the arguments which raged. Modern immunologists realize the essential contribution made by both cellular and humoral products to the effective response to infection and can recognize specific disorders which develop when one or more components of the system becomes deficient or abnormal (Chapter 17).

Specific and non-specific immunity

The initial dichotomy of thought reflects the dual nature of the immunological response. There is some merit in acknowledging the existence of two phases to the response: first, to recognize and 'capture' foreign material, e.g. bacteria, immobilize it and present it, secondly, to the effector system which may then use the appropriate means to neutralize or remove the offending substance. The first phase has often been described as 'non-specific', implying a generalized and undefined role, contrasting it with a 'specific' response, tailor-made for the actual organism, toxin or other foreign product. It should be understood that there are close links between both phases of the response and major interactions between the cells and soluble factors of both components. Instead of assuming separate functions for each, it is best to see one as the essential adjunct of the other in the development of an effective response.

Antibody versus cell-mediated responses

It has commonly been the practice to describe the specific immunological response to a given foreign antigen as if it were composed of two separate parts, one mediated by antibody, as in toxin neutralization or bacterial killing and the other a cell-mediated response, as in the elimination of viruses or the response to *M. tuberculosis*. It is wrong to separate the events which take place after antigen challenge in this artificial fashion. It is now obvious that most responses involve both types of reaction; the relative contribution of each to the successful elimination of antigen or infection may vary, but mainly as a reflection of the type of antigenic challenge rather than as the result of separation of two kinds of response. The separation of functional properties is initially useful in developing a simple explanation of how specific responses might be induced but some elaboration is required when considering the complex system of recognition and response which can be observed after procedures such as organ transplantation or immunosuppression by cytotoxic drugs or irradiation.

Phagocytic cells

Types of cell

The phagocytic cells of mammalian blood and tissues are divided into two classes, the *myeloid cells* and the *mononuclear phagocytes*. Mature and functioning forms of myeloid cells are known as granulocytes because of their prominent cytoplasmic granules. The two major variants are the *neutrophil granulocyte* (*neutrophil leucocyte*) and the *eosinophil granulocyte* (*eosinophil leucocyte*). Although these cells resemble one another morphologically, they are probably each derived from different committed bone marrow stem cells. Neutrophils are also often known as polymorphonuclear leucocytes (usually abbreviated to PMN), though, since other leucocytes sometimes also show a lobed nuclear morphology, the term *neutrophil* is probably preferable. The mononuclear phagocyte system includes the blood *monocytes* and their marrow precursors, as well as the *macrophage/histiocyte* series found in many tissues either as a resident population in the steady state or as a newly recruited 'elicited' population in inflamed tissues. It includes a variety of tissue-specific cells including hepatic Kupffer cells, splenic macrophages and osteoclasts *inter alia*. Whether cells such as the Langerhans cells of skin, the interdigitating cells of lymphoid tissue, and other similar cells whose major function seems to be antigen-presentation rather than direct disposal of pathogens should be included in the mononuclear phagocyte system, is debatable. The term *mononuclear phagocyte system* was introduced (van Furth et al. 1972) to replace the older term *reticuloendothelial system* (RES), though the latter term has certainly not gone into desuetude. Histopathologists prefer the term macrophage/histiocyte series. In this section, the formation, functions and fate of each of the cell-types mentioned above will be outlined.

The neutrophil leucocyte

Formation and fate

The neutrophil is the characteristic cell of acute inflammation and of acute infections. General reviews of neutrophil function include those of Klebanoff and Clark (1978) and of Murphy (1976). Neutrophils can be mobilized rapidly and can reach inflamed sites quickly and in large numbers. In contrast, mononuclear phagocytes are mobilized more slowly, but, since these are long-lived cells, they persist longer in the inflamed site and are thus most typically seen in chronic inflammation.

In the adult animal, neutrophils originate from the bone marrow. It is probable that there are pluripotential stem cells from which all the blood cells originate. A later stage of development is represented by committed stem cells which in *in vitro* culture form colonies containing mature neutrophils and mononuclear phagocytes, but not other cell-types. Still later stages (myeloblasts and myelocytes) can be identified by histological procedures as granulocyte precursors in normal marrow. During these stages, granulocytes begin to acquire some of the properties of the mature cell, i.e. cytoplasmic granules, cell-surface receptors for chemotaxis and phagocytosis, and deformability and motility. At the metamyelocyte and later stages, neutrophils are able to leave the bone marrow. Egress from marrow requires active deformation, and locomotion of the neutrophil through pores in the cytoplasm of the endothelial cells that line the lumen of the marrow sinuses (De Bruyn 1981). This migration is probably regulated by chemical signals that control the rate at which specific cell-types leave the marrow. In the case of neutrophils under steady-state conditions, there is a considerable reserve of mature cells which remain in the marrow and which can be mobilized when required. In man this reserve comprises about 2×10^9 granulocytes/kg body weight and is larger than the total blood neutrophil population (7×10^8 cells/kg body weight).

Neutrophils circulate in the blood for a brief period, possibly only a few hours, though estimates vary. They leave the blood stream rapidly, but their subsequent fate is uncertain. Many may migrate into the gut. For example, it has been estimated that about 6 per cent of the circulating neutrophils enter the oral cavity per day. Neutrophils are short-lived end cells with no capacity for replication and little or no capacity for protein synthesis, so their turnover must be very rapid—in man, about 10^{11} cells are formed and destroyed each day.

Despite this huge turnover, the blood level of neutrophils remains remarkably steady in normal persons, but it can be altered very rapidly to produce a neutrophil leucocytosis, as is characteristically seen in acute inflammations. In the blood at any time, about half of the neutrophils are circulating, and the other half are marginated on endothelia, especially in the pulmonary capillary bed. The marginated cells can be mobilized rapidly to rejoin the circulating pool by stimuli as simple as adrenaline or exercise. Likewise, a neutropenia can be induced by increasing the proportion of marginated cells. Circulating bacterial endotoxin may do this, and transient neutropenias are seen in patients undergoing renal dialysis (Craddock *et al.* 1977*a*), since the dialysis material activates complement to release C5a into the circulation. C5a causes a transient increase in neutrophil adhesiveness resulting in greater margination (Craddock *et al.* 1977*b*). As well as the marginated pool, the marrow reserve of mature granulocytes can also be called upon rapidly in inflammatory states. A slower mechanism for leucocytosis results from an increased rate of differentiation of marrow precursors, either by shortening of the cell cycle time or by increase in the number of cell divisions.

Neutrophils in inflammation

Exit from blood vessels

As mentioned above, about half of the blood neutrophils are in the 'marginated pool' at any one time. If one observes blood flowing through the microcirculation, for example in hamster cheek-pouch preparations, one can see neutrophils becoming attached to vascular endothelium, then detaching again and rejoining the circulation. This is most easily seen in post-capillary venules, in which blood flow is slowest. Introduction of an inflammatory stimulus to nearby tissues dramatically increases the number of cells sticking to endothelium. We do not really understand what has occurred to cause this change. The subject is reviewed by Lackie and Smith (1980). Either the neutrophils themselves may have become more sticky, as in the example mentioned above where intravascular C5a increased neutrophil margination or the vascular endothelial cells, or both, may have become more sticky. Possibilities include enhanced adhesiveness of endothelium generated by chemotactic factors, or shut-down of release by endothelial cells of a substance that normally prevents leucocytes adhering to them (prostacyclin has been suggested as a possibility).

Once leucocytes are marginated on endothelium, it is quite possible that their egress from the vessels is determined by chemotactic factors diffusing from the extravascular tissues. *In vitro* leucocytes overlaid on monolayers of endothelium, or of related tissues such as renal glomerular epithelium, can penetrate the monolayer in response to a chemotactic factor, as can be shown by culturing such monolayers on the surface of micropore filters, then allowing a chemotactic factor to diffuse from below the filter (Cramer *et al.* 1980). Locomotion, both *in vitro* and *in vivo*, is through the junctional spaces between endothelial cells. *In vivo* the leucocytes must then penetrate the vascular basement membrane to reach the extravascular tissue. At this stage, it is reasonably certain that subsequent locomotion is chemotactic, though this has been much easier to study *in vitro* than *in vivo*.

Neutrophil locomotion and chemotaxis

Neutrophils move *in vitro* at speeds of about 10–25 μm per minute. On planar surfaces they move by crawling along the surface, and require to be adherent enough to it to gain traction. Typically the moving cell shows a polarized morphology (Fig. 10.1) with a ruffled,

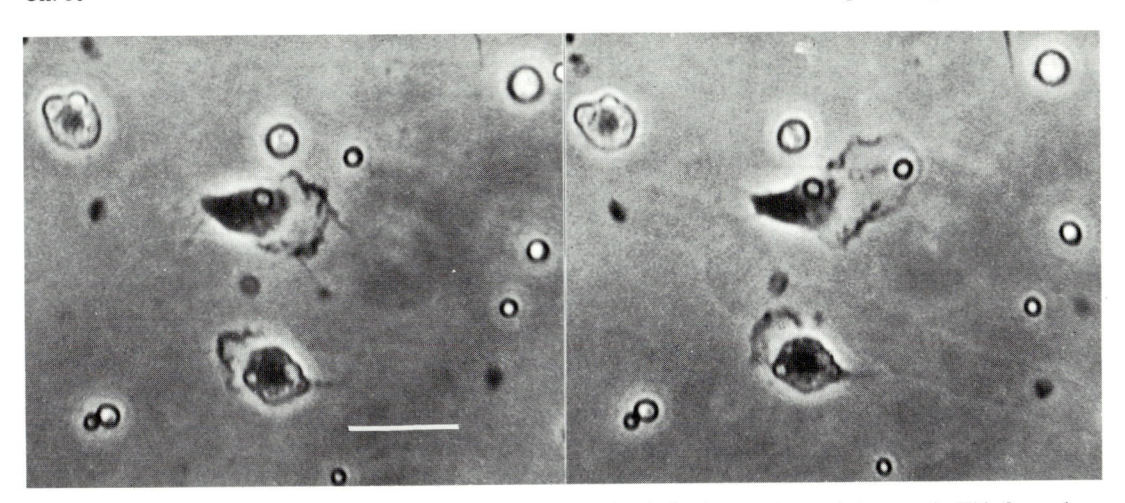

Fig. 10.1 Neutrophils and macrophages show similar morphologies during locomotion and chemotaxis. This figure shows mouse peritoneal macrophages in locomotor morphology; the two pictures were taken 2 minutes apart. The upper cell is moving towards, and ingests, a spore of *Candida albicans*. Both moving cells already contain ingested candida spores. Note the broad ruffled anterior lamellipodium, and the phase-dark, tapered, organelle-rich cell body behind it. A posterior retraction fibre can just be made out in the lower cell. (Bar = 20 μm.) From Wilkinson (1982b).

hyaline, anterior veil (or lamellipodium) rich in cytoplasmic actin and myosin, but excluding organelles. Behind this is the cell body, which is typically tapered, and which contains the organelles and nucleus. There may be a narrow posterior tail, or uropod, with retraction fibres. In three-dimensional matrices such as collagenous tissues, cells may be able to use the matrix as a 'climbing frame' and there may be a less strict requirement for adhesion. Also, in 3-D tissues, the cells may show considerable variation in shape as they move forward.

The speed at which neutrophils move can be modified by various environmental factors. Reactions by which cell speed (or rate of turn) is modified by the environment are called *kineses* (Keller *et al.* 1977). When the kinesis is due to a chemical substance, it is called a *chemokinesis*.

Movement through the tissues towards sites of inflammation and infection is almost certainly directed by the reaction named *chemotaxis* (Keller *et al.* 1977). In this, cell locomotion is oriented relative to the axis of a chemical gradient, the result being an unidirectional locomotion of cells. When the leucocytes are observed during their response to the gradient, they can be seen to assume an oriented morphology such that the anterior lamellipodia face the gradient source (Zigmond 1974, 1977). This morphological orientation may take place before any movement is evident, and the cells then move toward the gradient source, where they accumulate. Chemotaxis is at present under intensive study in many laboratories and there are a number of recent reviews (Gallin and Quie 1978; Wilkinson and Lackie 1981; Wilkinson 1982b). This reaction obviously requires that the cells possess a sensory detector system to monitor differences in concentration of attractant chemicals in their environment. It has been estimated that a neutrophil can detect a difference in attractant concentration of 1 per cent across its own length (Zigmond 1977). The sensor must be connected to the motor apparatus of the cell, i.e. the actin-myosin microfilament system, by transduction mechanisms that are not fully understood but are under study.

Chemotactic factors

A large series of chemotactic factors for neutrophils and other leucocytes have been documented and a list is given in Table 10.1 (see p. 300). Some of these which are of current interest are discussed below.

$C5a$ is the major chemotactic factor generated by complement activation. It is a 74-residue glycosylated peptide cloven from C5 by enzymes with trypsin-like activity, including C5 convertase. The C-terminal arginine of C5a may itself be cloven, by a carboxypeptidase B (anaphylatoxin inactivator) present in normal serum, to form $C5a_{des\ Arg}$. C5a has biological activity both as chemotactic factor and as anaphylatoxin. $C5a_{des\ Arg}$ is inactive as anaphylatoxin but retains chemotactic activity, though somewhat weaker than C5a. The chemotactic activity of $C5a_{des\ Arg}$ may be enhanced by a 'helper factor' peptide found in normal serum. Specific receptors for C5a have been demonstrated on the surfaces of neutrophils (Chenoweth and Hugli 1980). The importance of C5a is almost certainly that it is generated when complement is activated at the surfaces of micro-organisms or damaged cells, and is released to form a gradient which is detected by nearby leucocytes. This allows leucocytes to reach and ingest pathogens, even when these organ-

Table 10.1 Some chemotactic factors for leucocytes

Source	Factor
Complement activation	C5a
Cell-derived	Leukotriene B$_4$
lymphocytes	Lymphokines
mast cells	Eosinophil chemotactic factors
tumour cells	Macrophage chemotactic factors
neutrophils	Various
macrophages	Various
Tissue damage	Denatured proteins and other factors
Micro-organisms	
Bacteria (putative)	Formyl-methionyl peptides
	Lipids (undefined)
Viruses	No directly-acting factors, but virus-infected cells release chemotactic factors.
Products of specific immune reactions	Lymphocyte migration to antigen or ? to antigen-primed cells
	Release products of neutrophils in contact with immune complexes attract other neutrophils.

isms are themselves producing no chemotactic factors. For example, chemotaxis to *Staphylococcus aureus* occurs largely by a complement-dependent mechanism (Russell *et al.* 1976). Patients with C5 deficiency suffer from severe pyogenic infections (Leddy *et al.* 1978; Snyderman *et al.* 1979).

Formyl methionyl peptides are strongly active chemotactic factors for which receptors have been demonstrated on neutrophils (Schiffmann and Gallin 1979). The type-peptide is formyl-Met-Leu-Phe. These are synthetic peptides, though it has been conjectured that they may be analogues of bacterial products, since *Escherichia coli* uses formyl-methionine as a starting sequence for protein synthesis, whereas protein synthesis on eukaryotic cell ribosomes does not take this pathway. Thus neutrophils may use recognition of similar prokaryote-specific peptides as a mechanism for specifically homing in on to bacterial cells.

Leukotriene B$_4$ is a lipoxygenase-derived product of arachidonic acid which is released by neutrophils and other cells upon activation by a variety of stimuli. It has recently been shown to be chemotactic for neutrophils and eosinophils. Its physiological role is unknown but release of chemotactic products by cells in inflammatory lesions may provide a mechanism for amplifying the influx of leucocytes into such lesions.

A variety of *denatured proteins* have been shown to be chemotactic for neutrophils, though not as strongly so as the factors quoted above. Recognition of these proteins may represent a mechanism for non-immune clearance of altered body constituents which may not require a highly specific recognition system (Wilkinson 1978).

Lymphokine chemotactic factors may play a role in attracting leucocytes into sites of immune inflammation (Altman 1978), though more work is required to confirm this. The best studied is the macrophage chemotactic factor, though other lymphokines that attract neutrophils and eosinophils have also been reported.

These and other chemotactic factors have multiple effects on neutrophil function. At low concentrations, such as the cell encounters when it first enters the lesion, their main effect is chemotactic. However, at higher concentrations, such as the cell might encounter in the centre of the lesion, they stimulate granule release, increase adhesiveness and cause a metabolic burst in neutrophils (Gallin, Wright and Schiffmann 1978). These effects may enhance the microbicidal function of the cells.

Phagocytosis

The phagocytic event can be dissociated into two phases, *attachment* and *ingestion* (Rabinovitch 1967). Attachment requires recognition of the particle at the plasma membrane of the phagocyte, and can take place in the presence of metabolic inhibitors. Ingestion requires the action of actin-myosin microfilament networks and thus requires metabolic events.

Phagocytic recognition can be divided into two broad categories. The more primitive of these phylogenetically is *non-immune recognition*. Gelatin-stabilized carbon particles are cleared as rapidly by coelomocytes in the sea urchin as they are by neutrophils or macrophages in mammals. There is a widely diverse collection of particles that may be phagocytosed in the absence of antibody or complement, of which polystyrene latex beads are probably in the widest experimental use. It is difficult to visualize specific receptors for such objects, and this is a poorly understood, but probably non-specific, form of phagocytic recognition. A more physiological example of non-immune phagocytosis is the recognition and ingestion of physically or chemically altered red cells—glutaraldehyde-treated red cells are frequently used experimentally—which may be analogous to the physiological mechanism for clearance of old red cells from the body. Furthermore, many bacteria, particularly those of low virulence, may be cleared by non-immune phagocytic recognition. However, many of the more successful pathogens have evolved mechanisms for avoiding non-immune phagocytosis. For example, it has frequently been observed that capsulated bacteria resist phagocytosis, whereas non-capsulated variants are readily ingested. The pneumococcus is the classical example of this. To deal successfully with these organisms, the phagocytic cells require to use *immune phagocytosis*.

Neutrophils, mononuclear phagocytes and eosinophils possess plasma membrane receptors for the Fc region of IgG (the Fc receptors) and for the activated complement component, C3b. During the course of an immune response to a micro-organism or other particle, the particle becomes coated with antibody, bound to surface antigens of the particle, so that its Fc region is free, and with C3b. This process was named *opsonization* by Sir Almroth Wright (Wright and Douglas 1903, 1904). The particle-bound IgG and C3b can then interact with the appropriate receptors on the surface of the phagocytic cell and the opsonized particle thus becomes attached to the phagocyte. This rapidly activates the ingestion process and the microbicidal pathways of the phagocyte. It is probable that the majority of pyogenic bacteria are disposed of by opsonic phagocytosis. The Fc and C3b receptors of phagocytic cells are probably more complex than was initially realized. For example, on mouse macrophages there are different Fc receptors for different IgG subclasses (Unkeless 1977). In man, IgG1 and IgG3 are the major opsonizing immunoglobulins, and it is possible that phagocytic cells bear different receptors for each.

Particle ingestion results from the formation of a cup-like process around the particle; the latter eventually becomes fully enclosed by fusion of the edges of the advancing process. It has been suggested, with some firm experimental evidence, that this is a 'zippering' process (Silverstein *et al.* 1977) in which there is sequential circumferential attachment of particle-bound ligand molecules to phagocytic plasma-membrane receptors, so that the advancing edge of the 'cup' remains in close apposition to the particle as ingestion proceeds. As the phagocyte encircles the particle, microfilamentous actin networks are formed actively in its advancing tip, the cytoplasm of which is probably gelated, while further back the networks are broken down and the cytoplasm is isolated. This breakdown is probably necessary to allow the next step, once ingestion is completed. The 'phagosome'—the vesicle containing the ingested particle—is drawn through the cytoplasm towards the Golgi region, where fusion of cytoplasmic granules with the phagosome takes place.

Neutrophils contain two major classes of granule, the azurophil granules which are similar to lysosomes and contain a wide variety of acid hydrolases, and the specific granules, which contain lactoferrin, lysozyme, a vitamin B_{12} binding protein, collagenase and other neutral proteases. Release of these granules is sequential. First the specific granules fuse with the phagosome, then the azurophil granules (Bainton 1973). Granule contents may be released, not only into phagosomes, but also to the exterior by fusion of granule membranes with the plasma membrane. This happens when the neutrophil is activated by contact with a surface which is too large to be phagocytosed. For example, neutrophils on immune-complex-coated surfaces flatten on the surface, bound to it by their Fc receptors, and release enzymes extracellularly (Henson 1971*a*, *b*); this may contribute to immune-complex mediated tissue damage, but granular enzyme release may also have a protective function as discussed later in reference to eosinophils.

Microbicidal activity

Phagocytic cells probably have several mechanisms for killing bacteria and parasites. These include (a) oxidative microbicidal systems, (b) cationic proteins, (c) membrane-damaging enzymes including phospholipases. Of these, the first is probably the most important in the neutrophil. Normal neutrophils which are allowed to ingest a bacterium such as *Staphylococcus aureus* kill the majority of the bacteria within 20 minutes or so. In patients in whom a defect of the oxidative microbicidal system can be detected, for example, children with chronic granulomatous disease, the bacteria are killed much more slowly and the children suffer from severe recurrent infections. Contact with phagocytosable particles or with chemotactic factors activates a rapid metabolic burst in neutrophils and monocytes with increased hexose monophosphate shunt activity and increased generation of superoxide anion (O_2') and hydrogen peroxide. These latter molecules are believed to play an important role in the microbicidal pathway (Babior 1980; Roos and Balm 1980). H_2O_2, in the presence of myeloperoxidase released from azurophil granules into the phagosome, and halide, forms a rapidly acting microbicidal system (Klebanoff 1975, 1980).

Neutrophil granules contain cationic proteins with chymotrypsin-like activity, and other cationic, but non-enzymic, proteins. These have been shown to kill *Esch. coli in vitro* (Elsbach and Weiss 1981), and phospholipases may also contribute by disorganizing bacterial membranes. However, the importance of these non-oxidative mechanisms has not been fully evaluated. Once the intracellular bacteria have been killed, degradation can proceed more slowly, and it is probable that the armamentarium of hydrolases released into the phagosome from azurophil granules is instrumental in this process.

The importance of normal neutrophil function is emphasized by studies of patients in whom this function is defective. Leucocyte mobilization may be defective owing to defects of locomotion and chemotaxis because of an intrinsic defect in the cells themselves, or because of a complement deficiency, or because the serum contains inhibitors that depress either cell locomotion or the generation of chemotactic factors from serum. The other defects of neutrophil function that have been studied in detail are those of microbicidal activity, including chronic granulomatous disease, referred to above, and a genetic deficiency in which the neutrophils lack myeloperoxidase. The reader is referred to recent reviews (Gallin 1981; Clark 1978; Klebanoff and Clark 1978) for further discussion of these defects.

Eosinophils

There are recent reviews of eosinophil function by Beeson and Bass (1977) and Weller and Goetzl (1979). Though resembling neutrophils morphologically and, to some extent functionally, eosinophils are probably derived from a separate lineage of bone marrow precursors, and perform specialized functions in two major immunological reactions, namely immediate hypersensitivity and the response to metazoan parasites. A rise in the blood eosinophil count may be seen in both conditions and, in one experimental parasitic infection, that of rats with *Trichinella*, the eosinophilia has been shown to be T-cell dependent and may be part of a cell-mediated immune reaction (Basten and Beeson 1970). Primed lymphocytes in experimental metazoan infections release a factor which is chemotactic for eosinophils. In immediate hypersensitivity, however, there is no evidence that eosinophils are recruited by lymphocytes. Much more emphasis has been placed on products released from mast cells. Various mast-cell-derived eosinophil chemotactic factors have been described, including peptides, and, more recently, lipid derivatives of arachidonic acid, possibly leukotrienes. These are released when mast cells bearing cell-bound IgE come into contact with specific antigen. It is still not clear what role, if any, eosinophils play in immediate hypersensitivity reactions. It has been suggested that they are repair cells, phagocytosing mast cell granules, and neutralizing mast cell products, e.g. eosinophils release a histaminase which inactivates histamine.

The most obvious morphological feature of the eosinophil is the presence of prominent cytoplasmic eosinophilic granules—the specific granules. These granules contain large quantities of cationic proteins, the major one being present in the crystalloid bar, typically seen in transmission electron micrographs. These proteins are non-enzymic with the exception of a cationic lysophospholipase (Charcot-Leyden crystal protein). Eosinophil granules also contain hydrolases, though in smaller quantities than in neutrophils. There is a peroxidase, non-identical with the myeloperoxidase of neutrophils. A second series of small granules contains enzymes such as aryl sulphatase.

Eosinophils are phagocytic cells, but not as effectively so as neutrophils. Recent work suggests that their primary function may be secretion and extracellular, rather than intracellular, killing. This suggestion derives from experimental studies of the eosinophil response to the schistosomula stage of *Schistosoma mansoni* (Butterworth *et al.* 1979; Glauert *et al.* 1978). IgG antibody to schistosomal coats the parasite and acts as an opsonin. Eosinophils become bound by their Fc receptors to the parasite-bound IgG, and flatten on the parasite surface. They then release their granule contents into the space between the eosinophil and the parasite. The parasite is killed by co-operation between many eosinophils on its surface, which are thus able to deal with a pathogen much larger than themselves and which cannot be phagocytosed. It is thought that secreted cationic proteins play a major role in this killing. Other phagocytic cells, such as neutrophils or macrophages, may perhaps respond to other large parasites in a similar way, by opsonic binding followed by secretion of toxic products. However, it is still not certain that the various *in vitro* models for parasite killing reflect accurately what happens in immune responses to the same parasites *in vivo*.

Mononuclear phagocytes: monocytes, macrophages and tissue histiocytes

It was customary in the past to divide the mononuclear phagocyte system into two categories, the 'fixed' and the 'free' macrophages. The former were regarded as permanent features of the architecture of various tissues. The Kupffer cells of the liver and the splenic and lymph node macrophages were important examples. In contrast to these were motile tissue macrophages or 'wandering histiocytes' found in many tissues. However, it is by no means clear that 'fixed' macrophages are static populations; for example, the Kupffer cell population can increase considerably after challenge with a macrophage activator such as *Corynebacterium parvum*. Nowadays, it is more common to use the term *resident macrophage* or histiocyte to refer to those cells present in a tissue without any evident stimulus, and *elicited macrophage* to refer to the macrophages which are recruited into a tissue after an eliciting stimulus. For example, the peritoneal cavity of the mouse normally contains a population of resident macrophages whose biological activity, e.g. motility, secretion, microbicidal capacity, is fairly low. After injection of a variety of eliciting agents such as casein, thioglycollate or glycogen *inter alia* into the peritoneal cavity, a new population of inflammatory macrophages is recruited which are more active as eliciting agents than the resident populations.

Similarly, the pulmonary alveoli normally contain a resident population of macrophages; a new population of alveolar macrophages can be recruited by stimuli such as BCG. It is now generally accepted that elicited macrophages are derived from the blood monocytes and thus from bone marrow precursors. Majority opinion has it that resident populations, including Kupffer cells etc., are derived ultimately by the same route from the bone marrow, but there are dissenters from this view.

It is clear from this account that long lived mononuclear phagocyte populations are considerably more diverse than the short lived neutrophils and eosinophils discussed above. After development in the bone marrow, monocytes circulate in the blood for a short time, then leave it for the tissues where they may develop in a variety of ways. The inflammatory macrophages described above are actively chemotactic and phagocytic; these functions are similar to those described earlier in this chapter for neutrophils. Following an immunological stimulus, macrophages may follow a rather different pathway of development. Products of immune reactions cause differentiation to form populations of 'activated' macrophages, with considerably enhanced microbicidal and secretory powers. These cells may themselves release factors called 'monokines' which influence lymphocyte function. The differentiation pathways of macrophage populations have been followed by studying cell-surface markers such as Ia-antigens (see later). Resident and inflammatory macrophage populations contain a majority of Ia-negative cells. However, macrophage populations elicited by immunologically-active stimuli such as BCG or *Corynebacterium parvum* contain a high proportion of Ia-positive cells (Unanue 1981). Ia positivity or negativity has come to be seen as occupying a central position as a determinant of macrophage function in immune responses. Ia-positive macrophages or macrophage-like 'accessory' cells which have taken up antigen interact physically with helper T cells; this interaction is essential for T-dependent immune responses to the antigen. Ia-negative cells cannot perform this function. There is some debate whether these 'accessory' cells are part of the mononuclear phagocyte system at all, since they may be non-functional as phagocytes, and may constitute an independent antigen-presenting population.

Macrophage functions in clearance and inflammation

Mononuclear phagocytes play an important role in clearance of particles or of damaged cells from the body. The Kupffer cells of the liver and the macrophages that line the splenic sinusoids are well placed to act as filters of the blood stream. Intravenous injection of fluorescein-labelled bacteria or of radiolabelled denatured proteins is followed by rapid and effective clearance of these materials by hepatic macrophages. Aged red cells are removed from the circulation by macrophages, chiefly in the spleen, and uptake of abnormal or antibody-coated red cells by splenic or hepatic macrophages is much accelerated in haemolytic anaemias. Under normal conditions pulmonary alveolar macrophages probably remove large quantities of the surfactant that lines the alveolar spaces.

The recruitment of macrophages into inflammatory sites probably proceeds by mechanisms similar to those described earlier for neutrophils. Blood monocytes leave the blood stream by passing through gaps between vascular endothelial cells and are attracted to the site of the lesion by chemotaxis. This recruitment is much slower than that of neutrophils, owing not to slower locomotion of the monocytes but probably because there are fewer monocytes in the blood stream than neutrophils, and there is no marrow reserve of mature monocytes which can be released rapidly, as there is of neutrophils. Monocytes respond to many of the same chemotactic factors as neutrophils, including most of those listed earlier. However the lymphokine chemotactic factors that attract monocytes are reported to be different from those that attract other blood cells. Little is known about defects of monocyte chemotaxis in clinical states. However, one important defect has been documented in animals bearing experimental tumours. These tumours release factors that inhibit locomotion of monocytes and macrophages (Snyderman and Pike 1976; Normann and Sorkin 1977; Otu *et al.* 1977). Macrophages are believed to play an important role in preventing tumour growth by infiltrating the tumour and killing tumour cells; this defect may prevent access of macrophages to the tumour and thus may be of considerable importance. Studies of patients with malignant melanoma and breast carcinoma suggest that similar inhibitors are released by human tumours (Snyderman *et al.* 1977).

Many changes in function are seen in macrophages elicited by inflammatory stimuli compared with resident macrophages or blood monocytes (Cohn 1978). Elicited macrophages have a larger surface area of plasma membrane, which is extensively folded, and show more ruffling and spreading than resident macrophages. One of the most obvious morphological features of macrophages is pinocytosis. Plasma membrane is rapidly internalized to form intracellular vacuoles and membrane may be recycled by re-fusion of these vacuoles with the plasma membrane, a process known as the 'membrane shuttle' (Silverstein *et al.* 1977). This process is accelerated by inflammatory stimuli, which also induce numerous biochemical changes including increased synthesis and secretion of products such as proteases, collagenase, elastase, plasminogen activator, complement components, prostaglandins and leukotrienes. Superoxide generation is increased. There is an enhanced capacity for C3b receptor-mediated phagocytosis compared to resident cells. Many of these changes can be inhibited by corticosteroids.

Macrophages, like neutrophils, can kill many of the pyogenic bacteria and probably use similar oxidative microbicidal systems. However, there are a number of pathogens, both bacterial and protozoal, that can survive inside macrophages and are able to evade these microbicidal systems. These include *Mycobacterium tuberculosis*, *Listeria monocytogenes*, *Bordetella pertussis*, *Legionella pneumophila*, *Leishmania* spp, *Trypanosoma cruzi* and *Toxoplasma* spp. (see reviews in van Furth 1980). Killing of these species by macrophages requires a specific immune response and macrophage activation, a process that is considered below.

Macrophage activation. The importance of immunological activation for microbicidal activity of macrophages was first emphasized by experiments of Mackaness (1962, 1969), who reported the inability of unactivated macrophages to kill listeria or BCG, but who showed that macrophages taken several weeks after infection were able to kill these organisms. This enhanced microbicidal activity was dependent on formation of a cell-mediated immune response and could be transferred to control animals with T lymphocytes, but not with serum. The microbicidal effect itself was not specific, since activated macrophages could kill not only the organism originally injected, but also other organisms. These studies have since been extended to many other systems. Lymphokine 'macrophage activating factors' released by effector T lymphocytes after antigen-driven clonal expansion have been identified. Activated macrophages show many of the biochemical changes that inflammatory macrophages show, including increased protein synthesis, increased enzyme secretion, etc. They release hydrogen peroxide in much greater quantities than inflammatory macrophages; it is possible that this high level of H_2O_2 is necessary to kill organisms that are resistant to the microbicidal activity of unactivated macrophages.

The concept of macrophage activation has now extended considerably beyond the response to pathogenic micro-organisms. One of the most active areas of research has been in tumour cell killing, where, again, macrophages are effective killers of many tumour cell-types but, to perform this function, require to be activated by products of immune reactions (Keller 1980; Hibbs *et al.* 1980). The activation pathways for tumoricidal function may not be the same as for microbicidal function, and macrophages efficient in one may not be efficient in the other. Macrophages may also contribute as killer cells in other immunological responses to tissue cells, so that the term 'activated macrophage' has now become confusingly vague. It is necessary to specify which particular function is activated in reference to any particular study of these cells.

Macrophages are prominent in chronic inflammation and it seems likely that inflammation becomes chronic when the inflammatory stimulus is persistent, as is the case where pathogens are able to evade normal cellular microbicidal systems. A simple example of a persistent, but non-antigenic, stimulus is the foreign body granuloma in which an indigestible object becomes surrounded by macrophages. This is a low-grade inflammation and the macrophages present are probably relatively inactive. In contrast, when the inflammatory agent is both persistent and antigenic, as with the pathogens cited above, a 'high-turnover' granuloma evolves, with influx of both macrophages and lymphocytes. Effector lymphocytes release lymphokines, including macrophage activating factor, and the activated macrophages in turn release factors that increase T and B cell activity, and enhance T cell differentiation in the thymus.

Accessory or 'antigen-presenting' cells. Immune responses to thymus dependent antigens require the presence of antigen-bearing, Ia-positive macrophage-like cells (Unanue 1981). These 'antigen-presenting' or 'accessory' cells can be observed to interact physically with T lymphocytes in culture, and the T cells appear to form rosettes around them. Many of these T lymphocytes undergo blast transformation during a period of culture of two or three days. This is presumably the morphological counterpart of antigen recognition by T lymphocytes. Several different sorts of accessory cell have now been described. In some experimental models, these cells have both phagocytic and degradative function and accessory cell function (Unanue 1981), and the cells concerned appear to be true macrophages. Other accessory cells appear not to function as phagocytes, and there may be intermediate forms. Table 10.2 shows the properties of some of the different types of cell involved.

The 'dendritic' cells described in the spleen by Steinman and colleagues (Steinman and Nussenzweig 1980)—not to be confused with germinal centre dendritic cells, which are probably entirely different—are non-phagocytic, have no Fc or C3b receptors, but are Ia-positive. These are somewhat smaller than macrophages, with several dendritic cytoplasmic processes, many mitochondria and a nucleus that shows pulsatile movements in culture. They are good inducers of mixed leucocyte reactions *in vitro*. Possibly related to these are the skin Langerhans cells, but the latter possess Fc and C3b receptors. These cells take up contact-sensitizing agents. After contact sensitization, similar cells are found in the afferent lymph as 'veiled cells', so-called because of their extensive veil-like extensions. Possibly a later stage is represented by the 'interdigitating cells' found in lymph nodes. These cells are reviewed by Hoefsmit *et al.* (1982). Cells of dendritic morphology are found in many other sites after antigenic stimulation, and are frequently seen surrounded by a cluster of lymphocytes; thus formation

Table 10.2 Properties of mononuclear phagocytes and accessory cells

	Phagocytosis	Chemotaxis	Fc Receptor	C3b Receptor	Ia antigen
Blood monocytes	+ +	+ + +	+	+	Few cells
Resident peritoneal macrophages	+	±	+	+	Few cells
Elicited peritoneal macrophages (thioglycollate)	+ + +	+ +	+	+	Few cells
Elicited peritoneal macrophages (BCG or *C. parvum*)	+	?+	+	+	+ +
Langerhans and veiled cells	Poor	?	+	+	+ +
Splenic 'dendritic cells' (Steinman)	−	?	−	−	+ +

of cells of this type may be characteristic of many types of immune response.

Specific immunological responses

Lymphocytes and other mononuclear cells

Of the many cells concerned in the specific response to foreign antigens, lymphocytes are the most important. They are found in blood, lymph and lymphoid tissues. The lymphocyte is a mononuclear cell, approximately 7–12 μm in diameter, and in its mature state has a rim of cytoplasm and a large nucleus. The major and most obvious source of cells of the lymphoid series are those organs known to contain large numbers of lymphocytes (Table 10.3), but small local collections of lymphocytes are also found at many other sites in the body. In some organs, there are relatively few cells, e.g. as with the dendritic cells of the kidney. Elsewhere, the cells may be scattered and do not always form separate recognizable foci e.g. Langerhans cells in the skin. In association with those cells which are morphologically recognizable as lymphocytes there are many other kinds of cell with important functions in the recognition and handling of foreign antigens (Table 10.4).

Within the lymphoid organs, separate populations of lymphocytes can be recognized, each with well defined functions, detectable by reason of cell surface molecules or 'markers', characteristic of each type of cell. Antibody, i.e. immunoglobulin, production is the function of one lineage (B cells), and the second major population of lymphocytes (T cells) carry out the so-called 'cell-mediated' events of the immune response.

Lymphocyte recirculation

The lymphoid tissues which are ultimately responsible for the responses to antigen, both antibody and cell mediated, may be regarded as being divided into a series of interconnecting 'compartments' (Fig. 10.3). Lymphocytes move from one compartment to another during maturation and after activation. Various cell surface markers and functions are expressed as a result of passage through the separate tissues; the thymus, for example, produces hormones which affect the subsequent expression of surface markers. Each compartment provides an environment, sometimes known as the haemopoietic micro-environment, relevant to its role in the dynamic processes involved in immunological response. Lymphocytes are in constant transit through these micro-environments; their route is determined by their functional properties and by whether or not they have encountered antigen.

The major traffic of lymphocytes occurs between peripheral blood and lymphoid tissue, more particularly the spleen and lymph nodes (Ford 1975). Certain types of lymphocyte move in defined ways, depending on their tissue of origin, and activated lymphocytes, sometimes known as lymphoblasts, are described, which 'home' on specific tissues (Parrott and Ferguson 1974; de Sousa 1981). Thus, cells concerned in the response to antigens in the gut or gut wall tend to move from mesenteric nodes to the lamina propria and Peyer's patches. This 'homing' property is relevant to local immunity and is probably very important in the protection of the alimentary and respiratory tracts from local infection.

Table 10.3 Major lymphoid tissues in man

Spleen
Lymph nodes
Gut associated lymphoid tissue
    Tonsils, adenoids (Waldeyer's ring)
    Peyer's patches, appendix

Table 10.4 Specific immunological responses

Lymphocytes: T cells and subsets
                T_H—helper
                T_C—cytotoxic
                T_S—suppressor
            B cells
            Natural killer cells (NK)
Plasma cells: Mature B cells

Associated cells
Macrophage/monocyte/histiocyte
e.g. Langerhans cells
    Kupffer cells
    Dendritic cells
    Blood monocytes

306 *The normal immune system*

(a)

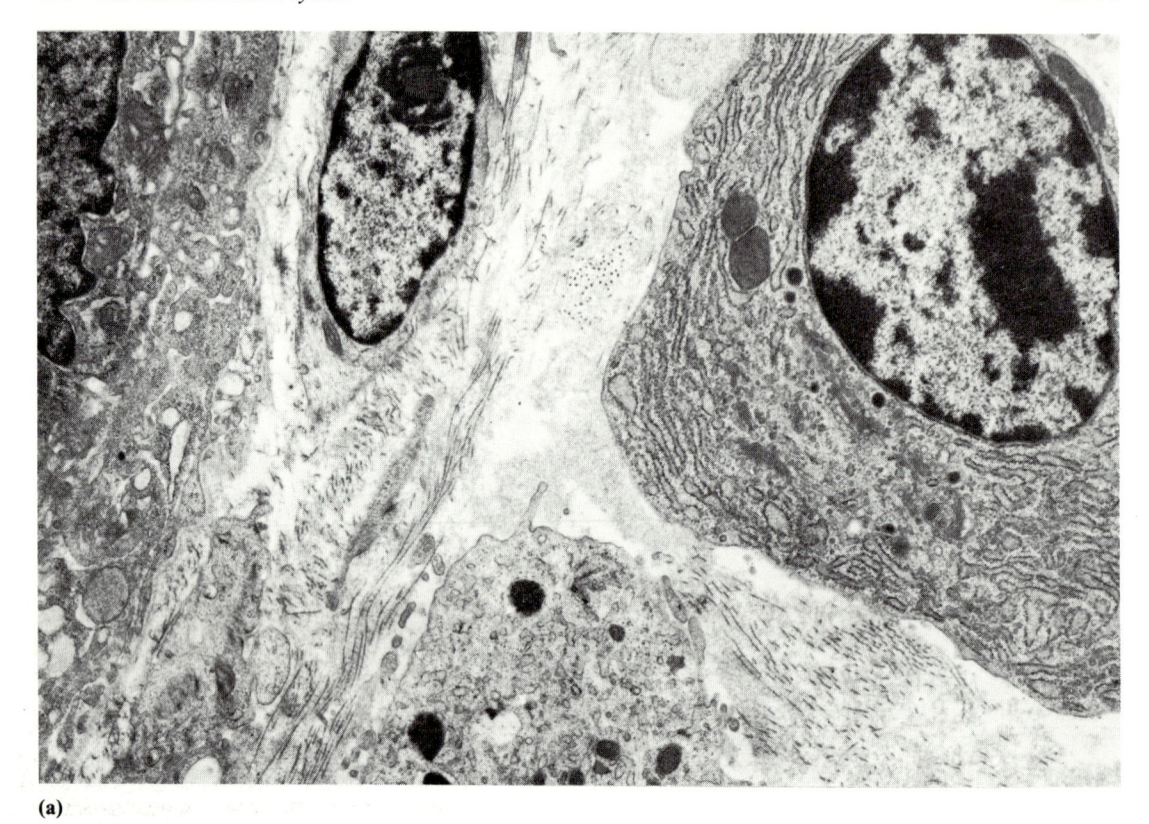

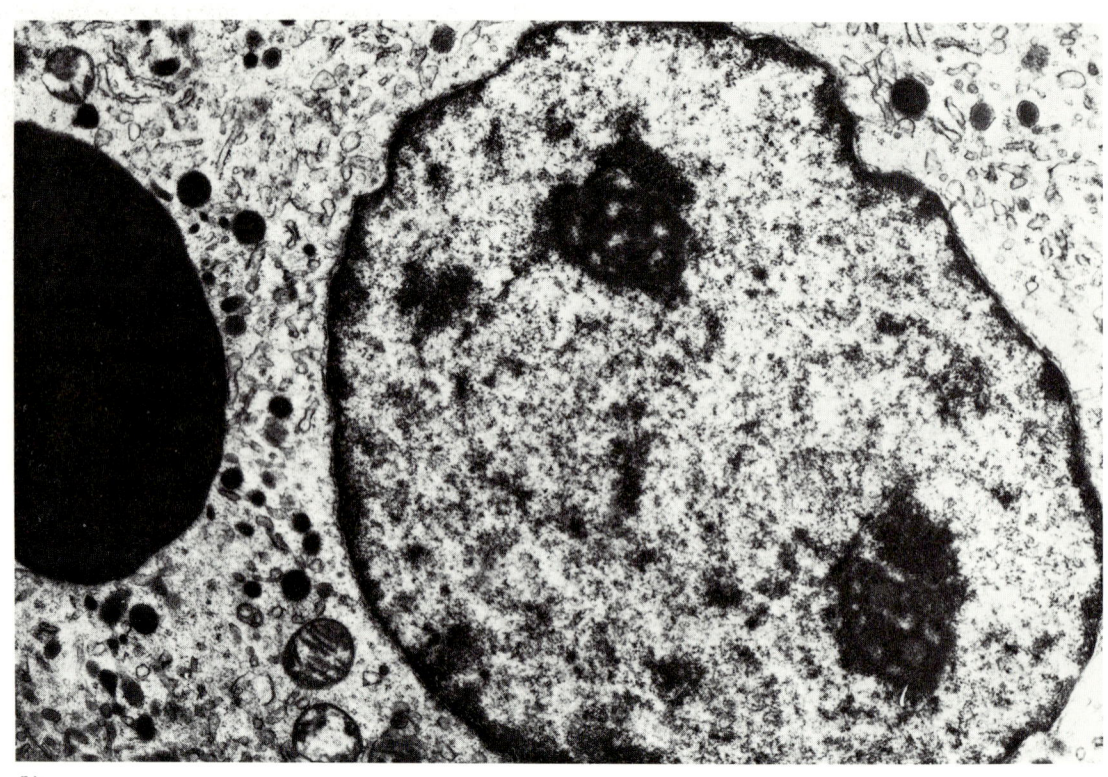

(b)

Fig. 10.2 (a)(b) See caption opposite.

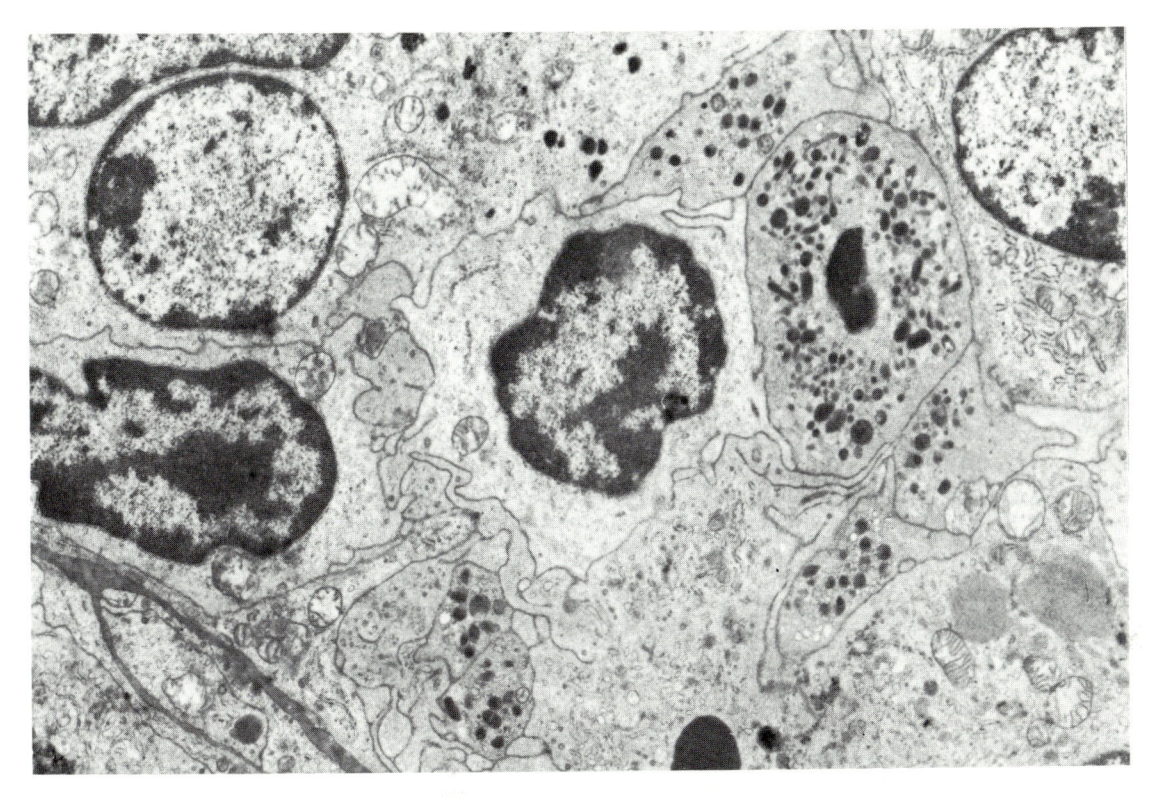

Fig. 10.2(a) Electronmicrograph of plasma cell (R) from small intestinal lamina propria, showing abundant endoplasmic reticulum and nucleus with typical chromatin distribution. The small cell on the left is probably a fibroblast ($\times 11\,500$). Electronmicrographs prepared by Dr A. L. C. McLay, Department of Pathology, Glasgow Royal Infirmary.
(b) Electronmicrograph of macrophage in spleen engulfing red cell (L). Note the numerous lysosomes in the surrounding cytoplasm (c 15 500). Electronmicrographs prepared by Dr A. L. C. McLay, Department of Pathology, Glasgow Royal Infirmary.
(c) Electronmicrograph of lymphocyte (centre) showing paucity of organelles in cytoplasm (compare Fig. 10.2(a) ($\times 12\,500$). Electronmicrographs prepared by Dr A. L. C. McLay, Department of Pathology, Glasgow Royal Infirmary.

The origin of T and B lymphocytes

The existence of different functions for lymphocytes was seen to be related to the micro-environment through which the cells passed or where they were found. Two distinct kinds of lymphocyte, T for thymus-derived and B for bursa or bone-marrow derived, can be distinguished. The properties of T cells arise from their exposure to the micro-environment of the thymus at a critical period in their early maturation. B cells have a different lineage; their role as precursors of the antibody-producing plasma cell was first detected in the chicken.

The thymus is a bilobed organ which arises early in fetal development from the third and fourth pharyngeal pouches and eventually migrates to the anterior mediastinum. It is large during fetal and neonatal life but later diminishes in size. In the early 1960s several observations pointed to an important role for this organ in the functioning of the immune system. Patients with a thymoma, a tumour of the thymus, were observed to develop concomitant immunological disorders such as hypogammaglobulinaemia and auto-immune disease. Moreover, thymectomy of neonatal mice resulted in animals which could not reject skin grafts. Thymectomy of adult mice, however, did not result in any abnormality (Good and Gabrielson 1964).

A small posterior appendage of the avian cloaca, the Bursa of Fabricius, was also found to play a key role in the immune system. Chickens which had been bursectomized during foetal life could not produce antibody, although their ability to reject allografts was retained. Bursectomy of neonatal chickens, however, had no detrimental immunological effect. The mammalian bursal equivalent is thought to be initially the fetal liver and, later, the bone marrow (Cooper et al. 1966).

Subsequent experiments soon clarified these observations. Mice with immune tissues destroyed by

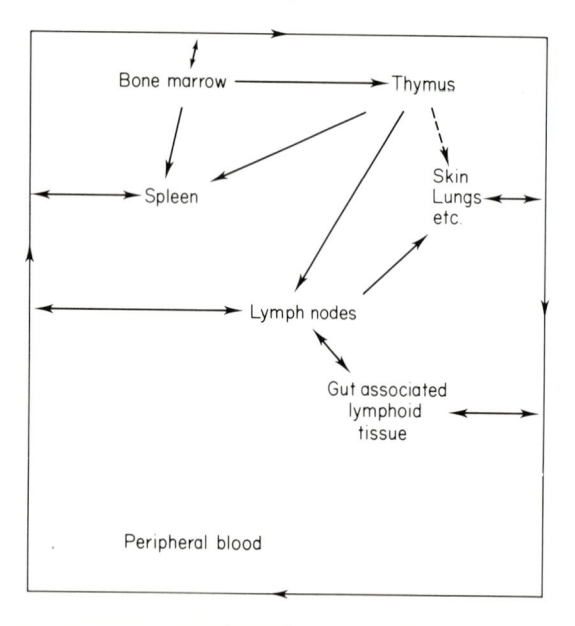

Fig. 10.3 Compartments of lymphocyte circulation.

sublethal irradiation were 'reconstituted' by grafts of bone marrow and of thymus cell suspensions from syngeneic animals. Antibody was not produced when either population was injected alone, but only when both were injected together. This was the first indication that cooperation between cell populations existed within the immune system. Further studies revealed that it was cells of the bone marrow which actively produced antibody and not those of the thymus. Hence lymphocytes which differentiate in the thymus, called T_H cells (T helper), help lymphocytes that differentiate in the bone marrow, called B cells, to produce antibody (Claman et al. 1966).

A third population of cells is necessary for antibody production. Murine spleen cells can be divided into two populations according to their ability to adhere to plastic. In the presence of antigen, antibody is produced only when both adherent and non-adherent cells are cultured together, but not when either is cultured alone. The adherent cell population consists predominantly of macrophages, whereas the non-adherent population consists predominantly of lymphocytes (Mosier 1967).

B cells (lymphocytes)

B lymphocytes are characterized by the presence of cell surface immunoglobulin molecules (sIg) which may be detected by fluorescent-labelled anti-immunoglobulin. This surface immunoglobulin can bind antigen molecules. The binding can be shown to be specific by using radiolabelled antigen (Lefkovits 1974). When the antigen used emits sufficient radioactivity to destroy any cell to which it becomes bound, then specific populations (clones) of B cells are eliminated. B cells possess the capability of producing antibody of the same specificity as the immunoglobulin present on their surface. Sequence analysis has shown that sIg differs from secreted antibody only in a small number of C-terminal aminoacids close to the point where the immunoglobulin molecule is inserted into the cell membrane. In the presence of antigen and with the cooperation of other cell populations, B cells can be triggered through sIg receptors to proliferate and differentiate into plasma cells. The mature plasma cell has an extensive endoplasmic reticulum since its function is to produce large amounts of secreted antibody of the same class and specificity as that of the B cell from which it arose. Plasma cells express little sIg, presumably because the cell is now committed to its final antigen-driven form.

Differentiation of B lymphocytes

The total B cell population in the adult may consist of 10^5–10^7 separate clones, each clone having a unique sIg specificity for antigen. A clone may range in size from 1–10^7 cells, depending on whether the cells have been exposed to antigen and triggered to develop into antibody secreting cells. The majority of B-cell clones are committed, i.e. genetically programmed, to expressing a particular antigen binding receptor before the appearance of sIg.

B cells can be subdivided according to the class of sIg which they express during differentiation (Fig. 10.4). The first recognizable cell in the pathway is the pre-B cell, a large lymphocyte, which contains intracytoplasmic immunoglobulin heavy chains, but lacks sIg. The chains are of the IgM class. During fetal life small numbers of pre-B cells are present in the spleen, liver and bone marrow. In the liver and bone marrow they may be almost as numerous as B cells with sIgM. In adults, pre-B cells are found only in the bone marrow where they constitute less than 1 per cent of nucleated cells (Gathings et al. 1981).

Pre-B cells give rise to small resting lymphocytes which express firstly IgM alone and then sIgM together with sIgD. These cells are particularly sensitive to tolerogenic stimuli and it is probably at this stage of development that reactivity to many self antigens is regulated or eliminated. sIgM$^+$ D$^+$ cells differentiate further to produce sIgG$^+$ and sIgA$^+$ cells of the same antigen specificity. This occurs through a process termed 'heavy chain switching', involving rearrangement of immunoglobulin genes. B cell clones may therefore consist of cells expressing a variety of immunoglobulin classes. During B cell differentiation, mutations may occur in that part of the immunoglobulin gene coding for the antigen binding site.

Hence IgG$^+$ B cells possess a greater range of specificities for antigenic determinants than do IgM$^+$ cells. This mechanism allows for a further expansion of the B cell clonal repertoire (Gearhart *et al.* 1981).

In addition to plasma cells, antigen-triggered B$^+$ cells may also form expanded clones of memory cells which retain their ability to produce further plasma cells or memory cells of the same antigen specificity. Whereas the plasma cell lives for only a few days, memory cells may exist for decades. It is this mechanism which accounts for the phenomenon known as immunological memory; a secondary response to antigen is both quicker and greater than a primary response, because the memory cell clone is stimulated by re-exposure to the antigen and proliferates rapidly to give rise to plasma cells and perhaps a further generation of memory cells.

B cells also possess several cell surface antigens other than immunoglobulin. The *Fc receptor* is specific for the Fc portion of immunoglobulin and its presence is demonstrated by showing that erythrocytes coated with immunoglobulin bind to and form 'rosettes' with Fc$^+$ cells. B cells possess a receptor which binds complexes of erythrocytes, antibody and complement. This receptor is specific for C3b, a cleavage product of the complement pathway (Katz 1977).

T cells (lymphocytes)

Experiments with radiolabelled antigen have demonstrated that T cells possess an antigen-specific receptor. However, this receptor is not immunoglobulin and its molecular structure is at present unknown.

Our knowledge of T cells has mostly been derived from the study of murine differentiation antigens— antigens which are expressed only at certain stages of a cell's development. Many of these exist as alloantigens and can therefore be investigated with alloantisera produced by immunization of inbred mouse strains with allogeneic cells. An inbred strain is produced after 20 generations of brother-sister mating, by which time the resulting offspring are 98.6 per cent genetically homozygous (Festing 1979). Mice of the same inbred strain all possess the same alloantigens on their cell surfaces; thus injection of the cells into another inbred strain possessing a different allelic form of the same antigen will result in the production of a specific alloantiserum, which can be used to examine the cellular distribution of the antigen under investigation.

One such antigen studied in this way is theta (θ) or Thy-1 which exists in two allelic forms, Thy-1.1 and Thy-1.2; 95 per cent of thymic lymphocytes express θ, whereas it is not expressed by bone marrow cells or B cells. Theta has therefore become a definitive murine T cell marker.

Murine T cells can further be subdivided into three populations by use of the alloantigens Lyt 1, Lyt 2 and Lyt 3, each of which exists in two allelic forms. T cells with different functional properties can be distinguished by their phenotype. Thus, T-helper cells (T$_H$), which cooperate or 'help' B cells to produce antibody

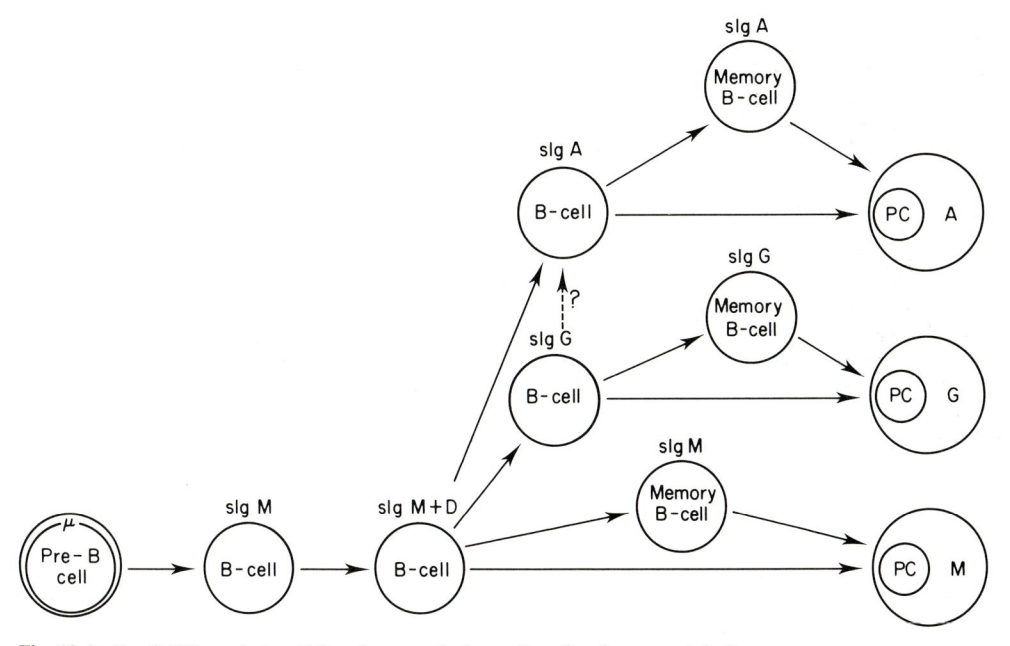

Fig. 10.4 B cell differentiation. PC = plasma cell. sIg = cell surface immunoglobulin.

are recognized as Lyt $1^+2^-3^-$, whereas T-suppressor cells (T_S) which, as the name indicates, have controlling functions, are Lyt $1^-2^+3^+$. A population of cytotoxic T cells (T_C) which are responsible for cell killing events, e.g. elimination of virus infected cells, also exists, and sub-populations with further fine distinctions in function may also be recognizable.

Human T-cell differentiation antigens are harder to identify simply because human beings constitute an outbred population. Until recently, only one reliable marker existed for human T cells, a receptor for sheep erythrocytes, demonstrable by rosette formation when the two cell populations were mixed *in vitro* (E rosetting) (Coombs *et al.* 1970; Bentwich *et al.* 1973). However, with the advent of monoclonal antibodies, a wide range of human T-cell differentiation antigens has now been identified (Reinherz and Schlossman 1981) with many similarities to those already defined in the mouse. The pathway of human T-cell differentiation has now been defined by the use of these antigens and is shown in Fig. 10.5. Murine monoclonal antibodies prepared from mice immunized with human T cells can be used to identify specific cell surface markers, which are currently named by reference to the original monoclonal antibody used to identify the sub-population (e.g. antigens T1, T4 etc: for details see Table 10.5).

Natural killer cells

Ten per cent of murine lymphocytes are 'null' cells, i.e. cells which express neither B-cell nor T-cell markers. Within this population is a group of cells called natural killer (NK) cells which spontaneously lyse a number

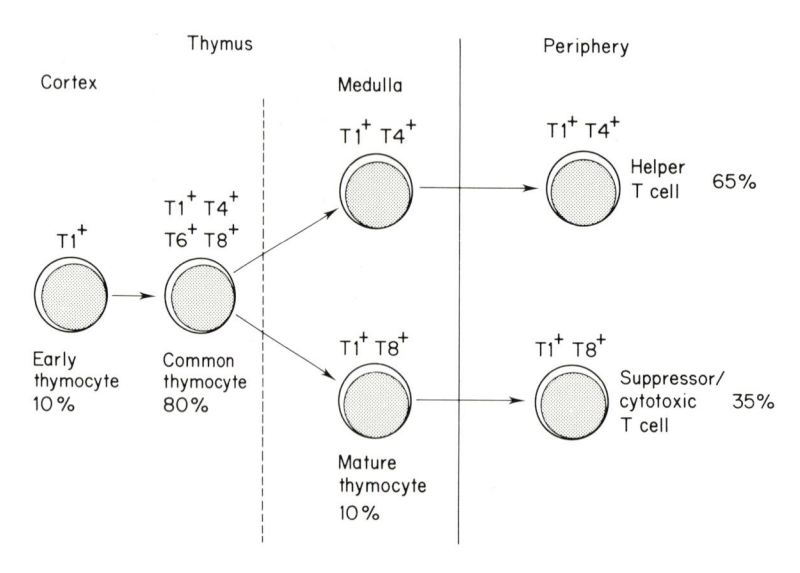

Fig. 10.5 Human T cell differentiation. Cell surface markers recognized by murine monoclonal antibodies.

Table 10.5

Antigen*	Molecular weight Non-reduced	Reduced	Distribution
T1	69 K	69 K	Expressed on all T cells and thymocytes, although only in low density on cortical thymocytes.
T4	62 K	62 K	Expressed on the majority of thymocytes and on 65% of peripheral T cells.
T6	49 K	49 K	Expressed on 70–80% of thymocytes. (Homologous to murine T1a).
T8	76 K	30 K + 32 K	Expressed on the majority of thymocytes and on 35% of peripheral T cells.

*Many other antigens have also been described in this series but are not represented here.

of *in vitro* tumour cell lines. Natural killer cells are activated by interferon and may have an important role in antitumour and antiviral immune mechanisms (Warner and Dennert 1982; Bloom 1982).

Recognition of antigen. The macrophage/histiocyte series of cells is crucial to the processes which follow the encounter with foreign antigen, e.g. in the early stages of bacterial infection. Metchnikoff (1899) had noted the phagocytic properties of wandering mononuclear cells, and commented on this essential event in activating the host defence mechanisms. Neutrophils and macrophages both participate in the phagocytosis and killing of micro-organisms, but the macrophage series of cells (see Table 10.2) seems to have additional functions in the presentation of antigen to lymphocytes. Macrophages, activated by encounter with antigen, also produce soluble mediators of the inflammatory response which act as chemotactic, opsonic and toxic factors.

Surface receptors on macrophages

The recognition and processing of antigenic material is faciliated by the presence of several types of receptor molecule displayed by neutrophils and macrophages. These receptors help to bind antigen, particularly when it is complexed with antibody, and make it easier for the phagocytic cells to fix the antigen and digest it. Cells of the macrophage lineage express receptors for the Fc region of immunoglobulin, particularly for IgG. The complex of antibody fixed to antigen is even more readily bound than antibody alone, and multiple sites increase the stability of the bound complexes. Other membrane receptors for part of the C3 component of complement are also crucial in the recruitment of cells to participate in uptake and removal of antigen. The close relation between the other receptors (Fc, C3 etc) and the sites which are involved in lymphocyte interaction has been demonstrated by the use of monoclonal antibodies (Adams 1982). It is likely that after antigen is internalized major rearrangement of surface membrane molecules precedes the presentation of antigen to lymphocytes and is accompanied by the production of a soluble factor by the macrophage. Antigen may persist in macrophages for prolonged periods, sometimes in uncatabolized form, as with *M. tuberculosis*, for example, but is frequently catabolized with great rapidity. Retained antigen appears to exist in a separate phase within the cell, associated with small pinocytic vacuoles. Some antigen may eventually be released, undegraded.

The mechanism for the transfer of either antigen or messenger information about antigen to the lymphocyte is not clear. Early suggestions that RNA was concerned proved unfounded (Haurowitz 1960; Roelants *et al.* 1971) and it has recently become apparent that the presentation of antigen to T lymphocytes is intimately involved with the expression of antigens of the major histocompatibility complex (MHC), the so-called immune-associated or Ia antigens (see below).

A typical antigen consists of many different determinants, each of which may be bound by specific antibody. When T cells and B cells cooperate in antibody production each cell population recognizes different determinants on the stimulating antigen. This phenomenon is called the 'carrier effect'. A hapten is a small molecular weight substance which can combine with antibody but cannot initiate an immune response unless bound to a larger molecule called a carrier (Landsteiner 1936). The recognition of hapten and carrier by the immune system has been studied in the following series of experiments (Ovary and Benaceraf 1963). Injecting a rabbit with the hapten dinitrophenyl (DNP) conjugated to the carrier bovine gammaglobulin (BGG) stimulated the production of antibody specific for DNP. A secondary response, with its characteristic increase in DNP-specific antibody titre, resulted only when the animal was further injected with exactly the same hapten-carrier combination. When DNP was injected bound to another carrier, such as egg albumin (EA), or BGG was injected alone, no increase in DNP-specific antibody was observed.

To reveal whether two populations of cells were involved in this phenomenon, one animal was immunized with BGG-DNP and another with EA alone. When the spleen cells from both these animals were transferred together to an irradiated syngeneic recipient which was then injected with DNP-EA, a specific anti-DNP response occurred. As a control, transfer of cells from the DNP-immunized animal alone followed by injection of DNP-EA resulted in no anti-DNP response. Subsequently it was shown that those cells recognizing the carrier were T cells and those recognizing the hapten B cells (Raff 1970). The carrier effect is an important feature of many immune responses, when antigen, whether soluble protein, viral or bacterial requires T cells and B cells to cooperate in stimulating antibody production by recognizing different determinants present on the same antigen.

T-dependent and T-independent antigens. Antigens which require T cell help to trigger antibody production are classed as T-dependent antigens and those that do not as T-independent antigens. T-independent antigens are characteristically large molecules with repeating antigenic determinants which stimulate an antibody response consisting predominantly of IgM. The lipopolysaccharide of gram-negative bacteria and pneumococcal polysaccharide are two such T-independent antigens. The majority of microbial antigens are, however, T dependent.

Cell cooperation: cell-mediated immunity

In addition to their role in antibody production T cells also possess important effector mechanisms of their own. These are the traditional 'cell-mediated' responses, formerly distinguished from that part of the immune response due to antibody. Cell-mediated responses can characteristically be transferred from one animal to another by lymphocytes but not by serum. Several *in vivo* and *in vitro* responses have been recognized which are predominantly cell mediated in nature.

Delayed hypersensitivity: This is classically demonstrated by the immune response to tuberculin. Intradermal injection of a small amount of this antigen leads to an inflammatory response at the site of injection 48 hours later. Once again, this response is mediated by T cells as confirmed when histological examination of biopsied skin is performed (see Chapter 16).

Allograft rejection: An allograft is tissue transferred between two genetically unrelated members of the same species. Some time after transfer (10–14 days) the graft becomes necrotic and eventually sloughs off. This is predominantly the result of a T-cell mediated immune response, although specific antibody may also be involved.

Graft versus host reaction (GVH): The graft versus host reaction is a T-cell mediated response with importance in the field of bone marrow transplantation. In a normal immune response, host lymphocytes reject a graft which is recognized as non-self. In GVH, however, immunocompetent lymphocytes present in the graft recognize host antigens as non-self and mount their own immune response.

Mixed lymphocyte reaction: When lymphocytes from one individual are cultured with those from a genetically unrelated individual, each population recognizes the other as non-self. A proliferative response results which can be assessed after 5 days by measuring the cellular uptake of tritiated thymidine for incorporation into DNA. This is known as a mixed lymphocyte reaction (MLR). The cells responding in the MLR are T_H cells.

Experiments showing cooperation in cell-mediated immunity are similar in design to those demonstrating cooperation in antibody production. Two populations of cells are mixed in order to determine whether their combined effects are greater than the sum of their individual effects, e.g. when this was done to study GVH reactions, it was found that a mixture of murine thymocytes and peripheral blood cells resulted in a response many times more effective than that produced by the individual populations.

T-cell suppression. In addition to the helper mechanisms already described, the immune system is also regulated by a sub-population of T cells which suppress the response to antigen (T_S cells). Suppression was first observed by Gershon and Kondo (1971), who were studying the murine immune response to sheep red blood cells (srbc). When lymphocytes from a mouse which was tolerant to srbc were transferred to a mouse capable of producing srbc-specific antibody, a state of tolerance resulted in the recipient. Suppression of antibody production could therefore be transferred from one animal to another. In man, suppressor cells express the T8 antigen.

The major histocompatibility complex (MHC)

The mechanisms underlying cellular interaction within the immune response are intricately linked to a set of antigens encoded in the genes of the major histocompatibility complex (MHC). This is a closely linked cluster of genes located on chromosome 17 in the mouse, in which it is referred to as the H-2 complex, and on chromosome 6 in man (the HLA system). The discovery of the murine H-2 system was the result of work to study allograft rejection and tumour immunity (Gorer 1936; Gorer *et al.* 1948). The subsequent development of strains of genetically homozygous inbred mice has been the cornerstone on which the study of MHC immunogenetics has been founded (Snell 1948; Festing 1979). The existence of these strains has, of course, meant that the complex is much better understood in the mouse than in man, in whom experimental findings tend to lag several years behind analogous findings in the mouse.

The MHC is divided into several main regions which code for three major classes of antigens, and some of these regions can be further divided into subregions. It should be stressed that each region or subregion probably contains many more genes encoding for products than those which have already been identified in the MHC.

The K and D/L regions in the mouse and the A and B/C regions in man code for class I histocompatibility antigens. They consist of a 44 K transmembrane glycoprotein which is present on the cell surface non-covalently bound to β-2 microglobulin, a 12 K protein coded for by a gene on a separate chromosome. The I region of the mouse and the D/DR region in man contain genes which code for class II histocompatibility antigens which consist of two glycoprotein chains (28 K and 32 K molecular weight) non-covalently bound and inserted into the cell membrane. Class II antigens are also known as Ia antigens, referring to the '*i*mmune-*a*ssociated' functions of these molecules, first described in the inbred mouse (McDevitt and Chinitz 1969). Studies of the genetic control of antibody responses to simple synthetic polypeptides, e.g. poly-L-lysine, had revealed that the relevant genes were closely linked to the H-2 antigens, and a major region of the MHC was defined which included several genes concerned in the immune response (Ir region). Thus, it was demonstrated that not only graft rejection was linked to the MHC, but that many aspects of the immunological response to antigens, not exclusively to transplantation antigens, were controlled by the

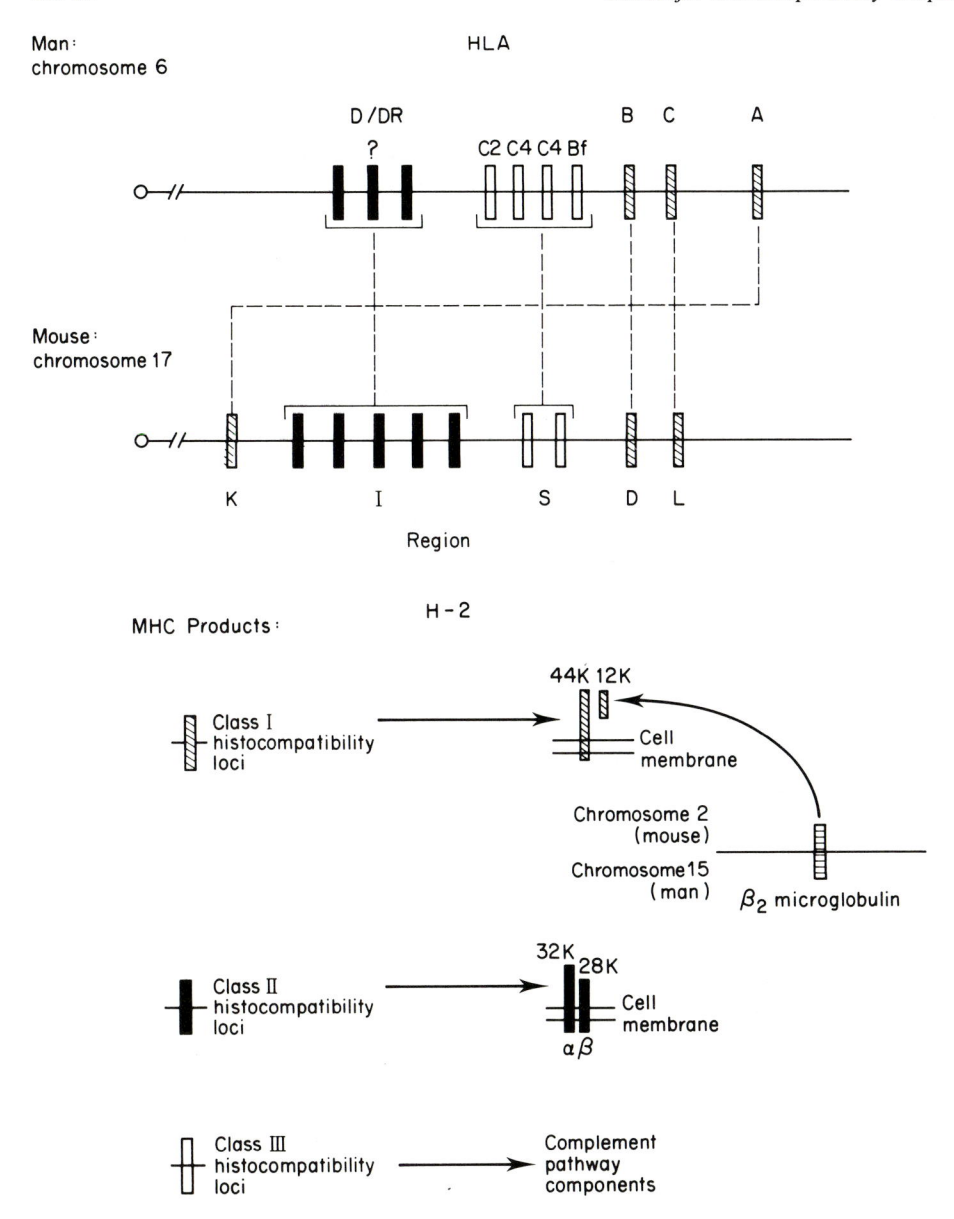

Fig. 10.6 The major histocompatibility complex in man and mouse.

products of genes in the Ir region of murine MHC. The Class II antigens of mice have their counterpart in the immune-associated products of the human MHC, known as HLA-D and DR (D-Related, in recognition of the close relation between the products of the D and DR loci). Class II antigens are expressed predominantly on B cells and cells of the macrophage/monocyte/histiocyte series, whereas Class I antigens may be detected on both B and T cells, and on many other nucleated tissue cells and platelets. A third group of cell products is coded for by genes in a further major region of the MHC: these include C_2, C_4 and Factor B of the complement components in man.

The MHC regions in man and mouse are diagrammatically illustrated and compared in Fig. 10.6. The precise number of loci in the D/DR region has not yet been ascertained: if this region is comparable with its Ia/Ir counterpart in the mouse, we may expect 3 or 4 loci, each coding for separate cellular products with different functional properties (Klein et al. 1981). The chromosomal arrangement of the major loci seems to be rather different in man and mouse. Biochemically and functionally HLA-A and H-2K products seem to be very similar; the apparent discrepancy in the siting of the loci (see Fig. 10.6) at opposite ends of the MHC region has been explained by invoking a chromosomal rearrangement (or 'loop') which must have occurred early after the divergence of the evolutionary pathways of the two species (Bodmer 1981). Certainly, evidence for the structural similarities and closely comparable functions would suggest that the MHC products are similar in mammalian species and that their role as membrane molecules is fundamentally the same. The antigens of the loci of the MHC occur in many different forms (alleles), i.e. the system is highly polymorphic. Individuals within an outbred species such as man are rarely identical for MHC products. What purpose is served by this remarkable molecular diversity and how is this related to the presence of MHC antigens on cells, most particularly, cells of the lymphoid series?

MHC and the recognition of antigen

Antigen recognition by cells of the immune system, especially by T cells, must be regarded as a critical event in the generation of an immunological response. B cells can bind antigen to their surface immunoglobulin. T cells do not seem to have such receptors, at least not recognizable immunoglobulin. However, it is likely that MHC antigens are the important factor in the recognition events involving T cells (Bach *et al.* 1976).

MHC Restriction

It has been demonstrated with murine cells *in vitro* that T cells will recognize foreign antigen only when it is presented to them in association with MHC antigens (Zinkernagel and Doherty 1974). Mice of a selected H-2 type were infected with lymphocytic choriomeningitis virus (LCMV). Several weeks later, the spleens of these animals were removed and a suspension of T cells prepared. The T cells were then assayed for their ability to lyse LCMV-infected target cells in tissue culture. Target cells from a variety of H-2 types were used: lysis and cytotoxic killing occurred only when the target cells and the T cells were of identical H-2 haplotype. Target cells not infected with virus were not lysed even when of suitable matched H-2 type. Detailed experiments were conducted to show that the cytotoxic T cells 'recognized' Class I MHC antigens on the target cells. A similar series of experiments later demonstrated the same phenomenon by using human T cells and influenza-virus infected target cells, when it was human HLA-A, B (Class I) determinants which had to be shared before effective cytotoxic killing took place (McMichael *et al.* 1977). This requirement for MHC matching has been christened MHC 'restriction' (for review, see Zinkernagel and Doherty 1979; also Munro and Brenner 1982).

Other types of T cells are also subject to MHC restriction in their interactions. T-helper cells respond only to macrophages which share MHC antigens; for such T-cell macrophage interaction, Class II MHC products must be shared between the primed T_H cell and the antigen-presenting macrophage (see Rosenthal and Shevach 1976) for murine model and Bergholtz and Thorsby (1978) for human D/DR restriction of response to PPD. The ability to transfer delayed hypersensitivity in guinea-pigs by means of lymph node cells is also dependent on shared MHC region antigens: this was demonstrated before the nature of the MHC had been elucidated. Landsteiner and Chase (1942) showed that cellular transfer of delayed hypersensitivity was most effective and strongly expressed when the lymphocytes used were injected into syngeneic, i.e. MHC identical, animals rather than into outbred strains, which would not necessarily share MHC antigens with the sensitized donor lymphocytes. The precise explanation for these early results is only now available, when the phenomenon of MHC restriction of the cellular interactions has been demonstrated. In the murine model, Class II compatibility is necessary before successful transfer of delayed hypersensitivity is realized when testing cellular and protein antigens (Miller *et al.* 1976). In exceptional cases, e.g. with LCM virus, Class I antigen restriction operates, presumably through a different sub-population of T cells (Zinkernagel 1976).

Two hypotheses have been put forward to explain the association between MHC products and the interaction of T cells with other cells in the development of immunological responses. The 'altered self' hypothesis states that the T cell possesses a single receptor site which recognizes a new antigenic determinant: this receptor is formed between the foreign (non-self) antigen and MHC (self) antigens (Matzinger 1981). An alternative view, the 'dual recognition' hypothesis postulates two different T-cell receptors, one of which recognizes the foreign antigen and the other, the MHC antigen, and states that both must be present for effective triggering of the T-cell response. So far, there is no clear-cut experimental evidence to exclude either hypothesis (Williamson 1980). MHC products provide the framework within which T cells interact with other cells in the system. The surface receptors of the T cell are acquired during the maturation process and, although the genetic information for the development of these membrane molecules is carried by the differentiating cell, the regulation or switching on of the expression of such receptors may well be brought about by external influences, in particular, the effect of hormones and soluble factors encountered in the thymus (van Boehmer *et al.* 1978).

The thymus

The thymus has both lymphoid and epithelial elements, the latter arising from the third and fourth pharyngeal pouches and the lymphoid cells as stem cells from the marrow and possibly even locally. Recent evidence derived from studies of thymectomized and grafted mice suggests that at least one class of pre-T cell originates in the thymus itself (Le Dourain and Joterau 1975; Gregoire et al. 1979). Within the mature thymus, the lymphoid tissues are arranged in lobular form: epithelial cells are found throughout the thymus; in the outer layers beneath the capsule the epithelial strands are densely infiltrated with thymic lymphocytes, the thymocytes. Deeper in the tissues in the medulla, epithelial cells predominate and have long dendritic processes interdigitating with each other. These dendritic cells are positive for Class II MHC (Ia) antigens and are important in antigen presentation. Hassall's corpuscles, whose function is unknown, lie in the medulla and are round or elongated groups of epithelial cells forming the focal point of an extended network of such cells, with a patchy infiltration of lymphocytes interspersed between the epithelial cells.

The thymic environment is essential to the maturation of cells of the T-lymphocyte series, and the acquisition of the specific membrane markers and of functional properties depends on residence in the thymus. There are differences in the expression of various alloantigens between cortical and medullary thymocytes, those in the latter site resembling mature peripheral blood T cells (Mathieson et al. 1979; McConnell et al. 1981).

The mechanisms which operate in the differentiation processes occurring in T cells in the thymus are poorly understood. Some workers have postulated a role for thymic hormones or soluble factors in altering cells and stimulating maturation. Several such soluble products have been isolated (for review, see Trainin et al. 1983). Most are small peptides, synthesized in thymic tissue, but some have been identified in other tissues—serum, bacteria, yeasts etc. One, thymosin, is synthesized by thymic epithelial cells and may alter the expression of surface markers by peripheral T cells and thymocytes (Leino et al. 1980; Elfenbein et al. 1980). Another small (9 amino acids) peptide has been demonstrated which is present in fresh serum and in thymus, but disappears from the circulation after thymectomy. This molecule has been synthesized and its effect on immune competence has been studied showing that T lymphocytes are increased in the presence of the factor (Factor Thymique Sérique—FTS) (Bach 1980). Even B lymphocytes may be affected in their differentiation pathways by selective effects of thymic factors on T cell precursors in bone marrow (Pahwa et al. 1980; Palacios et al. 1982). Other molecules with putative hormonal effects have been isolated from bovine thymus and used in attempts to correct immunological defects—thymic deficiency etc.—but none has been shown to substitute effectively for the intact thymus (Incefy et al. 1981).

Soluble factors and T-cell activation

Many of the interactions which occur between cells of the immune response are mediated by soluble factors. When these are produced by lymphocytes they are called lymphokines and when produced by cells of the macrophage series they may be called monokines. Many such factors have been described which carry out the helper and suppressor functions of T cells, but few have been extensively characterized. Recently, however, a lymphokine and a monokine have been identified which appear to play a central role in the initiation of the immune response.

Macrophages activate antigen-specific T cells through two distinct mechanisms. Firstly, antigen is presented in association with class II histocompatibility antigens. Secondly, after encounter with antigen, macrophages produce a low molecular weight monokine called interleukin 1 (IL1) or lymphocyte activating factor (LAF) (Mizel 1981). Interleukin 1 induces a set of T cells to produce the lymphokine, interleukin 2 (IL2) or T-cell growth factor (TCGF), an antigen non-specific immuno-enhancing factor (Watson et al. 1982). In addition, IL1 also induces another set of T cells to express cell surface receptors for IL2. Interaction between IL2 and its receptors on these cells leads to cell division. A proliferative response is therefore initiated from which progeny cells, also possessing IL2 receptors, are further stimulated to divide by IL2.

IL2 is also capable of supporting the long-term *in vitro* growth of human or murine activated T cells. This occurs through a continuous proliferative response in which progeny are further stimulated to divide by IL2. These continuous T-cell lines may be cloned to obtain monoclonal cell populations which

Table 10.6 Lymphokines and monokines

Lymphokines
1. Affecting macrophages
    Macrophage chemotactic factor
    Macrophage activating factor
    Migration inhibitory factor
2. Affecting polymorphonuclear leucocytes
    Chemotactic factors
    Leucocyte inhibitory factor
3. Affecting other cells
    Lymphotoxin
    Interferon
    Interleukin 2 (also known as TCGF)

Monokines
Interleukin 1 (also known as LAF)
B cell activating factor
Complement proteins

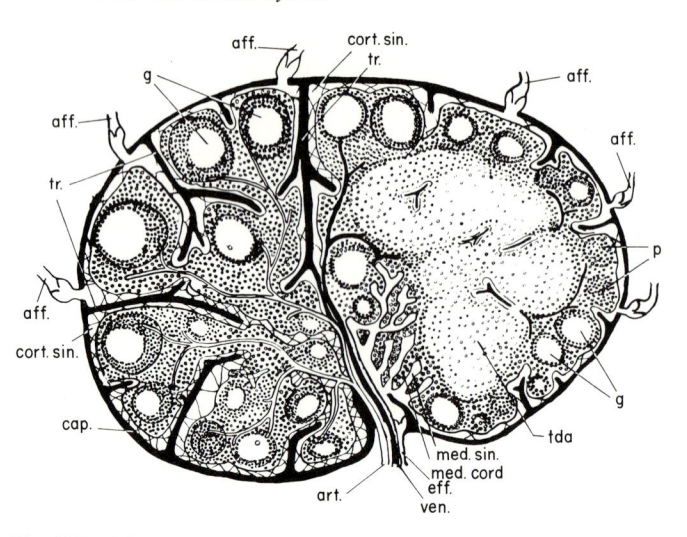

Fig. 10.7 Diagram of a human lymph node to illustrate the main compartments: cortex, paracortex or thymus-dependent area (tda) and medulla. On the left side of the diagram the arteries (art) are shown as they enter the hilum, travel in the trabeculae (tr) to enter the dense lymphoid tissue of the cortex. On the right side of the diagram the veins (ven) are shown, and a section through cortex, lymphoid nodules with germinal centres (g), thymus-dependent area and medulla.

express a single antigen specificity, an important experimental tool in the study of T cell antigen recognition and cellular interaction.

Other lymphokines and monokines, characteristically produced by T cells capable of transferring delayed hypersensitivity, are of importance in inflammatory responses (Rocklin *et al.* 1980). These act by attracting non-specific cells such as macrophages, neutrophils, eosinophils and basophils to the site of antigen deposition, activate them and inhibit their departure. As most lymphokines and monokines are produced in very small amounts, they have proved extremely difficult to purify and characterize. Therefore many activities have been described which may well be attributable to the actions of a single or relatively few molecules. Table 10.6 lists some of the most commonly described lymphokines and monokines.

Lymph nodes and spleen

The other major organs which are involved in the interactions of lymphocytes with other cells of the immune system are the lymph nodes and spleen. Both contain accumulations of T and B cells, often in close juxtaposition to antigen presenting cells of the macrophage series, the latter cells characterized by the presence on their surface of Class II (Ia) MHC antigens. Other collections of T and B cells may be found in the upper respiratory tract, concentrated mainly in the tonsils and adenoids (Waldeyers ring).

The lymphocytes in these organs tend to be organized into zones in which one or other class of cell predominates. B cells form the majority population of the germinal centres, e.g. in the cortex of lymph nodes, which are often not readily distinguishable as distinct areas until antigen exposure has stimulated cellular proliferation. Around the germinal centres (Fig. 10.7), T cells tend to predominate, for example, in the paracortical areas of lymph nodes. These cells proliferate when challenged with T-dependent antigens e.g. PPD, or the synthetic contact sensitizer, dinitrochlorobenzene. In the spleen, the germinal centres are found at the margins of the white pulp, whereas T cells tend to be arranged close to the central arteriole (McConnell *et al.* 1981). Exchange of cells between the different zones or compartments is common, and a continuous stream of lymphocytes passes through the tissues, although many cells may spend the greater part of their life cycle in the same site. The entry of antigen, either in soluble form, e.g. bacterial polysaccharides or toxins, or transported by macrophages, is the stimulus which initiates proliferation of the appropriate cells. Splenic B cells are particularly important in generating plasma cells producing antibodies to many bacterial carbohydrate antigens. Splenectomy is known to impair IgG antibody production and may be followed by fatal septicaemia, pneumococcal infection being a particular hazard (Hosea *et al.* 1981). Within T-cell regions, there appears to be a mixture of most of the different sub-classes (T_C, T_S etc.). T_H cells are also found scattered in B-cell areas as might be expected from their role as helper cells in antibody production. T_H cells seem to associate particularly with Ia-positive dendritic cells, reflecting the importance of the latter type of cell in the presentation of antigen (Janossy *et al.* 1980). No doubt T-B cell interactions as well as T-cell macrophage interaction take place as cells migrate in and out of these microenvironments. Defects in one or more of the cellular or structural components of the major lymphoid organs can produce severe and often fatal defects in the ability to develop an immunological response to microbial infection (see Chapter 17).

References

Adams, D. (1982) *Immunology Today* **3**, 285.
Altman, L. C. (1978) In: *Leukocyte chemotaxis*, p. 267. Ed. by J. I. Gallin and P. G. Quie. Raven, New York.
Babior, B. M. (1980) In: *The Reticuloendothelial System: A comprehensive treatise*, Vol. II, Biochemistry and Metabolism, p. 339. Ed. by R. Strauss and A. Sbarra. Plenum, New York.
Bach, F. H., Bach, M. L. and Sondel, P. M. (1976) *Nature* **259**, 273.
Bach, J. F. (1980) *Ann. intern. Med.* **131**, 177.
Bainton, D. F. (1973) *J. Cell Biol.* **58**, 259.
Basten, A. and Beeson, P. B. (1970) *J. exp. Med.* **131**, 1288.
Beeson, A. and Bass, D. A. (1977) *The eosinophil*. Saunders, Philadelphia.
Bentwich, Z., Douglas, S. D., Skotelsky, E. and Kunkel, H. G. (1973) *J. exp. Med.* **137**, 1532.
Bergholtz, B. and Thorsby, E. (1978) *Scand. J. Immunol.* **8**, 63.
Bloom, B. R. (1982) *Nature* **300**, 214.
Bodmer, W. F. (1981) *Tissue Antigens* **17**, 9.
Boehmer, H. van, Haas, W. and Jerne, N. K. (1978) *Proc. nat. Acad. Sci. Wash.* **75**, 2439.
Butterworth, A. E., Wassom, D. L., Gleich, G. J., Loegering, D. A. and David, J. R. (1979) *J. Immunol.* **122**, 221.
Chenoweth, D. E. and Hugli, T. E. (1980) *Molec. Immunol.* **17**, 151.
Claman, H. N., Chaperon, E. A. and Triplett, R. F. (1966) *Proc. Soc. exp. biol. Med.* **122**, 1167.
Clark, R. A. (1978) In: *Leukocyte chemotaxis*, p. 329. Ed. by J. I. Gallin and P. G. Quie. Raven, New York.
Cohn, Z. A. (1978) *J. Immunol.* **121**, 813.
Coombs, R. R. A., Gurner, B. W., Wilson, A. B., Holm, G. and Lindgren, B. (1970) *Int. Arch. Allergy appl. Immunol.* **39**, 658.
Cooper, M. D., Peterson, R. D. A., Smith, M. A. and Good, R. A. (1966) *J. exp. Med.* **123**, 75.
Craddock, P. R., Fehr, J., Dalmasso, A. P., Brigham, K. L. and Jacob, H. S. (1977a) *J. clin. Invest.* **59**, 879.
Craddock, P. R., Hammerschmidt, D., White, J. G., Dalmasso, A. P. and Jacob, H. S. (1977b) *J. clin. Invest.* **60**, 260.
Cramer, E. B., Milks, L. C. and Ojakian, G. K. (1980) *Proc. nat. Acad. Sci. Wash.* **77**, 4069.
De Bruyn, P. P. H. (1981) *Semin. Haematol.* **18**, 179.
De Sousa, M. (1981) *Lymphocyte circulation*. John Wiley & Sons, Chichester.
Ehrlich, P. (1900) *Proc. Roy Soc., London* **66**, 424.
Elfenbein, G. J., Goldstein, A. L., Adams, J. S. and Ravlin, H. M. (1980) *Transplantation* **29**, 113.
Elsbach, P. and Weiss, J. (1981) *Advanc. Inflamm. Res.* **2**, 95.
Festing, M. F. W. (1979) *Inbred Strains in Biomedical Research*. Macmillan, London.
Ford, W. L. (1975) *Prog. Allergy* **19**, 1.
Furth, R. van (Ed) (1980) *Mononuclear phagocytes: Functional aspects*. Nijhoff, The Hague.
Furth, R. van, Cohn, Z. A., Hirsch, J. G., Humphrey, J. H., Spector, W. G. and Langevoort, H. L. (1972) *Bull. WHO* **46**, 845.

Gallin, J. I. (1981) *Rev. infect. Dis.* **3**, 196.
Gallin, J. I. and Quie, P. G. (Eds) (1978) *Leukocyte chemotaxis*. Raven, New York.
Gallin, J. I., Wright, D. G. and Schiffmann, E. (1978) *J. clin. Invest.* **62**, 1364.
Gathings, W. E., Kubagawa, H. and Cooper, M. D. (1981) *Immunol. Rev.* **57**, 107.
Gearhart, P. J., Johnson, N. D., Douglas, R. and Hood, L. (1981) *Nature* **291**, 29.
Gershon, R. K. and Kondo, K. (1971) *Immunology* **21**, 903.
Glauert, A. M., Butterworth, A. E., Sturrock, R. F. and Houba, V. (1978) *J. Cell Sci.* **34**, 173.
Good, R. A. and Gabrielson, A. E. (Eds) (1964) *The thymus in immunobiology*. Harper & Row, New York.
Gorer, P. A. (1936) *Brit. J. exp. Pathol.* **17**, 42.
Gorer, P. A., Lyman, S. and Snell, G. D. (1948) *Proc. R. Soc. Lond. (Biol.)* **135**, 499.
Gregoire, K. E., Goldschneider, L., Barton, R. W. and Bollum, F. J. (1979) *J. Immunol.* **123**, 1347.
Haurowitz, F. (1960) *Annu. Rev. Biochem.* **29**, 609.
Henson, P. M. (1971a) *J. Immunol.* **107**, 1535; (1971b) *Ibid.* **107**, 1547.
Hibbs, J. B., Chapman, H. A. and Weinberg, J. B. (1980) In: *Mononuclear Phagocytes: Functional Aspects*, p. 1681. Ed. by R. van Furth. Nijhoff, The Hague.
Hoefsmit, E. C. M., Duijvestijn, A. M. and Kamperdijk, E. W. A. (1982) *Immunobiology* **161**, 255.
Hosea, S. W., Burch, C. G., Brown, E. J., Berg, R. A. and Frank, M. M. (1981) *Lancet* **i**, 804.
Incefy, G. S., O'Reilly, R. J., Kapoor, N., Iwata, T. and Good, R. A. (1981) *Transplantation* **32**, 299.
Janossy, G., Tidman, N., Selby, W. S., Alerothomas, J., Granger, S., Kung, P. C. and Goldstein, G. (1980) *Nature* **288**, 81.
Katz, D. H. (1977) *Lymphocyte differentiation. Recognition and regulation*. Academic Press, New York.
Keller, H. U., Wilkinson, P. C., Abercrombie, M., Becker, E. L., Hirsch, J. G., Miller, M. E., Ramsey, W. S. and Zigmond, S. H. (1977) *Clin. exp. Immunol.* **27**, 377.
Keller, R. (1980) In: *Mononuclear Phagocytes: Functional Aspects*, p. 1725. Ed. by R. van Furth, Nijhoff, The Hague.
Klebanoff, S. J. (1975) *Semin. Haematol.* **12**, 117; (1980) In: *The Reticuloendothelial System*, Vol. II, Biochemistry and Metabolism, p. 279. Ed. by R. Strauss and A. Sbarra. Plenum, New York.
Klebanoff, S. J. and Clark, R. A. (1978) *The neutrophil: Function and clinical disorders*. North Holland, Amsterdam.
Klein, J., Juretic, A., Baxevanis, C. N. and Nagy, A. (1981) *Nature* **291**, 455.
Lackie, J. M. and Smith, R. P. C. (1980) In: *Cell Adhesion and motility*, p. 235. Ed. by A. S. G. Curtis and J. D. Pitts. Cambridge University Press, Cambridge.
Landsteiner, K. (1936) *The Specificity of Serological Reactions*. Charles C. Thomas, Ill.
Landsteiner, K. and Chase, M. W. (1942) *Proc. Soc. exp. Biol. Med.* **49**, 688.
Leddy, J. P., Baum, J. and Rosenfeld, S. I. (1978) In: *Leukocyte chemotaxis*, p. 389. Ed. by J. I. Gallin and P. G. Quie. Raven, New York.
Le Dourain, N. M. and Joterau, F. V. (1975) *J. exp. Med.* **144**, 79.
Lefkovits, I. (1974) *Curr. Top. Microbiol. Immunol.* **65**, 21.

Leino, A., Hirvonen, T. and Soppi, E. (1980) *Clin. Immunol. Immunopathol.* **17,** 547.
McConnell, I., Munro, A. and Waldmann, H. (1981) *The Immune system. A course on the molecular and cellular basis of immunity.* Blackwell Scientific Publications, Oxford.
McDevitt, H. O. and Chinitz, A. (1969) *Science* **163,** 1207.
McMichael, A. J., Ting, A., Zweerink, H. J. and Askonas, B. A. (1977) *Nature* **270,** 524.
Mackaness, G. B. (1962) *J. exp. Med.* **116,** 381; (1969) *ibid.* **129,** 973; (1970) In: *Mononuclear phagocytes in Immunity, Infection and Pathology,* Ed. by R. van Furth. Blackwell Scientific Publications, Oxford.
Mathieson, B. J., Sharon, S. O., Campbell, P. S. and Asofsky, R. (1979) *Nature* **277,** 478.
Matzinger, P. (1981) *Nature* **292,** 497.
Metchnikoff, E. (1899) *Ann. Inst. Pasteur* (*Lille*) **13,** 737.
Miller, J. F. A. P., Vadas, M. A., Whitelaw, A. and Gamble, J. (1976) *Proc. nat. Acad. Sci. Wash.* **73,** 2486.
Mizel, S. B. (1981) *Immunol. Rev.* **63,** 52.
Mosier, D. E. (1967) *Science* **158,** 1573.
Munro, A. J. and Brenner, M. K. (1982) In: *Clinical Aspects of Immunology,* 4th edn. p. 187. Ed. by P. J. Lachmann and D. K. Peters. Blackwell Scientific Publications, Oxford.
Murphy, P. (1976) *The Neutrophil.* Plenum, New York.
Normann, S. J. and Sorkin, E. (1977) *Cancer Res.* **37,** 705.
Otu, A. A., Russell, R. J., Wilkinson, P. C. and White, R. G. (1977) *Brit. J. Cancer* **36,** 330.
Ovary, Z. and Benaceraf, B. (1963) *Proc. Soc. exp. Biol.* **114,** 72.
Pahwa, S., Pahwa, R., Goldstein, G. and Good, R. A. (1980) *Cell Immunol.* **56,** 40.
Palacios, R., Liorente, L., Ruiz-Arguelles, A. and Alarcon-Segovia, D. (1982) *Immunol. Lett.* **4,** 35.
Parrott, D. M. V. and Ferguson, A. (1974) *Immunology* **26,** 571.
Rabinovitch, M. (1967) *Exp. Cell Res.* **46,** 19.
Raff, M. C. (1970) *Nature* **226,** 1257.
Reinherz, E. L. and Schlossman, S. F. (1981) *Immunology Today* **2,** 69.
Rocklin, R. E., Bendtzen, K. and Greineder, D. (1980) *Advanc. Immunol.* **29,** 56.
Roelants, G. E. (1977) In: *B and T cells in immune recognition,* p. 103. Ed. by F. Loor and G. E. Roelants. John Wiley & Sons, Chichester.
Roelants, G. E., Goodman, J. W. and McDevitt, H. O. (1971) *J. Immunol.* **106,** 1222.

Roos, D. and Balm, A. J. M. (1980) In: *The Reticuloendothelial System,* Vol. II, *Biochemistry and Metabolism,* p. 189. Ed. by R. Strauss and A. Sbarra. Plenum, New York.
Rosenthal, A. S. and Shevach, E. M. (1976) *Curr. Top. Immunobiol.* **5,** 47.
Russell, R. J., Wilkinson, P. C., McInroy, R. J., McKay, S., McCartney, A. C. and Arbuthnott, J. P. (1976) *J. med. Microbiol.* **9,** 433.
Schiffmann, E. and Gallin, J. I. (1979) *Curr. Topics in Cell Regulation* **15,** 203.
Silverstein, S. C., Steinman, R. M. and Cohn, Z. A. (1977) *Annu. Rev. Biochem.* **46,** 669.
Snell, G. D. (1948) *J. Genetics* **49,** 87.
Snyderman, R., Durack, D. T., McCarty, G. A., Ward, F. E. and Meadows, L. (1979) *Amer. J. Med.* **67,** 638.
Snyderman, R. and Pike, M. C. (1976) *Science* **192,** 370.
Snyderman, R., Seigler, H. F. and Meadows, L. (1977) *J. nat. Cancer Inst.* **58,** 37.
Steinman, R. M. and Nussenzweig, M. C. (1980) *Immunol. Rev.* **53,** 127.
Trainin, N., Pecht, M. and Handzel, Z. T. (1983) *Immunology Today* **4,** 16.
Unanue, E. R. (1981) *Advanc. Immunol.* **31,** 1.
Unkeless, J. C. (1977) *J. exp. Med.* **145,** 931.
Warner, J. F. and Dennert, G. (1982) *Nature* **300,** 31.
Watson, J., Frank, M. B., Mochizuki, D. and Gillis, S. (1982) *Lymphokines* **6,** 95.
Weller, P. F. and Goetzl, E. J. (1979) *Advanc. Immunol.* **27,** 339.
Wilkinson, P. C. (1978) In: *Taxis and Behavior,* p. 293. Ed. by G. L. Hazelbauer. Chapman-Hall, London; (1982*a*) *Chemotaxis and Inflammation,* 2nd edn. Churchill Livingstone, Edinburgh; (1982*b*) *Immunobiology* **161,** 376.
Wilkinson, P. C. and Lackie, J. M. (Eds) (1981) *Biology of the chemotactic response.* Cambridge University Press, Cambridge.
Williamson, A. R. (1980) *Nature* **283,** 527.
Wright, A. E. and Douglas, S. R. (1903) *Proc. R. Soc. B.* **72,** 364; (1904) *Ibid.* **73,** 136.
Zigmond, S. H. (1974) *Nature* **249,** 450; (1977) *J. Cell Biol.* **75,** 606.
Zinkernagel, R. M. (1976) *J. exp. Med.* **144,** 776.
Zinkernagel, R. M. and Doherty, P. C. (1974) *Nature* **248,** 701; (1979) *Advanc. in Immunology* **27,** 51.

11

Antigen-antibody reactions—*in vitro*

Heather M. Dick

Introductory	319	Species differences in antigen-antibody complex formation	324
Types of antigen-antibody reaction	319	Formation of precipitates	324
Modified assays	320	Agglutination	324
Precipitin reactions	320	Bacterial agglutination	324
Intermolecular forces in antigen-antibody binding	320	Co-agglutination	324
Van der Waal's bonds	320	Antiglobulin reaction	325
Hydrostatic bonds (coulombic bonds)	320	The 'prozone' in bacterial agglutination	325
Hydrogen bonding	321	Applications of immunochemical principles	325
The effect of hydrophobic properties of antigen-antibody aggregates	321	Immunofluorescence (IMF)	325
		Immunoelectron-microscopy	326
Effect of electrolytes	321	Radioimmunoassays (RIA)	326
Precipitin tests	321	Radioimmunoassay of antibody	326
The Lattice hypothesis	322	Enzyme-linked Immunoassay (ELISA)	326
Optimal proportions: the ratio of antigen to antibody	322	Complement Fixation *in vitro*	327
		The complement cascade	327
Valency of antibody	323	Lysis of sheep red cells	327
Antibody affinity	323		

Introductory

Types of antigen-antibody reactions

The interaction between antigen and specific antibody molecules forms the basis of the humoral arm of the response of the immune system to antigenic stimulation. The reaction between antigens and antibody may be recognized by the use of one or more techniques which are designed to allow the two components to bind together in controlled conditions such that a visible reaction ensues. The result may be a qualitative or quantitative assay of the combined molecules but does not always reflect the biological (*in vivo*) circumstances under which the antigen and antibody would react, nor will *in vitro* tests necessarily reflect the mechanisms leading to tissue damage after *in vivo* reactions. However, since antigen-antibody reactions are fundamentally examples of chemical reactions, it is possible to use such reactions to make specific observations on the result of antigen and antibody combination. In some instances, this interaction is directly relevant to the biological role of antibody, as in toxin-antitoxin reaction or virus neutralization; other reactions are useful in quantitation or isolation of antigen-antibody complexes, as with precipitation or agglutination. The microbiologist will make most use of those techniques which rely for their endpoint on one of the four principal types of recognizable antigen-antibody reaction (Table 11.1).

Various modifications of these primary tests have been developed, which extend the usefulness of the simple assays. Some of these modifications use techniques which are based on the biochemical or physical properties of the antibody molecule, e.g. immunoelectrophoresis, which uses surface charge differences between the major classes of immunoglobulin molecule to separate individual antibody-

320 *Antigen-antibody reactions—in vitro* Ch. 11

Table 11.1 Antigen-antibody reactions

Reaction	Test(s)	Modified tests
Precipitation	Oudin tube	Immunoelectrophoresis
	Ouchterlony test	Immunoprecipitation
	Double diffusion (Mancini)	
Agglutination	Simple	Latex (and other inert particles)
	Mixed (cell type)	Indirect haemagglutination
	Haemagglutination	Co-agglutination
		Antiglobulin (Coombs) test
Complement fixation	Haemolytic assay	Conglutination
		Plaque assay
	Cytotoxicity	
	Cell lysis	
Neutralization	Toxins, lysins etc.	Measurement of LD (Chap. 19)
	Viral cytopathic effect	

antigen reactions within a mixture, and may also be adapted to make use of the difference in charge of isolated protein antigens.

Modified assays

Second generation assays are now available which often exploit the basic principles of these 'primary' methods, as well as making use of a variety of new procedures to identify specific reactions, to quantify antigen or antibody, or to measure complexes of both (Table 11.2). In some assays, use is also made of pre-

Table 11.2 Antigen-antibody reactions ÷ second generation tests

Reaction/Test	Modified test
Immunofluorescence (IMF)	Indirect Immunofluorimetric assays
Enzyme immunoassay (EIA)	Indirect e.g. peroxidase staining
Enzyme linked immunosorbent assay (ELISA)	
Radioimmunoassay (RIA)	Immunoradiometric assays (IMRA)

liminary procedures designed to separate antigen components (gel electrophoresis, for example) or to immobilize soluble antigen or immunoglobulins on columns of inert material e.g. Sepharose, thus allowing interaction between antigen and antibody. From such columns, elution of the various components by simple procedures such as pH shift may follow, simplifying the separation of individual molecules or of complexes of antigen and antibody.

Precipitin reactions

Serological precipitation occurs when an antigen in solution reacts with specific antibody (Landsteiner 1945). The appearance of visible precipitate depends on the presence of electrolytes; and the amount of the precipitate, which is made up of antigen-antibody complexes, will vary with the concentration of each constituent and with other factors, including time and temperature of incubation. The forces binding the antigen-antibody complexes are believed to be relatively weak, and derive from attractions between oppositely charged ions in each molecular structure. These forces include van der Waal's bonding, hydrogen bonding and hydrostatic forces, as well as the hydrophobic properties of some protein aggregates.

Intermolecular forces in antigen-antibody binding

Van der Waal's bonds. These arise from close juxtaposition of two molecules or atoms, the force rising with increased proximity. The force exerted is inversely proportional to the seventh power of the distance ($F \propto 1/d^7$) and is exerted at its optimum strength between two atoms which form part of molecules with close 'fit' i.e. spatial configuration is crucial. The complementary spatial relation between antigen and antibody molecules thus contributes to the exertion of such forces and increases the tendency of the two molecules to bind together. This 'lock-and-key' concept of antigen-antibody interactions also applies in the development of other intermolecular forces which contribute to the formation of complexes between specific antibody and antigen.

Hydrostatic bonds (coulombic bonds). These occur between two oppositely charged groups e.g. amino (NH_2^+) and carboxyl (COO^-) groups in their ionized form. In this case, the force varies with the square of the distance ($F \propto 1/d^2$). Similar forces can be generated where there are possibilities for transfer of charge between the two molecules.

Hydrogen bonding. A hydrogen bond can occur between a covalently bound hydrogen atom, which has a positive charge, and an adjacent negatively charged atom (e.g. N—H and C=O groups). The force exerted is relatively weak (F α $1/d^6$) and is direction dependent. The H^+ must be opposite to and close to the negatively charged acceptor atom, a condition which is most obviously fulfilled when there is a close spatial relationship between the antigen and antibody molecules, resulting from the basic configuration of each molecule.

The effect of hydrophobic properties of antigen-antibody aggregates. Antibody molecules are hydrophilic, bearing surface groups which attract water, thus tending to maintain the molecules in solution. When antigen and antibody bind there is a tendency for such groups to be less accessible to water: at the same time, hydrophobic bonding increases as two protein molecules become closely opposed. The latter event has the effect of reducing the number of water molecules associated with the surfaces in contact and leads to a reinforcement of the binding between the molecules of antigen and antibody.

Effect of electrolytes. The effect of antibody on the surface of bacteria is to make the bacteria behave as strongly hydrophobic colloids with a relatively high surface charge: in the presence of electrolytes, the bacteria agglutinate. The presence of the free ions in the solution reduces the charge with the result that the bacteria no longer remain in the dispersed state but tend to aggregate together. In the absence of electrolytes, antigen-antibody binding may occur, but the formation of visible aggregates depends on the presence of an appropriate concentration of salts.

Precipitin tests

The visible result of antigen-antibody combination in the form of a precipitate forms the basis of several widely used tests (Table 11.3). Careful analysis of the relative proportions of antigen and antibody present in precipitates and determination of the optimal conditions for the formation of precipitate forms the basis of an understanding of antigen-antibody reactions in general. Quantitative studies of precipitin reactions were developed by Heidelberger and Kendall (1935a, b) and were examined in detail by Kabat and Mayer (1961: see also Heidelberger 1956; Kabat 1976).

Formation of Precipitates. Precipitation occurs rapidly after mixing antibody (as serum) and soluble antigen e.g. pneumococcal polysaccharide. The visible result contains complexes of both, together with uncomplexed molecules of antigen and antibody, and, depending on the conditions under which precipitation has taken place, the precipitated material may

Table 11.3 Tests derived from the precipitin reaction

Test	Reaction chamber	Antigen (Ag)	Antibody (Ab)	Diffusion	Examples of application
Ring	Test tube Capillary	in solution layered over Ab	whole serum, beneath Ag	Simple Ag⇌Ab	Streptococcal grouping
Oudin	Test tube Capillary	in solution layered over Ab	serum, incorporated in agar, beneath Ag	Single, one dimension	
Oakley-Fulthorpe variation of Oudin	Test tube Capillary	in solution over agar, which is layered over Ab	serum, in agar	Double, one dimension	
Ouchterlony	Petri dish Glass plate Microscope slide with thin layer of clear agar	in well, cut in agar layer	in well, cut in agar layer	Double, two dimensions	Aspergillosis Farmers' lung
Mancini	As for Ouchterlony	In well, cut in agar + Ab layer	In agar layer on reaction plate	Double, one dimension	Assay of immunoglobulins, C' components
Immunoelectrophoresis (Grabar and Williams)	Agar layer, Agarose layer	In well or slot in agar layer	In trough in agar layer after electrophoresis	(1) Electrophoresis (to separate components) followed by (2) Double diffusion of Ag's and Ab	Characterization of Ag's Detection of paraproteins

322 *Antigen-antibody reactions—in vitro* Ch. 11

also include some other serum components e.g. complement molecules. The latter do not contribute specifically to the union of antigen with antibody but are involved as 'bystanders' in the overall reaction.

The lattice hypothesis

The arrangement of molecules in the precipitate may be likened to a lattice of alternating antigen and antibody (Marrack 1934; Marrack *et al.* 1951). The number of antibody molecules binding to each antigen molecule will depend on the number of separate antigenic determinants recognized by the antibodies. The precise conformation assumed is thought to depend on the relative proportions of antigen and antibody in the mixture (Humphrey and White 1970). The content of precipitate and supernatant may be studied by combining antigen and antiserum in a series of dilutions, when the effect of antibody excess and antigen excess can be observed (Fig. 11.1). Theoretically there is a precise point in the curve, the equivalence zone, where neither free antigen nor free antibody can be detected in the supernatant, provided sufficiently close points are selected for assay. In practice, there is a 'plateau' effect where equivalence appears to be present across a range of antigen concentrations, which is probably due to the polyclonal, i.e. heterogeneous nature of most animal antisera (see below, optimal proportions). It will also be noted that in the region of antigen excess, there is a decrease in the amount of detectable precipitate which may be interpreted as an inhibition of precipitate formation, or may be due to the dissolution of existing complexes in the presence of excess antigen. If the lattice arrangement is accepted as the basis of the precipitin reaction, then these phenomena may be represented as in Fig. 11.2, where the antibody is pictured as bivalent, i.e. with two binding sites for a specific antigenic determinant. The antigen is multivalent according to this hypothesis, presenting several binding sites which have the same specificity for antibody, i.e. identical or closely similar antigenic determinants.

Optimal proportions: the ratio of antigen to antibody

Semi-quantitative studies with a constant amount of antiserum (antibody) and increasing amounts of antigen show that a point is reached where maximum opacity appears soonest, and after which the amount of precipitate decreases; this demonstration of the solubility of antigen-antibody complexes in excess antigen also gives us the *optimal proportions* for antigen and antibody combination (Ramon 1922; Dean and Webb 1926). This ratio (Ag:Ab) is a useful basis for quantitative studies on the behaviour of antisera, including antitoxins.

The method is applicable to the study of the antibody content of two or more antisera, in terms of a fixed amount of specific antigen, or may be used to determine the concentration of an antigen in different solutions with fixed amounts of antibody. The former method was used to demonstrate that diphtheria antitoxin could be titrated against a constant amount of toxin, the so-called constant-antigen optimal ratio. When using sera from hyperimmunized animals it is

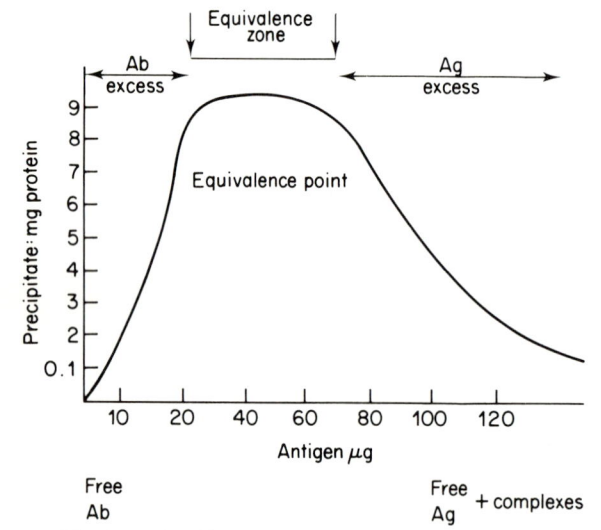

Fig. 11.1 Ag-Ab precipitate curve (hypothetical)

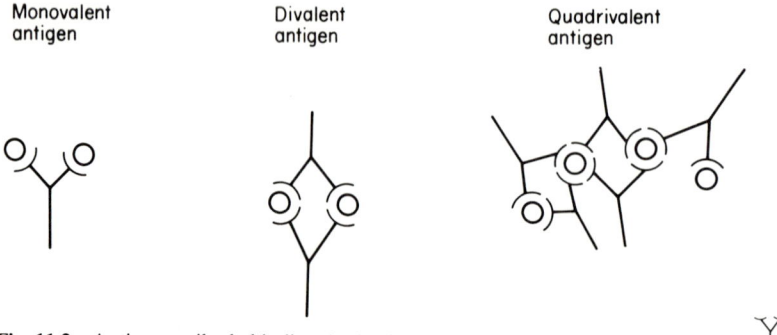

Fig. 11.2 Antigen-antibody binding: lattice formation with bivalent antibody e.g. IgG Y.

common to observe a relatively 'broad' zone of equivalence where the range of antigen concentration is wide, but the amount of precipitate formed is very similar across this range. This 'plateau' effect (Fig. 11.1) is attributed to the presence in hyperimmune sera of many classes of antibody molecule, all binding to antigen but with slightly differing properties and affinities for the antigen binding site. Many microbial antigens are also complex and will exhibit multiple binding sites, thus contributing to the heterogeneity of antigen-antibody binding. Only the use of highly purified antigen and monoclonal antibody can circumvent this problem, allowing the study of highly selected, specific antigen-antibody interaction. Even with relatively pure antigen, the presence of several different antigenic determinants on the antigen molecule may complicate the results obtained in semiquantitative studies of precipitation curves, because of the heterogeneity of the antigen-antibody complexes that are formed, not all of which are necessarily included in the precipitate under study.

Valency of antibody

Even before the structure and function of immunoglobulins was elucidated, simple observations had indicated that antibody might have multiple binding sites for antigen. With hyperimmune sera, the majority of antibody (immunoglobulin) is likely to be IgG with a molecular weight of 150 000. In optimal precipitates it was observed that the ratio of antibody to antigen varied inversely with the molecular weight of the antigen (Boyd and Hooker 1934). Antigen of mol. wt around 4-5000 e.g. type III pneumococcal polysaccharide, gave a ratio of 60:1 (Ag:Ab) whereas a larger molecule such as serum albumin (mol. wt 70 200) produced the value of 7:1, and a very large antigen, the haemocyanin pigment of keyhole limpet (KLH) with mol. wt of 3×10^6, produced a ratio of 1.5:1. Given the antibody size of 150 000, each of these molecules must be presenting several similar reactive groups of determinants for combination with the antibody molecules.

In conditions of antibody excess, every antigenic determinant group is likely to combine with antibody: in contrast, at high antigen concentration, all the antibody combining sites are likely to be occupied. By means of estimates of the antigen-antibody ratios formed in antigen excess, it is possible to derive the number of binding sites for each molecule. From these relatively simple early observations, we have now reached the point where the individual antigen combining sites of antibody can be described in terms of both structure (amino-acid sequence and crystallographic evidence) and functional behaviour (affinity). For a bivalent antibody (IgG) the two antibody combining regions (Fab$_2$: see Chapter 10) are identical in both structure and affinity (Nisonoff et al. 1960).

Antibody affinity

The relatively static observations which can be made by analysis of the content of antigen-antibody precipitates in the standard precipitin reactions have been followed by kinetic studies designed to examine the functional properties of different antibody molecules when combining with their specific antigenic determinants. The *affinity* of a given antibody reflects the goodness-of-fit between antigenic determinant and antibody (immunoglobulin); the better the correspondence between the two structures, the closer the 'fit' in terms of structure and of the contribution made to binding by several kinds of intermolecular forces. In particular, the term *affinity* is applicable to the strength of interaction between a single antibody-combining site (Fab) and a simple antigenic determinant such as a small hapten, with a single antigenic site. When such conditions prevail, the reaction between antibody and the complementary determinant site (or *epitope*) is governed by the Law of Mass Action. Thus

$$Ab + Ag \underset{Kd}{\overset{Ka}{\rightleftharpoons}} Ab.Ag$$

where Ka and Kd are the association and dissociation constants respectively. The affinity is given by the equilibrium constant K, which is derived from the equation:

$$K = \frac{Ka}{Kd} = \frac{[Ab.Ag]}{[Ag][Ab]}$$

As K increases, so does the amount of antigen-antibody complex formed i.e. high affinity antibody tends to produce more, as well as more tightly bound, complexes. Several equations which describe the behaviour of antigen-antibody complexes under various conditions can be derived from these two equations, giving a mathematical basis for the study of the interactions (Steward 1978).

When considering the interactions between multivalent antibodies and multiple combining sites—much more common state of affairs in the processes of infection than the theoretical reaction between monovalent antibody and a single epitope—the calculations become more complex because of the difficulties of determining true values of K for a mixture of antibodies, but relevant values can be derived which are particularly useful when applied for the purpose of standardizing radioimmunoassays and enzyme linked assays. The calculations are based on the theory of the events which occur in so-called displacement assays. Such assays apply physico-chemical techniques to the study of antigen and/or antibody (for details, see Otterness and Karush 1982).

With reference to the strength of interactions between multiple complex antigenic determinants and the more usual polyclonal antiserum, it is customary to use the term *avidity* (Glenny and Barr 1932; Glenny et al. 1932). Many events contribute to avidity; because of the very multiplicity of interactions which are involved, the dissociation constant of each antigen-antibody complex will be different from that of all the others, and the likelihood of complete dissociation is much less than for a single antibody-single epitope complex. The effect of multivalent binding is to produce a range of complexes with more likelihood of stability and deposition, either as precipitate *in vitro*, or as bound complex *in vivo*. The site of complex deposition *in vivo* may be critical in determining the result of antibody binding. When the relevant antigen is tissue bound or forms part of the cellular structure of an organ, so-called 'immune' complexes with antibody may be retained as for example, in the glomerular basement membrane, and exogenous antigens such as bacteria can become cell associated by reason of opsonizing antibody complexed on the surface.

Species differences in antigen-antibody complex formation. Different species produce antibodies which may vary in their properties. Horse antisera tend to give precipitates which are relatively soluble in both antigen and antibody excess. Rabbit antisera are liable to give no soluble complexes until the zone of antigen excess is reached (the curve in Fig. 11.1 is skewed to the left with rabbit serum). Chicken antisera precipitate antigen fully in the presence of a vast excess of antibody. *In vivo*, in response to antigenic challenge, the outcome may depend on the relative affinities of the different classes of immunoglobulin produced at each stage of the response, as much as on the nature of the antigen or the actual quantity of antibody produced. In the laboratory, selection of appropriate test conditions and suitable antisera are both important when studying new antigen-antibody systems or when designing new assay procedures.

Agglutination

Agglutination may be regarded as a special case of the precipitin reaction, taking place at the surface of large particles such as bacteria, erythrocytes or artificial, e.g. carbon or latex, particles (Kabat and Mayer 1948). The antigenic determinants on the surface of the cell or of an inert particle coated with (soluble) antigen are linked together by the multivalent antibody (Fig. 11.2). In principle, any type of cell might be agglutinated by appropriate antibody; in practice, the test is most often applied to suspensions of bacteria or of red cells. The latter may be agglutinated in their own right, for example by antibodies directed against blood group antigens, but are also frequently used as inert carriers of soluble antigens e.g. thyroglobulin, the red cell merely acting as a vehicle for presentation of the soluble antigen, with a visible outcome in the form of 'indirect' haemagglutination when antibody reacts with the antigen. Numerous bacteria, viruses and plants also produce molecules which are capable of agglutinating red cells (lectins): there is considerable variety in the species of red cell which are susceptible to these haemagglutinins, and the phenomenon is not directly attributable to antigen-antibody interaction, but depends on the presence of suitable receptors on the red cell surface.

Bacterial agglutination. Agglutination is frequently used in tests for the detection and typing of bacterial antigens e.g. in salmonella typing (q.v.). The test is adaptable to tube or slide techniques, the choice often depending on the volume of specific antiserum which is available. The bacterial suspension, in the presence of electrolytes, will be clumped or agglutinated by the relevant antiserum. The reaction is complex, depending as it does on multiple antigenic determinants on the surface of the bacteria and on polyclonal antibodies. IgM antibodies are generally the most effective antibacterial agglutinins, but agglutinating properties are not confined to this class of immunoglobulin.

Co-agglutination

An effective and adaptable modification of the principle of agglutination has been developed, making use of the separate binding sites on the IgG molecule (see Chapter 14). Immunoglobulin will bind to the protein A of *Staphylococcus aureus*, by the Fc terminal portion, leaving the antigen combining Fab regions free. Thus, *Staph. aureus* coated with IgG antiserum can be used in agglutination tests for bacterial antigens. The use of *Staph. aureus* as the 'carrier' allows the test to be adapted to the detection of a wide range of bacterial antigen. The test is economical in the amount of antibody required because the large surface area offered by the Ig-coated bacteria provides an ideal distribution of antigen combining sites to allow the appropriate lattice formation. The test has been used for streptococcal grouping (Christensen *et al.* 1973), as well as for typing of *N. gonorrhoeae* (Shanker *et al.* 1981), for mycobacterial grouping (Jublin & Winbald 1976) and for the identification of antibaterial antibodies.

Instead of employing whole organisms extracts of antigen may be used, for example, streptococcal antigens prepared by the use of enzymes (Maxted 1948, Castle *et al.* 1982). This technique seems to have the advantage, for streptococcal grouping at least, of reducing cross-reactivity which may be troublesome in tests with whole bacteria.

The rapidity and simplicity of the co-agglutination technique make it an acceptable alternative to other methods, particularly since it is economical in the amount of antiserum needed. Staphylococcal Protein A may be obtained commercially and individual tests prepared in the laboratory suited to the particular requirements of the work. Many commercial 'kits' for rapid bacterial identification by serology now use this method, but it may not be suitable for precise typing, because of the problems of cross-reactivity and false positivity (Shanker et al. 1982).

Antiglobulin reaction

A widely used modification of the haemagglutination test, the Coomb's reaction (Coombs, Mourant and Race 1945), was designed to detect the presence of so-called 'incomplete' antibodies to red cell antigens, but was quickly appreciated as a sensitive technique for the detection of the presence of antibody (immuno-globulin) when bound to cells. The test is widely used, both in its original form and in various adaptations and is suited to a variety of methods for detecting the endpoint of antigen-antibody interaction (Coombs and Roberts 1959). The test uses heterologous antiserum directed against the immunoglobulin(s) under study. Thus, human immunoglobulin bound to a cell as a result of specific reactivity with a cellular antigen is then detected by an antihuman globulin e.g. rabbit antihuman globulin (indirect antiglobulin test). The binding of the rabbit globulin label may be observed by one of several methods e.g. by the agglutination of the cells, or the rabbit globulin may be 'tagged' with a fluorescent marker, (immunofluorescence) or enzyme (ELISA) which can be used to render the reaction visible.

Antiglobulin reactions are widely exploited in many branches of laboratory medicine e.g. in red cell serology, in the Rose-Waaler test (for rheumatoid factor), in the fluorescent treponemal antibody (FTA) test for the diagnosis of syphilis, and for the detection of antibodies in brucellosis (for references, see Gell and Coombs 1975). The test may be adapted to a wide variety of cells, in tissues and in suspension, and can also be modified to detect the binding of a specific class or sub-class of immunoglobulin. The application of the technique is most often on a qualitative basis—the detection of the presence or absence of a specific bacterial antigen—but it should not be forgotten that the quantitative test was widely applied to the study of the antibody content of antibacterial sera (see Kabat and Mayer 1948).

The 'prozone' in bacterial agglutination. Some bacterial suspensions may fail to show agglutination in high concentrations of antiserum; the inhibition may be partial or complete, varying with the concentration of the bacterial suspension. The effect has been variously attributed to non-agglutinating (so-called 'incomplete' antibodies), to the presence of 'blocking' antibody, or to antigenic determinants deep below the cell surface. The most likely explanation is that the binding of antibody and subsequent lattice formation is prevented by steric factors which hinder the proper physical juxtaposition of immunoglobulin and antigenic determinant. Steric hindrance might occur in the presence of excess antibody (every antigen site would be instantaneously bound to an antibody combining site) or where the antigen sites were relatively inaccessible (i.e. below the immediate cell surface). The latter explanation is supported by the observation that methods which alter the surface or which depend on the development of successive layers of antigen-antibody interaction (as in the Coomb's reaction) may produce agglutination.

Applications of immunochemical principles

Using the principles derived from these basic techniques, numerous modifications have been introduced which are widely applied in the study of cell biology: in many instances, immunochemical procedures are used for purposes far removed from microbiology, but the originating ideas were derived from observations on the behaviour of bacteria and other micro-organisms when exposed to specific ('immune') sera. Many of the modified procedures are useful in the microbiology laboratory, in the detection and characterization of antigens, in the purification of antigens for vaccine preparation, and in the identification of microbial materials in fixed and living culture cells as well as more directly in the diagnosis of disease.

Immunofluorescence (IMF)

The binding of specific antibody to cellular antigen can be made visible if the antibody is tagged with a marker molecule or fluorochrome which can be rendered fluorescent by exposure to light of suitable wavelength, generally in the blue or ultraviolet wavelength zone of the spectrum, but varying for different fluorochromes (Coons et al. 1941). The labelled antibody can be used to locate cellular products on the surface or even inside cells, either in tissues or in cultured material. The test is sensitive and highly specific when well characterized reagents are used, but is subject to various pitfalls, some of which are peculiar to the fluorescence procedure. Non-specific staining can be a problem unless care is taken in the selection of the antiserum to be conjugated. The test may be used in direct or indirect form. Direct immunofluorescence uses fluorochrome conjugate of a selected specific antiserum e.g. group specific anti-streptococcal antibody and is useful for typing or for location of microbial antigen in tissues or on cells. Indirect immunofluorescence makes use of the antiglobulin reaction: instead of conjugating each individual specific serum, an antiglobulin serum is conjugated to the fluorochrome, providing a universal reagent for the detection of bound (unconjugated) globulin. Thus, in the FTA for treponemal antibody, the patient's serum is incubated with the treponemal suspension or layered over a smear of organisms, and this is followed, after washing, by the fluorochrome conjugated-antiglobulin. Where specific treponemal antibody has bound, the fluorescent marker will in turn bind to the antibody and render it 'visible' when the smear is exposed to a suitable light source.

This so-called 'sandwich' technique is widely applicable in microbiology for the detection of antibodies in human and animal antisera, for the location of infected cells, e.g. in sputum, urine or biopsy material, and for typing cultured organisms isolated from patients (for details of IMF techniques and their application the reader is referred to the numerous reviews including Nairn (1968) and De Luca (1982).

Immunoelectron microscopy

By using suitable electron-dense markers, linked to specific antibody, the visualization of antigen-antibody reactions can be extended beyond the range of the light microscope and applied to cellular and subcellular constituents. The principle is derived directly from that of immunofluorescence, using appropriate markers. The most widely used in early work was ferritin (Morgan et al. 1961), which was conjugated to antibody and employed to trace intracellular viral antigen. Subsequently, several other markers with similar properties were introduced, including gold as colloidal particles (Horisberger and Rossett 1977) to study yeasts, and horse radish peroxidase, the latter proving remarkably useful in the study of lymphocyte membrane antigens (for details, see Andres et al. 1973; Hoyer and Bucana 1982).

Radioimmunoassays (RIA)

Although this technique was originally applied to the measurement of hormone levels (specifically, insulin) in blood (Yalow and Berson 1959), it has proved to be an extremely versatile tool for the detection of specific antigenic determinants and measurement of antibody levels. The test lends itself to automation and is therefore useful where large numbers of samples must be assayed. Radioimmunoassays employ the principle of competitive binding, where labelled antigen, mixed with a fixed amount of specific antibody, is then allowed to mix with unlabelled antigen. Competition occurs between the unlabelled and labelled antigen molecules in the dynamic events of antibody-antigen binding, and the final amount of labelled antigen fixed is inversely proportional to the amount of unlabelled antigen present in the original mixture. Standard curves for antigen-antibody binding are derived by use of a known amount of pure antigen, and the 'dose-response' curve of a series of dilutions of the antigen under assay is compared. The test is relatively easy to perform, but does require more or less pure antigen for the preparation of the standard activity curves, and is also affected quite critically by the avidity of the antibody used. The radiolabelling of antigen is most often accomplished by iodination (using ^{125}I), and this chemical process may affect the properties of the antigen, including its binding sites. In an attempt to avoid this problem and perhaps to increase the sensitivity of the assay, a newer test has been developed, using labelled antibody, the immunoradiometric assay. This procedure requires relatively large amounts of antibody and should be particularly suited to the use of monoclonal antibodies which ought to be available in large quantities.

Radioimmunoassay of antibody. Initially exploited for the assay of antigens (hormones, enzymes etc.), RIA is proving equally useful in the measurement of specific antibody. Bacterial antigens are very suitable as a substrate for the assay because they can be prepared in relatively large amounts, purified where necessary and are readily bound to solid phase material when soluble. The solid phase used is generally some form of synthetic polymer or cross-linked dextran. Antigen coupled to the solid phase support is stable and easily handled and can often be stored for long periods. Many antigens of microorganisms are soluble or are produced as exotoxins or exoenzymes, making RIA tests readily adaptable to the assay of the specific antibody. By use of the antiglobulin reaction, it is also possible to develop a range of assays for different antibacterial or antiviral antibodies and to detect both the class and subclass of antibody. The test is thus particularly valuable where it is desired to measure specific IgM and IgG and to quantitate both in a single serum e.g. in rubella or brucellosis. Equally, the test can be adapted to measure specific IgE, directed against selected antigens (a test known as RAST—radioallergosorbent test—a convenient abbreviation for an ungainly 'portmanteau' description) including protozoal, bacterial, fungal and non-organic materials (for fuller details, see Parratt et al. 1982).

Enzyme-linked Immunoassay (ELISA)

This is also known as enzyme-linked immunosorbent assay (Engvall and Pesce 1978).

The use of enzyme reactions as markers for antigen-antibody binding developed out of RIA techniques. Instead of radioisotope as a marker to trace the antibody or antigen molecules, an enzyme coupled to the appropriate molecule is identified when it reacts with its appropriate substrate. The technique avoids the use of radiolabelled compounds with their attendant hazards of ionization and potential alteration in the structure of the bound molecule. The method was first applied to the measurement of specific proteins, one of which was immunoglobulin (Engvall and Perlmann 1971), but has since been applied to the detection of bacterial, viral, fungal, and protozoal antigens and, conversely, using known antigen, often coupled to solid phase materials, to the measurement of specific antibody (Johnson and Nakamura 1980). The ELISA uses enzyme molecules linked to antibody, frequently in the form of antiglobulin, to detect antigen-antibody reactions. Various suitable enzymes are available e.g. horse radish peroxidase, β-galactosidase, alkaline phosphatase and urease. The coupled enzyme is detected by virtue of its reaction with an appropriate substrate, when a colour change is observed. Thus, antibody fixed to a suspension of virus particles or bacteria can be detected as in the diagnosis of rubella, hepatitis B infection or Legionnaires' disease. The test is easy to perform, fast (1–3 hours) and relatively cheap (apart from the cost of conjugated antibody when purchased commercially) and is easy to automate for large numbers.

There are inherent problems in ELISA, as with most immunologically based tests. A peculiar feature of ELISA is the relative lack of stability of the chromagens used in the end-assay for the enzymes: these organic compounds, e.g. o-phenylenediamine, have a very short half-life (24 hours or less) and must be prepared fresh for each assay. Non-specific binding of serum proteins, including immunoglobulin, to the plastic used for reaction plates is a second major problem, leading to high background levels in the final readings. The precise reason for this binding is not known; it can be reduced by selecting plastic with low binding properties and by adding high molecular weight protein (e.g. bovine albumin) to the diluent used for titrating the serum.

In tissue sections, indirect labelling procedures, generally using peroxidase, can be usefully employed to increase the sensitivity of the assay, by building up layers of antigen + antibody + anti-antibody with a consequent increase in the amount of labelled immunoglobulin which is bound in the final layer (Nakane 1980).

Complement fixation *in vitro*

Complement is the collective name given to a series of twenty or more proteins found in human plasma. The effect of this collection of interacting molecules was first noted as early as 1895 when Pfeiffer remarked on the lytic property of fresh serum on cholera vibrios, and Bordet (1898) demonstrated that part of the lytic activity was heat-labile. At first, the property was merely identified as a serum component and it was not for many years that the complexity of the system was understood. Although the earliest observations which were made concerned the lysis of bacteria, the emphasis shifted on to haemolytic activity with Bordet's demonstration of the effect of fresh serum on red cells. For many years, the study of haemolytic mechanisms *in vitro* eclipsed the view of complement as an essential factor in many antibody-mediated reactions *in vivo*. Complement-fixation tests were developed, using the haemolytic end point, for the detection of antibodies in a wide variety of diseases. Only in recent years has there been a return of attention to the biological properties of complement and the role of complement components in inflammatory and allergic tissue damage. This section deals principally with the *in vitro* properties of complement. For a detailed description of the individual components, the reader is referred to Chapter 15 and for a discussion of the role of complement *in vivo* to Chapter 12.

The complement cascade. A simplified diagram of the components of complement which are relevant to the study of *in vitro* reactions is given in Chapter 15. The sequential interaction of the various proteins (many of them enzymes) which make up the system has been likened to a cascade, each event acting as the trigger or catalyst to the next. The normal activation and degradation of complement components is controlled by a finely balanced series of activators and inhibitors so that *in vivo* activity is neither excessive nor lacking. The *in vitro* study of red cell haemolysis, as mediated by antibody in the presence of complement, has been used to dissect the system into its individual components, but the application of the result to diagnostic tests is best understood in the simplest of terms, by a consideration of the properties of fresh whole serum.

Lysis of sheep red cells

Bordet (1898) was testing the properties of rabbit serum, from animals immunized with sheep cells, and observed that the haemolytic effect of the serum was lost if the serum was aged or heated to 55°C. He noted that the addition of *fresh* normal i.e. non-immune rabbit serum would restore the lytic properties of the 'depleted' serum (see Topley and Wilson 1975, pp. 272–275). The combination of red cells, haemolytic antibody (haemolysin or amboceptor) and complement leads to haemolysis only when conditions of temperature, concentration of salt ions (especially Ca^{++} and Mg^{++}) and of complement components are optimal.

This fundamental property of complement to participate in reactions between antigen and antibody has been exploited in the development of tests which are designed to identify specific antigen-antibody interactions e.g. in the diagnosis of many diseases, bacterial, viral, fungal and non-infectious. The *fixation* of complement to the antigen-antibody complex at the terminal region of the immunoglobulin heavy chains (the so-called 'Fc' region—see Chapter 15) effectively removes the complement proteins from the reaction mixture. *In vivo*, the biological effects of complement arise largely from this event; *in vitro*, we can measure the haemolytic activity—or the lack of it, when fixation has occurred—and, by standardizing the conditions of the test, obtain a qualitative estimate of whether antigen has combined with specific antibody. Complement-fixation tests are designed to detect specific antigen-antibody reactions by measuring the uptake of complement i.e. the fixation to the complex. To reveal the presence or absence of free complement, an indicator system is necessary. Sheep red cells sensitized with appropriate amounts of rabbit anti-sheep cell serum are an excellent and sensitive means of assaying the residual complement activity. The test is standardized by using a calculated amount of complement, most frequently in the form of guinea-pig serum. The haemolysis of sensitized sheep red cells by diminishing amounts of complement forms the basis of a standard lytic curve which is sigmoid in shape, revealing that lysis is not represented as a linear function. The central portion of the curve is relatively steep, indicating that here the reaction is very sensitive to small changes in the amount of complement present, and it is usual to employ 50 per cent lysis as the end point to ensure maximum sensitivity of the test. It should be noted that this test is at best semi-quantitative and is not a very suitable assay for the activity of complement itself, except as a measure of function in terms of haemolysis. There is no discrimination between the relative activities of each of the numerous complement components. These can be assayed separately by either immunochemical or functional assays (Thompson 1981).

This form of the diagnostic complement-fixation test has been widely used in diagnostic microbiology but is gradually being replaced by assays with less inherent variability which can be automated. It is also possible to recognize the individual fixed complement components by other procedures e.g. IMF, now that the exact nature of the complement proteins is known and antibodies to each are available. This much broader application of the study of 'complement fixation' indicates the revival of greater interest in the biological importance of complement in the immunological reactions which take place in the tissues.

(For a fuller description of antigen-antibody reactions, with a greatly extended list of references, see Chapter 7 in the 6th edition of this book.)

References

Andres, G. A., Hsu, K. G. and Seegal, B. C. (1973) Immunological techniques for the identification of antigens or antibodies by electron microscopy. In: *Handbook of Experimental Immunology*. 2nd edn, Vol. 2. Ed. by D. M. Weir. Blackwell, Oxford.

Bordet, J. (1898) *Ann. Inst. Pasteur* **12**, 688.

Boyd, W. C. and Hooker, S. B. (1934) *J. gen. Physiol.* **17**, 341.

Castle, D., Kessock-Philip, S. and Easmon, C. S. F. (1982) *J. clin. Path.* **35**, 719.

Christensen, P., Kahlmeter, G., Jonsson, S. and Kronvall, G. (1973) *Infect. Immun.* **7**, 881.

Coombs, R. R. A., Mourant, A. E. and Race, R. R. (1945) *Brit. J. exp. Path.* **26**, 255.

Coombs, R. R. A. and Roberts, F. (1959) *Brit. med. Bull.* **15**, 113–118.

Coons, A. H., Creech, H. J. and Jones, R. N. (1941) *Proc. Soc. exp. Biol. Med.* **47**, 200.

Dean, H. R. and Webb, R. A. (1926) *J. Path. Bact.* **29**, 473.

Engvall, E. and Perlmann, P. (1971) *Immunochemistry* **8**, 871.

Engvall, E. and Pesce, A. J. (1978) *Scand. J. Immunol.* Suppl. 7.

Gell, P. G. H. and Coombs, R. R. A. (1975) Basic immunological methods. In: *Clinical aspects of Immunology*. 3rd edn, pp. 3–53. Ed. by P. G. H. Gell, R. R. A. Coombs and P. J. Lachmann. Blackwell, Oxford.

Glenny, A. T. and Barr, M. (1932) *J. Path. Bact.* **35**, 91.

Glenny, A. T., Barr, M. and Stevens, M. F. (1932) *J. Path. Bact.* **35**, 495.

Heidelberger, M. (1956) *Lectures in Immunochemistry*. Academic Press, London.

Heidelberger, M. and Kendall, F. E. (1935a) *J. exp. Med.* **61**, 559; Ibid. (1935b) **62**, 697.

Horisberger, M. and Rosset, J, (1977) *J. histochem. Cytochem.* **25**, 295.

Hoyer, L. C. and Bucana, C. (1982) Principles of Immunoelectron microscopy. In: *Antibody as a Tool*, pp. 233–271. Ed. by J. J. Marchalonis and G. W. Warr. John Wiley, Chichester.

Humphrey, J. H. and White, R. G. (1970) *Immunology for Students of Medicine*. Blackwell Scientific Publications, Oxford.

Johnson, R. B. and Nakamura, R. M. (1980) Improved techniques in ELISA for viruses and IgG and IgM-type viral antibodies. In: *Immunoassays: Clinical laboratory techniques for the 1980s*, pp. 141–156. Ed. by R. M. Nakamura, W. R. Dito and E. S. Tucker. Alan R. Liss Inc., N.Y.

Jublin, J. and Winbald, S. (1976) *Acta pathol. microbiol. scand* (B) **81**, 179.

Kabat, E. A. (1976) *Structural Concepts in Immunology and Immunochemistry*. Holt, Rinehart & Winston Inc., N.Y.

Kabat, E. A. and Mayer, M. M. (Editors) (1948) *Experimental Immunochemistry*. 1st edn, pp. 67–95. Charles C Thomas, Springfield, USA; Ibid. (1961).

Landsteiner, K. (1945) *The specificity of serological reactions*. Harvard University Press, Cambridge.

De Luca, D. (1982) Immunofluorescence Analysis. In: *Antibody as a Tool*, pp. 189–231. Ed. by J. J. Marchalonis and G. W. Warr. John Wiley, Chichester.

Marrack, J. R. (1934) *Spec. Rep. Ser. med. Res. Coun., Lond.* No. 194.

Marrack, J. R., Hoch, H. and Johns, R. G. S. (1951) *Brit. J. exp. Path.* **32**, 212.

Maxted, W. R. (1948) *Lancet* **ii**, 225.

Morgan, C., Rifkind, R. A., Hsu, K. C., Holden, M., Seegal, B. C. and Rose, H. M. (1961) *Virology* **14**, 292.

Nairn, R. C. (1968) *Fluorescent Protein Tracing*. 3rd edn. Churchill Livingstone, Edinburgh.

Nakane, P. K. (1980) New developments in Tissue Enzyme-labelled Immunoassays. In: Immunoassays: Clinical Laboratory Techniques for the 1980s, III pp. 157–169. Ed. by R. M. Nakamura, W. R. Dito and E. S. Tucker. Alan R. Liss Inc., N.Y.

Nisonoff, A., Wissler, F. C. and Woernly, D. L. (1960) *Archs. Biochem. Biophys.* **88**, 241.

Otterness, I. and Karush, F. (1982) Principles of antibody reactions. In: *Antibody as a Tool*, pp. 97–137. Ed. by J. J. Marchalonis and G. W. Warr. John Wiley, Chichester.

Parratt, D., McKenzie, H., Nielsen, K. H. and Cubb, S. J. (1982) *Radioimmunoassay of antibody and its clinical applications.* John Wiley, Chichester.

Pfeiffer, R. (1895) *Z. Hyg. InfektKr.* **19,** 75.

Ramon, G. (1922) *C. R. Soc. Biol.* **86,** 661, 711, 813.

Roitt, I. (1980) *Essential Immunology.* 4th edn. Blackwell, Oxford.

Shanker, S., Daley, D. A. and Sorrell, T. C. (1981) *J. clin. Path.* **34,** 420.

Steward, M. W. (1978) Introduction to methods used to study antibody-antigen interactions. In: *Handbook of Experimental Immunology.* Volume 3. Ed. by D. M. Weir. Blackwell, Oxford.

Thompson, R. A. (Ed.) (1981) *Techniques in clinical immunology.* 2nd edn. Blackwell, Oxford.

Topley and Wilson's *Principles of Bacteriology, Virology and Immunity* (1975). 6th edn. Ed. by G. S. Wilson and A. Miles. Edward Arnold, London.

Yalow, R. S. and Berson, S. A. (1959) *Nature Lond.* **184,** 1648.

12

Antigen–antibody reactions—*in vivo*

Heather M. Dick

Introductory	330	Natural antibodies	333
Antibodies and inflammation	330	Maternal transfer	333
Antibodies and infection	330	Inapparent or subclinical infection as a	
Bacterial killing: the role of antibodies	331	source of natural antibody	333
Complement and bactericidal antibodies	331	Heterophile antibodies: Forssman antigen	334
Antibodies and opsonization	332	Natural antibodies and immunity	334
Antibodies in viral and parasitic infections	332	Immune complexes	335
Non-specific opsonins: C-reactive protein	332		

Introductory

Antibodies and inflammation

The study of the response to infection made by the various tissues which together form the so-called 'immune system' reveals a complicated series of events, in which both antibody and lymphocytes participate. It is necessary, when examining antibody-mediated processes, to see these in the context of cellular events, because in many kinds of infectious disease, both types of response contribute to eventual recovery from infection. The historical separation of humoral from cellular factors, dating from the late nineteenth century, should now be recognized for what it was: a wholly artificial division between physiological processes which are inextricably mixed. However, the immediate response to infection, particularly the primary encounter with a specific micro-organism, depends in great part on the development of antibodies which appear rapidly, with ready access to the site of infection via the blood and tissue fluids. The combination of these antibodies with antigens present on the invading organisms forms the background to many well recognized features of inflammation. The use of the phrase 'humoral response' tends to suggest that cells play no part but a moment's thought will make it obvious that a wide variety of cells must be involved from the earliest stages. Antigen-antibody reactions contribute positively by increasing the migration of cells, especially macrophages and polymorphonuclear leucocytes. The binding of complement is of critical importance in this process of 'recruitment'.

The dynamic nature of the events which follow infection are most conspicuous when inflammation is present in the classical localized form, with obvious 'rubor' (redness) 'calor (heat), 'tumor' (swelling) and 'dolor' (pain), as so graphically described in early texts on disease. Although not all infections appear in this readily recognizable form, the same underlying processes are present, whether buried deep in the tissues, as in pneumonia or appendicitis, or widely disseminated, as in brucellosis, malaria or influenza. The initial events following the invasion of tissues by microorganisms are dictated by the interaction between the products of the organisms (enzymes, toxins etc.) and the general 'first line' defence mechanisms. The subsequent events will be affected by the development of antibody directed against the pathogenic products of the organisms. Different types of immunoglobulin and the ease with which they can gain access to the organisms will also determine the eventual outcome, together with the contribution made by chemotactic and other soluble inflammatory products (Chapter 10).

Antibodies and infection

The various classes of immunoglobulins and their physicochemical properties and the mechanisms of hypersensitivity as mediated by antibodies are detailed in Chapter 14. These processes are generally harmful to the normal functions of the tissues where they develop. In the response to infection, we can observe the *beneficial* effects of antibody-antigen interaction, although it should be recognized that elements of hypersensitivity may also contribute to the pathological lesion e.g. in post-streptococcal glomerulonephritis, in endotoxic shock or in many chronic bacterial and parasitic infections.

The role of antibodies may be classified in terms of their effects on organisms or tissues (Table 12.1).

Table 12.1 Antimicrobial antibodies and their primary effects

Descriptive name	Effective properties
Opsonins	Promote phagocytosis
Cytotoxins	Cause cell death
Cytolysins	Cell lysis
Cytophilic antibody	Bind to tissue cells, and promote killing by lymphocytes (K cells).

Secondary events, which are the consequence of antigen-antibody reactions *in vivo* also contribute substantially to the response to infection (Table 12.2). The

Table 12.2 Secondary events due to antibody-antigen reactions *in vivo*

Immune complex deposition
Activation of complement
Promotion of chemotaxis
Release of soluble mediators

events that follow are influenced by the general health of the infected person and by genetic factors, which are now understood to play a major role in determining both the quality and magnitude of the immunological response (see Chapter 10).

Bacterial killing: the role of antibodies

The production of specific immunoglobulin directed against the antigenic determinants of invading microorganisms is accepted as one of the early events by which the immunological defence mechanisms can deal with infection. In the nineteenth century, it was believed that the presence of antibody or antitoxin was essential to recovery from infection and that antibody was primarily responsible for the death and even dissolution of bacteria. Pfeiffer (1894: see Wilson and Miles 1975*a*) demonstrated the killing and lysis of cholera vibrios *in vivo* and by serum from animals immunized against cholera; von Behring and Kitasato (1890) used serum from an immunized animal to confer immunity against tetanus toxin. These discoveries led to the search for antisera for the treatment of many killing infections: successful though this search was in some conditions (diphtheria, tetanus), it proved an unattainable goal in many other diseases, either because the necessary antibody could not be produced in sufficient quantity or because of the failure to realize that cell-mediated immunological responses were needed to eliminate or contain many infectious organisms. The relative eclipse of interest in the *in vivo* use of antibacterial antibodies that followed the introduction of chemotherapy and antibiotics has now been reversed by the introduction of monoclonal antibodies, produced *in vitro* by hybridomas between murine and/or human antibody-producing cells, with potential for unlimited quantities of specific Ig.

Direct bactericidal effects of antibody are not easy to confirm *in vivo*, where the contribution of other factors such as phagocytosis, prevention of adhesion, and toxin neutralization may be considerable. Lysis of bacteria *in vitro* is generally complement-dependent and the same is probably true *in vivo*. Antibody and complement are both necessary for the clearance of *Esch. coli* from the blood stream (Medhurst and Glynn 1970) in experimental infections in the mouse. Lysis of cholera vibrios in the peritoneal cavity is largely mediated by antibody, possibly because of damage to the microbial cell wall, rendering the organism more susceptible to the effect of lysozyme released by phagocytic cells.

Complement and bactericidal antibodies

Complement-dependent antibody-mediated lysis of bacteria is probably brought about by mechanisms very similar to those observed in studying red cell lysis. The activation of the complement cascade can be initiated by the presence of antibody fixed to the cell wall. Successive components of the complement cascade, beginning with C_1 (primarily an esterase in activity) are fixed to the Fc portion of the bound immunoglobulin. The final steps of complement fixation result in the formation of an enzymatically active complex which leads to irreparable cell wall damage. The subsequent loss and/or penetration of salts causes disorganization of the cell cytoplasm, with disruption of lysosomes and cell death (Humphrey and Dourmashkin 1969) (Fig. 12.1). (See Chapter 15.)

Gram-negative bacteria exhibit varying degrees of susceptibility to the lethal effects of antibody and complement. Organisms with smooth colonial forms due to the presence of a complete surface 'coat' of lipopolysaccharide are relatively or completely resistant e.g. freshly isolated strains of salmonellae (Nelson and Roantree 1967). The lipopolysaccharides are however, much more important as endotoxins which can themselves activate complement via the alternative pathway, even in the absence of antibody; and it is unlikely that the effects attributed to them in offering resistance to attack by complement plus antibody are either

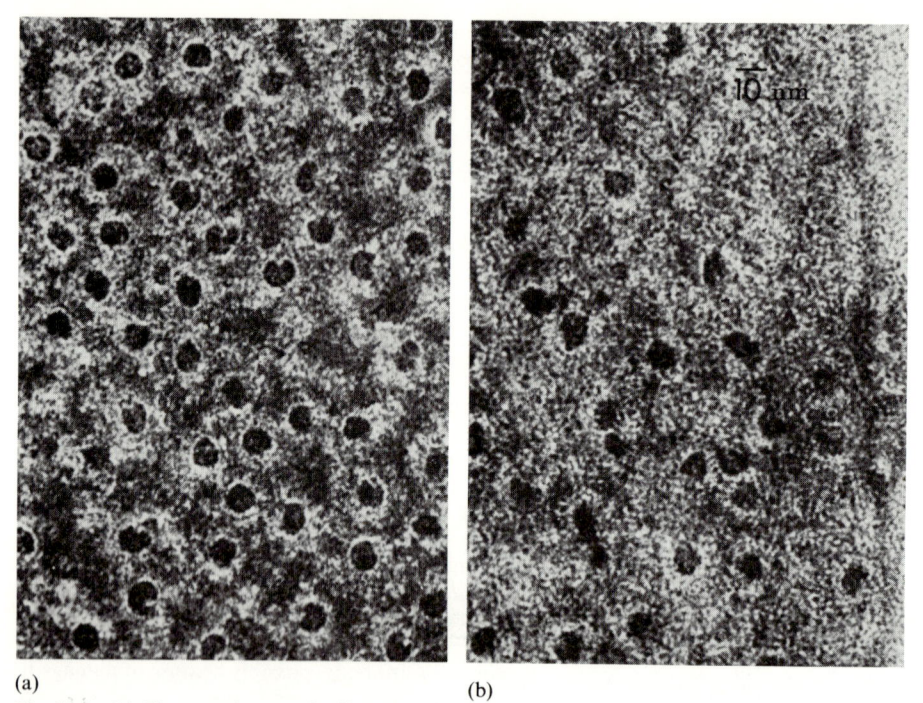

Fig. 12.1 (a). Electronmicrograph of human red cell membrane with lesions produced by antibody and human complement. (b). Electronmicrograph of *Esch. coli* cell wall with lesions produced by antibody and human complement. (× 320 000.) (From electronmicrographs kindly supplied by Drs R. Dourmashkin and J. H. Humphrey.)

major or significant. Other surface components, present in fresh isolates, may be lost after repeated culture, rendering previously resistant organisms susceptible to antibody (Ward *et al.* 1970).

Complement-dependent killing is most often produced through the classical pathway of activation by IgM antibodies, the first class of immunoglobulins to be produced in primary infections. Defects in IgM production are accompanied by repeated bacterial infections with septicaemia, attributing to the efficiency of IgM antibodies in controlling bacterial growth (Hobbs 1975).

The lytic functions of complement do not always require the fixation of antibody. The alternative pathway of complement activation can be triggered by several other means, including release of leucocyte proteases and the presence of endotoxins from bacteria, in particular, lipopolysaccharides (LPS) with the lipid A component playing the major role. Endotoxins have complex effects in inflammatory lesions: the molecules are themselves antigenic (T-cell independent) and stimulate specific antibodies of all three major classes (IgG, IgM and IgA) (Turner & Rowe 1964). Stimulation of both B-cells and macrophages also occurs, with production of interleukin 1 by the latter cells, promoting recruitment of cells and exacerbating the inflammatory response. The activation of complement also promotes the production of soluble chemotactic factors, and the combined effect of all these events undoubtedly contributes to bacterial killing within the lesion.

Antibodies and opsonization

Phagocytosis is an important mechanism for the removal and killing of bacteria (Chapter 10). Antibody contributes to the process, promoting the uptake of bacteria, often with the additional involvement of complement, although promoters of phagocytosis also exist which are non-specific. The presence of antibody is crucial for ingestion of capsulated strains and virulent strains of non-capsulated organisms. In the absence of opsonic antibody, phagocytosis is less rapid and intracellular death slowed down.

Antibodies in viral and parasitic infections. The relevance of specific antibodies in resistance to reinfection and recovery from infection has been regarded as of little consequence, with most emphasis being placed on the role of cell-mediated (lymphocyte) immunity. However, it is clear that antibodies may be important at several stages although effective immune responses require both lymphocytes and antibody (Chapter 16).

Non-specific opsonins: C-reactive protein

Evidence exists for opsonizing activity which is not dependent on the production of specific antibodies (see Wilson and Miles 1975*b*). Some of these effects which are detectable with fresh normal serum may be due to complement activity, since some of the effect is heat labile, but others must be attributed to soluble factors. C-reactive protein (CRP), one of the serum proteins associated with inflammation, may be one such molecule. CRP was discovered in the serum of patients with acute lobar pneumonia and was found to precipitate pneumococcal C polysaccharide (Tillett and Francis 1930). Measurement of CRP levels was used at one time as a screening test for possible inflammatory lesions and to monitor disease activity (e.g. in rheumatic fever) after it was noted that the rise and fall of CRP activity generally paralleled the progress of organic disease. CRP was regarded as one of the 'acute phase reactants' and until recently, its role in inflammation was ill defined. The biochemical structure and functions of the molecule have now been more fully described, and it can be seen to play an important part in the resolution of the inflammatory process (Pepys 1982). CRP binds to many soluble materials, including bacterial polysaccharides, and may then activate the classical pathway, bringing complement components into play to potentiate bacterial killing. CRP has opsonizing activity, both through its interaction with complement and of itself, perhaps increasing phagocytosis by polymorphonuclear cells of a wide variety of organisms, including pneumococci and *Esch. coli* (Kindmark 1971, Mold *et al.* 1982). It is currently suggested that the function of CRP is primarily to act as a binding mechanism for many foreign molecules, pneumococcal C polysaccharide being an example of a bacterial product which is most effectively precipitated (or agglutinated, when present on whole bacteria) in the presence of CRP. The binding of such ligands renders them accessible to phagocytosis, and consequently to clearance and metabolic breakdown. Numerous micro-organisms produce similar products, all with phosphorylcholine residues which have a particular affinity for CRP (Gotschlich *et al.* 1982).

Natural antibodies

It is possible to demonstrate the presence of various kinds of antibodies in the serum of individuals who have apparently not suffered from clinical infection with the organism(s) against which these so-called 'natural' antibodies are directed. Agglutinating, antitoxic and bactericidal antibodies have been described, directed against both gram-negative and gram-positive species. The phenomenon has also been observed with the sera of other mammals and of birds. The origin of these antibodies has been disputed and several likely sources have been suggested.

Maternal transfer

Passively transferred immunoglobulin (mainly IgG in man) is present in the serum of newborn infants owing to transfer *in utero* from the maternal circulation. As the mammalian infant matures and its own immunoglobulin-producing cells develop, the maternal antibody level gradually declines and although there may be a short period of hypogammaglobulinaemia, this is transient and usually not harmful (Brambell 1966). Maternal immunoglobulin (mainly IgA) also reaches the newborn infant via colostrum and/or breast milk, and probably contributes to the protection of the upper respiratory tract and gut. Serial studies of animals delivered in germ-free conditions have shown that the appearance of intrinsic immunoglobulin production coincides with the introduction of bacterial antigens in the diet, either as living or dead organisms, and at the same time, the lymphoid tissues of spleen and nodes show an increase in size and maturity, with plasma cells and immunoglobulin-bearing B cells increasing in number (Miller and Davies 1964).

The presence of maternal 'natural' antibody must be considered when measurements are made of antibody response following immunization in very young infants. It has been claimed that such pre-existing antibodies can interfere with the response to vaccines, e.g. with the capsular polysaccharides of meningococci of Type A and C (Gold *et al.* 1975) and pneumococcal vaccines (Borgono *et al.* 1978).

Inapparent or subclinical infection as a source of natural antibody

The presence in 'normal' serum of antibodies directed against both bacteria and viruses has been attributed to previous exposure with little or no clinical evidence of infection. Epidemiological surveys of pre-existing immunity to diphtheria and scarlet fever usually reveal a significant number of individuals, with no history of infection, who nevertheless give the appropriate skin reaction. The frequency of such reactors has been related to the likelihood of exposure to the organism depending on the known local incidence of disease; for example, the observation of a substantial increase in schick-negative reactors in a semi-isolated community following episodes of clinical diphtheria (Dudley 1923 and 1926, quoted in Wilson and Miles 1975c). Antibacterial antibodies have also been detected in the sera of a wide range of animal species, by means of both agglutination, complement-fixation and bactericidal assays. Some of these antibodies were found to react with bacterial species normally regarded as human pathogens (e.g. *V. cholerae*, *Bruc. abortus*, *Salm. typhi*), whereas others were directed against the antigens of species widely distributed in animals (e.g. *Esch. coli*, *Salmonella* sp. etc.). It seems likely that some of this antibody reactivity is the result of infection, probably of the gut in the first instance, with bacteria expressing cross-reactive antigens. Thus, the normal presence in the gut of numerous gram-negative bacteria leads to the exposure of the cells of the gut-associated lymphoid tissue to the antigenic determinants on these organisms, stimulating the production of antibody and memory cells. Although the antibodies formed may not have the precise specificity of the pathogenic strains, there are enough similarities (particularly in polysaccharide determinants) to give rise to serological cross-reactions.

Additional stimuli may be present in ingested food, both animal and vegetable. Many plants contain carbohydrate constituents which mimic antigenic determinants on human cells, bacteria and viruses. The isoagglutinins of the human AB blood groups are believed to arise from exposure to such cross-reactive determinants (Wiener 1951).

Heterophile antibodies: Forssman antigen

Forssman antigen is widely distributed in many animal species and may also constitute a source of antigenic stimulation, leading to the development of 'natural' antibodies. The antigen-antibody reaction which detected the presence of this ubiquitous lipid-associated polysaccharide material was first described by Forssman, who observed that the injection of guinea-pig tissues into rabbits stimulated the production of a haemolysin for sheep red cells (Forssman 1911). The name 'Forssman' has subsequently been applied to any material (generally of animal origin) which would stimulate the production of sheep cell haemolysin. The general characteristic of antigens of the Forssman type is that they stimulate antibody production in an animal in response to an antigen derived from one species, which antibody is found to react with antigen in yet another species (heterophile antibody). Thus, guinea-pig cells stimulated the rabbit to produce antibodies reactive with sheep red cells. Forssman antigens derived from different species are

Table 12.3 Some examples of the distribution of Forssman antigen

Red cells and other tissues
  horse
  cat
  mouse
  sheep
  chicken
Kidney cells of guinea-pig (not red cells)

Bacteria with Forssman antigen
  Pneumococci
  Shig. dysenteriae
  Pasteurella
  Neisseria catarrhalis
  Some *Salmonella* spp.
  Clostridium perfringens

similar, if not identical, and have many similarities to known antigens e.g. human blood group A_1 substances (Franks and Coombs 1969, Sweeley and Dawson 1969). Forssman antigens are present on the red cells of many species and are also found in the cell walls of several bacterial species (Table 12.3).

Other examples of cross-reacting antigens which stimulate heterophile antibody production have been identified. The exact structure of this type of antigen is not known, although most have polysaccharides as a major constituent. Some cross-react with bacteria and may confuse serological diagnosis, although others are put to good use in diagnostic tests where their cross-reactivity can be interpreted, e.g. the Weil-Felix reaction, where antibodies against rickettsia causing typhus also agglutinate a suspension of *Proteus*, strain OX19. Some examples of heterophile antibody reactions are listed in Table 12.4.

Table 12.4 Cross-reactive antigens revealed by heterophile antibody reactions

Blood group B substance and Esch. *coli* O86
Blood group A substance and pneumococcus type 14 capsular polysaccharide
Blood group P_1 antigen and hydatid cyst fluid
Blood group A substance and streptococcal extracts
Paul-Bunnell reaction (sheep and ox RBCs) with human serum

Natural antibodies and immunity

Although the existence of antibodies can be demonstrated, apparently without previous overt clinical infection stimulating specific immunity, it is less easy to be certain whether such antibodies can be protective against subsequent infection. Epidemiological studies might suggest that such antibodies are the result of subclinical infection; it is difficult to demonstrate correlations for all the so-called natural antibodies and we must assume that some, if not all, of such immunoglobulins are the consequence of response to cross-reactive antigens. The most clear-cut example of specific natural antibody is the immunoglobulin which is transferred to the neonate from the maternal tissues, either by the transplacental route, or via colostrum or milk. The human infant receives a significant proportion of its early protection against infectious organisms from the maternal immunoglobulin (IgG) transferred across the placenta. Human milk has been shown to contain secretory IgA which may also protect the infant gut directly (Michael *et al.* 1971). Such antibodies help to tide the infant over the critical period immediately after birth; the maternal immunoglobulins are slowly metabolized over the next few months and the maturing immune system of the child responds to external antigenic stimuli, beginning to produce specific immunoglobulins.

Immune complexes

Antibody bound to antigen may give rise to complexes, which may be precipitated out or solubilized, depending on the relative proportions of antigen and antibody (Chapter 11). *In vivo*, such antigen-antibody complexes are detectable in the circulation and in the tissues, where they are believed to cause damage by their interaction with complement components, and the subsequent generation of numerous soluble substances which contribute to the development of inflammation at the site of deposition of the immune complexes. Such complexes have been detected in a wide variety of diseases including infections with bacteria, viruses and parasites. Some forms of autoimmune disease are known to be associated with complexes, including systemic lupus erythematosus (SLE) (Lambert 1978, Hughes 1979); and complexes have also been detected in rheumatoid arthritis (Winchester *et al.* 1969, Hay *et al.* 1979), glomerulonephritis (Cochrane 1971) and malignant disease (Hellström *et al.* 1971).

Circulating complexes or their presence in tissues may be detected by a wide variety of tests—perhaps such a wide variety as to raise the suspicion that no one test is ideal. The techniques may be based on the physical properties, or may use indirect methods to identify the presence of antibody (immunoglobulin) and/or complement components (Table 12.5) (see also Levinsky 1981, 1982).

It is perhaps significant that in very few diseases has the specific *antigen* been clearly identified by elution from the complexes. In SLE, antigen has been identified as DNA and various breakdown products of DNA. In subacute bacterial endocarditis, it has not yet been convincingly demonstrated that the antigen is whole bacteria or bacterial products. In post-streptococcal glomerulonephritis, direct evidence for the presence of streptococcal antigens on the glomeruli is lacking although the presence of both immunoglobulins and complement may be detected by direct immunofluorescence.

The study of antigens present in circulating immune complexes can be facilitated by the separation of the complex from serum by physico-chemical methods, followed by enzymic digestion or other methods designed to separate antigen from the bound antibody. Complexes may be removed by one of the techniques used for detection (Table 12.5) and purified by ultracentrifugation. Subsequent elution of antibody can be secured by pepsin digestion or by pH shift to acid levels (pH 3) when bound antibody tends to elute from the complex. Both methods may alter the structure of antigen and care is necessary to provide the correct conditions. By such methods, the presence of Epstein-Barr virus has been demonstrated in complexes from patients with Burkitt's lymphoma (Lachmann *et al.* 1981). The same workers also detected microbial antigen in complexes from the serum of patients with nephritis. In a detailed study of immune complex glomerulonephritis Gamble *et al.* (1982) successfully eluted antibody from renal biopsy material and demonstrated that it had specificity for *Klebs. pneumoniae* although they did not obtain antigenic material from the kidney tissues.

Many virus infections are associated with the presence of immune complexes, including measles, rubella, influenza and hepatitis B infection (WHO Report 1977). Hepatitis B infection may be followed by a variety of complications which are attributed to the deposition in the tissues of complexes which produce chronic inflammatory changes and lead to permanent damage. These sequelae include glomerulonephritis and periarteritis nodosa (Combes *et al.* 1971). Several animal models exist, where immune complex formation, viral infection and the development of the features of autoimmunity occur together e.g. Aleutian mink disease, murine leukaemia (Fudenberg and Wells 1976) and infection with LDH virus (Notkins *et al.* 1966).

Many parasitic and tropical infections have been reported as having a close association with the presence of complexes, either as an integral factor in the lesions which develop or as an important feature of the response to infection. Schistosomiasis has been particularly well studied; the presence of both circu-

Table 12.5 Some methods for the detection of immune complexes

1. Based on physical properties of complexes
    Ultracentrifugation of serum
    Cryoprecipitation (cryoglobulins)
    Precipitation by polyethylene glycol (PEG).
2. Detection of complexes or fixed immunoglobulin and/or C′
    Antiglobulin techniques (e.g. rheumatoid factors)
    C1q binding
    Anticomplementary effect of serum
3. Effect of complexes on target cells by binding to receptor on cell
    IgG   Fc receptors   e.g. B-lymphocytes
                                RAJI cells
                                K cells
    C_3b receptors   e.g. Macrophages
                                B-lymphocytes etc.

lating antigen and soluble complexes has been detected; the antigens reported include both DNA and carbohydrate components of the organism (for reviews, see Colley 1977 and Houba 1980). The presence of complexes has been linked to renal lesions in schistosomiasis (Hillyer and Lewert 1974) but, more recently, the importance of complexes of specific IgE with schistosomal antigen in stimulating eosinophils has been noted as a crucial event in the destruction and clearance of the parasites (Butterworth 1979, Cox 1979). Malaria is also complicated by lesions which can be attributed to the deposition of immune complexes formed between antigens derived from parasite (both the schizont and merozoite forms give rise to specific antigens) and precipitating antibody. Malarial antigen has been identified in glomerular lesions in both acute glomerulonephritis (associated with *P. falciparum*) and in the chronic nephrotic lesions characteristic of *P. malariae* infection (Houba 1977).

References

Behring, E. von and Kitasato, S. (1890) *Dtsch. med. Wschr.* **16**, 1113.
Borgono, J. M., McLean, A. A. and Vella, P. P. (1978) *Proc. Soc. exp. Biol. Med.* **157**, 148.
Brambell, F. W. R. (1966) *Lancet* **ii**, 1087.
Butterworth, A. et al. (1979) *J. Immunol.* **122**, 221.
Cochrane, C. B. (1971) *J. exp. Med.* **134**, 75.
Colley, D. G. (1977) In: *Recent Advances in Clinical Immunology*. p. 101. Ed. by R. A. Thompson. Churchill Livingstone, Edinburgh.
Combes, B. Statsny, P., Shorey, J., Barrera, A., Carter, W. W., Hull, A. and Eigenbrot, E. H. (1971) *Lancet* **ii**, 234.
Cox, F. E. G. (1979) *Nature* **278**, 401.
Forssman, J. (1911) *Biochem Z.* **37**, 78.
Franks, D. and Coombs, R. R. A. (1969) In: *Infectious mononucleosis*. Ed. by R. L. Carter and H. G. Penman. Blackwell Scientific Publications, Oxford.
Fudenberg, H. H. and Wells, J. V. (1976) In: *Recent Advances in Rheumatology*. Ed. by W. Watson Buchanan and W. Carson Dick. p. 171. Churchill Livingstone, Edinburgh.
Gamble, C. N., Kinchi, A., Depner, T. A. and Christensen, D. (1982) *Amer. J. clin. Pathol.* **77**, 347.
Gold, R., Lepow, M. L., Goldschneider, J., Draper, T. L. and Gotschlich, E. C. (1975) *J. clin. Invest.* **56**, 1536.
Gotschlich, E. C., Liu, T-Y. and Oliveira, E. (1982) *Ann. N.Y. Acad. Sci.* **389**, 163.
Hay, F. C., Nineham, L. J., Perumal, R. and Roitt, I. M. (1979) *Ann. rheum. Dis.* **38**, 1.
Hellström, I., Sjögren, H. O., Warner, G. A. and Hellström, K. E. (1971) *Int. J. Cancer* **7**, 226.
Hillyer, G. V. and Lewert R. M. (1974) *Amer. J. trop. Med. Hyg.* **23**, 404.
Hobbs, J. R. (1975) In: *Immunodeficiency in Man and Animals*. Ed. by D. Bergsma, R. A. Good and J. Finstad. Sinauer Associates, Inc., Sunderland, Mass.
Houba, V. (1977) *Amer. J. trop. Med. Hyg.* **26**, 233; (1980) In: *Immunological investigation of tropical parasitic diseases*. p. 130. Ed. by V. Houba. Churchill Livingstone, Edinburgh.
Hughes, G. R. V. (1979) *Connective Tissue Diseases*, 2nd edn. Blackwell Scientific Publications, Oxford.
Humphrey, J. H. and Dourmashkin, R. R. (1969) *Advanc. Immunol.* **11**, 75.
Kindmark, C-O. (1971) *Clin. exp. Immunol.* **8**, 941.
Lachmann, P. J., Macanovic, M., Harkiss, G. D., Oldroyd, R. G. and Habicht, J. (1981) *Clin. exp. Immunol.* **46**, 250.
Lambert, P. H. (1978) *J. clin. Lab. Immunol.* **1**, 1.
Levinsky, R. J. (1981) *Immunology Today* **2**, 94; (1982) In: *Clinical Aspects of Immunology*. 4th edn. p. 398. Ed. by P. J. Lachmann and D. K. Peters. Blackwell Scientific Publications, Oxford.
Medhurst, F.A. and Glynn, A. A. (1970) *Brit. J. exp. Path.* **51**, 498.
Michael, J. G., Ringenback, R. and Hottenstein, S. (1971) *J. infect. Dis.* **124**, 445.
Miller, J. F. A. P. and Davies, A. J. S. (1964) *Annu. Rev. Med.* **15**, 23.
Mold, C., Duclos, T. W., Nakayama, S., Edwards, K. M. and Gewurz, H. (1982) *Ann. N.Y. Acad. Sci.* **389**, 251.
Nelson, B. W. and Roantree, R. J. (1967) *J. gen. Microbiol.* **48**, 179.
Notkins, A. L., Mahar, S., Scheele, C. and Goffman, J. (1966) *J. exp. Med.* **124**, 81.
Pepys, M. B. (1982) In: *Clinical Aspects of Immunology*. 4th edn. Ed. by P. J. Lachmann, D. K. Peters. Blackwell Scientific Publications, Oxford.
Pfeiffer, R. (1894) *Z. Hyg. InfektKr.* **16**, 268.
Sweeley, C. C. and Dawson, G. (1969) In: *Red cell membrane: structure and function*. Ed. G. A. Jamieson and T. J. Greenwalt. Lippincott, Phil.
Tillett, W. S. and Francis, T. (1930) *J. exp. Med.* **52**, 561.
Turner, M. W. and Rowe, D. S. (1964) *Immunology* **7**, 639.
Ward, M. E., Watt, P. J. and Glynn, A. A. (1970) *Nature, Lond.* **227**, 382.
WHO Scientific Group (1977) *WHO Technical Report Series* 606. WHO, Geneva.
Wiener, A. S. (1951) *J. Immunol.* **66**, 287.
Wilson, G. S. and Miles, A. (Eds) (1975a) *Topley and Wilson's Principles of Bacteriology, Virology and Immunity*, 6th edn., Vol. 1, p. 17, Edward Arnold, London; (1975b) Ibid. Vol. 2, p. 1330; (1975c) Ibid. Vol. 2, p. 1424.
Winchester, R. J., Agnello, V. and Kunkel, H. G. (1969) *Arth. Rheum.* **12**, 343.

13

Bacterial antigens

J. P. Arbuthnott, Peter Owen and R. J. Russell

1 Introductory	337	3.3.2 Staphylococcal Protein A	349
2 Major antigens of medically important bacteria	338	3.4 Surface Carbohydrate antigens	349
		3.4.1 General comments	349
3 Cell-associated antigens	345	3.4.2 Capsular polysaccharides	351
3.1 Flagella and axial filaments	345	3.4.3 Teichoic acids and lipoteichoic acids	354
3.2 Fimbrial antigens	345	3.4.4 Lipopolysaccharides	355
3.2.1 Fimbriae of enterobacteria	346	3.4.5 Enterobacterial common antigen	360
3.2.2 Fimbriae in other bacteria	347	3.4.6 Other glycolipids	361
3.3 Other surface proteins of Gram-positive bacteria	348	3.5 Outer membrane proteins	361
3.3.1 Streptococcal M, R and T proteins	348	4 Major extracellular antigens	367
		5 Vaccines and vaccine development	369

1 Introductory

The bacterial cell is a rich source of antigenic determinants. The major antigens comprise the macromolecules that are found on the bacterial cell surface and are excreted into the environment.

The degree to which any component may become an antigen depends on a number of different factors. The genetic background of an individual may affect the recognition of a particular antigenic determinant. The form which the antigenic determinant may take can vary and it rarely, if ever, occurs in pure state. It may be either synthesized or revealed to the immune system in varied quality and/or quantity. The determinant may occur alone or in association with other carrier or adjuvant substances derived from the bacterial cell, giving it the properties of tolerogen, immunogen or hapten separately or simultaneously. Evidence also exists that bacteria *in vivo* may envelop themselves with host-derived material. In addition, many species of bacteria exhibit antigenic variation. Antigenic determinants may undergo further changes or modifications as degradation products of phagocytosis, while yet more variation may be introduced in the way they are presented to the lymphoid cells by the mononuclear phagocytes. Intracellular bacterial components are also antigenic but these factors become involved in interactions between the host's immune surveillance system only when bacteria lyse or are degraded by phagocytic cells. This chapter will deal mainly with cell surface antigens and excreted (extracellular) antigens. From the standpoint of bacterial physiology the most important antigens are those associated with the cell surface. Such components form part of the structure of the bacterial plasmamembrane, periplasmic space, wall or envelope and are concerned in protection of the bacterial cell from antagonists such as: (a) adverse physiological conditions—including high salt concentrations, extremes of pH, temperature changes and the effect of dryness; (b) bacteriophages; (c) antibacterial agents and (d) antibodies.

In addition, external structures such as capsular or slime layers, fimbriae and flagella interact with the environment in such a way as to allow bacteria to establish themselves in a particular ecological niche. This is important whether the niche be the gingival crevice or a rock crevice in a river bed.

In terms of interaction between bacteria and man, these surface structures are factors that determine the distribution of the many bacterial types that make up the so-called normal flora; these interactions are vital to the success of the bacterial species involved.

In the relatively few cases where the equilibrium between man and his bacterial flora is disturbed, or where man encounters a true pathogen, with the consequent initiation of infection, the bacterial surface often determines the initial events in pathogenesis.

Where this property is well recognized and where the antigen has been identified, surface antigens have been used in the development of protective vaccines. With the growing limitations and costs of antibiotics, new possibilities for controlling infection by prophylactic immunization are being assessed. Thus the search continues for surface antigens that are major or minor determinants of pathogenicity and that might be candidates for new or improved vaccines.

Extracellular antigens are also important in host/pathogen interactions. Sometimes, tissue destructive toxins (membrane-damaging cytolysins) aid the establishment of infection often creating conditions where secreted enzymes such as phospholipases and proteases can act to provide the pathogen with essential precursors of biosynthesis. Other toxins including those of diphtheria, tetanus, botulism, cholera and *Escherichia coli* gastro-enteritis are responsible for the symptoms of disease and in such cases antibodies are protective against the symptoms of the disease. Thus established vaccines and vaccines under development often contain toxoid antigens. The major extracellular antigens will be dealt with in this chapter.

Antigenic analysis has proved extremely important in providing the microbiologist with serological methods for identifying and classifying bacteria. Diagnosis is therefore often based on serological tests for group- or type-specific antigens. Such tests are also invaluable in epidemiological investigations.

Progress continues towards an understanding, at the molecular level, of the organization of the macromolecular complexes that make up the surface components of bacterial cells. The high specificity of immunological labelling methods has played a major part in analysing the distribution and function of individual antigenic components within this mosaic. It is fascinating to note how a sensitive method such as crossed immunoelectrophoresis (CIE) reveals a complex pattern of 20-40 antigens where only 2 or 3 antigens were detected previously in double diffusion tests.

Several recent developments including the use of ultrasensitive immunological techniques, the application of recombinant DNA technology and the use of monoclonal antibodies have advanced knowledge of bacterial antigens. These developments separately or in conjunction are likely to have a profound influence on identification of virulence-associated antigens, on the separation and purification of antigens, on standardization of antigen preparations and detecting antibodies and on the production of better vaccines through an improved content of protective antigens.

Although traditional agglutination, precipitation, neutralization and complement-fixation tests are still used they have been complemented by the more sensitive methods of immunoelectrophoresis and its variants, enzyme-linked immunoassay (ELISA) and radioimmunoassay (RIA) which allow the detection of nanogram or picogram quantities of antigen or antibody. The use of monoclonal antibodies in the isolation, analysis and identification of biological substances has added a new dimension to the degree of resolution and reproducibility made possible in these areas.

The application of monoclonal antibodies to the study of bacterial antigens is yet in its infancy when compared with that of viruses. However, their effectiveness has adequately been demonstrated by Macario and colleagues (1982) in studying the antigenic structure and relatedness of the methane-producing Archaebacteria. Other studies have used monoclonal antibodies as immunological probes in the analysis of steric structure in lipopolysaccharides (LPS), and for discovering previously undetected antigens. The introduction of monoclonal antibodies for the detection of trace amounts of extracellular products, e.g. alginate from *Pseudomonas aeruginosa* and cholera toxin in human body fluids and for serodiagnosis of tuberculosis also shows a great deal of promise. In time, it should be possible to supplement the standard biochemical methods for definition of a bacterial antigen by reference to a panel of standardized monoclonal antibodies.

Recombinant DNA technology has made it possible to isolate and manipulate antigens at genetic level, allowing their investigation clear from other interfering antigens. This approach is also revealing the complex genetic control of antigen production as in, for example, pilus production in *Neisseria gonorrhoeae* (Meyer *et al.* 1982).

2 Major antigens of medically important bacteria

The major antigens of a number of medically important bacteria are described in summary form in Table 13.1 which documents the location, chemical nature and biological importance of these components. For a more detailed coverage of the serological relationships in the enterobacteria the reader is referred to Kwapinski (1974). Other useful reference sources covering the biological and immunochemical properties of antigens are: Ghuysen, Strominger and Tipper (1968); Jann and Westphal (1975); Steele, Jenkin and Rowley (1977); Sutherland (1977); Rogers, Perkins and Ward (1980); Beachey (1980) and Good and Day (1981).

Table 13.1 Major antigens of medically important bacteria

Species	Cell-associated antigens	Extracellular protein antigens	Remarks
Gram-positive cocci *Staphylococcus aureus*	**Peptidoglycan** (glycine interpeptide bridge) **Teichoic acid,** cell wall (glucosamino-ribitol) **Lipoteichoic acid,** membrane (glycerol phosphate backbone) **Protein A** present as bound and free forms **Clumping factor/bound coagulase** (protein) Reacts directly with fibrinogen. **Capsule** (contains aminoglucuronic acid, N-acetyl-L-alanine	**Coagulase** several isotypes exist **Leucocidin** F and S components both required for activity **Membrane-damaging cytolysins** $\alpha, \beta, \gamma, \delta$ **Enterotoxins** A, B, C$_1$, C$_2$, D, E **Epidermolytic toxin** serotypes A and B exist **Pyrogenic exotoxin** associated with Toxic Shock Syndrome	Serological typing not used extensively. Antibody to α-lysin, γ-lysin and leucocidin rise in chronic infections e.g. osteomyelitis. Teichoic acids act as species-specific antigens. No overall protective antigen though α-lysin antibody protects against necrosis.
Streptococcus pneumoniae	**M protein** type specific, unrelated to M protein of *Streptococcus pyogenes* **Capsule** (polysaccharide) 83 serological types **C-substance** (glucose,2-acetamido,2-4-6-trideoxygalactose, galactosamine, ribitol phosphate, choline phosphate)—species-specific	**Membrane-damaging cytolysin** oxygen-labile, serologically related to streptolysin O of *Str. pyogenes*	Antibody to capsule is protective and cross-reacts with *Klebsiella* and *Esch. coli* capsules. Capsular rough to smooth variation exists.
Streptococcus pyogenes	**Group-specific** (rhamnose, N-acetyl glucosamine) in cell wall **M Protein** type-specific, approx. 60 serotypes **R Protein** **T Protein** **Capsule** (hyaluronic acid) **Lipoteichoic acid,** membrane involved in attachment to mammalian cells	**Membrane-damaging cytolysins** O and S types **Streptokinase** **Erythrogenic toxins** A, B, C **NADase** **DNAase** **Leucocidin** **Hyaluronidase** **Fibrinolysin** **Serum opacity factor**	M protein is a poor antigen and exists in close physical association with serum opacity factor and M-associated protein. M protein antibody confers type-specific protection. M, T and R antigens are used in epidemiological typing. Additional antigens cross-react with myocardium, synovial fluid, skeletal muscle, brain and skin. Antibody levels to extracellular proteins are used to monitor streptococcal infection.
Streptococci of other Lancefield groups	**Group polysaccharide** **Lipoteichoic acid,** membrane		Teichoic acids constitute the group-specific antigens of group D and group N. Groups B, C, D and G are the important human pathogens.
Oral streptococci	**Lipoteichoic acid,** membrane **Glucosyl transferase** (glycoprotein) various isotypes exist	**Glucosyl transferase** (glycoprotein) various isotypes exist.	Oral streptococci are implicated in dental caries. Glucosyl transferase makes adhesive polysaccharide polymers in dental plaque. Teichoic acid is group-specific antigen of *Str. mutans*.

Table 13.1 (continued)

Species	Cell-associated antigens	Extracellular protein antigens	Remarks
Gram-positive bacilli			
Bacillus anthracis	**Somatic polysaccharide** (N-acetyl-glucosamine-galactose) **Capsule** (poly D-glutamic acid)	**Exotoxin** { Factor I (lipoprotein) / Factor II (protein) / Factor III (protein) }	Antibody to capsule is protective. Combination of exotoxin factors is required for maximum toxicity, antibody to the complex of factors is protective. Somatic polysaccharide antigen cross-reacts with type XIV pneumococcal polysaccharide.
Clostridium botulinum	**Flagella** **Spore** **Somatic heat-labile agglutinin**	**Neurotoxin** 8 serotypes: A, B and E important in humans	Neurotoxin antibody is protective. Antibody to spore antigens is more specific than antibody to vegetative cells.
Clostridium difficile	Poorly defined	**Cytotoxin**	Cytotoxin cross-reacts with toxin of *Cl. sordelli*.
Clostridium perfringens	**Somatic** no clearcut grouping	**Lethal toxins** $\alpha, \beta, \gamma, \delta, \varepsilon, \iota, \eta$	Serotyped on extracellular antigen pattern—types A–F; α-toxin is the major lethal toxin.
Clostridium tetani	**Flagella** 10 serotypes **Somatic** species-specific **Spore**	**Membrane-damaging cytolysin** oxygen-labile **Tetanospasmin** major lethal neurotoxin	Antibody to tetanospasmin is protective.
Corynebacterium diphtheriae	**Somatic** (arabinogalactan) group-specific **K** (protein) heat-labile, used for serological typing **Cord factor** (trehalose corynemycolate) **Fimbriae**	**Exotoxin** major lethal toxin	Antibody to exotoxin is major protective antigen. Toxoid is used as an immunogen. Somatic antigen cross-reacts with mycobacteria and nocardiae.
Listeria monocytogenes	**O-somatic polysaccharide** (contains galactose, glucosamine, rhamnose, glucose) **H-flagellar** **Rantz** (probably teichoic acid)	**Membrane-damaging cytolysin** O-labile	11 serotypes based on somatic polysaccharides.
Propionibacterium acnes	**Cell wall polysaccharides** (hexoses and hexosamines)—TCA-extractable	**Haemolysin**	Two serogroups based on extractable polysaccharides.
Gram-negative cocci			
Neisseria gonorrhoeae	**O(LPS)** strong cross-reactivity with *N. meningitidis*. Glucosamine and galactose are the major sugars in O-specific side-chains of virulent Kellog type 1 organisms. **Fimbriae** **Outer membrane proteins** Protein I is major antigen, Protein II is heat-modifiable and has five forms.		Serological typing is not widely used, but fluorescent antibody tests useful in identification. LPS, fimbriae and outer membrane proteins are all implicated in adhesion.

Neisseria meningitidis

O(LPS) strong serological cross-reactivity with *N. gonorrhoeae*
Outer membrane proteins type-specific protein antigens; 15 known protein serotypes
Fimbriae
Capsule (polysaccharide) group-specific antigen (for chemical composition see Table 13.3) Groups A, B, C, D, X, Y, Z, 29E, W135.

Vaccines containing A and C polysaccharide are protective. Group B polysaccharide is weakly immunogenic but outer membrane (serotype 2) protein antigen is a potential vaccine component.

Enterobacteriaceae
Enterobacter species

O(LPS) typical of enterobacteria, 68 types
H-flagellar 34 types
K capsular on some strains
Fimbriae type 1 on some strains
Enterobacterial common/ECA/Kunin

Escherichia coli

O(LPS) LPS architecture and sugar content similar to salmonellae. 162 O-serogroups known
H-flagellar 52 H-types known
K capsular designated A, B, L on basis of heat sensitivity. 100 serotypes known
Fimbriae Types 1, K88, K99, 987, CFAI, CFAII
Outer membrane proteins 3 heterogeneous groups corresponding approx. to omp A, omp F and omp C of *Esch. coli* K12.
Slime/mucous/M (colanic acid)
Enterobacterial common/ECA/Kunin heteropolymer, see text

Heat-labile enterotoxin/LT
Heat-stable enterotoxin/ST
Membrane-damaging cytolysin haemolysin

K1 antigen is associated with meningitis and cross-reacts with meningococcal group B polysaccharide. Antibody to K88, K99 and 987 protective against diarrhoeal disease in animals. M antigen cross-reacts with antigen K30. Antigenic formula has the form e.g. 0111:K58:H2.

Klebsiella species

O(LPS) O-side chains mainly homopolysaccharides. 120 antigenic serogroups.
K capsular (charged monosaccharide e.g. glucuronic acid) Over 80 types known.
Fimbriae Types 1 and 3.
Enterobacterial common/ECA/Kunin.

Proteus species

O(LPS) several interesting features; contains D and L glyceromannoheptose and hexuronic acids. Lysine also reported. Hexose and hexosamine prominent in O-side chains. 117 O-serogroups
H-flagellar 44 types
Fimbriae Types 1, 3, 4
Enterobacterial common/ECA/Kunin

O-antigen of *P. vulgaris* shares determinants with rickettsial group antigen.

Table 13.1 (continued)

Species	Cell-associated antigens	Extracellular protein antigens	Remarks
Salmonella species	**O(LPS)** structure well defined, archetype of enterobacteria. Specificity of O-side-chains forms basis of identification scheme **H-flagellar** also used in serotyping. Phase variation exists (see text). Altogether approx. 1700 serotypes **Vi** (N-acetyl, O-acetyl, galacturonic acid) present in *S. typhi*, *S. paratyphi C*—major virulence antigen **Fimbrial antigen** type 1 present on many species **Enterobacterial common/ECA/Kunin**	**Enterotoxin** produced by *S. typhimurium*, immunologically related to cholera toxin	Antigenic formula has the form: O antigens:H (phase 1):H (phase 2) e.g. 1,9,12:a:1,5. Human vaccine consists of heat-killed *S. typhi* and *S. paratyphi* A, B, C organisms. Acetone-treated vaccine retains Vi antigen and is more effective.
Serratia marcescens	**O(LPS)** typical of enterobacteria—150 antigenic types **Capsule** (polysaccharide) at least two heteropolysaccharides present **Enterobacterial common/ECA/Kunin** **H-flagellar** 16 serotypes **Fimbriae** Type 1 and type 5		Serotyping is useful in epidemiological studies.
Shigella species	**O(LPS)** both core and O-side chains vary according to species but generally typical of enterobacteria. Numerous serotypes. **Enterobacterial common/ECA/Kunin** **Cell-associated toxin** **Fimbriae** Type 1	Neurotoxic, cytotoxic and enterotoxic activities found in supernatants; all three activities thought to be associated with a single protein	There are 39 serotypes divided into 4 species or serologic serogroups A–D. Lysogenic conversion may alter antigenic patterns. *Sh. sonnei* shows S–R phase variation of O-antigens. 2-amino-2-deoxyhexuronic acid is major determinant of phase 1 specific polysaccharide. Type-specific protection through oral administration of attenuated organisms.
Yersinia pestis	**O(LPS)** rough type **F1 protein** heat-labile **V protein** **W lipoprotein** **Cell-associated toxin** additional to endotoxin **Enterobacterial common/ECA/Kunin**		Surface antigen pattern complex. Several antigens are required for full virulence. Some are found only at 37°. Killed/attenuated vaccines available for those at risk.
Other gram-negative bacilli *Bacteroides fragilis*	**O(LPS)** heptose and KDO absent, contains unusual fatty acids Monosaccharide composition of *B. fragilis* and *B. melaninogenicus* similar **Surface-layer polysaccharide** present in *B. fragilis* subsp. *fragilis* (contains hexose, hexosamine, methylpentose)	**Neuraminidase** **Fibrinolysin**	Numerous serotypes have been reported, but no single classification scheme yet adopted.

Bacteroides melaninogenicus	**O(LPS)** similar to *B. fragilis* **Bound collagenase**	
Bordetella pertussis	**Agglutinogens** 1, 2, 3 are predominant, 4, 5, 6 minor **O(LPS)** **Heat-labile dermonecrotoxin** **Histamine-sensitizing factor/ lymphocytosis-promoting factor**	**Fimbrinolysin** **RNAase** **DNAase** Certain cell-associated factors e.g. heat-labile toxin and histamine-sensitizing factor are also present in supernatants
Brucella species	**O(LPS)** Chemical composition still doubtful 'A' and 'M' shared determinants present on *Br. abortus*, *Br. melitensis* and *Br. suis*. Antigen A predominates in *Br. abortus* and *Br. suis*. Antigen M predominates in *Br. melitensis*. Antigens A and M consist of protein-LPS complexes	4 serogroups (A, B, C, C$_1$) by immunofluorescence studies. Phase variation S to R affects antigenic structure. Phase I organisms contain protective antigens. Smooth to rough variation occurs. Smooth type organisms should be used for serology (agglutination, complement fixation, antihuman globulin tests).
Campylobacter foetus	**O** type-specific antigens **H-flagellar** non-specific **Surface protein**	Surface antigens not yet clearly defined.
Francisella tularensis	**O(LPS)** **Polysaccharide** **Protein** cross-reacts with agglutinating antigens of *Brucella*	Only 1 serotype described. Attenuated vaccine available for persons at risk.
Fusobacterium species	**O(LPS)** resembles that of enterobacteria rather than bacteroides **Glycoprotein**	No serotype classification.
Haemophilus influenzae	**O(LPS)** similar to enterobacteria **Capsular polysaccharide** 6 chemically distinct capsular types a, b, c, d, e and f. Type b capsule (phosphodiester ribose-ribitol copolymer) is a major virulence factor **'M' substance** (protein)	Antibody to type b capsular polysaccharide protective but active immunization failed to protect infants. Capsular antigen and its antibody detected by a variety of methods e.g. CIE, ELISA, latex agglutination. Type b antigen cross-reacts with numerous organisms (e.g. *S. aureus*, *Str. pyogenes*).
Pseudomonas aeruginosa	**O** (O-side chains rich in amino sugars; phosphorus content of LPS higher than in enterobacteria). 17 serotypes of O-antigen **H-flagellar** 10 types **Fimbriae** **Slime (polysaccharide)** 7 antigenic types **Leucocidin**	**Exotoxin A** **Phospholipase C haemolysin** Heat-stable phospholipase glycolipid **Elastase** **Proteases**

Table 13.1 (continued)

Species	Cell-associated antigens	Extracellular protein antigens	Remarks
Vibrio cholerae	**O(LPS)** lacks KDO. 3 antigenic types based on shared group antigen A and type-specific antigens B and C **H-flagellar** one serological type **Outer membrane protein** heat-stable, present on Inaba and Ogawa serotypes **Adhesins** not yet fully characterized	**Choleragen/Choleragenoid** (natural toxoid) major diarrhoeagenic toxin **Neuraminidase** **Mucinase**	Serotype AB = Inaba, AC = Ogawa, ABC = Hikojima. Killed vaccines partially protective in endemic areas. Though cholera antitoxin neutralizes *Esch. coli* LT enterotoxin, *Esch. coli* antitoxin does not neutralize cholera toxin. Active stimulation of antitoxic immunity presents difficulties.
Various other bacteria *Borrelia* species	**Axial filaments** **Lipid complexes**		Antigenic variation exists.
Chlamydia trachomatis	**Group-specific** 2 types: lipopolysaccharide and protein **Species-specific** **Type-specific** not fully characterized		12 Serotypes exist: A, B, Ba, CJ, Ed, GF, H, I, K, L1, L2, L3. Antibody to type-specific antigen protective antigen in mice.
Legionella pneumophila	**Common surface** **Type-specific** 4 known		6 serogroups exist.
Leptospira interrogans	**Common somatic** (lipopolysaccharide) **Sheath** **Axial filaments**	**Membrane-damaging cytolysin/haemolysin**	Large number of serogroups and serotypes (serovars) by agglutination.
Mycobacterium leprae	**Common mycobacterial**		
Mycobacterium tuberculosis	**Common mycobacterial** (D-arabino-D galactan) **Cord factor** (trehalose dimycolate) **Sulphatides**. **Purified protein derivative (PPD)**		Serotyping not yet reliable. Cord factor is protective antigen in mice. PPD used in Mantoux test.
Mycoplasma species	**Membrane glycolipids** (galactosyl and glycosyl diglycerides) **Membrane proteins** **Capsule-like** (galactan and hexosamine polymer)		Species classification based on serology. Field trials have been conducted with inactivated vaccine.
Nocardia species	**Purified protein derivative (PPD)** **Polysaccharide** species-specific antigens		Most antigens cross-react with mycobacteria.
Rickettsia species	**Group-specific** **Cell envelope** type-specific **Common** (polysaccharide)		Antibodies to common polysaccharide antigen cross-react with *Proteus* (e.g. strain OX19).
Treponema pallidum	**Cardiolipin (Wassermann)** **Sheath** **Axial filament (Reiter)** (protein) **Slime** (polysaccharide) **Envelope** (protein) site for immobilizing antibody		Numerous serological tests available for diagnosis.

3 Cell-associated antigens

3.1 Flagella and axial filaments

The flagellar (H) antigens of motile bacterial species consist of protein subunit polymers of flagellin. The structure and function of bacterial flagella have been reviewed recently by Doetsch and Sjoblad (1980). The designation H antigen derives from the German word 'Hauch' meaning breath and signifies the similarity in appearance of motile bacterial species on solid media to breath misting over cold glass. The protein subunit of flagellin has been shown to be approximately 300 amino acids in length. Subunits originating from different species may differ slightly in molecular weight e.g. 33 kilodaltons (K) in *Bacillus subtilis*, 60 K in *Escherichia coli* and 51–57 K in salmonella species (Kondoh and Hotani 1974) but all are thought to have the same 'wedge-shaped' configuration (Calladine 1976). Flagellins from different bacterial species are antigenically distinct and are often characteristic of a particular species. For example, flagellin produced by salmonella species contains a novel amino acid, ε-N-methyl lysine, not encountered in any other flagellins (Joys and Kim 1978). All flagellins possess a low aromatic amino acid content but have increased contents of glutamic and aspartic acids.

The protein of the hook structure to which the flagellar filament is attached does not cross-react antigenically with flagellin. It has been shown that within salmonella species the hook protein remains antigenically homogeneous, whereas the flagellin is antigenically very heterogeneous (Kagawa et al. 1973).

Some bacterial genera e.g. *Pseudomonas*, *Vibrio* and *Proteus* possess a sheath structure surrounding the flagellar filament. This sheath is antigenic, and for *Vibrio cholerae* there have been conflicting reports as to whether it is immunologically related to the filament or not (Yang et al. 1977, Mulholland et al. 1978).

Flagellin as an immunogen is active only in polymerized or aggregated form. Monomeric subunits do not induce an immune response. In its immunogenic form, parenteral flagellin induces predominantly IgM antibodies in a T-independent response which tends to be long lasting. Unusually for a T-independent antigen, some IgG is also induced. In acetoacetylated form, in very low or very high doses, flagellin can induce specific immunological tolerance in experimental mice and rats (Shellan and Nossal 1968). If this is true also for man a possible mechanism for lack of resistance to infection is obvious.

It has long been held that flagella antigens do not stimulate the production of protective antibodies. Evidence for this originally came from studies using parenteral administration. Tannock and colleagues (1975) showed that O and Vi antigens of *Salmonella* were protective in chimpanzees whereas H antigens were not. However, recent research on mucosal immunity has shown that the H antigens can be protective. Flagellin of *V. cholerae*, for example, can induce the secretion of IgA along the mucosal surfaces and block colonization (Steele et al. 1975). Similarly, oral administration of attenuated *Salmonella typhi* has been shown to produce protective local immunity in the mucosa in human beings.

Flagellar antigens of the *Salmonella-Arizona* groups of bacteria are subject to a phenomenon known as phase variation, where the flagellin alternates between two antigenic types or phases. Flagellins of phase 1 are termed specific and are not shared by many species, whereas those of phase 2 are shared quite widely. Lederberg and Edwards (1953) showed there to be two separate non-allelic structural genes designated H1 and H2 coding for flagellins of phases 1 and 2 respectively. The H2 gene is now known to be closely linked to a repressor of the H1 gene. Under normal circumstances flagellin of phase 2 is produced and flagellin of phase 1 is repressed. However, at a frequency of 10^{-3} to 10^{-5} an inversion event stops synthesis of both phase 2 flagellin and the phase 1 repressor, thus allowing production of phase 1 flagellin (Silverman et al. 1979).

Structures called axial filaments, which resemble flagella, are found in treponemes, leptospires and borreliae. Little is known of the antigenic properties of this group or their appendages. The filaments of treponemas are flagella-like, composed of protein monomeric subunits of 37 K whereas leptospiral filaments are much more complex. The axial filaments are, except those of borreliae, enclosed by a sheath. The outer sheath of leptospires is highly antigenic and contains lipopolysaccharides which elicit both protective and bactericidal antibodies.

3.2 Fimbrial antigens

Fimbriae (Duguid et al. 1955) are adhesive filamentous appendages consisting of polypeptide subunits having molecular weights of approximately 20–30 K. The term 'fimbriae' will be used in preference to 'pili' because, as argued by Duguid and Old (1980), the former has priority, and because it seems appropriate to reserve the term pili to describe the sex fimbriae that are directly concerned with conjugative transfer of DNA (Ottow 1975, Jones 1977, Ørskov et al. 1977). Fimbrial antigens were classified into types (Duguid et al., 1966) based on their haemagglutinating properties and have recently been reviewed in detail (Duguid and Old 1980, Levine 1981, Pearse and Buchanan 1980). The general properties of fimbriae of types 1–4 are summarized in Table 13.2.

Table 13.2 Properties of Fimbriae types 1–4[a]

Type	Width (nm)	Length (μm)	Haemagglutination pattern	Sensitivity to mannose[b]	Organism
1	7.0	0.2–2.0	Broad range including guinea-pig, horse, fowl, human, rabbit and sheep	MS	*Esch. coli* *Klebsiella* species *Serratia marcescens* *Shigella flexneri* *Salmonella* spp.
2	7.0	0.2–2.0	Non-adhesive, non-haemagglutinating	—	*Salmonella gallinarum* *Salmonella pullorum*
3	4.8	0.2–0.15	Human, guinea-pig, ox, (acts only on tannic acid-treated cells)	MR	*Klebsiella aerogenes* *Serratia marcescens*
4	3.0–4.0	0.2–0.6	Sheep, fowl, guinea-pig, horse, human, rabbit, ox	MR	*Proteus* species

[a] Data compiled from Pearse and Buchanan (1980). [b] MS = mannose-sensitive; MR = mannose-resistant.

3.2.1 Fimbriae of enterobacteria

Investigations into the relation of fimbriae to haemagglutinins showed that although these substances often occurred together there was no strict relation between them; strains with certain haemagglutinating specificities lacked fimbriae and some strains with fimbriae were unable to cause haemagglutination (see review Duguid and Old 1980).

The filamentous surface antigens associated with enterotoxigenic strains of *Esch. coli*. viz. K88 antigen, K99 antigen, the colonization factor antigens CFAI and CFAII, and antigen 987 were not included in the classification of fimbrial types proposed by Duguid *et al.* (1966). Jones (1977) described these antigens non-committally as 'fibrillae'. It is now known that CFAI and CFAII and 987P have the appearance of typical fimbriae when examined by electron microscopy whereas K88 and K99 appear as flexible fibrils with a diameter of approximately 3–5 nm.

Type 1 fimbriae responsible for mannose-sensitive (MS) haemagglutination The haemagglutinating activity of fimbriae can be readily demonstrated by mixing a dense suspension of bacteria grown in static broth culture (48 hr at 37°) with a suspension of washed erythrocytes; the test is easily read on a white tile that is rocked continuously for 10 min. Most strains of *Esch. coli* belong to group I as defined by Duguid *et al.* (1955) and cause haemagglutination of erythrocytes of most animal species in a manner which is strongly inhibited by D-mannose (0.01–0.5 per cent w/v) or closely related compounds. This type of agglutination is known as mannose-sensitive (MS) haemagglutination. MS adhesins of this type have been found in many genera and species of enterobacteria (Table 13.2). The fimbriae responsible for MS haemagglutination are known as type 1 or common fimbriae. They are typically 7–10 nm in diameter, are produced in static broth culture, forming surface pellicles but not on solid medium, and cause mannose-sensitive agglutination of erythrocytes at temperatures between 0° and 55°. The presence of type 1 fimbriae confers on bacteria the ability to adhere to many kinds of cells other than erythrocytes and the haemagglutinin is referred to generally as MS adhesin.

The high specificity of the inhibitory action of mannose and related compounds together with other evidence suggests that the sites with which D-mannose reacts are located on the bacterial fimbriae rather than on the cell to which they adhere.

Although there is a partial sharing of type 1 fimbrial antigens among the species *Esch. coli*, *Shigella flexneri*, and *Klebsiella aerogenes* and among the genera *Salmonella*, *Arizona* and *Citrobacter*, there is no antigenic similarity between the type 1 fimbriae of these two groups of organisms. Other genera, such as *Edwardsiella*, *Enterobacter*, *Hafnia*, *Providencia* and *Serratia* possess still different antigenic forms. Type 1 fimbriae from haemagglutinating strains of *Salmonella paratyphi B* are antigenically related to non-haemagglutinating fimbriae (type 2) present on a minority of strains of this organism. Thus the mannose-sensitive site on the fimbrial structure does not seem to contribute significantly to the observed cross-reaction.

Fimbriae associated with mannose-resistant eluting (MRE) haemagglutination Strains of *Esch. coli* belonging to haemagglutination groups II and III (Duguid *et al.* 1955) show a variety of patterns of agglutination reactions with different species of erythrocytes. Unlike group I strains, haemagglutinins are produced optimally in cultures incubated at 37° on solid medium, and agglutination tests are best demonstrated at 3–5°. The reactions are unaffected by addition of D-mannose, and haemagglutination is reversed when tests are warmed. These haemagglutinins are therefore described as mannose-resistant and eluting

(MRE). Such MRE haemagglutinins have been found in several members of the enterobacteria and some strains possess both MS and MRE haemagglutinins (Table 13.2).

MRE activity is not always associated with the presence of morphologically typical fimbriae. Indeed, haemagglutination groups II and III can be distinguished from one another in that organisms in the former are fimbriate whereas those in the latter are not. Moreover, in a study of 19 MRE$^+$ *Esch. coli* strains, Duguid *et al.* (1979) found 12 strains to be fimbriate and 7 to be non-fimbriate. In fimbriate MRE$^+$ strains there was a strong association between the presence of fimbriae and MRE haemagglutination; fimbriae were absent under conditions that prevented formation of haemagglutinin. The problem of the structural location of MRE haemagglutinins in apparently non-fimbriate strains remains to be solved though it must be noted that MRE$^+$ strains that do not form fimbriae had haemagglutination patterns that differed from those of fimbriate MRE$^+$ strains. The filamentous antigens K88 and K99 of enterotoxigenic *Esch. coli* strains are examples of adhesive MRE haemagglutinins that do not have typical fimbrial structure.

A mannose-resistant haemagglutinin has been described in *Klebsiella aerogenes* which differs from other MRE haemagglutinins in that it acts only on tanned erythrocytes and does not elute when warmed to 50°. This haemagglutinin is found in strains of *K. aerogenes* that possess fimbriae termed 'type 3' (Duguid 1968) which are thinner (diameter 4.8 nm) than the common type 1 fimbriae that cause MS haemagglutination. Most *K. aerogenes* strains produce both type 1 and type 3 fimbriae.

A fimbrial antigen designated MR/P, which causes mannose-resistant non-eluting haemagglutination, is found in strains of *Proteus mirabilis*, *Proteus vulgaris*, *Morganella morgani*, and *Proteus rettgeri*. Many different patterns of haemagglutination and presumably fimbrial types (MS, MR/K, MR/P), however, have been found in individual strains of *Proteus* and *Providencia* (Duguid and Old 1980).

Role of fimbriae in adhesion Fimbrial antigens confer the ability on bacteria to adhere to the surfaces of cells possessing the appropriate receptors. For instance, enterobacteria with common type 1 MS fimbriae adhere to many kinds of cells other than erythrocytes. Most animal, plant and fungal cells so far tested serve as substrates for such organisms. The function of type 1 fimbriae is not known; but the ability of MS$^+$ bacteria to adhere to a wide variety of biological surfaces suggests that it might assist in colonization, for instance, of mucosal surfaces in animals. As pointed out by Duguid and Old (1980), non-fimbriate members of the enterobacteria, e.g. certain salmonellae and shigellae and some strains of *Esch. coli* are pathogenic, whereas conversely fimbriate strains can be isolated from normal healthy subjects; so there is no strict relation between possession of type 1 fimbriae and pathogenicity. Probably type 1 fimbriae also contribute to colonization of sites in the environment through adherence to plant or fungal tissues or through promotion of pellicle formation.

The role of certain *Esch. coli* MRE adhesins including K88, K99, 987, CFAI and CFAII is well established and has been thoroughly investigated and reviewed (e.g. Levine 1981). The K88 antigen (Ørskov and Ørskov 1966) is an essential virulence determinant of K88-positive strains of enterotoxigenic *Esch. coli* that cause diarrhoea in newborn piglets. Pathogenesis requires adherence to the upper small intestine and production of either heat-labile (LT) or heat-stable (ST) enterotoxin. The same holds true for the K99 antigen associated with enterotoxigenic strains of *Esch. coli* that cause diarrhoea in calves and lambs (Ørskov *et al.* 1975). Both K88 and K99 are filamentous protein surface antigens which are preferentially expressed after cultivation on solid medium at 37° but not at 18°. The genes encoding these adhesins reside in transferable plasmids.

In these strains the absence of K88 or K99 as a result of plasmid loss or mutation results in loss of ability to adhere and to cause diarrhoea. However, there are many strains isolated from piglets with diarrhoea that lack K88 or K99 antigens but are clearly pathogenic. These strains adhere to intestinal epithelium and cause diarrhoea when fed to piglets. In some such strains an additional fimbrial antigen (987) has been identified (Nagy *et al.* 1977). Evans and colleagues (1978a, b) have also described for human strains MRE fimbrial antigens termed colonization factor antigens I and II (CFAI and CFAII) which may be analogous to the adhesins of animal strains.

The identification of particular surface antigens responsible for adherence of enterotoxigenic *Esch. coli* raised the possibility that such antigens might be of value in the development of vaccines. Several studies have shown that immunization of sows provides sufficient antibody in colostrum to protect suckling piglets from diarrhoea and death following challenge with strains of enterotoxigenic *Esch. coli* (ETEC) bearing the homologous antigen. In six separate studies protection ranged from 80-100 per cent. Vaccines against strains of ETEC possessing K88, K99 or 987 are now commercially available.

3.2.2 Fimbriae in other bacteria

In addition to the enterobacteria, fimbriae have been observed in several other groups of bacteria, including *Neisseria*, *Vibrio*, *Bordetella*, *Pseudomonas*, and *Corynebacterium* (see review of Pearse and Buchanan 1980).

Fimbriae are universally present on fresh clinical isolates of *Neisseria gonorrhoeae* and *Neisseria men-*

ingitidis but the fimbrial character is lost after several subcultures *in vitro*. The fimbriae of *N. gonorrhoeae*, produced mainly by T1 and T2 colony types, have been studied in detail; they resemble type 1 fimbriae of enterobacteria, being 7 nm in diameter and 0.5–4.0 μm in length with 100–200 organelles per cell arranged peritrichously. A mannose-resistant haemagglutinin is associated with the fimbriae. Fimbriae facilitate attachment of gonococci to a wide range of eukaryotic cells with the exception of polymorphonuclear leucocytes (PMNs) where fimbriae serve an antiphagocytic function. The role of fimbriae in adherence may be to penetrate the electrostatic barrier between the gonococcus and the host cell. It must be emphasized, however, that non-fimbriate strains of *N. gonorrhoeae* also adhere, suggesting that fimbriae are not the only factors responsible for attachment.

Purified gonococcal fimbriae block attachment of *N. gonorrhoeae* to host cells, and immunization of human volunteers with purified fimbriae give rise to antibody capable of blocking attachment (Tramont 1981). However gonococcal fimbriae are antigenically heterogeneous and this is recognized as posing problems in developing an effective vaccine.

Less is known about the properties of fimbriae of other organisms (see Pearse and Buchanan 1980). Fimbriae resembling type 1 common fimbriae have been detected in strains of cholera and non-cholera vibrios (see review of Jones 1980), though several investigators have failed to find fimbriae on vibrio cells. The fimbriae of members of the Pseudomonadaceae are often polar or bipolar rather than peritrichous. Where fimbriae are polar in pseudomonads, e.g. in *Ps. aeruginosa*, the number of organelles per cell is low (e.g. 10), whereas in *Pseudomonas multivirons* and *Pseudomonas fragi* where the fimbriae are peritrichate the number per cell is of the order of 100–300. Fimbriae in pseudomonads tend to be flexible and narrower than type 1 fimbriae and often have retractile properties. The filamentous haemagglutinin of *Bordetella pertussis* acts on a broad range of erythrocytes. Recent work, however, suggests that this haemagglutinin is not associated with the fimbriae present on this organism.

The only genus of gram-positive bacteria reported to possess fimbriae is *Corynebacterium*. In *Corynebacterium renale* there are three serotypes (I, II, III) of which type II possesses most fimbriae. Bundles of fimbriae are common in these organisms but isolated organelles measure 2.5–3.0 nm in diameter and are up to 5–10 μm in length. They cause mannose-resistant haemagglutination of trypsinized sheep erythrocytes. Interestingly, organisms of type I and III are more pathogenic than type II but possess fewer fimbriae, so that the association of fimbriae with pathogenicity is doubtful. Several other species of *Corynebacterium* including *Corynebacterium diphtheriae* are fimbriate, possessing up to 100 organelles per cell.

3.3 Other surface proteins of gram-positive bacteria

Apart from flagella and fimbriae several species of bacteria possess additional surface protein antigens. Most of these are poorly characterized but two, streptococcal M protein and staphylococcal protein A, have been thoroughly studied and these are described in the following sections.

3.3.1 Streptococcal M, R and T proteins

Streptococcal M protein is found on organisms belonging to the group A streptococci and is located as fimbriae-like projections which cover the entire coccal surface (Swanson *et al.* 1969). Recent evidence, however, suggests that M-protein differs from pili and fimbriae in being based upon a highly extended structure rather than on assemblies of globular subunits. Thus, M-protein molecules are supposed to be anchored via their carboxy termini in the cell wall and to extend about 50 nm out from the cell in two-stranded α-helical coiled structures (Philips *et al.* 1981).

M protein is thought to have two main biological roles, namely mediating attachment to epithelial cells and providing an anti-phagocytic mechanism for group A streptococci. More recently, lipotechoic acid has been found to be concerned in attachment, and the role of M protein has been challenged (see Section 3.4.3). M protein is also of use for classification purposes owing to possession of type-specific antigenic determinants. Usually, only one type is present per strain, though some strains do have more. Despite numerous variations in extraction and purification methods, M protein has always appeared to be heterogeneous with regard to molecular weight and biological properties. Several other antigens are found in close association with M protein, notably serum opacity factor and M-associated protein. These have physico-chemical characters similar to those of M protein and hinder its purification. No definite role in virulence has been ascribed to either of these substances, although high levels of M-associated protein are invariably detected in rheumatic fever (Widdowson *et al.* 1971). M-associated protein is not type-specific.

Extraction of M protein yields components of differing molecular weight. Extracts of low molecular weight contain only the type-specific determinants, while those of higher molecular weight (27 K or over) also possess antiphagocytic properties and thus more closely resemble the characters of M protein in intact organisms. Variation in molecular weights is probably a reflection of proteolysis and might account in part for differing reports of antigenicity (Widdowson *et al.* 1971). In general, M protein is classed as a relatively poor immunogen and less than 50 per cent of strains having physically detectable M protein are serologically typable in precipitin tests with HCl extracts. Antibody to M protein is slow to appear in streptococcal infection. It may take several months to reach significant levels and is more regularly associated with throat infections than with skin infections (Bergner-Rabinowitz et al. 1969, Dajani and Wannamaker 1970). A surface protein, similar to M protein, has been described for pneumococci but this lacks antiphagocytic properties.

R and T protein antigens are also found on the surface of streptococci—R in streptococcal groups A, B, C and G, and T in group A only. R protein is moderately antigenic and is apparently not related to virulence (Lancefield and Perlman 1952). It occurs in two antigenically distinct forms, R28 and R3, R28 being trypsin-stable and R3 trypsin-labile (Lancefield 1958). T proteins occur in many different antigenic forms, are highly antigenic but do not contribute to streptococcal virulence (Lancefield 1962). Predictable T antigenic types are usually associated with the occurrence of certain M antigenic types. They are useful in serological typing and are detected by agglutination tests (Williams and Maxted 1955).

3.3.2 Staphylococcal protein A

Protein A occurs as a surface or extracellular antigen in most strains of *Staphylococcus aureus*. Strain Cowan I has been used by most workers as the prototype protein A producer. In this strain protein A may form up to 6 per cent of cell-wall dry weight of which up to 8 per cent of total cell protein A may be excreted during growth (Sjöquist et al. 1972). Structurally, protein A consists of an extended polypeptide chain (mol. wt 42 K) having five distinct domains. Four of these domains are highly similar to one another and have the property of non-specifically binding to the Fc region of IgG from many mammalian species. In some species they also bind to the Fc regions of IgA and IgM. Interaction of protein A with immunoglobulins contained in soluble immune complexes results in the formation of insoluble precipitates. Protein A can induce conventional specific antibodies and will react with the Fab antigen-binding sites. The fifth domain appears to be linked covalently to the peptidoglycan of the cell wall (Sjöquist et al. 1972). There is no clear evidence of the role of protein A in pathogenicity although one might envisage an anti-opsonic effect produced by blocking of the Fc region of specific antibody. The non-specific interactions with antibody Fc regions may also cause a variety of other biological effects e.g. hypersensitivity, histamine release, B cell mitogenicity and complement activation (Grov 1977).

Protein A has become a very important immunological tool through its Fc binding property; it has been used in isolation and purification of Ig, as a cytochemical probe when coupled with specific antibody and it has many other varied uses (Surolia et al. 1982).

3.4 Surface carbohydrate antigens

3.4.1 General comments

Surface antigens which possess carbohydrate as a major structural feature are widespread in bacteria. Included in this broad category are the capsular polysaccharides and S-layer glycoproteins of the bacterial glycocalyx (Costerton et al. 1981), the lipopolysaccharides and enterobacterial common antigen of gram-negative bacteria, the teichoic acids and lipoteichoic acids of gram-positive bacteria, the complex glycolipids found in mycoplasmas and the *Corynebacterium-Mycobacterium-Nocardia* group of organisms, and peptidoglycan. The structural and immunological properties of peptidoglycan and S-layer glycoproteins have been the subject of numerous reviews and the interested reader is directed to recent articles by Sleytr (1978), Stewart-Tull (1980) and Rogers et al. (1980). The remaining paragraphs in this Section (3.4) will thus concentrate on properties of the other carbohydrate antigens listed above. First, a few general comments.

Glycolipids and glycoproteins apart, most bacterial carbohydrate antigens have an ordered structure in which the sugar moieties are organized into relatively short repeating units. Homopolysaccharides are composed of a single structural unit e.g. glucose, and can be divided into linear polymers exemplified by bacterial cellulose and branched polymers such as dextrans. Other examples of homopolymeric carbohydrate moieties in bacterial antigens can be found in Tables 13.3 and 13.6. Most bacterial polysaccharides are heteropolysaccharides, however, and contain a repeating unit which can vary in complexity from a disaccharide to an oligosaccharide composed of over six sugars. In many instances, the oligosaccharide repeats possess a short side-chain. In general, bacterial polysaccharides differ in structure from their counterparts in plants and higher organisms, which are typified by complicated and irregular structures.

The repeating units of bacterial carbohydrate antigens may be neutral, as for example, in the transient T antigens (galactans and ribosans) of *Salmonella* and in many of the O-antigen repeats of lipopolysaccharide (see Table 13.6); or they may be negatively charged as is the case, for example, in the teichoic acids and in the Vi antigen of *Salm. typhi* and K1 antigen of *Esch. coli* (see Table 13.3 and Fig. 13.1). The negative charge may be conferred by uronic acids or by non-carbohydrate moieties e.g. phosphate, pyruvic acid and succinic acid. Other structural features of bacterial polysaccharides may be found in useful reviews by Jann and Jann (1977) and Sutherland (1979).

Numerous studies of antigenic specificity (see re-

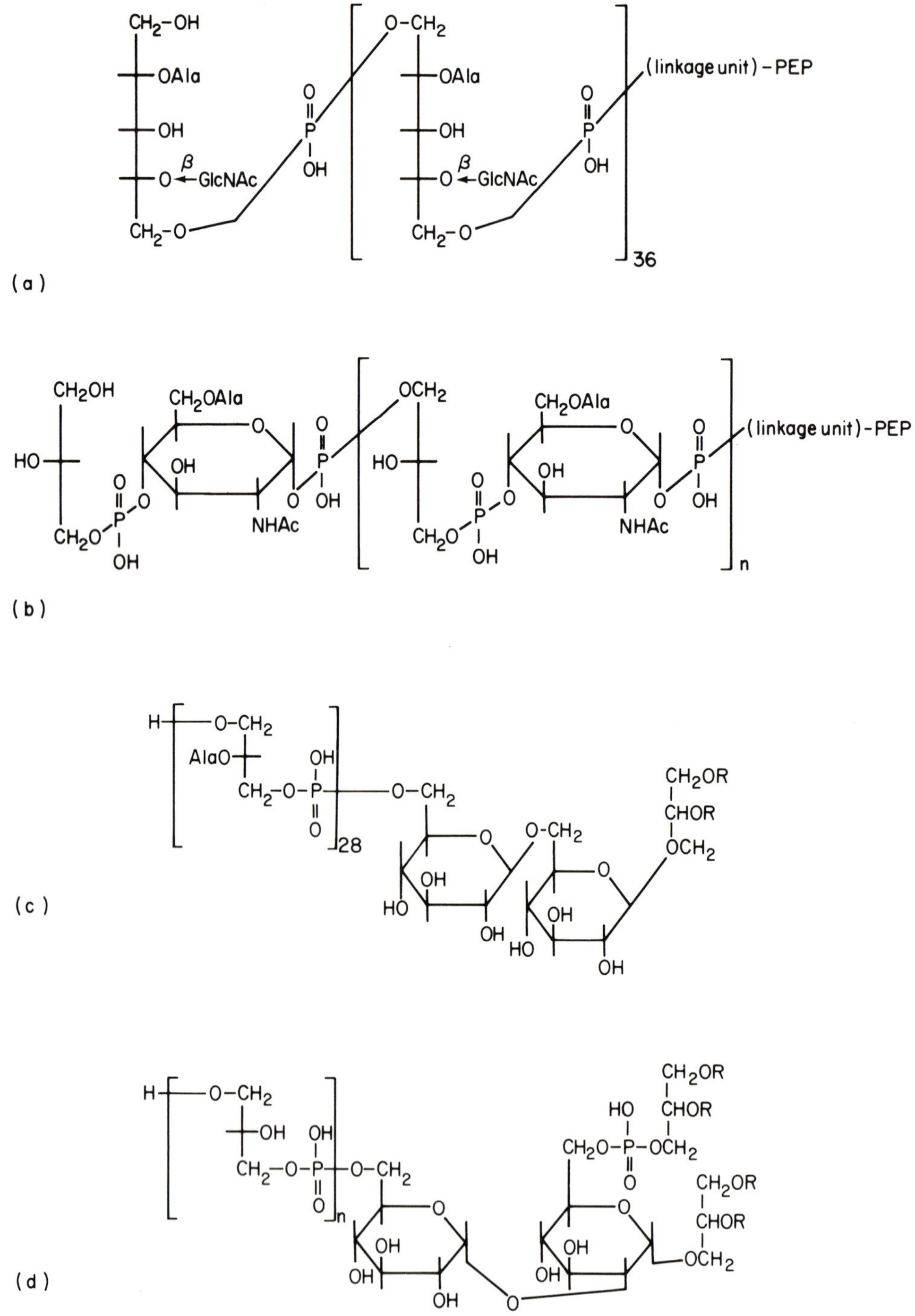

350   *Bacterial antigens*   Ch. 13

Fig. 13.1 Proposed structures for some teichoic acids and lipoteichoic acids. (a) The ribitol teichoic acid of *Staph. aureus* H; (b), the wall teichoic acid of *Staph. lactis* (*Micrococcus varians*) 13 in which the repeating unit is N-acetylglucosaminyl-1-phosphate glycerol phosphate; (c), the lipoteichoic aid of *Staph. aureus* H for which the polyglycerol phosphate repeats are linked to the glycolipid, gentiobiosylydiglyceride; (d), the lipoteichoic acid of *Str. faecalis* for which the polyglycerol phosphate backbone is linked to the phosphoglycolipid, phosphatidylkojibiose diglyceride. **Abbreviations:** Ac, acetyl; Ala, alanine; GlcNAc, N-acetyl-glucosamine; PEP, peptidoglycan; R, long chain fatty acids.
Taken from Duckworth (1977), Fischer *et al.* (1978) and Rogers *et al.* (1980).

views Jann and Westphal 1975, Lüderitz et al. 1966, 1968, 1971) have aimed at defining immunologically important sites on carbohydrate antigens. The sugar unit which contributes most to the serological specificity is the immunodominant sugar; this may be terminal or within the oligosaccharide chain. One polysaccharide can have different antigenic determinants owing to recognition by the immune system of different residues and linkages.

In linear polysaccharides, determinants can be at the non-reducing end and in the middle of the chain. On the other hand, in branched polysaccharides side-chains are usually immunodominant and sequences within the backbone of the repeating structure are usually weakly antigenic. In acidic polysaccharides e.g. the capsular polysaccharides of *Str. pneumoniae*, *Klebsiella* and *Esch. coli*, the charged constituents are usually the immunodominant sugars. Non-carbohydrate moieties e.g. acyl, formyl, succinyl groups, phosphate, glycerol phosphate, acetaldehyde, formaldehyde or pyruvate residues may also function as antigenic determinants.

Identical oligosaccharide regions are found in many polysaccharides (i.e. the same determinant is present in several polysaccharides). Thus polysaccharides bearing such identical determinants are immunologically related and cross-react serologically. Antibodies directed against common determinants are removed from an antiserum by reaction with polysaccharide bearing these determinants. The other antibodies remain in the antiserum which is now developed as an absorbed antiserum.

Cross-reactions formed the basis of the Kauffmann and White scheme for *Salmonella* typing (Kauffmann 1954, 1961) and were used by Heidelberger (1960, 1973) and Heidelberger and Nimmich (1976a, b) for quantitative description of immunological reactions and for structural analysis. For instance, antibody against a known structure can be used to probe for the presence of this structure in unknown polysaccharides. Conversely, oligosaccharides of known structure may be used as inhibitors of serological reactions with polysaccharides that are only partly characterized.

Cross-reactions occur between certain bacterial polysaccharides and carbohydrate components, e.g. blood group substances, human HLA antigens and other glycoproteins, found in mammalian tissue and body organs. A well documented cross-reaction is that between galactosyl diglycerides of *Treponema reiteri* and galactosyl ceramides present in nervous tissue. Such cross-reactions can be important in pathogenesis due to tissue damage by autologous antibodies. They may perhaps also render a host unresponsive to infection, although evidence for this is lacking.

3.4.2 Capsular polysaccharides

Capsular polysaccharides are important antigenic features of numerous gram-positive and gram-negative bacteria. Although not essential for cell viability, they do appear to provide the bacteria with an (additional) defence against desiccation, phagocytosis and infection by bacteriophage. For several bacteria, notably *Streptococcus mutans* which produces a highly branched insoluble glucan and some plant pathogens like *Xanthomonas* which have a xanthan glycocalyx, the capsular polysaccharides also appear to have a central role in adhesion to surfaces (Costerton et al. 1981, McGhee and Michalek 1981). Indeed, capsules may be regarded as important virulence determinants for a number of bacterial pathogens.

As outlined in the previous section, capsular polysaccharides are based upon a fairly simple oligosaccharide repeat. The chemical nature of this repeat is known in many instances and Table 13.3 lists the structures found for capsular polysaccharides in pathogenic bacteria (see also reviews by Dudman 1977, and Robbins 1978). A relatively recent finding is the presence of phospholipid in covalent linkage to some capsular polysaccharides. The capsular antigens of meningococcal groups A, B and C, and the K12, K82 and K92 polysaccharides of *Esch. coli*, have all been shown to possess at their reducing termini a phospholipid, most likely phosphatidic acid, linked to the reducing sugar via a labile phosphodiester bridge (Gotschlich et al. 1981, Schmidt and Jann 1982). The prevalence and biological significance of these structures is not clear at the present time.

In general, the immunogenicity of polysaccharides depends on molecular weight. This emerged from the work of Kabat and Bezer (1958) on dextrans. Dextrans with a molecular weight of 90 K and above were found to be good immunogens; those with a molecular weight of < 50 K were not immunogenic. The immune response to a polysaccharide antigen depends on the dose. Large doses of pneumococcal capsular polysaccharide when administered to mice failed to induce antibody production, whereas small doses (0.5–1 μg) were effective. This so-called immunological paralysis results from persistence of the polysaccharide in the tissues and fluids which serves to remove antibody.

Bacterial polysaccharides and lipopolysaccharides are notable because they stimulate antibody production without the mediation of T-cells. T-cells normally serve to present antigenic determinants in a spatially concentrated form to the B-cell membrane. Bacterial polysaccharides have determinants which are repeated at regular intervals and are close together. For this reason they presumably do not require the intervention of T-cells. However, large amounts of polysaccharide may interfere with the membrane alterations on the B-cell, resulting in tolerance.

For a number of human pathogens, notably pneu-

Table 13.3 Repeating structures for capsular polysaccharides for bacteria commonly associated with infection[a]

Bacterial species	Polysaccharide	Structure of oligosaccharide repeat[b]
Streptococcus pneumoniae	Type 1[c]	---- GalUA $\xrightarrow{1,3}$ GlcNAc $\xrightarrow{1,3}$ GalUA -----
	Type 6	$\xrightarrow{2}$ Gal $\xrightarrow[1,3]{\alpha}$ Glc $\xrightarrow[1,3]{\alpha}$ Rha $\xrightarrow[1,3]{\alpha}$ Rib–P $\xrightarrow[1]{}$ $\xrightarrow[1]{}$
	Type 8	$\xrightarrow{4}$ Glc $\xrightarrow[1,3]{\beta}$ Glc $\xrightarrow[1,4]{\alpha}$ Gal $\xrightarrow[1,4]{\alpha}$ GlcUA $\xrightarrow[1]{\beta}$
Neisseria meningitidis	Group A[d]	$\xrightarrow{6}$ ManNAc-1-P $\xrightarrow[1]{\alpha}$; $\|3$ O-Ac
	Group B	$\xrightarrow{8}$ NeuNAc $\xrightarrow[2]{\alpha}$
	Group C[d]	$\xrightarrow{9}$ NeuNAc $\xrightarrow[2]{\alpha}$; $\|7/8$ O-Ac
Haemophilus influenzae	Type b[e]	$\xrightarrow{3}$ Rib $\xrightarrow[1,1]{\beta}$ Rib–P $\xrightarrow[5]{}$
Escherichia coli	K1[d,f]	$\xrightarrow{8}$ NeuNAc $\xrightarrow[2]{\alpha}$; $\|$ O-Ac
Salmonella typhi	Vi[d]	$\xrightarrow{4}$ GalNAcUA $\xrightarrow[1]{\alpha}$; $\|3$ O-Ac

[a] Data compiled from Jann and Westphal (1975), Lindberg (1977), Sutherland (1977), Robbins (1978), Ørskov et al. (1977) and De Voe (1982).
[b] Abbreviations are: Ac, acetyl; Gal, galactose; GalUA, galacturonic acid; Glc, glucose; GlcNAc, N-acetylglucosamine; GlcNAcUA, N-acetylgalactosaminuronic acid; GalUA, galacturonic acid; ManNAc-P, N-acetylmannosamine phosphate; NeuNAc, N-acetylneuraminic acid; Rha, rhamnose; Rib, ribose; Rib-P, ribitol phosphate.
[c] Partial structure only.
[d] Not all the sugar residues are O-acetylated.
[e] An example of a teichoic acid polymer in a gram-negative organism.
[f] Internal ester bridges are also present.

mococci, meningococci, *Haemophilus influenzae* type b, group B streptococci, *Esch. coli, Salm. typhi, Klebsiella pneumoniae* and *Staph. aureus*, capsules are important virulence determinants (Table 13.3, Robbins 1978). However, they are not the only surface-associated virulence factors in these organisms. The virulence role of capsules is based on several lines of evidence:

(i) Freshly isolated strains from pathological sources are capsulated, whereas non-capsulated strains are less obviously associated with disease.

(ii) Loss of capsule due to mutation or enzymic degradation results in lower virulence.

(iii) The amount of capsular polysaccharide influences virulence, e.g. for pneumococcus types II, III and VII virulence in mice is related to the amount of polysaccharide formed *in vitro*. Similar relations between capsule size or the amount of extracellular surface material and virulence or resistance to phagocytosis has been found with *Klebs. pneumoniae, Staph. aureus, Salm. typhi, Salmonella typhimurium* and *Esch. coli* (see Dudman 1977).

In addition, certain serotypes or serogroups of capsulated organisms are more commonly associated with disease, e.g. 14 of the 83 serotypes of *Str. pneumoniae* account for 85 per cent of clinical infection. Similarly, of the 40 serogroups of *Salmonella*, five account for 90 per cent of all isolates (Robbins 1978).

Contact and engulfment of capsulated bacteria by phagocytes depends to some extent on the hydrophobicity of the surface of the bacterial cell (van Oss and Gillman 1972). Organisms with surfaces more hydrophobic than that of human neutrophils appear to be more readily phagocytosed and do not require opsonins. The more hydrophilic the bacterial cell surface, the more resistant is the organism to phagocytosis. Polysaccharide capsules are therefore antiphagocytic by virtue of their hydrophilic nature. Surface polysaccharides also protect bacterial cells from the bactericidal action of complement and antibody. This is well established for the Vi antigen of *Salm. typhi* and certain K antigens of *Esch. coli* (see Dudman 1977).

Not all capsular polysaccharide antigens are species specific. For instance the M or mucus antigen, first described by Kauffmann (1935), is a capsular non-type-specific antigen common to all naturally mucoid strains of the Enterobacteriaceae. Indeed growth under certain conditions, e.g. low temperature/high salt concentration will induce its production in any strain (Anderson 1961).

Structurally, the M antigen consists of colanic acid composed of glucose, fucose, glucuronic acid and galactose (Goebel 1965). The 3-O-β-D-glucuronosyl-galactose element of the M antigen has been shown to cross-react with an identical residue in the K30 type-specific capsular antigen of *Esch. coli* (see review by Lüderitz, Jann and Wheat 1968).

Control of infection by encapsulated pathogens is becoming more difficult owing to the problem of antibiotic resistance. This together with other factors such as high morbidity and mortality, and constant or increasing incidence of disease has prompted a renewed interest in prevention of infection by immunization with highly purified capsular polysaccharides. An encouraging feature of such vaccines is the lack of reactogenicity in man: approximately 130 million people have been injected with these antigens with little evidence of adverse side-effects (Robbins (1978).

The protective effect of antibody to capsular polysaccharide from pneumococcal types 1, 2, 5 and 7 against pneumonia caused by these types has been known since the work of MacLeod and colleagues (1945). The success of antibiotics led to a loss of serious interest in the protective effects of anticapsular antibodies. However, through the efforts of Austrian and others (see Austrian 1981) a polyvalent vaccine containing purified capsular polysaccharide from the 14 serotypes responsible for most serious pneumococcal infections 1, 2, 3, 4, 6A, 7F, 8, 9N, 12F, 14, 18C, 19F, 23F and 25) has been tested in a number of trials. This vaccine proved to be effective in protecting populations with high attack rates (e.g. South African miners). However considerable controversy exists as to the efficacy of the vaccine on other populations (see for example Bentley 1981, Broome 1981).

Prevention of meningococcal meningitis caused by groups A, C and Y has been successful with capsular polysaccharide vaccines (WHO 1976, Peltola *et al.* 1977, Farquar *et al.* 1977). These vaccines are now being manufactured in at least five different countries. Problems however have been encountered in the protection of infants, the group at greatest risk, owing to the poor immune response in this age group (see reviews Robbins 1978, Sippel 1981). Prevention of infection against group B meningococci also presents a problem. Because the group B polysaccharide is a poor immunogen, an alternative approach is being followed. Another group of surface antigens, the outer membrane proteins have been identified and studied in some detail (see Section 3.5), and it has been suggested that a combination of membrane protein and group B polysaccharide may prove to be efficacious as a vaccine against group B infections.

Of the six serotypes of *Haemophilus influenzae*, the most important cause of serious disease is type b. The type b polysaccharide has been purified and is non-toxic. The material is immunogenic in children older than 14 months (Makela *et al.* 1977) but the protective effect appears to be short lived (see Robbins 1978 and Section 3.5).

The efficacy of a typhoid vaccine was for a long time in dispute. However, controlled trials in several countries have now shown that a heat-killed phenolized vaccine or an acetone-extracted vaccine confers a considerable degree of immunity against the disease (see 6th edition, p. 2027); but the part played in this result

Table 13.4 Teichoic acids as bacterial antigens[a]

Organism	Serological group	Teichoic acid Cellular location	Polyol	Immunodominant sugar(s)[b]
Staphylococcus aureus Wood 46		Wall	Ribitol	β-GlcNAc
Staphylococcus aureus 263		Wall	Ribitol	α-GlcNAc
Staphylococcus epidermidis T1		Wall	Glycerol	α-Glc
Staphylococcus epidermidis T2		Wall	Glycerol	β-Glc
Streptococcus	D	Membrane	Glycerol	Glc $\xrightarrow[1,2]{\alpha}$ Glc
	N	Membrane	Glycerol	Gal-P
Lactobacillus	A	Membrane	Glycerol	α-Glc
	D	Wall	Ribitol	α-Glc
	E	Wall	Glycerol	α-Glc
	F	Membrane	Glycerol	Gal $\xrightarrow[1,2]{\alpha}$ Glc

[a] Date from Knox and Wicken (1973) and Rogers *et al.* (1980).
[b] Abbreviations are defined in legend to Table 13.3; Gal-P, galactose-phosphate.

by the Vi antigen or other constituents of the bacillus is still undetermined. There is reason to believe that the somatic antigen is chiefly responsible and that the flagellar antigens are devoid of protective action.

Studies of *Esch. coli* K1 strains and group B streptococci, both of which are important causes of neonatal bacterial meningitis, are gaining momentum. The role of capsular polysaccharides in virulence is well established in each case. It is to be hoped that the difficult problem of conferring anticapsular immunity in the neonate will be given high priority.

3.4.3 Teichoic acids and lipoteichoic acids

The general term teichoic refers to a family of polymers possessing 'phosphodiester groups, polyol and/or sugar residues, and usually, but not always, D-alanine residues' (Baddiley 1970). These acidic macromolecules are generally regarded as being confined to gram-positive bacteria and can be divided into two broad classes depending on their cellular location *viz.* wall teichoic acids and membrane-bound lipoteichoic acids. Wall teichoic acids are covalently attached to the peptidoglycan via a short linkage unit and are based upon simple repeats of (a) glycerol phosphate or (b) ribitol phosphate or upon (c) slightly more complex repeats with additional sugars or sugar phosphates, e.g. glycosylglycerol phosphate repeating units. Polymers containing simple sugar phosphate repeats are also found in some micro-organisms (see Fig. 13.1).

In contrast to wall teichoic acids, all lipoteichoic acids have a backbone consisting of glycerol phosphate repeating units. Furthermore, cell-bound lipoteichoic acids are not linked to the peptidoglycan but are integrated into the outer leaflet of the plasma membrane. They owe their association with the bilayer to the presence of a covalently bound glycolipid or phosphoglycolipid and have their repeating units extending out from the cell (see Fig. 13.1). Many micro-organisms also appear to excrete lipoteichoic acids into the external medium, either in the fully acylated form or as deacylated derivatives (Knox and Wicken 1973, Duckworth 1977, Wicken and Knox, 1980, Wicken 1980).

A primary function of teichoic acids appears to be related to their ability to bind divalent cations. It has been suggested that wall and membrane teichoic acids form a continuous anionic matrix capable of assimilating and concentrating cations for subsequent release at the membrane surface (Lambert *et al.* 1977, Duckworth 1977). Lipoteichoic acids have also been implicated in the control of cellular autolysins and as possible carrier lipids in the biosynthesis of teichoic acids (Rogers *et al.* 1980, Ward 1981).

Both wall and membrane teichoic acids are immunogenic, the latter being a T-cell dependent immunogen giving rise to both IgM and IgG antibodies (Wicken and Knox 1980). Immunologically, the dominant structural features of the polymers are the glycosyl substituents which occur on the glycerol and ribitol moieties. Unfortunately, the limited range of sugar substituents restricts the usefulness of teichoic acids in classification of gram-positive bacteria. Nevertheless, they have been shown to represent the grouping antigens for some strains of streptococci, staphylococci and lactobacilli, and to play an important role in the serological classification of these bacteria (see Table 13.4).

Antibodies can also be elicited to the polyglycerol phosphate backbone of teichoic acids as well as to their glycosyl substituents. For this reason lipoteichoic acids have been implicated in the cross-reactions observed between diverse groups of gram-positive bacteria. Anti-polyglycerolphosphate antibodies will also react with the phospholipid cardiolipin, which carries a phosphorylglycerol-phosphate determinant. The significance of this reaction is that it might account for some of the false positive reactions observed in the Kahn test for syphilis.

Although lipoteichoic acids differ from the lipolysaccharides of gram-negative bacteria in being non-toxic and non-pyrogenic, they have been shown to share some biological properties in common with this molecule e.g. anticomplementary activity, positive Shwartzman reaction, stimulation of non-specific immunity, release of macrophage lysosomal enzymes, stimulation of bone resorption, positive *Limulus* lysate tests and phage receptor properties (Lindberg 1977, Wicken and Knox 1980). A further property shared with other surface amphiphiles is an ability to bind to eukaryotic cell membranes (Wicken 1980). Indeed, some workers have proposed that the adherence of group A streptococci to the human pharyngeal epithelium may be due to a fimbrial network which is composed of M-protein and both membrane-bound and secreted lipoteichoic acids and which is organized in such a way as to allow the fatty acyl groups of the glycerol phosphate polymers to interact with protein receptors in the mammalian membrane. The intercalated network might thus act as a biological 'glue' sticking the bacterium to the cell surface (Beachey 1981, Ofek *et al.* 1982). There is also a body of opinion which maintains that cationic bridges formed between the phosphate groups of lipoteichoic acids and pellicle-coated tooth enamel may play some role in the ability of *Str. mutans* to colonize the oral cavity and thus to cause dental caries (Wicken 1980, Hamada and Slade 1980). Little appears to be known about the potential importance of anti-lipoteichoic antibodies in immunity to infection caused by these bacteria.

3.4.4 Lipopolysaccharide

It is now well established that the O-somatic antigen of gram-negative bacteria is an endotoxic lipopolysaccharide (LPS), which is located in the outer leaflet of the outer membrane in high copy number ($\sim 3.5 \times 10^6$ molecules/cell). For the cell, LPS plays a major role in stabilization of the protective outer membrane. It also appears to be essential for the biological activity of several outer membrane proteins, notably the omp A and omp F/C proteins. In addition, it serves as the receptor for numerous bacteriophages and some bacteriocines.

A variety of mild procedures are available for the extraction and purification of LPS (Wilkinson 1977, Wicken and Knox 1980), the most popular being based upon the use of 45 per cent aqueous phenol at 65°. In aqueous solution, LPS forms large micellar aggregates which may be dispersed with detergents and EDTA. The molecule is thermostable, in the sense that it retains its antigenicity and immunogenicity after boiling.

The importance of the LPS antigen cannot be overemphasized. As might be anticipated from a consideration of its abundance and cellular location, it is the major surface antigen for many gram-negative bacteria. It forms the basis of the serological classification of numerous bacterial species and is an indispensible diagnostic aid. It also is capable of inducing an extensive range of important biological effects, some harmful, some beneficial, in man and experimental animals.

The structure of LPS has been extensively studied and numerous reviews on this topic are available (Jann and Westphal 1975, Wilkinson 1977, Jann and Jann 1977, Lindberg 1977, Ørskov *et al.* 1977, Rogers *et al.* 1980). The molecule itself is amphiphilic and can be thought of as being composed of three distinct but covalently connected regions, viz. a hydrophobic lipid A region and two hydrophilic areas called the core oligosaccharide and the O-specific polysaccharide (or O-antigen) regions (see Figs. 13.2 and 13.3). Of the lipopolysaccharides studied to date, that of *Salm. typhimurium* is understood in greatest structural detail. This will be described below as will an indication of how much the structure can be expected to vary in other organisms.

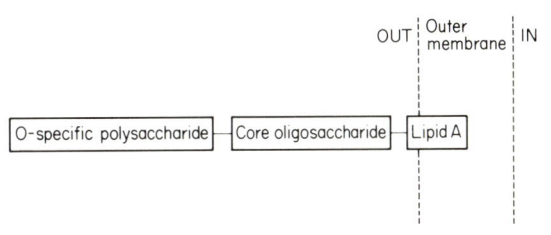

Fig. 13.2 Schematic diagram illustrating the general structure and location of bacterial LPS.

A common feature of the lipid A moieties of almost all bacterial lipopolysaccharides appears to be a central β (1-6)-linked glucosamine disaccharide bearing β-hydroxy fatty acids in N-acyl linkage and phosphate groups at positions 1 and 4'. Under some cultural conditions these terminal phosphates are replaced by 4-amino-L-arabinose and phosphorylethanolamine (see Fig. 13.4). Three O-acyl fatty acids can also be present and these, together with the N-acyl hydroxy fatty acids, serve to anchor the LPS molecule in the outer membrane. In *Salm. typhimurium*, the acyl substituents are usually β-hydroxymyristic and myristoxymyristic acids together with lauric, myristic and palmitic acids (Fig. 13.4). The type of fatty acids can vary with the organism as can substitution on the phosphate groups (Wilkinson 1977, Rogers *et al.* 1980, Qureshi *et al.* 1982).

In contrast to the fairly constant nature of bacterial lipid A, the core oligosaccharide region, to which lipid A is linked, shows more variability in structure. For example, all salmonellae appear to have the same core structure, but this differs from that in the Pseudomonadaceae and from that found in strains of *Esch. coli*, for which five different core types (designated R1–R4, and K12) have been observed (see Table 13.5). Core structures similar to those observed in *Esch. coli* also occur in other gram-negative bacteria, notably *Shigella*.

The chemistry of this oligosaccharide region has been elaborated mainly through the use of rough (R) mutants defective in various stages of core biosynthesis (Fig. 13.3). Some of the better characterized cores are shown in Table 13.5. A characteristic feature of these regions is the presence of two somewhat unusual sugars *viz.* 2-keto-3-deoxy-D-mannulosonic acid (KDO) and heptose (such as L-glycero-D-mannoheptose). The former sugar appears to be unique to LPS and is often used as a convenient molecular marker for this polymer. The core region is highly charged by virtue of carboxyl groups on KDO, as well as phosphate, pyrophosphate and ethanolamine residues. These appear to be important in the interactions of LPS with other outer membrane components (proteins) and with bacteriophages.

Although the core regions do possess immunological determinants, the R-specificities are seldom expressed since (a) R-mutants do not occur frequently in nature and (b) the expression becomes cryptic in wild type (S-)strains where the relatively lengthy O-specific polysaccharide becomes immunodominant.

The structural variability of LPS, which is evident to some extent in the bacterial core regions, increases dramatically in the O-specific polysaccharide region. This part of the molecule is made up of oligosaccharide repeating units. The repeats themselves usually consist of a short linear main chain with or without single hexose branch substituents. In almost all cases non-reducing termini are presented. The potential for antigenic variability is immense. For example, there are currently 169 recognized O-serogroups of *Esch. coli* (see also Table 13.1). The immunological specificity can be dictated by the nature of the non-reducing termini, the nature of the branch substituents, the anomeric configuration of the linkages and secondary substituents such as O-acetyl groups. The distinctive dideoxyhexoses, colitose, abequose, tyvelose and paratose which occur as branch substituents in the LPS of enterobacteria can exert a major influence in this respect. Table 13.6 gives a flavour of the variety of structural repeats observed for bacterial LPS.

Evidence is also accumulating for the presence in some bacteria of O-antigen repeats which are not covalently linked to the lipid A-core region of LPS (Goldman *et al.* 1982). For example, in *Esch. coli* serogroups O100, O111 and O113 up to half of the O-antigen may be present in this form. The number of repeating units observed per molecule of free O-antigen is over 30-times that found in native LPS (where it is about 12). Consequently, these O-antigen polymers have molecular weights over 300 K *cf.* about 16 K for the LPS monomer. They seem to be located on the bacterial surface and behave in a similar fashion to K-antigens in the sense that, unless first extracted by heat treatment at 100°, they effectively mask antigenic determinants expressed on the native LPS (Goldman *et al.* 1982). It will be interesting to ascertain the structural basis for their apparent inability to

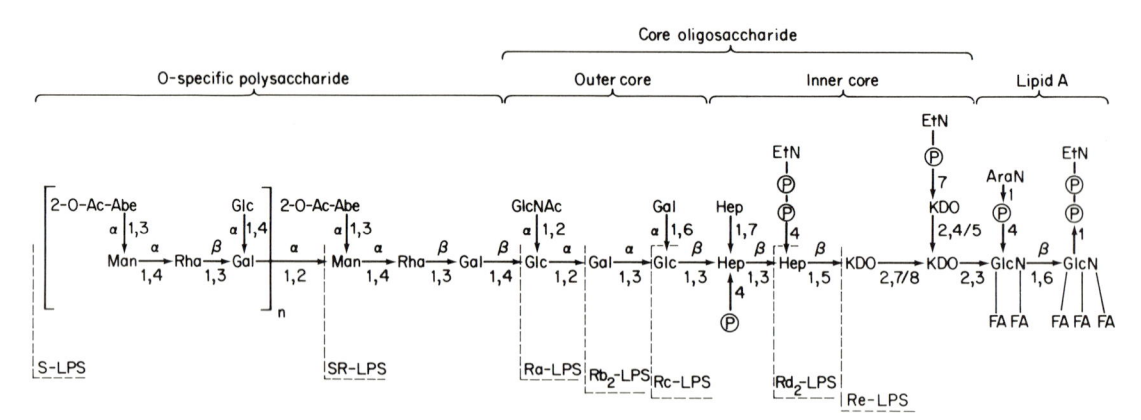

Fig. 13.3 Structure of the lipopolysaccharide from *Salm. typhimurium*. Also indicated are the structures of lipopolysaccharides produced by mutants defective at various stages in LPS biosynthesis. **Abbreviations:** Abe, abequose; Ac, acetyl; AraN; L-arabinosamine, EtN, ethanolamine; FA, fatty acids; Gal, D-galactose; Glc, D-glucose; GlcN, D-glucosamine; GlcNAc, N-acetyl-D-glucosamine; Hep, L-glycero-D-manno-heptose; KDO, 2-keto-3-deoxy-D-*manno*-octulosonic acid; Man, D-mannose; P, phosphate; Rha, L-rhamnose. Heterogeneity in LPS composition may be caused by variations in the value of *n* and in the degree of substitution at key points in the molecule, *viz.* the 2-O-acetylation of Abe and the glucosyl substitution of Gal in the O-antigen region; the N-acetyl glucosaminyl and galactosyl substitution of Glc in the outer core; the heptosyl, phosphoryl and pyrophosphorylethanolamine substitution of Hep in the inner core; and the arabinosaminyl and phosphorylethanolamine substitution of GlcN and GlcN-P respectively in the Lipid A region.

Data from Jann and Westphal (1975), Rogers *et al.* (1980) and Palva and Makela (1980).

Table 13.5 Proposed structure of some core oligosaccharides[a,b]

Organism	Structure[c]

Esch. coli (R1):

$$\text{Glc} \xrightarrow[1,3]{} \text{Glc} \xrightarrow[1,3]{} \text{Glc} \xrightarrow[1,3]{} \text{Hep} \xrightarrow[1,3]{} \text{Hep} \longrightarrow (\text{KDO})_3 —$$

with Gal $\xrightarrow{1,2}$ Gal $\downarrow_{1,2}$ above second Glc, and Hep $\downarrow_{1,7}$ above first Hep.

Esch. coli (R2):

$$\text{Glc} \xrightarrow[1,2]{\alpha} \text{Glc} \xrightarrow[1,3]{\alpha} \text{Glc} \xrightarrow[1,3]{} \text{Hep} \xrightarrow[1,3]{} \text{Hep} \longrightarrow (\text{KDO})_3$$

with GlcNAc $\alpha|_{1,2}$ above first Glc; Gal $\alpha|_{1,6}$ above third Glc; Hep $\downarrow_{1,7}$ above first Hep; Gal $\alpha\vdots 1,7/8$ above second Hep.

Esch. coli K12[d]:

$$\text{Glc} \xrightarrow[1,2]{} \text{Glc} \xrightarrow[1,3]{} \text{Glc} \xrightarrow[1,3]{} \text{Hep} \longrightarrow \text{Hep} \xrightarrow[1,3]{} \text{Hep} \longrightarrow (\text{KDO})_3 —$$

with GlcNAc $\downarrow_{1,6}$ above first Glc; Hep $\downarrow_{1,7}$ above second Hep.

Esch. coli B[d,e]:

$$\text{Glc} \xrightarrow[1,3]{\alpha} \text{Glc} \xrightarrow[1,3]{\alpha} \text{Hep} \xrightarrow[1,3]{} \text{Hep} \xrightarrow[1,5]{} \text{KDO} \xrightarrow[2,7/8]{} \text{KDO} ——$$

with Hep $\downarrow_{1,6}$ above first Hep; KDO $\downarrow_{2,4}$ above last KDO.

Salm. typhimurium:

$$\text{Glc} \xrightarrow[1,2]{\alpha} \text{Gal} \xrightarrow[1,3]{\alpha} \text{Glc} \xrightarrow[1,3]{\beta} \text{Hep} \xrightarrow[1,3]{\beta} \text{Hep} \xrightarrow[1,5]{\beta} \text{KDO} \xrightarrow[2,7/8]{} \text{KDO} —$$

with GlcNAc $\alpha|_{1,2}$ above first Glc; Gal $\alpha|_{1,6}$ above second Glc (Glc after Gal); Hep $\downarrow_{1,7}$ above first Hep; KDO $\downarrow_{2,4/5}$ above last KDO.

Ps. aeruginosa:

$$\text{Glc} \xrightarrow[1,2]{\beta} \text{Rha} \xrightarrow[1,6]{\alpha} \text{Glc} \longrightarrow \text{GalN} \xrightarrow[,3]{} \text{Hep} \xrightarrow[1,3]{} \text{Hep} \xrightarrow[1,4/5]{} \text{KDO} \xrightarrow[2,4/5]{} \text{KDO} ——$$

with Glc $\xrightarrow[1,6]{\alpha}$ Glc branch attached above GalN; L-Ala$_{\frac{1}{2}}$ below GalN.

[a] Data from Ørskov *et al.* (1977) and Wilkinson (1977).
[b] Abbreviations are as listed for Fig. 13.3. In addition: Ala, alanine, GalN, galactosamine.
[c] Most core oligosaccharides are substituted with phosphate, phosphorylethanolamine and pyrophosphorylethanolamine. In this Table, these have been omitted for clarity. However, see Fig. 13.3 for an example of full substitution.
[d] This organism has a rough LPS which lacks an O-antigen side chain.
[e] *Esch. coli* B is an example of an organism with an incomplete core whose ancestor is unknown.

Table 13.6 Structure of some bacterial O-antigen repeats[a]

Organism	O-Group	Repeating unit[b]
Salm. paratyphi A	A	→2) Man (α1,4) Rha (α1,3) Gal (α1)→ with Par α1,3 on Man, Glc α1,4 on Gal, OAc on Rha
Salm. typhi	D₁	→2) Man (α1,4) Rha (α1,3) Gal (α1)→ with Tyv α1,3 on Man, 2-O-Ac-Glc on Gal
Salm. milwaukee	U	→Gal (β1,3) GalNAc (1,3) GlcNAc (1,4) Fuc→ with Gal α1,3 on Gal
Sh. boydi[c]	6	→3) GalNAc (β1,3) Gal (α1,6) Man (α1,2) Man (α1)→ with GlcUA α1,4 on Gal
Esch. coli	8	→3) Man (α1,2) Man (α1,2) Man (α1)→
Esch. coli	111	→GlcNAc (β1,2) Glc (α1,4) Gal (α1)→ with Col α1,6 on GlcNAc, Col on Glc
Klebsiella	7	→2) Rha (α1,2) Rib (β1,3) Rha (α1,3) Rha (α1)→
Ps. aeruginosa[c]	3a, b	→4) ManImUA (β1,4) Man(NAc)₂UA (β1,3) FucNAc (β1)→
Y. pseudotuberculosis	IB and IIB	→2) Man (α1,3) Fuc (α1)→ with Par α1,3 on Man

[a] Data from Jann and Westphal (1975), Wilkinson (1977), Ørskov et al. (1977), Rogers et al. (1980) and Knirel et al. (1982).
[b] Abbreviations not listed in Fig. 13.3 are as follows: Col, colitose; FucNAc, N-acetyl-D-fucosamine; GalNAc, N-acetyl-D-galactosamine, GlcUA, glucuronic acid; ManImUA, 2,3-(1-acetyl-2-methyl-2-imidazolino-5,4)-2,3-dideoxy-D-mannuronic acid; Man(NAc)₂UA, 2,3-diacetamido-2,3-dideoxy-D-mannuronic acid; Par, paratose; Rib, D-ribose; Tyv, tyvelose.
[c] An example of an acidic O-antigen repeat.

react competently with O-antiserum and to determine whether they are covalently linked to phospholipid in a manner similar to that recently demonstrated for meningococcal surface polysaccharides (see Section 3.4.2).

Virulence can be determined in part by the chemical nature of the lipopolysaccharide. In general, rough mutants are less virulent than their smooth (wild type) counterparts. Thus, Medearis et al. (1968), investigating the relation of O-antigen composition to virulence, found that a parent wild-type *Esch. coli.* (O111:B4) was 100 times more virulent than a mutant lacking colitose and was 1000 times more virulent than a mutant with a LPS deficient in galactose, glucose, N-acetyl glucosamine and colitose. As endotoxin prepared from all three antigens was equally toxic and pyrogenic, it appears that altered virulence was due to alterations in the structure of the polysaccharide region of the O-antigen. Furthermore, in experiments in which the composition of O-antigen side-chains was altered by genetic manipulation (Valtonen 1970) it was shown that different O side-chains were related to different degrees of virulence in mice. A detailed appraisal of the prevalence of serological groups of *Esch. coli* in various pathological conditions of men and animals has been published (Ørskov et al. 1977) and is recommended for a fuller appreciation of this particular topic.

Lipopolysaccharides are T-independent immunogens that produce polyclonal B cell activation with a predominantly IgM humoral response. As is outlined in Section 3.5, LPS will also act as adjuvants for other immunogens, notably outer membrane proteins.

The biological properties of LPS are many and varied and have been reviewed recently (Bradley 1979, Wicken and Knox 1980). It has been suggested that the classical endotoxic properties of pyrogenicity and lethal toxicity displayed by LPS result from an interaction of the amphiphile with the inner mitochondrial membrane leading to inhibition of NADH-Q reductase and ATPase activities. The resultant accumulation of NADH and ADP in the cytosol leads to enhanced glycolytic metabolism and lactate formation, and to raised levels of superoxide radicals and H_2O_2. This in turn results in autophagy and enhanced lysosomal enzyme activity. It is further suggested that:

'These cellular events that occur in hepatocytes and phagocytic cells set the stage for the interlocked, multifactoral alterations that lead to death: Acute hypoglycaemia precipitates an insulin-type shock; disseminated intravascular coagulation precipitates renal failure; and released vasoactive mediators produce severe circulatory hypotension, and secondarily produce hypoxia and acidosis. If the animal cannot regain homeostasis quickly, the vascular changes exacerbate the cellular injuries' (Bradley 1979).

Other biological activities attributable to LPS include complement activation via the classical and/or alternate pathways, macrophage stimulation, production of a Shwartzman reaction, stimulation of bone resorption, gelation of *Limulus* lysates at low concentration (ng/ml) and tumour necrotizing potential (Bradley 1979, Wicken and Knox 1980).

The lipid A region has been shown to be responsible for most of the endotoxic activities of native LPS. More precise definition of the basic toxic moiety is not possible at the present time, although it may be relevant to note that fully acylated monophosphoryl lipid A (Fig. 13.4) is *non-toxic* but retains tumour regression activity (Takayama et al. 1981, Qureshi et al. 1982). Clearly, it is feasible to dissociate some biological properties of LPS from others.

The toxicity of lipid A poses obvious problems in the use of LPS as a protective vaccine. This is not an academic point; LPS is a good immunogen and has long been implicated in immunity conferred by some of the cruder vaccine preparations i.e. heat killed or attenuated vaccines. More recent studies with mutants of *Esch. coli* and *Salmonella minnesota* with defined LPS structures have clearly shown that both the core polysaccharide and the O-antigen repeat are excellent antigens and generate, in a variety of experimental animals, antibodies which protect against both the toxic manifestations of LPS and bacteremia (Braude 1980). Attention, however, has focused on the use of core polysaccharides, since their spectrum of anti-endotoxin activity is much broader and is not restricted to a single species or serotype. Interestingly, although anti-core antibodies are extremely effective against homologous and heterologous wild type (smooth) organisms, the bulky O-antigen chains do appear to interfere with the *production* of protective antibody to the shared core region containing KDO and heptose. For this reason, immunization is most effective with the rough (Rc-type) LPS (see Fig. 13.4 and Table 13.5).

Active immunization with core endotoxin from *Esch. coli* or passive immunization with anti-core antiserum has consistently protected rabbits against a wide range of gram-negative bacteria e.g. *Ps. aeruginosa, Klebs. pneumoniae* and *Esch. coli.* Encouraging preliminary results have also been obtained in the use of anticore antiserum in the treatment of patients severely ill with bacteraemia caused by gram-negative micro-organisms (Braude 1980).

The toxicity of lipid A remains a problem for active immunization; however, one possible approach would be to construct non-toxic conjugates of inner core carbohydrate and (outer membrane) protein, in a manner similar to that used for the O-antigen of *Salm. typhimurium* (see Section 3.5). An alternative, and probably more realistic approach from the point of view of the immunologically deficient patient, would be the production, for passive immunization, of monoclonal antibodies directed towards determinants in the inner core region.

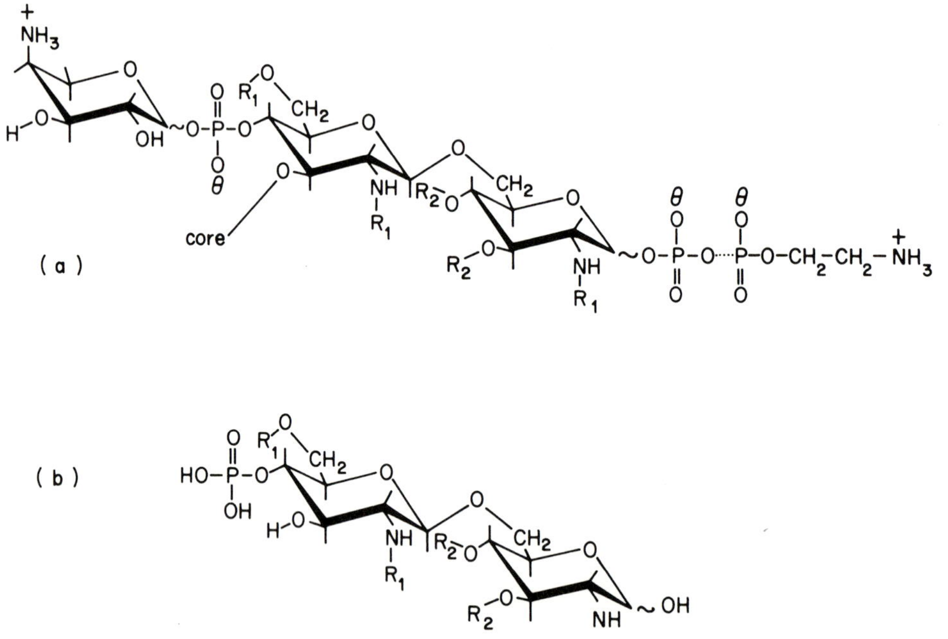

$$R_1 = CH_3\text{-}(CH_2)_{10}\text{-}\underset{}{\overset{OH}{\overset{|}{C}}}H\text{-}CH_2\text{-}CO\text{-}$$

$$R_2 = \underset{CH_3\text{-}(CH_2)_{12}\text{-}CO\text{-}O}{CH_3\text{-}(CH_2)_{10}\text{-}CH\text{-}CH_2\text{-}CO\text{-}} \quad \text{and} \quad CH_3\text{-}(CH_2)_{10}\text{-}CO\text{-} \;\text{or}\; CH_3\text{-}(CH_2)_{14}\text{-}CO\text{-}$$

Fig. 13.4 Structure proposed for (a) the Lipid A moiety of *Salm. typhimurium* LPS and (b) a non-toxic derivative. Data from Rogers *et al.* (1980) and Qureshi *et al.* (1982).

3.4.5 Enterobacterial common antigen

Enterobacterial common antigen (ECA) was first described in 1962 by Kunin and his colleagues (Kunin *et al.* 1962) and is an amphiphilic glycophospholipid located in the outer membrane of almost all wild type strains of the Enterobacteriaceae. Its expression on the cell surface varies, being readily available to homologous antibody in non-encapsulated rough strains and poorly so, or not at all, in smooth strains carrying complete O-antigen chains. The antigen has been purified from *Salmonella montevideo* and has been shown to be an acidic linear heteropolysaccharide composed of alternating 1–4 linked N-acetyl-D-glucosamine and N-acetyl-D-mannosaminuronic acid. Phosphate and ester-linked palmitate are also present, possibly as covalently bound phospholipid. ECA can occur in two distinct forms: a commonly occurring and poorly immunogenic free form, and a bound form which is highly immunogenic. The latter is restricted to a few rough strains where it appears to be covalently bound to the core region of the LPS. Both free and bound forms share the same antigenic determinant, which is based upon the D-mannosaminuronic acid moiety. In smooth strains, the molecule can show considerable heterogeneity with respect to its molecular weight and solubility characters—features which probably reflect differing degrees of acylation. The soluble monomer has an apparent molecular weight of 2.7 K. Striking differences have also been noted in the mitogenicity and immunogenicity of different preparations of the antigen and these have been attributed to varying amounts of associated membrane protein which the polymer seems capable of binding avidly. ECA does not appear to be either toxic or pyrogenic (Wicken and Knox 1980, Acker et al. 1982, Gannon et al. 1982).

ECA should not be confused with the so-called common protein antigen. This is a soluble, acidic cytoplasmic protein of subunit molecular weight 60 K which shows even broader cross-reactivity than ECA. Antibodies to common protein antigen can be demonstrated in preimmune serum from most mammals and frequently complicate the results of immunological tests performed with bacterial extracts (Sompolinsky et al. 1980, Owen 1983).

3.4.6 Other glycolipids

Several of the lipids commonly found in microbial membranes are immunologically active. These include acidic phospholipids such as cardiolipin, phosphatidylglycerol and phosphatidylinositol and various glycolipids and phosphoglycolipids. Generally, the purified lipids do not elicit antibody responses in laboratory animals unless they are adsorbed to a 'carrier' protein which assists in the formation of immunologically active micellar structures. They should thus be regarded as haptens, whose antigenic properties are determined largely by the nature of their hydrophilic headgroups.

For many bacteria, membrane lipids are not major surface antigens and do not appear to play a significant role in the immunogenicity of the organism. However, in wall-less micro-organisms such as streptococcal L-forms and mycoplasmas, where the head groups of (glyco)lipids are surface expressed, they become significant antigenic determinants (Feinman et al. 1973, Owen 1981). This is especially true in mycoplasmas where glycolipids can account for a considerable percentage of the total membrane lipid (Razin 1978, 1981). The role of glycolipids in the serology of *Mycoplasma pneumoniae*, for example, is well established. They react with specific antibody to fix complement and are intimately involved in the mechanism of immune lysis. The serologically active lipids, which include di- and tri-galactosyl diglycerides and di- and tri-glycosyl diglycerides containing both glucose and galactose, also react with antibodies inhibiting growth and metabolism of the organism (Owen 1981, Razin 1981). Serologically active glycolipids and phosphoglycolipids have been demonstrated in several other species of mycoplasmas and many have been shown to cross-react with each other and to possess determinants in common with glycolipids from gram-positive bacteria and from plants.

Glycolipids with more extended oligosaccharide chains have also been detected in some species of *Acholeplasma*. These lipids were originally called lipopolysaccharides. However, this name is confusing since the molecules show little structural resemblance to classical LPS of gram-negative bacteria lacking lipid A, KDO and heptose. The alternative term 'lipoglycan' has been proposed and this would seem more appropriate in the circumstances. Unlike other glycolipids, the lipoglycans are immunogenic *per se* and it has been suggested that they may play a role in the interaction of the parasite with its host (Razin 1981).

Glycolipids of somewhat unusual structure are important in the immunology of the so-called CMN group of micro-organisms (*Corynebacterium*, *Mycobacterium* and *Nocardia*). For example, the chief antigenic polysaccharide of the mycobacterial cell wall, a D-arabino-D-galactan, is covalently linked to mycolic acids. These are complex long chain fatty acids which are branched at the α-position and hydroxylated at the β-position (Fig. 13.5 (a) and Ch. 24). Other important constituents of the wall of organisms in the CMN group are haptenic glycolipids based on the disaccharide trehalose. One class, termed 'cord-factors', are dimycolates of trehalose (Fig. 13.5 (b)) and are toxic for mice by virtue of damage inflicted to the mitochondrial membrane. Toxicity appears to depend upon certain free hydroxyl groups on the trehalose moiety (Barksdale and Kim 1977). A structurally related class of lipids are the sulphatides. These acidic multiacylated trehalose sulphates (Fig. 13.5 (c)) are not by themselves toxic but appear capable of potentiating the toxicity of cord factor. Both cord factor and the sulphatides have been considered virulence characteristics for *Mycobacterium tuberculosis* (Barksdale and Kim 1977, Lefford 1981).

A third group of complex glycolipids found in mycobacteria are the mycosides (see Chapter 24). These can be either (a) lipidic aglycones consisting of methylated sugar residues glycosidically linked to phenol *para* substituted with a branched alkyl chain whose hydroxyl groups are further esterified with fatty acids or (b) peptidoglycolipids consisting of a sugar moiety, a short peptide and fatty acids (Fig. 13.5 (d)). Mycosides appear to be located at the cell surface and act as receptors for some bacteriophages. Their precise function is not clear, but they may serve to prevent desication in a non-aqueous environment.

3.5 Outer membrane proteins

Owing in part to the development of techniques for the separation of inner and outer membranes and for resolution of membrane polypeptides, the past fifteen years have witnessed a remarkable advance in our understanding of the structure, function, biosynthesis, genetics and immunology of outer membrane proteins (Osborn and Wu 1980, Hall and Silhavy 1981, Owen 1981, Lugtenberg 1981). It is now clear that gram-negative bacteria possess in their outer membrane a highly characteristic complement of proteins which

Table 13.7 Properties of *Esch. coli* outer membrane proteins

Protein[a]	(Apparent) subunit molecular weight	Map position of structural gene[b] (min)	Estimated no molecules/cell[c]	(Part of) receptor for
83K protein	83 000			
Fep A protein	81 000	13.0		Colicine B
Fec A protein	80 500	7		
Fhu A protein	78 000	3.4	$\sim 10^3$	Phage T1, T5, ϕ80 Colicine M and albomycine
Cir protein	74 000	41		Colicine I
74K protein	74 000	Encoded by ColV plasmids		Cloacin DF13
Btu B protein	60 000	89	~ 250	Phage BF23, E-type colicines
Lam B protein	47 392	91	$\sim 10^5$	Phage λ, K10, TP1
Protein a	40 000	12.5	$\sim 4 \times 10^4$	Phage LP81
Pho E protein	36 782	5.9	$\sim 5 \times 10^4$	Phage TC23, TC45
Protein K	40 000			
Omp F protein	37 205	20.7	$\sim 10^5$	Phage TuIa, T2, TF1 and Colicine A
Omp C protein	36 000	47		Phage TuIb, Mel, PA-2 and 434
Lc protein	36 000	Encoded by prophage PA-2		
Nmp C protein	36 000	12		
Omp A protein	35 159	21.5	$\sim 10^5$	Phage K3 and TuII*
31K protein	31 000			
O-13	31 000		$\sim 1.6 \times 10^4$	
Phospholipase A	28 000		~ 500	
Tsx protein	26 000	9.2	$\sim 10^4$	Phage T6 and Colicine K
Tra T protein	25 000	Encoded by F and F-like plasmids	2.1×10^4	
21K lipoprotein (PAL)	21 000		$\sim 10^4$	
Lom protein	20 500	Encoded by prophage λ		
Protein III	18 000		$\sim 10^4$	
LPS-binding protein[f]	15 000			
Murein lipoprotein (lpp protein)	7 200	36.5	7.5×10^5	
LT-toxin	25 500(A) 11 780(B) Stoichiometry $A_1 B_{5/6}$	Encoded by *Ent*-plasmid		

[a] The currently favoured nomenclature is one in which the proteins are ascribed the name established for their structural gene in current linkage maps. For other nomenclatures see Osborn and Wu (1980) and Lugtenberg (1981).
[b] According to 6th Edition of *Esch. coli* K12 linkage map (Bachman and Low 1980).
[c] Values refer to full expression of gene product.
[d] Refers to apparent increase (↑) or decrease (↓) in molecular weight observed on SDS-polyacrylamide gels after heating of sample to 100°C in SDS. The latter is usually indicative of the presence in the membrane of homopolymers (e.g. trimers).

Conditions for expression	Heat modifiability[d]	Peptidoglycan associated[e]	Function
Fe^{3+}-limitation			Uptake of Fe^{3+}?
Fe^{3+}-limitation			Uptake of Fe^{3+}-enterochelin
Fe^{3+}-limitation in presence of citrate			Uptake of Fe^{3+}-citrate
Fe^{3+}-limitation			Uptake of ferrichrome
Fe^{3+}-limitation	+↑		Uptake of Fe^{3+}?
In plasmid-bearing strains under Fe^{3+}-limitation			Uptake of Fe^{3+}-aerobactin
Vitamin B_{12}-limitation			Uptake of Vitamin B_{12}
Induced by maltose	+↓	+	1.6-nm diameter porin for small hydrophilic solutes; shows specificity for maltose and maltodextrins.
Synthesis decreased under conditions (30°) favouring production of mucoid colonies			Protease; involved in regulation of capsular polysaccharide biosynthesis.
Phosphate limitation	+↓	+	1.2-nm diameter porin for small hydrophilic solutes; shows preference for anions.
Observed in encapsulated (K) strains			?
Major outer membrane proteins observed under most growth conditions with omp F protein favoured by growth on nutrient broth and omp C by high osmolarity	+↓	+	1.4-nm diameter general porin for small hydrophilic solutes with mol. wt <700
	+↓	+	1.3-nm diameter general porin for small hydrophilic solutes with mol. wt. <700.
	+↓	+	Replaces omp C protein as new general porin which surface excludes and prevents superinfection by PA-2.
Found in strains lacking omp F/C proteins			
Major outer membrane protein observed under most growth conditions	+↑		Role in acceptor cell in stabilization of mating aggregates during F-pilus-mediated conjugation; influence on cell shape.
			Mediates attachment of DNA (ori C locus) to membrane.
		+	?
	+↑		Possesses phospholipase A_1, A_2 and lysophospholipase activities. Function unclear.
Expression favoured at 30°. Coregulated with nucleoside uptake			Porin for nucleosides, deoxynucleosides and amino acids.
In strains carrying F and some F-like plasmids	+↓	+	Produces surface exclusion and inhibits formation of stable mating aggregates with other F-factor carrying donors.
		+	?
In λ-infected cells			?
Repressed by c-AMP	+↓		?
			Basic protein with high affinity for LPS. Role unclear.
Major outer membrane protein expressed constitutively		+(1/3)[g]	Anchors outer membrane to peptidoglycan and influences cell shape and stability of outer membrane.
In plasmid bearing enterotoxigenic Esch. coli	+↓(B)		B binds to G_{M1} ganglioside receptors on mucosal surface and facilitates entry of A which catalyses NAD-dependent ADP-ribosylation of membrane proteins thus stimulating adenylcyclase activity.

[e] With the exception of the murein lipoprotein, where the linkage is known to be covalent, the association of proteins to the peptidoglycan is thought to be noncovalent in nature.
[f] Data relate to the salmonella analogue.
[g] 1/3 of the lipoprotein molecules are covalently linked to the peptidoglycan, the remaining 2/3 are not.

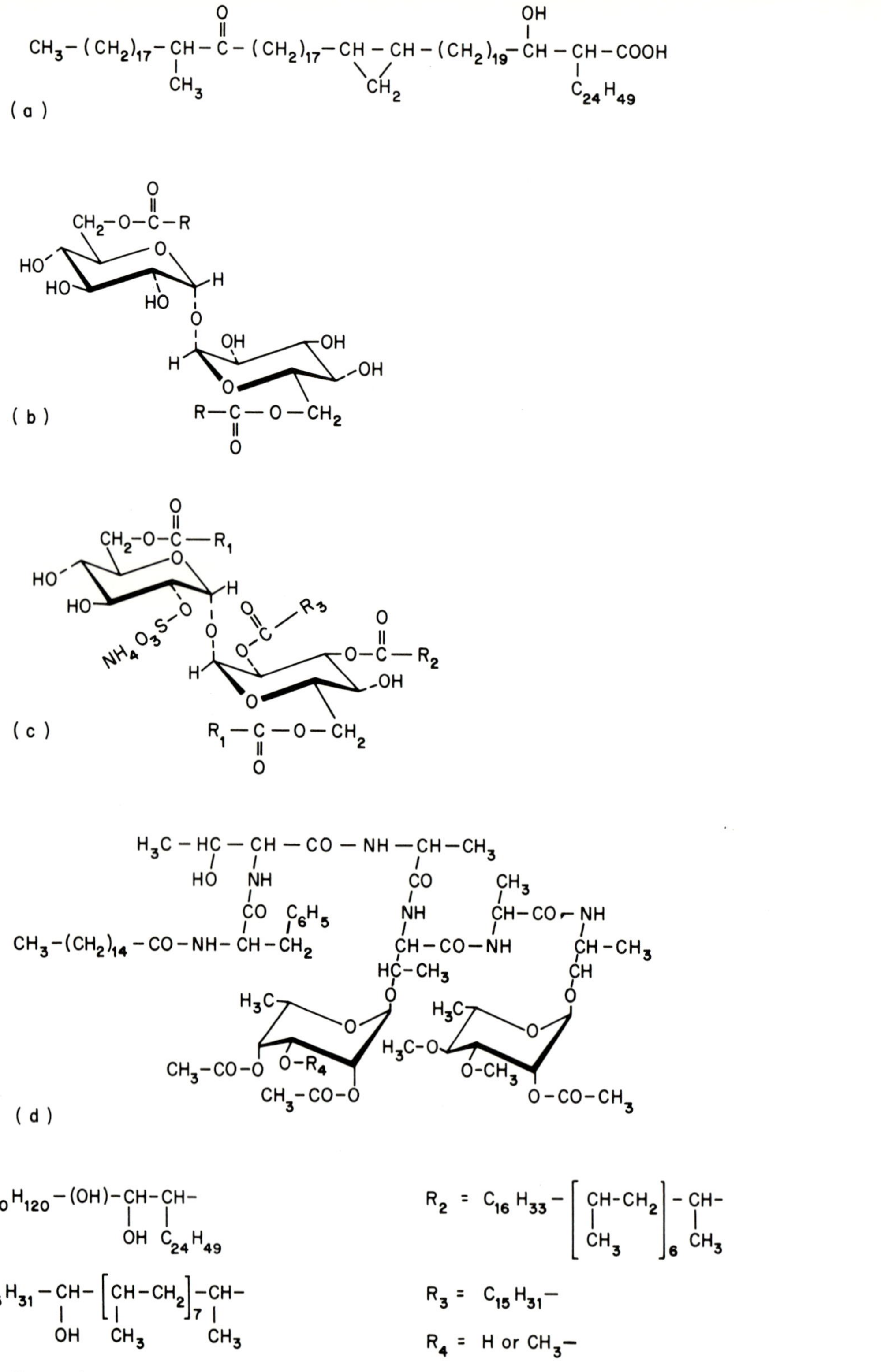

(a) $CH_3-(CH_2)_{17}-\underset{\underset{CH_3}{|}}{CH}-\overset{\overset{O}{\|}}{C}-(CH_2)_{17}-CH-\underset{\underset{CH_2}{|}}{CH}-(CH_2)_{19}-\overset{\overset{OH}{|}}{CH}-\underset{\underset{C_{24}H_{49}}{|}}{CH}-COOH$

(b) Cord factor structure

(c) Sulpholipid I structure

(d) Mycoside C_2 structure

$R = C_{60}H_{120}-(OH)-\underset{\underset{OH}{|}}{CH}-\underset{\underset{C_{24}H_{49}}{|}}{CH}-$

$R_1 = C_{15}H_{31}-\underset{\underset{OH}{|}}{CH}-\left[\underset{\underset{CH_3}{|}}{CH}-CH_2\right]_7-\underset{\underset{CH_3}{|}}{CH}-$

$R_2 = C_{16}H_{33}-\left[\underset{\underset{CH_3}{|}}{CH}-CH_2\right]_6-\underset{\underset{CH_3}{|}}{CH}-$

$R_3 = C_{15}H_{31}-$

$R_4 = H \text{ or } CH_3-$

Fig. 13.5 Proposed structures of some important mycobacterial glycolipids: (a) The principal mycolic acid (β-mycolic acid) of *M. tuberculosis* var. *hominis*. This is found ester-linked through its carboxyl group to the aribinogalactan of the cell wall. (b) Cord factor (6,6'-dimycolyl-α-α'-D-trehalose) and (c) sulpholipid I (2,3,6,6'-tetraacyl-α-α'-trehalose-2-sulphate) from strains of the same organism. (d) Mycoside C_2 from *Mycobacterium avium*.
Data from Voiland *et al.* (1971) and Goren (1972).

Table 13.8 Apparent correlation between outer membrane proteins of various bacteria[a]

Outer membrane protein in *Escherichia coli*	Analogous proteins in		
	Pseudomonas aeruginosa[b]	*Neisseria gonorrhoeae*	*Neisseria meningitidis*[c]
Lam B protein	Protein D1		
Pho E protein	Protein P		
Omp F/C proteins	Protein F	Protein I	Class 2 proteins
Omp A protein		Protein II	Class 5 proteins
21K lipoprotein (PAL)	Protein H2		
Murein lipoprotein	Protein I		

[a] A similar nomenclature to that employed for *Esch. coli* outer membrane proteins is used to describe analogous proteins in *Salm. typhimurium*.
[b] Nomenclature of Hancock and Carey (1979).
[c] Nomenclature of Tsai *et al.* (1981).

differ qualitatively and functionally from their counterparts in the cytoplasmic membrane. The outer membrane of *Esch. coli*, for example, is well characterized and contains upwards of two dozen discrete protein species (Table 13.7). There is also increasing evidence that proteins performing apparently analogous functions exist in the outer membrane of many other gram-negative bacteria (Table 13.8). Of the major outer membrane proteins, some, like the murein lipoprotein, clearly have a structural role. Many others, including several inducible and derepressible proteins, are concerned in the transport into the periplasm of solutes of low molecular weight (Table 13.7). Several can be considered as potential virulence determinants for the cell (Table 13.9).

As befits molecules with major roles in transmembrane diffusion and as receptors for various bacteriophages and colicines (Table 13.7), most outer membrane proteins, with the notable exception of the murein lipoprotein, are surface expressed. In addition, almost all appear to be antigenic. In this context it should be stressed that outer membrane proteins are notoriously difficult components to purify in undenatured form free of contaminating lipopolysaccharide or lipoprotein. Consequently, antiserum to 'purified' outer membrane proteins frequently contains precipitating antibodies directed against immunogenic contaminants and/or can be shown to be directed primarily against denatured forms of the antigen.

Despite these and other problems (see Hofstra and Dankert 1981), it now seems probable that the major porins of *Esch. coli* (*viz.* the pho E, omp F and omp C

Table 13.9 Outer membrane proteins as potential virulence determinants

Organism	Protein	Plasmid encoded	Putative virulence characteristics
Escherichia coli	74 K protein	+	Outer membrane receptor for highly efficient iron uptake system (Fe^{3+}-aerobactin). Found in certain invasive strains of colicinogenic (ColV⁺) *Esch. coli*.
	Tra T protein	+	Responsible for serum resistance in strains bearing Inc FII-group plasmids.
	LT-toxin	+	Heat-labile enterotoxin produced by enterotoxigenic (ETEC) *Esch. coli*.
Neisseria gonorrhoeae	Certain class(es) of Protein I		Implicated in resistance to complement-dependent serum killing.
	Certain class(es) of Protein II		Implicated in increased adhesion to human buccal epithelial cells and to leukocytes.
Yersinia pseudotuberculosis	Protein I	+	Temperature-inducible 140 K protein (W-antigen?) associated with virulence plasmid and probably required for intracellular survival.
Aeromonas salmonicida	54 K protein		Forms an ordered layer (A-layer) on external surface of outer membrane of virulent strains. Has been implicated in adhesion to fish epidermis.
Vibrio anguillarium	86 K protein	+(?)	Probably outer membrane receptor for plasmid-encoded iron uptake system. Observed under conditions of iron-limitation for strains of this fish pathogen harbouring a virulence plasmid.

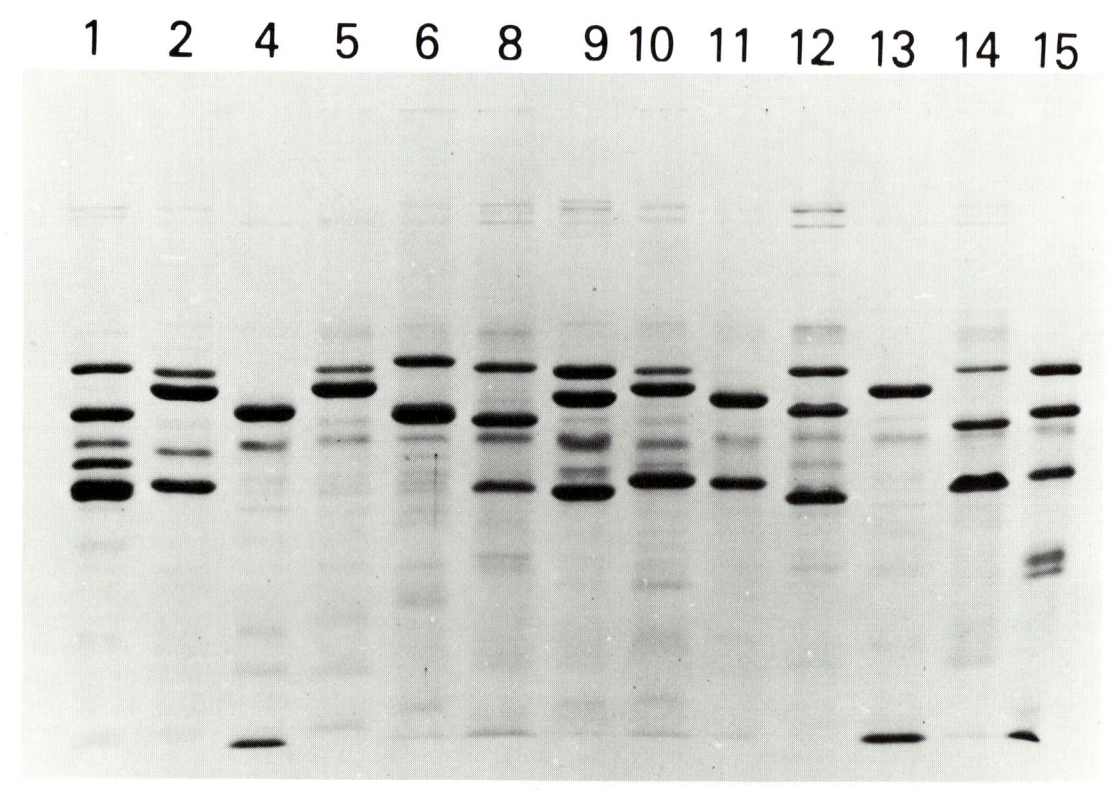

Fig. 13.6 Polypeptide profiles of outer membranes from different serotypes of *N. meningitidis* as resolved by SDS-polyacrylamide gel electrophoresis. Numbers at the top of each lane relate to the protein serotype in question. Serotypes 3 and 7 cross-react with serotypes 8 and 2 respectively and have been omitted. Note the uniqueness of the profiles obtained for each serotype. The four polypeptides resolved for outer membranes of serotype 2 correspond to (from top to bottom) proteins of Classes 1, 2, 4 and 5 and possess molecular weights of 46 K, 41 K, 32 K and 28 K respectively.

Reproduced with kind permission of the authors from Tsai *et al.* (1981). Copyright by the American Society for Microbiology. (Photograph kindly supplied by Dr Carl Frasch.)

proteins) are immunologically related; also the omp A protein, the 21 K lipoprotein (PAL), an LPS-binding protein, the murein lipoprotein and perhaps the porins may represent common antigens for the Enterobacteriaceae (Nakamura *et al.* 1979, Geyer *et al.* 1979, Beher *et al.* 1980, Overbeeke *et al.* 1980, Hofstra *et al.* 1980, Mizuno 1981). Immunological cross-reactivity does not appear to extend outside this family of microorganisms despite the presence of functionally similar proteins in a number of other bacteria (see Table 13.8).

The major protein of the *Esch. coli* outer membrane, the murein lipoprotein, is an excellent immunogen and a potent B-mitogen. Major determinants have been demonstrated for this molecule at the C-terminal linkage region. In contradistinction, the N-terminal glycerylcystene residue with attached fatty acids appears to be the biologically important part of the molecule with regard to its ability to stimulate B-lymphocytes (Braun *et al.* 1976, Bessler *et al.* 1977). Antibodies to the murein lipoprotein can be detected in sera from rabbits with experimental pyelonephritis induced by encapsulated strains of *Esch. coli* and from patients with severe enterobacterial infections, this despite the fact that the molecule is not surface expressed (Smith 1977, Griffiths *et al.* 1977, Owen 1981). However, to date there is no evidence that anti-lipoprotein antibody has any pronounced protective effect.

Several other outer membrane proteins, including the omp A protein and omp F/C proteins of *Esch. coli*, and proteins F, H2 and I of *Ps. aeruginosa* are also established B-mitogens (Bessler and Henning 1979, Chen *et al.* 1980). Furthermore, the omp F/C proteins, the *omp A* gene product and murein lipoprotein have recently been shown to be major constituents of the so-called endotoxin protein of *Esch. coli*, to which has been ascribed such diverse biological properties as mitogenic and polyclonal activation of B-lymphocytes, release of vasoactive amines from mast cells and stimulation of macrophages to differentiate into cytotoxic cells capable of lysing tumour cells (Goldman *et al.* 1981). Studies with the major 39 K pore protein of *Pr. mirabilis* have also clearly established the fact that some outer membrane proteins can dramatically stimulate the host's antibody response to lipopolysaccharide and convert it from a predominantly IgM to a predominantly IgG response (Karch and Nixdorff 1981). The reciprocal type of reaction i.e. the adjuvant effect of lipopolysaccharide is well documented.

Different strains of a particular bacterium often show pronounced differences in the pattern of major outer membrane proteins observed after sodium dodecyl sulphate-polyacrylamide gel electrophoresis (Fig. 13.6). This variability has been successfully utilized in the subtyping of strains of *H. influenzae* type b (Barenkamp *et al.* 1981*a*, *b*) and in the serotyping of pathogenic neisseriae. One of the earliest uses of this technique, which appears to have considerable epidemiological potential, was in the study of *N. gonorrhoeae*. Here it was shown that the gonococcus could be subdivided into 16 different serotypes based upon the antigenic specificity of an outer membrane complex, for which variations in the molecular weight of a principal membrane protein (Protein I) could be related in large part to antigenic differences (Johnston *et al.* 1976). There are now indications, however, that other antigens (possibly lipopolysaccharide and Protein III) also play a significant role in this typing system. For example, a recent survey indicated that 99 per cent of gonococcal strains tested could be serotyped by an ELISA assay based upon a family of only nine different purified Protein Is. A further significant feature of this latter study relates to the observation that almost all invasive strains could be typed by means of just two of the nine Protein I serotypes *viz.* the closely related serotypes 1 and 2 (Buchanan and Hildebrandt 1981, see Table 13.9). This raises the possibility of using these (Protein I) serotype proteins as vaccines against invasive gonorrhoea.

Meningococci of serogroup B can also be subdivided into protein serotypes based upon antigenically different sets of outer membrane proteins (see Fig. 13.6). Of the fifteen distinguishable serotypes, only serotypes 2 and 15 appear to be associated with outbreaks of meningococcal disease. Since purified group B polysaccharide is poorly immunogenic (see Section 3.4.2), considerable attention has focused upon the use of outer membrane proteins of these serotypes as potential vaccines. Initial results were encouraging inasmuch as purified preparations of the class 2 protein from serotype 2 (see Table 13.8 and Fig. 13.6) were found to protect experimental animals against challenge with homologous (type 2) strains of the meningococcus (Frasch and Robbins 1978). Unfortunately, although this vaccine was non-toxic, it proved to be poorly immunogenic in adults and children. The problem can be circumvented, at least in part, by use of a more soluble but cruder vaccine composed of non-covalent complexes of group B polysaccharide and LPS-depleted outer membrane proteins. Results of recent trials have shown the improved vaccines to cause no serious reactions and to produce in both adults and infants bactericidal antibody against serotype 2 protein. Bactericidal antibodies against the polysaccharide component are also elicited, although these are largely of the IgM type and of short duration (Zollinger *et al.* 1979, Sippel 1981, Zahradnik *et al.* 1981). At present, the most efficacious vaccine against all pathogenic meningococci would appear to be one which contains both outer membrane proteins and the capsular polysaccharides.

These conclusions parallel to some extent those obtained in studies of experimental salmonellosis in mice. Here vaccination of mice with outer membrane porins which had been covalently coupled to an octasaccharide derived from the salmonella O-antigen afforded protection against an approximately tenfold higher challenge dose of *Salm. typhimurium* than was given by either porin or octasaccharide alone. Furthermore, antibodies elicited by this non-toxic conjugate possessed specificity for both antigenic moieties and were active in passive immunization tests (Svenson *et al.* 1979).

The failure of the purified type b capsular polysaccharide of *H. influenzae* to protect the age group (infants under 18 months of age) most at risk from this serious invasive pathogen has also stimulated investigations into the immunogenic potential of outer membrane proteins. In a series of illuminating papers, Hansen and his colleagues have shown that young rats and, more significantly, infants recovering from systemic *Haemophilus* type b meningitis can form antibodies to a number of outer membrane proteins including a major one of molecular weight 39 K, which is surface expressed and accessible to homologous antibody. Baby rats passively immunized with monoclonal antibody against this protein are protected against systemic disease caused by virulent (homologous) strains of *H. influenzae* type b (Hansen *et al.* 1982, Gulig *et al.* 1982). These observations are important and encouraging since they reinforce the notion that antibodies against outer membrane proteins can be protective even in encapsulated pathogens, and suggest that a proteinaceous vaccine may soon be a reality for an organism which is the single most important cause of endemic bacterial meningitis in infants.

4 Major extracellular antigens

Several human pathogens release extracellular toxins and enzymes, which are antigenic (Table 13.1). These antigens are a particular feature of gram-positive bacteria but, increasingly, important extracellular products are being identified in gram-negative organisms. Some bacterial species, e.g. *Staph. aureus*, *Str. pyogenes* and *Clostridium perfringens* secrete a large num-

Table 13.10[a] Exotoxins with known role in pathogenicity

Disease	Causative organism	Toxin
Diphtheria	*Corynebacterium diphtheriae*	Diphtheria toxin
Tetanus	*Clostridium tetani*	Tetanus toxin (Tetanospasmin)
Botulism	*Clostridium botulinum*	Botulinum toxins
Enterotoxaemia in domestic animals	*Clostridium perfringens* (types B, C, D)	β and ε Toxins
Cholera	*Vibrio cholerae*	Cholera enterotoxin
Diarrhoeal disease	*Escherichia coli* (enteropathogenic strains)	Heat labile (LT) and heat stable (ST) enterotoxin
Food poisoning	*Staphylococcus aureus*	Staphylococcal enterotoxin
	Clostridium perfringens (type A)	Clostridial enterotoxin
	Bacillus cereus	*B. cereus* enterotoxin
Scalded skin syndrome	*Staphylococcus aureus*	Epidermolytic toxin
Scarlet fever	*Streptococcus pyogenes*	Erythrogenic toxin
'Opportunistic' infections in debilitated patients	*Pseudomonas aeruginosa*	Exotoxin A

[a] Data from Arbuthnott (1978).

ber of extracellular antigens; for example supernatants of *Staph. aureus* contain up to 30 distinct exoproteins. (For Symposia on Toxins, see Reports 1975, 1978, 1982.)

In several instances extracellular antigens play an important role in pathogenesis (Table 13.10). The pioneering studies at the turn of the century led to the discovery of diphtheria toxin, tetanus toxin and botulinum toxin. It was soon realized that the pathogens concerned were unique in that the causative organisms were non-invasive, each elaborating a single powerful exotoxin that accounted for the symptoms of disease. Antitoxic immunity confers a high degree of protection against the overt disease; toxoid vaccines against diphtheria and tetanus are among the commonest and most successful in present day use.

Unfortunately, despite the hopes of early microbiologists, it became clear that prophylactic 'antitoxic' immunization against other bacterial infections was at best only partly successful. Indeed, around the 1940s interest in bacterial toxins declined for two main reasons, (i) increasing reliance on antibiotics and (ii) the failure to appreciate the complexity of pathogenic mechanisms. It is now realized that with few exceptions pathogenicity is multifactorial, often involving numerous cell-associated and extracellular virulence factors. This can be even more so in mixed infections.

Toxins are classically associated with symptomatic tissue damage, e.g. tetanus and dysentery. However, extracellular products of invasive pathogens can have a more subtle role. The membrane-damaging toxins of staphylococci, streptococci and clostridia (see Table 13.1) can kill phagocytic cells or, at very low doses, impair the ability of these cells to respond to a chemotactic stimulus (see review, Arbuthnott 1981). These toxins also act on lymphocytes and mast cells and the responses of these cells are important in determining the outcome of host-pathogen interactions.

Where the mechanism of pathogenesis is complex it is difficult to evaluate the role of individual exotoxins and exoenzymes. Convincing evidence for the role of staphylococcal α-toxin came from studies of experimental rabbit mastitis by Adlam et al. (1977), who showed that the introduction of small numbers of certain strains into the mammary tissue caused a spreading lesion in which the tissue became oedematous and haemorrhagic, but that other strains induced the formation of pus-filled abscesses. In rabbits immunized with highly purified α-toxin the lethal, haemorrhagic oedematous form of the disease was reduced to a localized abscess. No such protection was obtained by immunization with the other major staphylococcal cytolysin, β-toxin. Thus, availability of purified extracellular antigens and of specific antisera are valuable tools in the study of pathogenesis.

Some generalizations can now be made about structure-function relation of exotoxins which have implications for these antigens as immunogens. Whatever the tissue affected, the potent biological effects of toxins involve toxin-membrane interactions. These can be of two types (i) binding of the toxin (through the B or binding region of the molecule) to a specific membrane receptor enables the active (or A) region to start off membrane-associated or intracytoplasmic events that lead to impaired metabolism of host cells e.g. diphtheria toxin, pseudomonas exotoxin A, cholera toxin and *Esch. coli* heat-labile enterotoxin: (ii) interaction with targets in the membrane, usually lipids or phospholipids, causes impairment of normal permeability properties (e.g. α-toxin and θ-toxin of *Cl. perfringens*, streptolysins O and S and staphylococcal α, β, γ and δ-toxins). For certain bipartite toxins consisting of A and B regions, e.g. cholera toxin and *Esch. coli* LT toxin, the non-toxic B subunits could perhaps be used as natural toxoids or subunit vaccines. Indeed the B region of cholera toxin termed choleragenoid, was purified along with the whole toxin (Finklestein and Lo Spolluto 1969); choleragenoid induces antibodies that neutralize the intact toxin. Recently an alternative approach has been followed by Finkelstein and his colleagues (Honda and Finkelstein 1979) who isolated a variant of *V. cholerae* with a mutation in the gene for toxin production. This mutant produces only choleragenoid and no A subunit. Such a variant, designated 'Texas-Star', might be the forerunner of a live vaccine which would stimulate a specific IgA response in the gut.

By contrast with cholera toxin it is the A fragment of diphtheria toxin that is associated with immunological specificity. The amino acid sequence associated with toxicity and immunological specificity has been identified and the first successful active immunization with a synthetic diphtheria toxin antigen was described recently (Audibert et al. 1982). A synthetic tetradecapeptide equivalent to part of the A fragment was linked covalently to two different carriers and elicited in guinea-pigs antibodies which bound specifically with toxin and neutralized its effects.

In some cases, where toxins share membrane receptors or target sites there appears to be an element of homology between such toxins which is revealed in immunological cross-reactions. This is seen in the B regions of cholera toxin and *Esch. coli* heat labile toxin, both of which bind to ganglioside receptors. The B regions of these two toxins exhibit substantial amino acid homology (Dallas and Falkow 1981). A well known and still unexplained relation exists for a group of membrane-damaging cytolysins, known as the thiol-activated cytolysins, such as streptolysin O, tetanolysin, pneumolysin and *Cl. perfringens* θ-toxin which have among other common properties cholesterol as a membrane target site (see review of Arbuthnott 1982). The different thiol-activated toxins exhibit cross-neutralization with hyperimmune horse sera; for example tetanolysin is neutralized by antistreptolysin O and vice versa. This implies the presence of a degree of homology among the extracellular proteins of unrelated organisms.

In addition to toxins that have a known or at least a putative role in disease, there are numerous exoenzymes, some of which are important in diagnosis. For example, coagulase production is of help in identification of *Staph. aureus*. Numerous other exoenzymes are of value in identification and taxonomy. Typing of *Cl. perfringens* is based on the pattern of production of exoenzymes and lethal exotoxins. Another application of exoenzymes is their use in serodiagnosis. Raised antibody levels to DNAase and NADase are useful diagnostic indices of preceding streptococcal infection and are often used along with antistreptolysin O measurement.

The isolation of highly purified extracellular antigens from culture filtrates containing numerous exoproteins has presented problems. Success has paralleled advances in protein purification technology. While techniques of high resolving power, such as preparative isoelectric focusing, ion exchange chromatography and gel filtration, have proved useful, the availability of high-speficity antibodies has provided a powerful tool. Immobilized antibody acts as an ideal substrate for affinity chromatography; antigen binds to the antibody and can be eluted subsequently by manipulation of pH and salt concentration. The wider availability of monoclonal antibodies will no doubt facilitate the important task of antigen purification and vaccine development.

5 Vaccines and vaccine development

The knowledge that prior exposure to an infectious agent can protect against subsequent infection has led to the search for suitable effective safe vaccines. The search lost impetus with the introduction of antibiotics for treatment of infections but with the emergence of antibiotic-resistant bacterial strains, notably among *Shigella* and *Salmonella* species, *V. cholerae*, *N. gonorrhoeae* and *Bord. pertussis* the search has regained its momentum. As can be seen from Table 13.11 the number of currently available, proven and effective bacterial vaccines is disturbingly small. A number of problems have hindered vaccine development.

For example, even at present, the identities of protective antigens and mechanisms of pathogenicity are not known for many species of bacteria. It is only recently that any detailed knowledge of the host immune responses *in vivo* have emerged, particularly an understanding of the regulation, genetic basis or localized nature of these responses. The latter feature is particularly important in relation to mucosal immunity, where it is evident that stimulation of immunity must be localized to be effective. Previously used parenteral vaccines often failed to protect against colonization of the mucosa because they acted at the wrong location and did not stimulate the correct type of immune response. Other problems besetting effective vaccine production are those of antigenic variation and mutation whereby alteration of the antigenic constitution of the product from that of the vaccine leads to evasion of the expected specific immunity. In order to keep adverse side effects of vaccines to a minimum, it has been standard practice to extract and purify protective antigens. These may no longer truly resemble the stimulus presented to the host *in vivo*; generally, isolated antigens are of lower molecular weight than natural antigens and may be readily dissociated from subsequently formed immune complexes. This may be the direct cause of low antigenicity or short lived immunity seen with many vaccine preparations, e.g. early meningococcal capsular vaccines.

The most effective vaccines to date have been those directed against single extracellular products, e.g. tetanus toxin and diphtheria toxin where the toxoided vaccines closely resemble the natural state of the antigen *in vivo*. Vaccines which are based on major anti-

Table 13.11 Bacteriological vaccines

Vaccine	Form of immunogen	Remarks
Anthrax	Protein fractions from *Bacillus anthracis* (adsorbed form)	Efficacy up to 80 per cent. Used on persons at risk only. Short duration.
Cholera	Whole phenol-killed *Vibrio cholera* of Ogawa and Inaba serotypes in equal numbers	Efficacy: poor and short-lived. WHO standard: vial containing 4×10^{10} Inaba or Ogawa serotype organisms.
Diphtheria	Cell-free formaldehyde toxoid (adsorbed form)	Efficacy: almost complete for 10 years. WHO standard: vial containing 132 IU = 75 mg toxoid. Given alone, with pertussis or pertussis and tetanus vaccines.
Dental caries	1. Dextran-synthesizing enzymes (glycosyl transferases) 2. Whole killed *Streptococcus mutans* organisms	Experimental
Enterotoxigenic *Esch. coli*	1. Fimbrial adhesins adsorbed 2. Enterotoxin	Experimental
Gonorrhoea	1. Fimbrial adhesins adsorbed	Experimental. Induces opsonizing antibody.
	2. Outer membrane proteins	Experimental
	3. IgA protease	Experimental
Haemophilus influenzae b	1. High molecular weight polysaccharide from capsule, not completely purified	Experimental. Effective but short-lived protection. Non-effective in children under 18 months.
	2. Protein-polysaccharide complexes	Experimental
	3. Live *Esch. coli* K100 microorganisms	Experimental. Ribose-ribitol phosphate capsule reacts with *Haemophilus* components.
Meningococcal A	Pure capsular polysaccharide from group A organisms	Efficacy: almost complete and long duration
Meningococcal C	Pure capsular polysaccharide from group* C organisms	Efficacy: 95 per cent in persons 2 yr or older. Long duration. *WHO recommendation: incorporate high molecular weight fractions from vaccines A, C, Y and W135 as a single vaccine.
Meningococcal groups Y, 29E and W135	Pure capsular polysaccharide from respective serotype organisms	Experimental
Meningococcal group B	1. Serotype-specific outer membrane protein	Experimental — Group B capsular polysaccharide alone is not immunogenic
	2. O-acetylated K-polysaccharide	Experimental
Pertussis	Whole heat/chemically-killed phase 1 *Bordetella pertussis* organisms	Efficacy: c. 90 per cent for 10 years. WHO standard: vial containing 46 IU. Given alone, with diphtheria or with diphtheria and tetanus vaccines. Short duration.
Plague	Whole formalin-killed *Yersinia pestis* organisms	
Pneumococcal	Pure capsular polysaccharide from 14 serotypes	Efficacy: 80–90 per cent for short duration Usual dose 50 µg/serotype/dose
Shigella	Live attenuated type-specific *Shigella* organisms	Experimental. Administered orally.
Streptococcal Group B	Type III$_a$ capsular polysaccharide	Experimental
Tetanus	Cell-free formaldehyde toxoid (adsorbed form)	Efficacy: almost complete for 10 years. Administered alone, with pertussis or with pertussis and diphtheria vaccines. WHO standard: vial containing 340 IU = 27.5 mg toxoid
Tuberculosis	Live, attenuated *Mycobacterium tuberculosis* (B.C.G. strain)	Efficacy: undetermined, probably of permanent duration. WHO standard: vial containing 2.5 mg semi-dry weight.
Tularensis	Live, attenuated *Franciscella tularensis* organisms	Experimental. Given to those at risk only.
Typhoid	1. Whole heat/phenol-killed *Salmonella typhi* organisms	Efficacy, Good, of doubtful duration. Vi antigen is not preserved. WHO standard: vial containing 34 mg heat/phenol-killed organisms.
	2. Whole acetone-killed *Salmonella typhi* organisms	Efficacy: c. 90 per cent for several years. Vi antigen remains intact. WHO standard: vial containing 11 mg acetone-killed organisms.
	3. Isolated Vi polysaccharide	Experimental
	4. Live attenuated Ty21$_a$ streptomycin-dependent mutant	Experimental. Efficacy: 85–95 per cent for over 3 years. Usual dose 10^9–10^{10} organisms orally.
Typhus	1. Aqueous extract from organic solvent phase of killed rickettsia organisms grown in chick embryo	Efficacy: short duration
	2. Live attenuated *Rickettsia prowazeki*	Efficacy: short duration
	3. Formalin-killed *Rickettsia prowazeki*	Efficacy: short duration

gens of certain bacteria, e.g. pneumococcal capsular polysaccharides, where numerous serotypes exist should reflect the current serotype incidence in disease.

To more fully appreciate which antigens bacteria produce during infection, greater emphasis has to be placed upon the analysis of *in vivo*-grown organisms. Practical problems associated with this type of study have been largely overcome and preliminary investigations have already revealed differences between *in vivo*-grown and *in vitro*-grown *Staph. aureus* (Adlam et al. 1970, Day 1983) and between *in vivo*-grown and *in vitro*-grown *Esch. coli* (Finn *et al.* 1982).

References

Acker, G., Schmidt, G. and Mayer, H. (1982) *J. gen. Microbiol.* **128**, 1577.
Adlam, C., Pearse, J. H. and Smith, H. (1970) *J. med. Microbiol.* **3**, 147.
Adlam, C., Ward, P. D., McCartney, C., Arbuthnott, J. P. and Thorley, C. M. (1977) *Infect. Immun.* **17**, 250.
Anderson, E. S. (1961) *Nature, Lond.* **193**, 501.
Arbuthnott, J. P. (1978) *Companion to Microbiology* Ch. 6, Longman, London; (1981) *Microbial Perturbation of Host Defences*, p. 97, Academic Press, London; (1982) *Molecular Actions of Toxins and Viruses*, p. 107, Elsevier Biomedical Press, Amsterdam.
Audibert, F., Jolivet, M., Chedid, L., Arnon, R., Sela, M. (1982) *Proc. nat. Acad. Sci. Wash.* **79**, 5042.
Austrian, R. (1981) *Rev. infect. Dis.* **3**, S1.
Bachman, B. J. and Low, K. B. (1980) *Microbiol. Rev.* **44**, 1.
Baddiley, J. (1970) *Account. chem. Res.* **3**, 98.
Barenkamp, S. J., Munson, R. S. and Granoff, D. M. (1981a) *J. infect. Dis.* **143**, 668.
Barenkamp, S. J., Granoff, D. M. and Munson, R. S. (1981b) *J. infect. Dis.* **144**, 210.
Barksdale, L. and Kim, K-S. (1977) *Bacteriol. Rev.* **41**, 217.
Beachey, E. H. (1980) *Bacterial Adherence*, Chapman and Hall, London; (1981) *J. infect. Dis.* **143**, 325.
Beher, M. G., Schnaitman, C. A. and Pugsley, A. P. (1980) *J. Bact.* **143**, 906.
Bentley, D. W. (1981) *Rev. infect. Dis.* **3**, S61.
Bergner-Rabinowitz, S., Beck, A., Ofek, I. and Davis, A. M. (1969) *Israel J. med. Sci.* **5**, 285.
Bessler, W. and Henning, U. (1979) *Z. Immunitaeforsch.* **155**, 387.
Bessler, W., Resch, K., Hancock, E. and Hantke, K. (1977) *Z. Immunitäbsforsch.* **153**, 11.
Bradley, S. G. (1979) *Annu. rev. Microbiol.* **33**, 67.
Braude, A. I. (1980) *Advanc. intern. Med.* **26**, 427.
Braun, V., Bosch, V., Klumpp, E. R., Neff, I., Mayer, H. and Schlecht, S. (1976) *Eur. J. Biochem.* **62**, 555.
Broome, C. V. (1981) *Rev. infect. Dis.* **3**, S82.
Buchanan, T. M. and Hildebrandt, J. F. (1981) *Infect. Immun.* **32**, 985.
Calladine, C. R. (1976) *J. molec. Biol.* **118**, 457.
Chen, Y-H. U., Hancock, R. E. W. and Mishell, R. I. (1980) *Infect. Immun.* **28**, 178.
Costerton, J. W., Irwin, R. T. and Cheng, K-J. (1981) *Annu. Rev. Microbiol.* **35**, 299.
Dajani, A. S. and Wannamaker, L. W. (1970) *J. infect. Dis.* **122**, 196.
Dallas, W. S. and Falkow, S. (1980) *Nature, Lond.* **288**, 499.

Day, S. E. J. (1983) *PhD thesis* Dublin.
De Voe, I. W. (1982) *Microbiol. Rev.* **46**, 162.
Doetsch, R. N. and Sjoblad, R. D. (1980) *Annu. Rev. Microbiol.* **34**, 69.
Duckworth, M. (1977) *Surface of Carbohydrates of the Prokaryotic Cell*, p. 177. Academic Press, London.
Dudman, W. F. (1977) *Surface Carbohydrates of the Prokaryotic Cell*, p. 357. Academic Press, London.
Duguid, J. P. (1968) *Arch. Immun. Ther. exp.* **16**, 173.
Duguid, J. P., Anderson, E. S. and Campbell, I. (1966) *J. Path. Bact.* **92**, 107.
Duguid, J. P., Clegg, S. and Wilson, M. I. (1979) *J. med. Microbiol.* **12**, 213.
Duguid, J. P. and Old, D. C. (1980) *Bacterial Adherence*, p. 185. Chapman and Hall, London.
Duguid, J. P., Smith, I. W., Demster, G. and Edmunds, P. N. (1955) *J. Path. Bact.* **70**, 325.
Evans, D. G. and Evans, D. J., Jr. (1978a) *Infect. Immun.* **21**, 638.
Evans, D. G., Evans, D. J., Jr., Tjoa, W. S. and Du Pont, H. L. (1978b) *Infect. Immun.* **19**, 727.
Farquar, J. D., Hankins, W. A., Sanctis, A. N. de, Meio, J. L. de and Metzgar, D. P. (1977) *Proc. Soc. exp. Biol. Med.* **155**, 453.
Feinman, S. B., Prescott, B. and Cole, R. M. (1973) *Infect. Immun.* **8**, 752.
Finklestein, R. A. and Lo Spolluto, J. J. (1969) *J. exp. Med.* **130**, 185.
Finn, T. M., Arbuthnott, J. P. and Dougan, G. (1982) *J. gen. Microbiol.* **128**, 3083.
Fischer, W., Laine, R. A. and Nakano, M. (1978) *Biochim. biophys. Acta* **528**, 298.
Frasch, C. E. and Robbins, J. D. (1978) *J. exp. Med.* **147**, 629.
Gannon, P. J., Jacobs, D. M., Marx, A., Mayer, H., Romanowska, E. and Neter, E. (1982) *Infect. Immun.* **35**, 193.
Geyer, R., Galanos, C., Westphal, O. and Golecki, J. R. (1979) *Eur. J. Biochem.* **98**, 27.
Ghuysen, J-M., Strominger, J. L. and Tipper, D. J. (1968) *Comprehensive Biochemistry* Vol. 26A, p. 53. Elsevier, Amsterdam.
Goebel, W. F. (1965) *15th Coll. Ges. Phy. Chem., Mosbach. 1964*, p. 135. Springer, Berlin.
Goldman, R. C., White, D. and Leive, L. (1981) *J. Immunol.* **127**, 1290.
Goldman, R. C. et al. (1982) *J. Bact.* **151**, 1210.
Good, R. A. and Day, S. B. (1981) *Comprehensive Immunology* Vol. 8. Plenum, New York.
Goren, M. B. (1972) *Bact. Rev.* **36**, 33.
Gotschlich, E. C., Fraser, B. A., Nishimura, O., Robbins, J. B. and Liu, T.-Y. (1981) *J. biol. Chem.* **17**, 8915.
Griffiths, E. K., Yoonessi, S. and Neter, E. (1977) *Proc. Soc. exp. Biol. Med.* **154**, 246.
Grov, A. (1977) *Microbiology 1977*, p. 350. Amer. Soc. Microbiol., Washington, D.C.
Gulig, P. A., McCracken, G. H., Frisch, C. F., Johnston, K. H. and Hansen, E. J. (1982) *Infect. Immun.* **37**, 82.
Hall, M. N. and Silhavy, T. J. (1981) *Annu. Rev. Genet.* **15**, 91.
Hamada, S. and Slade, H. D. (1980) *Microbiol. Rev.* **44**, 331.
Hancock, R. E. W. and Carey, A. M. (1979) *J. Bact.* **140**, 902.
Hansen, E. J., Robertson, S. M., Gulig, P. A., Frisch, C. F. and Haanes, E. J. (1982) *Lancet* **i**, 366.

Heidelberger, M. (1960) *Fortschritte der chemie Organischer Naturstoffe* **18**, 503; (1973) *Research in Immunochemistry and Immunobiology*, Vol. 3, p. 1. University Park Press, Baltimore.

Heidelberger, M. and Nimmich, W. (1976a) *Immunochemistry* **13**, 67; (1976b) *J. Immunol.* **109**, 1337.

Hofstra, H. and Dankert, J. (1981) *J. gen. Microbiol.* **125**, 285.

Hofstra, H., Van Tol, M. J. D. and Dankert, J. (1980) *J. Bact.* **143**, 328.

Honda, T. and Finklestein, R. A. (1979) *Proc. nat. Acad. Sci. Wash.* **76**, 2052.

Jann, K. and Jann, B. (1977) *Surface of Carbohydrates of the Prokaryotic Cell*, p. 247. Academic Press, London.

Jann, K. and Westphal, O. (1975) *The Antigens*, Vol. III, p. 1. Academic Press, London.

Johnston, K. H., Holmes, K. K. and Gotschlich, E. C. (1976) *J. exp. Med.* **143**, 741.

Jones, G. W. (1977) *Microbial Interactions*, Vol. 3, p. 139. Chapman and Hall, London; (1980) *Bacterial Adherence*, p. 219. Chapman and Hall, London.

Joys, T. M. and Kim, H. (1978) *Microbios. Lett.* **7**, 65.

Kabat, E. A. and Bezer, A. E. (1958) *Arch. Biochem. Biophys.* **78**, 306.

Kagawa, H., Asakura, S. and Lino, T. (1973) *J. Bact.* **113**, 1474.

Karch, H. and Nixdorff, K. (1981) *Infect. Immun.* **31**, 862.

Kauffmann, F. (1935) *Z. Hyg. Infektkr.* **116**, 617; (1954) *Enterobacteriaceae*. Munksgaard, Copenhagen; (1961) *Die Bakteriologie der Salmonella Species*. Munksgaard, Copenhagen.

Knirel, Y. A. et al. (1982) *Eur. J. Biochem.* **128**, 81.

Knox, K. W. and Wicken, A. J. (1973) *Bact. Rev.* **37**, 215.

Kondoh, H. and Hotani, H. (1974) *Biochim. biophys. Acta* **336**, 117.

Kunin, C. M., Beard, M. V. and Halmagyi, N. E. (1962) *Proc. Soc. exp. Biol. Med.* **111**, 160.

Kwapinski, J. B. G. (1974) *CRC Handbook of Microbiology*, Vol. IV, p. 717. CRC Press, Cleveland.

Lambert, P. A., Hancock, I. C. and Baddiley, J. (1977) *Biochim. biophys. Acta* **472**, 1.

Lancefield, R. C. (1958) *J. exp. Med.* **108**, 329; (1962) *J. Immunol.* **89**, 307.

Lancefield, R. C. and Perlman, G. E. (1952) *J. exp. Med.* **96**, 83.

Lederberg, J. and Edwards, P. (1953) *J. Immunol.* **71**, 323.

Lefford, M. J. (1981) *Immunology of Human Infection*, Pt. I, p. 345. Plenum Publishing Corp., New York.

Levine, M. M. (1981) *Adhesion and Microorganism Pathogenicity*, p. 142. Pitman Medical, Tunbridge Wells.

Lindberg, A. A. (1977) *Surface of Carbohydrates of the Prokaryotic Cell*, p. 289. Academic Press, London.

Luderitz, O., Jann, K. and Wheat, R. (1968) *Comprehensive Biochemistry*, Vol. 26A, p. 105. Elsevier, Amsterdam.

Luderitz, O., Staub, A. M. and Westphal, O. (1966) *Bact. Rev.* **30**, 192.

Luderitz, O., Westphal, O., Staub, A. M. and Nikaido, H. (1971) *Microbial Toxins*, Vol. IV, p. 145. Academic Press, London.

Lugtenberg, B. (1981) *Trends biochem. Sci.* **6**, 262.

Macario, E. C. de, Marcario, A. J. L. and Wolin, M. J. (1982) *J. Bact.* **149**, 320.

McGhee, J. R. and Michalek, S. M. (1981) *Annu. Rev. Microbiol.* **35**, 595.

MacLeod, C. M., Hodges, R. G., Heidelberger, M. and Bernhard, W. G. (1945) *J. exp. Med.* **82**, 445.

Makela, P. H. et al. (1977) *J. infect. Dis.* **136**, 543.

Mendearis, D. N. Jr., Camitta, B. M. and Heath, E. C. (1968) *J. exp. Med.* **128**, 399.

Meyer, T. F., Mlawer, N. and So, M. (1982) *Cell* **30**, 45.

Mizuno, T. (1981) *J. Biochem., Tokyo* **89**, 1039.

Mulholland, A., Larson, A. D., Hart, L. T. and Eubanks, E. R. (1978) *Abstr. Annu. Meet. Amer. Soc. Microbiol.* **78**, 40.

Nagy, B., Moon, H. W. and Isaacson, R. E. (1977) *Infect. Immun.* **16**, 344.

Nakamura, K., Pirtle, R. M. and Inouye, M. (1979) *J. Bact.* **137**, 595.

Ofek, I., Simpson, W. A. and Beachey, E. H. (1982) *J. Bact.* **149**, 426.

Ørskov, I. and Ørskov, F. (1966) *J. Bact.* **91**, 69.

Ørskov, I., Ørskov, F., Jann, B. and Jann, K. (1977) *Bact. Rev.* **41**, 667.

Ørskov, I., Ørskov, F., Smith, H. W. and Sojka, W. J. (1975) *Acta Path. Microbiol. scand. B.* **83**, 31.

Osborn, M. J. and Wu, H. C. P. (1980) *Annu. Rev. Microbiol.* **34**, 369.

Oss, C. J. van and Gillman, C. F. (1972) *J. Reticuloendothelial Soc.* **12**, 283.

Ottow, J. C. G. (1975) *Annu. Rev. Microbiol.* **29**, 79.

Overbeeke, N., Scharrenburg, G. van and Lugtenberg, B. (1980) *Eur. J. Biochem.* **110**, 247.

Owen, P. (1981) *Organization of Prokaryotic Cell Membranes*, Vol. 1, p. 73, CRC Press Inc., Boca Raton, Florida; (1983) *Electroimmunochemical Analysis of Membrane Proteins*, Ch. 19. Elsevier, Amsterdam.

Palva, E. T. and Makela, P. H. (1980) *Eur. J. Biochem.* **107**, 137.

Pearse, W. A. and Buchanan, T. M. (1980) *Bacterial Adherence*, p. 289. Chapman and Hall, London.

Peltola, H. et al. (1977) *New Engl. J. Med.* **297**, 686.

Philips, G. N. Jr., Flicker, P. F., Cohen, C., Manjula, B. N. and Fischetti, V. A. (1981) *Proc. nat. Acad. Sci. Wash.* **78**, 4689.

Qureshi, N., Takayama, K. and Rivi, E. (1982) *J. biol. Chem.* **257**, 11808.

Razin, S. (1978) *Microbiol. Rev.* **42**, 414; (1981) *Organization of Prokaryotic Cell Membranes*, Vol. 1, p. 165. CRC Press, Boca Raton, Florida.

Report (1975) *Jap. J. med. Sci. Biol.* **28**, 55; (1978) *Ibid.* **31**, 155; (1982) *Ibid.* **35**, 105.

Robbins, J. B. (1978) *Immunochemistry* **15**, 839.

Rogers, H. J., Perkins, H. R. and Ward, J. B. (1980) *Microbial Cell Walls and Membranes*. Chapman and Hall, London.

Schmidt, M. A. and Jann, J. (1982) *FEMS Microbiol. Lett.* **14**, 69.

Shellan, R. G. and Nossal, G. J. V. (1968) *Immunology* **14**, 273.

Silverman, M., Zieg, J., Hilman, M. and Simon, M. (1979) *Proc. nat. Acad. Sci. Wash.* **76**, 391.

Sippel, J. E. (1981) *CRC crit. Rev. Microbiol.* **8**, 267.

Sjöquist, J., Movitz, J., Johansson, I. B. and Hjelm, H. (1972) *Eur. J. Biochem.* **30**, 190.

Sleytr, U. B. (1978) *Int. Rev. Cytol.* **53**, 1.

Smith, J. W. (1977) *Infect. Immun.* **17**, 366.

Somplolinsky, D., Hertz, J. B., Hoiby, N., Jensen, K., Bendt, M. and Samra, Z. (1980) *Acta Path. Microbiol. scand. B* **88**, 143.

Steele, E. J., Chaicumpa, W. and Rowley, D. (1975) *J. infect. Dis.* **132**, 175.
Steele, E. J., Jenkin, C. R. and Rowley, D. (1977) *The Antigens*, Vol. IV, p. 247. Academic Press, London.
Stewart-Tull, D. E. S. (1980) *Annu. Rev. Microbiol.* **34**, 311.
Surolia, A., Pain, D., Khan, M. I. (1982) *Trends biochem. Sci.* **7**, 74.
Sutherland, I. (1977) *Surface Carbohydrates of the Prokaryotic Cell*, p. 27. Academic Press, London; (1979) *Trends biochem. Sci*, **4**, 55.
Svenson, S. B., Nurminen, M. and Linberg, A. A. (1979) *Infect. Immun.* **25**, 863.
Swanson, J., Hsu, K. C. and Gotschlick, E. C. (1969) *J. exp. Med.* **130**, 1063.
Takayama, K., Ribi, E. and Cantrell, J. L. (1981) *Cancer Res.* **41**, 2654.
Tannock, G. W., Blumershine, R. V. H. and Savage, D. C. (1975) *Infect. Immun.* **11**, 365.
Tramont, E. C. (1981) *Adhesion and Microorganism Pathogenicity*, p. 188. Pitman Medical, Tunbridge Wells.
Tsai, C-M., Frasch, C. E. and Mocca, L. F. (1981) *J. Bact.* **146**, 69.
Valtonen, V. V. (1970) *J. gen. Microbiol.* **64**, 255.
Voiland, A., Bruneteau, M. and Michel, G. (1971) *Eur. J. Biochem.* **21**, 285.
Ward, J. B. (1981) *Microbiol. Rev.* **45**, 211.
WHO report (1976) *Techn. Rep. Ser. Wld. Hlth. Org.* No. 588. WHO, Geneva.
Wicken, A. J. (1980) *Bacterial Adherence*, p. 137. Chapman and Hall, London.
Wicken, A. J. and Knox, K. W. (1980) *Biochim. biophys. Acta* **604**, 1.
Widdowson, J. P., Maxted, W. R. and Pinney, A. M. (1971) *J. Hyg. Camb.* **69**, 553.
Wilkinson, S. G. (1977) *Surface of Carbohydrates of the Prokaryotic Cell*, p. 97. Academic Press, London.
Williams, R. E. O. and Maxted, W. R. (1955) *Atti VI Congr. int. Microbiol. Roma 1953*, Vol. 1, p. 46.
Yang, G. C. H., Schrank, G. D. and Freeman, B. A. (1977) *J. Bact.* **129**, 1121.
Zahradnik, J. M., Cate, T. R. and Frasch, C. E. (1981) *Absts. 21st Intersci. Conf. Antimicrobial Agents and Chemotherapy*, p. 689.
Zollinger, W. D., Mandrell, R. E., Griffiss, J. M., Altieri, P. and Berman, S. (1979) *J. clin. Invest.* **63**, 836.

14

Immunity to infection—immunoglobulins

Heather M. Dick and Eve Kirkwood

Antibodies and infection	374	IgM	377
Immunoglobulins	375	IgG	378
Immunoglobulin structure	375	IgA	378
Heavy and light chains	375	IgE	379
Constant and variable regions	375	IgD	380
Structural features of immunoglobulins	377	Antibody diversity	381
Immunoglobulin classes and subclasses	377	Synthesis of immunoglobulins: genetics	381
Polymeric immunoglobulins	377	Generation of diversity	382
Properties of immunoglobulin classes	377		

Introductory

The challenge presented by infectious micro-organisms is met by specific responses which are mediated by both antibodies and cellular reactivity. The physiological basis for these events has been outlined in Chapter 10 (q.v.). For a proper understanding of the processes involved in combating infection, it is necessary to look in more detail.

Antibodies are specific for the antigens which stimulate their production: the means by which such specificity is achieved and the type of antibody produced are genetically predetermined in the B lymphocytes. Cell-mediated responses, involving several different types of function in T lymphocytes are activated and controlled by membrane antigens and soluble products which are also under genetic control. Superimposed on these dual pathways of response and conveyed by surface membrane markers and soluble factors is a controlling system exhibiting many features of a network, using feedback inhibition to switch responses on and off (Jerne 1974).

The concept of the recognition of foreign antigens by individual populations of cells was developed by Burnet (1959), who postulated that antigenic determinants were recognized as foreign by the cells of the immune system, because of the unique ability of such cells to distinguish between molecular structures as self and non-self. This ability arose because of the presence on the surface of lymphocytes of preformed receptors, capable of binding to any antigenic structure which was not already expressed on the tissues of the individual. The process of recognition and binding automatically selected that cell (lymphocyte), which then proliferated in response to antigen challenge, forming a clone of cells. This clone actively responded to the specific antigen, and its cells were capable of producing antibody or of taking part in some form of cellular activity designed to eliminate the foreign antigenic material. The clonal selection theory still forms the basis of modern immunology, although much elaborated and modified as with all good theories, to take account of the many detailed observations which have since been made. The definition of lymphocyte classes, the role of the macrophage and the production of a wide variety of soluble factors have all been integrated into the picture of clonal selection. The production of specific antibody is an interesting example of the way in which the simple concept of selection and subsequent clonal expansion of a cell can be elaborated to explain the complex mechanisms which operate in the immunological response.

Antibodies and infection

The presence of specific neutralizing substances in

serum was postulated to explain the observation that bacteria and bacterial toxins could be affected by the presence of serum from an infected individual. Bordet, and subsequently other workers, delineated the natural of the response to the introduction of antigenic material, and showed that, in the course of infection, antibodies could be produced which were important to the process of recovery, or which were merely incidental to that process (Bordet 1920, Wilson and Miles 1975).

Immunoglobulins

Serum proteins can be separated electrophoretically into α_1, α_2, β and γ-globulin fractions. Antibody activity is found predominantly in the γ-globulin region, residing in a widely heterogeneous group of glycoproteins known as immunoglobulins (Tiselius and Kabat 1939a,b). These molecules secreted by plasma cells circulate freely in serum and other body fluids where their role is to combine specifically with antigen. Membrane bound immunoglobulin can also be found on B lymphocytes, where it is concerned in the recognition and binding of antigen and in stimulating the ultimate maturation of B lymphocytes into antibody secreting plasma cells (Chapter 10).

Immunoglobulin structure (Fig. 14.1)

Heavy and light chains

The basic immunoglobulin unit consists of four polypeptide chains—two heavy chains and two light chains bound together by non-covalent hydrophobic forces and by disulphide bonds. This monomeric unit has a molecular weight of approximately 150 000. The heavy chains, in particular, show major structural and antigenic differences and on this basis they are identified by the Greek letters gamma (γ) mu (μ) alpha (α) delta (δ) and epsilon (ε). Only two types of light chain have been identified i.e. kappa (κ) and lambda (λ). An immunoglobulin molecule is classified according to its heavy chain type, for example, an IgG molecule contains gamma heavy chains and an IgA molecule alpha heavy chains. Kappa and lambda light chains are found throughout all the immunoglobulin classes although any one molecule will contain one type of light chain only. Molecules containing kappa light chains are approximately twice as common as lambda-containing molecules.

Constant and variable regions

The immunoglobulin molecule can be split easily near the disulphide bond linking the two heavy chains (the *Hinge Region*) by the enzyme action of *pepsin* and *papain* (Porter 1967). These substances have been useful experimentally in determining the properties of the various parts of the molecule. It has been shown that antibody combining activity resides at the end of the molecule which contains both light chains and heavy chains. These are known as the *Fab fragments*. The opposite end of the molecule which contains heavy chains only does not exhibit antibody combining activity; however, secondary functions such as complement fixation and the ability to bind to cells are determined by this region. Because this fragment of the

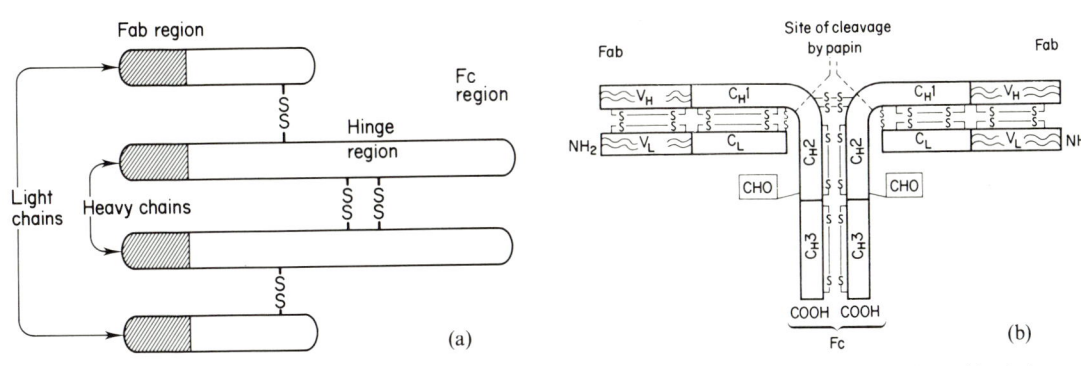

Fig. 14.1 (a) Diagram of immunoglobulin molecular structure. (b) Diagram of the arrangement of peptide chains and disulphide linkages in the IgG_1 molecule. NH_2 = amino terminal. COOH = carboxyterminal. CHO = carbohydrate. The site of cleavage by papain into Fab and Fc fragments is indicated.
  V_H and V_L are the homologous variable regions of the heavy and light chains.
  C_H1, C_H2, C_H3 = homology regions from constant region of heavy chain.
  C_L = constant region from light chain (homologous to C_H1, C_H2 and C_H3).

(After Edelman 1971.)

Table 14.1 Biological and physical properties of immunoglobulin classes

Immuno-globulin class	Biological properties	Mol. wt	Conc. in normal serum	% of total immunoglobulin	Valency for antigen binding
IgG	Most abundant class in serum and body fluids. Fixes complement via classical pathway. Crosses placenta. Binds to macrophages and polymorphs.	150 000	8–16 mg/ml	80	2
IgA	Major secretory immunoglobulin. Particularly high levels in colostrum. Can fix complement but only via alternative pathway. Binds macrophages and polymorphs.	160 000 (monomer)	1.4–4 mg/ml	13	2,4
IgM	Produced early in immune response. Efficient agglutinator. Fixes complement effectively via classical pathway.	900 000	0.5–2 mg/ml	6	5(10)
IgD	Function unknown. Mostly bound to lymphocytes.	185 000	0–0.4 mg/ml	0–1	2
IgE	Elevated in parasitic disease. Responsible for atopic symptoms. Fixes to mast cells and basophils. Does not fix complement.	200 000	17–450 mg/ml	0.002	2

Table 14.2 Subclasses of immunoglobulins

Class	Subclasses	Comment
IgG	IgG_1	Predominant circulating IgG subclass. Fixes complement.
	IgG_2	
	IgG_3	Fixes complement.
	IgG_4	Shorter half-life than other subclasses of immunoglobulin.
IgA	IgA_1	Predominant subclass of IgA in circulation.
	IgA_2	Predominant IgA subclass in secretions.
IgM	IgM_1	Biological differences between these subclasses not yet defined.
	IgM_2	
IgD		No subclasses have as yet been defined.
IgE		No subclasses have as yet been defined.

molecule can be crystallized under certain circumstances, it has become known as the *Fc fragment* (Cohen and Porter 1964).

Structural features of immunoglobulins

Studies of the amino acid sequence of immunoglobulin molecules demonstrate that heavy and light chains can be subdivided into regions or units on the basis of interchain disulphide bridges. These are known as *domains*. Both light and heavy chains have *variable* (*V*) and *constant* (*C*) domains (Edelman and Gall 1969). The variable regions are responsible for the antibody diversity of immunoglobulin molecules and it is in this region that antigen-antibody combining occurs. Both kappa and lambda light chains have only one constant domain; the number of heavy chain constant domains varies depending on the class of immunoglobulin. Gamma and α heavy chains have three constant domains, whilst μ and ε appear to have four, as well as having the largest number of carbohydrate residues.

There are three regions within the variable domains known as *hypervariable areas* which are particularly important in antigen binding. In its fully assembled form, the immunoglobulin molecule consists of a series of 'pleats' and the V domains of the light and heavy chains interact with each other in such a way that the hypervariable regions form the actual antibody combining site (Feinstein and Beale 1977). It is possible to raise antisera known as *anti-idiotypic antisera* specific for these areas. The term *idiotype* refers to the antigenic determinant formed by the amino acid residues in the hypervariable region which together represent the antibody combining site. Each clone of antibody-secreting plasma cells produces a unique idiotype (Oudin 1966). Since the original description of anti-idiotype antibodies, the definition of what constitutes an idiotype has been extended to include sequences of amino acids which are unique to particular classes of immunoglobulin, the sequences being dictated by a series of genes in the germ line. Such idiotypes can occur on antibody molecules with different specificities because of the presence of the appropriate amino acids in their combining site. These are referred to as cross-reacting idiotypes; they can be detected in all individuals of a species who have the appropriate gene(s). These markers have proved immensely useful in the study of the mechanisms of immunoglobulin synthesis, since the idiotype acts as a genetic marker which can be identified on the appropriate antibody or myeloma protein (Capra and Nisonoff 1979), thus, for example, pinpointing identical V regions on different heavy chains.

It is now thought that naturally occurring *anti-idiotypic* antibodies are produced during the normal immune response (Oudin and Michel 1969). They can be regarded as anti-antibodies and play an important role in the regulation of the immune response.

Immunoglobulin classes and subclasses

The heavy chain structure of the immunoglobulin confers very different properties on each class of immunoglobulin. These properties are summarized in Table 14.1.

In each class of immunoglobulin, minor antigenic differences can be detected within the Fc region of the heavy chains. On this basis, immunoglobulins are still further divided into subclasses (Table 14.2).

Polymeric immunoglobulins

Not all immunoglobulins exist in monomeric form. IgM for example, almost always exists as a polymer (*pentamer*) and IgA frequently as a *dimer*. Polymers contain an additional structure known as *joining chain* or *J-chain*, a 15 000 mol. wt. polypeptide molecule rich in cysteine, which is synthesized by antibody producing cells but is coded for by a separate gene from those which determine Ig heavy and light chain production.

Properties of immunoglobulin classes

For a review of this topic, see Spiegelberg (1974).

IgM

This high molecular weight immunoglobulin is found in relatively small amounts, approximately 6-7 per cent of total Ig in the serum, but is also bound to the membrane of early cells in the B lymphocyte series where it has an important function as an antigen receptor. In this role it is present in the monomeric form, and may have additional amino acids at the C-terminal end of the μ chain (McCune et al. 1980) as well as having a large part of the molecule buried in the cell membrane (Fu and Kunkel 1974). The pentameric structure of five basic Ig units gives rise to special antigen combining sites, which makes it a particularly effective agglutinating antibody (Fig. 14.2). IgM is also complement fixing, making it very effective in bacterial killing. IgM is produced early in the response to antigen (see below) and its optimum effect is reached in the circulating plasma probably because its large size reduces the chance of the molecules entering the extravascular compartment. Many of the so-called 'natural' antibodies are IgM, including the blood group iso-agglutinins (anti-A, B etc.) and heterophile antibodies (see Chapter 12). IgM has a relatively short half-life of five days in the circulation, but increased synthesis may occur in infection and in some autoimmune disorders, e.g. rheumatoid arthritis; in this, the exact stimulus for increased IgM production is unknown, although the reason may lie in defective control of biosynthesis rather than specific induction of increased metabolism (Allison 1973) since no single

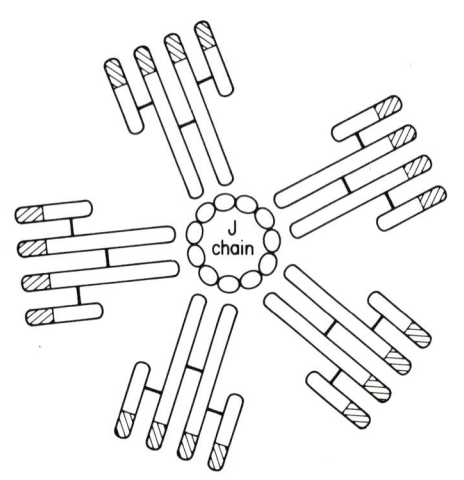

Fig. 14.2 The IgM pentamer.

microbial cause for the disease has ever been clearly demonstrated. The IgM produced in rheumatoid arthritis is generally regarded as anti-IgG in specificity, but the mechanism for the loss of tolerance to a self-product is not clear (Johnson 1979).

IgG

This immunoglobulin is the major component of the serum γ globulin fraction, forming 70-80 per cent of the total immunoglobulin material. The relatively small size of the molecule (15 K) makes it possible for IgG to reach most sites in the tissues. IgG_1 and IgG_3 bind readily to cells which have the receptor for the Fc terminal component of immunoglobulin, including macrophages and neutrophils, potentiating antigen uptake and phagocytosis of bacteria. These two subclasses may also activate complement by the classical pathway, whereas all four subclasses are effective in the alternative pathway (see below, Complement) with IgG_2 being the least active and thus less responsible for bacterial killing (Table 14.3). Staphylococcal protein A substance is reactive with $IgG_{1, 2}$ and $_4$ through the Fc portion (Forsgren and Sjoquist 1966), a reaction which is used in serological tests when protein A is employed as a material to immobilize IgG antibodies. The precise significance of this property of protein A *in vivo* is not clear, although it has been shown to have antiphagocytic effects *in vitro* (Dossett et al. 1969). However, IgG and protein A will activate complement, and this may lead to the production of chemotactic factors which are important mediators of inflammation.

The IgG subclasses differ from each other in the primary structure of the C-terminal portion of the γ chain and in the number and sites of disulphide chains, both between the heavy chains and between light and heavy chain. They also have different heavy chain determinants known as Gm allotypes (Natvig and Kunkel 1973). The concentrations of each subclass may vary; infants tend to have most IgG_1, and there is a continuous increase of IgG_3 with increasing age (Morrell et al. 1976) for reasons which are unknown.

Antibodies may be limited to one or more of the subclasses; bacterial carbohydrates stimulate IgG_2 antibodies (Yount et al. 1968), whereas cell membrane antigens, eg. MHC molecules, stimulate IgG_1 and IgG_3 antibody production. Antibodies to Factor VIII of the coagulation cascade are exclusively IgG_4 (Anderson and Terry 1968), whereas platelet antibodies in idiopathic thrombocytopenic purpura often are of the IgG_3 subclass (Mollison 1972).

IgG molecules have a half-life of approximately 21 days in plasma, with synthesis of up to 40 mg/kg body weight taking place. IgG_3 seems to have a much shorter life span, 7 days or less, probably because of rapid catabolism of the γ chain of this subclass (see Waldmann and Strober 1976 for review on metabolism of immunoglobulins).

IgA

This class of immunoglobulin is detectable both in plasma and in many secretions; in the latter fluids, it forms the major source of immunoglobulin. IgA occurs in all the mucous secretions of man, i.e. in the alimentary, respiratory and genital tracts. It is also found in colostrum in man, mouse and rat. In plasma, IgA occurs as a dimer of two immunoglobulin units (mol. wt 160 K) joined by the J-chain and forms approximately 13 per cent of the circulating immunoglobulin.

In secretions, IgA also occurs as the dimeric molecule, but with an additional fragment, the secretory piece, or secretory component—SC, which is added to the dimer during the processes of IgA secretion. SC is produced in the epithelial cells of the gut lining and is

Table 14.3 Properties of IgG subclasses*

	IgG_1	IgG_2	IgG_3	IgG_4
Concentration (mg/ml)	10	3	1	0.5
Binding to Fc receptors	+	−	+	−
Activation of complement—classical pathway, via C_1q	+	(+)	+	−
Placental transfer	+	(+)	+	+

*For fuller details, see Spiegelberg (1974).

united with the dimeric IgA, produced by submucosal plasma cells, when the whole Ig-SA molecule is transported across the gut wall into the lumen. The IgA-Sc complex passes through the enterocytes to the lumen in small endocytic vesicles, which are subsequently discharged into the gut lumen. IgA dimers may reach the circulation via draining lymphatics (Heremans 1974, Tomasi 1976). Most IgA is produced by cells in the submucosal layers of the gut, respiratory tract etc., and plasma IgA may be derived predominantly from the plasma cells in the alimentary tract. Transport of IgA formed in this site probably takes place so that organs which have relatively few IgA secreting plasma cells may be adequately protected. The migration of precursor B lymphocytes in lymph and the blood stream is well documented and it may be that it is only after antigenic stimulation that such cells move into the submucosal region and mature to the final immunoglobulin secreting plasma cell (Parrott 1976). An additional source of IgA-SC is found in the liver where hepatocytes also produce SC and actively transport IgA into the biliary secretions (Birbeck et al. 1979, reviewed in Hall and Andrew 1980).

IgA-SC is also found in large amounts in breast milk, where the activity of the antibody is largely directed against bacterial and viral antigens. In the presence of other antibacterial substances known to occur in breast milk, the IgA probably potentiates the local defence offered to the newborn infant against intestinal micro-organisms (Adinolfi et al. 1966, Goldman and Smith 1973). Much of the IgA in milk may be formed locally; the human mammary gland has appreciable numbers of IgA secreting plasma cells (Pumphrey 1977).

Antibacterial antibodies in milk IgA appear to arise in response to gut colonization, and for some pathogenic species there is evidence that alimentary exposure to antigens precedes the formation of milk IgA eg. against *V. cholerae* and enterotoxin producing strains of *Esch. coli* (for review, see Hanson 1982), and against live poliomyelitis virus vaccine (Ogra and Karzon 1969; Smith and Sweet; see Chapter 84). IgA has been shown to have a protective effect *in vivo* in infection with the helminth *T. taeniformis* in mice (Lloyd and Soulsby 1978). It is suggested that the protective effect of IgA may be the result of preventing or reducing the adherence of organisms to gut epithelium.

IgA occurs in two subclasses IgA$_1$ and IgA$_2$, the latter being more abundant in secretions than the former, although both types combine with SC. The two can be distinguished by amino acid differences in the α-chain, and by differences in disulphide bonding. IgA$_2$ heavy chains are linked to the light chains by non-covalent forces and have disulphide bonds between the light chains (Grey et al. 1968). In addition, IgA$_2$ has a deletion of the enzyme susceptible peptide (prolythreonyl) on one chain, which makes it peculiarly resistant to degradation in the gut. IgA$_1$ may be destroyed by some bacterial proteases, rendering it less suitable as a protective antibody than IgA$_2$, perhaps accounting for the predominance of the latter subclass in secretions.

The protective effects of IgA antibodies are attributed to the ability to neutralize or prevent adherence of micro-organisms to epithelial surfaces. There is some argument as to whether IgA can have opsonizing effects; it is known not to bind the C$_1$ component of complement, although it is stated that activation of the alternative pathway may occur when IgA is aggregated (Glynn 1982). IgA has a relatively short half-life as measured in plasma (5–6 days), but detailed evidence for its persistence as the secreted form is lacking.

Selective deficiency of IgA is fairly common in man, as measured by circulating levels of the dimer, with a reported rate of 1 in 700 in white Caucasians (Buckley 1975). However, such individuals frequently have IgA bearing B cells (Lawton et al. 1972); the defect may be one of T helper cells rather than a total absence of the capacity to produce IgA, or of a selective defect of B cells relating to the inability to secrete whole IgA (Cassidy et al. 1979). IgA deficiency may be accompanied by other significant abnormalities, including the syndrome of ataxia telangiectasia, and structural abnormalities of chromosome 18. However, many individuals with selective IgA deficiency may have no symptoms or signs attributable to lack of this immunoglobulin; IgM and IgG antibodies may provide adequate protection (Ogra et al. 1974).

IgE

The presence of antibodies which had an affinity for skin and which occurred in the presence of allergic states was demonstrated by Prausnitz and Küstner (1921) before the isolation of IgE from serum γ-globulin (see also Gell and Coombs 1968). Küstner had noted that he developed an acute reaction after eating cooked fish. Tests with his serum failed to reveal the presence of precipitins to fish antigens, nor would his serum transfer sensitivity to guinea-pig skin, when the technique known as passive cutaneous anaphylaxis (PCA) was used to reveal the vascular changes which accompanied acute allergic reactions (Feldberg and Miles 1952). However, when his serum was injected into the skin of a normal person, followed by an injection of fish antigen 24 hours later, the classical weal and flare reaction developed—the P-K reaction, in honour of its discoverers. Attempts to isolate the serum fraction responsible were initially unsuccessful; the name 'reaginic antibody' was coined for the mediator. It was not until the 1960s that a unique immunoglobulin was identified (Ishizaka et al. 1966) as the carrier of reaginic activity. The presence of increased amounts of this immunoglobulin in sera from patients with asthma was noted (Johansson 1967). A chance finding of a myeloma (plasma cell) tumour secreting

IgE proved an invaluable source of IgE, which appeared in normal sera in minute amounts; hence the long delay in identifying it (Bennich and Johansson 1971).

IgE occurs in plasma in monomers (mol. wt 190 K). In the tissue it is bound to mast cells and basophils, by the Fc terminal portion of the ε chains. This immunoglobulin is heat-labile, losing its ability to sensitize skin after being heated at 56° for 30 minutes. The damage occurs in the Fc portion of the molecule, leaving the antibody combining site (Fab) intact. IgE persists for long periods on the mast cells in skin but in the plasma seems to have a short half-life of around 2 days. The amount of IgE in normal serum is minute, being less than 0.005 per cent of the total Ig.

When IgE bound to mast cells or basophils is challenged by specific antigen, histamine and other vasoactive enzymes are released from the granules within these cells. The response is immediate, hence the acute allergic reaction which occurs in the affected tissues, skin and respiratory tract mucous membrane being the sites mainly affected. Although the term immediate hypersensitivity is applied to this reaction, long term effects may also be invoked by some of the mediators other than histamine. The features of this form of hypersensitivity, also known as type I, are discussed later in Chapter 16.

Although type I reactions mediated by specific IgE are commonly associated with exposure to pollens, animal hair, and some food proteins, this form of immunological reactivity is also a feature of exposure to both fungal spores and particularly to antigens present in association with a mite which proliferates in house dust. The mite *Dermatophagoides pteronyssimus* is found in vast numbers in house dust; the exact antigen(s) which induce hypersensitivity has not been identified, although it appears that the major antigenic component, found in the faecal material of the mite, is a glycoprotein (Chapman and Platts-Mills 1980). Many parasitic and helminthic infections in man and animals are characterized by raised serum levels of IgE, and animal models have been used to study the biosynthesis and function of IgE. The IgE found in helminth infections is believed to arise as a result of polyclonal stimulation of B cells by the complex antigens of the worms, with a selective effect on IgE production (Jarrett 1978). The exact role of IgE in such infections is still not clearly established, but the importance of eosinophils, mast cells and probably T lymphocytes is indicated by the presence of such cells in fairly large numbers in the infected tissues. Evidence that IgE helps to eliminate or reduce helminth infestation is lacking (Ogilvie and Jones 1973). It has been suggested that there may be a link between helminth infection and the acute disorders of type I class, but the evidence is conflicting. Other workers have failed to show any relation between IgE levels and subsequent response to intestinal parasites (Turner *et al.* 1978). Jarrett *et al.* (1980), using rats infected with *Nippostrongylus brasiliensis*, demonstrated that the non-specific IgE induced by helminth infection did not inhibit hypersensitivity reactions due to actively produced specific IgE, although the same non-specific IgE could block passive sensitization as measured by PCA techniques.

The relation between high IgE levels in serum and the severity of Type I hypersensitivity reactions is poor; specific IgE can be identified by radioimmunoassay techniques and purified antigen. The levels of specific IgE to given antigens, e.g. grass pollen, house dust etc., may show a better relation than do total IgE levels, but affected individuals often have raised levels of antibody to multiple antigens, suggesting some form of polyclonal stimulus which is selective for IgE. In one study in the tropics it was noted that asthma was relatively rare despite the frequent finding of increased IgE levels; this might be explained if non-specific IgE, stimulated by chronic parasitic infestation, could block the binding of specific antibody to mast cells in the respiratory tract (Godfrey and Gradidge 1976).

IgD

IgD is found principally as a surface membrane-bound immunoglobulin on the majority (90 per cent) of B lymphocytes (Sjoberg 1980) where it is expressed simultaneously with surface IgM. Only traces of IgD are found in normal serum. IgD or similar molecules have been detected in primates including man, rats, frogs and chickens.

The molecule was first examined in detail when it was found as a myeloma protein (Rowe and Fahey 1965). The exact function of IgD is still a matter for argument; it is generally agreed that membrane IgD acts as a surface receptor for antigen. In structure, the molecule is a monomer immunoglobulin. The δ chain has three constant domains and one variable domain. The membrane form has a larger δ chain than the secreted molecule, the extra sequence taking the form of a hydrophobic peptide, anchoring the Ig in the membrane. The Fab fragment of IgD is very susceptible to proteolysis, in contrast to the behaviour of other immunoglobulins when subjected to enzyme degradation (Jefferis 1981). A predominance of λ light chains is found in IgD, nearly 90 per cent, whereas most other immunoglobulins are found to have a ratio of 3:2 for $\kappa:\lambda$. The reason for this difference is unknown.

Although specific IgD forms a significant amount of some antibody responses, the function of the secreted molecule is unexplained: only trace amounts of IgD are found in normal serum (Jefferis and Mathews 1977). Some patients with raised IgE levels are also found to have increased amounts of serum IgD (Josephs and Buckley 1980).

It would seem that the membrane bound IgD is the important functional molecule. On murine lymphocytes, both monomer and half molecules have been identified (i.e. $\delta_2 L_2$ and δL) (Eidels 1979). IgM is almost always present. IgD$^+$ M$^+$ B cells are not committed to respond to a specific antigen (Cooper *et al.* 1976). The simultaneous expression of the two classes of immunoglobulin appears to mark a special stage in the differentiation of B cells (Dosch *et al.* 1979). After encounter with antigen and proliferation, the cells cease to express IgD, and are thus believed to have matured to give rise to memory cells, capable of a secondary response on subsequent challenge with the same antigen (Black *et al.* 1980). It was suggested that IgD$^+$ and IgD$^-$ B cells differed in their ability to bind T cell dependent and T cell independent antigen but this hypothesis was disproved in experiments using murine spleen cells, stimulated by polymerized flagellin and human γ globulin. B cells with surface IgM alone and those which were also IgD$^+$ responded to the T cell independent antigen (Layton *et al.* 1979). Calvert and Jefferis (1981) have recently reviewed the evidence for the possible mechanism of action of IgD as a receptor for antigen, suggesting that it may act in concert with sIgM as a binding site for T-dependent antigens which have less repeating antigenic determinant sequences than T-independent antigens (e.g. bacterial polysaccharides). The latter could probably bind quite adequately to IgM alone, with its two binding sites, whereas the more polymorphic T dependent antigen would require to cross-link receptors of IgD and IgM before generating a suitable 'signal' to the cell to initiate proliferation and antibody production. Certainly, removal of IgD from B cells by papain treatment abolishes the response to T dependent antigens; treatment with anti δ serum can abolish responses to both types of antigen under certain conditions (Calvert and Jefferis 1981). Pernis (1977), using monkeys, injected rabbit anti δ serum and demonstrated increased Ig production of anti-rabbit antibodies, in comparison with monkeys given normal rabbit serum, as a primary response, suggesting that IgM binding was adequate to initiate the response. It is possible that the presence of antibodies to the δ chain could affect the binding of antigen by steric hindrance but without IgD binding; memory cells were not produced.

Antibody diversity

Antibody formation is stimulated by a selective process; the cells which proliferate to secrete specific immunoglobulin are driven to do so by the presence of antigen (Landsteiner 1945).

A wide range of immunoglobulin molecules, with specificity for the stimulating antigen is produced by a complex of genes, coding for a series of proteins and peptides which together form the heavy and light chains. Major segments of immunoglobulin movecules have close similarities in the amino acid sequences between the various classes and subclasses and are presumably coded for by the same or closely related genes. The structural basis of specificity lies in the arrangement of amino acid residues at the antibody combining site and most particularly, in those residues in the *variable* regions of both light and heavy chains in the Fab region.

Variable region sequences can be grouped into three recognizable areas, relevant to each of the chains i.e. κ, λ and heavy chains all have variable regions. Within each of these there are particular amino acids which are invariable, that is, identical in the V regions of light and heavy chains. Some of these invariant amino acids are concerned with the folding of the peptide chain and, hence, with the tertiary structure of the immunoglobulin. Those amino acids which show the greatest variation within the variable region are found close together in a limited number of *hypervariable* regions. By means of antibodies against haptens, e.g. 2,4 dinitrophenyl (DNP), methods have been developed for attaching marker molecules to side chains in close proximity to the antibody combining site, permitting the identification of amino acid sequences in or near the site. The anti-hapten antibodies used are selected to cross-react with various amino acid residues, e.g. tyrosine, and will thus show affinity for such sites within the antibody combining site. From a knowledge of the residues to which the antihaptens bind it has proved possible to identify the amino acids which together form the hypervariable region and thus, the major position of the antibody combining site (reviewed in Nisonoff *et al.* 1975). A relation between variability of amino acid sequence in these sites has been demonstrated by the use of murine myeloma proteins as the antigen (Cebra *et al.* 1974).

Synthesis of immunoglobulins: genetics

By markers, such as idiotype specificity, the inheritance of the genes responsible for the immunoglobulin proteins has been studied in great detail in the inbred mouse and rabbit (Mage *et al.* 1971, Eichman 1975). There is a cluster or 'family' of genes coding for the λ chains, one for κ and a third for each of the class and subclass specific heavy chains (for review, see Williamson 1976). The clusters are unlinked and those for light and heavy chains are present on separate chromosomes. Separate genes code for V and C regions. The V region genes are obviously crucial to the determination of the final arrangement of amino acids in the antibody combining site. By DNA hybridization techniques it can be shown that in immature cells, the V and C genes are separated by a lengthy stretch of DNA, whilst in the mature plasma cell they are placed close together (Hozumi and Tonegawa 1976). Two other segments of DNA are essential to the final assembly of the chain, the J and D segments. The J segment encodes for a group of amino acids at the N-terminal end of heavy chains and the κ light chain. The existence of a J-region in the genome for the λ chain is unconfirmed. Although the J_H genes code for the terminal amino acids, a stretch of perhaps 7-10 amino acids immediately preceding this is accounted for by the short fourth segment, D (for diversity) which shows the greatest amount of variability in the heavy chain sequence.

During maturation of the immunoglobulin-producing cell, these segments are joined to each other to determine the final genetic arrangement for mRNA, which is the basis for the assembly of the final product. The intervening stretches of DNA (introns) which are present between V, J, D and the C region (exons) genes must be excised and the proper order of genes attained by a complex group of enzymes which break the chains. Most crucial is the mechanism whereby the combination of these genes can give rise to specific antibody, with the appropriately fashioned antibody combining site.

Generation of diversity

Two hypotheses have been advanced to explain how the synthesis of immunoglobulins is arranged in order to realize the almost infinite repertoire of specific antibody molecules. One model, the germ-line hypothesis (Hood et al. 1975) postulates the presence of all antibody genes in germ line cells, i.e. the zygote, and accounts for diversity by postulating either gene interactions, e.g. different combinations of genes, or a very large 'library' of separate V region genes, which would expand with evolutionary development. The alternative model, the somatic gene hypothesis (Jerne 1971) invokes the presence of a minimal number of V genes in the germ line; during somatic differentiation and maturation, successive mutations within these genes occur followed by antigen-driven selection. The fundamental differences between the two hypotheses have not yet been reconciled. The germ line hypothesis requires the presence of many thousands of V region genes, even allowing for interactions between them. The somatic gene model, on the other hand, might be encompassed by a library of less than 10^3 genes, but requires a very high rate of mutation within the maturing cell. The production of antibodies binding to cross-reactive molecules must also be accounted for as well as the means by which only one set of chromosomal genes is transcribed in each cell, to give the final immunoglobulin molecule with two pairs of matching heavy and light chains (allelic exclusion). Readers who wish to study this topic in depth are recommended to consult Hood et al. (1975) and for the evolution of the selective hypothesis of antibody formation, Landsteiner (1945).

References

Adinolfi, M., Glynn, A. A., Lindsay, M. and Milne, C. M. (1966) Immunology 10, 517.

Allison, A. C. (1973) Ann. rheum. Dis. 32, 283.

Anderson, B. R. and Terry, W. D. (1968) Nature 217, 174.

Bennich, H. and Johansson, S. G. O. (1971) Advance. Immunol. 13, 1.

Birbeck, M. S. C., Cartwright, P., Hall, J. G., Orlans, E. and Peppard, J. (1979) Immunology 37, 477.

Black, S. J., Tokuhisa, T. and Herzenberg, L. A. (1980) Eur. J. Immunol. 10, 846.

Bordet, J. (1920) Traité de l'immunité dans les maladies infectieuses. Masson, Paris.

Buckley, R. H. (1975) In: Immunodeficiency in man and animals. Ed. by D. Bergsma et al. Sinauer Assoc. Inc., Stamford, Conn.

Burnet, F. M. (1959) The Clonal Selection Theory of Acquired Immunity. Cambridge University Press, London.

Calvert, J. E. and Jefferis, R. (1981) Trends biochem. Sci. 6, 125.

Capra, J. D. and Nisonoff, A. (1979) J. Immunol. 123, 279.

Cassidy, J. T., Oldham, G. and Platts-Mills, T. A. E. (1979) Clin. exp. Immunol. 35, 296.

Cebra, J. J., Koo, P. H. and Ray, A. (1974) Science 186, 263.

Chapman, M. D. and Platts-Mills, T. A. E. (1980) J. Immunol. 125, 587.

Cohen, S. and Porter, R. R. (1964) Advanc. Immunol. 4, 287.

Cohn, M., Blomberg, B., Geckeler, W., Raschke, W., Riblet, R. and Weigert, M. (1974) In: The Immune System: genes, receptors, signals, p. 89. Ed. by E. E. Sercarz, A. R. Williamson and C. F. Fox, Academic Press, London.

Cooper, M. D., Kearney, J. F., Lawton, A. R., Abney, E. R., Parkhouse, R. M. E., Preud'Homme, J. L. and Seligman, M. (1976) Ann. Immunol. (Inst. Pasteur) 127c, 573.

Dosch, H-M., Kwong, S., Tsui, F., Zimmerman, B. and Gelfano, E. W. (1979) J. Immunol. 123, 557.

Dossett, J. H., Kronvall, G., Williams, R. C. Jr. and Quie, P. G. (1969) J. Immunol. 103, 1405.

Edelman, G. M. and Gall, W. E. (1969) Annu. Rev. Biochem. 38, 699.

Eichmann, K. (1975) Immunogenetics 2, 491.

Eidels, L. (1979) J. Immunol. 123, 896.

Feinstein, A. and Beale, D. (1977) In: Immunochemistry: an advanced textbook, p. 263. Ed. by L. E. Glynn and M. N. Steward. John Wiley & Sons, Chichester.

Feldberg, W. and Miles, A. A. (1952) J. Physiol. 120, 205.

Forsgren, A. and Sjoquist, J. (1966) J. Immunol. 97, 822.

Fu, S. M. and Kunkel, H. G. (1974) J. exp. Med. 140, 895.

Gell, P. G. H. and Coombs, R. R. A. (Eds) (1968) Clinical Aspects of Immunology, 2nd edn. Appendix B. Blackwell Scientific Publications, Oxford.

Glynn, A. A. (1982) In: Clinical Aspects of Immunology, 4th edn. p. 1393. Ed. by P. J. Lachmann and D. K. Peters. Blackwell Scientific Publications, Oxford.

Godfrey, R. C. and Gradidge, C. F. (1976) Nature 259, 484.

Goldman, A. S. and Smith, C. W. (1973) J. Paediatrics 82, 1082.

Grey, H. M., Abel, C. A., Yount, W. J. and Kunkel, H. C. (1968) J. exp. Med. 128, 1223.

Hall, J. G. and Andrew, E. (1980) Immunology Today 1, 100.

Hanson, L. Å. (1982) Immunology Today 3, 168.

Heremans, J. F. (1974) In: The Antigens, Vol. 2, p. 365. Ed. by M. Sela. Academic Press, London.

Hood, L., Campbell, J. H. and Elgin, S. C. R. (1975) *Annu. Rev. Genet.* **9,** 305.

Hozumi, N. and Tonegawa, S. (1976) *Proc. nat. Acad. Sci. Wash.* **73,** 3628.

Ishizaka, K., Ishizaka, T. and Hornbrook, M. M. (1966) *J. Immunol.* **97,** 75.

Jarrett, E. E. E. (1978) *Immunol. Rev.* **41,** 52.

Jarrett, E. E. E., Mackenzie, S. and Bennich, H. (1980) *Nature* **293,** 302.

Jefferis, R. (1981) *Trends biochem. Sci.* **6,** 111.

Jefferis, R. and Matthews, J. B. (1977) *Immunol. Rev.* **39,** 25.

Jerne, N. K. (1971) *Eur. J. Immunol.* **1,** 1; (1974) *Ann. Immunol.* (*Inst. Pasteur*) **125C,** 373.

Johansson, S. G. (1967) *Lancet* **ii,** 951.

Johnson, P. M. (1979) In: *Immunopathogenesis of Rheumatoid Arthritis*, p. 45. Ed. by G. S. Panayi and P. M. Johnson. Reedbooks, Chertsey.

Josephs, S. H. and Buckley, R. H. (1980) *J. Paediatrics* **96,** 417.

Landsteiner, K. (1945) (Reprinted 1962) *The specificity of serological reactions*, 2nd edn. Harvard University Press. Reprinted edition Dover Publications, New York.

Lawton, A. R., Royal, S. A., Self, K. and Cooper, M. D. (1972) *J. Lab. clin. Med.* **80,** 26.

Layton, J. E., Pike, B. L., Battye, F. L. and Nossal, G. J. V. (1979) *J. Immunol.* **123,** 702.

Lloyd, S. and Soulsby, E. J. L. (1978) *Immunology* **34,** 939.

McCune, J. M., Lingappa, V. R., Fu, S. M., Blobel, G. and Kunkel, H. G. (1980) *J. exp. Med.* **152,** 463.

Mage, R. G., Young-Cooper, G. O. and Alexander, C. (1971) *Nature, New Biol.* **230,** 63.

Mollison, P. L. (1972) *Blood Transfusion in Clinical Medicine*, 5th edn. Blackwell Scientific Publications, Oxford.

Morrell, A., Skvaril, F. and Barandun, S. (1976) In: *Clinical Immunobiology*, Vol. 3, p. 37. Ed. by F. H. Bach and R. A. Good. Academic Press, London.

Natvig, J. B. and Kunkel, H. G. (1973) *Advane. Immunol.* **16,** 1.

Nisonoff, A., Hopper, J. E. and Spring, S. S. (1975) *The Antibody Molecule*. Academic Press, New York.

Ogilvie, B. M. and Jones, V. E. (1973) *Progr. Allergy* **17,** 93.

Ogra, P. A., Coppola, P. R., MacGillivray, M. H. and Dzieba, J. L. (1974) *Proc. Soc. exp. Biol. Med.* **145,** 811.

Ogra, P. L. and Karzon, D. T. (1969) *J. Immunol.* **102,** 1423.

Oudin, J. (1966) *Proc. Roy. Soc. Lond.* **B166,** 207.

Oudin, J. and Michel, M. (1969) *J. exp. Med.* **130,** 619.

Parrott, D. M. V. (1976) *Clin. Gastroenterol.* **5,** 211.

Pernis, B. (1977) *Immunol. Rev.* **37,** 210.

Porter, R. R. (1967) *Sci. American* **217,** No. 4, 81.

Prausnitz, C. and Küstner, H. (1921) *Zentrabl. Bakteriol. Orig A* **86,** 160.

Pumphrey, R. S. H. (1977) *Symp. Zoo. Soc. Lond.* **41,** 261.

Rowe, D. S. and Fahey, J. L. (1965) *J. exp. Med.* **121,** 171.

Sjöberg, O. (1980) *Scand. J. Immunol.* **11,** 377.

Smith, H. and Sweet, C. (See Chapter 84).

Spiegelberg, H. L. (1974) *Advanc. Immunol.* **259,** 19.

Tiselius, A. and Kabat, E. A. (1939*a*) *Science* **87,** 416; (1939*b*) *J. exp. Med.,* **69,** 119.

Tomasi, J. B. Jr. (1976) *The Immune System of Secretions*. Prentice Hall, New Jersey.

Turner, K. J., Quinn, E. H. and Anderson, H. R. (1978) *Immunology* **35,** 281.

Waldmann, T. A. and Strober, W. (1976) In: *Clinical Immunobiology*, Vol. 3, p. 71. Ed. by F. H. Bach and R. A. Good. Academic Press, London.

Williamson, A. R. (1976) *Annu. Rev. Biochem.* **45,** 467; (1981) In: *Antibody Production*, p. 141. Ed. by L. E. Glynn and M. W. Steward. John Wiley & Sons, Chichester.

Wilson, G. S. and Miles, A. A. (Eds) (1975) *Topley and Wilson's Principles of Bacteriology, Virology and Immunity*, 6th edn. Edward Arnold, London.

Yount, W. J., Dorner, M. M., Kunkel, H. G. and Kabat, E. A. (1968) *J. exp. Med.* **127,** 633.

15

Immunity to infection—complement

Heather M. Dick and Eve Kirkwood

Introductory	384	Breakdown of C3: the effector mechanism for	
Terminology	385	complement activity	386
Classical pathway activation	385	Membrane attack	387
Alternative pathway activation	385	Control mechanisms	387
		Effects of complement lysis	387

Introductory

Since before the beginning of the century it has been known that red blood cells or certain bacteria could be lysed by serum which contained two 'factors':

1. A 'thermostable' factor which was highly specific and appeared following immunization, later identified as specific antibody to the red blood cell or bacteria.
2. A 'thermolabile' factor which was non-specific and appeared to be easily destroyed by heating to 56°. This was named 'complement' because it 'complemented' the action of antibody (Bordet 1898; Ehrlich and Sachs 1905).

A few years after its discovery, the complement-fixation test for the detection of antibody was developed (Bordet 1898) and this for many years proved to be an indispensable technique for the detection of antibody in many bacterial and later viral infections (Bordet and Gengou 1901; Bradstreet and Taylor 1962; Humphrey and White 1970).

It is now clear that complement is not a single substance but is made up of as many as twenty proteins all of which play a part in the complement system. These proteins act sequentially in a cascade fashion similar to the coagulation and fibrinolytic systems, i.e. as the result of one single chemical reaction involving an early complement component later components can be activated leading to an amplification of the response. Some of the components act as inhibitors of the process, so that uncontrolled activation cannot proceed under normal physiological conditions.

The complement proteins constitute the major effector system for the elimination of foreign antigenic material (see Chapter 12). The production of irreversible lesions on cell membranes causes bacteriolysis of gram-negative bacteria such as enteric bacilli or meningococci. Although gram-positive organisms such as pneumococci are not lysed in the presence of complement and specific antibody, complement does exhibit an opsonizing effect in that it enhances phagocytosis of these organisms.

Certain micro-organisms when coated with specific antibodies and complement adhere to the surface of red blood cells, neutrophils, platelets and macrophages. This immune adherence results in immediate phagocytosis of the micro-organism if it happens to be bound to a phagocytic cell. If the organism is bound to platelets or red blood cells, these too are ingested by macrophages. This is beneficial to the host, provided red cell and platelet destruction is not too great. During complement activation, breakdown products are generated which have a direct chemotactic effect on neutrophils. In addition to these effects on foreign proteins of an infective nature, it is likely that the complement cascade exerts similar effects leading to the destruction of malignant cells in the host, although the tumour cell killing is mediated by lymphocytes coated with antibody, rather than by antibody and complement alone.

Unfortunately, the consequences of complement activation are not always beneficial. Just as the coagulation system can be goaded into unchecked activation leading to disseminated intravascular coagulation, the

Table 15.1 Complement terminology

Abbreviation	Interpretation
C	The complement system.
C1–C9	Components of classical pathway. (NB. The early components do not follow strict numerical order, C4 precedes C3, therefore the sequence of activation is C1→C2→C4→C3- - - - -→C9.)
C1qrs	C1 is a trimolecular complex which requires calcium ions Ca^{++} to function.
E	Erythrocyte.
A	Antibody.
EAC	Signifies a complex of antibody and complement bound to an erythrocyte.
$EAC_{1,4}$	Further detail is represented by numbers indicating the complement components involved.
$EAC_{1,4,2,3}$ or (EAC_{1-3})	
$\overline{C1}$	Enzymatically active C1 i.e. a component or group of components with enzyme activity is indicated by a bar above.
$\overline{C4,2}$	Enzymatically active form of C4,2 complex. This is also known as the C3 convertase of the classical pathway.
$C3_a$, $C3_b$	These are examples of low molecular weight proteins which result from cleavage by other enzymatically active components. These fragments are also known as conversion products. (3a is also known as an anaphylotoxin (as is C5a).)

complement system can similarly be activated in a way which can prove to be extremely damaging to the host.

Terminology

The complement system can be activated by either the classical pathway or the alternative pathway. These are two distinct enzyme cascades which both culminate in cleavage at the level of the complement component known as C3 (see Table 15.1: complement terminology).

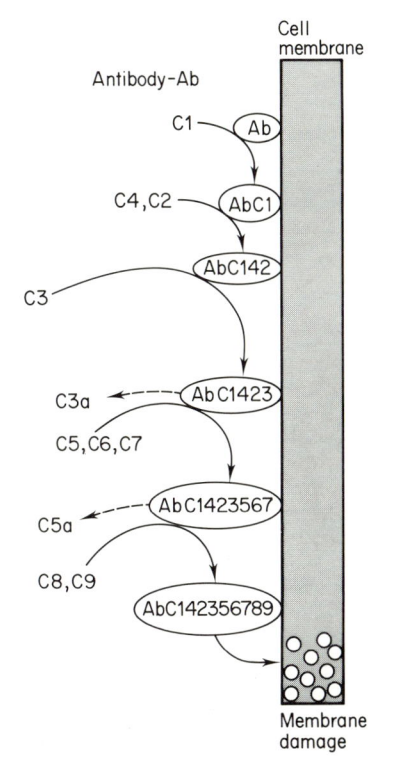

Classical pathway activation (Fig. 15.1)

The complement sequence may be activated by the binding of its first major active component, C1 to immunoglobulin, mainly IgM and IgG, which is bound to antigen. C1 is actually a complex made up of three proteins, C1q, C1r and C1s. C1q has been studied in detail: it has six identical subunits which are joined together to form a structure resembling a daisy head on a stalk (Reid and Porter 1976). The head forms the site for binding to antibody. A single molecule of IgM is sufficient to bind C1, and the fixation results in bending of the 'arms' of the immunoglobulin. With IgG, two molecules of antibody, fixed to closely placed antigenic sites, are required before effective C1 binding occurs (Feinstein and Richardson 1981). Ca ions are also an essential addition for effective C1 binding. After C1 fixation, the breakdown of C1r and C1s leads to the production of active enzymes (esterases) which then act on the C4 and C2 proteins in the serum in the presence of magnesium ions (Mg^{++}). The end result is the production of an unstable active complex of C42, which is an enzyme capable of splitting C3, for which reason it is often called C3 convertase.

Alternative pathway activation

The second route for complement activation is effective immediately at the point where cleavage of C3 occurs, with the help of an additional protein, Factor B. Thus, alternative pathway activation can occur both through the action of the same activators which are effective in the classical pathway and, in addition,

Fig. 15.1 Classical pathway activation.

386 *Immunity to infection—complement* Ch. 15

Table 15.2 Synonyms in complement terminology

Name	The classical pathway	
	Synonym	Principal breakdown fragments
C1q	11S-component	–
C1r	–	–
C1s	C1 esterase	–
C4	β_1E-globulin	C4a, C4b
C2	–	C2a, C2b
C3	β_1C-globulin	C3a, C3b (C3c, C3d)
C3c	β_1A	–
C3d	α_2D	–
C5	β_1F-globulin	C5a, C5b
C6	–	–
C7	–	–
C8	–	–
C9	–	–
Inhibitors		
C1 INH	C1 esterase inhibitor	–
	C1$_s$ inactivator	
C3b INA	KAF	–
C3b INA accelerator	A. C3b INA	–
	β_1H	
C6 INA	–	–
Anaphylotoxin INA	Carboxy peptidase B	–

in any circumstance where C3b is produced. This may occur without antigen-antibody complex formation, and is thus effective early in infection.

Several molecules take part only in the alternative pathway. The nomenclature of these proteins has been complicated in the past by the existence of several synonyms for many of them (see Tables 15.2 and 15.3).

A protein known as initiating factor (IF) is capable of binding directly to the cell membrane without the help of specific antibody. In the presence of Mg^{++} and factor D, factor B becomes activated to interact with IF and C3b to form a complex C3bB which has transient C3 convertase activity. This forms a positive feedback amplification loop in which the product of C3 cleavage forms more C3 convertase. This C3 convertase activity probably 'ticks over' slowly in the normal state and is held in check by the action of an inhibitor known as C3b inactivator. Substances such as endotoxin enhance C3 convertase activity and the active complex of C3 and Factor B is stabilized by properdin, an additional serum protein.

Once the C3 stage, which is common to both classical and alternative pathways, is reached C5 enters the system and the membrane attack unit is assembled in the way previously described in classical pathway activation.

Breakdown of C3: the effector mechanism for complement activity

C3 is the most abundant of the complement components in serum, approximately 1 g/ml, and is the key

Table 15.3 Synonyms in complement terminology

Name	The alternative pathway	
	Synonym	Principal breakdown fragments
Initiating factor	IF	–
	Nephritic factor	
	Factor 1	
Properdin	P	–
C3	β_1C-globulin	C3a, C3b etc.
	Factor A	
Factor B	C3 proactivator	GAG, factor B
	C3PA	
	GBG	
Factor $\overline{\text{B}}$	C3 activator	–
	C3A	
	GGG	
Factor D	C3 proactivator Convertase	–
	C3 PA-ase	
	GBG-ase	

molecule in both classical and alternative pathway activation. It is synthesized in the liver and by macrophages (see Colten 1976 for a review on the biosynthesis of complement).

The cleavage of C3 is the reaction which initiates the remainder of the complement pathways. C3 is a two-chain molecule and can be digested by proteolytic enzymes at several points. C3 convertase enzymes are produced in both classical and alternative pathways. Other enzymes, not belonging to the complement system, can also act on C3, including some bacterial proteases. The active breakdown product (C3b) has an acceptor site for C5, which is then cloven by the same enzymes. One of the breakdown products C5b then forms the site on the cell membrane to which the remaining components bind (C6, 7, 8, 9). Binding of C5 to C3b is very active on membrane surfaces, such as those of bacteria and virus-infected cells.

Membrane attack

The complex of C5b, 6, 7 etc. forms the lytic unit which finally leads to disruption of the cell membrane. This it does by penetrating the cell membrane. Complement lysis produces holes in the membrane which are seen on electron micrographs and are approximately 10 nm in diameter (Humphrey and Dourmashkin 1969 (see Fig. 12.1, p. 332)). The lesion develops in the lipid bilayer as a cylindrical channel or pore, the outside layer of which is hydrophobic and the inner one hydrophilic; low molecular weight molecules and ions leak out of the cell, and water enters, swelling the cell until eventually the membrane ruptures (Mayer 1977).

Control mechanisms

The complex cycle of events that occur in alternative pathway activation includes both amplifying and controlling products that lead to the steady production of the C3 convertase enzymes (Lachmann and Peters 1981). Factors H and I form the basis of the control mechanism for increased breakdown of C3b, whereas events which decrease C3b destruction will tend to amplify the process by feedback inhibition. This can arise in the absence of H or I (genetic defects) or where C3b is protected from breakdown by binding to a suitable surface, e.g. bacterial polysaccharide (Fearon and Austen 1977).

Effects of complement lysis

The classical concept of lysis of the red blood cells or bacteria following complement activation can be extrapolated to many other types of cell. The consequences of a damaging membrane lesion depend on the properties of the cell concerned.

Nucleated cells are more resistant to lysis than red blood cells. Bacterial lysis can be antibody-dependent as with gram-negative cocci antibody-dependent or with some viruses antibody-independent. *Neisseria spp.* appear to be relatively susceptible to complement lysis; and congenital defects of some complement components have been reported to make infections such as gonorrhoea and meningococcal meningitis more frequent in persons with such defects (Lachmann and Rosen 1978).

It appears that complement is not as effective in lysing autologous cell membranes as it is on cell membranes from another species. This concept of 'self-recognition', however, is far from complete and at times damage to autologous cells does occur. For example, complement-mediated platelet lysis occurs in the course of normal clotting mechanisms and at inflammatory sites. When excessive, this can initiate intravascular coagulation by the release of platelet Factor 3. Autologous cells infected by viruses can be lysed by complement; this is likely to be an important mechanism in the defence against infection by intracellular organisms.

Certain cells, e.g. polymorphs, macrophages, B lymphocytes and platelets have receptors for one or more of the products, C3a, C3b, C4b and C5a and will therefore bind to other cells coated with these components, bringing about the subsequent release of cytotoxic lysosomal enzymes from polymorphs, histamine release from mast cells and platelets, and initiation of macrophage activation. In addition, factors such as C3a and C5a have chemotactic activity for polymorphs, eosinophils and macrophages (see Chapter 10); and a kinin-like substance is liberated after the interaction of C1s and C2, with a direct effect on vascular permeability.

The effect of complement activation can be seen as much wider than simply lysis of an offending bacterium. It is unfortunate that, while the protective role of complement is undisputed, the same reactions can be at times extremely damaging to the host tissues (see Chapter 16).

References

Bordet, J. (1898) *Ann. Inst. Pasteur* **12**, 688.
Bordet, J. and Gengou, O. (1901) *Ann. Inst. Pasteur* **15**, 290.
Bradstreet, C. M. P. and Taylor, C. E. D. (1962) *Mon. Bull. Min. of Health & PHLS* **21**, 96.
Colten, H. R. (1976) *Advanc. Immunol.* **22**, 67.
Ehrlich, P. and Sachs, H. (1905) *Collected Studies on Immunity*. John Wiley, New York.
Fearon, D. T. and Austen, K. F. (1977) *J. exp. Med.* **146**, 22.
Feinstein, A. and Richardson, N. E. (1981) In: *Endocytosis and Exocytosis in Host Defence* (Monographs in Allergy Series, Vol. 17). Ed. by L. La Eidero and O. Stendhal. Karger. Basel.
Humphrey, J. H. and Dourmashkin, R. R. (1969) *Advanc. Immunol.* **11**, 75.

Humphrey, J. H. and White, R. G. (1970) *Immunology for Students of Medicine*, 3rd edn. Blackwell Scientific Publications, Oxford.

Lachmann, P. J. and Peters, D. K. (1981) In: *Clinical Aspects of Immunology*, 4th edn, p. 18. Ed. by P. J. Lachmann and D. K. Peters. Blackwell Scientific Publications, Oxford.

Lachmann, P. J. and Rosen, F. S. (1978) *Springer Sem. Immunopathol.* **1**, 339.

Mayer, M. M. (1977) *Monogr. in Allergy* **12**, 1.

Reid, K. B. M. and Porter, R. R. (1976) *Biochem. J.* **155**, 19.

16

Immunity to infection—hypersensitivity states and infection

Heather M. Dick and Eve Kirkwood

Introductory	389	Antigen excess	394
Antibody-mediated hypersensitivity	390	Serum sickness	395
Type I hypersensitivity	390	Chronic immune complex disease	395
Specific IgE production	390	Type III hypersensitivity and infection	395
Attachment of specific IgE to mast cells	390	Dust diseases of the lung	395
Mechanism of type I, IgE mediated hypersensitivity	391	Tests for type III hypersensitivity	396
		Cell-mediated hypersensitivity	396
Antigens producing type I hypersensitivity	391	Type IV hypersensitivity	396
Type I hypersensitivity and infection in man	391	Production of type IV hypersensitivity reaction in skin	396
Other causes of type I hypersensitivity	393	The cellular response in the skin	397
Type II hypersensitivity	393	The role of T lymphocytes in type IV hypersensitivity	397
Immunoglobulin class and type II hypersensitivity	393	Mechanism of T-cell cytotoxicity	397
Mechanisms of type II hypersensitivity	393	T-cell immunity and infection	398
Type II hypersensitivity and infection	394	Contact hypersensitivity	399
Type III hypersensitivity	394	Autoimmune disease	399
Immunoglobulin class and type III hypersensitivity	394	Transfer factor	399
Mechanisms of type III hypersensitivity	394	Genetic factors in the response	400

Introductory

Although the immunological responses we have been describing are directed primarily towards the elimination of foreign antigenic material, we must also be aware of the potentially harmful effects that can be produced in the tissues both by antibodies and by sensitized lymphocytes. Bacteriologists have long been aware of these damaging events. The acute and sometimes lethal effect of intravenous injection of bacteria or bacterial toxins was recognized after injection of pneumococci and streptococci (Zinsser and Grinnell 1927) and intravenous tuberculin (Zinsser 1921). The name 'anaphylaxis' or 'anaphylactic shock' was used for the reaction which ensued (Portier and Richet 1902, Richet 1913). The animal might have local or generalized signs, or a combination of both. Locally, vasodilation was present around the injection site; generalized reaction included shock, respiratory distress due to bronchospasm and perhaps asphyxia; and, in man, asthma and sometimes generalized urticaria (Parish and Oakley 1940). The reaction was shown to be specific, and repeated exposure to the same antigen might produce the anaphylactic symptoms even after a lapse of many months or years. The features of anaphylaxis could be transferred between animals of the same species in systemic or local form by using serum from sensitized animals or man (Otto 1907, Prausnitz and Küstner 1921—see Chapter 14). Non-bacterial antigens were also implicated in man, when therapeutic horse serum was found to give rise to a similar set of symptoms (von Pirquet and Schick 1905).

Subsequent investigation has revealed that the

Table 16.1 Classification of hypersensitivity states (see Coombs and Gell 1975)

Type I	Anaphylactic: immediate: IgE mediated
Type II	Cytotoxic
Type III	Immune complex: Arthus: serum sickness
Type IV	Delayed: T-cell mediated

tissue damage which occurs in hypersensitivity may take several forms, and that each type of lesion has a different time of onset in relation to challenge and in the precise nature of the damage which is produced. Microbial infections may be accompanied by one or more of the features of these hypersensitivity reactions.

A classification of hypersensitivity states was suggested by Gell and Coombs (1963) who also made a plea for the use of the alternative term 'allergy'. They based their case for the choice of allergy as a suitable general name following the suggestion of von Pirquet (1906) who had coined the word to indicate a state of awareness of antigen by the tissues, which had neither the connotation of protective nor harmful responses von Pirquet derived it from the Greek 'allos'—other, implying 'deviation from the original state, from the behaviour of the normal individual' (see translation in Gell et al. 1975). However rational this usage may be, the word 'hypersensitivity' has continued to be employed by immunologists, and the use of the word 'allergy' has remained mainly in the clinical field, referring principally to the anaphylactic or immediate response, mediated by IgE.

Although the classification of hypersensitivity states into four separate divisions (types I-IV) (Gell and Coombs 1963) is a useful basis for simplifying the events which occur in the sensitized individual, it is now increasingly obvious, both from clinical and experimental evidence, that such clear cut separation does not fully allow for the proper description of the tissue damage which occurs in hypersensitivity reactions. A recent edition of the same text (Lachmann and Peters 1982) suggests that these types (I-IV) be dropped, and that a more realistic view be taken of the considerable overlap between the events which follow re-exposure to antigen in a sensitized individual. However, the types I-IV as originally described make it possible to dissect out the various effects of hypersensitivity in most reactions which follow re-exposure to antigen, particularly in infectious disease (Table 16.1).

Antibody-mediated hypersensitivity

Immediate hypersensitivity is used as a general description for reactions which follow very rapidly after re-exposure to antigen. The term should probably be reserved for the most acute anaphylactic response, but is sometimes also applied to reactions which have a longer time scale.

Type I hypersensitivity (Fig. 16.1)

This is the most rapidly developing form of response and encompasses the acute anaphylactic shock observed by many early workers (see above). The role of serum in the response was recognized both by *in vivo* and *in vitro* transfer of sensitivity (Prausnitz and Küstner 1921, Ovary 1961). They ascribed the effect to 'reaginic' antibodies in serum which were subsequently identified as IgE immunoglobulins (Ishizaka et al. 1966; see Chapter 14, IgE).

Specific IgE production When antigens (allergens) cross the skin or epithelium of the respiratory or gastro-intestinal tracts, they are processed by macrophages, T and B lymphocytes. The latter transform into plasma cells and synthesize IgE which is liberated into the circulation (Chapter 10).

Attachment of specific IgE to mast cells IgE antibodies—also known as homocytotropic antibodies are able to bind to mast cells and basophils throughout the tissues. These cells are found in relatively large numbers in the submucosal and mucosal layers of the respiratory tract and gut wall, and also in skin (Riley 1959), all of which tissues are the major targets for IgE mediated reactions. Mast cells and basophils have a high affinity for the Fc receptor of IgE and bind the immunoglobulin very effectively. Eosinophils may also have this receptor. The IgE molecules on the cell surface have the Fab regions exposed, ready to bind antigen (Kay 1981). Some mucosal mast cells also contain intracytoplasmic IgE and in rats infected with *N. brasiliensis*, these cells are also found in the epithelial layer of the gut, migrating towards the lumen (Mayrhofer et al. 1976). Mast cells probably reach the tissues from the draining lymph nodes and there is good evidence that IgE production takes place in the nodes, where the mast cell acquires its IgE before migrating to the peripheral sites.

On second and subsequent exposure to antigen, IgE antibody production proceeds more quickly and efficiently. Antibody for the primary response may still be bound to mast cells (see Fig. 16.1).

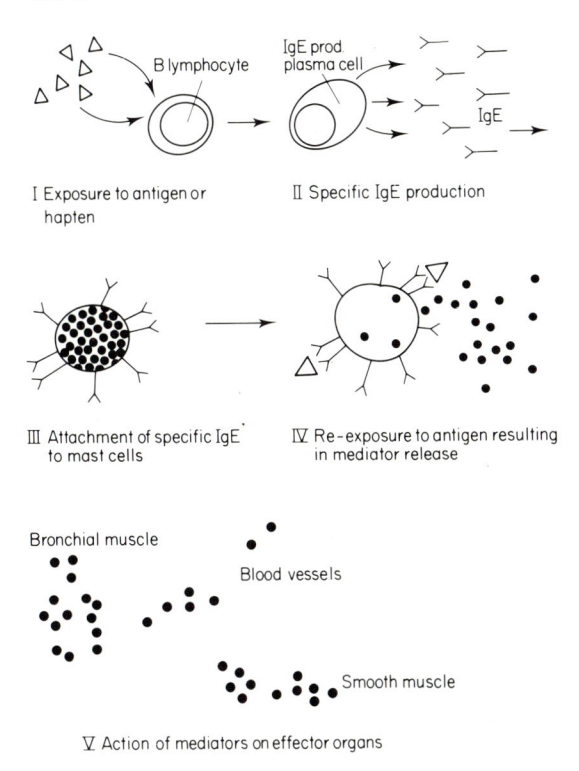

Fig. 16.1 Type I hypersensitivity.

I Exposure to antigen or hapten
II Specific IgE production
III Attachment of specific IgE to mast cells
IV Re-exposure to antigen resulting in mediator release
V Action of mediators on effector organs

Mechanism of type I, IgE mediated hypersensitivity

IgE bound to mast cells by the Fc portion can bind antigen by its two Fab combining sites. Mast cells and basophils are both noted for their content of basophilic granules, rich in histamine and several other small molecular weight products and highly active amines. These products include slow-reacting substances (SRS) (Feldberg and Kellaway 1938), and a series of metabolites of arachidonic acid, called leukotrienes. Several leukotrienes have been described, including two which are active in producing smooth muscle contraction and increased vascular permeability (Leukotrienes C_4 and D_4) while others are potent chemotactic factors for leucocytes, especially eosinophils (Leukotriene B_4) (Borgeat and Samuelsson 1979, Samuelsson 1981). Leukotrienes and SRS-A—the slow reacting substance of anaphylaxis—are derived from fatty acids such as arachidonic acid and belong to a group of short chain metabolites. Leukotrienes C_4 and D_4 are both constituents of SRS-A and both are substantially more active than intact SRS-A in producing smooth muscle contraction and oedema.

Further active products are also released from mast cell granules, from the oxygenation and breakdown of polyunsaturated fatty acids: the prostaglandins, which on further metabolism give rise to prostacyclin and thromboxanes, are also derived from arachidonic acid, but vary in their tissue distribution. PGI_2, for example, is produced in vascular endothelial cells and to a lesser extent, macrophages (Moncada et al. 1977). PGE_2 and PGF_2 are produced in many tissues, but are particularly important in the smooth muscle of the respiratory tract (Christ and Van Dorp 1972). The release of histamine generates large amounts of these highly active compounds (Lewis et al. 1981) together with PGD_2. It has been suggested that histamine release and PGD_2 synthesis are linked in a cycle of enzyme activity, involving both adenylate cyclase in the membrane and various PG synthetase and oxygenase enzymes in the cytoplasm, with the consequent breakdown of arachidonic acid (Holgate et al. 1980, Austen 1981).

In addition to histamine, mast cells contain substantial amounts of heparin, arylsulfatase, tryptase and several other proteases. The release of these products by degranulation of the mast cell has profound effects on the microenvironment of that cell. Vasoconstriction, chemotactic attraction of eosinophils and other leucocytes, platelet activation (PGD_2 also possess this activity), increase in vascular permeability and smooth muscle contraction are all involved and contribute to the production of the signs and symptoms of acute anaphylactic hypersensitivity. The tryptases and acid hydrolases, including β-glucuronidase and arylsulfatase, can have damaging effects but may also help in scavenging the products of that damage by destroying cellular products such as collagen (Goetzl and Kay 1982). Several other substances released by mast cells and other tissues are implicated in the acute reaction. The role of some (serotonin, bradykinin) is not well defined and they probably play a minor part (Table 16.2, see p. 392).

Antigens producing type I hypersensitivity Pollens (tree and grass, predominantly) animal dust and the house mite (*D. pteronyssimus*) are the major antigenic stimuli inducing type I hypersensitivity reactions in man (for review, see Platts-Mills 1982, Goetzl and Kay 1982). Such antigens are not the primary concern of microbiologists.

Type I hypersensitivity and infection in man

Infection with fungi and parasites is frequently associated with the effects of type I hypersensitivity. The presence of antigen-specific IgE in the serum of patients is detectable by the 'prick' test. Antigen in solution is applied to the skin surface and a small prick is made with a suitable sharp needle point, sufficient to break the superficial layers of the dermis. The IgE bound to the skin mast cells binds antigen and the resultant degranulation produces the classical urticarial weal and flare within 10–15 minutes (for full details, see Platts-Mills 1982). *Aspergillus sp.* may stimulate type I responses, and patients with asthmatic symptoms due to exposure to this organism give positive

Table 16.2 Possible mediators in Type I hypersensitivity

Histamine	Found in mast cells, basophils, platelets. Acting via H_1 receptor leads to increased vascular permeability, contracts smooth muscles, increases goblet cell secretion of mucus. Acting via H_2 receptors leads to increased gastric secretion and decreased mediator release from mast cells and basophils. Chemotactic for eosinophils.
Serotonin (5-hydroxytryptamine)	Found in mast cells, platelets, brain, gastric mucosa. Leads to increased vascular permeability, capillary dilatation, smooth muscle contraction. Role in human anaphylaxis not clear.
Kinins (including bradykinin)	Found in plasma, tissues and organs, e.g. pancreas. Increase vascular permeability, lower blood pressure, contract smooth muscle. Levels increase during anaphylaxis but probably play only a minor role.
Slow-reactive substances of anaphylaxis (SRS-A)	Break down to form leukotrienes C4 and D4. Produce slow contraction of smooth muscle. Not inhibited by antihistamines. May contribute to prolonged bronchospasm seen in asthma. Released by lung tissue.
Eosinophil chemotactic factor of anaphylaxis (ECF-A)	Released from mast cells and basophils. Selectively chemotactic for eosinophils.
Platelet Activating Factor (PAF)	Induces platelet aggregation and secretion. May amplify anaphylactic reactions.
Prostaglandins	Wide tissue distribution. Relase during anaphylaxis. Some contract, e.g. $F_{2\alpha}$ and others relax, e.g. E_1 and E_2, smooth muscle. Some intermediates of prostaglandin synthesis are important mediators of platelet aggregation and smooth muscle contraction.
Leukotrienes	Metabolites of arachidonic acid. Several molecular species (LTA, LTB, LTC, LTD, LTE). (see Samuelsson and Hammarstrom 1981).

skin reactions. *Aspergillus* infection, e.g., in a pulmonary aspergilloma—a late event in cavitational tuberculosis—is accompanied by both type III and type I reactions (Longbottom and Pepys 1964). It has been suggested that type IV (delayed) reactions also form part of the picture in infection with *C. albicans* (Holti 1966). Cross-reactivity in skin testing with fungal extracts has been noted and the method used for antigen preparation may be important (Longbottom and Pepys 1975). Other fungal infections, including histoplasmosis and coccidioidomycosis are considered in the section below, on type IV cell-mediated hypersensitivity.

Infection with various parasitic worms and helminths stimulates the production of specific IgE and positive prick tests can be obtained. *Ascaris* antigen has been used to sensitize the skin experimentally but the IgE produced was short lived (Fisherman 1962). IgE production is both greater and more persistent in natural infections with the parasite (Johansson *et al.* 1968). Filarial infection may also be accompanied by IgE antibody, e.g., infection with *Onchocerca volvulus* (Buck *et al.* 1973). The stage of infection and the species of filaria are critical. No antibody is detectable when microfilariae are circulating in filariasis due to *W. bancrofti* (Chandra *et al.* 1974).

The role of IgE antibody in cure and/or protection from parasitic infection is not clear (see Chapter 14, IgE). In many cases where gut infection or invasion via the gut are predominant in parasitic lesions, then it would be reasonable to ascribe a more active role to IgA antibodies (Lloyd and Soulsby 1978); and T-cell mediated killing may also be involved (Dobson and Soulsby 1973).

Rupture of hydatid (*Echinococcus*) cysts in the liver may precipitate acute anaphylaxis, presumably because of the sudden release of a very large amount of antigen. The fluid also activates complement which is involved in type III hypersensitivity; the clinical features of the reaction show evidence for this as an additional factor (Schantz 1977). IgE levels are raised in infection and a positive skin test can be elicited (Casoni 1911) by intradermal injection. However, the test gives a high rate of false positives and antigen preparations vary greatly in their specificity. It is not recommended as a diagnostic test for this reason (Kagan 1968).

The condition known as 'Job's Syndrome' where patients present with recurrent skin sepsis and staphylococcal abscesses has been described. These patients also have high levels of IgE (>2000 IU/ml) and may have evidence of phagocyte defects. The IgE antibodies may be directed against staphylococcal cell wall antigens (Davis *et al.* 1966, Hill and Quie 1974).

Bacterial infections may well involve some element of type I hypersensitivity but specific examples are not well documented as being due to IgE antibody.

Other causes of type I hypersensitivity Parish (1970) has reported that IgG antibody may sometimes be involved in type I skin reactions; this type of antibody may be responsible for some of the urticarial rashes which occur after the use of materials such as tetanus toxoid and therapeutic antiglobulin or blood transfusion. Urticaria and other acute dermal manifestations after penicillin are IgE-mediated (Juhlin and Wide 1972) whereas the rash which may follow administration of ampicillin is not (Kraft and Wide 1976).

Hypersensitivity to the enzymes of *Bac. subtilis*, used for the preparation of so-called 'biological' washing powders, has also been reported to give rise to type I skin reactions, in addition to asthmatic problems (Belin and Norman 1977). Dermatitis, in contrast, is due to type IV reactions and is T-cell mediated. Sensitivity to laboratory animals is not uncommon in scientists and technicians who are exposed to epithelial scales (dander) and probably also to antigenic proteins in excreta, and is IgE mediated (Stechschulte 1982).

Bee stings are prone to give rise to anaphylactic reactions in predisposed individuals who have IgE antibodies. In fact, the venom of several members of the genus *Hymenoptera* which includes bees, hornets and wasps, may give rise to hypersensitivity. The manifestations may be local or generalized and can be fatal. Immunization with suitable venom preparations is highly effective in producing productive antibodies, which are likely to be IgG, although this has never been clearly demonstrated (Sobotka 1977, Adkinson et al. 1979, Golden et al. 1979).

Type II hypersensitivity

Antibody mediated cytotoxic damage is the principal feature of type II hypersensitivity. The antigen forms part of the membrane of the cell under attack, or is firmly bound to the membrane, e.g. haptens, drugs (Fig. 16.2). The binding of antibodies by their combining sites (Fab_2) is followed by activation of the complement pathways (Chapter 15). The effect is to produce damage to the cell wall which becomes 'leaky', loses high molecular weight products and takes up water, ultimately lysing.

The effect of antibody directed against cellular antigen was noted after the injection of serum containing Forssman antibody into guinea-pigs. The animals developed not only symptoms of anaphylactic shock but also had multiple haemorrhages and haemolysis, when the antibody was fixed to the tissue cells which contained the heterophile Forssman antigen (Doerr and Pick 1913, Redfern 1926). Immune haemolysis of transfused blood by preformed antibodies to red cell antigens, and haemolytic disease of the newborn (HDN), due to incompatibility for the Rhesus D antigen, are the most obvious examples of type II reactions in current clinical practice.

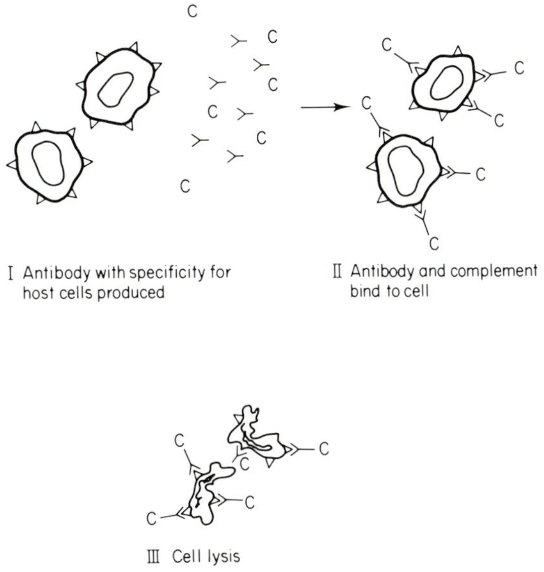

I Antibody with specificity for host cells produced

II Antibody and complement bind to cell

III Cell lysis

Fig. 16.2 Type II hypersensitivity.

Immunoglobulin class and type II hypersensitivity

IgG and IgM antibodies are involved; both bind complement (C1q) and activate the classical pathway. IgG antibodies are important in HDN because IgM does not cross the placenta. IgG_1 and IgG_3 antibodies must be the main subclasses, because of their ability to fix complement. No precipitating or skin fixing antibody is present and consequently serological diagnosis may require special techniques. These have been well developed in red blood cell studies. Immunofluorescence is useful for the detection of fixed Ig by means of suitable animal anti-Ig sera. In Goodpasture's syndrome, for example, IgG can be detected fixed to the basement membrane of glomeruli and pulmonary tissue (but, see below, type III, immune complex diseases).

Mechanisms of type II hypersensitivity

Cell surface antigens of many types of cell can be affected. The cells of the blood are most commonly involved, including red blood cells (RBCs), platelets, polymorphonuclear leucocytes and lymphocytes. The antigens may be either alloantigens, isoantigens or haptens bound to the cell membrane. Auto-antibodies directed against RBC isoantigens produce auto-immune haemolytic anaemia, and antibodies may also lyse RBCs to which drugs or drug metabolites have become bound.

The method of cell lysis is similar to that described in Chapter 15 (Complement). The bound complex of antigen, antibody and complement components leads to the production of enzymes which effectively 'puncture' the cell membrane, causing osmotic changes and eventual lysis.

Where the antibodies are directed against self constituents, tissue damage may produce destructive effects. Placental transfer of maternal IgG can produce neonatal disease, e.g. myasthenia gravis, thyrotoxicosis and thrombocytopenic purpura. Not all autoantibodies are destructive; some are concerned in the regulation of the immunological response (Jerne 1974) and others appear to be secondary to the pathological changes in affected organs or reflect the known increase in the incidence of such antibodies in ageing persons (see Glynn and Holborow 1964 for full details; also Roitt 1980).

Type II hypersensitivity and infection

It has long been suggested that virus infection may be an initiating event in autoimmune disease, but the evidence to support the hypothesis is weak. Some viral infections in animals have features which resemble those found in human disease. Lymphocytic choriomeningitis infection in mice and equine haemolytic virus infection both include autodestructive features. However, it is when we look closely at such diseases that the overlap between type II and type III hypersensitivity, immune complex mediated, becomes most obvious. Deposition of antigen-antibody complexes occurs in many autoimmune disorders, and although this may sometimes be due to virally induced antigens on the cell membrane, it is clear that we are making semantic distinctions in separating the effect of antibody binding to (normal) membrane constituents and that binding to antigens induced by infection or other cellular pathology.

T-cell mediated damage as well as that produced by antibody bound to mononuclear killer (K) cells may also be involved in lesions where type III and possibly type II damage is occurring. The relevance of all these mechanisms must be considered in post-streptococcal rheumatic heart disease and glomerulonephritis.

Type III hypersensitivity

The first observations of this form of hypersensitivity were made by Arthus (1903) who noted that multiple subcutaneous injections of horse serum into rabbits produced local reactions which increased in severity with successive injections. He described local oedema, hyperaemia and erythema, proceeding to actual necrosis of the skin. The features exhibited vary a little in other species; the guinea-pig frequently has simultaneous type I anaphylactic reactions such as may occur in man, when specific IgE antibodies are present in patients given anti-lymphocyte globulin (Salaman 1979) for the treatment of kidney graft rejection. The Arthus reaction is mediated by precipitating antibodies which combine in the intravascular space and in the tissues with soluble antigen. Arthus sensitization can be passively transferred by serum; the amount of antibody required is large, presumably to saturate the tissues and provide sufficient combining power. Only precipitating antibody can mediate passive transfer and large amounts of antigen must be injected locally and the antibody intravenously to produce the passive phenomenon (cf. PCA of Ovary, 1958). The reverse experiment (local antibody, IV antigen) is also positive, given that the proportions of antibody to antigen are such as to produce local aggregates (Culbertson 1935, Benacerraf and Kabat 1949).

Immunoglobulin class and type III hypersensitivity

The formation of antigen-antibody complexes by precipitating antibodies in serum is the key event in type III hypersensitivity: the *in vivo* tissue damage is produced by such complexes in association with complement components. The ratio of antigen to antibody is important. In the presence of moderate antigen excess, the complexes remain soluble rather than precipitating out (Chapter 11). In this state, toxic effects are produced in the tissues; depletion of complement, by the use of cobra venom factor, for example, prevents the damaging effects (Henson and Cochrane 1974). IgG antibodies are the main source of precipitating antibody in most complexes in disease. IgA has also been implicated in some diseases. IgM and IgE appear to be little involved (Levinsky 1981). The detection of circulating complexes is relatively easy; the identification of specific antigen in the complexes in many diseases has not been so successful, with the exception of DNA and rheumatoid factor in systemic lupus erythematosus and rheumatoid arthritis respectively.

Mechanisms of type III hypersensitivity (Fig. 16.3) The sequence of events which occurs depends on the site of complex formation and the relative amounts of antigen and antibody. In experimental animals, serum proteins have been used to study the results of antigen-antibody complex deposition.

Antigen excess The introduction of a large dose of antigen either intravenously or by inhalation into an already sensitized animal will produce the formation of complexes in vascular endothelium or in the alveolar lining of the lungs. Complement activation releases chemotactic factors and induces local inflammatory changes with vessel wall destruction, vasculitis, oedema and platelet and fibrin deposition. Sites such as the glomeruli are damaged, particularly by smaller aggregates. The tissues are infiltrated by polymorphonuclear leucocytes, eosinophils and neutrophils in particular (Henson 1971, Cochrane and Koffler 1973).

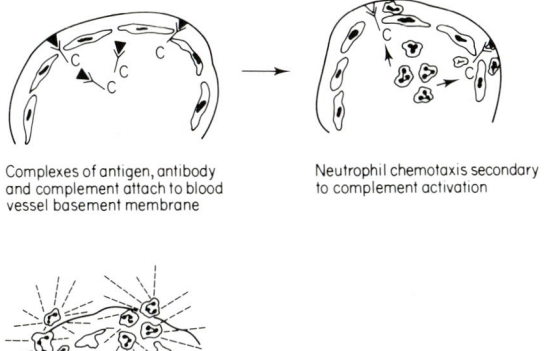

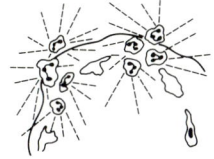

Fig. 16.3 Type III hypersensitivity.

Serum sickness (von Pirquet and Schick 1905) The intravenous injection of a large single dose of antigen, such as therapeutic horse serum, is followed by a period of 7–10 days when no ill effects are obvious. As antibody begins to appear, the antigen is being catabolized by the tissues; vascular and perivascular lesions may appear in the joints, muscles and kidneys. As the amount of antibody increases and antigen levels decline, the lesions resolve. Urticarial rashes are probably due to type I IgE mediated hypersensitivity (Davis 1938, Smith and Gell 1975).

Chronic immune complex disease The continued presence of antigen, whether endogenous, e.g. autoantigen such as DNA, or exogenous, e.g. in parasitic infections, leads to the persistent deposition of complexes. Small vessels in the kidneys, skin and joints are particularly affected. Immunofluorescent staining with anti-Ig and anti-complement antisera will show the deposits, often in linear pattern along the basement membrane in kidney and lung, in the so-called 'lumpy-bumpy' granular deposits. Eventually, the basement membrane thickens and renal function is impaired. In other tissues, the vasculitis and inflammatory processes may destroy vital cells e.g. the synovial lining.

Type III hypersensitivity and infection

In many infectious diseases, the lesions induced by immune complex deposition are central to the pathology of the disorder. Parasitic, bacterial and viral infections may all give rise to immune complexes. The rate of formation of the antigen-antibody complex and the size of complexes formed are both important. The phagocytic cells, macrophages, histiocytes etc., readily take up larger size complexes, but small ones may evade the process and reach vascular endothelium etc. to cause damage (Leslie 1980). The relative paucity of Fc receptors exposed on the surface of small complexes may make them less likely to bind to the macrophage, at least in a stable configuration (Knutson et al. 1979). Leprosy of the acute lepromatous form, where lesions with polymorphonuclear leucocyte infiltration are found, is associated with the deposition of both Ig and complement components, C3 in particular (Wemambu et al. 1969). The tuberculoid lesions in leprosy are, of course, granulomatous and involve cell mediated (type IV) hypersensitivity. Borderline lesions may exhibit features of both types of immunologically mediated tissue damage.

Protozoal and helminth infections are often accompanied by the production of large quantities of IgM and IgG antibody, and both circulating immune complexes and deposits of immunoglobulin and C3 are present. In CNS trypanosomiasis due to *T. brucei*, IgG, IgM, C3 and the antigens of the trypanosomes can be detected by the use of appropriate fluorescent antibodies for staining brain tissue of experimentally infected mice (Poltera et al. 1979).

The renal lesions of malaria are related to the presence of immunoglobulins and C3 in the glomerular basement membrane (Houba 1975). Circulating immune complexes occur in schistosomiasis, and are probably related to the lesions in man occurring in the kidneys and liver (Jones et al. 1977, Le Bras et al. 1980).

Dust diseases of the lung Type III hypersensitivity reaction may be precipitated by the inhalation of fine dusts or particulate organic matter. Farmers' lung, bagassosis, maltworkers lung and bird fanciers lung are all examples of this form of pulmonary damage, extrinsic alveolitis. As the names of the syndromes suggest, most are due to occupational exposure. Where micro-organisms are known to be the precipitating factor (Table 16.3, see p. 396), the damage to the alveolar membrane is not due to infection and multiplication of the organisms but to the formation of soluble complexes. The particle size is important, as is the dose of antigen inhaled. The symptoms are 'flu-like', with pyrexia, shivering, muscle pains as well as respiratory distress due to poor oxygen exchange. This form of hypersensitivity is not true 'asthma' with smooth muscle constriction, mediated by IgE. These disorders are accompanied by the development of IgG, and sometimes IgM, precipitating antibodies. Not all persons who are exposed to the dust develop such precipitins—perhaps only 10–20 per cent—and the predisposition to develop lesions may depend on genetic factors, although familial patterns of susceptibility are not wholly convincing. Successive exposures lead to increasingly severe symptoms, and a change of occupation may be essential. Proven cases are entitled to compensation for industrial injury.

Table 16.3 Some microorganisms associated with extrinsic alveolitis

Organism	Disease and source
Micropolyspora faeni	Farmers' lung—exposure to mouldy hay Mushroom workers' lung—exposure to mouldy fibre Ventilation pneumonitis—orgainsims in ventilation systems
Thermoactinomyces sacchari	Bagassosis—exposure to dust from mouldy sugar 'cake'
Aspergillus clavatus *Aspergillus fumigatus*	Malt workers' lung—exposure to mouldy grain
Penicillium casei	Cheese washers' lung—mould on cheese
Miscellaneous proteins in other materials	Bird fanciers' lung—pigeon, budgerigars Fish meal extracts Bovine pituitary extract

Tests for type III hypersensitivity

The presence of circulating immune complexes is not always associated with type III tissue damage and positive tests should be interpreted with caution (see Chapter 11). Precipitins may be detectable, e.g. in Farmers' lung and aspergillosis by means of the Ouchterlony test (Edwards and Davies 1981): radioimmunoassay is probably more sensitive and has good specificity (Parratt *et al.* 1982).

The presence of deposits of immunoglobulins and complement, especially C3, in tissues can be detected by immunofluorescent antibody staining in biopsied material e.g. skin, kidney, liver and is often the main diagnostic criterion for making a diagnosis of immune complex mediated tissue damage (for methods, see Johnson *et al.* 1978).

Skin tests may be employed but care should be taken to use low concentrations of antigen, because a strong positive response may proceed to actual necrosis. The reaction is delayed in its appearance and it is generally some 8 hours before it becomes obvious (cf. the type I reaction in 15 minutes, and type IV, taking 24 to 48 hours). Resolution of the skin test can take over 24 hours. Type IV reactions can occur after this, and in some patients, type I immediate responses can precede it, e.g. with aspergillus. The histology of the lesion includes many polymorphonuclear cells, especially granulocytes with some basophils and eosinophils. Vasculitis is a prominent feature, developing as a result of the presence of soluble complexes of antigen and antibody on the endothelium of small vessels. Bronchial provocation with inhaled antigen has been observed but is not recommended, because of the risk of severe systemic symptoms.

Cell-mediated hypersensitivity

Type IV hypersensitivity

The tissue damaging events in this form of hypersensitivity are mediated by T lymphocytes, not by antibody. The process is also referred to as 'cell-mediated' hypersensitivity, or 'delayed' hypersensitivity. The use of the term 'delayed' reflects the very different time scale of this type of response when compared with immediate hypersensitivity, mediated by antibodies. After primary antigen exposure, a period of 7–10 days or longer elapses before the production of the sensitized T lymphocytes, which migrate to the site of antigen deposition. Secondary responses take 24 to 72 hours to appear, again because of the time taken for proliferation of the relevant T cells and for their migration into the affected tissues. For many bacterial and viral infections the presence of multiple antigenic determinants will stimulate both antibody production and the proliferation of specific T cells. Examples of immunological responses which involve both major mechanisms are discussed later in this section. The tuberculin skin test is the best known example of an evoked delayed hypersensitivity reaction but the complex antigenic structures of *Myco. tuberculosis* and of tuberculin or purified protein derivative (PPD) must be recognized as introducing responses which are not 'pure' T cell.

Production of type IV hypersensitivity reaction in skin

The original observations of delayed skin responses were made by Koch (1891) in experiments with tubercle bacilli and with his extract of the bacilli, which he called tuberculin. Injecting this material subcutaneously into guinea-pigs previously infected with tubercle bacilli, he noted that an inflammatory reaction appeared in the skin at the injection site within one or two days. This lesion became necrotic within the next day or two and the skin sloughed off leaving a shallow ulcer which rapidly healed. The guinea-pig would eventually succumb to the original infection, but the skin reaction reflects the hypersensitivity state present in the infected animal. The 'Koch phenomenon' was subsequently confirmed, and other infectious agents and antigens were shown to have a similar effect, i.e. the primary exposure to antigen or infection sensitizes the animal, which may develop a measurable degree of immunity. Subsequent challenge with the same antigen produces the delayed, type IV response, which is detected by the skin test.

The transfer of this form of hypersensitivity is by cells, not by serum (Landsteiner and Chase 1942). The same workers also demonstrated that even small synthetic molecules, such as haptens, could stimulate this type of cellular intervention. Using dinitrochlorobenzene (DNCB) and other halogenated nitrobenzene haptens, they demonstrated that leucocytes could transfer reactivity from immunized to non-immunized guinea-pigs, resulting in positive skin tests with the general pattern described by Koch. The response is specific, just as with antibody-mediated hypersensitivity reactions (McCluskey et al. 1963). Many bacterial and viral infections result in type IV hypersensitivity; the complex antigens such as proteins are more likely to stimulate than the simpler repeating structures such as bacterial polysaccharides.

The cellular response in the skin After an initial mild inflammatory reaction at the injection site, there is generally a time lag of some 24 to 72 hours before the appearance of the reaction. Biopsy of the full-blown lesion will reveal the presence of large numbers of mononuclear cells in the skin, with some polymorphonuclear leucocytes and evidence of the release of lymphokines (Spector 1967). The lymphocytes in and around the lesion include many 'by-standers' which have been attracted by the lymphokines; in studies where type IV hypersensitivity was transferred to non-sensitized animals the host lymphocytes labelled with radioisotope may be seen in large numbers at the site of antigen injection (Turk and Oort 1963).

The role of T lymphocytes in type IV hypersensitivity

The presence of lymphocytes in the skin lesion induced by antigen in the previously sensitized individual is an indication of their central role in the production of type IV hypersensitivity. Although such reactions are a feature of many bacterial, viral and parasitic infections, it was by the study of a most unnatural process that their damaging effects were elucidated. Transplantation of cells or organs between animals of the same species is also followed by activation of the cellular response. Gowans (1965) removed lymphocytes from rats by thoracic duct cannulation and showed that such animals were incapable of graft rejection. The replacement of the cells would bring about rejection. The lymphocytes from a rat that had already rejected a graft could be transferred to a second rat, inducing a very rapid rejection of grafted tissue from the same donor, as if the recipient of the transferred cells had been sensitized against the donor antigens (Billingham et al. 1954). The histological appearances of rejected tissue are very similar to those which can be seen in type IV reactions, and rejection is clearly the result of an immunological reaction (Medawar 1958).

It is now possible to identify the types of lymphocyte which participate in type IV reactions. The use of monoclonal antibodies to identify surface markers has revealed the presence of numerous T lymphocytes, with many cytotoxic cells (Tc). Other sub-sets are present in lesser numbers, including T-suppressor and T-helper, some of the latter presumably involved with B cells in response to T-dependent antigen. As with other kinds of hypersensitivity reaction, the type IV response is often complicated by the presence of one or more of the antibody-mediated responses, type III reactions being prominent. It has been suggested that a proportion of T_H cells when faced with antigen proliferate to form a separate functional type of T cell, the T-delayed-type hypersensitivity cell (T_{DTH}). The existence of this population of specialized cells has been partly supported by experiments with skin grafts, incompatible for minor transplantation antigens, but has not been confirmed by the detection of specific markers for a further sub-class of T cells. It may well be that both T_H and T_C cell can function in more than one mode, i.e. behave as T_{DTH} cells, the change in activity being produced by soluble factors present in such reactions, with interleukin 2 as a strong candidate for such a modifying influence (Bianchi et al. 1981, Lin Yun Lu and Askonas 1981; for review, see Liew 1982). The T cells which participate in type IV delayed reactions are MHC restricted, i.e. the antigen-presenting cell and the responding T cell must have the same cell membrane antigens of the major histocompatibility complex (Chapter 10), the Ia or DR antigens being the relevant markers. The other types of cells in the lesion are non-specific and are attracted to the site purely by the chemotactic factors liberated by T cells; only the initial stages are antigen specific.

Mechanism of T-cell cytotoxicity (Fig. 16.4, see p. 398)

The cytotoxic effect of T cells is important both in local type IV reactions and in the host response to viruses and many bacteria. The effect is specific, almost as much so as antibody activity, and the cytotoxicity directed against target cells can occur independently, in the absence of antibody or complement (Cerrottini and Brunner 1974, Henney 1977). Cell contact is necessary, and the cytotoxic T cell must be viable; the

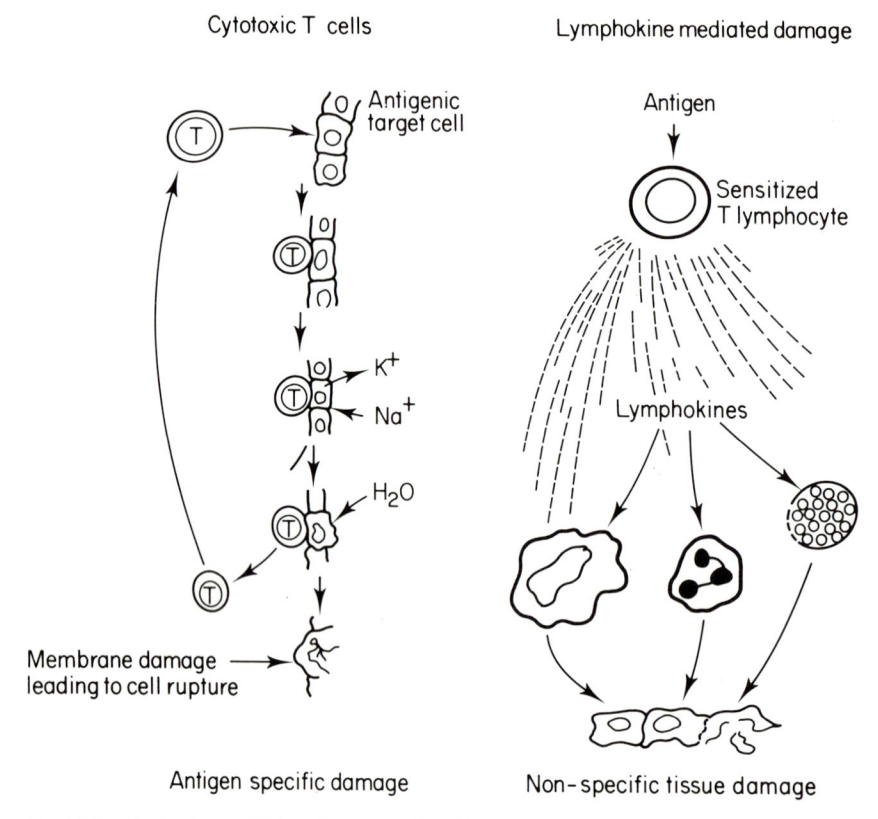

Fig. 16.4 Mechanisms of T-lymphocyte mediated hypersensitivity.

presence of Mg^{++} is essential before membrane damage occurs in the target cell. The mechanism by which T_C cells actually lyse the target cells is still poorly understood, although it has been suggested that a phospholipase or even trypsin-like enzymes may be involved (for review of current evidence, see Henney 1980). The target cell, whether it be virus infected or grafted tissue or the object of an auto immune attack, undergoes osmotic lysis and death.

T-cell immunity and infection

The role of T cells in the response to virus infections is supported by the greatly increased incidence of such infections in patients who have defects in their T-cell immunity (Chapter 17). In experimental infections in mice, influenza virus, lymphocytic choriomeningitis and ectromelia have all been demonstrated to stimulate cytotoxic T cell production, and these cells have been shown to lyse virus-infected target cells. The response to viruses is complicated by the presence of other mononuclear cells in virus-infected tissues, including the natural killer (NK) cells where the effect is non-specific and may be very important in the early stages of virus elimination. Interferons are also present, at least one of which is released by T cells (Doherty et al. 1974, Zweerink et al. 1977).

T cells may be responsible for some of the adverse reactions to infecting micro-organisms, in addition to their role in defence. The cerebral lesions in mice infected with LCMV are abrogated by T-cell depletion; the rash in measles is T-cell mediated; hepatitis-B virus liver damage is probably also due to T-cell activity. In parastic infections, much of the tissue damage may be produced by this mechanism, e.g. the cardiac lesions in South American trypansomiasis (Teixeira et al. 1978). In fungal infections, T-cell immunity is important to recovery, and patients with defects in their T-cell responses are peculiarly susceptible to serious infections with fungi not normally regarded as major pathogens (Chapter 17). The possible role of T cells in the development of the lesions in fungal diseases is problematic. Some of the generalized features, known to dermatologists as 'id', which occasionally develop, with eruptions in skin at sites distant to the primary lesion, could be attributed to type IV or even a mixture of type III and type IV events, due to the action of antibodies of IgG and IgM classes and T_{DTH} cells in the presence of soluble antigen released from the focus of fungal growth. Vasculitic features are observed in skin lesions associated with infection and may well be immunologically produced. Firm evidence of the presence of specific fungal antigens in such lesions is not available.

Contact hypersensitivity Type IV reactions may develop in skin after contact with a wide range of sensitizing materials, including nickel, chrome, penicillin, hair dyes, plasticizers, formaldehyde, and many industrial chemicals, oils and dyes. The lesions are infiltrated with lymphocytes and are very similar in histological appearances to those of the tuberculin test. The inflammatory changes in the skin may progress to frank vesiculation with scaling and exudation. The specific nature of the sensitization may be determined by the use of the 'patch' test, when a solution of the chemicals under examination is applied to an area of healthy skin in the form of a small patch of material (not injected). After 24–48 hours a localized reaction will be observed where sensitized T cells have migrated to the epithelial surface layers. Many of the antigens which produce contact hypersensitivity, also known as industrial dermatitis, are in fact, haptens, and are immunogenic only after becoming bound to proteins in the tissues.

Autoimmune disease

A detailed description of the pathogenesis and a description of the pathological processes involved in this large and complex group of diseases is outside the scope of the present text. Readers should consult the major texts on pathology, immunology and specialized monographs (e.g. Glynn and Holborow 1965, Anderson *et al.* 1967, Bellanti 1978, Anderson 1980, Roitt 1980, Lachmann and Peters 1982). It will be clear from the preceding description of hypersensitivity states that the pathological processes which follow the loss of self-tolerance are basically determined by immunological events. The loss of integrity of the tissues and the breakdown of the barrier to recognition of self-antigens lead to both humoral and cellular responses, with antibodies and lymphocytes—T cells and others—directed against body constituents. It has been suggested that infection may precipitate this loss of self-tolerance by allowing or facilitating interactions between cell surface immunoglobulin receptors and antigen without producing cross-linking of determinants, which is regarded as essential for the induction of B cell responses. A tolerogenic effect might also be brought about by a change in the nature of the cell-surface determinants of the individual, i.e. the alteration of self-constituents because of the presence of the organism in or on the surface of the cell. That many changes do occur in and on the surface of cells which are associated with infecting organisms is not denied; the unsolved problem is whether these metabolic and destructive processes actually produce new antigenic structures. The resemblance between such new antigens and normal tissue antigens might be close and infected cells might initially evade the normal immunological response directed against the specific antigens of the infecting organism by not being sufficiently 'foreign' to be destroyed. This could lead to persistent infection; some viruses might be strong candidates for such a contingency. Continued growth and tissue destruction might then produce the breakdown in normal regulatory processes.

Several autoantibodies directed against specific hormones or cell membrane receptors have been described, e.g. against the acetycholine receptor in myasthenia gravis, and against insulin and islet cells in diabetes. However, the involvement of infectious micro-organisms in such diseases has not been directly demonstrated. Similarly, in systemic lupus erythematosus where anti-DNA antibodies are formed, no positive proof of infection has been available. Viruses have been blamed but, apart from some suggestive evidence in inbred animals, not directly identified. LCMV and some other murine viruses are associated with autoimmune phenomena in some strains of mice; the tendency to develop lesions seems to be strain dependent and linked to the MHC.

Transfer factor

The ability to transfer type IV delayed hypersensitivity skin reactions is apparently limited to living lymphoid cells, with one rather unusual exception which was reported by Lawrence and co-workers (Lawrence 1955). Using a cell free extract of human leucocytes, they demonstrated the transfer of type IV skin reactions against tuberculin, streptococcal M substance and *Candida* sp. The dialysable extract, which they called 'transfer factor', produced local and distant skin reactivity, and the effect persisted for long periods after the injection of the material into the skin. The reaction produced in the recipient of transfer factor was antigen-specific. Lymphocytes from a tuberculin sensitized donor gave transfer factor which converted a previously tuberculin negative recipient (Lawrence 1969).

Transfer factor is prepared by the treatment of frozen and thawed human leucocytes; the cells are digested with ribonuclease and the extract is dialysed to give a purified fraction which is non-antigenic and does not contain immunoglobulins. The major constituents of the extract are polypeptides and polynucleotides and the transfer factor appears to have a molecular weight of 12 K. That such a small molecule should mediate type IV reactivity seems to preclude the possibility that antigen fragments are transferred. The mechanism of action of transfer factor, even its very existence, has long been the subject of dispute. It has been suggested that it merely heightens an otherwise undetectable 'hidden' state of hypersensitivity and, further, that the transfer is non-specific and not truly antigen-specific. Recently, an *in vitro* test for activity has been reported, by the use of human lymphocytes in tissue culture, tested in the presence of transfer factor and antigen for the production of lymphokines (Borkowsky et al. 1981).

Therapeutic trials have been conducted with transfer factor for the treatment of patients suffering from defects of T-cell immunity, e.g. chronic mucocutaneous candidiasis (Chapter 17), but the results have not shown any improvement in T-cell function.

Genetic factors in the response

The possible role of genetic factors in the development of hypersensitivity states is now a matter for experimental study. Our increasing knowledge of the role of MHC antigens would lead us to suppose that genes within this region, especially those of the Ia (Immune-associated) group, known as HLA-DR in man, might have some part to play in the regulation of the balance between stimulation and suppression of both T- and B-cell mediated events (McDevitt and Chinitz 1969, Benacerraf 1981). The finding that many autoimmune disorders show a close association with the presence of one or more HLA-DR antigens, especially HLA-DR3, would implicate the human MHC (Bodmer 1972, Dick 1981).

The importance of infectious disease in past centuries as a selective force has been recognized; survival from killing infections may be genetically determined. Known facts on the epidemiology of many serious infections would support this observation, but the mechanism by which selection is brought about has not been elucidated. The HLA system, with its central role in cellular interaction, in maintaining regulatory functions within the immune system, may be an important factor in disease processes, particularly those where resistance to infection and the survival to reproductive age could be related (Bodmer 1980).

References

Adkinson, N. F. Jr., Sobotka, A. K. and Lichtenstein, L. M. (1979) *J. Immunol.* **122**, 965.
Anderson, J. R. (Ed.) (1980) *Muir's Textbook of Pathology*, 11th edn. Edward Arnold, London.
Anderson, J. R., Buchanan, W. W. and Goudie, R. B. (1967) *Auto-immunity: Clinical and Experimental.* Charles C Thomas, Springfield, Ill.
Arthus, M. (1903) *C.R. Soc. Biol. (Paris)* **55**, 817.
Austen, K. F. (1981) *Immunology Today* **2**, 146–148.
Belin, L. and Norman, P. S. (1977) *Clin. Allergy* **7**, 55.
Bellanti, J. A. (Ed.) (1978) *Immunology II.* W. B. Saunders, Philadelphia.
Benacerraf, B. (1981) *Science (Wash)* **212**, 1229.
Benacerraf, B. and Kabat, E. A. (1949) *J. Immunol.* **62**, 517.
Bianchi, A. T. J., Hooijkaas, H., Benner, R., Tees, R., Nordin, A. A. and Schreier, M. H. (1981) *Nature* **290**, 62.
Billingham, R. E., Brent, L. and Medawar, P. B. (1954) *Proc. Roy. Soc. B.* **143**, 58.
Bodmer, W. F. (1972) *Nature (Lond.)* **237**, 139 (1980) *J. exp. Med.* **152**, 353s.
Borgeat, P. and Samuelsson, B. (1979) *Proc. nat. Acad. Sci. Wash.* **76**, 3213.
Borkowsky, W., Suleski, P., Bhardwaj, N. and Lawrence, H. S. (1981) *J. Immunol.* **126**, 80.
Buck, A. A., Anderson, R. I. and Macrae, A. A. (1973) *Z. Tropenmed. Parasitol.* **24**, 21.
Casoni, T. (1911) *Fol. clin. Chem. Micro.* **4**, 5.
Cerrottini, J-C. and Brunner, K. T. (1974) *Advanc. Immunol.* **18**, 67.
Chandra, R., Govila, P., Chandra, S., Katiyar, J. D. and Sen, A. B. (1974) *Indian J. med. Res.* **62**, 1017.
Christ, E. J. and Van Dorp, D. A. (1972) *Biochim. biophys. Acta* **270**, 537.
Cochrane, C. G. and Koffler, D. (1973) *Advanc. Immunol.* **16**, 185.
Coombs, R. R. A. and Gell, P. G. H. (1975) In: *Clinical Aspects of Immunology*, p. 761. Ed. by P. G. H. Gell, R. R. A. Coombs and P. J. Lachmann. Blackwell Scientific Publications, Oxford.
Culbertson, J. T. (1935) *J. Immunol.* **29**, 29.
Davis, H. M. (1938) *Lancet* **i**, 193.
Davis, S. D., Schaller, J. and Wedgewood, R. J. (1966) *Lancet* **i**, 1013.
Dick, Heather M. (1981) In: *Recent Advances in Medicine, No. 18*, p. 1. Ed. by A. M. Dawson, N. Compston and G. M. Besser. Churchill Livingstone, Edinburgh.
Dobson, D. and Soulsby, E. J. L. (1973) *Exp. Parasitol.* **35**, 16.
Doerr, R. and Pick, E. P. (1913) *Z. Immun Forsch.* **19**, 251.
Doherty, P. C., Zinkernagel, R. M. and Ramshaw, K. A. (1974) *J. Immunol.* **112**, 1548.
Edwards, J. H. and Davies, B. H. (1981) *Allergy clin. Immunol.* **68**, 58.
Feldberg, W. and Kellaway, C. H. (1938) *J. Physiol. (Lond.)* **94**, 187.
Fisherman, E. W. (1962) *J. Allergy* **33**, 12.
Gell, P. G. H. and Coombs, R. R. A. (Eds.) (1963) *Clinical Aspects of Immunology*, 2nd edn. Blackwell Scientific Publications, Oxford.
Gell, P. G. H., Coombs, R. R. A. and Lachmann, P. J. (Eds.) (1975) *Clinical Aspects of Immunology*, 3rd edn., p. 1723. Blackwell Scientific Publications, Oxford.
Glynn, L. E. and Holborow, E. J. (1964) *Autoimmunity and Disease.* Blackwell Scientific Publications, Oxford; (1965) *Autoimmunity and Disease.* Blackwell Scientific Publications, Oxford.
Goetzl, E. J. and Kay, A. B. (1982) In: *Current Perspectives in Allergy*, p. 1. Ed. by E. J. Goetzl and A. B. Kay, Churchill Livingstone, Edinburgh.
Golden, D., Valentine, M. D., Sobotka, A. K. and Lichtenstein, L. M. (1979) *J. Allergy clin. Immunol.* **63**, 180.
Gowans, J. L. (1965) *Brit. med. Bull.* **21**, 106.
Henney, C. S. (1977) *Contemp. Top. Immunobiol.* **7**, 245; (1980) *Immunology Today* **1**, 36.
Henson, P. M. (1971) *J. exp. Med.* **134**, 1145.
Henson, P. M. and Cochrane, C. G. (1974) *Ann. N.Y. Acad. Sci.* **256**, 426.
Hill, H. R. and Quie, P. G. (1974) *Lancet* **i**, 183.
Holgate, S. T., Lewis, R. A. and Austen, K. F. (1980) *Proc. nat. Acad. Sci., Wash.* **77**, 6800.

Holti, G. (1966) In: *Symposium on Candida Infections*, p. 73. Ed. by H. L. Winner and R. Hurley, Churchill Livingstone, Edinburgh.
Houba, V. (1975) *Bull. Wld. Hlth. Org.* **52**, 199.
Ishizaka, K., Ishizaka, T. and Hornbrook, M. M. (1966) *J. Immunol.* **97**, 75.
Jerne, N. K. (1974) *Ann. Immunol. Inst. Pasteur* **125C**, 373.
Johansson, S. G. O., Mellbin, T. and Vahlquist, B. (1968) *Lancet* **i**, 1118.
Johnson, G. D., Holborow, E. J. and Dorling, J. (1978) In: *Handbook of Experimental Immunology*, 3rd edn., Vol. 1. Ed. by D. M. Weir. Blackwell Scientific Publications, Oxford.
Jones, C. E., Rachford, F. W., Ozcel, M. A. and Lewert, R. M. (1977) *Exp. Parasitol.* **42**, 261.
Juhlin, L. and Wide, L. (1972) In: *Mechanisms in Drug Allergy*, p. 139. Ed. by C. H. Dash and H. E. H. Jones, Churchill Livingstone, Edinburgh.
Kagan, I. G. (1968) *Bull. Wld. Hlth. Org.* **39**, 25.
Kay, A. B. (1981) In: *The Inflammatory Process*, p. 293. Ed. by P. Venge and A. Lindblom, Almquist and Wiskell International, Stockholm.
Knutson, D. W., Kijlstra, A. and Van Es, L. A. (1979) *J. Immunol.* **123**, 2040.
Koch, R. (1891) *Dtsch. med. Wschr.* **17**, 1189.
Kraft, D. and Wide, L. (1976) *Brit. J. Derm.* **94**, 593.
Lachmann, P. J. and Peters, D. K. (Eds.) (1982) *Clinical Aspects of Immunology*, 4th edn., Blackwell Scientific Publications, Oxford.
Landsteiner, K. and Chase, M. W. (1942) *Proc. Soc. exp. Biol. Med.* **49**, 688.
Lawrence, H. S. (1955) *J. clin. Invest.* **34**, 219, (1969) *Advanc. Immunol.* **11**, 195.
Le Bras, M., Dupont, A., Longy, M. and Delmas, M. (1980) *Med. trop. (Mars.)* **40**, 67.
Leslie, R. G. Q. (1980) *Europ. J. Immunol.* **10**, 317.
Levinsky, R. J. (1981) *Immunology Today* **2**, 94.
Lewis, R. A., Holgate, S. T., Roberts, L. J. II, Oates, J. A. and Austen, K. F. (1981) In: *Biochemistry of the Acute Allergic Reactions*, p. 239. Ed. by E. L. Becker, A. S. Simon and K. F. Austen, Alan R. Liss Inc., New York.
Liew, F. Y. (1982) *Immunology Today* **3**, 18.
Lin, Yun, Lu and Askonas, B. A. (1981) *J. exp. Med.* **154**, 225.
Lloyd, S. and Soulsby, E. J. L. (1978) *Immunology* **34**, 939.
Longbottom, J. L. and Pepys, J. (1964) *J. Path. Bact.* **88**, 141; (1975) In: *Clinical Aspects of Immunology*, 3rd edn., p. 99. Ed. by P. G. H. Gell, R. R. A. Coombs and P. J. Lachmann, Blackwell Scientific Publications, Oxford.
McCluskey, R. T., Benacerraf, B. and McCluskey, J. W. (1963) *J. Immunol.* **90**, 466.
McDevitt, H. O. and Bodmer, W. F. (1974) *Lancet* **i**, 1269.
McDevitt, H. O. and Chinitz, A. (1969) *Science (Wash)* **163**, 1207.
Mayrhofer, G., Bazin, H. and Gowans, J. L. (1976) *Europ. J. Immunol.* **6**, 537.
Medawar, P. B. (1958) *Proc. Roy. Soc. B.* **149**, 145.
Moncada, S., Higgs, E. A. and Vane, J. R. (1977) *Lancet* **i**, 18.

Otto, R. (1907) *Münch med. Wschr.* **55**, 1665.
Ovary, Z. (1958) *Progr. Allergy* **5**, 459; (1961) *C.R. Acad. Sci.* **253**, 582.
Parish, J. and Oakley, C. L. (1940) *Brit. med. J.* **i**, 294.
Parish, W. E. (1970) *Lancet* **ii**, 591.
Parratt, D., McKenzie, H., Nielsen, K. H. and Cobb, S. J. (1982) *Radioimmunoassay of antibody*. John Wiley, Chichester.
von Pirquet, C. (1906) *Münch. med. Wschr.* **53**, 1457.
von Pirquet, C. and Schick, B. (1905) *Die Serumkrankheit*, Leipzig.
Platts-Mills, T. A. E. (1982) In: *Clinical Aspects of Immunology*, 4th edn., p. 598. Ed. by P. J. Lachmann and D. K. Peters, Blackwell Scientific Publications, Oxford.
Poltera, A. A., Lambert, P. H. and Miescher, P. A. (1979) In: *The Menarini Series on Immunopathology*, Vol. 2, Ed. by P. A. Miescher. A. G. Schwabe, Basel.
Portier, P. and Richet, C. (1902) *Comptes rendues Soc. Biol.* **54**, 170.
Prausnitz, C. and Küstner, H. (1921) *Zentrabl. Bakteriol (Orig A)* **86**, 160. Translation in *Clinical Aspects of Immunology*, 2nd edn., p. 1298. Appendix B, Ed. by P. G. H. Gell and R. R. A. Coombs (1963). Blackwell Scientific Publications, Oxford.
Redfern, W. W. (1926) *Amer. J. Hyg.* **6**, 276.
Richet, C. (1913) *Anaphylaxis*. Trans. by J. M. Bligh. Constable, London.
Riley, J. F. (1959) *The Mast Cells*. Churchill Livingstone, Edinburgh.
Roitt, I. (1980) *Essential Immunology*, 4th edn. Blackwell Scientific Publications, Oxford.
Salaman, J. R. (1979) In: *Kidney Transplantation. Principles and Practice*, p. 163. Ed. by P. J. Morris, Academic Press, London.
Samuelsson, B. (1981) In: *Biochemistry of the Acute Allergic Reactions*, p. 1. Ed. by E. L. Becker, A. S. Simon and K. F. Austen, Alan R. Liss Inc., New York.
Samuelsson, B. and Hammarstrom, S. (1981) *Immunology Today* **2**, 3.
Schantz, P. M. (1977) *Exp. Parasitol.* **43**, 268.
Smith, J. M. and Gell, P. G. H. (1975) In: *Clinical Aspects of Immunology*, 3rd edn., p. 903. Ed. by P. G. H. Gell, R. R. A. Coombs and P. J. Lachmann, Blackwell Scientific Publications, Oxford.
Sobotka, A. K. (1977) *J. Allergy clin. Immunol.* **60**, 213.
Spector, W. G. (1967) *Brit. med. Bull.* **23**, 35.
Stechschulte, D. J. (1982) In: *Current Perspectives in Allergy*, p. 113. Ed. by E. J. Goetzl and A. B. Kay, Churchill Livingstone, Edinburgh.
Teixeira, A. R. L., Teixeira, G., Macedo, V. and Prata, A. (1978) *Amer. J. trop. Med. Hyg.* **27**, 1097.
Turk, J. L. and Oort, J. (1963) *Immunology* **6**, 140.
Wemambu, S. N. C., Turk, J. L., Waters, M. F. R. and Rees, R. J. W. (1979) *Lancet* **ii**, 933.
Zinsser, H. (1921) *J. exp. Med.* **34**, 495.
Zinsser, H. and Grinnell, F. B. (1927) *J. Bact.* **14**, 30.
Zweerink, H. J., Courtneidge, S. A., Skehel, J. J., Crumpton, M. J. and Askonas, B. A. (1977) *Nature* **267**, 354.

17

Problems of defective immunity. The diminished immune response

J. Douglas Sleigh and Heather M. Dick

Introductory	402	Chronic granulomatous disease	405
Immunodeficiency. Syndromes with a (congenital) genetic basis	403	Chediak – Higashi syndrome	406
		Miscellaneous disorders of phagocyte function	406
Immunodeficiency: recognized syndromes and main defects in immunity	403	Treatment of phagocytic and bacterial killing defects	406
Absence of stem cells: severe combined immunodeficiency (SCID)	403	Defects of complement components	406
Swiss-type agammaglobulinaemia	403	Problems of acquired defects in immunological response	406
Nezelof's syndrome	404	Immunosuppression: action of drugs	406
Adenosine deaminase deficiency	404	Corticosteroids	406
Reticular dysgenesis	404	Azathioprine and related compounds	407
T-cell defects	404	Alkylating agents	407
Di George's syndrome	404	Cyclosporin A	407
Purine nucleosidase phosphorylase deficiency	404	Immunodepression produced by trauma, surgery, burns etc.	408
B-cell defects	404	Defects in the immune response in malignant disease	408
Hypogammaglobulinaemia	404		
Other defects of immunoglobulin production	404	Infection in patients with defective immunological responses	408
IgA deficiency	404	Pulmonary infections	409
IgM deficiency	405	CNS infections	409
Physiological hypogammaglobulinaemia of newborn	405	Septicaemia	409
Immunodeficiency affecting antibody formation	405	Screening of the patient with defective immunity	410
Partial defects of T- and B-cell function	405	Precautions to reduce the risk of endogenous infection	410
Wiskott-Aldrich syndrome	405		
Chronic mucocutaneous candidiasis	405	Precautions to diminish the risk of exogenous infection	410
Ataxia telangiectasia	405		
Defects of phagocyte function	405	Acquired immunodeficiency syndrome (AIDS)	410

Introductory

Most people develop infections at some time in their life, yet the majority of us suffer relatively little from such infections, largely because of the efficiency of the body's defence mechanisms. These defences are both non-specific and specific, ranging from the skin and mucous membranes acting as excellent physical barriers to invading micro-organisms, to the highly specific immunological responses leading to the development of both antibody and lymphocyte defences. Defects in these mechanisms may increase the chance of acquiring infection or increase the severity of the eventual lesion. Such defects may be genetically determined (congenital) and present from birth or may be acquired by reason of some disturbance in the internal

Table 17.1 Immunodeficiency syndromes with a genetic basis (congenital immune deficiency disorders)

Physiological or anatomical defect	Main clinical features and presentation
1. Absence of stem cells (lack of cells capable of differentiating into T or B cells).	Failure to thrive from birth. Bacterial, fungal and protozoal infections, especially candidiasis, pneumonia due to *Pneumocystis carinii*.
2. Failure of development of thymus (4th branchial pouch)—T-cell defect.	Persistent and recurrent fungal and viral infections. Cardiac abnormalities. Tetany (absence of parathyroids).
3. Deficient plasma cells: impaired antibody synthesis—B-cell defect.	Recurrent bacterial infections, especially of respiratory tract and skin (staphylococcal, streptococcal or due to gram-negative organisms). Meningitis. Otitis media.
4. Partial defects of T- and B-cell functions.	Recurrent infections—mixed pattern, with other features depending on type of defect e.g. skin rashes (purpuric), candidiasis, diarrhoea, encephalitis, anatomical and neurological defects.
5. Defects in phagocytic function.	Bacterial infections. Skin lesions (staphylococci, candida). Respiratory problems (fungi especially *Aspergillus sp.*).
6. Defects in complement system.	Recurrent infections, gonococcal or meningococcal. Commonest syndrome is SLE (Systemic lupus erythematosus) or other autoimmune disease.

or external 'milieu' of the individual. The microbiologist needs detailed knowledge about the type of defect which may arise, and more particularly, of the likely infectious hazards of such disorders.

Congenital immunodeficiency diseases are rare disorders; most bacteriologists may not encounter a single case, unless they are involved in specialist paediatric care. Like the inborn errors of metabolism which threw such light on the metabolic pathways of normal physiology, the immunodeficiencies of neonatal life have afforded unique opportunities to understand the normal functions of the tissues which make up the immune 'system'. A proper understanding of these rare defects forms the basis of an informed approach to diagnosis and treatment.

Acquired defects of greater or less severity are far commoner than the neonatal disorders, but their eventual outcome can best be understood in the context of the congenital defects. Most of the acquired defects develop as a result of the forms of treatment which have profound widespread effects on cellular function and metabolism. Chemotherapy, radiotherapy, steroids and immunosuppressive drugs all contribute to the subsequent morbidity and mortality from infectious disease. The type of infection, the predominant organisms and the organs or systems affected can all be predicted, given a few facts about the kind of cell likely to have been destroyed or suppressed by a particular form of treatment.

Immunodeficiency. Syndromes with a (congenital) genetic basis

The principal forms of these rare syndromes may be identified by the basic cellular defect which has arisen during development. The precise reason for most of these defects is unknown; some few can be identified as the result of defective genes, with consequent metabolic enzyme defects e.g. severe combined immunodeficiency due to lack of adenosine deaminase, which disrupts normal DNA synthesis in stem cells, affecting both the red and white cell series. Others may be due to intrauterine infection or unidentified gene mutations. The outcome of each type of defect is directly related to the normal functions which affected cells subserve. For example, absence of B cells means that there can be no maturation of B cells to immunoglobulin synthesizing plasma cells and, inevitably, little or no antibody production. Those micro-organisms which are usually controlled by antibody are thus free to multiply, and recurrent bacterial infections occur with staphylococci, particularly *Staphylococcus aureus*, and pneumococci predominating.

The major defects encountered are detailed in Table 17.1, where the physiological or anatomical abnormality and its consequent effects are listed. A fuller description of those disorders that have been more thoroughly characterized is appended (for reviews, see MRC Report 1971, Buckley 1977, Hayward 1979, Hayward 1982).

Immunodeficiency: recognized syndromes and main defects in immunity

Absence of stem cells: severe combined immunodeficiency (SCID).

Several variants of this defect occur, with differing degrees of severity. The well-recognized syndromes include Swiss type agammaglobulinaemia and reticular dysgenesis. SCID may present as one of the following disorders:

(a) Swiss-type agammaglobulinaemia, which may be due to an X-chromosome linked recessive defect, occurring in males, or may be the result of an autosomal recessive inherited defect which can occur in either sex. The thymus is absent and there are few lymph nodes. Multiple infections with viruses and fungi occur as well as gastro-enteritis due to gram-negative organisms. The local lesion after BCG vaccination may become generalized. These infants generally die before 6–12 months of age (see Hitzig 1973) unless treated. Bone marrow transplant may correct the defect (Gatti *et al.* 1968, Dupont *et al.* 1974) together with fetal thymus grafting (Biggar *et al.* 1975).

(b) Nezelof's syndrome, where there is cellular immunodeficiency but immunoglobulins are present (Nezelof *et al.* 1964, Lawlor *et al.* 1974) although specific antibody responses are poor. Majority of Ig is IgM. Presentation similar to (a) and respiratory infections are very common.

(c) Adenosine deaminase deficiency, which affects DNA synthesis. Immunoglobulin synthesis is depressed and the child may also have skeletal defects (Giblett *et al.* 1972). The enzyme defect leads to the accumulation of intracellular adenosine which affects not only lymphocytes but also neutrophils. This condition is inherited as an autosomal recessive; heterozygotes for the gene are normal.

(d) Reticular dysgenesis, where SCID is accompanied by polymorphonuclear leucocyte defects. Affected children often die early from fulminating septicaemia and have infiltration of tissues with histiocytes and eosinophils. The exact defect is unknown (Ochs *et al.* 1974).

T-cell defects

The main syndromes which involve defects in T-cell production or maturation generally characterized by recurrent viral infections and recurrent or chronic fungal disease. Infection with the protozoan, *Pneumocystis carinii*, may also occur.

(a) Di George's syndrome (thymic hypoplasia) results from a failure of tissue development *in utero*, affecting the 3rd and 4th branchial pouches. As a result the thymus and parathyroids are absent or vestigial. Abnormalities in the aortic arch may also be present (it is derived from the same embryonic tissues). There are low numbers of circulating T cells. The mechanism responsible for the developmental failure is unknown.

Infants present with tetany, due to the absence of parathyroid tissue, cardiac problems and infections. The latter include pneumonia, gastro-enteritis and candidiasis of the skin. The disorder responds dramatically to thymus grafting with fetal thymic tissue (Biggar *et al.* 1975).

(b) Purine nucleosidase phosphorylase deficiency (PNP deficiency) is due to a structural gene defect on chromosome 14, which results in a lack of the enzyme (PNP) which is closely concerned in the metabolism of cytosine and inosine to purines. T-cell proliferation is decreased. Children may not present with symptoms until after infancy. They have low T-cell levels and normal immunoglobulins. The predominant infections are viral, especially varicella, and those due to the viruses of the herpes group, Auto-antibodies may appear, presumably owing to lack of proper regulation by T cells, and haemolytic anaemia may develop (Edwards *et al.* 1978).

B-cell defects

Hypogammaglobulinaemia may occur in X-linked and autosomal recessive forms. The former may include a history of affected male relatives or siblings dying early in infancy from infection. This defect generally manifests itself as severe pneumonia in very young infants—under 6 months of age (Bruton 1952). The B-cell counts are low and many of the cells lack surface immunoglobulin: there may be sIg$^+$ cells in the marrow. Antibody production is diminished or absent. Bacterial infections predominate—staphylococcal, pneumococcal and infections with gram-negative organisms. T-cell function is relatively normal: skin tests of type IV hypersensitivity are positive and skin grafts are rejected. The main defect may be in the maturation of B cells (Hayward and Greaves 1975). The condition is eminently treatable by the use of regular monthly intramuscular injections of pooled human immunoglobulin, mainly IgG in content, with a little IgA and IgM (MRC Report 1971, Buckley 1977).

Other defects of immunoglobulin production

Several partial defects of one or more Igs have been described (see Hayward 1982).

IgA deficiency (less than 5 mg/100 ml) is probably the commonest selective immunoglobulin defect. It is found in approximately 1 in 700 'normal' Caucasian adults, some of whom have no apparent problems due to the lack of these protective antibodies. Tests for the defect should include the measurement of secretory IgA, e.g. in saliva. IgA$^+$ B lymphocytes are generally present in blood, but there appears to be a block in their maturation into IgA secreting plasma cells. Patients may suffer from respiratory tract infections, including sinusitis and otitis media and there is an increased incidence of asthma and other forms of type I hypersensitivity; the link between IgA and IgE production is unknown, but might be due to a central regulating mechanism for B cells. Auto-immune disorders are relatively common in IgA deficient adults e.g. myasthenia gravis, thyroiditis and rheumatoid arthritis (Amman and Hong 1970, Behan *et al.* 1976). IgA deficient individuals are liable to form anti-IgA antibodies after blood transfusion or plasma administration and may subsequently be subject to anaphylactic (type I) hypersensitivity reactions to further blood or blood products.

IgM deficiency is usually characterized by very low levels rather than absence of IgM. The patients tend to present with bacterial infection, meningitis, pneumonia and septicaemia being reported (Hobbs 1975).

Physiological hypogammaglobulinaemia of newborn The newborn infant relies for its antibody on maternal immunoglobulins, acquired transplacentally, and on IgA present in breast milk (Chapter 14). As the maternal immunoglobulins are slowly catabolized in the infant's circulation, the level of Ig falls, reaching a nadir at about 4 to 6 months of age. The production of immunoglobulins by the infant's own B cells and plasma cells should begin to increase at this time and levels slowly rise to normal by about 12 months of age. No treatment by replacement therapy is required. The process may very occasionally be delayed, perhaps owing to an inherited defect, but most infants have an excellent capacity to synthesize specific antibody when challenged with infection (Buckley 1977).

Immunodeficiency affecting antibody formation, also known as 'variable' immunodeficiency.

A heterogeneous collection of immunologically deficient individuals is classified under this heading, generally because their symptoms and signs do not fall clearly into one of the recognized syndromes. Here, immunoglobulin production is deficient or perhaps suppressed (Feldmann et al. 1975, Siegal et al. 1978). These patients suffer from bacterial infections and, even in some cases, virus and fungal diseases.

Partial defects of T- and B-cell function There are three well described forms of partial defect involving both T and B cells. Other less well defined patterns of defect probably occur: the results will depend on the relative lack of one or other functional subset.

Wiskott–Aldrich syndrome This syndrome is characterized by the presence of eczema, purpura and recurrent infections (Aldrich et al. 1954) and is X-chromosome linked. The purpura develops as a result of poor platelet aggregation and thrombocytopenia. The IgE levels are often raised, and poor IgM and IgG response to antigens is a feature of these cases. Infections include gastro-enteritis, often with blood staining, and the eczematous skin may become infected. Cerebral haemorrhage due to the defective platelets or bacterial pneumonia are the commonest causes of death. Bone marrow transplantation from a suitable HLA-matched donor has been used successfully to treat this disorder (Bach et al. 1968).

Chronic mucocutaneous candidiasis This disfiguring and intractable disease is characterized by infection of skin and mucous membranes with *Candida albicans*, often with large granulomatous lesions affecting the face and scalp. The defects in immunological response include reduced but not absent T-cell function, normal or even high levels of antibody response to *Candida* together with autoimmune disorders. The condition probably includes a spectrum of defects, varying in severity and age of onset, although it generally appears in infancy. Numerous attempts to treat it have been made, with little success. These have included transfer factor (Valdimarsson et al. 1972) and bone marrow or fetal thymus grafts (Buckley et al. 1968, Levy et al. 1971).

Ataxia telangectasia This inherited disorder is frequently accompanied by IgA deficiency. The presence of multiple telangiectases and neurological defects is accompanied by the features associated with defective T-cell function and a tendency to recurrent infection of the upper respiratory tract. Decreased T cells and absent or poor type IV skin hypersensitivity to common antigens—tuberculin, *Candida*—is present. The condition is not always severe and may be diagnosed only with the finding of telangiectases in the mouth or nasal mucous membrane, accompanied by IgA deficiency (Peterson et al. 1966).

Defects of phagocyte function These are found in numerous forms, some of which are reasonably well understood as far as the mechanism is concerned, and other forms where the functional defect can be demonstrated but the exact mechanism of which has not yet been elucidated.

Chronic granulomatous disease Yet another disorder which may occur either as an autosomal recessive gene defect in either sex, or more often in a form which is X-chromosome linked in males, the principal defect in chronic granulomatous disease appears to be the lack of ability of the polymorphonuclear granulocytes to effect bacterial killing of phagocytosed organisms. This is due to a specific defect in the enzymes which are central to the oxidation processes involved in bacterial killing. The primary oxidase, which is essential to superoxide production, is defective, and the lack of this dehydrogenase reduces the ability of superoxide and peroxide (H_2O_2) to interact. In these circumstances, the metabolic 'burst' which is necessary for bacterial killing is not produced (Segal 1979).

Patients with this disorder, generally male infants, usually present with severe infections, often with organisms not otherwise regarded as pathogens—the 'opportunists'—which may include *Candida albicans*, *Serratia marcescens*, as well as *Staphylococcus aureus* and the more usual pathogenic species of *Shigella*. Infections with catalase negative organisms are not often a problem (streptococci, for example), because their own production of peroxide acts within the phagocytic vacuole to mediate bacterial killing (Kaplan et al. 1968). Thus, catalase-positive organisms are the main infecting organisms (cf. Chediak–Higashi syndrome, see below). The lesions are those of chronic inflammation with granuloma formation; persistent deep-seated abscess formation can occur in organs such as the liver and long bones. Serum immunoglobulin levels are high and leucocytosis is common (Berendes *et al.* 1957, Johnston and Newman 1977). The cells of the affected infants can be shown to lack a special cytochrome *b*, which is unique to neutrophils (Segal and Jones 1980), and tests for killing of phagocytosed bacteria are abnormal (for methods see Segal 1981).

Chediak-Higashi syndrome Multiple defects are present in this rare syndrome, with severe defects in phagocytic function, albinism, platelet defects and the tendency to recurrent severe bacterial infections (Blume and Wolff 1972). The phagocytic cells contain large granules which stain for peroxidase. Both catalase-positive and catalase-negative bacteria cause infections; and chemotaxis as well as phagocytosis and killing of bacteria are abnormal. There appears to be a primary defect in the cells concerned in the assembly of cytoplasmic microtubules which are crucial to degranulation and the release of lysosomal enzymes (Boxer *et al.* 1976). This disorder is unusual in that an exact counterpart has been detected in several animal species, including mice, cats and mink (Oliver *et al.* 1975, Baehner and Boxer 1979).

Miscellaneous disorders of phagocyte function Several other rare disorders, principally caused by defects in phagocyte function, have been described, testifying to the importance of this activity in protection from bacterial infection. Some are associated with specific enzyme defects in the phagocytic cells, including defects in myeloperoxidase, pyruvate kinase and glutathione reductase enzymes.

Treatment of phagocytic and bacterial killing defects Currently, no specific treatment is available to remedy the metabolic defects in phagocytes, and the clinician must therefore rely on appropriate antibiotic or chemotherapeutic agents. It is possible that the development of new techniques for grafting stem cells or for correcting genetic defects by appropriate transfer of the relevant DNA might be applied at some time in the future to the treatment of these rare but life-threatening disorders.

Defects of complement components Rare deficiencies of complement components have been reported, mostly associated with abnormal regulation of immunological responses, leading to the development of autoimmune disorders, most frequently systemic lupus erythematosus (SLE). Deficiencies in the production of C1, C2 and C4 are associated with this and other immune complex disorders—polyarteritis nodosum, rheumatoid arthritis, nephritis, etc. (Agnello 1978).

Lack of the later components which mediate cell lysis (C5, 6, 7, 8 etc.) tends to be associated with recurrent infections. *Neisseria* sp. are a particular problem, suggesting that complement plays an important part in the elimination of these organisms by immunological means. Recurrent meningococcal meningitis and disseminated gonococcal infection have both been reported in patients with defects of one or other of these terminal components (Petersen *et al.* 1979, Lachmann 1982) as well as lupus-like disease (Zeitz *et al.* 1981). Lack of the inhibitor for C3b production, the activated form of C3, leads to susceptibility to both gram-positive and gram-negative bacterial infection (Thompson and Lachmann 1977). Lack of C3 itself also predisposes to bacterial infections, which may be recurrent and relate to the lack of normal opsonic activity in the absence of complement (Alper *et al.* 1972).

Problems of acquired defects in immunological response

Patients with acquired immunodeficiency are unusually susceptible to infection either because of the nature of their disease or as a result of their treatment.

Patients may also have single or multiple defects in their specific responses. In many diseases, both humoral and lymphocyte-mediated, responses may be seriously disturbed, e.g. in advanced malignant disease, and in other disorders very specific defects may be recognizable, e.g. in acute lymphocyte leukaemia of the T-cell type, where there is a lack of B cells and of only some subsets of T cells.

The history of the illness, together with information about treatment used, may provide valuable clues which indicate the likelihood of a relative defect in one component rather than another. In general, the same pattern prevails as in the congenital defects. Bacterial sepsis, particularly with gram-positive organisms and septicaemia, would suggest B-cell depression, possibly with diminution in antibody production or phagocytic defects. Fungal and viral infections may indicate a relative lack of T cells, or depression of both B cell and T cell, especially T-helper cell activity.

Sometimes it is possible to associate the disease or treatment with the defect it produces in the immune system. Patients with Hodgkin's disease suffer a significant impairment of their cell-mediated immunity, whereas after splenectomy, patients are unable to mount a normal acute antibody response to infection with encapsulated bacteria such as *Streptococcus pneumoniae*. More often, however, it is difficult to pin-point the site of the deficiency, and defence against infection may be diminished in a variety of ways. The action of drugs as well as the effects of underlying disease must be recognized. The main predisposing conditions are listed in Tables 17.2 and 17.3 (p. 407).

Immunosuppression: action of drugs

Corticosteroids The most commonly used corticosteroids, methyl prednisolone and hydrocortisone, affect the circulation of lymphocytes from nodes and spleen to the periphery (Ernström and Larssön 1967)

Table 17.2 Diseases in which a diminished immune response may be present

Diabetes.
Renal failure.
Disseminated malignant disease especially those of the lymphoreticular system, e.g. leukaemias, lymphomas, myeloma.
Diseases causing severe neutropaenia, including aplastic anaemia, leukaemia.
Autoimmune diseases, e.g. SLE, rheumatoid arthritis.
Acquired immunodeficiency syndrome (AIDS).

Table 17.3 Treatment which may result in a diminished immune response

Extensive radiotherapy, i.e. total nodal or whole body irradiation.
Drugs*
(i) Corticosteroids
(ii) Immunosuppressive/Cytotoxic drugs (e.g. cyclophosphamide, azathioprine, chlorambucil, cyclosporin A, methotrexate, 6-mercaptopurine).
Antilymphocyte globulin.
Splenectomy.

*Those drugs given to treat malignant or autoimmune diseases or to prevent organ transplant rejection.

and also many of the functions of specific lymphocyte populations, including cytotoxicity and some aspects of T-helper activity (Dougherty et al. 1964, Claman 1972). Lymphocytes bear receptors for steroids on their membranes, and it is likely that endogenous glucocorticosteroids are important in some aspects of the control of the interaction between cells during immunological responses (Fauci et al. 1976). Administration of steroids may therefore interfere with normal controlling mechanisms. T-helper function is diminished in mice given high doses of steroids, whereas in man, T-suppressor activity is preferentially inhibited (Knapp and Posch 1980). Long term steroid therapy, e.g. after renal transplantation, may be accompanied by a drop in immunoglobulin levels, but this is seldom large, and bacterial infections due to defective antibody response are not often a problem (Gabrielsen and Good 1967).

Azathioprine and related compounds Many other immunosuppressive drugs are used in clinical practice today, for example, azathioprine which is used to diminish graft rejection and in the treatment of some autoimmune disorders; one of its derivatives, 6-mercaptopurine, is widely used in haematological malignancies. Both act as purine analogues and are metabolized to produce products which inhibit the synthesis of DNA, RNA and protein, particularly in cells of the lymphoid series. In animals, antibody synthesis is reduced, the primary response being obviously diminished or suppressed, and even secondary responses are reduced in magnitude (for details, see Bach 1975). In man, at the therapeutic doses usually administered, humoral immunity is not dramatically affected, although antibody production can be affected if the drug is given at the same time as antigen exposure (Rowley et al. 1969). IgG production is more affected than is that of IgM. The activity of cytotoxic T cells in antibody dependent killing is also diminished (Campbell et al. 1976).

Alkylating agents In the treatment of malignant disease, several forms of cytotoxic drugs are employed, including cyclophosphamide and chlorambucil. These drugs (or alkylating agents) affect the accuracy of DNA reduplication in the cell nucleus. Both T and B cells are affected by the subsequent loss of membrane markers, and functional activity ceases even before cell death. Cyclophosphamide is used for the suppression of rejection after bone marrow transplantation (Santos et al. 1971). Patients treated with such drugs will have combined defects of antibody and cell-mediated immunity; phagocytic functions are not grossly affected. Fungal and protozoal (*P. carinii*) infections as well as viral infections, herpes zoster, varicella etc., are common; bacterial infections with poor antibody production may be a problem when high doses of the drugs are used (Mackay et al. 1973).

Cyclosporin A The newest immunosuppressive drug, Cyclosporin A is a metabolite extracted from the fermentation broth of fungi, either *Trichoderma polysporum* or *Cylindrocarpon lucidum* (Borel et al. 1976). The molecule, which has now been synthesized, is a cyclic endecapeptide, and is one of a series of cyclosporins, many of which have antibiotic activity. Cyclosporin A (CS-A) has profound immunosuppressive activity, by its action on T cells. There is a lesser but appreciable effect on B cells in man. Myeloid cells appear to be unaffected by CS-A. The action on lymphocytes occurs at an early stage in differentiation perhaps affecting the ability of these cells to respond to growth factors, interleukins, for example.

The drug has been used to suppress rejection of both bone marrow and renal grafts with remarkable success

(Powles et al. 1980, Report of European Multicentre Trial 1982). There have been problems with nephrotoxicity and an early report of a disturbingly high incidence of lymphomas (Calne et al. 1979), which appeared to be the result of using CS-A in combination with other immunosuppressive agents such as azathioprine and steroids. The same combination was also linked to the development of serious infections in many patients. The infections were predominantly bacterial, including both gram-negative and gram-positive organisms, but viral and fungal infections also occurred. Respiratory tract infections were commonest and gave most problems in treatment and diagnosis. A particular feature of some cases treated with CS-A was the detection of rising antibody titres to Epstein-Barr virus, probably the result of the absence of cytotoxic T cells which help to eliminate the EB-infected B lymphocytes (Crawford and Edwards 1982). Most, if not all, of the toxic effects of CS-A can be avoided by using minimal doses of the drug without other potent immunosuppressives simultaneously. Nephrotoxicity, in particular, may be reversed by lowering the dose of CS-A and by avoiding the use of aminoglycosides, the presence of which leads to increased blood levels of CS-A.

Immunodepression produced by trauma, surgery, burns etc. Numerous studies have attempted to pinpoint the exact defects in the immunological responses which are believed to accompany major trauma, surgical treatment or burns, leading to the increased risk of infection. In most of these studies, defects of both phagocytic and specific immune responses have been reported, but the precise time-cause relation between such defects and susceptibility to infection is not always clear. Most defects are temporary, and resolve during recovery. Probably the most significant effects are the depression of macrophage function leading to less than optimal phagocytosis, which together with the general inflammatory changes should stimulate the development of the immunological response. The production of lymphokines, monokines and other chemotactic factors may be affected adversely, for example in burned patients (Ninnemann 1981). The presence of increased numbers of T-suppressor cells, and of monocytes with suppressive functions has been reported, after both splenectomy and burns (Miller and Baker 1979a, b). Infection with gram-negative bacteria may introduce the additional complication of the presence of large amounts of bacterial lipopolysaccharide in the circulation, which has been shown to depress cytotoxic T-cell function (Vallera et al. 1980).

Defects in the immune response in malignant disease Advanced malignancy is frequently accompanied by profound depression of the immunological response to antigen. Most attention has been focused on the possible relevance of such reduced responses to the control of the malignant proliferation in the early stages of disease or after surgical removal of the main tumour. However, the bacteriologist is more interested in the consequences of the diminished immunological response in so far as it may predispose the patient to infection. Infective problems seem to be restricted more to the patient with advanced disease, who is cachetic and debilitated. T-cell responses are particularly depressed, but widespread secondary deposits of tumour may reduce myeloid proliferation so that polymorphonuclear cell function is also interfered with. Antibody production is often little affected as far as secondary responses are concerned, e.g. to flagellin, (Whittingham et al. 1978) but lymphoid tumours, replacing normal marrow, may reduce the ability to produce antibody by decreasing the number of available B cells (Steinherz et al. 1980), as demonstrated by the poor anti-haemagglutinin responses to influenza vaccine in children with leukaemia and other lymphoid neoplasms.

Infection in patients with defective immunological responses Many of the infections are due to well recognized pathogens, but others are caused by micro-organisms that rarely, if ever, produce serious disease in the healthy individual (Table 17.4, p. 409). The infections, colloquially called 'opportunistic' are often unusually severe with a tendency to disseminate and they may present unfamiliar clinical manifestations. Knowledge of the clinical syndromes and awareness of their possible causes is essential for patient management; although accurate laboratory diagnosis may be difficult, it must be made so that appropriate specific treatment can be prescribed.

Infections of the skin and soft tissues are common. Cellulitis usually begins in areas of moist skin such as the axilla, groin or perineum and, especially in patients with neutropaenia, the infection may fail to localize and form pus. A particular problem is created by venous access sites containing a foreign body which has to remain *in situ* for a long time, e.g. indwelling plastic lines, shunts for haemodialysis. These are necessary for the management of the patient and even the most careful local aseptic and antiseptic measures may fail to prevent infection. *Staphylococcus aureus* is the usual cause, but the infection may be due to other bacteria such as *Pseudomonas aeruginosa* or to fungi. *Ps. aeruginosa* may produce a characteristic black lesion surrounded by erythema that exudes serous fluid known as ecthyma gangrenosum. Common warts, usually a trivial virus infection, may be more exuberant and more numerous than usual. The vesicles of shingles normally restricted to a dermatome may spread by viraemia to involve other sites on the body surface in a generalized varicella-zoster infection.

In patients with defective immunity, thrush of the oropharynx is extremely common. This infection, due to *Candida albicans*, is often unusually severe, tends to be persistent and may spread to involve the oesophagus. Another troublesome infection of the mouth is

Table 17.4 Micro-organisms responsible for infection in the host with defective immunity

Bacteria
  Gram-negative aerobic bacilli, e.g. *Escherichia coli, Klebsiella species, Pseudomonas aeruginosa*
  Streptococcus pneumoniae and other streptococci,
  Staphylococcus aureus, Staphylococcus epidermidis
  Mycobacterium tuberculosis and atypical mycobacteria
  Listeria monocytogenes
  Legionella pneumophila
  Legionella micdadei ('Pittsburgh pneumonia agent')
  Nocardia asteroides

Viruses
  Herpes simplex virus
  Cytomegalovirus
  Epstein-Barr virus
  Varicella-zoster virus
  Vaccinia virus
  Measles virus
  Papovaviruses: JC (human polyoma) virus Papillomavirus

Fungi
  Candida albicans
  Cryptococcus neoformans
  Histoplasma capsulatum
  Coccidioides immitis
  Blastomyces dermatitidis
  Aspergillus species esp. *A. fumigatus*
  Phycomycetes esp. *Rhizopus species, Mucor species*

Protozoa
  Pneumocystis carinii
  Toxoplasma gondii

Helminths
  Strongyloides stercoralis

intra- and peri-oral herpes simplex; a feature of the lesions is abnormal local extension to the face and oesophagus.

Pulmonary infections Determination of the cause of pulmonary infiltrates detected by chest x-ray poses a special problem in patients with defective immunity particularly as some are not the result of infection (for review, see Williams *et al.* 1976). Acute infections due to *Streptococcus pneumoniae, Haemophilus influenzae* and other acknowledged respiratory pathogens should be easily detected by routine examination of sputum, although in tuberculosis direct microscopy and culture may fail to demonstrate acid-fast bacilli until the disease has produced lung destruction and caseation. Aerobic gram-negative bacilli, such as *Klebsiella species* or *Ps. aeruginosa*, are often cultured and this, usually but not always, indicates colonization rather than infection of the oropharynx or bronchial tree, particularly when the patient has been receiving broad-spectrum antibiotics. Caution must be exercised in interpreting the laboratory results, however, because these organisms are sometimes the cause of pneumonia. Other possible pathogens including the common opportunists, are difficult or impossible to isolate. *Aspergillus species, Pneumocystis carinii* and *Legionella pneumophila* are rarely shed in sufficient numbers in sputum to permit diagnosis by microscopy or culture, and the isolation of *Nocardia asteroides*, when it is accomplished, requires prolonged incubation. Virus infection, almost always due to cytomegalovirus, is another cause that has to be considered. In some suspected pulmonary infections serological tests may be of value, but a definite diagnosis will often require lung biopsy by the transbronchial, percutaneous or open route.

CNS infections Meningitis is not a common problem and is usually due to the conventional bacterial pathogens. *Listeria monocytogenes* is a recognized bacterial opportunist and *Cryptococcus neoformans* the commonest fungal cause. Meningoencephalitis is the result of infection with viruses of the herpes group or, occasionally, the protozoan *Toxoplasma gondii*. A rare demyelinating disease found only in the immuno-deficient host is multifocal leucoencephalopathy caused by JC virus (Narayan *et al.* 1973).

Septicaemia Septicaemia is the most serious complication and may be rapidly fatal in the presence of severe neutropaenia (granulocyte count of less than 500/μl). Rapid diagnosis is essential and treatment must be started on clinical suspicion before the results of blood culture are available. Fever in such patients may also be due to cytotoxic drug therapy; its significance in terms of infection is difficult to assess. The infecting organism does not always come from a septic focus, it may originate from the colonic flora.

Screening of the patient with defective immunity
The source of the infecting micro-organisms may be endogenous or exogenous. Endogenous infection is due to a micro-organism which is part of the body's normal flora, e.g. *Candida*, or to the reactivation of a healed but latent, infection that has remained asymptomatic, e.g. tuberculosis, herpes simplex, varicellazoster, cytomegalovirus infection. Exogenous infection is acquired from the environment, sometimes the hospital environment.

Endogenous and exogenous infection cannot always be distinguished, because sometimes a potential pathogen will colonize the body before infection. This observation is the rationale for repeated cultures taken from a variety of sites, throat, mouth, nose, axilla, rectum etc., which some believe will enable pathogens to be recognized before they are able to establish infection. For example, Schimpff *et al.* (1972) showed the value of detecting colonization with *Ps. aeruginosa* in patients with leukaemia, since two-thirds of those colonized subsequently developed septicaemia; and Aisner *et al.* (1979) showed that the presence of *Aspergillus flavus* in the nose preceded invasive pneumonia. Kramer *et al.* (1982), however, in a study of granulocytopaenic patients with fever argued against routine body surveillance cultures, which they found confusing. Often more than one potential pathogen was isolated and the results were difficult to assess and so of little value in patient management. Cultures failed to allow prediction with any certainty of the cause of the patient's septicaemia which was subsequently established by blood culture.

Precautions to reduce the risk of endogenous infection A major source of endogenous infection, especially in patients with neutropaenia, is the enormous and varied bacterial flora of the colon, and it is recommended that oral non-absorbable antibiotics should be given to some groups of patients (Storring *et al.* 1977). Favoured regimens are FRACON (framycetin, colistin and nystatin) and GVN (gentamicin, vancomycin and nystatin). Although the administration of prophylactic antibiotics systemically is contraindicated because it will result in the emergence of a highly antibiotic-resistant normal flora, prophylaxis with cotrimoxazole has been notably successful in the prevention of pneumocystis pneumonia in patients who are intensively immunosuppressed (Hughes *et al.* 1977). The epidemiology of pneumocystis infection is not clear. There are reports of person-to-person spread in patients being treated with cytotoxic drugs, but the alternative explanation that disease in those patients is the reactivation of latent infection brought about by their treatment is thought to be the cause of most cases (Jameson 1980). Shepherd *et al.* (1979) on the basis of a serological study deduced that many apparently healthy persons must have had mild undetected respiratory infection with *P. carinii*, possibly in early childhood.

Precautions to diminish the risk of exogenous infection Because some of the infections suffered by patients with immunodeficiency are exogenous, attempts have been made to prevent this route of infection. Treatment in a single room with reverse barrier nursing precautions is a relatively simple measure, but some advocate strict isolation with the use of laminar-flow beds and the consumption of sterilized food. *Aspergillus* spores are ubiquitous and, in an attempt to prevent pulmonary aspergillosis, the installation of sophisticated ventilation systems with air filtration has been recommended when the disease is common. Hospital epidemics have been reported (Rose 1972). Unusual outbreaks of pneumonia in renal transplant recipients due to aerosol-spread *Legionella pneumophila* (Tobin *et al.* 1980) and dustborne *Nocardia asteroides* (Houang *et al.* 1980) have also been reported.

A particular risk in organ recipients is acquisition of infection from the transplanted tissues. Many agents may be transmitted in this way but the best documented and most important is primary cytomegalovirus infection acquired by a sero-negative recipient from a sero-positive donor.

Acquired immunodeficiency syndrome The acquired immunodeficiency syndrome (AIDS) in which previously healthy homosexual men develop opportunistic infections, usually pneumocystis pneumonia, or Kaposi's sarcoma, was first reported from California and New York in 1981 (Lancet 1981). Since then there has been a minor epidemic of AIDS in the USA and cases have been reported in Europe, although it appears to be uncommon in the UK. Three-quarters of the patients are male homosexuals and most of the remainder are intravenous drug abusers. Of the rest, who are neither homosexual nor drug abusers, a few are haemophiliacs receiving factor VIII and a larger number Haitian immigrants.

The fully developed syndrome carries a high mortality: patients present with severe, often repeated, opportunistic infections or unusual rapidly progressive neoplasms, classically Kaposi's sarcoma, but other tumours such as Burkitt-like lymphomas have been reported. Neurological complications are important, the most common being encephalitis due to infection with either cytomegalovirus or *Toxoplasma gondii* (Gapen 1982). In the same population another milder syndrome of fever, weight loss and generalized lymphadenopathy has been recognized. Cell-mediated immunity is depressed but, in contrast, humoral immunity remains intact. Polyclonal hypergammaglobulinaemia may be present, perhaps as a result of B-cell stimulation by replicating virus (Fauci 1982). Evidence of infection with cytomegalovirus is almost always found in affected homosexuals. Although it has been suggested that cytomegalovirus infection may contribute to the cell-mediated immune depression it is not considered to be the cause of the syndrome. A causal agent has yet to be identified, but AIDS is apparently transmissible by intimate mucosal contact and parenterally by the intravenous route (Lancet 1983). Recent speculation has incriminated a retrovirus but full confirmation of its role in pathogenesis is still lacking.

References

Agnello, V. (1978) *Medicine* **57**, 1.
Aisner, J., Murillo, J., Schimpff, S. C. and Steere, A. C. (1979) *Ann. intern. Med.* **90**, 4.
Aldrich, R. A., Steinberg, A. G. and Campbell, D. C. (1954) *Paediatrics* **3**, 133.
Alper, C. A., Colten, H. R., Rosen, F. S., Rabson, A. R., Macnab, G. N. and Gear, J. S. S. (1972) *Lancet* **ii**, 1179.
Ammann, A. J. and Hong, R. (1970) *Clin. exp. Immunol.* **7**, 833.
Bach, F. H., Albertini, R. J., Joo, P., Anderson, J. L. and Bortin, M. M. (1968) *Lancet* **ii**, 1364.
Bach, J. F. (1975) *The Mode of Action of Immunosuppressive Agents*. Elsevier-North Holland, Amsterdam.
Baehner, R. L. and Boxer, L. A. (1979) In: *Inborn Errors of Immunity and Phagocytosis*, p. 201. Ed. by F. Güttler, J. W. T. Seakins and R. A. Harkness, MTP Press, Lancaster.
Berendes, H., Bridges, R. A. and Good, R. A. (1957) *Minn. Med.* **40**, 309.
Behan, P. O., Simpson, J. A. and Behan, W. M. H. (1976) *Lancet* **i**, 593.
Biggar, W. D., Park, B. H., Stutman, O., Gaji-Peczalska, K. and Good, R. A. (1975) In: *Immunodeficiency in Man and Animals*, p. 361. Ed. by D. Bergsma, R. A., Good, J. Finstad and N. W. Paul, Sinauer Press, Sunderland.
Blume, R. S. and Wolff, S. M. (1972) *Medicine* **51**, 247.
Borel, J. F., Feurer, C., Gubler, H. U. and Stoehelin, H. (1976) *Ag. Actions* **6**, 468.
Boxer, L. A., Watanabe, A. M., Rister, M., Besch, H. R. Jr., Allen, J. and Baehner, R. L. (1976) *New Engl. J. Med.* **295**, 1041.
Bruton, O. C. (1952) *Paediatrics* **9**, 722.
Buckley, R. H. (1977) In: *Recent Advances in Clinical Immunology*, No. 1, p. 219. Ed. by R. A. Thompson, Churchill Livingstone, Edinburgh.
Buckley, R. H., Lucas, Z. J., Hattler, B. G., Zmijewski, C. M. and Amos, D. B. (1968) *Clin. exp. Immunol.* **3**, 153.
Calne, R. Y. C., White, D. J. G. and Thiru, S. (1979) *Lancet* **i**, 1033.
Campbell, A. C. et al. (1976) *Clin. exp. Immunol.* **24**, 249.
Claman, H. N. (1972) *New Engl. J. Med.* **287**, 388.
Crawford, D. H. and Edwards, J. M. B. (1982) *Lancet* **i**, 1469.
Dougherty, T. F., Berliner, M. L., Schneebeli, G. L. and Berliner, D. L. (1964) *Ann. N.Y. Acad. Sci.* **113**, 825.
Dupont, B., O'Reilly, R. J., Jersild, C. and Good, R. A. (1974) In: *Progress in Immunology II*, Vol. 5, p. 203. Ed. by L. Brent and J. Holborow, North-Holland, Amsterdam.
Edwards, N. L., Gelfand, E. W., Biggar, D. and Fox, I. H. (1978) *J. Lab. clin. Med.* **91**, 736.

Ernström, U. and Larssön, B. (1967) *Acta path. microbiol. scand.* **70**, 371.
Fauci, A. S. (1982) *Ann. intern. Med.* **96**, 777.
Fauci, A. S., Dale, D. C. and Balow, J. E. (1976) *Ann. intern. Med.* **84**, 304.
Feldman, G., Koziner, B., Talamo, R. and Bloch, K. G. (1975) *J. Pediatr.* **87**, 534.
Gabrielsen, A. E. and Good, R. A. (1967) *Advanc. Immunol.* **6**, 91.
Gapen, P. (1982) *J. Amer. med. Ass.* **248**, 2941.
Gatti, R. A., Meuwissen, H. J., Allen, H. D., Hong, R. and Good, R. A. (1968) *Lancet* **ii**, 1366.
Giblett, E., Anderson, J., Cohen, F., Pollara, B. and Meuwissen, H. J. (1972) *Lancet* **ii**, 1067.
Hayward, A. R. (1979) In: *Immunological aspects of infectious diseases*, p. 151. Ed. G. Dick, MTP Press Lancaster; (1982) In: *Clinical aspects of Immunology*, 4th edn., p. 1658. Ed, by P. J. Lachmann and D. K. Peters, Blackwell Scientific Publications, Oxford.
Hayward, A. R. and Greaves, M. F. (1975) *Clin. Immunol. Immunopathol.* **3**, 461.
Hitzig, W. H. (1973) In: *Immunologic Disorders in infants and children*, Ed. by E. R. Stiehm and V. Fulginiti. W. B. Saunders, Philadelphia.
Hobbs, J. R. (1975) In: *Immunodeficiency in Man and other Animals*. Ed. by D. Bergsma, R. A., Good, J. Finstad and N. W. Paul, Sinauer Press, Sunderland.
Houang, E. T., Lovett, I. S., Thompson, F. D., Harrison, A. R., Joekes, A. M. and Goodfellow, M. (1980) *J. Hosp. Infect.* **1**, 31.
Hughes, W. T. et al. (1977) *New. Eng. J. Med.* **297**, 1419.
Jameson, B. (1980) *J. Hosp. Infect.* **1**, 103.
Johnston, R. B. and Newman, S. L. (1977) *Ped. Clin. N. Amer.* **24**, 365.
Kaplan, E. L., Laxoal, T. and Quie, P. G. (1968) *Paediatrics* **41**, 591.
Knapp, W. and Posch, B. (1980) *J. Immunol.* **124**, 168.
Kramer, B. S., Pizzo, P. A., Robichaud, K. J., Witesbsky, F. and Wesley, R. (1982) *Amer. J. Med.* **72**, 561.
Lachmann, P. J. (1982) In: *Clinical Aspects of Immunology*, 4th edn. p. 36. Ed. by P. J. Lachmann and D. K. Peters, Blackwell Scientific Publications.
Lancet (Editorial) (1981) **ii**, 1325; (1983) *Ibid.*, 162.
Lawlor, G. J., Ammann, A. J., Wright, W. C., Lafranchi, S. H., Bilstrom, D. and Stiehm, R. (1974) *J. Paediatr.* **84**, 183.
Levy, R. L. et al. (1971) *Lancet* **ii**, 896.
Mackay, I. R., Dwyer, J. M. and Rowley, M. J. (1973) *Arth. and Rheum.* **16**, 455.
Miller, C. L. and Baker, C. C. (1979a) *Trans. Proc.* **11**, 1460; (1979b) *J. clin. Invest.* **63**, 202.
MRC Report. (1971) In: *Special Report Series No. 310*. HMSO, London.
Narayan, O., Penney, J. B. and Johnson, R. T. (1973) *New Eng. J. Med.* **289**, 1278.
Nezelof, C., Jammet, M. L., Lortholary, P., Labrune, B. and Lamy, M. (1964) *Arch. fr. Pediat.* **21**, 897.
Ninnemann, J. L. (1981) *The Immune Consequences of Thermal Injury*. Williams & Wilkins, Baltimore.
Ochs, H. D., Davis, S. D., Mickelson, E., Lerner, K. G. and Wedgwood, R. J. (1974) *J. Pediatr.* **85**, 463.
Oliver, J. M., Zurier, R. B. and Berlin, R. D. (1975) *Nature (Lond.)* **253**, 471.

Petersen, B. H., Lee, T. J., Snyderman, R. and Brooks, G. F. (1979) *Ann. intern. Med.* **90,** 917.
Peterson, R. D. A., Cooper, M. D. and Good, R. A. (1966) *Amer. J. Med.* **41,** 342.
Powles, R. L. *et al.* (1980) *Lancet* **i,** 327.
Report of European Multicentre Trial. (1982) *Lancet* **ii,** 57.
Rose, H. D. (1972) *Amer. Rev. resp. Dis.* **105,** 306.
Rowley, M. J., Mackay, I. R. and Mackenzie, I. F. C. (1969) *Lancet* **ii,** 708.
Santos, G. W., Sensenbrenner, L. L., Burke, P. J., Colvin, M. and Owens, A. H. (1971) *Transplant. Proc.* **3,** 400.
Schimpff, S. C., Young, V. M. and Greene, W. H. (1972) *Ann. intern. Med.* **77,** 707.
Segal, A. W. (1979) In: *Inborn Errors of Immunity and Phagocytosis*, p. 247. Ed. by F. Güttler, J. W. T. Seakins and R. A. Harkness, MTP Press, Lancaster; (1981) In: *Clinics in Immunology and Allergy*, Vol. 1, p. 581. Ed. by A. D. B. Webster, W. B. Saunders, Philadelphia and Eastbourne.
Segal, A. W. and Jones, O. T. G. (1980) *FEBS Letters* **110,** 111.
Shepherd, V., Jameson, B. and Knowles, G. K. (1979) *J. clin. Path.* **32,** 773.
Siegal, F. P., Siegal, M. and Good, R. A. (1978) *New Eng. J. Med.* **299,** 172.
Steinherz, P. G. *et al.* (1980) *Cancer* **45,** 750.
Storring, R. A., Jameson, B. and McElwain, T. J. (1977) *Lancet* **ii,** 837.
Thompson, R. A. and Lachmann, P. J. (1977) *Clin. exp. Immunol.* **27,** 23.
Tobin, J. O'H. *et al.* (1980) *Lancet* **ii,** 118.
Vallera, D. A., Pflugfelder, U. and Schmidtke, J. R. (1980) *J. Immunol.* **124,** 635.
Valdimarsson, H., Wood, C. B. S., Hobbs, J. R. and Holt, P. J. L. (1972) *Clin. exp. Immunol.* **11,** 151.
Whittingham, S., Buckley, J. D. and Mackay, I. R. (1978) *Clin. exp. Immunol.* **34,** 170.
Williams, D. M., Krick, J. A. and Remington, J. S. (1976) *Amer. Rev. resp. Dis.* **114,** 359.
Zeitz, H. J., Miller, G. W., Lint, T. F., Ali, M. A. and Gewurz, H. (1981) *Arth. Rheum.* **24,** 87.

18

Herd infection and herd immunity
Ashley Miles

Herd structure in relation to herd immunity	413	Effect of variation in the epidemic strain on the course of herd infection	422
The epidemiology of some microbial infections	414	Bacterial infections	423
Herd immunity in diphtheria	414	Variation in viral infections in mice	424
Herd immunity in scarlet fever	416	Effect of changes in herd structure on herd infection and herd immunity	424
Herd immunity in other microbial diseases	416	Continuous contact	424
Heterogeneity of herds at risk	417	Discontinuous contact	425
Genetics and herd structure	418	Dispersal of a closed herd	425
The experimental study of herd infection and herd immunity	419	Dispersal of an increasing herd	425
Spread of infection and natural immunization in mice	419	Herd infection and herd immunity in nature	425
		The control of herd infection	426
Effect of artificial immunization on spread of infection in mice	421	Quarantine	426
		Isolation	426

Introduction

In the preceding chapters our host unit has been the individual man or animal. For the study of the epidemic spread of infection, and the factors that favour or prevent it, it is with the larger unit, the herd, that we must deal.

As we relate in Chapter 48, epidemiology may be considered from a variety of standpoints, ranging from the rigorously mathematical to general social studies. For our immediate purpose we shall adopt the definition of epidemiology proposed by Evans (1979) after a survey of over twenty definitions offered during the last half-century. 'Epidemiology is the quantitative analysis of the circumstances under which disease processes, including trauma, occur in population groups, the factors affecting their incidence, distribution and host responses, and the use of this knowledge in prevention and control.' In this chapter we consider the analysis of immunity as a characteristic of communities as a whole. This herd immunity may be considered as being dependent on resistance to attack by a disease to which a large population of the individual members are immune, thus lessening the likelihood of an individual with the disease coming into contact with a susceptible individual.

Herd structure in relation to herd immunity

The herd, like each of its members, has a characteristic structure; and this structure, from our present point of view, includes not only the hosts belonging to the herd species, and their spatial relationships to one another, but the presence and distribution of alternative animal hosts and possible insect vectors of infection, as well as all those environmental factors that promote or inhibit the spread of infection from host to host. This herd structure, apart altogether from the susceptibility or resistance of the individual hosts, may play a decisive part in the immunity of the herd as such. A herd may be free, or almost free, from a particular disease although each of its members is fully susceptible, and would fall an easy victim if he strayed to a herd with a structure that allowed an endemic prevalence of the disease in question. There is for example in the British herd no such prevalence of plague because the association of man, the rat and the flea is not now of a kind to allow spread of the bacillus. It is likewise with cholera, as the result of an adequate system of water purification; much in the way that as a result of quarantine measures, the British dog is free from rabies, and British cattle, though very much less effectively, from foot-and-mouth disease. The human herd in Britain is not, and does not seem likely to become, immune to any of those diseases that are disseminated from the upper respiratory tract and acquired by inhalation. It would take us altogether beyond our present scope to consider the known or problematical effects on herd resistance of such changes in environmental conditions; but we may at least note that many of the most striking successes of preventive medicine have been attained by altering herd structure without inducing any increased resistance in its individual members. By attacking insect vectors of infection such as the mosquito, by preventing the frequent passage of bacteria from one person's intestine to another person's mouth by way of water and food, and by a general improvement in environmental conditions, we have succeeded in eliminating, or reducing to negligible proportions, diseases that formerly took a heavy toll of lives, and still take that toll in areas where such measures are not applied.

The type of herd immunity with which we are here concerned is that in which this freedom from the spread of infection has not yet been attained, so that contact with the bacterial parasite is at least an occasional event in the normal experience of the host species. Under such conditions the course of events in any infected herd will be determined mainly by the distribution of the parasite and the distribution of persons susceptible to the diseases in question, which in turn may be modified by persons with an innate or acquired immunity, both specific and non-specific. In practice, any decrease in the prevalence of an infective disease is largely the end result of specific immunization, whether it is natural or artificial. Intelligent interference with the course of events is impossible without a clear idea of what is actually happening.

We may recognize, in theory, at least six categories of hosts among any infected herd. Four of these categories include individuals who are themselves infected: (1) the typical case, (2) the atypical case—usually a milder and less lethal form of the disease, (3) the latent infection, and (4) the healthy carrier; the other two categories are not themselves infected, or infective, (5) the uninfected immune and (6) the uninfected susceptible. The division between (2), (3) and (4) is formal rather than actual—these conditions shade into each other by imperceptible degrees. It may be noted that the individuals falling in categories (3) and (4) are in general, though in a very varying degree, resistant to further infection from without, having made some degree of immune response to the infecting organism. The only fully susceptible hosts at risk are those in class (6).

Fig. 18.1 illustrates the kind of distribution, both of infection and of immunity, that may be met with in infected herds under different epidemic conditions. The schemes are deceptively simple in that all are based on the assumption of a homogeneous randomly mixing population whose individuals differ only in their reaction to the infecting agent; whereas in real life, open populations are made up of innumerable definable but often interlocking subgroups, which differ in respect to proportions of immune and intimacy of contact (Fox *et al.* 1971). No distinction is made between latent infections and healthy carriers and between typical and atypical cases. The arrows indicate the direction of effective spread—effective in the sense of producing new cases of disease or in converting susceptibles to immunes. As we shall see, an epidemic of an infective disease is usually accompanied by an epidemic of symptomless immunization.

A is an example of an epidemic phase in an endemic-epidemic prevalence, that is, during an outbreak of an infective disease from which the affected herd is never completely free, epidemics of varying severity recurring at intervals. B may be regarded as a later stage of A, or as a small epidemic wave occurring in a herd in which susceptibles are few, and carriers are frequent. C represents a severe epidemic occurring in a herd with little initial immunity. An extreme example of this catastrophic type of prevalence occurs with the introduction of a disease like measles into an isolated island community that has either never experienced the infection before, or has been free from it for many years. D represents a stage of relative quiescence between two outbreaks of the type depicted in A. It will be noted that the proportion of susceptibles is higher than in A or B, and with such a distribution as this a fresh outbreak of the A type is likely to occur.

The epidemiology of some microbial infections

Herd immunity in diphtheria

Two infections associated with an exotoxin—diphtheria and scarlet fever—are no longer prevalent in the UK but as subjects of extensive epidemiological study in the past inducing a readily measurable immune response, they serve as excellent material for discussion.

The epidemiology of diphtheria in the United Kingdom, as studied during the last few decades before it virtually disappeared as a clinical disease, exemplifies the conditions pictured in A, B and D. The disease is essentially a toxaemia; and an effective antitoxic immunity will protect the host against clinically detectable infection. By means of the schick test we can divide the members of any herd into susceptibles and immunes. Except in very young children a negative reaction indicates that an individual has over 0.01 AU of antitoxin per ml of circulating blood, and that he will produce further antitoxin briskly and effectively in response to any entry of toxin into his tissues. By swabbing throats and noses, and testing the virulence of any morphologically typical diphtheria bacilli cultivated, we can determine with some approach of accuracy the distribution of the parasite among the hosts at risk; and we are thus in a position to describe the course of events in a community in which diphtheria occurs.

The diphtheria bacillus is isolable not only from those suffering from the clinical disease in its typical form. It is frequently isolated from cases of mild sore throat associated with an epidemic of typical diphtheria, less frequently from healthy contacts, and still less frequently from non-contacts.

Kober (1899) isolated diphtheria bacilli from 70 per cent of 139 contacts suffering from mild sore throat, and from 8 per cent of 123 contacts with apparently normal throats. Closeness and continuity of contact, here as in other diseases, have a considerable influence on the carrier rate. A carrier rate of virulent diphtheria bacilli was found among 610 contacts in barracks or hospital wards of 15 per cent, 7 per cent among 10 883 home contacts, and 0.6 per cent among the general non-contact population (Monograph 1923).

The response to virulent diphtheria bacilli by a non-infected person will depend on his immunological condition. Susceptible (schick-positive) persons will respond by producing antitoxin and becoming immune (schick-negative), or get clinical diphtheria. Hence schick-positive carriers of virulent diphtheria bacilli are very rare. We may take it that a person carrying virulent bacilli in sufficient numbers to be detectable by an ordinary swabbing is (a) immune (schick-negative), or (b) undergoing rapid immunization to the schick-negative level, or (c) incubating the disease.

If, then, we take a relatively isolated community, such as a large boys' school, and trace the spread of infection and the development of immunity during an epidemic of diphtheria, we shall observe the following changes of immunological status in a population about 60 per cent immunes, including a few per cent of carriers, and 40 per cent susceptibles.

Among these boys there will be a certain number of carriers of virulent diphtheria bacilli; these bacilli may be acquired by a susceptible or an immune; and, depending on the number acquired and the degree of immunity, they may alter the status of the uninfected boys as follows.

Frequently:
(a) schick-positive susceptible→case (mild or severe).
(b) schick-positive susceptible→schick-negative immune
    (a) without detectable carrying,
    (b) with detectable carrying.
(c) schick-negative immune→schick-negative immune carrier.

Very rarely:
(d) schick-positive susceptible→schick-positive carrier.
(e) schick-negative immune→mild 'bacteriological' case.

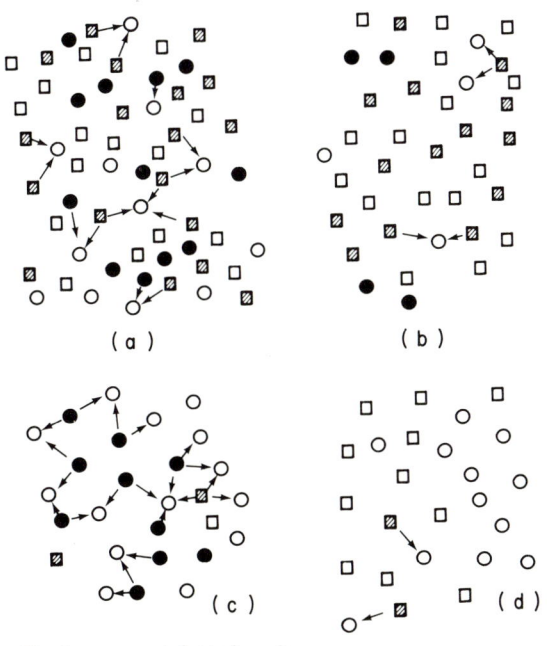

□ Immune uninfected
▨ Latent infection or healthy carrier (usually immune)
● Clinical case. Typical and atypical
○ Susceptible, uninfected
→ Direction of effective spread

Fig. 18.1

As an example we may cite the studies of Dudley (1923, 1926, 1932) on diphtheria prevalence in the Royal Naval School at Greenwich, a school of some 1000 boys during 1919–22. There was one epidemic in early 1919 and another in the second half of 1921. The results (Table 18.1) indicate a clear relation between experience and immunity, the immunity rate varying according to the boys' length of stay in

Table 18.1 Showing relation between Schick reaction and previous experience among boys tested in 1922

Previous residence in school	Experienced 1919 epidemic	Experienced 1921 epidemic	Per cent immune (schick-negative)
None (new boys)	No	No	58
6 months–2 years	No	Yes	85
Over 3 years	Yes	Yes	95

the school. Of entrants to the school before the first epidemic, between 98 and 100 per cent had become schick-negative by 1922; among entrants who had experienced only the second epidemic, the schick-negative rates lay between 82 and 90 per cent; whereas among boys joining after the second epidemic the percentage of immunes was 58. There had apparently been two peaks of immunization in the school, coinciding with the peaks of clinical infection.

During the later part of the 1921 epidemic there were 28 carriers of virulent diphtheria bacilli. In the first 6 months after the outbreak the carrier rate was 2.2 per cent, rising in the second 6 months to 4.5 per cent; and during this period cases of diphtheria were still occurring. But immunization of susceptibles was also taking place. Of 88 schick-positive boys, 24 were negative 3 months later. In subsequent years, the carrier rates varied between 1.0 and 4.5 per cent, with sporadic recurrence of clinical diphtheria; and here again there was a coincident natural immunization of susceptibles.

The general significance of such studies is clear enough; but we must remember that the two categories—immunes and susceptibles—that are divided from one another by simple schick-testing present a very incomplete picture of the graduations in resistance that actually exist in a herd at risk. Glenny (1925) defines five grades of immunity, the most immune of which are schick-negative and two of which are schick-positive but either immune or readily immunizable; the fourth is positive and sub-immune and the fifth is positive and fully susceptible.

A point of major interest is the reaction of schick-negative persons to the acquisition of virulent diphtheria bacilli, who are less liable to develop clinical diphtheria than their schick-positive companions. When they are exposed to highly virulent strains of diphtheria bacilli, they may contract the disease, though usually it is mild (see Hartley, Tulloch et al. 1950 and Chapter 53). Many of them become symptomless carriers. Is a point reached in natural immunization at which a person resists carrier infection? When a human herd is rendered 100 per cent schick-negative, either by natural immunization or by prophylactic inoculation, will infection with the diphtheria bacillus be eliminated in the bacteriological sense, or will a freedom from clinical diphtheria be associated with a persistent carrier rate not much below that of the population at large, and perhaps with carrier epidemics? Is complete immunity, of the kind that would render a herd entirely free from infection, likely to be attained by immunization of this kind, however far it is pushed?

These questions have been answered to some extent. Wherever extensive and continuous immunization has been practised, the carrier rate among the normal population has ultimately fallen. Once a sufficiently high proportion of persons in a community are immunized—naturally or artificially—and kept immune, diphtheria in both its clinical and its latent form will gradually disappear. Diphtheria immunization eliminates the disease from a given herd, not, presumably, by making each individual wholly resistant to infection but by decreasing the number of fully susceptible members to the point at which the diphtheria bacillus cannot maintain a continued existence within the community. These issues in relation to tuberculosis and diphtheria are more fully discussed in Chapters 51 and 53, and in relation to the biology of epidemics as a whole in a paper by Topley (1942).

Herd immunity in scarlet fever

Scarlet fever is another disease in which, by the aid of the Dick test and the examination of throat swabs for haemolytic streptococci, we can obtain information of the same kind as that with diphtheria. The production of the erythrogenic toxin forms a relatively small part of the pathogenic potentialities of haemolytic streptococci; and an antitoxic immunity that will suffice to prevent the clinical syndrome of scarlet fever will not suffice to prevent the spread of tonsillitis or other streptococcal infections (see Okell 1932). For effective immunity against all the clinical manifestations of infection with haemolytic streptococci we need an antibacterial as well as an antitoxic immunity; and this antibacterial immunity will almost certainly be type-specific. (For an alternative view, see Chapter 59.)

Herd immunity in other microbial diseases

With bacterial diseases in which antitoxic immunity plays no part, we have no simple tests, like the Schick and Dick tests, by which to determine immunes and susceptibles. There are measures like the estimation of circulating specific antibody. Antibody, however, is not an unequivocal index of immunity. Even when it is known to be protective, we are with man in no position to establish by direct test any relation between antibody concentration and resistance, as determined by response to known doses of the infective agent. When specific immunity is predominantly cell-mediated, the exaggerated hypersensitivity reaction of a tissue like the dermis to the pathogen in question, or its antigens, is used in a number of diseases as an index of immunity, as is the reaction of these antigens with host cell preparations *in vitro*; indeed they are often regarded as synonymous with immunity. It cannot be too strongly emphasized however that the demonstration of reactivity of this kind does not *per se* necessarily signify a potentially protective reactivity.

Some general deductions however can be made. Thus, in certain diseases, such as cerebrospinal meningitis, the ratio of carriers to cases is very high (see for instance Glover 1918), so that it is probable that the immunes greatly outnumber the susceptibles. In other diseases, such as typhoid fever, the ratio of carriers to cases is not so high (see Chapter 68). But atypical cases of typhoid fever occur during an epidemic, and many carriers of typhoid bacilli give no history of ever having suffered from a typhoid-like disease. It is a fairly safe assumption that an endemic or epidemic prevalence of typhoid fever is associated with the occurrence of sub-clinical immunizing infections. Such a prevalence leaves behind it infected healthy carriers who may be a potent source of further spread.

The same general relationships hold for virus diseases. Immunity of the individual is a direct result of infection, either inapparent or clinically recognizable. In certain diseases, such as poliomyelitis or rubella, there may be many more infections than clinical cases; in others, such as smallpox or measles, most of the infections result in clinically manifest disease. In either instance a decrease in the number of susceptibles, by making the spread of virus less easy, tends towards a stage at which the infection dies out. The rate at which this happens will depend partly on the period during which virus is broadcast from infected to healthy persons. With many virus infections, the diseased patient is maximally infective for others at the end of the incubation period and at the onset of illness; and susceptibles are at risk for relatively short periods. With others, like poliomyelitis, the infected individual, whether ill or a healthy carrier, excretes infective virus for long periods; and sporadic infection of susceptibles is thus likely to continue for longer periods.

Heterogeneity of herds at risk

The various epidemic circumstances illustrated in Fig. 18.1 were based on the theoretical existence of at least six categories of reactor to the infecting agent, categories ranging from immune to susceptible, uninfected to infected hosts. It will be clear from the few examples of epidemics cited above that this assumption of herds, the resistance of whose members form a graded series from nil to total, is often unjustified.

The possibility that an apparently homogeneous herd may contain subgroups of substantially different reactivity—i.e. the herd is grossly heterogeneous as regards, say, structure and environmental conditions, or in innate immunological constitution—is clearly important in any epidemiological analysis. It is equally important, as related in Chapter 19, in immunological measurement involving animals, where accuracy is dependent on the degree of homogeneity that can be assigned to the test populations used. In this chapter we are particularly concerned with the effect first of heterogeneity of epidemiological conditions, and second of discontinuous variations of innate resistance in populations at risk.

The first is usually exemplified in mathematical models that have been devised to describe the course of epidemics of infection. For example, using mathematical models based on that of Reed and Frost (see Serfling 1952, Abbey 1952) in which it is postulated that the progress of an epidemic is determined when the initial set of conditions is established, Fox and his colleagues (1971) conclude that this progress is primarily regulated by the number of susceptibles and the rate of contact between infectious and susceptible persons. If these are held constant, changes in population size, and therefore in the proportion of immunes, does not influence the probability of spread.

In a series of 100 computer simulations of epidemic spread under various conditions from one infected among 100 susceptible children, with the assumption of no contacts but those occurring in a randomly mixing (i.e. a homogeneous) population, the epidemic's size was limited to a median value of 1, with a mean number of 1.2 case per epidemic; in 82 per cent of the trials there was no epidemic. When heterogeneity was introduced in the form of added high contact rates within families, the median epidemic size rose to 3, and mean number of cases to 3.3. With added contacts in play groups outside the family, the corresponding figures were 4 and 5.6; when contacts in a nursery attended by 40 of the 100 susceptibles were added, the median size of the epidemics was 58 with a mean number of 45 cases; and in only 23 per cent of the trials was there no epidemic.

One conclusion of importance is that, with a fixed number of susceptibles the epidemic potential increases with opportunity for contact between infected and susceptibles; and that the proportion of immunes is not *per se* a deciding factor. A simple relation between increasing the proportion of immunes and decreasing probability of spread is valid only for randomly mixing populations, which in nature occur only in small closed communities. The typical community is to be regarded as heterogeneous, made up of many interrelated subgroups—families, living conditions, schools, places of works, economic status, race and so forth—of different sizes, proportions of susceptibles and frequency of contact of susceptibles. Consequently, the occurrence of infections is not uniform; and an assumption of uniformity may well invalidate any attempt to specify the proportion of the community as a whole that should be immunized to ensure a decrease in the number of susceptibles sufficient to prevent outbreaks of an infection.

The best immunization programme must aim at decreasing the supply of susceptibles in all definable subgroups so as to leave no group with enough susceptibles for a high epidemic potential. The identification of such high risk subgroups is an important part of such programmes; as is the aim, for example, of immunization in the U.S. against poliomyelitis and measles. Alternatively it may be more effective to decrease the probability of effective exposure, as when foci of smallpox infection are contained by intensive vaccination.

The hypotheses upon which models of this kind are based may be manipulated to accommodate a wide variety of observed features of epidemics. It may be shown, for example, given a sufficient number of susceptibles, that when the probability of exposure is high epidemics are short and explosive; when low, not only is the epidemic long lasting, but many susceptibles remain when it is finished.

The underlying hypotheses are clearly invalid when a model fails to correspond with observation; but a good correspondence, it is to be emphasized, is not evidence of their validity—i.e. that the postulated mechanisms are operative. As Lilienfeld and Lilienfeld (1980) emphasize in their discussion of the subject, mathematical models may indicate factors determining epidemic spread and their interrelations and methods of exploiting them. Models should be judged mainly by their ability to change the concepts or methods in epidemiological disciplines; and little of this nature has as yet emerged.

Genetics and herd structure

The analysis of the epidemiological disposition to infection in a community in terms of specific immunity, as in the time-honoured classification into fully immune, partially immune and susceptibles, must clearly include innate differences in resistance. We are concerned mainly with the heterogeneity that might divide populations at risk into subgroups that differ substantially in innate resistance. But it will be useful first to indicate what is known of innate resistance in general. Innate resistance and its heritability has been explored rather haphazardly over the last forty years. Most of the evidence in animals concerns the selection by inbreeding of strains of differing resistance to given infections or of differing immunological responsiveness; and of differing efficiencies of non-specific mechanisms, like the bactericidal action of complement. Deliberate inbreeding is not practicable in man; and analysis of inheritance is accordingly restricted. The relevance of this kind of evidence to herd immunity lies in the demonstration of discontinuous heterogeneity within the herd. When susceptibility or responsiveness varies continuously presumably many genes are involved and there is no clear-cut Mendelian segregation among the progeny. When single factors are involved susceptibility may have a bi- or tri-modal distribution, and resistance segregates among the progeny of mating between resistance and susceptibility (Allison 1965). A similar discontinuity is inferred in man from the association of response with a marker like an HLA histocompatibility antigen possessed by only part of a community.

Selective inbreeding in mice for susceptibility to a wide variety of pathogenic bacteria, viruses and both protozoan and metazoan parasites indicates a large number of genes controlling the immune responses.

Thus Webster (1933) bred one strain of mice relatively susceptible to, and another relatively resistant to *Past. septica*, *Klebsiella aerogenes* and the pneumococcus. He later (1937) obtained three strains, one resistant to *Salm. typhimurium* but not the virus of St Louis encephalitis, a second resistant to the virus but not the bacterium, and the third resistant to both; in accord with the hypothesis that two distinct sets of factors regulated the inheritance of susceptibility. Genetic differences are demonstrable too, in terms of antibody production and cell-mediated immune responses; and in selective capacity to respond to certain antigens. In some instances, the genetic determination is precise, as in Odaka's (1969) demonstration of a single autosomal gene for susceptibility to the Freund leukaemia virus. Again, a single dominant gene within the Ir region determines the response of inbred guinea-pigs to a family of synthetic poly-L-lysine conjugates (Levine and Benacerraf 1964); and Ir autosomal genes in the mouse linked to the H-2 histocompatibility loci likewise determine the responses to certain multichain polymers of amino acids (Sela 1966, Dorf 1981).

A strain of mice susceptible to *Salm. typhimurium* inbred by Robson and Vas (1972), in comparison with a strain resistant to the organism, was not immunizable by vaccine. Resistance to the salmonella proves to be under the control of a gene, *Ity* (Plant and Glynn 1979). It is of interest that this locus may be identical with *Lsh*, controlling resistance of the mouse to an unrelated pathogen, *Leishmania donovanii*; both are near the H-2 histocompatibility locus (Bradley 1982). With *L. tropica*, on the other hand, a gene independent of the H-2 locus appears to determine a susceptibility in which a population of T-suppressor cells develops and macrophages fail to be activated (Howard *et al* 1982). (For further references to earlier work on the genetics of resistance see Chapter 41 of the sixth edn and 16 of the present edn.)

Susceptibility to schistosomiasis is associated with certain HLA haplotypes in man, and with several non-H2 genes in mice (see Fanning *et al*. 1981). Haverkorn and colleagues (1975) describe possible linkage between HLA Cw3 and antibody production after attenuated measles vaccine. Coovadia and colleagues (1981) provide cogent evidence of a distinct subgroup of differing responsiveness in their analysis of the link between HLA and various indices of humoral and cell-mediated immunity in measles. On the average about 9 per cent of the community studied constituted a subgroup, associated with HLA Aw32, tending to severe measles characterized by lymphocytopaenia. It is to be noted however, that the clinical outcome, including survival, was not HLA-linked.

The association of the HLA complex and susceptibility is reviewed by Zinkernagel (1979). As examples of associations with the immune responses to various microbial antigens we may note those with response to streptococcal antigen (Greenberg *et al*. 1975); to various viral infections (McDevitt *et al*. 1974); to *Myco. leprae* (de Vries *et al*. 1981); to live influenza vaccine (Spencer *et al*. 1976); to tetanus toxoid (Sasazuki *et al*. 1978) where a low response was associated with HLA B5; and to vaccinia virus, where a low immune response was associated with HLA Cw3 (de Vries *et al*. 1977). In this last instance there was in the population studied evidence of bimodality late in the responses.

The implied immunological heterogeneity of populations in which these associations are demonstrable, whether it is of susceptibility to infection or response to infection and immunizing agents, is clearly important in any epidemiological context. But it is also important in the first place to realize that most of the genes reported as determining a response emerge only as a result of inbreeding or other genetic manipulations. In the second, association with a histocompatibility complex—however striking, is seldom 100 per cent, and there may be substantial degrees of association with other complexes; and interaction between a complex and genes controlling some immune process is usually a matter of conjecture. In the third place the associations, in man particularly, are often restricted to *in vitro* reactions which do not unequivocally indicate immunity—as distinct from an 'immune' responsiveness—or are significant only in interreaction with unknown factors.

In summary, there is obvious immunological heterogeneity in both the human and animal herds studied, but in only a few instances are subgroups distinguishable with substantially different responses from the main herd. There is little epidemiological justification for treating the remainder other than as examples of herds varying continuously in the individual immunological disposition.

The experimental study of herd infection and immunity

It is clearly possible, by selecting a convenient host species and a parasite that spreads naturally among them, to submit problems of the kind we have considered above to direct experimental study—initiating an epidemic of a particular disease among our test population, and studying the reactions of new entrants, of old members of the herd, or of migrants from one herd to another, by any available means (see Greenwood 1935; Greenwood *et al.* 1936). We can thus observe the effects of any intentional interference we choose, splitting a herd into smaller units, reaggregating these units after any selected interval, varying the rate of immigration of susceptibles, actively immunizing some or all of our animals by vaccines of different kinds administered by different routes, and so on, as new problems suggest themselves for attack. In this way we gain enormously in an increased control over many of our unknown variables and may if desired approach a simulation of the model of a randomly mixing population; but we lose, in so far as we wish to argue from our experimental herds to happenings in the natural world outside our cages, by having to rely on analogies that are certainly incomplete, and may well be misleading. Because our control is still far from complete we must use large numbers of animals and make constant use of statistical methods to assess the meaning of our results. We must, therefore, use a small, easily controlled and relatively inexpensive animal, and the mouse fulfils these requirements. Experiments of this kind were carried out during a number of years, in England and the United States, on mouse typhoid—caused by *Salm. typhimurium* or *Salm. enteritidis*—on mouse pasteurellosis, and on the virus disease ectromelia. Further investigations on ectromelia (mouse-pox) were also made in Australia. In addition, the American workers studied the spread of infection among rabbits and fowls housed in laboratories, or maintained as breeding stock by dealers. It will be convenient to consider the results thus obtained in direct relation to the observations on human herds, noting the points at which particular caution must be exercised in applying the argument from analogy and in generalizing from an isolated experience.

Is there, in the spread of infection among mice, the same distribution of typical cases, atypical cases, latent infections and carriers, and the same natural immunization associated with natural infection, that occurs among human herds?

In mouse typhoid—and almost certainly in other epidemic infections—it is quite certain that most of the mice at risk are infected by the bacterium at some period or other during a prolonged epidemic prevalence.

In a particular experiment (Topley, Ayrton and Lewis 1924) 5 mice were fed on a culture of *Salm. typhimurium*. As soon as they were found to be excreting the bacillus in their faeces, 20 normal mice were added, and thereafter 1 normal mouse was added to the cage each day for 115 days. The faeces of every mouse were examined on 6 days a week; in each mouse that died, and in all survivors (55 in number), the spleen was examined for *Salm. typhimurium* and the blood for agglutinins. The existence of non-fatal, and apparently mild or trivial, infections was clearly demonstrated. Some mice that excreted *Salm. typhimurium* on several occasions remained in apparent health during the whole period of their residence in the cage. At the end of the experiment some had negative spleen cultures and no agglutinins in their blood, others had agglutinins and negative cultures and others positive spleen cultures with or without serum agglutinins. With death from the infective disease, and in surviving mice, either presence of the organism in the faeces or spleen, or of agglutinins in the blood, as criteria of infection, of 128 mice which had resided in the cage for periods of 14 to 115 days, 122 proved to have become infected. Of these infected mice 46 were surviving in apparent health on the day when the experiment terminated, but 25 of them (54.3 per cent) had *Salm. typhimurium* in their spleen. The collected figures for eleven different epidemics of mouse typhoid, lasting from 60 to 117 days, showed that of 267 surviving mice 190 (71.2 per cent) were latently infected (Topley 1926).

Similarly (Greenwood et al. 1936) with ectromelia, some 80 per cent of the mice entering the herd were infected with the virus within three weeks of entry (see also Fenner 1949b).

The distribution of the bacteria among the hosts at risk is then of the same general kind as obtains in natural epidemics in man, differing mainly in the fact that fewer hosts escape infection. Is there any evidence that the spread of infection results in immunization as well as, or in place of, the production of overt disease? Survivors from an epidemic are, on the average, more resistant than newcomers to an infected herd; they live longer—usually much longer—during a subsequent period of high mortality (Topley 1921, Amoss 1922).

This increase in resistance with increasing herd experience under epidemic conditions can be studied in greater detail by constructing life tables for the mice submitted to risk of infection during a long continued prevalence of such a disease as mouse typhoid. It is convenient in calculating the expectation of life to limit it to 60 days.

Employing the value ($_{60}E_x$) we find, in an experiment in which three normal mice were added to an infected herd each day for 116 days, that the expectation of life then rises, at first slowly, but more steeply after the 15th day, until by the 33rd day it reaches the figure of 34.75 days, and fluctuates round this figure up to the 100th day of herd residence. In general, then, we may say that a mouse that has survived about 30 days' exposure to risk will live more than half as long again as a newcomer to the cage (Greenwood and Topley 1925, Greenwood et al. 1936).

In another type of test with mouse pasteurellosis, survival through a testing period of 30 days or more added some 50 per cent to a mouse's expectation of life under severe epidemic conditions.

But clearly we cannot assume from such results that the mice have been actively immunized. We are dealing with fatal diseases, and natural selection has certainly been at work. In the mouse typhoid experiment, of every 1000 mice living on day 0 (their day of entry to the herd) only 256 survived for 30 days. Death had by then removed some 75 per cent of those at risk; and, if no active immunization had occurred, the surviving 25 per cent would have a higher average resistance than newcomers to the herd, on the safe assumption only that the natural resistance of mice differs *inter se*, and that this difference plays some significant part in determining survival during the first 30 days. Webster and his colleagues (see Webster 1923a, c, d, 1924a, b, 1926, Pritchett 1925, 1926) consider that this natural, inborn resistance is the dominant factor in determining death or survival during an epidemic. It is clear from our discussion of the genetics of resistance to infection that non-specific resistance may vary within a herd, sometimes in a graded, often in a discontinuous manner. But it is not necessarily a dominant factor; there is good evidence that active immunization has an effect similar in kind, if less in degree, to that exerted in the spread of diphtheria in man.

The data for mouse typhoid and mouse pasteurellosis (Greenwood et al. 1926, 1928, Newbold 1927) showing a positive correlation between the length of afterlife and length of previous exposure, indicated that the longer a mouse survives under constant risk of death, the longer it survives under continued exposure to infection. But length of afterlife and death rate during previous exposure were not so correlated, indicating that with equal lengths of exposure to risk, the survivors from a period of high mortality have no significant advantage over the survivors from a period of lower mortality; clearly suggesting that the elimination by death of the more susceptible mice is not a very important factor in raising the average resistance of the survivors, whereas the advantage derived from prolonged exposure at a constant death rate is compatible with the view that active immunization is important.

It may be noted, as a further argument against selection of individuals with an innate non-specific resistance, that the increased resistance gained by survival in an infected herd, or after artificial infection, was specific for the original infecting agent, in that experience of *Salm. typhimurium* did not immunize against *Past. muriseptica* and vice versa (Topley et al. 1925, Greenwood and Topley 1925). This, of course, accords well with epidemiological experience in general. If differences in innate resistance are of major importance they must either be specific or must be differences of immunizability rather than of immunity.

Whether or not active immunization is the main factor that determines the increased resistance of surviving mice, the grade of antibacterial immunity produced is only moderate. When surviving mice are exposed to a continuous risk of heavy infection, the expectation of life in both mouse typhoid and mouse pasteurellosis is raised only from about half to two-thirds of the normal figure (Greenwood et al. 1931b).

This is a very different picture from that presented by our study of diphtheria in man—even allowing for the fact that in diphtheria we are dealing with a disease of relatively low fatality. Either mice differ in some fundamental way from men; or the conditions in mouse cages differ fundamentally from those in schools and institutions—an obvious possibility; or natural immunization against enteric infection or against pasteurellosis is less effective than natural immunization against diphtheria or scarlet fever—an instance, perhaps, of a more general hypothesis that natural antitoxic immunization is much more effective than natural antibacterial immunization.

There is no known disease of mice that presents any close analogy with such toxaemic human infections as diphtheria and scarlet fever; but a natural disease of mice—ectromelia which gives rise to severe and fatal epidemics under experimental conditions (Marchal 1930) affords an opportunity for the study of natural immunization against a virus infection. The results are in striking contrast to mouse typhoid or mouse pasteurellosis (Greenwood et al. 1936).

On the same plan of starting an epidemic in a herd of mice and thereafter adding 3 mice a day, two epidemics of this disease were observed for some 18 months. The limited expectation of life ($_{60}E_x$) on entry to the infected herd was about 31 days. It fell to about 26 days. From then, within a week, the $_{60}E_x$ value rose; but instead of reaching a maximum in some 30 days at a figure far below the normal expectation of life it continued to rise until, on the 100th day after entry, it reached a plateau at about 55 days. Unlike the two bacterial epidemics, where the expectation of life reached only two-thirds of the normal span, it reached almost a normal expectation in mice surviving up to 3 months during a severe epidemic.

It should be pointed out that ectromelia (mouse-pox) differs from mouse typhoid and mouse pasteurellosis in that the mode of infection is by skin abrasions; infectivity of the animals ceases as soon as the skin lesions heal; and potent protective antibodies are formed as the result of infection (Fenner 1949b).

A more prolonged experience of the same epidemic prevalence showed that waves of exceptionally high mortality occurred during which many of the old survivors succumbed; but this fact does not abolish the significant difference in behaviour between a bacterial and a virus infection over epidemic periods that last through a large part of the lifetime of a normal mouse. It is clear that, even under conditions of severe and continuous exposure to infection, a mouse can develop a relatively effective immunity to ectromelia. An analogy may be drawn with outbreaks in man of the related virus infection smallpox. Members of the community who are immune because of prior infection or vaccination suffer no illness during epidemic periods unless or until their immunity has waned to the point when they become susceptibles at risk. When the herd immunity is high, but incomplete, the disease—as is the case also with measles—will smoulder in the community as long as a proportion of susceptibles is maintained by new births.

How far may the course of events in an infected herd be modified by the artificial immunization of some or all of the hosts at risk?

We shall discuss in later chapters the evidence with regard to the efficacy of different kinds of prophylactic vaccination in human herds. So far as experimental epidemiology is concerned the answer has been sought in mouse typhoid and, with far scantier data, in ectromelia.

Active immunization with a killed culture of *Salm. typhimurium* consistently increases the resistance of mice to that organism (Topley 1929, Topley, Wilson and Lewis 1925).

Thus, in one experiment, 50 mice that were immunized by a heat- and formalin-killed vaccine of *Salm. typhimurium* were partially protected against a uniformly fatal dose of the bacillus; the time-to-death of some animals was prolonged; 24 per cent were saved, though left with a latent infection; and 18 per cent were apparently wholly protected.

As might be expected, an immunity of this order was not very effective in herds submitted to a continuous risk of infection.

An epidemic of mouse typhoid was started and maintained by adding three normal mice daily, and every 7th day groups of 10 mice that had been given certain vaccines were added (Greenwood et al. 1931a). The limited expectation of life ($_{60}E_x$) (Table 18.2) of the normal mice on entry was 26.26 ± 0.641 days; for the four immunized groups that had received vaccines containing the S surface antigen the $_{60}E_x$ figures ranged from 32.10 to 35.06. Thus by active immunization with a killed vaccine an increase in average resistance was obtained of the same order as that attained naturally by normal mice after living for some 30-50 days in an infected herd under the joint influence of active immunization and mortuary selection.

Very similar results were obtained in another experiment with a killed *Salm. typhimurium* vaccine given by mouth, or by intraperitoneal inoculation (Greenwood et al. 1931c).

The immunity to ectromelia induced by vaccinating mice with a formolized virus is far more effective than the immunity to mouse typhoid induced by the bacterial vaccine. Thus, in one test (Greenwood et al. 1936), 96 per cent survived an otherwise fatal dose of virus.

The expectations raised by these results were fulfilled, up to a point, in an experiment on active immunization under epidemic conditions (Greenwood et al. 1936).

A herd was selected in which ectromelia was actively spreading, and causing a high mortality. At weekly intervals 30 normal and 30 immunized mice were added; and 3 normal mice were added daily. For the 300 immunized mice, the 300 normal controls added with them and the 210 normal mice added daily over the 70-day period, the $_{60}E_x$ values for the day of entry to the herd were: immunized mice 49.1 days, normal controls added with immunized 20.8 days, normals added daily 21.4 days (Table 18.2). This shows a very great advantage to the immunized mice, and it was maintained throughout later cage-ages until the combined effects of natural immunization and selection, which are so effective in this disease, had raised the $_{60}E_x$ of the surviving normal entrants to the same high level.

The results provided no reliable guide to the permanence of the protection against ectromelia under these very severe conditions of exposure until further results were available; but it is noteworthy that mice which survived infection remained almost completely immune to reinfection by pad inoculation for a few weeks, and even after a year restricted the growth of virus to the site of inoculation. Further, mice protected as the result of previous infection had a higher degree of immunity than those protected by artificial vaccination (Fenner 1949a, b).

The evidence is summarized in Table 18.2. The $_{60}E_x$ figures for the surviving mice are recorded after that period of exposure (30–100 days) when the full advantage appears to have been attained, and the limited expectation of life has reached relative stability. For the uninfected standard herd the figure of 58 days is taken both for new entrants and for survivors at all periods. The two sets of figures for the vaccinated mice

Table 18.2 Showing limited expectation of life ($_{60}E_x$), in days, for various groups of mice submitted to epidemic infection

	Normal		Vaccinated			
			Intraperitoneally		*Per os*	
Infection	NE	S	NE	S	NE	S
None	58.00	58.00	—	—	—	—
Mouse Typhoid (*a*)	22.49	34.75	—	—	—	—
Mouse Typhoid (*b*)	26.26	36.47	32.10	40.17	—	—
			35.06	36.90		
Mouse Typhoid (*c*)	31.18	42.40	38.19	49.52	35.30	44.87
Ectromelia (*a*)	30.71	54.69	—	—	—	—
Ectromelia (*b*)	20.8	52.8	49.1	51.6	—	—

NE = New Entrants.   S = Survivors after 30–100 days.

in the second mouse typhoid experiment refer to the least and most effective of the four antigenically similar vaccines employed.

The significance of the figures is obvious. With such a bacterial infection as mouse typhoid residence and survival in an infected herd increase resistance, as measured by expectation of life, but the expectation is never increased to that of healthy animals. However long a mouse has lived in such a herd it is never indifferent to the risk that it continues to run. There is no solid immunity.

Active immunization via the peritoneal cavity increases resistance to the degree otherwise attained after long residence in the herd. This increased resistance on entry is augmented by life in the infected herd; but vaccination followed by exposure to risk never leads to prolongation of life equal to that of normal uninfected mice. Here, again, solid immunity is not attained. The effect of immunization *per os* is similar.

With ectromelia, natural immunization is far more effective, and though immunity is not complete, it is of a high order. In conformity with this, vaccination with a formolized virus greatly increases the expectation of life of the mice exposed to risk.

Will immunization of all entrants to an infected herd eliminate the disease, at least in its overt form? In mouse typhoid, it will not.

In a herd receiving three normal entrants daily, $_{60}E_x$ was 27.00 days. When three mice immunized daily were substituted over a period of 22 weeks, the limited expectation of life for these animals rose only to 34.92 days; and after a further 68 weeks of one immunized mouse daily, only to 36.38 days (Greenwood *et al.* 1931*c*). Mouse typhoid was spreading and killing as actively at the end as it was at the start of the experiment. The mice at risk lived a little longer, but that was all.

This does not mean that active immunization of such a kind is useless. As noted above, it is capable of completely protecting a proportion of mice against a single experimental infection with a dose of living bacilli that is fatal to unvaccinated controls. Under conditions in which exposure to risk is slight, transient or intermittent, a resistance of this order may make all the difference between escape and infection, or between life and death. But this degree of resistance does not allow its possessor to ignore the risk of infection. Universal immunization of such a kind will not stamp out a disease irrespective of the sanitary environment.

How far does variation of the epidemic strain of a microbe modify the course of herd infection?

In many descriptions of epidemics of microbial infection reference is often made to 'epidemic' strains of particular bacteria, to loss and gain of virulence and infectivity during an epidemic prevalence, and to secular changes in the pathogenic powers of a particular parasite. But nearly all such writing is really only a way of saying that the disease in question was more prevalent, or more fatal, or spread more rapidly at one time than at another. The difficulty is that a description of epidemic events is, of necessity, a description of a fluctuating equilibrium between parasite and host. We can describe an increase in the severity of an infection in terms either of an increase in bacterial virulence or of a decrease in host resistance; but in epidemiology, if our descriptions are to be used as evidence in solving this particular problem, the rise in virulence, or the fall in resistance, must be proved, not assumed; and the proof is no easy matter.

As is evident from our examples changes in host resistance are quite certainly of primary importance; this is obvious from the natural history of diphtheria in man, and the experiments with herds of mice. Although many epidemic diseases have displayed wide secular fluctuations in fatality, there is little firm evidence that changes in infectivity or virulence of the microbial parasite are responsible for the fluctuations and other changes commonly observed during the course of the epidemic.

Of indirect and suggestive evidence there is plenty. Selection of the more virulent variants within a microbial population may occur during the infection of an animal—the well established increase in virulence of a strain by animal passage. Strains less virulent or less toxigenic than usual are often isolated from man and animals, suggesting selection in the opposite direction. Again, the results of widespread use of an antibiotic in a community clearly show that one variety of a bacterial species prevalent in that community may be replaced by another. This particular change, which we discuss in Chapter 5 and the relevant chapters of Volume 3, is important too in that it may affect the control of epidemic disease—not necessarily because the new strain differs in virulence or infectivity from the original strain; but because it has become resistant to therapeutically useful agents.

Bacterial infections

Another change, not necessarily associated with a variation in virulence—namely a change in immunogenicity—directly affects the immune status of the community, since the specific immunity induced by infection with the originally prevalent variety may be less effective or even ineffective, against the newly established strain. The influenza virus is a case in point (see below).

The conclusion that bacterial variation probably plays an important part in epidemic happenings is a natural one. With human populations, proof is difficult because any independent test of virulence of the prevalent microbe is of necessity indirect and unreliable; indirect because it must be made in the experimental animal, and unreliable because the properties of the microbe determining its virulence for the experimental animal are not necessarily those determining it in man.

With the same species of animal as that in which the disease is prevalent, as in experimental epidemics, the results are more immediately applicable. But even when such tests demonstrate that microbial variation does, or does not, determine epidemic fluctuations, we must be cautious in generalizing from them to epidemics in general, because our findings may be peculiar to the system we have chosen to study.

In the hands of Webster and his colleagues (Webster 1923*a,b,c,d*, 1924*a,b,c,d*, 1925, 1926, 1927, 1928, 1930*a,b,c*, Webster and Burn 1926, 1927*a,b*,) in a given epidemic prevalence the virulence of the infecting bacterium remained substantially unchanged, even during a long continued epidemic; but differences in virulence were observed among strains of the same bacterial species isolated from different epidemic prevalences of the same infective disease, such as in the proportion of carrier infections to fatal cases, which appeared to determine differences between one epidemic and another (Webster 1930*d*, Hughes 1930, Hughes and Pritchett 1930, Pritchett, Beaudette and Hughes 1930*a,b*). Webster's main contention was that the course of any particular epidemic depended essentially on the degree of genetically determined resistance of the individuals composing the herd (Webster 1946).

Virulence may not be the only variable factor. Defining *infectivity* as the capacity to spread from one host to another under any specified conditions of infection, we may inquire whether infectivity is highly correlated with virulence.

As a rule, those strains of *Salm. typhimurium* which spread rapidly, giving rise to severe epidemics of mouse typhoid with a high mortality rate, had a relatively high virulence determined by parenteral injection; strains of low virulence did not spread widely and, though persisting for long periods in the originally infected hosts, gave rise only to a few latent infections and an occasional death among the normal mice at risk (Topley *et al*. 1928). This association between virulence of *Salm. typhimurium* and the capacity to induce severe epidemics by contact infection was not consistently observed. One strain, of moderate virulence, had little power of producing severe epidemics by contact infection (Topley *et al*. 1931).

With two strains of low virulence, the limited survival time of exposed mice was about 52 days—near the normal limit; and 9 to 10 per cent had become latently infected. With a more virulent strain, though the epidemics were mild—the limited survival time was also about 52 days—there was a considerable power of contact spread, since some 31 per cent acquired a latent infection. When tested with a highly virulent and infective strain, the survivors of these mild epidemics were much more resistant than normal mice, significantly more resistant than vaccinated mice and just as resistant as survivors from severe epidemics. The strain giving rise to these mild epidemics might be regarded as being a better natural inducer of immunity than of lethal infection. It was originally isolated from an infected mouse; had it appeared in, and been excreted from, a mouse during an epidemic prevalence, it might have significantly changed the course of events—a speculation, though a plausible one.

Studies on pasteurellosis, however, provide good evidence of changes in infectivity during the epidemic spread of a bacterial parasite. Certain strains of *Past. septica* of moderately high virulence when tested by direct inoculation, but having little power of contact infection, gave rise, during their slow and limited spread in an infected herd of mice, to variants with a high capacity for rapid and fatal spread (Greenwood *et al*. 1936).

Thus with *Past. septica* and *Salm. typhimurium* infections virulence and infectivity may vary independently. High virulence and high infectivity, moreover, appear to be necessary attributes of an 'epidemic strain', defined as one having the capacity to spread naturally among a herd at risk to give rise to a severe and fatal epidemic; with the loss of either character, the strain loses its epidemic character. Little is known of the properties of an organism that determine its infectivity; but they clearly must include those that enable it both to adapt to the environmental changes associated with transfer from host to host, and to establish, in or on the surfaces of the recipient host, a primary lodgment from which tissue invasion can occur.

Viral infections

Observations on the animal viruses parallel in most respects those on bacteria. The main difference appears to consist in the greater susceptibility of viruses to adverse environmental conditions and in the wider range of virulence among the varieties of a given virus observed under natural conditions.

Fenner (1949*b*), working with the ectromelia virus, confirmed the distinction, observed by Topley and Greenwood in *Salm. typhimurium*, between virulence and infectivity. Two strains, for example, had almost the same degree of virulence for mice as judged by direct pad inoculation, yet varied greatly in their power to infect mice under natural conditions.

Most viruses can be adapted to hosts different from those in which they are normally found. Sometimes only a few passages through the chick embryo, the mouse, or some other animal are necessary to modify their original characters. Moreover, owing to the mode of virus multiplication in the cell, periodic mutations can occur with the release of virus with altered virulence. With this degree of plasticity, it is not surprising that several varieties of the same virus are often met with under natural conditions. Good examples are those of foot-and-mouth disease and poliomyelitis; but the most striking example is the influenza virus. The variations of this virus are discussed in Chapter 96. Here we may note that in most outbreaks in recent years a strain or strains have been isolated differing in behaviour to a greater or lesser extent from each other and from strains met with previously. One major example is the appearance throughout the world of the A2 or 'Asian' subtype in 1957 and the Hong Kong variant in 1968. The readiness with which mutants of the influenza virus—or indeed of many other viruses—can be selected and established in laboratory conditions does not, however, imply that the successive appearance of variants in the world is due to such a change taking place in the prevailing virus population. The newly appearing variety may well have been latent in some parts of the world or in an animal reservoir, the previous one being at a selectional disadvantage, perhaps because it has immunized a sufficient proportion of the populations at risk to make its further spread difficult. Even in influenza, however, no conclusive evidence has yet been brought to show that the rise and fall of a given epidemic is determined by variations in the virulence or infectivity of the strain responsible.

Myxomatosis deserves special mention in this context, as an example of an epidemic virus disease in which the populations of both causative organism and host appear to have changed as a result of their mutual interaction. The European rabbit, also prevalent in Australia, was highly susceptible to this virus, which induced in it a severe, generalized and often fatal infection. The disease was artificially introduced into Australia and Europe in the hope of decreasing the rabbit population by epidemic spread of the infection via the normal vectors—fleas and mosquito. In the early years of the attempt this hope was fulfilled, the death rate being high; but as a result of natural selection, the progeny of surviving rabbits have proved to have higher innate resistance to the virus. This semi-experimental epidemic also provided evidence of a variation in virulence, because the virus isolated from animals in the late seasonal epidemic prevalences proved to be significantly less virulent in laboratory rabbits than the virus originally introduced. It is suggested (see Fenner and Ratcliffe 1965) that the virus of lesser virulence prevailed over the more virulent as the epidemic strain because it was less rapidly lethal, with the result that each diseased rabbit was infective for other rabbits—via the insect vectors—for periods up to five times that with the virulent strain (see Chapter 87).

In summary, variations in virulence in a microbial parasite during an epidemic prevalence have often been sought for, largely with negative results. There is experimental evidence that different strains of the same parasite, possessing different degrees of virulence or of infectivity, produce very different results when they are allowed to spread by natural infection among a susceptible herd; and that a microbial parasite may vary in infectivity during an epidemic, though there is as yet no satisfactory proof that the resulting variant influences the progress of the epidemic. The experimental evidence as a whole accords quite well with the view, expressed by many epidemiologists, that variants in the characters of the parasite are of major importance in the spread of human infections, and that an outbreak of disease may be initiated by the evolution, or importation, of an 'epidemic strain' of the causative organism. Nevertheless, it seems probable that, with the exception of certain viruses, the evolution within any parasitic species of a strain of high epidemicity or virulence is an occasional event, rather than part of a normal or periodic process determining the fluctuations in mortality and morbidity that may occur during a long continued epidemic.

What modification of herd infection and herd immunity can be made by changes in herd structure?

All those factors affecting the resistance of the individual will affect the average resistance of the herd of which he is a member. Those factors that have been adequately studied are considered in the relevant chapters in Volume 3.

There are special factors, such as the spatial distribution of the hosts at risk, the duration or the intermittency of contact, and other conditions of arrangement or environment, that affect herds but not individuals. Some of them have been studied experimentally in mouse typhoid (Greenwood *et al.* 1936, 1939).

Continuous contact. The addition of susceptibles at a low rate of 1 or 2 a day to an already infected herd resulted in irregular fluctuations in the death rate. With higher rates of addition, the death rates fluctuated in a minor way round an almost steady average value; in the circumstances the population increased rapidly at first, and later more slowly, suggesting that a still higher rate of addition would have produced a stable population in which the daily deaths equalled the daily additions. Thus, continuous contact of infected animals with susceptibles introduced into the herd at a steady rate did not produce a system with any inherent tendency to fluctuate, as epidemics fluctuate in natural conditions.

Discontinuous contact. Twenty-five mice were infected with *Salm. typhimurium*. These and 100 normal mice were kept each in a separate cage. Three times weekly, all the mice were herded in one large cage for 4 hours and at the same time two normal mice were added to the herd. An epidemic started among the normal mice, but soon died down, and after 70 days, there were no deaths from mouse typhoid. By the 149th day the population of mice had increased to 180, and there had been no deaths for 80 days. The population was then aggregated in a single cage. An epidemic wave resulted from this continuous contact, and in spite of the addition of six new mice per week, the total population had fallen by the 289th day from 180 to 44 mice. The resumption of discontinuous contact at this time was followed by a prompt response in the reverse direction. The death rate sank and remained at a low level, and the population rose to over 100 mice. These conditions were maintained by intermittent contact over a period of 15 months, at the end of which reaggregation again produced a major epidemic, which within 30 days reduced the population to 20.

Dispersal of a closed herd. The effect of dispersal depended on the degree of dispersal and the age of the epidemic. Thus the dispersal of 100 mice into four groups of 25 had little effect upon an epidemic that had just begun to spread. With dispersal, however, into ten groups of 10 mice, the subsequent mortality was much lower than that of a control group of 100 maintained as a single unit. Dispersal, even in small groups, had little effect if it was delayed until the epidemic was well under way, (Topley 1922, Topley and Wilson 1925, Greenwood *et al.* 1930).

Dispersal in an increasing herd. Large batches of mice were added at short intervals to an infected herd so as to maintain a continuous spread of infection. Mice were withdrawn to isolation in single cages, after varying intervals of exposure to infection, and the rates of infection and mortality compared with those of added mice left in the herd. The average mortality of mice subsequent to withdrawal was lower than that of their contemporaries left in the herd. The advantage of withdrawal was pronounced for all periods of exposure up to the longest tested, which was 35 days. The greatest effect, as would be expected, was observed in mice withdrawn after the shortest exposure (Greenwood *et al.* 1939).

The conclusions to be drawn from these experiments, though they must in some cases be tentative, add to the evidence from the study of natural epidemics in man and animals, that movements of susceptible and infected hosts in relation to one another, and the aggregation or dispersal of human or animal herds, apart from any introduction of new infection, are sufficient to induce major changes in the incidence of many infectious diseases (Topley 1942). Thus the addition of susceptibles to an infected population ensures the recurrence of epidemics; and a continuing high rate of addition is associated with a high death rate. (We should note however that the susceptibility to infection of normal entrants to an infected herd is not necessarily a reflection of a high risk of infection. Christian (1968) for example suggests it may, in part at least, be due to the stress to which entrants are subjected by reason of their socially inferior status, with a decrease of adrenal activity and a consequent decrease of non-specific immunity.) On withdrawal of susceptibles, the infection dies out, even when the survivors are not all immune. Dispersal of the herd into sufficiently small groups will shorten an epidemic, presumably by lessening the contacts between infected and susceptibles in the population. As to the undoubted innate variations in resistance of individuals in natural or experimentally assembled populations there are few instances where the responses are demonstrably clustered in two or more grossly different modes; and there are therefore as yet few analyses of the epidemiological consequences of multimodality of this kind.

Herd infection and herd immunity in nature

The results of experimental epidemiology must be applied with caution to natural epidemics of infections of both animals and man by agents other than the few that have been studied. These epidemic diseases are discussed in detail in the relevant chapters of Volume 3; here we need touch only on examples bearing directly on those features of herd immunity discussed above.

As a probably valid example of the effect of introducing susceptibles into a herd, we may cite the epidemic prevalence of measles every 2–3 years in urban communities at high risk. The inter-epidemic period is almost certainly due to the natural immunization of children that results from infection, and outbreaks to the accumulation, by birth, of a sufficient number of new susceptible children. At the other extreme we have the isolated community which does not experience measles for a generation or more. The resulting accumulation of susceptibles of all ages is such that, when measles is introduced, a rapidly spreading epidemic of severe disease is engendered (see Chapter 98).

The effects of changing the structure of the herd are evident in the variations in incidence of poliomyelitis. In primitive crowded communities, clinical poliomyelitis is uncommon, and both clinical and subclinical infections are largely confined to young children, most of whom are thereby immunized by their fifth years. The risk is diminished in more advanced communities with good sanitation, so that older children and adults escape the immunization of infection; and the disease when it develops tends to be severe and epidemic. In such communities, when artificial immunization has to some extent restored the insusceptibility of the older members of the population, the incidence of the disease may shift back to the younger children unless artificial immunization is instituted early in life (see Chapter 99).

No hard and fast rules can be formulated about the proportion of individuals in a population that must be immunized if the epidemic spread or the epidemic prevalence of an infection is to be prevented. Thus a general figure of 70 per cent is cited (Chapter 53) for diphtheria; but it may differ for a special subgroup of the population. A figure of more than 98 per cent is cited for measles, and the maintenance of herd immunity demands regular vaccination of all susceptible persons. A similar rigour of procedure is advocated in the United States for the elimination of rubella (Chapter 94) by vaccination of all young adolescents of both sexes in order to minimize rubella in pregnant women with its attendant risk of severe fetal damage during early pregnancy. In Britain reliance is placed on natural infection and on vaccination of girls aged 11–14, backed by vaccination of women liable to pregnancy who prove serologically to be susceptible. Cowpox vaccination has led to the eradication of smallpox from the whole world (Chapter 87)—probably the most successful to date of all mass immunization programmes. Pertussis in Britain affords an example of epidemic outbreaks occurring when the proportion of susceptibles rises with a decrease—in this case owing to unpopularity of the vaccine—in the proportion of the immunized (Chapter 67). In other instances the immunization of subgroups is sufficient. Thus a population of well immunized domestic animals is a barrier against human rabies in spite of a high incidence of disease in the wild life of the environment.

As we have seen, the changes in prevalence of a communicable disease, and in its morbidity and mortality, are largely explicable in terms of the population at risk; there is no need to postulate variations in infectivity or virulence of the infecting strain of the parasite, and in the few instances when tests have been made little direct evidence of such variation has been apparent. Nevertheless the possibility should be borne in mind that some substantial modification, distinct from replacement of one strain by another, might occur. For example, some non-toxigenic strains of the diphtheria bacillus, on lysogenation with a beta-prophage, become fully toxigenic (Chapter 25). Whether this is likely to occur in nature is at present a matter for speculation. An analogous change is postulated as occurring *in vivo* among prevalent strains of salmonellae, whereby resistance to certain antibiotics is conferred on susceptible strains by plasmids (Chapters 6 and 72); virulence is not affected in this instance; but the control of salmonella infection of a herd might thus be diminished.

The control of herd infection

We have discussed experimental epidemiology in some detail, not only for its bearing on the problems raised by observational epidemiology in man and animals but as an example of the way in which further evidence might be obtained for more effective means of controlling herd infection. In the last four decades the investigation of immunity to infection has centred largely on the individual animal and its cellular reactions (Chapters 10–17); little has been done to extend the pioneer work of Topley in Britain and Webster in the United States. Nevertheless limited studies of herd infection and herd immunity by controlled experiment and, in recent years, by intensive field work in some of the virus diseases of man, have modified many of the older conceptions about the effective administrative control of epidemic disease. Isolation and quarantine, for instance, have, in the past, played a major part in public health administration. A very cursory consideration of the various types of distribution of a bacterial parasite within an infected herd that have been described earlier in this chapter, and of the results obtained in experimental epidemics, will raise serious doubts about the efficacy of such measures.

Quarantine. In the case of quarantine, it is clear that it can succeed only where it is complete; and the history of sanitary control suggests that successful quarantine is possible only under very exceptional conditions, and over relatively short periods of time. The carrier and the atypical case defeat a sanitary barrier, just as they defeat isolation; and, under the conditions of modern transport, there is little possibility of preventing the introduction into any one part of the inhabited world of any infective parasite which is prevalent in another. Once the barrier is passed, the subsequent course of events will depend upon the conditions obtaining within the community into which infection has been introduced.

Isolation. The policy of the isolation of sick persons within a community is based on a similar failure to realize that the clinical picture provides a very incomplete description of the true state of affairs. If isolation removed from the community the whole, or even the great majority, of the infected individuals, it might considerably influence the prevalence of an infectious disease. But when the ratio of latent or atypical infections to clinically recognizable cases is high, we cannot hope to decrease the morbidity rate much by removing to hospital those cases which exhibit the typical stigmata of the disease. We may indeed decrease the total infective material less than the ratio of isolated to non-isolated, among infected individuals, would suggest; for the sick person would, in any case, move less freely among his fellows than the apparently healthy carrier. As with quarantine, we should expect a policy of isolation to be successful only in exceptional circumstances, when the recognizable cases form a very high proportion of the total infected; when the main period of infectivity coincides with the appearance of manifest symptoms of disease—and not, as with many infections, when it occurs in the prodromal stage of infection as well; and when the total mass of infection to be dealt with is small. Once a given infective disease has assumed an endemic-epidemic prevalence within a herd, we should expect no appreciable result from the isolation hospital, so far as a reduction in morbidity is concerned. It is interesting to find that the expectations based on bacteriological and experimental findings are borne out by administrative experience. Thus, all applications of the calculus of correlations wholly failed to bring out any connection whatever between the incidence rate of scarlet fever and the extent of isolation (see Greenwood and Topley 1925). The isolation hospital or the infectious disease unit seem to have little effect on the health of the community as a whole. Their chief value lies in convenience to both patient and staff and the ready availability of specialized nursing and treatment; and in minimizing the transmission to contacts of the infecting agent during the usually short period of the patient's infectivity. The number of microbial diseases in which such precautions are either necessary—or useful—is small. It includes some diseases, like pertussis, pneumonic plague and acute pulmonary tuberculosis transmitted via the respiratory tract; and others with stages in which the patient is broadcasting the microbe in the environment, like diphtheria, smallpox, cholera, hepatitis A, mumps and measles. Special isolation units are usually demanded for the isolation of patients with the highly contagious and lethal Marbug, Lassa and Ebola fevers; but even here surveillance rather than quarantine may prove to be a sufficient precaution for all but close contacts.

When animal communities are concerned, destruction of the infected or potentially infected members of the herd is practicable—as in the British practice for the eradication of bovine tuberculosis by slaughter of cattle with any evidence of infection (Chapter 51) or the containment of imported outbreaks of foot-and-mouth disease by slaughter of all cattle in the region of an outbreak (Chapter 89).

Turning to those administrative measures which bring about a general reduction in the opportunities for the transference of infection from host to host, theoretical considerations suggest that action along these lines is likely to be effective. Any improvement in sanitation, which ensures a clean water, or milk, or food supply, which reduces the frequency of insect vectors of infection, or which lessens the opportunities for close contact between individuals, may be expected to lead to a reduction in the incidence of those infections, with the transference of which that particular measure of sanitation interferes. This expectation has been fulfilled whenever an effective method of reducing the general opportunities for the transference of a particular infection has been adopted. With infections of the respiratory tract, however, it is particularly difficult under conditions of modern civilization to devise any generally applicable and effective means of preventing transit of the parasite from host to host, and it is these diseases that are least susceptible to attack by preventive measures.

Finally, the whole trend of our present knowledge stresses the importance of the susceptible host; because the proportion of susceptible hosts appears to be the main determinant of the course of epidemic diseases in a community where the infecting agent is a continuing risk. Any measure which decreases that proportion, or which increases the average resistance of a herd, whether it depends on specific immunization, or on some factor which confers an increased immunity less specific in its range, or on the destruction of the organism as it reaches the tissues by chemoprophylactic means, will directly influence the incidence of the corresponding infections. For most infectious diseases, the accurate determination of the conditions which must be fulfilled in order that induced variations in immunity should have their optimal effect, and the development of methods by which such immunity can be induced, are two of the most important problems in the control of herd infection.

References

Abbey, H. (1952) *Hum. Biol.* **24**, 201.
Allison, A. C. (1965) *Arch. ges. Virusforsch.* 280.
Amoss, H. L. (1922) *J. exp. Med.* **36**, 45.
Bradley, D. V. (1982) *Trans. roy. Soc. trop. Med. Hyg.* **76**, 143.
Christian, J. J. (1968) *Amer. J. Epidemiol.* **87**, 255.
Coovadia, H. M., Wesley, A., Hammond, M. G. and Kiepïela, P. (1981) *J. infect. Dis.* **144**, 142.
Dorf, M. E. (1981) Editor: *The Role of the Major Histocompatibility Complex in Immunobiology.* J. Wiley & Sons, Chichester.
Dudley, S. F. (1923) *Spec. Rep. Ser. med. Res. Coun., Lond.* No. 75; (1926) *Ibid.* No. 111; (1932) *J. Hyg., Camb.* **32**, 193.
Evans, A. S. (1979) *Amer. J. Epidemiol.* **109**, 379.
Fanning, M. M., Peters, P. A., Davis, R. S., Kagura, J. W. and Mahmoud, A. A. F. (1981) *J. infect. Dis.* **144**, 148.
Fenner, F. (1949a) *Aust. J. exp. Biol. med. Sci.* **27**, 1, 19, 31, 45; (1949b) *J. Immunol.* **63**, 341.
Fenner, F. and Ratcliffe, F. N. (1965) *Myxomatosis.* Cambridge University Press.
Fox, J. P., Elveback, L,, Scott, W., Gatewood, L. and Ackerman, E. (1971) *Amer. J. Epidemiol.* **94**, 179.
Glenny, A. T. (1925) *J. Hyg., Camb.* **24**, 301.
Glover, J. A. (1918) *J. Hyg., Camb.* **17**, 350.
Greenberg, L. J., Gray, E. D. and Yunis, E. J. (1975) *J. exp. Med.* **141**, 935.
Greenwood, M. (1935) *Epidemics and Crowd Diseases.* Williams and Norgate, London.
Greenwood, M., Hill, A. B., Topley, W. W. C. and Wilson, J. (1936) *Spec. Rep. Ser. med. Res. Coun., Lond.* No. 209; (1939) *J. Hyg., Camb.* **39**, 109.
Greenwood, M., Newbold, E. M., Topley, W. W. C. and Wilson, J. (1926) *J. Hyg., Camb.* **25**, 336; (1928) *Ibid.* **28**, 127; (1930) *Ibid.* **30**, 240.
Greenwood, M. and Topley, W. W. C. (1925) *J. Hyg., Camb.* **24**, 45.
Greenwood, M., Topley, W. W. C. and Wilson, J. (1931a) *J. Hyg., Camb.* **31**, 257; (1931b) *Ibid.* **31**, 403; (1931c) *Ibid.* **31**, 484.
Hartley, P., Tulloch, W. J. et al. (1950) *Spec. Rep. Ser. med. Res. Coun., Lond.* No. 272.

Haverkorn, M. J., Hofman, B., Masurel, N. and van Rood, J. J. (1975) *Transplantation Rev.* **22,** 120.
Howard, J. G., Hale, C. and Liew, F. Y. (1982) *Trans. roy. Soc. trop. Med., Hyg.* **76,** 152.
Hughes, T. P. (1930) *J. exp. Med.,* **51,** 225.
Hughes, T. P. and Pritchett, I. W. (1930) *J. exp. Med.* **51,** 239.
Kober, M. (1899) *Z. Hyg. InfektKr.* **31,** 433.
Levine, B. B. and Benacerraf, B.(1964) *J. exp. Med.* **120,** 955.
Lilienfeld, A. M. and Lilienfeld, D. E. (1980) *Foundations of Epidemiology.* Oxford University Press, Oxford.
McDevitt, H. O., Oldstone, M. B. A. and Pincus T. (1974) *Transplantation Rev.* **19,** 209.
Marchal, J. (1930) *J. Path. Bact.* **33,** 713.
Monograph. (1923) *Diphtheria. Med. Res. Coun., Lond.*
Newbold, E. M. (1927) *J. Hyg., Camb.* **26,** 19.
Odaka, T. (1969) *J. Virol.,* **3,** 543.
Okell, C. C. (1932) *Lancet* **i,** 761, 815, 867.
Plant, J. and Glynn, A. A. (1979) *Clin. exp. Immunol.* **37,** 1.
Pritchett, I. W. (1925) *J. exp. Med.* **41,** 195; (1926) *Ibid.* **43,** 161.
Pritchett, I. W., Beaudette, F. R. and Hughes, T. P. (1930*a*) *J. exp. Med.* **51,** 249; (1930*b*) *Ibid.* **51,** 259.
Robson, H. G. and Vas, S. I. (1972) *J. infect. Dis.* **126,** 378.
Sasazuki, T., Kohno, Y., Iwamoto, I., Tanimura, M. and Naito, S. (1978) *Nature, Lond.* **272,** 359.
Sela, M. (1966) *Advanc. Immunol.* **5,** 30.
Serfling, R. E. (1952) *Hum. Biol.* **24,** 145.
Spencer, M. J., Cherry, J. D. and Terasaki, P. I. (1976) *New Eng. J. Med.* **297,** 13.
Topley, W. W. C. (1921) *J. Hyg., Camb.* **20,** 103; (1922) *Ibid.* **21,** 20; (1926) *Lancet* **i,** 477, 531, 645; (1929) *Ibid.* **i,** 1337; (1942) *Proc. roy. Soc., B* **130,** 337.
Topley, W. W. C., Ayrton, J. and Lewis, E. R. (1924) *J. Hyg., Camb.* **23,** 223.
Topley, W. W. C., Greenwood, M. and Wilson, J. (1931) *J. Path. Bact.* **24,** 523.
Topley, W. W. C., Greenwood, M., Wilson, J. and Newbold, E. M. (1928) *J. Path. Bact.* **27,** 396.
Topley, W. W. C. and Wilson, J. (1925) *J. Hyg., Camb.* **24,** 295.
Topley, W. W. C., Wilson, J. and Lewis, E. R. (1925) *J. Path. Bact.* **23,** 421.
Vries, R. R. P. de, van Eden, W. and van Rood, J. J. (1981) *Lepr. Rev.* **52,** Suppl. **1.,** 109.
Vries, R. R. P. de, Kreeftenberg, H. G., Loggen, H. G. and van Rood, J. J. (1977) *New Eng. J. Med.* **297,** 692.
Webster, L. T. (1922) *J. exp. Med.* **36,** 71; (1923*a*) *Ibid.* **37,** 231; (1923*b*) *Ibid.* **37,** 781; (1923*c*) *Ibid.* **38,** 33; (1923*d*) *Ibid.* **38,** 45; (1924*a*) *Ibid.* **39,** 129; (1924*b*) *Ibid.* **39,** 843; (1924*c*) *Ibid.* **39,** 857; (1924*d*) *Ibid.* **40,** 117; (1925) *Amer. J. Hyg.* **5,** 335; (1926) *J. exp. Med.* **43,** 573; (1927) *Ibid.* **45,** 529; (1928) *Ibid.* **47,** 685; (1930*a*) *Ibid.* **51,** 219; (1930*b*) *Ibid.* **52,** 901; (1930*c*) *Ibid.* **52,** 909; (1930*d*) *Ibid.* **52,** 931; (1933) *Ibid.* **57,** 793; (1937) *Ibid.* **65,** 261; (1946) *Medicine, Baltimore* **25,** 77.
Webster, L. T. and Burn, C. G. (1926) *J. exp. Med.* **44,** 359; (1927*a*) *Ibid.* **45,** 911; (1927*b*) *Ibid.* **46,** 887.
Webster, L. T. and Clow, A. D. (1933) *J. exp. Med.* **58,** 465.
Zinkernagel, R. M. (1979) *Transplant. Proc.* **11,** 624.

19

The measurement of immunity
Frank Sheffield

Introduction	429	Estimations and comparisons of ImDs50	436
The concept of resistance	430	The quantal assay	436
The heterogeneity of experimental animals	430	Measurement of resistance by means of	
The dose-response curve	430	quantitative responses	438
Standard deviations and probabilities	432	Student's t test	439
Standard error of the mean	433	Assay with six treatment groups	439
Levels of significance	433	Estimations of protective potency	439
Calculation of the LD50	434	International standards	440
Probits	434	Field trials	441
Comparison of LDs50	434	Value of quantitative measures of error	442
Comparisons based on chance events	435		

Introduction

The measurement of the ability of laboratory animals and of man to resist infection and the harmful effects of microbial toxins is fundamental to the development, manufacture and application of all preparations intended to provide protection against microbial infections. Without a technique for comparing the resistance of a group of treated animals with that of a group of untreated animals it is impossible to assess the protective effect of an antiserum, a vaccine, an antibiotic or an antiviral agent. Without a technique for comparing the attack rate of an infection in vaccinated subjects with that in unvaccinated controls it is impossible to assess the efficacy of a vaccine in its target population. Furthermore, without assays of immunity one cannot compare two batches of the same prophylactic, or two prophylactics of different formulations, and thus judge which is more efficacious.

Many of the difficulties inherent in the measurement of immunity were largely overlooked by the pioneers of microbiology for reasons that are not altogether surprising. The effects demonstrated in many of the earlier experiments were so dramatic as to need no exact measurement to establish their significance. The need for standardizing antitoxic sera did, indeed, provide an obvious opportunity for developing adequate methods; but the need was not realized and the opportunity was allowed to pass. As new methods of immunization were developed and tested in the laboratory, in the ward and in the field, the publication of violently conflicting reports on their efficacy showed that something was amiss; but the remedy, the application of biometrical techniques to the design and analysis of the tests, was not obvious. It was natural enough that the laboratory worker and the clinician should show little eagerness to learn and apply them. But the result was a quite unnecessary amount of confusion.

The causes of the difficulties are now sufficiently obvious and some of them remain. They were set out very clearly by Greenwood (1924) and by Yule (1924). We cannot reduce our variables to a small and perhaps negligible residuum and so obtain constantly reproducible results. In experiments on living animals, and still more in assessing the value of therapeutic or prophylactic procedures in man, we cannot exclude the interplay of factors about which we know little, except that they are certainly very numerous and may be very important. We cannot avoid a large element of randomness in our observations; but, if we plan our experiments wisely and fortune is kind in eliminating major disturbing factors that we could not have forseen, this randomness will be of the sort which we can express numerically and whose effects can therefore be assessed with some accuracy in the interpretation of our results.

We therefore devote this chapter to the concepts that underlie the techniques that are used for the assessment of immunity, to the techniques themselves, and to an elementary consideration of the biometrical principles used in the assessment and in comparisons of immunity. We eschew, however, statistical detail and theory as inappropriate in a textbook of this kind and refer the reader to textbooks such as those of Fisher (1960, 1970), Finney (1971, 1978), Armitage (1971), Bailey (1981) and Harper (1971).

The concept of resistance

Resistance is the ability of an animal to withstand, either wholly or in part, the pathogenic effects of an infecting microbe or the toxic products of a microbe. Resistance may be a natural attribute of a particular species, it may be acquired by the passive transfer of antibody or lymphocytes or by active immunization, or it may be consequent upon treatment with an antibiotic. It may be meagre and readily overcome by a large challenge, or gross and virtually absolute. However, as an attribute of any animal it is measurable only in terms of the size of the challenge that it enables the animal to withstand or of the size of the challenge that is required to overcome it. Conversely, the pathogenicity of the challenge is measurable only in terms of the sizes of the doses that will or will not overcome the resistance of an animal. Thus, when we try to express resistance in terms of a test toxin or culture, or to express toxicity or infectivity in terms of a test animal, the raw data from which the expressions are derived are no more than the results of the interaction of biological materials of very variable activities. If we take the toxin or the culture as our point of reference, we can make a quantitative estimate of the resistance of the host; if we take the animal as the point of reference, we can make a quantitative estimate of the toxicity of the toxin or the infectivity of the culture. The *degree* of the interaction—i.e. the severity of intoxication or infection—tells us nothing about these quantities. The interaction would be as severe with a host and a parasite respectively of high resistance and high infectivity, as with a host of low resistance and a parasite of low infectivity. Again, a mild interaction might mean that the animal was highly resistant to a parasite; but it might equally well mean that the parasite had an exceptionally low infectivity.

The heterogeneity of experimental animals

From the accumulated experience of a century of measurement of immune responses in man and animals, it is obvious that the uniform susceptibility of a batch of test animals is a figment of our imagination. Aside from the demonstrably large differences between different strains of particular animal species due to the genetic determination of the response to antigens and to living microbes, the age, sex, nutritional and physiological state of the animals and the conditions in which they are maintained, may all affect the response of each individual to immunization and infection. An essential part of a good experimental technique is, as Fisher emphasized, that all possible means must be used to randomize the conditions of test, i.e. to distribute the test objects between the various groups so that differences between groups due to possible heterogeneity are kept to a minimum. Let us first, therefore, consider how a batch of animals of average heterogeneity responds to graded doses of a test material.

The dose-response curve

On dividing a large number of similar animals into groups of equal size, and administering to each group one of a series of increasing doses of a lethal toxin, we may observe, if the scale of doses was suitably chosen, that none of the animals dies in the group receiving the smallest dose. With each increase of dose a higher death rate is observed and in the group receiving the largest dose all the animals die. Table 19.1 shows a set of hypothetical percentage mortalities in 12 groups of 100 animals, each group being given one of a series of doses of a lethal toxin. Each dose differs by one unit from the preceding dose. When we plot death-rate against dose, as in Fig. 19.1 (A), we obtain a *dose-response* curve.

The third column of Table 19.1 records the values obtained by subtracting from the mortality rate for a given dose of a units, the mortality for the preceding dose, $(a-1)$ units. Inspection both of column 3 and Fig. 19.1 (A) reveals one of the most important features of many dose-response curves, namely, that the rate of change of percentage mortality is greatest in the middle range of the curve. At the extremes, a relatively large change in dose results in a small change in mortality. That is, for a certain grading of doses, we shall be able to determine doses that kill 50 per cent of a group of animals with much greater accuracy than those that kill 99 per cent. Trevan (1927, 1929, 1930) introduced a convenient notation for these various doses. The 50 per cent killing dose is symbolized by LD50, where LD = lethal dose; the 99 per cent killing dose by LD99, and so forth. When, as in Fig. 19.1 (B), the dose-response curve is steep, the accuracy of estimating the exact LD50 is clearly greater than in (A). In the past, with toxins like that of *C. diphtheriae* which have a steep dose-response curve, an estimate of the 100 per cent killing dose, namely the least amount of substance killing all the animals in the test group, has successfully been used in the measurement of resistance. With other toxins, and with cultures of

Table 19.1

Dose a in arbitrary units	Percentage mortality in groups of 100 animals	Increase in percentage mortality with each increment of dose
2	0	—
3	1	1
4	4	3
5	10	6
6	22	12
7	40	18
8	60	20
9	78	18
10	90	12
11	96	6
12	99	3
13	100	1

pathogenic bacteria, the dose-response curve is far less steep. The LD50 can be more accurately measured and, even with steep dose-response curves, it is preferable to an estimate of the minimal 100 per cent killing dose. For example, when a strain of *Salm. typhimurium* was given in graded doses, each dose into 25 mice, the rise in mortality with increasing numbers of bacilli injected was as follows: 10 bacilli, 24 per cent; 10^3 bacilli, 56 per cent; 10^5 bacilli, 88 per cent, and 10^7 bacilli, 96 per cent. For 100-fold increases in dose the change in percentage mortality in the region of 95 per cent was relatively small; it was greater in the region of the 50 per cent killing dose. In essentials, then, when resistance to the lethal agents, whether toxins or microbes, is expressed in terms of dose of a given lethality, that dose is often best expressed as the LD50 of those agents. In doing so we define the median lethal dose characterizing a given population of animals; and this, when the dose response is symmetrical, is also the *mean* of the range of doses between LD>0 and LD<100.

It is fallacious to equate the LD100 with the measure employed by the early immunologists—the minimal lethal dose, MLD. Nor has the term LD100 any precise meaning, since it includes all possible doses larger than the MLD. The LD0 is likewise imprecise, since it includes all doses smaller than the maximum non-lethal dose.

The LD50 is commonly used as a unit of infectivity, in the sense that the dose employed in a test is referred to as so many LDs50. But to say that ten LDs50 were used to test the immunity of a group of animals gives no clear idea of the severity of the test unless the slope of the dose-response curve is known; if the slope were steep, 50 × LD50 might well be a 100 per cent killing dose, but if it were very shallow, as sometimes happens with, e.g. oral infections, it might kill only some 10–20 per cent more than the LD50.

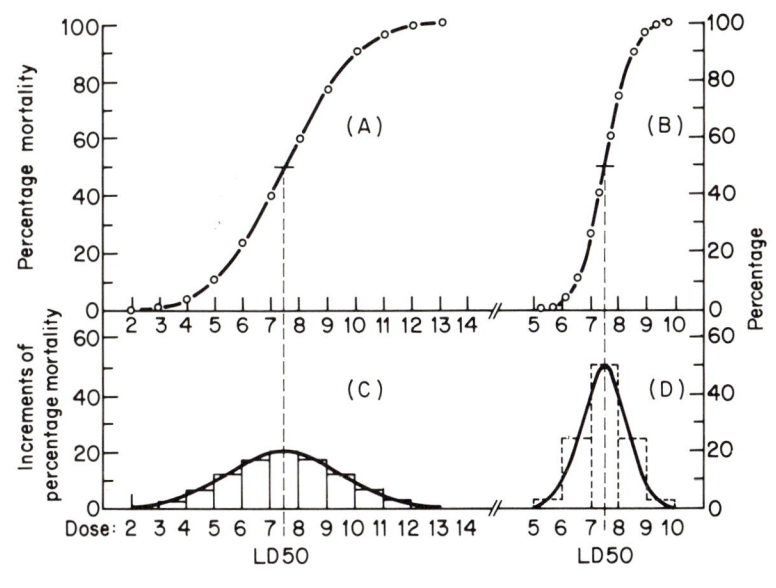

Fig. 19.1

Trevan's notation is now generalized to include wider concepts such as the infective dose (ID), the immunizing dose (ImD), the cell culture infective dose (CCID) or, more generally, the effective dose (ED).

The S-shape of most of the dose-response curves obtained in immunity measurements and the steepness of its slope characterizes the behaviour of the biological system from which it is obtained in a number of ways. One view is that the shape of the curve follows largely from the random nature of the events determining the outcome of inoculation. A second is that the curve reflects the distribution of individual resistances among the test animals. Thus, if in Table 19.1 we assume that each group of 100 animals is representative of the total of 1200 animals from which they were drawn, it is clear that the figures in column 3 represent the proportion of all the 1200 animals which are susceptible to a dose of a units, but not to a dose of $(a-1)$ units. In each case, then, we may say that the recorded proportion of animals is on the average susceptible to a dose half-way between these two values, that is $(a-0.5)$ units. Thus, the 3 per cent of the animals susceptible to 4 units but not to 3 units, are on the average susceptible to 3.5 units, and so forth. Plotting these proportions against doses as abscissae we obtain the histogram in Fig. 19.1 (C), which is in effect a frequency distribution of susceptibilities in the population of animals employed. The average, or mean resistance, corresponds to the LD50. The hypothetical curve B represents the response of another batch of animals to the toxin, estimated from a series of doses increasing by 0.4 unit. The histogram (D), the distribution of susceptibilities derived from it, has been drawn on a scale of unit increases in dose, for comparison with the histogram (C). It is clear that the steeper dose-response curve corresponds to a narrow distribution of susceptibilities obtained in the performance of the test. If we were to give very finely graded doses, each to 100 animals, we should obtain an almost continuous series of points, approximating to the continuous curves in C and D. These symmetrical curves are examples of the Gaussian curve (the so-called *normal curve*), a curve with well defined mathematical properties. Reasonable conformity with a normal distribution, however, is often evident in measurements of resistance only when frequencies are plotted against log dose, making the log-normal distribution the basis for statistical manipulation.

Standard deviations and probabilities

Figure 19.2 shows three normal curves such as might be obtained in experiments of the type just described but differing in the ranges over which resistances are spread. To compare the curves we need a measure of the spread that can be applied to all curves of this kind. This measure is the *standard deviation*, a value that can be estimated from the data and is represented

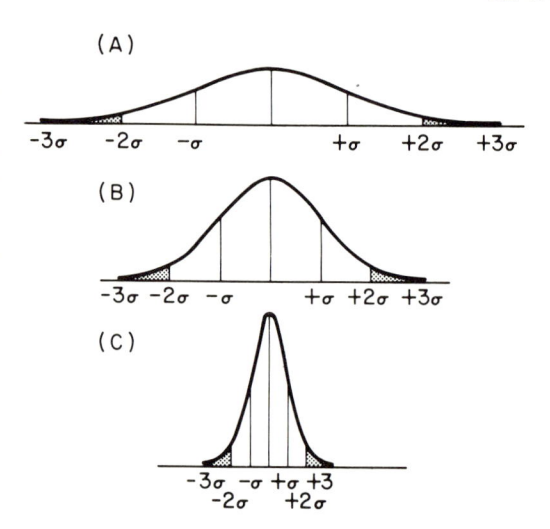

Fig. 19.2

by the symbol s and is thus distinguished from the corresponding value, σ, appropriate to the population as a whole.

It can have a positive and a negative value and is correspondingly measured to the right and left of the mean in graphic representations of the distribution. In Fig. 19.2B for example, the base line has been marked off at $\pm\sigma$, $\pm 2\sigma$ and $\pm 3\sigma$, and the verticals have been erected at these points. It will be seen that nearly all the events or observations summarized by the curve are included between the limits $\pm 3\sigma$. In fact only some 0.27 per cent of the observations lie outside $\pm 3\sigma$, and only about 4.6 per cent outside $\pm 2\sigma$. Figures A and C, representing normal curves in which the distribution of frequencies is respectively wider and narrower than that in B, are also divided in this manner, and it is obvious that the areas in A, B and C marked off by the verticals erected at multiples of the standard deviation in each case bear the same relation to the total area enclosed by the curve.

Since the frequencies of *all* the events or observations are represented by the total area enclosed by the curve, we may express the *probability* of observing one of a given set of these events by the ratio of the area representing the set of events to the total area. Suppose we wish to know the probability of observing events whose frequencies deviate from the mean by more than $+2\sigma$. In Fig. 19.2B it is clearly the ratio of the shaded area on the right of the vertical erected at $+2\sigma$ to the total area. This shaded area is 2.3 per cent of the total so that the probability, P, is 0.023, equivalent to odds of about 42 to 1. In practice it is found that, as probabilities are conceptually simple, it is convenient to deal in rounded values of P rather than whole number values of s or σ. Table 19.2 thus relates convenient rounded values of P to the corresponding values of the normal deviate, d. Each P value takes

Table 19.2

d	1.645	1.960	2.326	2.576	3.291
P	0.1	0.05	0.02	0.01	0.001
Odds to 1 against	9	19	49	99	999

into account both tails of the distribution and so indicates the probability that the observed deviation could occur in either the positive or the negative directions merely by reason of the inherent variability of the assay. Table 19.2 also contains for each value of P the odds against any particular observation occuring by chance.

Standard error of the mean

An interesting feature of observations that are normally distributed is that the means obtained from groups of such observations are themselves normally distributed. The standard deviation, or as it is usually called the *standard error* of the mean, is obtained by dividing the standard deviation of the observations by the square root of the number of observations contributing to the mean. Furthermore, when pairs of means are compared the *differences* between the pairs are normally distributed also, and with a standard error of the difference that can be calculated from the standard errors of the two means in the pair. As we shall see, this is a most important relationship as it provides a basis on which the significance of differences between pairs of observations can be assessed.

Levels of significance

The levels of significance conventionally used in biometrical work correspond to probabilities of 5 and 1 per cent (expressed as $P = 0.05$ and $P = 0.01$) or, more memorably if a little inaccurately, odds of 20 to 1 and 100 to 1. Unless we have good technical and statistical evidence of a reasonable degree of homogeneity among our experimental animals we shall be wise not to accept a single result of 20 to 1 against a chance effect as indicative of a real difference between two groups. On the other hand a requirement to reach 100 to 1 against a chance effect may be unreasonably high and cause us to ignore differences that are real. Of course, whatever the result, every experiment should be repeated, even when a single trial has given a result that passes all the conventional tests of statistical significance. We can never be quite sure that the conditions of a single trial do not include some determining factor other than that which we are trying to investigate. By repeating an experiment several times we are testing, in the only way we can, for the intrusion of unknown factors other than errors of random sampling. When each of several tests gives us a statistically significant answer, and each answer points in the same direction, we have some justification for stating our conclusion in general as opposed to particular terms; but when some tests give different answers from others we must solve the problem of why these differences occur before we can accept the indication given by the original experiment.

If, in a series of repeated experiments, we observe, despite all our efforts to ensure the uniformity of the strain, age, sex and diet of our experimental animals, a tendency to irregular responses, we must always consider the possibility that the host species is heterogeneous in a way that makes different animals respond to our procedure in different ways; some may succumb to a dose of toxin much smaller than others resist, some may respond to doses of a vaccine much smaller than that which evokes no measurable response in others. If heterogeneity appears to be gross the chances of obtaining results in which confidence can be placed are small but with populations of animals of no obvious heterogeneity, but the variability of whose response is nevertheless large enough to suggest heterogeneity, two courses are open to us. We may treat them as homogeneous, and compensate for the large variability by the use of large numbers of animals so as to increase the accuracy of our results. Alternatively, we may attempt by inbreeding to select strains whose response is less variable. These inbred strains, which may or may not be pure lines, may yield much steeper dose-response curves than those of the parent strains, and their use in biological assay permits either a greater increase in the accuracy of an assay for the expenditure of the same number of animals, or measurements of the same accuracy with smaller numbers of animals. For example, Prigge (1937) recorded that the ratio of the smallest to the largest dose of diphtheria toxoid required to immunize the individual animals in a mixed laboratory strain of guinea-pigs against a standard injection of toxin was 1:32,000. Similar tests of pure-line strains inbred by brother-sister matings revealed a strain for which the ratio was as small as 1:25. Lincoln and DeArmon (1959) in comparative infectivity tests of *Bac. anthracis* and *Fr. tularensis* in inbred and ordinary laboratory strains of mice were able to use up to 55 per cent fewer inbred than non-inbred mice for the same degree of precision in their measurements. Inbreeding, however, by no means always results in less variable responses; randomly bred strains may be less variable than inbred (see, e.g., Grüneberg 1954, Biggers *et al.* 1961); and the first generation offspring of two inbred lines less so than the parent strains (Brown and Dinsley 1962). When the animal is to be used for measuring the protective or therapeutic value of antisera or vaccines, the choice between random bred or inbred strains or between one species of laboratory animal or another may, however, be determined not on which strain gives the most accurate answer, but on which strain responds to the agent under test in a manner most resembling that of the animal species for the treatment of which the agent is intended. When a therapeutic agent for use in man is to be tested for efficacy the ultimate experimental animal is, inevitably, man himself, random bred and grossly heterogeneous.

Calculation of the LD50

The information in any sigmoid dose-response curve such as that in Fig. 19.1A can readily be used to calculate an estimate of the LD50. The observed mortalities between 25 and 75 per cent are plotted against the corresponding doses on arithmetic graph paper or on semi-log graph paper, respectively, for cases of normal and log-normal distributions. The dose-response curve is fitted to the plotted points, the intercept of the line of the 50 per cent response with the curve is marked and a perpendicular dropped to the abscissa. The LD50 is then read off at the point where the perpendicular and abscissa meet. Inevitably, however, the accuracy of graphical solutions depends on the efficiency with which the curve is fitted to the observations and, furthermore, the procedure is limited in that mortalities less than 25 and more than 75 per cent make no contribution to the estimate. Alternatives to graphical solutions include arithmetic procedures such as the cumulative response method of Reed and Muench (1938), the moving averages method of Thompson (1947), and the method of Tint and Gillen (1961).

Probits

Estimations of LDs50 using only the responses between 25 and 75 per cent can be wasteful of hard-gained data and methods for utilizing all observed mortalities, other than 0 and 100 per cent, are available. In essence the data are transformed by arithmetic manipulation so that a straight line can be fitted to the points representing the responses to all the doses. As all the doses and responses then contribute to the slope and the position of the dose-response curve, no data are wasted. The most satisfactory transformation differs from one biological system to another, so that for one a normal plot of doses may be satisfactory and for another a log plot may be required. A particularly useful treatment of the responses consists of their transformation to the *probits* of the percentages of the mortalities at each dose (Finney 1971). The values of the probits are such that points derived from sigmoid curves corresponding with a normal or log normal distribution fall on straight lines with probit values in the range of 0 to 10 and a 50 per cent response at the level of probit 5. Thus, when the percentage mortalities in Fig. 19.1A and B are transformed to probits and plotted against the corresponding doses, the sigmoid curves become the straight lines of Fig. 19.3 and LDs50 may be obtained by inspection. Moreover, any deviation of the plot from linearity is indicative of

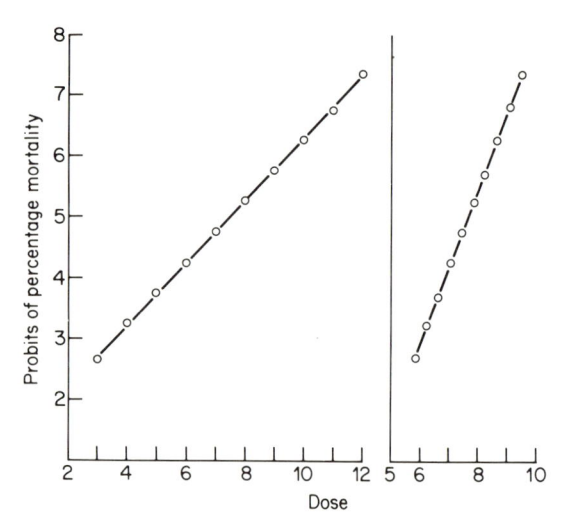

Fig. 19.3

non-conformity of the data with a normal, or, as the case may be, log normal distribution (Gaddum 1933, Bliss 1938, Irwin and Cheeseman 1939a, and b).

Simple and instructive as the manual methods for calculating LDs50 are, all are now rapidly falling into desuetude as computor programs for the purpose become ever more readily available.

Comparisons of LDs50

The assayist who has obtained an estimate of the LD50, x, of a pathogen or toxin for a treated group of animals and another estimate, y, for a control group of untreated animals, may reasonably wish to know whether any observed difference, $x-y$, in the LDs50 can be attributed to the treatment or was due merely to heterogeneity of the animals. To make the assessment it is first necessary to obtain the standard deviations, or, more properly, the standard errors, S_x and S_y, of the two LDs50 and, because the calculation of these values is beset with pitfalls and is rather more difficult than the calculation of the standard errors of means derived from quantitative data (see below), the assistance of a statistician can be very valuable. However, if such help is not available, appropriate methods are provided in the European Pharmacopoeia (1971), by Finney (1971) and by Irwin and Cheeseman (1939a, 1939b). When calculated the two standard errors are combined to

$$S_{x-y} = \sqrt{(S_x)^2 + (S_y)^2}$$

which is the standard error of the *difference* between the LD50 estimates, x and y. The difference may then be compared with its standard error in the form of the ratio $x-y/S_{x-y}$ and a normal deviate obtained. This will be a value akin to the values of d in Table 19.2

Comparisons based on chance events

In some circumstances, especially after immunization with a marginally effective dose of a vaccine, the resistances of the individual animals in a single treatment group may range from the low immunity of those that have responded poorly to the high immunity of those that have responded well. Thus, when a group of such animals is challenged with an appropriate dose of the infective agent or toxin against which the vaccine was intended to provide protection, the animals with low immunity are likely to die and those with high immunity are likely to survive. Further, when there are two, or more, groups of animals each group immunized with a different vaccine, a low proportion of survivors in a group will indicate a poor vaccine and a high proportion a good one. The extent to which it is justifiable to draw conclusions from the proportions of survivors about the potencies of two or more vaccines is based on our knowledge of the distribution of events that, like the immunizability of test animals, is, subject to a large variation inherent in biological systems.

We interpreted the dose-response curves in Fig. 19.1 as expressions of a frequency distribution of individual resistances in the population of animals tested. Dose-response curves of this kind can also be derived from probability arguments about batches of animals of quite uniform resistance; and since this derivation leads to simple and instructive methods of assessing the results of resistance tests, we shall briefly indicate its nature.

Suppose 30 untreated mice and 30 mice immunized against a given bacterium are tested by infection with that bacterium and the number and proportion surviving in each are as follows.

Vaccinated	24 = 80.0%
Controls	6 = 20.0%
Difference	18 = 60.0%

What is the probability that the observed result could occur by chance? We proceed on the assumption that the two groups were identical. The best index, therefore, we have of the resistance of the mice to the dose of the pathogen is the combined death rate in the 60 animals $(6+24)/60 = 0.5$. Since all the mice are by definition of equal resistance, then for each mouse, 0.5 represents the probability, p, of its dying; the probability of its survival is $q = (1-p) = 0.5$. For a group of n mice given this dose, the probabilities of $1/n$, $2/n$ ... n/n mice dying are given by the binomial expansion of $(p+q)^n$. For four mice it is p^4, $4p^3q$, $6p^2q^2$, $4pq^3$ and q^4 respectively. Since $p = 0.5$. these five terms are respectively 0.0625, 0.25, 0.375, 0.25, 0.0625. That is to say, in 100 repetitions of the test, 2/4 deaths would be observed 37.5 times, 1/4 and 3/4 each 25 times, and 0/4 and 4/4 each 6.25 times. These relative frequencies, and those for 8 mice $(p+q)^8$ and 12 mice $(p+q)^{12}$, are plotted in Fig. 19.4, A, B, C. As n increases, the figure become less step-like, and has a decreasing proportion of its total area at the end of each baseline. When n is large, say 1000, the corresponding distribution is an almost smooth curve approximating to the normal curve. The standard deviation of frequency distributions of this kind is $\sqrt{npq}$. Applying this to the distributions A, B and C in Fig. 19.4 and to the case of $(p+q)^{1000}$, we have

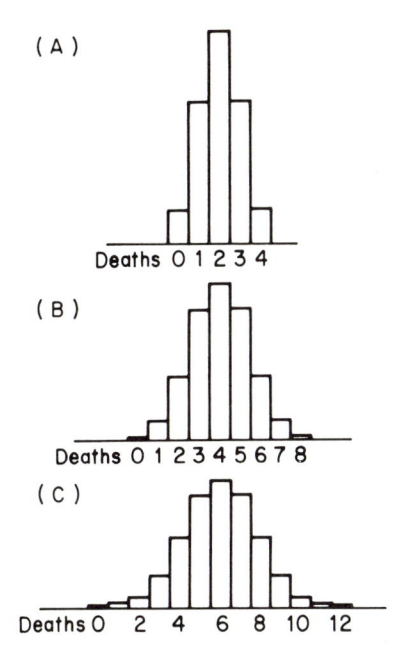

Distribution	Mean	Standard Deviation
A $(p+q)^4$	2 deaths	$\sqrt{1} = \pm 1$ death
B $(p+q)^8$	4 deaths	$\sqrt{2} = \pm 1.41$ deaths
C $(p+q)^{12}$	6 deaths	$\sqrt{3} = \pm 1.73$ deaths
D $(p+q)^{1000}$	500 deaths	$\sqrt{250} = \pm 15.49$ deaths

Or, if we regard mean and standard deviation as proportions, the latter becomes $\sqrt{pq/n}$ and the result is

Distribution	Mean	Standard Deviation
A $(p+q)^4$	0.5	± 0.250
B $(p+q)^8$	0.5	± 0.177
C $(p+q)^{12}$	0.5	± 0.144
D $(p+q)^{1000}$	0.5	± 0.016

As with the normal distribution, the differences between pairs of means are distributed normally, with a standard error equal to the square root of the sum of the standard errors of the two means from which the difference was derived. That is, if x_1 and x_2 are the numbers dying in the two groups of n_1 and n_2 animals, with respective expectation of death p_1 and p_2, the mean proportional death rate and its standard error in the first group are x_1/n_1 and $\sqrt{p_1 q_1/n_1}$ and in the second x_2/n_2 and $\sqrt{p_2 q_2/n_2}$. The standard error of the difference $x_2/n_2 - x_1/n_1$ is then $\sqrt{p_1 q_1/n_1 + p_2 q_2/n_2}$.

When, however, for the purpose of analysis, we assume, as in our example above, that the two groups do not differ significantly, the best measure of p is $(x_1 + x_2)/(n_1 + n_1)$, which we may call p_0. The standard errors of the two means are then $\sqrt{p_0 q_0/n_1}$ and $\sqrt{p_0 q_0/n_2}$ and of the difference between them

$$p_0 q_0 \left(\frac{1}{n_1} + \frac{1}{n_2} \right)$$

In the example $n_1 = n_2 = 30$, $p_0 = 0.5$, and the difference $= (24/30 - 6/30) = 0.6$, with a standard error of

$$p_0 q_0 \left(\frac{1}{n_1} + \frac{1}{n_2} \right) = \pm 0.1265;$$

The difference is $0.6/0.1265 = 4.74$ times its standard error which corresponds to $P = 0.000\,001$, a value to convince even the severest critic that the resistance of the mice to the test dose of bacteria was certainly increased by prior immunization.

The applicability of the arithmetic of the normal distribution to our test animals depends on how our experimental data approximate to the curve.

In the kind of experiments with which we are usually concerned there are two factors that affect this degree of approximation. One is the number of animals we use. We shall never use enough to give us a really smooth curve. In fact our distribution will usually be nearer to that shown at Fig. 19.4C than to the true normal curve. The other factor is that though the normal curve is symmetrical, corresponding to the expansion of $(0.5 + 0.5)^n$, the curve with which we are actually concerned will usually be asymmetrical. The symmetrical disposition of the frequencies round the mean in Fig. 19.4 A–C is, of course, the result of taking $p = q$. If the dose administered on the average kills 30 per cent of the animals, $p = 0.3$ and $q = 0.7$; and the relative frequencies of 4, 3, 2, 1 and 0 deaths for batches of four mice are given by the binomial expansion of $(0.3 + 0.7)^4$, and are respectively 0.0081, 0.0756, 0.2646, 0.4116 and 0.2401. The distribution is skew; one death in four will occur with the highest frequency, and the mean rate in four mice is 1.2. As p becomes progressively smaller or larger than 0.5, the distribution becomes more skew, and the use of means and standard deviations based on the mathematics of symmetrical distributions becomes less applicable— for example, to situations where death rates are either very low or high, or, if we are comparing two groups, where the overall average mortality is not in the region of 50 per cent. But when the number of animals in the group is not too small, say not less than 30, and the chance of the less likely event—life or death—is not less than 0.2, the odds calculated on the basis of the standard deviation are sufficiently accurate. Moreover, we can apply the calculation to differences between means with even greater confidence, since differences between means from relatively skew distributions tend to be normally distributed.

Estimations and comparisons of ImDs50

In the early stages of the development of a vaccine, an antiserum or an antimicrobial drug, a comparison of the resistance of a group of animals treated with the drug with that of an untreated group is often the only way in which biological activity can be recognized. Thus the methods such as those described in the foregoing sections which indicate whether or not animals treated with a particular preparation are significantly more resistant than similar animals treated with a placebo are of great value, as they inform the researcher that he is working with a truly active substance—or in less happy circumstances with an inactive one. Once a protective preparation has been identified, however, subsequent research is almost certain to require comparisons that are rather different— comparisons not of the resistances of treated and untreated groups of animals but comparisons of the protective potencies of two preparations. Such comparisons do not require estimations of LDs50 but are designed to estimate another value, the ImD50 or Immunizing Dose 50. This is the dose of a preparation, often a vaccine, which protects 50 per cent of the animals to which it is given from a challenge which, though given in the same amount to all animals in a particular test, is seldom the same from test to test. The procedure in its simplest form is the three plus three $(3 + 3)$ quantal assay.

The quantal assay

The $3 + 3$ quantal assay derives its name from the three graded doses of each of the two preparations that are used to treat six groups of experimental animals and the use of a quantal response, usually death, to indicate the effects of the treatments. Larger assays using more doses of each preparation and more than two preparations are feasible and take the forms $4 + 4$, $3 + 3 + 3$, etc. as appropriate to a particular investigation. In the $3 + 3$ form, as in the comparison of the potencies of two vaccines, the three graded doses of the two vaccines are prepared, usually in 2-fold to 5-fold dilution steps, so that the high doses immunize almost all of the animals to which they are given, the low doses immunize very few of the animals to which they are given, and the intermediate doses immunize about 50 per cent of the animals to which they are given. Thus when, after a period of 2 to 4 weeks during which time immunity develops, the animals are all challenged with the *same* amount of toxin or infectious agent, which need not be very precisely defined, most of those given the high doses of vaccine, only a few of those given the low doses, and about 50 per cent of those

and the numbers of animals surviving the challenge at each dose. As the doses are in a logarithmic series they need no transformation, but the numbers of survivors are transformed first to percentages and then to the probits of the percentages. Figure 19.5 shows the probits of the percentages of survivors plotted against the log-dose for each preparation and the resultant straight lines that may be described by the equation

Table 19.3

	Assay A						Assay B					
	Reference (R$_1$)			Test (T$_1$)			Reference (R$_2$)			Test (T$_2$)		
Vaccine dose in µl	50	10	2	50	10	2	50	10	2	50	10	2
Surviving mice of 32	26	9	1	29	17	5	24	8	1	24	19	13
Per cent survivors	81	28	3	91	53	16	75	2.5	3	75	59	41
Probits of per cent	5.9	4.4	3.1	6.3	5.1	4.0	5.7	4.3	3.1	5.7	5.2	4.8
Protective dose 50 (µl)		17.4			7.7			19.0			4.1	
Potency ratio (R/T)				2.2						4.7		
Potency ratio (computed)				2.224 30						4.684 30		
Fiducial limits of ratio				1.28–3.99						2.09–13.60		
Slopes		2.00			1.66			1.85			0.65	
Total deviations from parallel line model $-\chi^2$				0.7261						9.0420		

given the intermediate doses have sufficient immunity to resist the challenge and thus survive.

Results generated by two 3+3 quantal assays in which two test vaccines, T$_1$ and T$_2$, are assayed against two reference vaccines, R$_1$ and R$_2$, are given in Table 19.3. The Table shows the three doses of each vaccine

for linear regression, $y = a + bx$, in which y is the logarithm of the immunizing dose, x is the probit of the percentages of animals that survive the challenge at each dose, a is a constant, and b is the regression coefficient or 'slope'. At any probit value, therefore, the horizontal distance between the two dose-response

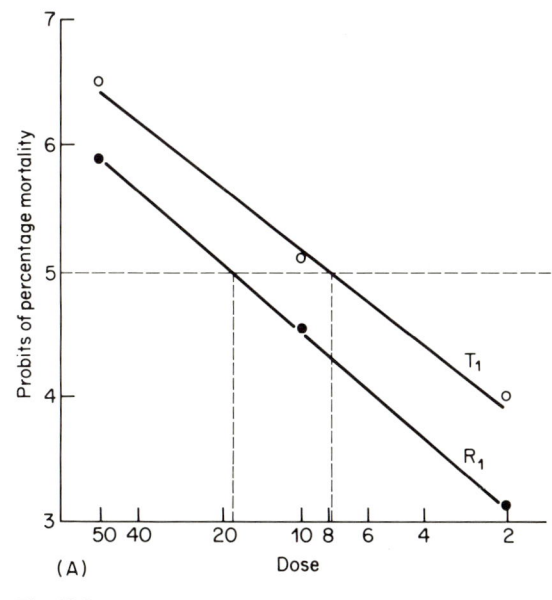

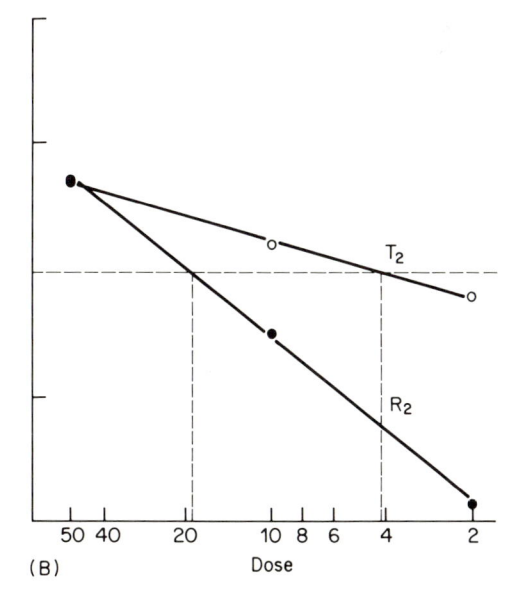

Fig. 19.5

lines on each graph is a measure of the relative potency of the preparations at that particular level of immunity, or response.

From Fig. 19.5A it is possible to read off on the abscissa the doses of both the reference material (R_1) and the test material (T_1) that protected 50 per cent of the animals (probit value = 5). The ratio of the doses, 17.4/7.7 is thus the measure of the potency of the test vaccine in terms of the reference one and shows that, volume for volume, the test vaccine was 2.2 times as active as the reference one. As the dose-response lines deviate little from parallelism this ratio may be considered to apply to all doses of the two preparations. From Fig. 19.5B it is possible to obtain the ED50s of the two preparations in the same way and to find that at the 50 per cent protection level the potency of the test vaccine (T_2) was as much as 4.7 times that of the reference one (R_2). However, as the two dose-response lines are far from parallel this particular relationship applies only at the ED50 level. At higher levels the protective efficacies of the two preparations approximate until at 50 µl both preparations protect the same percentage of animals, i.e. about 75 per cent. At lower doses the test vaccine, possibly due to the presence of an adjuvant, becomes relatively more and more potent than the reference one.

Although graphical treatment of the results of 3+3 quantal assays explains the principles that underlie the calculation of relative potency, it is today seldom used. Computerized probit analysis (Finney 1971) has almost completely superseded graphical and manual methods (European Pharmacopoeia, 1971) of calculation and can, in seconds, handle the data from experiments with several preparations, with inconsistent numbers of treatment groups of different size, and with irregular dose intervals. Such analyses provide not only very rapid estimates of potency with their fiducial limits but also the results of tests based on χ^2 of the validity of the assay. These warn the assayist of inadequate regression of response on log-dose and of significant deviations of the dose-response lines from linearity and parallelism.

The results of the analysis of the data in the upper part of Table 19.3 are given in the lower part of the table and there is little difference between the relative potencies of T_1 and T_2 as calculated graphically and as computed. In the assay of T_1 against R_1 the value of χ^2 at 0.7261 is clearly less than that (7.815 for three degrees of freedom) indicative of significant deviations of the log dose response lines from either parallelism or linearity and no evidence of invalidity is detected. In the assay of T_2 against R_2 the large value of χ^2, 9.0420, clearly exceeds the value indicative of significant deviations and the assay is accordingly invalid. An indication of the cause of the invalidity is to be found in the very considerable difference in the regression coefficients (slopes) of the two preparations R_2 and T_2.

The measurement of resistance by means of quantitative responses

So far in this chapter we have considered only those responses that are of the 'all-or-none' type—survival or death of the laboratory mouse, susceptibility or immunity of an animal after protective inoculation—and are known to statisticians as quantal. In some circumstances, however, it is advantageous to use some quantitative measure that is, in theory at least, infinitely variable. The size of the lesions that occur in the skin of guinea-pigs after intradermal injection of diphtheria toxin (Knight 1974), the agglutinating antibody that is evoked in mice by pertussis vaccine (Evans and Perkins 1954), and the time between injection of a lethal dose of toxin and death, are all examples of such measures. An advantage of some quantitative procedures, e.g. antibody titration, is that repeated measurements can be made on the same animals to provide an account of increasing or waning resistance over a period; a disadvantage is that the procurement and titration of a large number of sera can be both tedious and expensive. Moreover, despite much additional work the precision of the results is not greatly better than that provided by quantal assays (van Ramshorst 1972).

In its simplest form the quantitative assay contains only two groups of animals, but, because of the quantitative nature of the information from each animal, these two groups alone can provide convincing evidence of different responses, especially of responses to immunogens. Suppose that there are two groups of guinea-pigs, one containing n_1 and the other n_2 animals, and that the animals of the first group are immunized with a plain diphtheria toxoid and those of the second group with adsorbed diphtheria toxoid. A month after immunization all the animals are bled and the sera are titrated individually for diphtheria antitoxin and it is found that the animals immunized with plain toxoid have a mean titre of $\bar{x}_1$ and those immunized with the adsorbed toxoid a mean titre of $\bar{x}_2$. The standard deviations of the two groups of observations, s_1 and s_2, can be obtained, each as the square root of the sum of the squares of the deviations of the individual titres from the corresponding means divided by the number of observations less one,

$$\sqrt{\frac{\Sigma(x-\bar{x})^2}{n-1}}$$

Division of the standard deviations by the square roots of the numbers of animals in each group, respectively, provides estimates of the standard errors of the two means. Further the standard error of the distribution of the differences is the square root of the sum of the squares of the standard errors of the two means. We are now in a position to determine the ratio of the difference between the means to its standard error from the equation:

$$d = \frac{\bar{x}_1 - \bar{x}_2}{\sqrt{\left(\frac{s_1}{\sqrt{n_1}}\right)^2 + \left(\frac{s_2}{\sqrt{n_2}}\right)}}$$

By entering a table of normal deviates such as Table 19.2 at the value of d, we can find the probability that the observed difference between the means $\bar{x}_1$ and $\bar{x}_2$ could have occurred by chance.

Student's *t* test

Although the procedure just described has wide applicability in biological work the reader must be warned that it is unreliable when the sizes of the samples are smaller than 30 individuals. As group size is progressively reduced the distribution curve becomes flatter and flatter and the relation between d and probability characteristic of the normal curve no longer holds. Instead of the normal deviate, d, a rather different value known as *Student's t* is calculated:

$$t = \frac{\bar{x}_1 - \bar{x}_2}{s\sqrt{\frac{1}{n_1} + \frac{1}{n_2}}}$$

The values n_1, n_2, $\bar{x}_1$, and $\bar{x}_2$ are as in the previous calculation but s represents an estimate of the standard deviation based on both sets of observations. It is obtained by calculating the two sums of squares about the means, adding these sums, dividing by $n_1 + n_2 - 2$, and taking the root,

$$s = \sqrt{\frac{\sum(x_1 - \bar{x}_1)^2 + \sum(x_2 - \bar{x}_2)^2}{n_1 + n_2 - 2}}$$

A further difference with the *t*-test is that, because the distribution changes with the size of the samples, the relation of a particular value of t to its corresponding probability changes also. When a value for t is calculated, therefore, it is necessary to enter the table of t values at the point corresponding to the appropriate number of degrees of freedom, which are $n_1 + n_2 - 2$, in order to read off the relevant probability. Despite these minor complexities the t-distribution is of great value in biological assays, as it is applicable in so many situations in which, either by reason of availability or costliness, the numbers of possible observations are small.

Assay with six treatment groups

A more complex and informative version of the quantitative assay procedure uses the format of the six treatment groups—high, intermediate and low doses of two preparations—that features in the 3+3 quantal assay. The procedure is well suited to comparisons of the potencies of two products, such as vaccines, and the calculation of the relative potencies and their fiducial limits is based on the regression of the responses on the logarithms of the doses. Graphical representations of the results may be a sufficient guide to the relative potencies in some cases but procedures for calculating regression lines based on the least sum of squares of the deviations from the dose-response lines are often applied. Computerized analysis does the mathematics much more rapidly, however, and the analysis of variance usually built into such analysis indicates any sources of invalidity.

Although attractive in principle, quantitative assays have found few applications in microbiological research and standardization and their use today is limited to a few special situations, for example, the provision of supplementary information in the establishment of a new standard or reference preparation.

Estimations of protective potency

Important as are the techniques for measuring resistance and protective potency in the research laboratory, it is for the routine quality control of vaccines that they are most used today. A vaccine that contains killed bacteria, bacterial components, bacterial toxoids or killed viruses must generally have, in the recommended dose, sufficient immunologically active material to evoke immunity in every vaccinee to whom it is given. Obviously, even with poorly immunogenic substances, this may be ensured merely by increasing the amount of active material in each dose, but only with the disadvantage that each dose then becomes more costly to make and may approach or even exceed the level at which local and general side effects become unacceptable. Thus, for both economic and toxicological reasons, adequate immunogenicity must always be attained with the smallest possible dose of immunogen. It is the need to demonstrate adequate immunogenicity despite this severe constraint that predicates the need for routine assays of immunogenicity.

The simplest assays of immunogenicity are those which require a vaccine merely to evoke a stipulated response in a stipulated proportion of animals. As we have seen, the variability and heterogeneity of biological systems as complex as whole animals reacting with immunogens, toxins and living pathogens are quite beyond exact specification, and all we can hope to achieve is a broad similarity of circumstances, from the exploitation of which we shall arrive at approximately the same conclusion as before. When, however, such tests are made in order to specify an immunological product like a vaccine or an antitoxin for human use, the range of unpredictable variation within biological test systems may be so large that the answer we get from it is clinically unreliable. Suppose we are testing the immunizing potency of diphtheria toxoid on an apparently average batch of guinea-pigs. If in fact the animals were exceptionally easy to immunize, we should overestimate the potency of a given weight of the toxoid; and we should underestimate it with guinea-pigs of more than average resistance to immunization. Diphtheria toxoid happens to be relatively harmless in large doses, so that in this instance we can allow for overestimation of potency by using plenty of toxoid for routine immunization. This will mean that very large amounts of an underestimated toxoid may be unwittingly injected, which is wasteful, but not dangerous. With more toxic immunizing substances, obviously we cannot risk making such allowances for overestimated potency. Moreover, the allowances we make are in all cases largely guess-work. For these reasons this method of *direct* estimation of potency from the magnitude of the response in the test animal is to be avoided whenever possible.

The uncertainty about a biological test system may be eliminated, at least in part, by abandoning the concept of the standard animal, as typified by the 250–300 g Hartley guinea-pig, and taking advantage of the methods that are now available to preserve test materials in ways that stabilize attributes such as toxicity, infectivity and immunogenicity. Storage in glycerol-saline at low temperature, drying, lyophilization, are all used as appropriate to furnish test materials with biological properties that decline only slowly, if at all, with time. Although, by reason of the arguments rehearsed above, it is never possible to have absolute certainty that such materials are utterly stable, there is now overwhelming evidence that many biological substances, if correctly prepared, are sufficiently stable to maintain their properties for very long periods and so be suitable for use as reference preparations.

It might be thought from reading the earlier sections of this chapter that among reference preparations the most useful would be materials such as bacterial toxins and various infectious agents that might be used to test the immunity induced in experimental animals by vaccines. Such, however, is not the case. Toxins and infectious agents are generally less stable when preserved than are vaccines and antisera. Most assays of vaccines designed to include reference materials use as the reference a preserved form of the vaccine. This arrangement has the advantages that, the test material and the reference being generically similar, like is compared with like and the quantitation of the challenge is not critical as all the host animals receive the same dose.

For most comparisons we use the 3 + 3, or possibly larger, quantal assay. The responses that we measure are usually the death rates in the groups but can, as in the assay of diphtheria vaccine prescribed by the EP, be erythema after intradermal injection of toxin or, in the assay of tetanus vaccine, the appearance of paralysis after injection of toxin. It is, of course, important that the test is measuring the same kind of immunity as that we wish to induce for medical or veterinary purposes (see Miles 1954, Morrell and Greenberg 1954). There are, however, no general rules we can apply. We might expect that it would be best to test the immunized animal by infecting it in the most natural way. None the less, with whooping cough vaccine, the best laboratory index of the potency in man proves to be not that tested by infecting the lungs of mice but that tested by the highly artificial method of intracerebral infection. This is not to say that laboratory tests should not attempt to reproduce the circumstances of natural infection.

The results of such an assay may, as described previously, be treated graphically or subjected to probit analysis. Both methods allow the calculation of the potency of the test preparation or preparations in terms of the standard. If the standard preparation has not been assigned a unitage the potency of the test preparation can be expressed relative to a weight of standard, to a volume, or to the contents of an ampoule. When the standard has a unitage, usually today expressed in International Units (IU) per ampoule, the potency of a test preparation can be calculated in units per ml, per mg or per dose as is most convenient. Thus we find in the European Pharmacopoeia, for example, that each dose of pertussis vaccine must contain at least 4 IU of antigenicity calculated as the estimate of potency of the vaccine and, for rather confusing historical reasons, that each dose of adsorbed tetanus vaccine must contain not less than 40 IU calculated as the lower fiducial limit of the estimate of potency. Potency requirements stated in terms of an estimated potency have proved satisfactory in practice but can be misleading in that, depending on the confidence limits characteristic of the assay system used to estimate them, the true potency of a preparation that satisfies the requirement may be as little as one-half or even one-third of the estimated value. Potency requirements stated in terms of lower confidence limits of potency avoid such delusions and ensure that only very rarely is the true potency of a preparation less than the minimum required potency.

International standards

There is now a wide range of international standards available for the regulation of the potency of immunological products (WHO 1979) and, although such standards are not generally available for routine quality control purposes, they can often be obtained for the purpose of establishing national and laboratory standards.

The weight of each standard that has unit potency has no special significance, being a convenient and arbitrary value, peculiar to the preparation selected for each standard; and it is arrived at by international agreement. Nor is it important that no unit value is assigned to a standard; any stated amount of the standard or reference preparation serves for the purposes of comparative assay. Some of the standards are for diagnostic reagents, like agglutinating sera.

Not all immunological products can be standardized in this way, because in some instances no preparation stable enough for a standard has yet been devised. Vaccines composed of living organisms are, for obvious reasons, particularly difficult to standardize. Nevertheless, a beginning has been made with BCG, and with a number of viruses from which stable preparations can be made. But with products like yellow-fever vaccine we have to rely on direct tests on animals or tissue culture for an indication of potency.

(For discussions on the establishment and use of immunological standards, see Hartley 1935, 1945, Miles 1949, 1951, 1953, Maaloe and Jerne 1952.)

Field trials

The field trial of a vaccine or of an antiserum is merely a technique for assessing, usually in a quantal manner, the ability of one or more preparations to induce resistance to a particular infection in individuals of a target population. Deliberate selection of the test animal, particularly when that animal is man, and standardization of environmental factors are usually impossible, and efforts to ensure that a trial yields a significant result are largely restricted to the admission to the trial of adequate numbers of entrants. Furthermore, as pestilence always strikes unpredictably, we have to consider in detail the operation of chance. Because in almost all field trials the observed phenomena, immunity or susceptibility, death or survival, are of a quantal type, they may usually be considered as chance events with parameters defined by the binomial system. Thus, let us suppose that we are provided with the results of a trial of an influenza vaccine conducted in two groups of 1000 school children, one group being given placebo and the other the vaccine. Further, let us suppose that after one year in which 100 of the control children and 70 of the vaccinated children were attacked by 'flu' we are required to comment on the statistical significance of the 30 per cent reduction in the attack rate. Our task is, as before, to compare the difference in the attack rates, 100 per 1000 and 70 per 1000, in the two groups with the standard error of the difference of the attack rates using the formula:

$$\frac{x_p - x_v}{\sqrt{p_0 q_0 \left(\frac{1}{n_p} + \frac{1}{n_v}\right)}}$$

in which x_p and x_v are the attack rates, respectively, in the numbers of children receiving placebo (n_p) and vaccine (n_v). Substituting results for symbols we obtain:

$$\frac{0.1 - 0.07}{\sqrt{0.17 \times 0.83 \left(\frac{1}{1000} + \frac{1}{1000}\right)}} = 1.78$$

This value, being less than two standard deviations from the mean, fails to convince us that the observed results were due to the effects of the vaccine for they could have occurred by chance more often than, on average, once in 20 trials.

However, in vaccine trials of the type just described, the expectation is always that the vaccine will reduce the incidence of an infection, not predispose to it. If this be the case the attack rate in the control children will always be greater than in the vaccinated children and the value $x_p - x_v$ will always be positive. Thus only the right hand side of the distribution of the values of $x_p - x_v$ will apply. Now, as the tails of the distribution beyond two normal deviates each contain only 2.3 per cent of chance observations, the use of two normal deviates in a test in which we are considering only one tail provides us with a test of significance at about $P = 0.025$ rather than at about $P = 0.05$. If, therefore, we hold to the view that a result that could occur by chance only once in 20 trials is evidence of vaccine efficacy, and well we may for fear of erroneously discarding an effective product, we need to adopt a value of the normal deviate which leaves 5 per cent of chance observations in *one* tail. This value is that for 10 per cent in the two-tailed test, or 1.645. As the deviate obtained from the trial results was 1.78 we can be confident that, although the vaccine was not very effective in reducing the attack rate, the result that it provided would have occurred by chance less often than once in 20 trials.

The utility of the findings of a clinical trial are, inevitably, founded on the data that the trial provides and no amount of analysis can extract information of value from a trial inadequately designed or of inadequate size. Thus, in the planning of every trial it is necessary to consider the attack rate, the protection to be expected, and whether a comparison is to be made between a vaccine and a placebo (one tail) or between two vaccines of similar efficacy (two tails). Furthermore, it is essential to plan with a view to the confidence that is needed in the final result, that is, the need to design a trial in which an inactive product is unlikely to be shown to be active or an active product is unlikely to be shown to be inactive. In this connection we can fix the probability, say one in 20 or one in 100, that a significant result has occurred by chance, but the more we increase the odds to reduce the risk of declaring an inactive product to be active, the more we increase the risk of failing to recognize an active one. The solution to this problem is to endow each clinical trial at its inception with sufficient 'power' (Lachin 1981) to provide a result in which the required confidence can be placed.

The power of a field trial is the probability that, if treatment groups really differ, the trial will reveal a statistically significant difference. It is usually expressed as a percentage, say 80 or 90 per cent, and it is dependent upon the real difference in the treatments and the number of entrants to the trial. To calculate the numbers of entrants needed in any set of circumstances we use the following derivative of Lachin's

(1981) equation

$$N = \frac{4(Z_\alpha - Z_\beta)^2 \cdot p(1-p)}{(p_p - p_v)^2}$$

in which,

N is the number of entrants,

Z_α is the normal deviate (one tail) corresponding to the probability that is acceptable as evidence of efficacy,

Z_β is the normal deviate corresponding to the probability that is acceptable that although the two groups actually differ they are found by the trial not to differ. The converse, $1-\beta$, is the probability that the trial will show a difference when such a difference exists. It is the power of the trial.

p_p, p_v, and p are the attack rates in the placebo group, the vaccinated group and the combined groups, respectively.

In many infections the attack rate in unprotected individuals is well known and it is relatively easy to stipulate the confidence that is required in the result of a trial. Consequently, when vaccines of different presumed efficacies are tested against a placebo, the equation may be used to calculate the number of entrants needed for a trial of any degree of rigour. Table 19.4 shows the number of entrants necessary to provide an 80 per cent assurance that a true difference is not missed in the trials of vaccines with efficacies of 10 to 90 per cent against an infection with an attack rate of 1.0 per cent in the unprotected. Clearly, with a vaccine that is highly effective, e.g. 90 per cent, two groups of about 1700 entrants are large enough to ensure that an apparent lack of efficacy could have occurred by chance only once in five trials and any efficacy observed would have odds of at least 20 to 1 against being due to chance. Larger groups of entrants are needed for less effective vaccines and for comparisons of one vaccine with another, especially when the comparison is between two vaccines of similar potencies. Furthermore, it must be remembered that the number of entrants provided by the equation is merely the minimum required to obtain a statistically significant difference if the vaccine actually has some protective effect. To estimate the efficacy of a vaccine in terms of the percentage protection that it is likely to provide in wide scale use, a much larger number of entrants is required.

It should be emphasized that field trials in human populations raise important ethical problems which are discussed by Hill (1963). These include decisions whether the proposed treatment is likely to harm the test subject; whether a known effective treatment should be withheld from those in the control group; and what kind of subject may reasonably be brought into the trial. As to the possible harmful effects of the treatment, it must be noted that the use of immunological products is peculiarly beset with dangers that must be borne in mind when trials are made in man. They are fully discussed in Wilson's (1967) monograph on the hazards of immunization.

It is often desirable that no trial in man should be larger or take longer than is needed to obtain a significant answer. In some circumstances this economy of effort can be ensured by adopting a sequential method (see Armitage 1960). In essence, subjects are added to the trial in pairs, one of each pair being treated, the other untreated, and any difference in reaction between the members of the pair is recorded. By suitable analysis, it is possible to determine when, as pairs were added to the trial, an observed advantage conferred by the treatment becomes statistically significant, and the trial may accordingly be concluded.

The value of quantitative measures of error

The quantitative results of biological tests or observations can be used with confidence only if the error to which the estimation is subject is known with sufficient accuracy. Statistical analysis of the experimental data may reveal both the errors due to the variability of the material used, and the errors inherent in the technique of experiment; and with a measure of the total error we may assess the statistical significance of any numerical result. Without such an estimate of error, we can compare neither two experiments by one worker, nor the experiments of several workers. It is not, however, enough to submit finished data to statistical tests. Simple statistical principles of the kind we have been considering are not rules to be applied in drawing conclusions from an experiment that has been done—*they are part of the experiment itself, an essential factor in its proper planning.*

The statistical method does not consist merely in arithmetical analysis of numerical results. It includes the detection and elimination of bias, and the proper appreciation of the variables to be considered, both in the design and technique of the experiment, and in its analysis. It is often possible, for example, by a simple consideration of the numbers of animals available for a given test, to say that the experiment is unlikely to give a significant answer. In such a case, the test is much better made upon a larger scale, or abandoned altogether.

A word of warning is needed. No conviction of the

Table 19.4

Efficacy of vaccine in per cent	Number of entrants required to establish efficacy
10	4900069
20	1163115
30	109795
40	25820
50	15534
60	10064
70	6862
80	4867
90	3523

necessity for the proper design of experiments, and for the statistical analysis of the results obtained, should lead us to overestimate the support that a statistically significant result can give to the particular hypothesis the experiment was intended to test. As Fisher (1970) points out, all that a properly designed experiment can test is the '*null hypothesis*'; that is, the hypothesis that there is no difference between the control and test systems that we are studying. If, for example, $P = 0.25$ in the defined conditions of a therapeutic test of a fluid A in a fatal bacterial infection, no one could decide, without further experiment, whether the effect observed was due to 'chance' or to A. If $P = 0.001$, we can say that the effect is unlikely to be an example of a chance variation of the kind we expect from the materials used in the test; but we cannot say that it was *therefore* due to the administration of a few millilitres of the fluid A. The statistically significant result indicates that the test and control groups differ in some important respect and therefore permits us to consider the fluid A as a possible cause of that difference; but the decision that A *is* the cause of the observed reduction in mortality in the test group is independent of the statistical demonstration of significance, and in a final analysis depends on the experimenter's knowledge of his particular science and his appreciation of the consistency of his ideas of A's action with the general theory of that science. If A were normal saline, the conclusion, no matter how statistically significant the experimental result, that a small quantity of saline had cured the infection, would rightly be considered as suspect. If A, on the other hand, were a solution of a substance known to be antibacterial and non-toxic to the animals used in the experiment, then these pieces of scientific knowledge, coupled with the disproof of the null hypothesis, would go far towards justifying the conclusion that A was capable of curing a particular type of infection.

The aim, then, of the statistical method in experiments involving the comparison of two or more groups of events is to discover the probability that the null hypothesis is untrue; and only in those cases where the null hypothesis is disproved is there any justification for considering the possibility that an observed effect resulted from any of the deliberate variations in the conditions of the experiment.

The proper design and analysis of quantitative tests is essential for the efficient practice of a science. That the record of much immunological work contains few examples of statistical analysis is no proof that such analysis is unnecessary. Sometimes the results of an experiment are so clear cut that any arithmetical analysis is unnecessary; on many other occasions it will be found that the data are incapable of supporting the conclusions reached by the investigators. There is no doubt that much of value has been learned without any consideration of statistical design, but as much could have been learned with a great deal less expenditure of time and materials had the experiments been designed more skilfully.

The voyager who ignores tidal streams and compass deviations will arrive somewhere in the long run; but he is less likely to identify correctly the place at which he has arrived than he who allows for these variables by the established methods of elementary navigation.

References

Armitage, P. (1960) *Sequential Medical Trials*. Blackwell, Oxford; (1971) *Statistical Methods in Medical Research*. Blackwell, Oxford.
Bailey, N. T. J. (1981) *Statistical Methods in Biology*, 2nd edn. English Universities Press, London.
Biggers, J. D., McLaren, A. and Michie, D. (1961) *Nature, Lond.*, **190**, 891.
Bliss, C. I. (1938) *Quart. J. Pharm.*, **11**, 192.
Brown, A. M. and Dinsley, M. (1962) *Nature, Lond.*, **196**, 910.
European Pharmacopoeia (1971) Maisonneuvre S.A., 57-Sainte-Ruffin, France.
Evans, D. G. and Perkins, F. T. (1954) *J. Path. Bact.* **68**, 251.
Finney, D. J. (1971) *Probit Analysis*, 3rd edn. Cambridge University Press, London; (1978) *Statistical Methods in Biological Assay*, 3rd edn. Griffin, London.
Fisher, R. A. (1960) *The Design of Experiments*, 7th edn. Oliver and Boyd, Edinburgh: (1970) *Statistical Methods for Research Workers*, 14th edn. Blackwell, Oxford.
Gaddum, J. H. (1933) *Spec. Rep. Ser. med. Res. Coun., Lond.*, No. 183; (1945) *Nature, Lond.*, **156**, 463.
Greenwood, M. (1924) *Lancet*, **ii**, 153.
Grüneberg, H. (1954) *Nature, Lond.*, **173**, 674.
Harper, N. W. (1971) *Statistics*. 2nd edn. MacDonald and Evans Ltd, London.
Hartley, P. (1935) *Bull. Hlth Org. L.o.N.*, **4**, 735; (1945) *Ibid.*, **12**, 76, 98.
Hill, A. B. (1963) *Brit. med. J.*, **i**, 1043.
Irwin, J. O. and Cheeseman, E. A. (1939a) *J. roy. statist. Soc.*, **6**, 174; (1939b) *J. Hyg., Camb.*, **39**, 574.
Knight, P. A. (1974) *J. biol. Stand.* **2**, 69.
Lachin, J. M. (1981) *Controlled Clinical Trials* **2**, 93.
Lincoln, R. E. and DeArmon, I. A. (1959) *J. Bact.*, **78**, 640.
Maaloe, O. and Jerne, N. K. (1952) *Annu. Rev. Microbiol.*, **6**, 349.
Miles, A. A. (1949) *Bull. Wld. Hlth Org.*, **2**, 221; (1951) *Brit. med. Bull.*, **7**, 283; (1953) *J. Amer. pharm. Ass.*, **42**, 226; (1954) *Fed. Proc.*, **13**, 799.
Morrell, C. A. and Greenberg, L. (1954) *Fed. Proc.* **13**, 808.
Prigge, R. (1937) *Z. Hyg. InfektKr.*, **119**, 186.
van Ramshorst, J. D., Sundaresan, T. K. and Outschoorn, A. S. (1972) *Bull. Wld. Hlth. Org.* **46**, 263.
Reed, L. J. and Muench, H. (1938) *Amer. J. Hyg.*, **27**, 493.
Thompson, W. R. (1947) *Bact. Rev.*, **11**, 115.
Tint, H. and Gillen, A. (1961) *J. appl. Bact.*, **24**, 83.
Trevan, J. W. (1927) *Proc. roy. Soc.*, B, **101**, 483; (1929) *J. Path. Bact.*, **32**, 127; (1930) *Ibid.*, **33**, 739.
W.H.O. (1979) *Biological Substances*. International Standards, Reference Preparations and Reference Reagents. *Wld. Hlth. Org.* Geneva.
Wilson, G. S. (1967) *The Hazards of Immunization*. The Athlone Press, London.
Yule, G. U. (1924) *Med. Res. Coun. industr. Fatigue Board, Rep.* 28.

Index

The index should be read in conjunction with the chapter contents lists

Acholeplasms, 23
Acid-fast stain, 19
Acquired immunodeficiency syndrome (AIDS), 410
Aeration, effect of on growth, 57
Affinity of antibody, 323
Agglutination reactions, prozone in, 325
Air, bacteriology of, Chapter 9, 251 *et seq.*
Airborne infection, 252, 257
Aldehydes, disinfectant action of, 79
Algae, 222, 225, 228
Alkalies, effect of on bacteria, 79
Alleles, 146, 155
Allergy, 390
Allograft rejection, 312
Amino acids, biosynthesis of, 51
Anaphylaxis, 389
Anaphylatoxin, 385
Antibiotic agents, 97 *et seq.*
   therapy, 97 *et seq.*
Antibiotic resistance, plasmid-determined, 164–166
Antibodies, cytophilic, 331
   cytolytic, 331
   cytotoxic, 332
   opsonic, 332
Antibodies, diversity of, 381
Antibody immune responses, 297
Antifungal drugs, 136
Antigen-antibody reactions, 12
Antigen-antibody reactions *in vitro*, 319
Antigen-antibody reactions *in vivo*, Chapter 12
Antigen processing, 304, 311
Antigens. See Chapter 13
Antigens, bacterial, Chapter 13
   cell-associated, 345–367
   composition of, 339–344
   fimbrial, 345–348
      role of in adhesion, 347
   flagellar, 345
   glycolipid, 361
   immunogenicity of, 349–351
   major extracellular, 367–369
   polysaccharide, 349–361
   protein, 345 *et seq.*, 348
      gram-negative outer membrane, 361–367
      staphylococcal, 349
      streptococcal, 348
Antituberculous drugs, 134
Antiviral drugs, 139
Arthus reaction, 394
Ataxia telangiectasia, 379
Ataxia telangiectasia, 405
Autoclaves, 77
Automation, 10

Avidity in relation to antigen-antibody reactions, 323

Bacitracin, 111
Bacterial genetics, 8
Bacteriocines, 166, 247
Bacteriophage, 7
Bacteriophage, causing bacterial variation, 151 *et seq.*
Bacteriophages, 177–216
   as diagnostic agents, 177, 217
   assay of, 180
   burst size of, 182
   classification of, 186
   cosmid vectors of, 215
   genetics of, 208
   Löcher formed by, 180
   lysis by, 202
   lysogeny, 202
   mutation in, 209
   nucleic acids in particles of, 187
   plaque formation by, 180
   recombination of, 210
   transcription of, 193, 195, 196, 200
   transduction by, 206
   typing of bacteria by, 217
B-cell defects, 404–405
Bee stings, 393
Behring, Emil von, 11
Benthos, 225
Beta-lactam antibodies, 103, 105, 111
Beta-lactamase inhibitors, 131
Bifidobacterium in infant gut, 232
Blood, normal sterility of, 242
B lymphocytes, 308
Bordet, Jules, 11
Breast milk, human, flora of, 241
Breed smear for milk, 284

Cabinets, safety, 256
Capsules of bacteria, 21
Carbon cycle in nature, 226
Carbon dioxide requirements, 57
Casoni test, 392
Cell-mediated responses, 311, 396
Cephalosporins, 125
Chance in disinfection, 86
Chediak-Higashi syndrome, 406
Chemokinesis, 299
Chemostat, 62
Chemotaxis, 299
Chemotherapeutic agents, 78
Chemotherapy, 98 *et seq.*

Chick-Martin test, 89
Chlorhexidine, 89
Chromosomes, 146
    mapping of, 163
    transfer of, 159
Classification of bacteria, 6
Clonal selection, 374
Cloning, 173
Coagglutination, 324
Codon, 145
Colicines, 248
Colicins, 166
Coliform count in milk, 266
Competence, 156
Complement, 11
Complement, Chapter 15
    alternative pathway of, 385
    classical pathway of, 385
    components of, 386
    mode of action of, 386, 387
Complement activation, 299, 309
Complement and bactericidal antibodies, 331
Complement, defects of, 406
Complement-fixation reaction, 327
Conjugation, 148 et seq., 170
    determined by F, 157
    role of pili in, 157
Conjunctiva, bacterial flora of, 241
    effect of lysozyme on, 241
Convertase activity, 386
Coombs antiglobulin reaction, 325
Cosmetics, disinfection of, 91
Cotton-wool, antibacterial action of, 82
Counting of bacteria, 58
C-reactive protein, 332–333
Cream, bitty, 292
Cytochromes, 47
Cytotoxic T cell killing, 310, 314

Death of bacteria, 64
Dendritic cells, splenic, 304
Dick test and immunity, 416
Di Georges syndrome, 404
Diphtheria, epidemiology of, 414
Disinfectants, chemical, 78
    D_{10} value of, 77
    emulsified, 84
    gaseous, 90
    solid, 91
    sprays, 91
    standardization of, 88
    toxicity of, 90
Disinfection and disinfectants, 70 et seq.
Dispersal, effect of on spread of infection, 425
Distilled water, effect of on bacteria, 79
DNA, biosynthesis of, 54
DNA in relation to variation, 149 et seq.
Domagk, G., 9

Dormancy of bacteria, 64
Dose-response curve, 430
Dressings, sterilization of, 91
Droplets and droplet nuclei, 252, 253
Dust, 252
Dust and house mites, 391, 395
Dyes, bacteristatic effect of, 85

Ecology, microbial, 220
Ecosphere, 222
Ectromelia, experimental epidemiology of, 424
E_h, effect of on growth, 56
Ehrlich, Paul, 9
ELISA: enzyme-linked immunosorbentassay, 326
Enzymes in bacterial metabolism, 41
Epidemic infection, control of, 426
    by immunization, 421
    by isolation, 426
    by quarantine, 426
Epidemiology, definition of, 413
    experimental, 419
Epitope, 323
Esch. coli, lac operon, 42
Essential oils, bacteristatic effect of, 85
Ethanol, effect of on bacteria, 83
Ethers, effect of on bacteria, 83
Ethylene oxide, disinfection by, 90
Exotoxins, bacterial, 362–369
Experimental epidemics, spread of infection in, 419
    effect of artificial immunization on, 421
    effect of changes in microbial virulence on, 422
    effect of natural immunization on, 419

Faeces, microbial flora of, 233
Farmers' lung, 395
Fatty acids and lipids, biosynthesis of, 52
Fermentation, 44
Fermentation, nature of, 2
Field trials, 441
Fimbriae, 32
Flagella, 30
Fleming, Alexander, 9
Florey, Howard, 9
Fluctuation test, 146
Fluorine, disinfectant action of, 85
Forssman antigen, 334, 393
Frame-shift mutations, 148
Freezing, effect of on bacteria, 75

Genes,
    acquisition of new, 155
    plasmid-determined, 164 et seq.
Genetic engineering, 173, 214
Genetic homology, 155, 166 et seq.
Genetics, 208
Genetics in relation to herd structure, 418
Genome, 155
Genotype, 145, 160

Germ-free animals, 234, 237
Gonococci, serotypes of, 367
Goodpasture's syndrome, 393
Gram stain, 18
Granules, intracellular, 22
Growth of bacteria, 58–65

Haemagglutination, 324
Haemagglutination caused by fimbriae, 346
Hail, 261
Hands, disinfection of, 84, 91
Hassall's corpuscles, 315
Heat, disinfection by, 76
Helper cells, 308, 310, 316
Herd infection and herd immunity, Chapter 18
  in experimental epidemics, 419
  in nature, 425
Herd structure in relation to immunity, heterogeneity of, in epidemics, 413
d'Herelle, Félix, 178
Hershey-Chase experiment, 183
Heterophile antibodies, 334
Hexachlorophane, 84
HLA system, 312
Hodgkin's disease, 406
Hydrogen peroxide, toxicity of, 48, 57
Hydrostatic pressure, effect of, 78
Hypersensitivity, anaphylactic, 12
Hypersensitivity,
  antibody-mediated,
    type 1, 391
    type 2, 393
    type 3, 394
    type 4, 396
  cell-mediated, delayed, 396
  genetic factors in, 400
Hypersensitivity, delayed, 12, 312
Hypogammaglobulinaemia, 404–405

Ia-antigen, 303, 312
Ice, 261
Ideotype, 377
Iodine, disinfectant action of, 85
Iodophors, 85
Immune complexes, 335
Immune system, 296
Immune tolerance, 14
Immunity, antibody, 13
Immunity, cellular, 14
Immunity, herd, Chapter 18
Immunity, measurement of, Chapter 19
Immunity of lysogenic phage strains, 204
Immunity, specific and non-specific, 297
Immunodeficiency, Chapter 17
  acquired, 410
  IgA, 404
  IgM, 405
  severe combined (SCID), 403–404

Immunodepression, causes of, 408
Immunoelectron microscopy, 326
Immunofluorescence, 325
Immunoglobulins, 14, 375
  genes of, 381
  properties of, 376
  structure of, 375
  types of, 376
immunosuppression caused by drugs, 406–408
infants, microbial flora of, 232
Infection, subclinical, role of antibodies in, 333
Inflammation, 298, 303, 315
Inflammation, role of antibodies in, 330
Insertion sequences and transposons, 151 et seq.
Interferon, 10
Interleukin (lymphocyte activating factor), 315
Intestinal flora, 232
  effect of antibiotics on, 235
  effect on resistance to infection, 236
  in relation to nutrition, 237
  variations in, 233
Intestinal toxaemia, 235
Isolation in control of infection, 426

J-chain, 377, 378
Jenner, Edward, 10
Job's syndrome, 392

Killer cells, natural, 310
Kitasato, Baron Shibasaburo, 11
Koch, Robert, 5
Koch's phenomenon, 12, 397
Kupffer cells, 303, 305

Langerhans cells, 304, 305
Lattice hypothesis, 322
LD 50, 430 et seq.
Leeuwenhoek, Antony van, 1
Leukotriene B4, 300
Leukotrienes, 391
L-forms, 34
Light, effect of on bacteria, 71
Lister, Joseph, 4
Lupus erythematosus, systemic, 394, 399
Lymph nodes, 305, 316
Lymphocytes, 305
Lymphokines, 300, 315
Lysis medicated by complement, 387
Lysis of bacteria and red cells, 327

Macrolides, 107
Macromolecules, biosynthesis of, 53
Macrophage activation, 303
Major histocompatibility complex (MHC), 312
Mannose-sensitive fimbriae, 346
Mast cells, 302
Mathematical models of epidemics, 417–418
Measles, epidemic prevalence of, 425

Medical microbiology, 6
Membrane immunoglobulin, 375, 377, 378
Membranes of bacteria, 25, 27
Memory cells, 309
Meningococci, serotypes of, 367
Mesosomes, 27
Metabolism of bacteria, 39–58
    regulation of, 42
Metals, effect of on bacteria, 80
Metchnikoff, Élie, 10
Methylene blue reduction test, 286
MIC of antibiotics, 102
Microbial variation, effect of on epidemic spread, 422
Microscopy, 19
Milk, bacteriology of, 279
    bottles, 292
    diseases caused by, 188
    ropiness of, 280
Milk, human, flora of, 241
Mixed lymphocyte reaction (MLR), 312
Monoclonal antibodies, 331
Monoclonal antibodies, study of antigens by, 338
Monokines, 315
Morphology of bacteria, 16–38
Mouse pasteurellosis, experimental, 420, 423
Mouse typhoid, experimental, 419 et seq.
Mussels, 273
Mutagens, 147
Mutation and mutagenesis, 146–151
    fluctuation test for, 146
    gene transfer and recombination in, 164
    macro- and microlesions in, 147–150
    various types of, 148 et seq.

Nasopharynx, normal flora of, 237, 239
Natural antibodies and immunity, 333
Natural antibodies, maternal transfer of, 333
Natural killer cells, 310
Nezelof's syndrome, 404
Nitrofurans, 121
Nitrogen cycle in nature, 226
Normal flora of body, 230
    alimentary tract, 231
    auditory meatus, 241
    blood and internal organs, 242
    conjunctiva, 241
    hair, 241
    respiratory tract, 237
    saliva, 231
    skin, 240
    tonsils, 238
Nuclear apparatus of bacteria, 21
Nutrient limitations, 61

One-step growth experiment, 178, 182
Operons, 42, 145, 155, 157, 159
Opsonins, 11

Opsonins, role of, 332
    non-specific, 332
Opsonization, 301
Organic acids as disinfectants, 79
Ouchterlony test, Table 11.3
Oudin test, Table 11.3
Oxygen, effect of on growth, 57
    in water, 263
Oysters, 273
Ozone, effect of on bacteria, 90

Paralysis, immunological, 351
Parasitic infections, hypersensitivity due to, 392 et seq.
Pasteur, Louis, 2
Pasteurization of milk, 290
Patch test, 399
Penicillin, 9
Penicillins, 104, 105
Permeation of outer membrane, 50
pH, effect of on growth, 56
Phage conversion, 162
Phagocytes, 302
    eosinophils, 302
    macrophages, 302
    mononuclear, 302
    neutrophils, 301
Phagocytic defects, 405–406
Phagosomes, 301
Phenols and cresols, 83
Phenotypes, 145
Pheromones, 161
Phosphatase test, 290
Photodynamic sensitization, 73
Photoreactivation, 73
Photosynthesis, 225
Photosynthetic activity in water, 221
Pili, 32
    conjugative, 157, 167, 169, 170
    sex, 33
Pinocytosis, 303
Plankton, 225
Plasma cells, Fig. 10.2(a), 309
Plasmids, 163 et seq.
    mapping of, 167
Plate count in milk, 284
    in water, 268
Polymyxins, 119
Prausnitz-Küstner reaction, 379
Precipitin reactions, optimal proportions in, 322
Pressure, effect of on bacteria, 78
Prick test, 391
Probits, 434
Properdin, 386
Prophages, 203 et seq.
Prostaglandins, Table 16.2
Protein biosynthesis, 54
Protoplast fusion, 163

Protoplasts, 33
Prozone in agglutination tests, 325
Purines and pyrimidines, 52
Putrefaction, nature of, 3
Pyocines, 249

Quarantine, effect of in control of infection, 426
Quaternary ammonium compounds, 82

Radiation, effect of on bacteria, 71
    electromagnetic, 73
    gamma, 74
    infra-red, 74
    ionizing, 73
    ultraviolet, 71
    X-rays, 73
Radioimmunoassays, 326
Reagin antibody, 379
Recombination, 160
    genetic, 210
Recombination in bacteria, 8
Replica plating, 146
Resazurin test, 286
Respiration of bacteria, 46
Respiratory disease, air-borne, 254, 257
Reticular dysgenesis, 404
Reticuloendothelial system, 297
Rhapidosomes, 22
Ribosomes, 22
Rideal-Walker test, 89
Ringer's solution, 81
Rivers, self-purification of, 264
RNA, biosynthesis of, 54
Rose-Waaler test, 325
Rumen, microbial flora of, 237

Salts, effect of on bacteria, 80
Sandwich technique, 326
Scarlet fever, epidemiology of, 416
Schick test and immunity, 415
Schlieren photography of air movement, 252
Sea-water, 262, 268
Semmelweis, Ignatius, 70
Sewage, 260, 274
Sexuality in bacteria, 8
Shell-fish, 272
Skin, bactericidal action of, 82
    disinfection of, 82, 91
Skin, normal flora of, 240
    sampling of, 241
    self-disinfection of, 241
    sweating, effect of on, 241
Slow reacting substances, 391
Sneeze, 252-258
Snow, 261
Soaps, effect of on bacteria, 81
Soil, ecology of, 221
    pathogenicity of bacteria in, 228

Sonic waves, effect of on bacteria, 74
Spectrum, diagram of, 72
Spheroplasts, 33
Spontaneous generation, 3
Spores of bacteria, 28
Sprays, disinfectant, 257
S → R variation, 156, 166
Staining of bacteria, 18
Staphylococcal protein A, 349, 378
Statistical methods, 429 et seq.
Stem cell defects, 403–404
Sterilization of milk, 290
Stomach,
    high acidity of, 234
    normal flora of, 232
Streptococcal M, R and T proteins, 348
Streptomycin, 9
Student's test, 439
Sulphonamides, 122
Sulphur cycle in nature, 224, 227
Sunlight, effect of on bacteria, 71
    effect of on culture media, 71
Survival of bacteria, 64
Swimming-bath water, 270
Symbiosis in soil microbes, 222, 227
Syringes, disinfection of, 91

T-cell alloantigens, 310
T-cell defects, 404
Teeth, microbial flora of, 231, 232
Tcichoic acid antigens, Table 13.1
Temperature, effect of on growth, 63
Thermal shock, 78
Thymus, 315
T lymphocytes, 310
Toxigenicity, prophage-determined, 162
Transduction in bacteria, 8, 162
Transfection, 216
Transfer factor, 399
Transformation in bacteria, 8, 156
Transitions, 147
Transport into bacterial cells, 48
Transposons, 151
Transversions, 147
Tuberculin reaction, 12, 396, 397
Turbidostat, 61
Twort-d'Herelle phenomenon, 177–178
Tyndallization, 4, 77

Ultrasonic waves, 74
Ultraviolet light, effect of on bacteria, 71
Urethra, microbial flora of, 240

Vaccination against smallpox, 10
Vaccines, bacterial types of, 369–371
    influenzal, failure of, 367
Vagina and vulva, normal flora of, 239

Valency of antibodies, 323
Variation, bacterial, and epidemiology, 172
Ventilation, 256
Vibriocines, 249
Viral infection, experimental study of, 424
Virulence factors, 166
Viruses, filtrable, 7
Virus-induced chemotactic factors, Table 10.1
Virus infection and T-cell immunity, 398
Viscera, bacteria in, 232
Vitamins, destruction of by gut bacteria, 237
  synthesis of by gut bacteria, 237

Waksman, Selman A., 9
Walls of bacteria, 23
Water, bacteriology of, Chapter 9, 260
  diseases due to, 270
Watercress, 272
Wells, 261
Wiskott-Aldrich syndrome, 405

X-rays, effect of on bacteria, 73

Ziehl-Neelsen stain, 19

Library, Phila. College of Pharmacy & Science
43rd St. And Woodland Ave., Phila., Pa. 19104